Handbook of membrane reactors

Related titles:

Advanced membrane science and technology for sustainable energy and environmental applications
(ISBN 978-1-84569-969-7)

Functional materials for sustainable energy applications
(ISBN 978-0-85709-059-1)

Materials for fuel cells
(ISBN 978-1-84569-330-5)

Details of these books and a complete list of titles from Woodhead Publishing can be obtained by:

- visiting our web site at www.woodheadpublishing.com
- contacting Customer Services (e-mail: sales@woodheadpublishing.com; fax: +44 (0) 1223 832819; tel.: +44 (0) 1223 499140 ext. 130; address: Woodhead Publishing Limited, 80, High Street, Sawston, Cambridge CB22 3HJ, UK)
- in North America, contacting our US office (e-mail: usmarketing@woodheadpublishing.com; tel.: (215) 928 9112; address: Woodhead Publishing, 1518 Walnut Street, Suite 1100, Philadelphia, PA 19102-3406, USA)

If you would like e-versions of our content, please visit our online platform: www.woodheadpublishingonline.com. Please recommend it to your librarian so that everyone in your institution can benefit from the wealth of content on the site.

We are always happy to receive suggestions for new books from potential editors. To enquire about contributing to our energy series, please send your name, contact address and details of the topic/s you are interested in to Sarah.Hughes@woodheadpublishing.com. We look forward to hearing from you.

The Woodhead team responsible for publishing this book:

Commissioning Editor: Sarah Hughes
Publications Coordinator: Adam Davies
Project Editor: Rachel Cox
Editorial and Production Manager: Mary Campbell
Production Editor: Mandy Kingsmill
Project Manager: Vedhapriya Badrinarayanan, Newgen Knowledge Works Pvt Ltd
Freelance Copyeditor: Dick Hill
Proof reader: Suma George, Newgen Knowledge Works Pvt Ltd
Cover designer: Terry Callanan

Woodhead Publishing Series in Energy: Number 55

Handbook of membrane reactors

Volume 1: Fundamental materials science, design and optimisation

Edited by
Angelo Basile

WP
WOODHEAD PUBLISHING

Oxford Cambridge Philadelphia New Delhi

Published by Woodhead Publishing Limited,
80 High Street, Sawston, Cambridge CB22 3HJ, UK
www.woodheadpublishing.com
www.woodheadpublishingonline.com

Woodhead Publishing, 1518 Walnut Street, Suite 1100, Philadelphia,
PA 19102-3406, USA

Woodhead Publishing India Private Limited, G-2, Vardaan House, 7/28 Ansari Road,
Daryaganj, New Delhi - 110002, India
www.woodheadpublishingindia.com

First published 2013, Woodhead Publishing Limited

British Library Cataloguing in Publication Data
A catalogue record for this book is available from the British Library.

Library of Congress Control Number: 2012954744

ISBN 978-0-85709-414-8 (print)
ISBN 978-0-85709-733-0 (online)
ISSN 2044-9364 Woodhead Publishing Series in Energy (print)
ISSN 2044-9372 Woodhead Publishing Series in Energy (online)

The publisher's policy is to use permanent paper from mills that operate a sustainable forestry policy, and which has been manufactured from pulp which is processed using acid-free and elemental chlorine-free practices. Furthermore, the publisher ensures that the text paper and cover board used have met acceptable environmental accreditation standards.

Typeset by Newgen Knowledge Works Pvt Ltd, India
Printed by MPG Printgroup, UK

Contents

Contributor contact details

(* = main contact)

Editor

Prof. Angelo Basile
Institute on Membrane Technology
ITM-CNR
c/o University of Calabria
Via P. Bucci Cubo 17/C
87030 Rende (CS)
Italy

Email: a.basile@itm.cnr.it

and

AST Engineering S.p.A.
via Adolfo Ravà 30
00142 Rome
Italy

Chapter 1

Joaquim Vital*
REQUIMTE, CQFB, Departamento de Química
FCT, Universidade Nova de Lisboa
Campus da Caparica
2829-516 Caparica
Portugal

Email: jsmv@fct.unl.pt

José M. Sousa
Escola de Ciências da Vida e do Ambiente – Departamento de Química
Universidade de Trás-os-Montes e Alto Douro
Apartado 1013, 5001-801-Vila Real Codex
Portugal

Email: jsousa@utad.pt

and

LEPAE – Departamento de Engenharia Química
Faculdade de Engenharia da Universidade do Porto
Rua Roberto Frias S/N
4200-465 Porto
Portugal

Email: jmsousa@fe.up.pt

Chapter 2

Prof. Angelo Basile*
Institute on Membrane Technology
ITM-CNR
c/o University of Calabria
Via P. Bucci Cubo 17/C
87030 Rende (CS)
Italy

Email: a.basile@itm.cnr.it

Prof. Jianhua Tong
Metallurgical and Materials Engineering
Colorado School of Mines
1500 Illinois Street
Golden, CO 80401
USA

Prof. Pierre Millet
Institut de Chimie Moléculaire et des Matériaux d'Orsay
UMR 8182 - Université Paris sud
Centre d'Orsay, Bâtiment 410
91405 Orsay Cedex
France

Chapter 3

Pietro Pinacci*
Research on the Energetic System (RSE) S.p.A.
Via Rubattino 54
20134 Milano
Italy

Email: pietro.pinacci@rse-web.it

Prof. Angelo Basile
Institute on Membrane Technology
ITM-CNR
c/o University of Calabria
Via P. Bucci Cubo 17/C
87030 Rende (CS)
Italy

Email: a.basile@itm.cnr.it

Chapter 4

Alessia Santucci and Silvano Tosti*
Italian National Agency for New Technologies, Energy and Sustainable Economic Development (ENEA)
Unità Tecnica Fusione
C.R. Frascati
Via E. Fermi 45
00044 Frascati (RM)
Italy

Email: silvano.tosti@enea.it

Prof. Angelo Basile
Institute on Membrane Technology
ITM-CNR
c/o University of Calabria
Via P. Bucci Cubo 17/C
87030 Rende (CS)
Italy

Email: a.basile@itm.cnr.it

Chapter 5

Annarosa Gugliuzza
Institute on Membrane Technology
ITM-CNR
c/o University of Calabria
Via P. Bucci Cubo 17/C
87030 Rende (CS)
Italy

Email: a.gugliuzza@itm.cnr.it

Chapter 6

C. Algieri*
Institute on Membrane Technology
ITM-CNR
c/o University of Calabria
Via P. Bucci Cubo 17/C
87030 Rende (CS)
Italy

Email: c.algieri@itm.cnr.it

Dr Antonio Comite and Prof. Gustavo Capannelli
Dipartimento di Chimica e Chimica Industriale
Università degli Studi di Genova
Via Dodecaneso, 31
16146 Genoa
Italy

Chapter 7

Dr X. Tan
Department of Chemical Engineering
Tianjin Polytechnic University
399, Bingshui West Raod
Xiqing District, Tianjin, 300397
China

Kang Li*
Department of Chemical Engineering and Technology
Imperial College London
South Kensington
London SW7 2AZ
UK

Email: kang.li@imperial.ac.uk

Chapter 8

Simon Smart* and João C. Diniz da Costa
School of Chemical Engineering
The University of Queensland
Brisbane
Queensland 4072
Australia

Email: s.smart@uq.edu.au

Jose M. Serra
Instituto de Tecnología Química
Consejo Superior de Investigaciones Cientificas
Universidad Politécnica de Valencia
Campus UPV - Building 6C av. los Naranjos s/n
E-46022 Valencia
Spain

Email: jmserra@itq.upv.es

Shaomin Liu
Department of Chemical Engineering
Curtin University
Perth, WA 6845
Australia

Email: shaomin.liu@curtin.edu.au

A. Iulianelli and Prof. Angelo Basile
Institute on Membrane Technology
ITM-CNR
c/o University of Calabria
Via P. Bucci Cubo 17/C
87030 Rende (CS)
Italy

Email: a.basile@itm.cnr.it

Chapter 9

Simon Smart, J. Beltramini and João C. Diniz da Costa*
School of Chemical Engineering
The University of Queensland
Brisbane
Queensland 4072
Australia

Email: j.dacosta@uq.edu.au

A. Harale, S. P. Katikaneni and T. Pham
Saudi Aramco
Saudi Arabia

Chapter 10

Dr Kelly Briceño*
Department d'Enginyeria Quimica
Universitat Rovira i Virgili
Av. Pasos Catalans, 26
43007 Tarragona
Spain

Email: kellybm2004@gmail.com; Kelly.briceno@urv.cat

Prof. Angelo Basile
Institute on Membrane Technology
ITM-CNR
c/o University of Calabria
Via P. Bucci Cubo 17/C
87030 Rende (CS)
Italy

Email: a.basile@itm.cnr.it

Prof. Jianhua Tong
Metallurgical & Materials Engineering
Colorado School of Mines
1500 Illinois Street
Golden, CO 80401
USA

Dr Kenji Haraya
National Institute of Advanced Industrial Science and Technology (AIST)
Research Institute for Innovation in Sustainable Chemistry
Membrane Separation Processes Group
AIST Tsukuba Central 5
Tsukuba 305-8565
Japan

Chapter 11

Mika Huuhtanen*, P. K. Seelam, T. Kolli, E. Turpeinen and R. L. Keiski
Mass and Heat Transfer Process Laboratory
Department of Process and Environmental Engineering
P.O. Box 4300
FI-90014 University of Oulu
Finland

Email: mika.huuhtanen@oulu.fi

Chapter 12

Marcello De Falco*
Faculty of Engineering
University Campus Bio-Medico of Rome
via Alvaro del Portillo 21
00128 Rome
Italy

Email: M.DeFalco@unicampus.it

Prof. Angelo Basile
Institute on Membrane Technology
ITM-CNR
c/o University of Calabria
Via P. Bucci Cubo 17/C
87030 Rende (CS)
Italy

Email: a.basile@itm.cnr.it

Chapter 13

Prof. Naotsugu Itoh*
Department of Material and Environmental Chemistry
Utsunomiya University
7-1-2, Yoto
Utsunomiya
321–8585
Japan

Email: itoh-n@cc.utsunomiya-u.ac.jp

Prof. K. Mimura
Engineering Solution Unit
Chiyoda Corporation
4-6-2 Minatomirai, Nishi-ku
Yokohama 220-8765
Japan

Chapter 14

Leonardo Roses, Stefano Campanari* and Giampaolo Manzolini
Politecnico di Milano
Dipartimento di Energia
Via Lambruschini 4
20156 Milano
Italy

Email: stefano.campanari@polimi.it

Chapter 15

Y. Wang and Gregory L. Leslie*
UNESCO Centre for Membrane Science and Technology
School of Chemical Engineering
University of New South Wales
Kensington 2052
Australia

Email: g.leslie@unsw.edu.au

T. D. Waite
Water Research Centre
School of Civil and Environmental Engineering
University of New South Wales
Kensington 2052
Australia

Chapter 16

Stefano Curcio* and Prof. Gabriele Iorio
Department of Engineering Modeling
University of Calabria
Ponte P. Bucci Cubo 36/C
87036 Rende (CS)
Italy

Email: stefano.curcio@unical.it

Chapter 17

G. De Luca
Institute on Membrane Technology
ITM-CNR
c/o University of Calabria
Via P. Bucci Cubo 17/C
87030 Rende (CS)
Italy

Email: g.deluca@itm.cnr.it

Chapter 18

S. K. Schnell* and T. J. H. Vlugt
Process & Energy Department
Delft University of Technology
Leeghwaterstraat 44
2628CA Delft
The Netherlands

Email: s.k.schnell@tudelft.nl

Signe Kjelstrup
Department of Chemistry
Norwegian University of Science and Technology
Høgskoleringen 5
7491 Trondheim
Norway

Woodhead Publishing Series in Energy

1 **Generating power at high efficiency: Combined cycle technology for sustainable energy production**
Eric Jeffs

2 **Advanced separation techniques for nuclear fuel reprocessing and radioactive waste treatment**
Edited by Kenneth L. Nash and Gregg J. Lumetta

3 **Bioalcohol production: Biochemical conversion of lignocellulosic biomass**
Edited by K. W. Waldron

4 **Understanding and mitigating ageing in nuclear power plants: Materials and operational aspects of plant life management (PLiM)**
Edited by Philip G. Tipping

5 **Advanced power plant materials, design and technology**
Edited by Dermot Roddy

6 **Stand-alone and hybrid wind energy systems: Technology, energy storage and applications**
Edited by J. K. Kaldellis

7 **Biodiesel science and technology: From soil to oil**
Jan C. J. Bart, Natale Palmeri and Stefano Cavallaro

8 **Developments and innovation in carbon dioxide (CO_2) capture and storage technology Volume 1: Carbon dioxide (CO_2) capture, transport and industrial applications**
Edited by M. Mercedes Maroto-Valer

9 **Geological repository systems for safe disposal of spent nuclear fuels and radioactive waste**
Edited by Joonhong Ahn and Michael J. Apted

10 **Wind energy systems: Optimising design and construction for safe and reliable operation**
Edited by John D. Sørensen and Jens N. Sørensen

11 **Solid oxide fuel cell technology: Principles, performance and operations**
Kevin Huang and John Bannister Goodenough

12 **Handbook of advanced radioactive waste conditioning technologies**
Edited by Michael I. Ojovan
13 **Membranes for clean and renewable power applications**
Edited by Annarosa Gugliuzza and Angelo Basile
14 **Materials for energy efficiency and thermal comfort in buildings**
Edited by Matthew R. Hall
15 **Handbook of biofuels production: Processes and technologies**
Edited by Rafael Luque, Juan Campelo and James Clark
16 **Developments and innovation in carbon dioxide (CO_2) capture and storage technology Volume 2: Carbon dioxide (CO_2) storage and utilisation**
Edited by M. Mercedes Maroto-Valer
17 **Oxy-fuel combustion for power generation and carbon dioxide (CO_2) capture**
Edited by Ligang Zheng
18 **Small and micro combined heat and power (CHP) systems: Advanced design, performance, materials and applications**
Edited by Robert Beith
19 **Advances in clean hydrocarbon fuel processing: Science and technology**
Edited by M. Rashid Khan
20 **Modern gas turbine systems: High efficiency, low emission, fuel flexible power generation**
Edited by Peter Jansohn
21 **Concentrating solar power technology: Principles, developments and applications**
Edited by Keith Lovegrove and Wes Stein
22 **Nuclear corrosion science and engineering**
Edited by Damien Féron
23 **Power plant life management and performance improvement**
Edited by John E. Oakey
24 **Electrical drives for direct drive renewable energy systems**
Edited by Markus Mueller and Henk Polinder
25 **Advanced membrane science and technology for sustainable energy and environmental applications**
Edited by Angelo Basile and Suzana Pereira Nunes
26 **Irradiation embrittlement of reactor pressure vessels (RPVs) in nuclear power plants**
Edited by Naoki Soneda
27 **High temperature superconductors (HTS) for energy applications**
Edited by Ziad Melhem

28 **Infrastructure and methodologies for the justification of nuclear power programmes**
Edited by Agustín Alonso

29 **Waste to energy conversion technology**
Edited by Naomi B. Klinghoffer and Marco J. Castaldi

30 **Polymer electrolyte membrane and direct methanol fuel cell technology Volume 1: Fundamentals and performance of low temperature fuel cells**
Edited by Christoph Hartnig and Christina Roth

31 **Polymer electrolyte membrane and direct methanol fuel cell technology Volume 2: *In situ* characterization techniques for low temperature fuel cells**
Edited by Christoph Hartnig and Christina Roth

32 **Combined cycle systems for near-zero emission power generation**
Edited by Ashok D. Rao

33 **Modern earth buildings: Materials, engineering, construction and applications**
Edited by Matthew R. Hall, Rick Lindsay and Meror Krayenhoff

34 **Metropolitan sustainability: Understanding and improving the urban environment**
Edited by Frank Zeman

35 **Functional materials for sustainable energy applications**
Edited by John Kilner, Stephen Skinner, Stuart Irvine and Peter Edwards

36 **Nuclear decommissioning: Planning, execution and international experience**
Edited by Michele Laraia

37 **Nuclear fuel cycle science and engineering**
Edited by Ian Crossland

38 **Electricity transmission, distribution and storage systems**
Edited by Ziad Melhem

39 **Advances in biodiesel production: Processes and technologies**
Edited by Rafael Luque and Juan A. Melero

40 **Biomass combustion science, technology and engineering**
Edited by Lasse Rosendahl

41 **Ultra-supercritical coal power plant: Materials, technologies and optimisation**
Edited by Dongke Zhang

42 **Radionuclide behaviour in the natural environment: Science, impacts and lessons for the nuclear industry**
Edited by Christophe Poinssot and Horst Geckeis

43 **Calcium and chemical looping technology for power generation and carbon dioxide (CO_2) capture: Solid oxygen- and CO_2-carriers**
P. Fennell and E. J. Anthony

44 **Materials' ageing and degradation in light water reactors: Mechanisms, modelling and mitigation**
Edited by K. L. Murty

45 **Structural alloys for power plants: Operational challenges and high-temperature materials**
Edited by Amir Shirzadi, Rob Wallach and Susan Jackson

46 **Biolubricants: Science and technology**
Jan C. J. Bart, Emanuele Gucciardi and Stefano Cavallaro

47 **Wind turbine blade design and materials: Improving reliability, cost and performance**
Edited by Povl Brøndsted and Rogier Nijssen

48 **Radioactive waste management and contaminated site clean-up: Processes, technologies and international experience**
Edited by William E. Lee, Michael I. Ojovan, Carol M. Jantzen

49 **Probabilistic safety assessment for optimum nuclear power plant life management (PLiM): Theory and application of reliability analysis methods for major power plant components**
Gennadij V. Arkadov, Alexander F. Getman and Andrei N. Rodionov

50 **The coal handbook Volume I: Towards cleaner production**
Edited by D. G. Osborne

51 **The coal handbook Volume II: Coal utilisation**
Edited by D. G. Osborne

52 **The biogas handbook: Science, production and applications**
Edited by Arthur Wellinger, Jerry Murphy and David Baxter

53 **Advances in biorefineries: Biomass and waste supply chain exploitation**
Edited by K. W. Waldron

54 **Geoscience of carbon dioxide (CO_2) storage**
Edited by Jon Gluyas and Simon Mathias

55 **Handbook of membrane reactors Volume 1: Fundamental materials science, design and optimisation**
Edited by Angelo Basile

56 **Handbook of membrane reactors Volume 2: Reactor types and industrial applications**
Edited by Angelo Basile

57 **Alternative fuels and advanced vehicle technologies: Towards zero carbon transportation**
Edited by Richard Folkson

58 **Handbook of microalgal bioprocess engineering**
Christopher Lan and Bei Wang

59 **Fluidized-bed technologies for near-zero emission combustion and gasification**
Edited by Fabrizio Scala

60 **Managing nuclear projects: A comprehensive management resource**
Edited by Jas Devgun

61 **Handbook of process integration: Minimisation of energy and water use, waste and emissions**
Edited by Jiří Klemeš

Preface

This handbook is dedicated in particular to those readers interested in emerging applications of membrane reactors in the field of energy and environment. The main motivation for this handbook is to give to the reader a panorama of the various aspects of research related to membrane reactors and their applications. The utilisation of membrane reactor technology on a larger scale could constitute a relevant enhancement of conventional systems already in existence. For example, in the field of reforming processes, the main benefit of a membrane reactor is the selective removal of a compound such as hydrogen from the reaction side, which may allow the thermodynamic equilibrium restrictions of the conventional fixed bed reactors to be overcome.

To this end, I invited an international team of expert scientists from the field of membrane science and technology to write about: the state-of-the-art of the various kind of membranes (polymeric, Pd- and non Pd-based, carbon, zeolite, perovskite, composite, ceramic, etc.) used in membrane reactors; modelling aspects related to all kinds of membrane reactors, the various applications of membrane reactors and, finally, economic aspects.

Due to the large amount of material available in the specialised literature, the handbook is composed of two volumes. It should also be mentioned that all the chapters are strictly interconnected. However, for practical use of the handbook, each volume is composed of different parts. In particular, in this first volume, the various arguments are conceptually split into three different parts.

In Part I the various aspects related to polymeric, dense metallic and composite membranes for membrane reactors are extensively considered. The volume starts with Chapter 1, in which the authors (Vital and Sousa) give an overview of the polymeric membranes used in membrane reactors. After introducing some basic concepts of polymer science and polymer membranes, two different types of polymeric membrane reactors (inert and catalytic) are discussed. Various examples of the main reactor types (extractors, forced-flow or contactors) are also given. Finally, the modelling aspects of membrane reactors with dense polymeric catalytic membranes are also presented in detail. It is followed by Chapter 2 (Basile, Tong and Millet), which provides an extensive overview of the inorganic membrane reactors related to the hydrogen production. Almost all the reactions used by scientists for this purpose are summarised and, for the ones considered of practical value, a deeper analysis of both catalysts and membranes is also undertaken. The characteristics as well as the preparation of dense Pd-based membranes, non Pd-based membranes, and also amorphous membranes are discussed, having also in mind both the governing equations and the laws of sorption, diffusion and permeation. The final part of the chapter is dedicated to the analysis of the degradation of dense metallic (pure, alloy, amorphous) membranes: embrittlement, oxidation, polarisation effect, interaction with support (composite membranes) and degradation due to interactions with catalysts, coke, sulphur and morphology changes. Chapter 3 (Pinacci and Basile) focuses attention on preparation methods for thin dense palladium layer deposition onto microporous supports in the field of inorganic composite membrane reactors. In particular, special emphasis is paid to electroless plating and magnetron sputtering techniques, followed by an analysis of the chemical and physical stability of the prepared composite membranes. The most important and recent performances and developments of these membranes are also discussed.

Chapter 4 (Santucci, Tosti, Basile) is mainly focused on the development of membranes based on metals other than Pd, such as Ni, Nb, V and Ti, which are considered today promising substitutes for the Pd-alloys. Particular attention is given to the synthesis of these membranes as well as to the effect of alloying on their chemical–physical properties. The chapter also provides a description of two porous (ceramic and glass) membranes used as a support for the new metal alloys, in gas separation and in membrane reactors, respectively. The objective of Chapter 5 (Gugliuzza) is to document what is known about nanocomposite polymeric membranes and the procedures of fabrication. Their potentialities in catalytic membrane reactors, bioreactors and membrane operations for alternative power production are highlighted.

After this chapter, Part II is dedicated to zeolite, ceramic and carbon membranes and catalysts used in membrane reactors. In Chapter 6 (Algieri, Comite and Capannelli) the remarkable properties of zeolite membranes are illustrated. Moreover, the key role of zeolite membrane reactors to improve the yield and the selectivity of reactions is particularly emphasised. Furthermore, the possibility of using zeolite membranes as micro-reactors and sensors is also discussed. Chapter 7 (Tan and Li) deals with dense ceramic membrane reactors, which are made from composite oxides usually having perovskite or fluorite structures with appreciable mixed ionic (oxygen ion and/or proton) and electronic conductivity. This chapter mainly describes the principles of various configurations (disc/flat-sheet, tubular and hollow fibre membranes) of dense ceramic membrane reactors and the fabrication of the membranes and membrane reactors. The commercialisation of dense ceramic membrane reactor technology is also discussed.

After the dense ceramic membranes, Chapter 8 (Smart, Liu, Serra, da Costa, Iulianelli and Basile) is entirely dedicated to the porous ceramic membranes used in membrane reactors. To summarise, this chapter discusses the most commonly used preparation techniques as well as the various characterisation procedures for porous ceramic membranes and their application as membrane reactors for gas and liquid phase reactions, permeation and separation. Chapter 9 (Smart, Beltramini, da Costa, Harale, Katikaneni and Pham) introduces silica membrane reactors by discussing the research and development of membrane reactors which incorporate microporous silica-based membranes specifically for hydrogen production. A discussion of relevant gas transport mechanisms, membrane performance parameters, membrane reactor designs and membrane reactor performance metrics is followed by an in-depth analysis of the various research investigations where silica membrane reactors are used to produce hydrogen and/or syngas from hydrocarbon reforming reactions.

Chapter 10 (Briceño, Basile, Tong and Haraya) provides an introduction to carbon-based membrane reactors, which contain new and very interesting membrane materials that can be integrated in a compact configuration. Even if carbon membranes are still in an infant stage, they are today considered promising candidates for porous membranes in membrane reactors because of their ease of use, low raw material cost, low fabrication cost, molecular sieve separation effect and relatively high gas permeance values. Some interesting applications of carbon membranes for both macro- and micro-reactors are also reviewed. This Part II ends with Chapter 11 (Huuhtanen, Seelam, Kolli, Turpeinen and Keiski) dedicated to the new catalysts used in membrane reactors. Since the catalyst is one of the main components of a membrane reactor, it is also important to understand its function in these systems. The chapter discusses the different ways in which the catalyst can be introduced or incorporated in the reactor as a catalytic membrane wall. Due to the need also for new catalysts for membrane reactors, the development of some new materials that can be used as novel catalyst supports is also presented.

Part III is dedicated to the modelling, simulation and optimisation of membrane reactors. Modelling of reactors is a crucial topic in process design. The development of reliable and powerful tools for the specific design of particular reactors allows both the optimal dimensions and optimal operating conditions to be defined. Some general aspects of inorganic membrane reactor modelling start to be considered in Chapter 12 (De Falco and Basile), where the main guidelines for membrane reactor modelling are also reported. After presenting the model categories and the procedures for the development of the reactor model, a natural gas steam reforming Pd-based membrane reactor is completely modelled as a case study. Chapter 13 (Itoh and Mimura), in which a computational fluid-dynamics analysis of membrane reactors is presented, follows in the same vein. In particular, the results of the simulation of single- and multi-tube palladium membrane reactors for hydrogen recovery are shown and discussed. The model developed takes into account the concentration, temperature and velocity distributions due to mass, heat transfer and flow resistances in the membrane reactor, and it is verified for the dehydrogenation of cyclohexane in a shell-and-tube type of palladium membrane reactor as well as a multi-tube type. It is demonstrated that the multi-tube model developed is applicable for the reactor design. Using the same tool, i.e. the same computational fluid-dynamics analysis, Chapter 14 (Roses, Campanari and Manzolini) presents a bi-dimensional simulation of a steam methane reformer coupled with a palladium-based hydrogen-permeable membrane. The simulation predicts the system performances by modelling the combined phenomena taking place in the reactor. The chapter shows a detailed analysis of various parameters (temperature, and so on) and discusses their impact on reactor performances. Furthermore, the same modelling technique is used in Chapter 15 (Wang, Waite and Leslie) to provide an overview of membrane bioreactor design for two-phase and three-phase flows. Various effects (reactor geometry, etc.) are discussed and particular emphasis is given to the importance of model calibration, results validation and the outlook for future work.

In Chapter 16 (Curcio and Iorio) it is shown how various kinds of advanced models, based either on artificial neural networks or on a hybrid approach, could be used to predict the behaviour of some typical membrane process. The obtained results demonstrate that the proper combination of a theoretical model with a straightforward neural model is capable of widening the applicability of pure neural models outside the training range, thus paving the way for the exploitation of hybrid neural models for process optimisation purposes and for the implementation of efficient on-line controllers operating in different kinds of membrane processes. Chapter 17 (De Luca) starts from the consideration that several nanostructured materials are used in the preparation of membrane reactors. For this reason, the evaluation of their key properties is very important. Quantum mechanics is seen as a reliable tool for obtaining these fundamental properties, avoiding the use of empirical parameters. The assessment of some fundamental quantities, such as the adsorption energies of gases in nano-porous materials or on metallic surfaces, is reviewed. Moreover, the calculation of hydrogen solubilisation in metal alloys is also presented. Chapter 18 (Schnell, Vlugt and Kjelstrup) considers that non-equilibrium thermodynamics offer a better and more precise way to describe transport of heat and mass over membranes than the simple Fick's and Fourier's laws. In deriving the equations for single component transport (of *n*-butane) across a zeolite membrane, the authors use non-equilibrium molecular dynamics data to show how phenomena are strictly related. The transport models across a zeolite membrane are shown to be in agreement with the second law of thermodynamics.

I wish to take this opportunity to thank all the authors of the chapters for their expert contributions.

A. Basile
ITM-CNR, Italy

Part I

Polymeric, dense metallic and composite membranes for membrane reactors

1

Polymeric membranes for membrane reactors

J. VITAL, Universidade Nova de Lisboa, Portugal and
J. M. SOUSA, Universidade de Trás-os-Montes e Alto Douro, Portugal e Universidade do Porto, Portugal

DOI: 10.1533/9780857097330.1.3

Abstract: The objective of this chapter is to give an overview of the use of polymeric membranes in membrane reactors. Since the study of polymeric membrane reactors is a multidisciplinary activity, the chapter begins with some basic concepts of polymer science and polymer membranes. In the following, the different types of polymeric membrane reactors, classified into two main groups – polymeric inert membrane reactors (PIMRs) and polymeric catalytic membrane reactors (PCMRs), are presented and discussed. For each of these groups, examples of the main reactor types are given: extractors, forced-flow or contactors. Finally, there is a discussion of the modelling aspects of membrane reactors with dense polymeric catalytic membranes reported in the literature.

Key words: catalytic/inert polymeric membranes, polymeric membranes preparation, membrane reactors, extractor-type, distributor/contactor-type, forced-flow-type, polymeric inert membrane reactors (PIMRs), polymeric catalytic membrane reactors (PCMRs), modelling.

1.1 Introduction: polymer properties for membrane reactors

The polymers used as membrane materials can be classified as either natural or synthetic. The vast majority of membranes today are made from synthetic polymers; however, polysaccharides and rubbers are significant examples of natural membrane materials. A membrane's barrier properties and other significant characteristics are determined by its macromolecular structure. Key factors include the chemical structure of a polymer's chain segments, its chain length (molar mass) and chain flexibility, as well as intra- and inter-molecular interactions.[1]

The chain flexibility and transport properties of a polymer are determined by its *chemical structure*.[1,2] For example, if unhindered rotation is possible around single bonds in the main chain of a macromolecule, it can be said to be flexible. Hydrogenation can be used to decrease unsaturation in a polymer backbone, which in turn decreases diffusivity.[2] This is because unsaturation in polymer chains provides segmental mobility, whereas saturation tends to restrict it. Bulky substituents on the polymer chains can influence the transport process due to steric hindrances, which decrease the segmental motion of the polymer chains.[2] As the cohesive energy of a polymer increases, the permeability of species interacting weakly with the functional groups present in that polymer can be expected to decrease.[2]

The *average molecular* weight and *polydispersity* of a polymer influences its chemical and physical properties. This is achieved via interactions between the chain segments. On the one hand the number of interaction sites in a polymer chain increases with the length of the chain. Therefore, as the molar mass increases, the stability of the polymer will increase. However, as a consequence of its stability increase, the polymer solubility will decrease.[1] On the other hand, the polymer molecular weight also influences its sorption and transport properties, and the number of chain ends decreases as the molecular weight of the polymer increases. Since these chain ends correspond to discontinuities forming sites where permeant molecules may be sorbed into glassy polymers, the sorption of those permeant species will decrease when the molecular weight of the polymer increases.[2]

A polymer in which chain segmental motion is very restricted is referred to as a '*glassy state*' polymer; that is, it has the properties of a glass in that only bond angle deformation and bending can take place.[3]

A polymer with local segmental mobility but with restricted total chain flow due to its physical or chemical network matrix structure is referred to as being in a *rubbery state*.[3]

The transition between the glassy and rubbery states occurs at the *glass transition temperature* (T_g).[1–3] If a polymer is heated to a higher temperature than its T_g, it will attain a more flexible and mobile state, with lower elastic modulus and higher permeability. The presence of structural features such as bulky pendant groups, chain symmetry, polar groups and cross-linking increases the T_g of a polymer, and the presence of additives such as plasticizers and structural features, such as flexible pendant groups, non-polar groups or dissymmetry, may decrease the T_g.[3] So-called glassy polymers have a T_g above room temperature, and rubbery polymers (or elastomers) have a T_g below room temperature. Membranes with a non-porous selective barrier are reliant on polymer selection, because flux and selectivity are dependent on the mass-transport mechanism (sorption–diffusion, described in Section 1.2.1). For membranes with a porous selective barrier, the mechanical stability of the polymer is crucial in preserving the shape and size of the pores.[1]

The transport properties of dense membranes are dependent on the segmental mobility of the polymer chains and on the *free volume* within the polymer (i.e. the space not occupied by polymer chains).[2,3] While the polymer in most polymeric membranes is amorphous, polymers with flexible chains of regular chemical structure can form crystalline domains, leading to higher chemical and temperature resistance, higher mechanical stability, but lower free volume (which leads to lower permeability), than the same polymer in an amorphous state.[1] Positive correlation has been observed between polymer transport properties and free volume (measured by positron annihilation techniques).[4]

Where high-performance polymeric membranes are required, *block or graft copolymers* are often used in place of homopolymers. These contain two or more different repeating units within the same polymer chain and provide synergy between the properties of the different components.[1]

Blending of polymers or copolymers is also used for the production of membranes. In this case both (co)polymers must be compatible and miscible in the same solvent, in order to form a homogeneous solution. The solid membrane so obtained is characterized by a single T_g value, usually lying between the T_g values of the two pure components. However, if the rule of compatibility and miscibility is not observed, a heterogeneous (phase-separated) polymer blend is obtained, characterized by two (or more) T_g values corresponding to the individual phases.[1] The composition, phase morphology and miscibility of a polymer blend affect diffusion and transport through the membrane. Interactions between the component polymers influence diffusion through homogenous blends, whereas diffusion through heterogeneous blends is affected by interfacial phenomena and the rubbery or glassy nature of the phases.[2]

Chemical or physical cross-linking of a polymer can be used to enhance the chemical stability and mechanical strength and control swelling of a membrane. The latter is particularly important in separations of organic mixtures. Cross-linking is usually carried out after membrane formation as it decreases polymer solubility.[1] The polymer mass-transport properties depend on the cross-linking degree; for the same polymer and the same cross-linking degree, they depend on the nature of the cross-linker.[2]

Another important parameter in selecting the correct polymer for a membrane is its *hydrophilic–hydrophobic balance*, which depends on the functional groups present in the polymer chains. For example, when the required non-porous membrane properties are high permeability and selectivity for water, hydrophilic polymers (with high affinity to water) are suitable as materials for membrane production.[1] The hydrophilic–hydrophobic balance can lead to dissolution (sorption) of molecules in the membrane, which can cause swelling when the membrane is in contact with a liquid feed. This effect (especially for the case of huge swellings) should be taken into account when selecting a polymer.

The polymer *film-forming* properties should also be taken into account in the selection process of the material for membrane production. The ability of a polymer to form a cohesive film depends on aspects of its macromolecular structure, such as molar mass and attractive interactions between chain segments. Examples of good film-forming materials include polyethersulfones (PES), polysulfones (PS), polyamides (PA) and polyimides (PI).[1]

1.2 Basics of polymer membranes

A membrane (usually a thin film) is a permeable or semi-permeable phase that acts as a barrier between two adjacent fluid phases and controls the exchange of materials between them. This can be achieved either by sieving or controlling the relative rate of transport of the materials.[5]

1.2.1 Transport phenomena in polymer membranes

Mass transfer through dense polymeric membranes is nowadays accepted to be described by the sorption–diffusion mechanism.[5] According to this, the species being transported dissolve (sorb) in the polymer membrane surface on the higher chemical potential side, diffuse through the polymer free volume in a sorbed phase, and pass into the fluid phase downstream of the membrane (lower chemical potential side).[6] In the case of dense polymeric membranes the polymer is an active participant in both the solution and diffusion processes. However, since in many porous membranes the mass transfer takes place mainly in the pores, the membrane material is not an active participant and only its pore structure is important.[6]

Mass transfer through membranes involves the concepts of permeation and diffusion. Permeation involves transferring components from an upstream fluid phase to a downstream fluid phase.[6] If a porous membrane is used, mass transfer takes place in the pores, by convection. In the case of dense membranes, the components being transferred are sorbed on the surface of the membrane (usually in equilibrium with the adjacent fluid phase) and diffuse all along the membrane thickness until reaching the other surface, as described above. Permeation is based on the driving force difference between the upstream and downstream fluid phases, while diffusion is based on the driving force difference between the two surfaces but within the membrane phase.[6,7]

1.2.2 Basic aspects in polymeric membrane preparation

A wide range of materials and physical structures can be used for the preparation of synthetic membranes, which can be categorized based on their structure:[9]

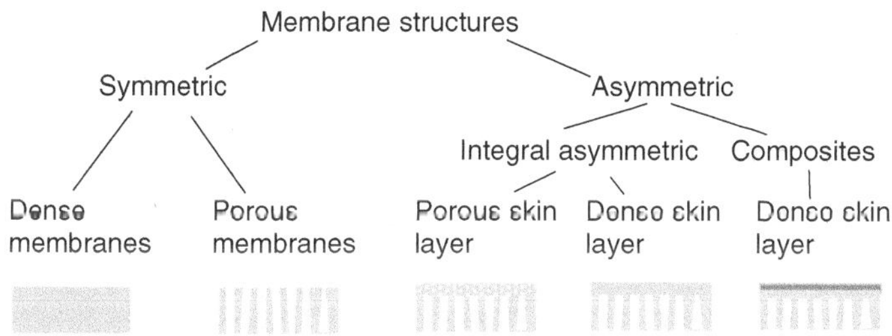

1.1 Main types of polymeric membrane structures. (Adapted from H. Stathmann *et al.*[9] Reprinted with permission from Elsevier, Copyright (2010).)

1. porous membranes,
2. homogeneous solid membranes,
3. solid membranes carrying electrical charges and
4. liquid or solid films containing selective carriers.

Synthetic membranes may be symmetric (with an identical structure over the cross-section of the membrane) or asymmetric, and may be flat, tubular or hollow fibre (see Fig. 1.1 for the main types of polymeric membrane structures). Table 1.1 shows the molecular structures of common polymers used in membrane preparation.

Symmetric and asymmetric membranes

In a symmetric membrane, the thickness of the entire membrane determines the flux of materials through it. An asymmetric membrane consists, for example, of a 0.1–1-μm-thick skin layer (the selective barrier) on a highly porous 100–200-μm-thick substructure. The separation characteristics of an asymmetric membrane are determined by the nature of the material (for the case of a dense skin layer) and size of the pores in the skin layer, whereas the mass flux is determined mainly by the thickness of the skin. Thin, fragile skin is usually supported by a porous sublayer which has little or no effect on the above properties.[9]

To prepare an asymmetric membrane, either the phase-inversion process (skin and support made of the same material) or a two-step process (barrier layer deposited on a porous substructure) is used. In the latter case, the barrier and support structures are usually made from different materials.[9]

Symmetric and asymmetric polymeric membranes can be prepared using the phase separation process.[1,9,10] A precipitation/solidification process is used to transform a polymer solution into two phases (a polymer-rich solid and a polymer-lean liquid phase). The following techniques can be used to solidify the polymer:

Table 1.1 Most common polymers used to prepare membranes

Molecular structure	Name	Abbreviation
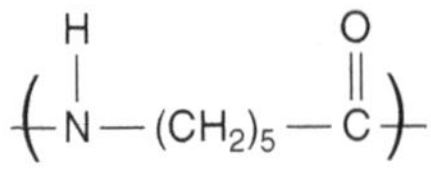	Aliphatic polyamide	PA
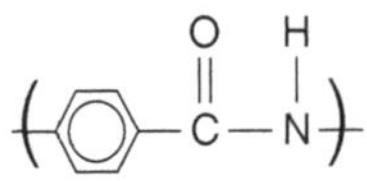	Aromatic polyamide	PA
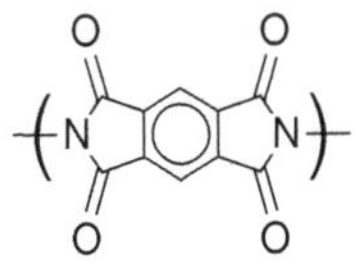	Polyimide	PI
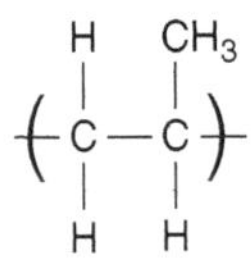	Polypropylene	PP
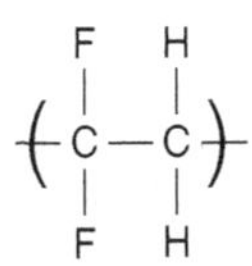	Polyvinylfluoride	PVDF
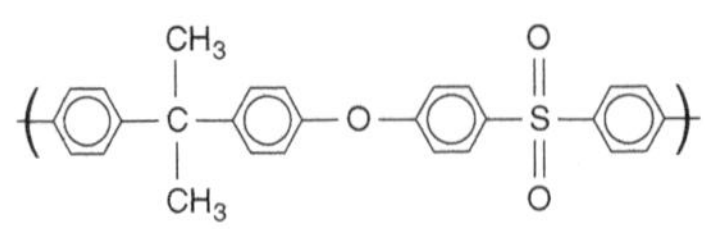	Polysulfone	PES

Source: Adapted from H. Stathmann *et al.*[9]

- Non-solvent induced phase separation (NIPS), or diffusion-induced phase separation, involves dissolution of the polymer in a good solvent in order to obtain a homogeneous solution, followed by the addition of a non-solvent miscible with the first solvent. This will cause precipitation of the polymer when the non-solvent concentration becomes significant.
- Vapour induced phase separation (VIPS) – in this process, the polymer solution is exposed to an atmosphere containing a non-solvent, which is absorbed and causes precipitation of the polymer.
- Thermally induced phase separation (TIPS) is used to create symmetric membranes and involves the cooling of a homogeneous polymer solution until it reaches a set temperature, at which point it will separate into two phases.
- Evaporation induced phase separation (EIPS) – where the homogeneous polymer solution contains two or more solvents of different dissolution capacities, the more volatile solvent can be evaporated.

Integral-asymmetric membranes by phase-inversion

Asymmetric phase-inversion membranes can be prepared from virtually all polymers by using the TIPS, EIPS and NIPS procedures, as long as they are soluble in an appropriate solvent. These three procedures result in the formation of membrane pores (the liquid phase poor in polymer) and structure (the solid phase rich in polymer), which may be either symmetric or asymmetric with a dense skin at one or both surfaces of a porous bulk phase.

The following steps are used in the NIPS process:[1,9]

1. A polymer is dissolved in an appropriate solvent to form a homogeneous solution.
2. The solution is cast into a film of 100–500 μm thickness (known as the proto-membrane).
3. Immersion in a non-solvent coagulation bath causes precipitation.
4. The resultant membrane may then undergo treatments such as rinsing, annealing or drying.

The pores at the surface of the film, where precipitation occurs first and most rapidly, are much smaller than those inside or on the base of the film, resulting in an asymmetric membrane structure.[9] Selection of different polymer solvents and non-solvents, additives, residence times and other parameters during NIPS can be used to fine-tune the pore structure.[10]

Mixed-matrix membranes

Inorganic materials such as metal oxides or zeolites can be embedded in an organic polymer matrix to form a mixed-matrix membrane. This is useful where a specific combination of properties is desired.[1]

The following steps are typically used to prepare mixed-matrix membranes:

1. Separate preparation of a polymer solution and a suspension of inorganic material.
2. The solution and the solid material are mixed to form a homogeneous mixed-matrix suspension.
3. This suspension is then cast (or spun).
4. The NIPS process is then used to induce phase separation.

Mixed-matrix membranes containing inorganic oxides (e.g., silica) can also be prepared by using the sol–gel method to synthesize nanoparticles *in situ* within a polymer solution, followed by phase separation.[1]

Hybrid inorganic–organic materials can be constructed with a variety of properties according to the desired application.[11] Some of the techniques used for the incorporation of inorganic building blocks into organic polymers include:[11]

- Preparation of interpenetrating networks by: (1) sol–gel processes in the presence of preformed polymers; (2) polymerization in sol–gel networks; (3) simultaneous formation of interpenetrating networks; (4) dual network precursors.
- Incorporation of metals and metal complexes in polymers by coordination interactions.
- Insertion of polymers in 2D layered materials.

Porous membranes

Porous membranes are made up of a solid matrix with pores of < 1 nm to > 10 μm in diameter. When the pore size is larger than 50 nm, the membrane is said to be macroporous. Mesoporous membranes have pore diameters between 2 and 50 nm, and microporous membranes have diameters between 0.1 and 2 nm. Where there are fluctuating free volumes rather than permanent pores, the membrane is described as dense.[9]

Symmetric porous membranes prepared by sintering, track-etching and leaching techniques

Sintering of organic or inorganic materials is a simple technique that can be used to obtain porous structures. A powder is pressed into a film or plate and sintered just below the melting point of the material, the particle size of the powder determining the pore size of the final membrane.[9] When preparing sintered membranes, the material selection is determined mainly by the chemical and thermal stability as well as the mechanical properties required for the membrane's application.

Porous membranes can also be prepared by stretching a partially crystalline homogeneous polymer film. This technique is mainly employed for polyethylene or polytetrafluoroethylene films. These films are extruded from a polymer powder at close to melting temperature coupled with a rapid drawdown[9] and the crystallites in the polymer are aligned in the direction of drawing. The extruded film is then annealed, cooled and stretched perpendicularly to the direction of the drawing, in order to obtain relatively uniform pores with diameters of 0.2–20 μm by partially fracturing the film.[9]

Track-etching is a two-step process used to create porous membranes with uniform, round cylindrical pores (see Fig. 1.2).[9,10] High-energy particle radiation is applied perpendicularly to a polymer foil or film, damaging the polymer matrix and creating tracks. The foil or film is then immersed in

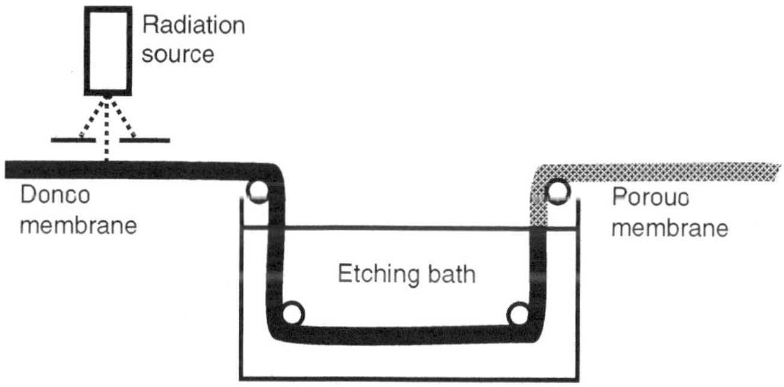

1.2 Track-etching method. (Adapted from H. Stathmann *et al.*[9] Reprinted with permission from Elsevier, Copyright (2010).)

an acid (or alkaline) bath and the polymeric material is etched away along the tracks to form uniform cylindrical pores with narrow pore distribution – pore size 0.2–10 μm and porosity 10%.

The length of time spent under irradiation controls the pore density of a track-etched membrane, while the residence time in the etching bath determines the pore diameter.[9]

Polycarbonate and polyester films are used to prepare capillary porous membranes, because those polymers are commercially available in very uniform films of 10–15 μm thickness. This is the maximum penetration depth of collimated particles obtained from a radiation source.[9]

Preparation of polymeric composite membranes

A typical polymeric composite membrane is composed of a porous film with a dense polymer barrier layer of a different material formed over the top.[9] Composite membranes have the advantage over integral-asymmetric structures that different polymers may be used for the different layers, depending on their properties (e.g. a polymer with the correct selectivity for a separation problem can be used over a support structure of a different porous material).

Membrane modification

Membranes for specific applications can also be prepared through the modification of pre-existing, for example, commercially available, membranes. Membrane functionalization can be carried out using different techniques including surface chemistry, polymer deposition, alternate adsorption of oppositely charged polyelectrolytes, plasma or radiation induced grafting and gold–thiol chemistry. Surface modification of membrane supports in order to obtain membranes with the desired properties is the most common technique used, which involves introducing functional groups through covalent or non-covalent attachment mechanisms. Porous membranes can be functionalized by depositing polymer inside the membrane pores, either by cross-linking the polymer chains or by *in situ* polymerization of corresponding monomers with simultaneous cross-linking. Relevant molecules carrying the desired functional groups can then be covalently attached to the polymer coating the pore walls. Examples of porous membrane supports which can be activated as described above are cellulose, cellulose acetate, alumina, polysulfone and poly(vinylidine fluoride) (PVDF).[8]

There are three different categories of application for functionalized membranes: separation, sorption and catalytic applications. For separation, the modified membranes must allow the selective permeation of the desired chemical species and can be prepared, for example, by layer-by-layer (LbL) assembling. In sorption applications, the modified membranes act as adsorbents, which can also lead to separation and capture. However, these membranes need to be regenerated before they can be reused. Finally, functionalized membranes for catalytic applications may include enzymes or immobilized nanoparticles that act as catalysts and convert the reactants into products as they pass through the membrane pore.[8] Porous membrane supported catalytic applications not only provide a way for catalyst immobilization, avoiding the need for its subsequent removal from the reaction mixture, but also lead to improved mass-transport conditions, since this transport is mainly done through the pores.

1.3 Membrane reactors

The main characteristic of membrane reactors is the integration of catalysis and membrane separation in a single unit operation.[12]

From the reaction point of view, the product removal reduces the flow rate of the reactant stream, while increasing the residence time; increases the reactant concentration, and hence the forward reaction rate; reduces product concentration, reducing the reverse reaction rate. The rate-determining steps of the reaction could change because, even though the species present are the same, they may have a very different concentration with respect to a traditional reactor.[13]

A possible classification of polymeric membrane reactors is based on the role of the membrane in the catalytic process:[5,13,14]

- *Polymeric catalytic membrane reactors (PCMRs)*, if the membrane is itself catalytically active.
- *Polymeric inert membrane reactors (PIMRs)*, if the membrane provides only a separation function.

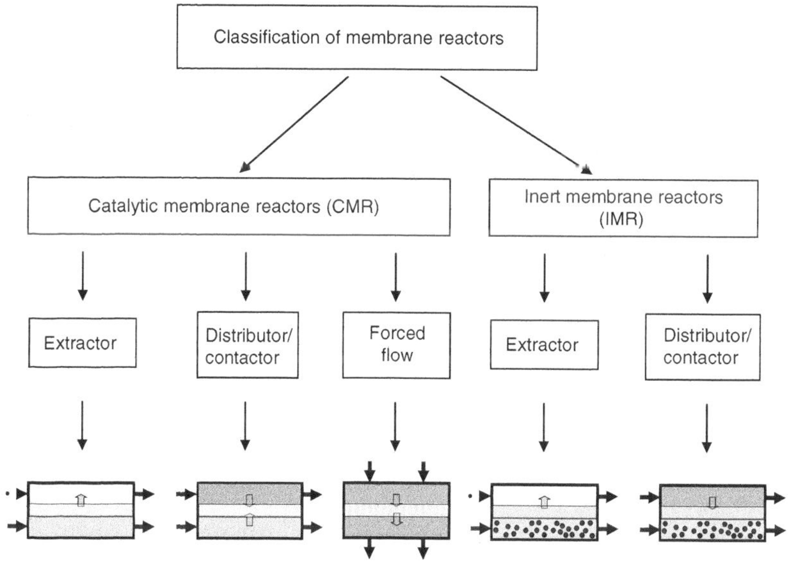

1.3 A classification of membrane reactors.

In this case two different reactor configurations are usually distinguished: the *packed-bed membrane reactor* (*PBMR*), and the *fluidized-bed membrane reactor* (*FBMR*).

Membrane reactors can also be classified according to the transport function of the membrane.[5,14,15] A possible classification may be that depicted in Fig. 1.3 with three different reactor types for PCMRs and only two for PIMRs:

- *Extractor-type membrane reactors*: Applied to PCMRs and PIMRs. This type of reactor is based on the selective removal of one or more reaction products, which could result in an increase of the conversion for equilibrium-limited reactions or in the improvement of the catalytic activity if the removed products are reaction-rate inhibitors. Dehydrogenation membrane reactors or pervaporation membrane reactors are examples of extractor-type membrane reactors.
- *Forced-flow membrane reactors*: Only applied to PCMRs. In this type of reactor a porous membrane is used. The flow is mainly convective, taking place through the membrane pores, where the catalytic active species are located. The effects of the permselectivity properties of the polymer are negligible and the membrane works mainly as a catalyst support.
- *Distributor/contactor-type membrane reactors*: Applied to PCMRs and PIMRs. This type of reactor finds application whenever the reactants'

addition must be controlled in order that the formation of undesired side products is avoided. Another possible application of contactor-type membrane reactors is when the membrane promotes the intimate contact of two immiscible reactants, fed on each side of the membrane, in such a way that solvents can be excluded, aiming environmentally sustainable processes.[12,16]. This last case requires the use of catalytically active membranes (CMRs).

1.3.1 Polymeric catalytic membrane reactors (PCMRs)

Polymeric catalytic membranes, when compared with inert membranes, offer the possibility of tuning the sorption of reactants and products in the close vicinity of the catalytic active sites, by a careful selection of the polymeric environment.[12] As was mentioned above, in porous catalytic membranes the choice of the polymer is of less importance, since permeation does not take place through the polymer matrix. However, in the case of dense membranes, sorption and transport properties are crucial for the catalytic performance and are strongly affected by the polymeric matrix.

Extractor-type membrane reactors

Pervaporation-assisted catalysis is a typical example of an operation efficiently carried out in extractor-type catalytic membrane reactors. Esterification is by far the most studied reaction combined with pervaporation.[14] Esters are a class of compounds with wide industrial application, from polymers to fragrance and flavour industries.[17] Esterification, a reaction between a carboxylic acid and an alcohol with water as a by-product, is an equilibrium-limited reaction. So, this is a typical reaction that can be carried out advantageously in a extractor-type membrane reactor. By selectively removing the reaction product water, it is possible to achieve a conversion enhancement over the thermodynamic equilibrium value based on the feed conditions.

The traditional process to carry out esterification reactions is reactive distillation. However, this process can only be applied if the difference of volatilities between the components is high enough and if azeotropes are not present. Pervaporation is an interesting alternative to reactive distillation, since it is not limited by the relative volatility of the components of the reaction mixture.[5,12,14,17]

Multilayer blends of poly(styrene sulfonic acid) with hydrophilic polymers such as poly(acrylonitrile) (PAN) or poly(vinyl alcohol) (PVA) on the top of commercial PVA membranes, for the esterification of propionic acid and propanol,[18] tungstophosphoric acid entrapped in PVA for the esterification of *n*-butanol and acetic acid,[19] or protonated Nafion membranes for the esterification of methanol and acetic acid,[20] are among the former successful attempts of PCMRs for pervaporation-assisted esterification. However, in the last case, the accumulation of reaction products in the membrane drastically changed its separation properties.[12,20] A catalytic polymer/ceramic composite membrane consisting of a PVA entrapped $Zr(SO_4)_2$ mixed-matrix supported on a porous ceramic plate was also reported to be effective in the pervaporation-assisted *n*-butyl alcohol/acetic acid esterification.[21] The main feature of these 'bifunctional' membranes is to combine the permselectivity for water of the dense polymer matrix with the catalytic activity for esterification of the entrapped material. In the last case of the *n*-butyl alcohol/acetic acid esterification, 95% conversion was achieved.

More recently, composite polymeric catalytic membranes consisting of a dense layer of a mixed-matrix of tiny particles of Amberlyst-35 dispersed in PVA cross-linked with maleic acid cast over a commercial PVA membrane (PERVAP 1000), were efficiently used in the pervaporation-assisted esterification of acetic acid and ethanol.[22] After 8 h of reaction, a 60% increase in conversion was observed for the catalytic membrane configuration, compared to an inert membrane/fluidized-bed configuration.

The role of the polymer cross-linking is an important issue in pervaporation-assisted esterification reactions, since it affects drastically the membrane's transport properties. In fact, it is in this feature that lies one of the advantages of polymeric membranes in relation to their inorganic counterparts: the possibility of changing the membrane's sorption and transport properties by small changes in polymer structure, such as cross-linking. PVA membranes bearing sulfonic acid functions were successfully used as catalysts in the acetic acid/isoamyl alcohol esterification.[23] The sulfonic functions were introduced in the PVA matrix by two different methods: via the sulfosuccinic acid cross-linker (PVA-SSA) or via sulfosalycilic acid anchored on PVA chains (PVASA), independently of the succinic acid cross-linker. Catalytic activity observed in batch experiments with membranes in which cross-linking is independent of the sulfonic acid functions, increases when cross-linking is increased from 1% to 10%. This increase in catalytic activity is explained by the improvement of the membrane's transport properties: the cross-linker acts as a spacer avoiding hydrogen bonding between different polymer chains and increasing free volume. The use of a PVASA membrane in a sweep gas pervaporation reactor configuration led to an increase in activity of about 16%.

The use of polymeric catalytic membranes in pervaporation-assisted catalysis applied to non-esterification reactions has also been referred in recent reviews.[14,16 24,25]

Composite HPA-polymer catalytic membranes were successfully used for the production of very pure isobutene by sweep gas pervaporation-assisted

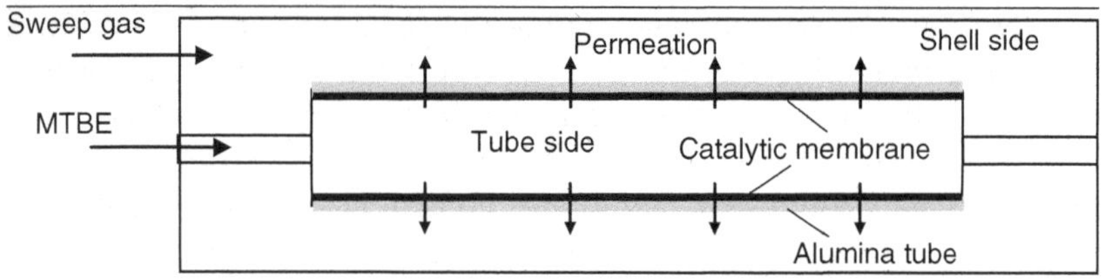

1.4 Shell-and tube-membrane reactor for pervaporation-assisted MTBE decomposition. (Adapted from Choi *et al.*[27] Reprinted with permission from Elsevier, Copyright (2001).)

methyl *tert*-butyl ether (MTBE) decomposition, in a shell-and-tube membrane reactor configuration[26,27] (Fig. 1.4). Three types of composite catalytic membranes comprising 12-tungstophosphoric acid (PW) and polypropyleneoxide (PPO) supported on alumina porous tubes, were tested. The best performance was achieved with the catalytic composite membrane PW–PPO/PPO/Al_2O_3, which was attributed to the intrinsic permselective capabilities of the PW–PPO catalytic membrane and the sub-layered PPO membrane. The selective removal of methanol through the catalytic membrane led to an equilibrium shift to the favourable direction in the decomposition of MTBE.

Chemical reaction with simultaneous pervaporation has also been used in non-equilibrium systems, such as in the works of D. Fritsch and co-workers on the hydrodehalogenation of chlorophenol and chlorobenzene.[28,29]

The reactions were carried out over polymeric catalytic membranes consisting of poly(ether block amide) (PEBA) or blends of PEBA and poly(vinylpyrrolidone) (PVP) loaded with palladium nanoclusters. These membranes were used to simultaneously hydrogenate and concentrate the organic compounds in the permeate side by catalytic pervaporation. The pervaporative enrichment of the organics was achieved by a factor of 100, and a significant conversion was only obtained when the catalytic membranes were employed in pervaporation mode.

A PEBA/PVA blend membrane filled with silica and loaded with 1.5 nm Pd-clusters exhibits the highest activity in the chlorophenol hydrodehalogenation, ten times higher than the activity obtained with a homogeneous membrane even with about six times higher Pd content.[29] This improvement was attributed to the smaller cluster size and the effect of the filler, known to elongate the diffusion path through the membrane, therefore increasing the contact rates with the catalyst.

The concept of simultaneous reaction and pervaporative concentration of the organic compounds was extended to the hydrogenation of acetophenone in diluted aqueous solutions.[30] In this study homogeneous films of PEBA and polydimethilsiloxane (PDMS) loaded with 2–5 nm Pd-clusters were used.

Since PDMS membranes exhibited high H_2 permeation, and acetophenone conversion demands at least stoichiometric availability of H_2 at the membrane enclosed catalyst, a much lower H_2 pressure was required with these membranes than with PEBA for identical catalytic performance.

The use of a non-pervaporative extractor-type catalytic polymeric membrane reactor has been reported for light alcohol/acetic acid esterifications.[31] A cross-linked poly(styrene sulfonic acid) (PSA)/PVA blend flat membrane was assembled in the reactor in a vertical configuration, separating two chambers. One of the chambers was loaded with an aqueous solution of ethanol and acetic acid, while the other chamber was filled with chlorobenzene. The esterification equilibrium is displaced to the product's side by the continuous extraction of the formed ester. In the esterifications of methanol, ethanol and *n*-propanol with acetic acid, the reactivity through the PSA/PVA membrane was higher than that with HCl as catalyst. In that of *n*-butanol with acetic acid, however, it was viceversa.

The use of an extractor-type polymeric catalytic membrane reactor has also been described by Wu *et al.*[32–34] for phenol allylation. Ion-exchange membranes, consisting of poly(styrene quaternary ammonium halide) cross-linked with divinylbenzene paste on polypropylene non-woven fabric, were assembled in a two-chamber flat membrane reactor, either in a horizontal configuration or in a vertical configuration. One of the chambers was filled with an aqueous solution of phenol and sodium hydroxide, while the other chamber was filled with a solution of allylbromide in dichloroethane, the membranes acting as phase transfer catalysts according to the mechanism depicted in Fig. 1.5.

Forced-flow membrane reactors

In forced-flow membrane reactors functionalized porous membranes are used instead of dense membranes. The advantages of this type of membrane reactor lie in the small resistance to mass transfer of the porous membranes, in comparison with their dense counterparts. However, this is also their

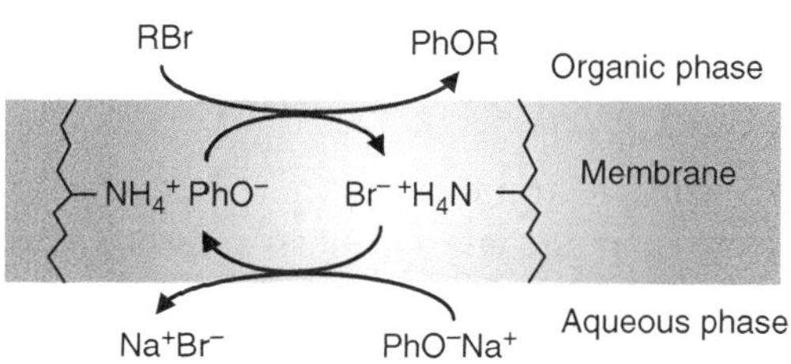

1.5 Mechanism for phenol allylation in the ion-exchange membrane reactor. (Adapted from Wu and Wu.[32] Reprinted with permission from American Chemical Society, Copyright (2005).)

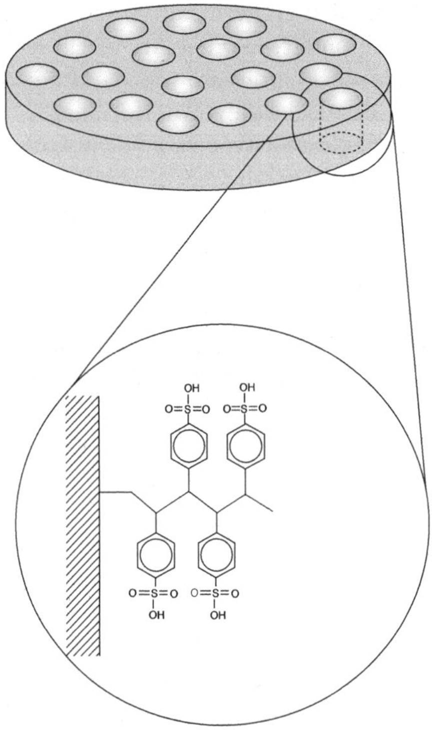

1.6 Scheme showing a poly(styrene sulfonic acid) graft on the wall of a membrane's pore. (Adapted from Shah *et al.*[36] Reprinted with permission from Elsevier, Copyright (2005).)

drawback: since the flow is mainly convective, though the membrane pores, the separation properties of the polymeric membrane are not used and the membrane acts basically as a catalyst support.

The use of PCMs consisting in commercial polyethersulfone (PES) microfiltration membranes functionalized by grafting sulfonated polystyrene in the pores (SPES, Fig. 1.6), has been reported for the esterification of acetic acid and ethanol to produce ethyl acetate.[35,36] Although the SPES membrane showed a catalytic activity comparable to that of Amberlyst-36 in experiments carried out in a batch reactor, significant improvements in apparent reaction rates were achieved using the catalytic membranes in flow-through experiments carried out in a microfiltration membrane reactor. These results were explained by the decrease of the internal mass-transfer resistance due to the achievement of convective flow in the solid phase. However, the use of a porous membrane puts pervaporation aside, since permeation takes place mainly through the membrane pores and the polymer permselectivity properties have negligible effects. Consequently, the esterification reaction remains equilibrium-limited in this reactor configuration.

The dimerization of isobutene carried out in a forced-flow polymeric catalytic membrane reactor was reported by D. Fritsch and co-workers.[37,38] The authors prepared composite porous membranes consisting of a catalytic layer made of solid acid catalysts, such as silica supported Nafion®, Nafion®NR50, Amberlyst™ 15 and silica supported tungstophosphoric acid dispersed in polymeric binders such as Teflon®AF, Hyflon®AD, polytrimethylsilylpropyne, or polydimethylsiloxane (PDMS), cast on microporous support membranes made of polyacrylonitrile (PAN) or Torlon®. The membranes were assembled in the membrane reactor into which isobutene was fed in the retentate side with a build-up pressure of 4 bar. The liquid product was collected on the permeate side.

Depending on the membrane used, high conversions of up to 98% at 22% selectivity to isooctane were achieved. The porous composite catalytic membranes provided the removal of the desired intermediate product isooctene, thus inhibiting secondary reactions to give trimer and oligomer compounds. In the membrane forced-flow reactor the by-products are purged from the catalyst's active sites before the reaction proceeds further, by the non-diffusive flow of educts and products. Therefore, catalyst deactivation by building up of higher oligomers cannot take place and no catalyst deactivation was found within operation for one week.

The selective hydrogenation of propyne to propene has been studied over porous ultrafiltration polymeric membranes of polyacrylonitrile (PAN), polyetherimide (PEI) or polyamideimide (PAI)[39] or over poly(acrylic acid) membranes,[40] treated with palladium acetate. Similarly, porous PVDF membranes, also loaded with palladium particles, were used in hydrogenation flow-through experiments of methylenecyclohexane to methylcyclohexanone.[41] The hydrogenation of viscous liquids such as vegetable oils could also be performed in flow-through experiments over porous polymeric membranes loaded with palladium or platinum particles.[42–44]

The use of a forced-flow polymeric membrane reactor has also been described for room temperature polychlorinated biphenyl (PCB) dechlorination.[45] Core/shell Fe/Pd nanoparticles were synthesized on the pore walls of PVDF microfiltration membranes functionalized with poly(acrylic acid) (PAA). PAA functionalization was achieved by *in situ* free radical polymerization of acrylic acid in the microfiltration membrane pores. Ferrous ions were then introduced into the membranes by the ion-exchange process. Subsequent reduction resulted in the *in situ* formation of 20–40 nm Fe nanoparticles. Bimetallic nanoparticles could be formed by post-deposition of Pd.

The membranes were assembled in a dead-end filtration module and the results obtained under convective flow conditions were compared with those obtained in a batch reactor. The PCB degradation rate under pressure-driven convective flow conditions benefited from narrow pore size, high intrinsic reaction rate and high diffusivity.

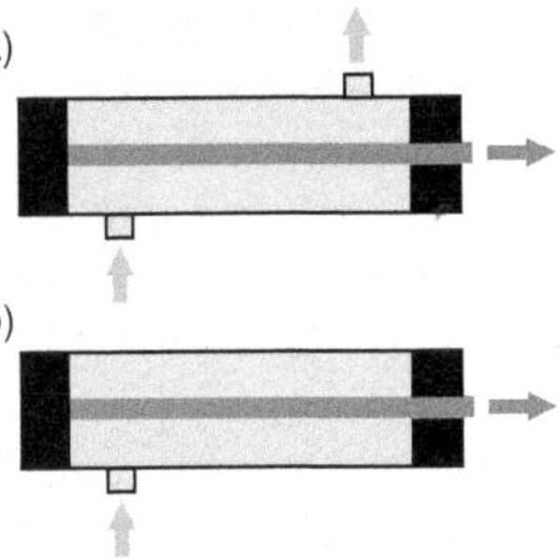

1.7 Forced-flow-type hollow fibre polymeric membrane reactor operated in the: (a) crossed-flow configuration; (b) dead-end configuration. (Adapted from Macanás *et al.*[49] Reprinted with permission from Elsevier, Copyright (2010).)

Forced-flow polymeric membrane reactors have also been successfully tested for the oxidation of benzene to phenol by Molinari and co-workers.[46–48] Mixed-matrix membranes consisting of CuO powder or CuO nanoparticles[46,47] dispersed in PVDF were prepared by the inversion phase method, by using dimethylacetamide (DMAc), dimethylformamide (DMF) or N-methyl-2-pyrrolidone (NMP) as solvents and water as non-solvent. The membranes were assembled in a ultrafiltration unit to which a solution of acetonitrile/benzene and hydrogen peroxide (H_2O_2) was fed. The best results were obtained with a PVDF membrane filled with CuO nanoparticles, with a phenol yield of 2.3% at 35°C and a contact time of 19.4 s in a single pass, in the presence of ascorbic acid.

More than twice the phenol yield (5.5%) at a much smaller contact time (3.5 s) in a single pass was obtained with a mixed-matrix membrane of Na-Fe-silicalite-1 dispersed in PVDF, which was obtained by a similar technique, by using DMAc as solvent and water as non-solvent.[48]

Porous composite polyethersulfone/poly(styrene sulfonate) hollow fibres containing palladium nanoparticles have been efficiently used in the liquid-phase reduction of nitrophenol to aminophenol by sodium borohydride in a forced-flow reactor.[49] The crossed-flow and dead-end reactor configuration were compared (Fig. 1.7).

The photocatalytic degradation of phenol and light alcohols in aqueous solution has also been carried out by using porous PVDF, PDMS or Hyflon® membranes loaded with polyoxomethalates (POMs).[50–53] The reactor was operated by feeding the aqueous solution saturated with molecular oxygen to the retentate side, as a forced-flow reactor or as a contactor, depending on the applied transmembrane pressure.[52] The best results were obtained in forced-flow operating mode. The forced-flow reactor configuration was also chosen for the photocatalytic mineralization of light alkenes, alcohols and carboxylic acids, in the liquid or gas phase, by using composite membranes (PHOTOPERM®) porous polypropylene membranes coated with a prepolymeric blend containing TiO_2.[54–56]

Distributor/contactor polymeric membrane reactors

In distributor/contactor membrane reactors the membrane can promote the controlled addition of reactants, therefore increasing the selectivity of kinetically-controlled reactions such as hydrogenations and selective oxidations, and decreasing the extension of parallel undesirable reactions.[14] In liquid-phase reactions the polymeric catalytic membranes can also be used to separate two immiscible reactants, promoting their contact in the catalytically active membrane phase and avoiding the use of solvents.[12, 14] As mentioned above, slight modifications in polymer functionalization can change the polymer hydrophilic/hydrophobic balance and, therefore, change the reactants concentrations near the catalyst active sites.[12]

The use of polymeric catalytic membranes in distributor/contactor-type reactors for hydrogenation or oxidation reactions has been widely described in several former and recent reviews.[12,14,16,57,58] Mixed-matrix membranes of PDMS filled with Pd particles,[59] composite membranes of ionic liquid-polymer gels filled with Pd/C,[60] ionic liquids containing rhodium complexes and supported in polystyrene sheets in a corrugated configuration,[61] have been used for the selective gas-phase hydrogenation of hydrocarbons in contactor-type membrane reactors.

More recently, the gas-phase hydrogenation of propene[62] and propyne[63] in contactor-type reactors assembled with polydimethylsiloxane (PDMS) membranes loaded with nanosized palladium clusters, has been reported. Since the catalytic membranes are composed of a dense polymer doped with a catalyst, the membrane acts as a catalyst support and the reaction occurs inside the polymer phase. PDMS is particularly suited for gas-phase hydrogenations carried out over dense membranes, because reactants and products are able to diffuse to and from the catalyst active sites, due to the polymer high gas permeability.

The hydrogenation of soybean oil has also been carried out in a distributer/contactor-type membrane reactor assembled with a composite catalytic polymeric membrane.[64–66] The metal/polymer composite catalytic membrane used consists of an integral-asymmetric membrane with high flux and selectivity for hydrogen and negligible permeability to vegetable oil. The polymeric membrane consists of a highly porous substructure with a thin (approximately 0.2 μm), dense and defect-free layer known as the membrane skin which results in high hydrogen selectivity and flux for defect-free membranes. Gases permeate through the thin skin by the well-known solution/diffusion mechanism while the porous substructure allows convection.

Hydrogen is supplied from the porous side of the membrane while the oil flows on the skin/platinum side.

Volkov *et al.*[67] describe the use of porous polypropylene hollow fibre membranes loaded with palladium particles on the external surface, as distributer/contactor membrane reactors for the catalytic removal of dissolved oxygen (DO) from water. Hydrophobic porous catalytic membrane serves three key functions:

1. A well-defined and easily controlled location of a gas–liquid interface.
2. Accessibility of the catalyst active sites for reactants (hydrogen and oxygen).
3. High hydrogen mass transfer.

Water containing DO flows over the outer surface of the Pd-loaded hydrophobic hollow fibre membrane, whereas hydrogen is supplied into the lumen side of hollow fibres and approaches the working surface of the catalyst through the membrane pores. Due to the catalytic activation of hydrogen adsorbed on the palladium surface, heterogeneous reduction reaction of the DO takes place, being formed water.

Espro *et al.*[68] reported the oxidation of propane to the corresponding oxygenated products (*n*-propanol, isopropanol, propionic aldehyde and acetone) mediated by the Fe^{2+}–H_2O_2 Fenton system in a distributor/contactor-type membrane reactor, under mild conditions. A composite Nafion/PEEKWC (a modified polyetheretherketone) catalytic membrane separates the gas (propane/helium) and liquid (Fe^{2+}–H_2O_2) phases. The flat and hollow fibre multi-tubular membrane reactor configurations were compared, being the best performances achieved with the hollow fibre multi-tubular configuration.

Contactor-type polymeric membrane reactors have also been used for the liquid-phase oxidation of hydrocarbons.[12,14,16,58] The polymeric membrane separates the two immiscible phases, the hydrocarbon and the oxidant reagent, usually *tert*-butylhydroperoxide or H_2O_2, acting as a 'solid solvent' common to both phases. In such a way the use of solvents is avoided with environmental and process economy gains. PDMS membranes loaded with metal phthalocyanine complexes encaged in the supercages of zeolite Y have been used for the liquid-phase oxidation of hydrocarbons such as cyclohexane or *n*-dodecane.[69–73] Contactor-type membrane reactors equipped with PDMS membranes embedding heterogeneous catalysts such as TS-1 or Ti-MCM-41 or homogenous catalysts such as TDCPP(Mn)Cl (tetrakis-[2,6-dichlorophenyl]porfyrino manganese chloride), have proven to be effective in several types of liquid-phase oxidation reactions.[74–77]

Buonomenna *et al.*[78] reported the liquid-phase oxidation of dibenzylamine to the correspondent nitrone using a polymeric catalytic membrane reactor.

The reactor was assembled with a plasma treated microporous PVDF membrane containing sodium tungstate as catalytically active species. The PVDF membranes were subjected to plasma treatments in NH_3 fed radio frequency glow discharges in order to generate amine functions on the surface of PVDF membranes prepared by a phase-inversion procedure, which are suitable for the immobilization of Na_2WO_4. The modes of reactor operation as forced-flow or contactor were compared, being the best results achieved with a transmembrane pressure $\Delta P = 0$ (contactor mode), since this mode of operation is the one that better favours the contact between the reactants and the catalytic active sites.

The use of chiral Schiff base complexes immobilized in PDMS membranes for the epoxidation of olefins,[79] has also been reported.

Contactor-type polymeric membrane reactors have been also applied to liquid-phase reactions other than hydrogenation or oxidation. The hydration of α-pinene has been carried out successfully over polymeric membranes consisting of mixed matrixes of PDMS embedded USY or beta zeolites or sulfonated activated carbon.[80,81] The membranes were assembled in a flat contactor-type reactor configuration, separating the aqueous and organic phases. Sulfonated PVA membranes were also reported to be effective in the acid catalysed methanolysis of soybean oil carried out in a flat contactor-type membrane reactor configuration.[82]

Electrochemical membrane reactors have already been the subject of several recent reviews.[14,83–85] Polymer electrolyte membranes (PEM) coated with catalytic layers have also been used in distributor/contactor membrane reactor configurations. Examples of electrochemical hydrogenation and oxidation reactions carried out in PEM membrane reactors in a distributor contactor configuration are the following: the electrochemical coupling of benzene hydrogenation and water electrolysis carried out on Nafion membranes coated with Pt-Rhfilms;[86] the electrochemical dehydrogenative oxidation of cyclohexane to benzene over Nafion membranes coated with Pt-Ni/C catalysts;[87] the total oxidation of oxalic[88] or formic[89] acids by using Nafion membranes as solid electrolyte between IrO_2/TiO_2 mesh and Pt/carbon fibre electrodes; the degradation of eosin B dye over Ti-Ru-Pt coated expanded TiO2 sheets pressed against a sulfonated poly(ether sulfone) membrane;[90] the use of perfluorosulfonic and perfluorocarboxylic polymers as PEMs in the electrochemical synthesis of chromium dioxide[91]; the reduction of nitrate in sandwiched type membrane reactors assembled with H^+ conducting Nafion membranes coated with Pt, Pd-Rh, Ag modified Pt, Pd, and Pt-Pd films;[92,93] the electrochemical hydrogenation of soybean oil over a composite membrane consisting of a Nafion membrane coated in both faces with polymer dispersed Pt-black and Pd-black powders, respectively;[94] the electrochemical hydrogenation of decene and acetone over a composite membrane consisting of a carbon cloth coated with a Nafion slurry of catalyst metal powder (Pt, Pd, Ni, Ag, Au or Cu) and pressed against a Nafion 115 sheet;[95] the reduction of carbon dioxide to formate over a membrane electrolyte assembly (MEA) consisting of a polymeric hydroxide-ion conducting membrane coated on both sides with Nafion inks of indium and lead powders, respectively, as electrodes.[96]

Electrochemical membrane reactors have also been used in fine chemistry reactions.[97] The electrochemical reduction of maleic acid to succinic acid has been reported by using a sulfonated poly(styrene divinylbenzene) membrane intercalated between copper and lead anodes, assembled in a continuous contactor-type membrane reactor.[98] Montiel *et al.*[99] have recently described the use of MEA technology for the synthesis of N-acetyl-L-cysteine by electroreduction of N,N-diacetyl-L-cystine.

1.3.2 Polymeric inert membrane reactors (PIMRs)

Inert membranes are by far the most widely used in polymeric membrane reactors. Since in PIMRs the catalytic function is absent from the membrane, in comparison with PCMRs this type of membrane reactor represents a lower level in process integration. However, because of this and because the membrane modulus can be separated from the vessel where the chemical reaction takes place, PIMRs allow a much wider range of reactor configurations than PCMRs.

Extractor-type PIMRs

Extractor-type membrane reactors and particularly pervaporation membrane reactors illustrate quite well the versatility of PIMRs. Van der Bruggen[17] distinguishes between pervaporation type reactors R1 and R2, wherein the extracted component is the main product or is a by-product, respectively. However, a variety of reactor configurations is possible for those two reactor types, as exemplified in Fig. 1.8.

The use of PIMRs in pervaporation has the advantage of the direct application of the well-established technology of the polymeric membranes used in solvent drying.[100] For R2 type reactors in which the component to be removed is usually water, hydrophilic polymers are usually used, such as PVA or chitosan. On the other hand, for R1 type reactors, in which the extracted component is the desired product, usually less polar than water, the membrane material is a hydrophobic polymer (e.g., PDMS).[14,17]

Zhu *et al.*[101] reported the esterification of acetic acid and ethanol in a tube-and-shell type reactor, using a polymer–ceramic composite membrane obtained by dip-coating a ceramic support tube with a hydrophilic polyetherimide. Reactants, along with sulfuric acid as homogeneous catalyst, were fed in the tube side while a vacuum was applied to the shell side.

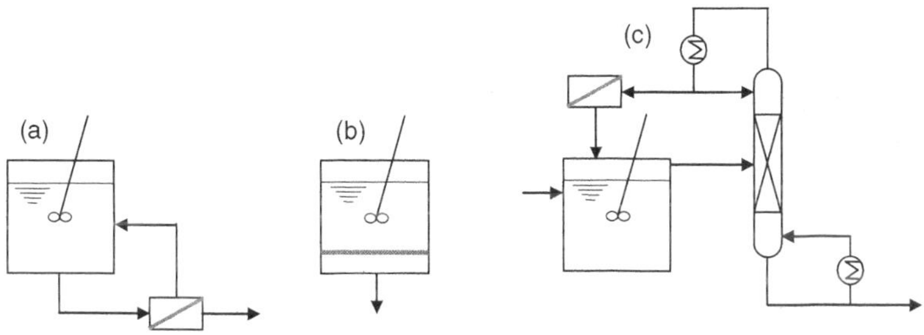

1.8 Examples of different configurations for pervaporation inert membrane reactors. (a) External membrane unit; (b) internal membrane unit; and (c) hybrid process with external membrane unit and distillation column. (Adapted from van der Bruggen.[17] Reprinted with permission from Elsevier, Copyright (2010).)

The same tube-and-shell-type reactor using a similar polymer–ceramic composite tube-membrane, was used in a hybrid configuration coupling pervaporation with adsorption, in which the shell side was loaded with $CaSO_4$ as water adsorbent, has been successfully applied to the acetic acid/ethanol esterification.[102]

A layout consisting of a semibatch tank reactor, loaded with a sulfonic acid functionalized poly(styrene divinylbenzene) copolymer (Amberlyst) as catalyst and connected to an external pervaporation module equipped with a commercial PVA membrane, has been used for esterification studies of acetic acid/isopropanol[103,104] and lactic acid/ethanol[105,106] systems. A similar arrangement was reported by Lauterbach and Kreis[107] for the propionic acid/propanol esterification.

Buchaly *et al.*[108] propose a hybrid process combining reactive distillation and pervaporation for the propionic acid/propanol esterification. The membrane module equipped with a commercial PVA membrane is located in the distillate stream in order to selectively remove the produced water. The desired product, *n*-propyl propionate is removed at the bottom of the distillation column, which is packed with Amberlyst 46 as catalyst, in a reactive zone.

An example of the use of extractor-type PIMRs in reactions other than esterification is the gas-phase decomposition of MTBE catalysed by tungstophosphoric acid. Lee *et al.*[109] reported the use of closed-loop recycle membrane reactors by using polycarbonate, polyarylate or cellulose acetate membranes to selectively permeate the formed methanol in a flat membrane reactor configuration, with the catalyst packed in the retentate side and by using helium as sweep gas in the permeate side. The authors also used a tube-and-shell reactor configuration with the catalyst packed in the shell side being the sweep gas fed to the tube side.

The same reaction was also performed in a tube-and-shell reactor configuration with polymer membranes of polyphenylene oxide, polysulfone, or cellulose acetate coated on a porous alumina tube packed with silica supported tungstophosphoric acid.[110]

Distributor/contactor-type PIMRs

Contrary to the extractor-type PIMRs, literature on distributor/contactor-type PIMRs is relatively scarce.

Buonomenna and Drioli[111] reported the liquid-phase selective oxidation of benzyl alcohol to benzaldehyde using a membrane contactor unit in a flat membrane configuration. Microporous hydrophobic polyvinylidene fluoride membranes are used to separate and to promote the contact between the aqueous phase containing H_2O_2 and ammonium molybdate or sodium tungstate as catalyst and the organic phase containing the substrate and the product.

The liquid-phase one-step production of phenol by direct hydroxylation of benzene has been carried out in a contactor-type membrane reactor, in which an inert polymeric membrane separates an aqueous phase consisting of H_2O_2 aqueous solution and metal salts as catalysts and an organic phase composed by benzene and the reaction products. Hydrophobic polypropylene and polytetrafluoroethylene or hydrophilic polyacrylonitrile porous membranes were tested,[112] achieving the best results with the polypropylene membrane as a barrier between the two immiscible phases. High selectivities to phenol were also obtained in the same reaction system, but using polydimethylsiloxane membranes as barrier.[113] The direct oxidation of benzene to phenol was also tested in a photocatalytic membrane contactor by using the same reactor composed of two chambers separated by a porous polypropylene membrane. In this case, the aqueous phase containing TiO_2 particles as catalyst is circulated between the membrane cell chamber and the reaction vessel subjected to UV irradiation.[114]

In inert polymeric membrane reactors, the membrane's role can be just the retention of the catalyst in the reactor volume, by using a membrane with the appropriate molecular weight cutoff.[14,25] In this configuration the catalyst can be heterogeneous, used as a powder suspension, as in the experiments of water purification by photocatalytic degradation of dyes[115] or pharmaceuticals,[116] in which nano-filtration membranes, acting at a molecular level, retain not only the catalyst but also a substantial part of the organic pollutants. But the catalyst can also be homogeneous, used in enlarged forms such as dendrimers or soluble hyperbranched polymers or bound to soluble polymers[14,25] as illustrated by the oxidation of alcohols catalysed by homogeneous polyurethane-supported 2,2,6,6-tetramethylpiperidine-1-oxyl (TEMPO) in a nano-filtration membrane reactor, by using a solvent-resistant nano-filtration membrane.[117]

1.4 Modelling of polymeric catalytic membrane reactors

Modelling and simulation are widely considered essential tools in many areas of science and engineering and they have proven to be invaluable in the study of simple or complex systems. These tools are important for a deep understanding of systems behaviour, for scale-up of lab-scale or pilot plants, optimization and control purposes. Modelling can also be important to conduct 'virtual experiments', used to propose and study new processes/ products at lower costs.

The modelling of CMRs is extensively reported in the literature. However, most of such works concern metallic or ceramic membranes. Only a few papers consider the modelling of membrane reactors with polymeric membranes, more specifically catalytic polymeric membranes.

Because of their intrinsic characteristics, mostly dense rubbery polymers have been considered for the preparation of polymeric catalytic membranes. The mass-transport mechanism considered has been the well-known sorption–diffusion model.[118] Modelling the kinetics of the reaction(s) occurring at the occluded catalyst level is a much more complex task. The reaction may be carried out under special operating conditions, for example in a batch reactor where the catalyst is dispersed in a support[62,63] or directly inside the catalytic membrane,[81,119–121] a reaction-rate model is assumed and the related parameters are determined by fitting the global model to the experimental data. In other cases, the kinetic models determined by other authors are used.[77,122] In some theoretical studies,[123–131] a hypothetical reaction-rate model and the respective model parameters are assumed.

The following discussion will focus exclusively on membrane reactors with polymeric catalytic dense membranes. Cases of membrane reactors with a polymeric non-catalytic membrane are not presented and discussed, as the membrane only performs some separation task and the models to be considered are independent of the type of membrane, considering that an adequate transport equation is considered. The cases of porous polymeric membranes with catalyst supported in the porous walls, typical in the biomembrane reactors, photocatalysis, amongst others, are also not considered here.

1.4.1 Membrane reactors with liquid/liquid or liquid/ vapour phases

Some of the modelling studies reported in the literature are conducted in liquid phase. Yawalkar *et al.*,[122] for example, studied the epoxidation of alkenes in the presence of peroxide to epoxides in liquid phase. The CMR consisted of two well-stirred chambers separated by the membrane, operated in a segregated mode (interfacial contact), that is, the reactants were fed separately to each side of the membrane. These conditions guarantee a homogeneous concentration of reactants and reaction products in both bulk organic and aqueous phases and no concentration polarization. To analyse the membrane reactor performance as a function of, namely, reactants concentration in the liquid phase and respective sorption coefficients and the catalyst particles loading and size, the authors developed a one-dimensional steady-state mathematical model, subjected to some main assumptions, namely:

- Homogeneous distribution of the cubic zeolite catalyst particles, all of the same size, in the homogeneous polymeric phase.
- Sorption coefficients for the peroxide and organic species independent of each other and described by the Henry's law, either for the interface liquid phase/membrane surface or polymer phase/catalyst surface.
- Diffusion of the reactant peroxide in both polymeric phase and catalyst particles described by Fick's law.
- The reaction scheme considered consisted of the main epoxidation (reaction between the peroxide and the alkene to give epoxide, considered of pseudo first-order relatively to the epoxide species) and the parallel undesirable peroxide decomposition.

Based on the model predictions, the authors concluded that the organophillic polymeric membrane with the built-in zeolite particles significantly limits the peroxide concentration at the catalyst surface, minimizing thus its decomposition and the deactivation of the catalyst. The authors concluded also that it would be possible to maximize the epoxide production and minimize the epoxide parallel decomposition by a proper selection of the model parameters.

The model by Yawalkar *et al.*[122] was extended by Nagy[123] some years later. The same geometrical distribution of the catalyst in the polymer phase and the same sorption–diffusion model for the mass transport, either in the polymer or in the catalyst, was considered. However, that author considered an irreversible first-order reaction in his model. From the study of the catalytic particles size and distribution, membrane thickness, membrane and catalyst diffusion coefficients, among other variables, the author concluded that the mass-transfer rate depends significantly on the size of the catalytic particles and the thickness of the membrane fraction between the surface and the first particles layer and, also, on the usually low diffusion coefficient through the catalytic particles.

Another catalytic polymeric membrane reactor operating in interfacial contact mode, but now for conducting the liquid/vapour phase oxyfunctionalization of *n*-hexane with H_2O_2 producing a mixture of hexanones and hexanols, was studied by Kaliaguine and co-workers.[77] The catalytic membrane was a PDMS polymer built-in with titanium silicalite zeoliteTS-1. Considering some main assumptions, namely isothermal conditions, Fickian transport with constant and independent diffusion coefficients and liquid phase/membrane surface equilibrium described by Henry's law, among others, the authors developed a steady-state model. The simulated results were shown to fit quite well to the experimental average formation rates of the oxygenate species. Simulation results showed also the strong effect of the diffusivity of the reactant H_2O_2 through the membrane on the rates of formation of oxygenates (the reaction was carried on in the presence of an excess of *n*-hexane). Both experimental and simulation results showed the suitability of using a catalytic membrane as interphase contactor in a biphasic reaction like the one studied. Despite the capacity of the system to produce hexanols and hexanones, it was also effective in the separation of these products from the organic feed.

Vital and co-workers developed also simple isothermal transient models to describe the hydration of α-pinene into α-terpineol and a series of other products.[81,120] either using zeolite USY 750 as catalyst immobilized in a PDMS polymeric membrane[81] or using molybdophosphoric acid as catalyst embedded in hydrophobic PVA membranes modified with acetic anhydride. The reaction was carried out in a membrane reactor working with total recycle (batch reactor), the reactant solution being a mixture of water and α-pinene dissolved in acetone. The authors developed their models based on some main assumptions, namely liquid phase/membrane surface equilibrium described by Henry's law and Fickian transport, with the diffusion coefficients of α-pinene and water depending on the α-terpineol concentration. The models were transient for the mass balances concerning the bulk solutions, but pseudo steady-state for diffusion and reaction inside the membrane. The reaction scheme was considered parallel for both studies. The secondary reaction was described by a first-order elementary equation in the reactant α-pinene.[81,120] The main reaction was described by a pseudo first-order elementary equation in the reactant α-pinene[120] and by a second-order elementary equation in the reactants' water and α-pinene.[81] These models predicted the increase of the reactants' permeability as a function of the catalyst loading and was proved to describe reasonably well the evolution of the reactants and reaction products average concentration in the liquid phase, as well as the reaction selectivity to the product α-terpineol.

Other authors studied the dehydrogenation of cyclohexane to benzene[119] and the *cis* to *trans* isomerization of piperylene.[121] Both reactions were carried out in a vapour phase batch reactor at low temperature with a catalytic polymeric membrane, a composite polyethylacrylate polymer built-in with zeolite 13X containing Ni or Ti[119] or Co[121] as catalyst. The reactor was fed with pure reactants in both cases. The authors developed a simple transient diffusion-reaction model, considering the reactions described by pseudo first-order kinetic equations. The necessary diffusion coefficients of the reactants in the membranes were obtained experimentally by the time-lag method. The authors were thus able to calculate the activation energies and reaction-rate constants in both studies, and showed that the kinetics was controlled by the surface diffusion.

1.4.2 Gas-phase membrane reactors

Mendes and co-workers have developed some modelling studies concerning dense PCMRs, both experimental[62,63] and theoretical.[124–131]

Some of these studies focused on the analysis of equilibrium-limited reactions, namely those in which the conditions of the respective conversion could be enhanced relatively to the value obtained in a 'conventional' reactor, the so-called thermodynamic equilibrium conversion.[124–130] The developed models considered generic equilibrium-limited reactions carried on in membrane reactors with perfectly mixed[124–126] or plug-flow patterns.[127–130] In all these studies, the main assumptions considered consisted in isothermal and steady-state operation, Fickian transport across a non-porous membrane with a homogeneously distributed nanosized catalyst with constant diffusion coefficients, Henry's law for describing the equilibrium condition at the interfaces membrane/gas, and equality of local concentrations at the interface polymer phase/catalyst surface.

Based on simulation results,[124–130] the authors concluded that it is possible to obtain conversions higher than the thermodynamic equilibrium value, for a given set of operating conditions related with the pressure difference between the retentate and permeate sides, Thiele modulus (ratio between a characteristic intramembrane diffusion time and a characteristic direct reaction time), and contact time (ratio between the maximum possible flux across the membrane for a reference component and the total molar feed flow rate) parameters. For a given stoichiometry, the authors found a possible conversion enhancement when, globally, the diffusion coefficients of the reaction products were higher than the ones of the reactants, or vice versa for the sorption coefficients (that is, the ones for the reactants higher than the ones for the reaction products). Also, it was found that an enhancement or a detriment of the conversion could be obtained based solely on the pressure difference between the two sides of the membrane.

In another set of studies,[62,63,131] this group studied the potentialities of the dense PCMRs for conducting consecutive-parallel reactions. In a theoretical study,[131] considering non-isothermal conditions and perfectly mixed flow pattern, beyond other main assumptions already assumed in previous works and referred above, the authors analysed the hydrogenation *propyne→propene→propane* in order to define in which conditions the propyne concentration in the outlet stream could be lowered to less than the levels obtained using a conventional catalytic reactor in a more efficient way, taking advantage of the flexibility provided by the effective diffusivity and sorption selectivities of the reaction species. From a set of simulation results, the authors analysed the concentration of propyne at the permeate stream, the selectivity and overall yield to the desirable intermediate product propene, and the conversion of the main reactants propyne and hydrogen along the parametric space describing the model.

Considering values reported in the literature for some of the variables, or defining values for others according to a qualitative knowledge of the reactions and catalytic membrane, and limiting the range of some of the parameters to interval values suitable for the system considered in the study, the authors concluded that:[131] considering a higher sorption and diffusion coefficients of hydrogen relative to those of hydrocarbons, the conversion of propyne and hydrogen is always enhanced relatively to the value obtained in a 'traditional' reactor, though in an extension depending on the Thiele modulus value; additionally, the key variable (concentration of propyne in the permeate stream) is also enhanced (lower concentration) in the parametric region analysed; on the other hand, selectivity and overall yield to the intermediate product are penalized or favoured depending on the Thiele modulus values region considered. Assuming now the lower sorption and higher diffusion coefficients of hydrogen relative to those of hydrocarbons, the performance of the reactor decreases globally: propyne and hydrogen conversions decrease relatively to the values attained in a 'traditional' reactor, in an extension depending on the Thiele modulus value. Also the concentration of propyne in the permeate stream is penalized in the parametric region analysed, in an extension depending on the Thiele modulus value; on the other hand, selectivity turned out to be favoured in the entire region of the Thiele modulus values, especially for the lower ones, while the overall yield to the intermediate product was enhanced only for an intermediate Thiele modulus region.

In a later work, the authors performed an experimental and simulation work studying the same reaction system carried out in a catalytic membrane reactor using a PDMS membrane occluded with nanoclusters of palladium,[62,63] with some of the variables used in the model obtained from independent experiments (for the transport[132] and for the kinetics[133,134]). With this model, the authors were able to describe reasonably well the experimental results obtained at 35°C, obtaining a good fitting.

1.5 Conclusions

Although polymeric membrane reactors can only be used with reactions taking place in a relatively low temperature range, the increasing variety of available polymer materials offers a large range of selection possibilities in terms of mechanical, chemical and temperature resistance, as well as in terms of sorption, transport and catalytic properties, when compared with their inorganic counterparts.

In PIMRs the polymer membrane may be used integrated in the chemical reactor or in a separated modulus, making the membrane reactor extremely versatile. In fact, the variety of possible reactor configurations allows not only the separate optimization of the membrane transport properties and the catalyst activity and selectivity, but also the integration with other separation operations such as distillation and adsorption.

Even when the inert membrane is used just for catalyst retention in ultra- or nano-filtration membrane reactors, the fine control of membrane porosity is a strong advantage of polymeric membranes when compared with inorganic ones.

PCMRs by combining, in the membrane phase, the catalytic and separation functions represent, when compared with PIMRs, a higher level of process integration. In extractor-type polymeric membrane reactors the membrane's transport properties can be easily changed by changing polymer's cross-linking. In contactor-type membrane reactors, polymeric membranes offer the possibility of fine-tuning the reactants' concentration in the close vicinity of the catalyst active sites by slight changes in the polymer structure, affecting its hydrophilic/hydrophobic balance. When the membrane's separation properties are not a key issue, a porous polymeric membrane can be used in a forced-flow reactor configuration. The pore walls can be functionalized in such a way that each membrane's pore behaves like a nanoreactor. The adjustment of the membrane's pore size allows the fine control of the reactants residence time. High conversions can be achieved due to the small resistance to mass transfer.

In spite of the growing research effort, with the exception of fuel cells, there are only a few examples of industrial applications of non-biocatalytic polymeric membrane reactors, such as the Remedia Catalytic Filter System for the destruction of dioxins and furans from industrial combustion sources or pervaporation-assisted esterification processes.[12] More research is required in order to find long-lasting high-performance and cheap polymeric materials and catalysts that can effectively compete with the traditional processes. On pursuing this quest, mathematical modelling and simulation are fundamental tools for the better understanding of membranes' behaviour and optimization.

1.6 References

1. Ulbricht M and Susanto H (2009), 'Polymeric membranes for molecular separations', in Drioli E and Giorno L, *Membrane Operations, Innovative Separations and Transformations*, Weinheim, Wiley-VCH, 19–43.

2. George S C and Thomas S (2001), 'Transport phenomena through polymeric systems', *Prog Polym Sci*, **26**, 985–1017.
3. Carraher C E Jr (2003), *Seymour/Carraher's Polymer Chemistry*, New York, Basel, Marcel Dekker.
4. Nagel C, Gunther-Schade K, Fritsch D, Strunskus T and Faupel F (2002) 'Free volume and transport properties in highly selective polymer membranes', *Macromolecules*, **35**, 2071–2077.
5. Marcano J G S and Tsotsis T T (2002), *Catalytic Membranes and Membrane Reactors*, Weinheim, Wiley-VCH.
6. Paul D R (2010), 'Fundamentals of transport phenomena in polymer membranes', in Drioli E and Giorno L, *Comprehensive Membrane Science and Engineering, Vol. 1, Basic Aspects of Membrane Science and Engineering*, Kidlington, Elsevier, 75–89.
7. Bruggen B (2009), 'Fundamentals of membrane solvent separation', in Drioli E and Giorno L, *Membrane Operations, Innovative Separations and Transformations*, Weinheim, Wiley-VCH, 45–61.
8. Ladhe A R, Xu J, Hollman A M, Smuleac V and Bhattacharyya D. (2010), 'Functionalized membranes for sorption, separation, and reaction: an overview', in Drioli E and Giorno L, *Comprehensive Membrane Science and Engineering, Vol. 1, Basic Aspects of Membrane Science and Engineering*, Kidlington, Elsevier, 13–26.
9. Strathmann H, Giorno L and Drioli E (2010), 'Basic aspects in polymeric membrane preparation', in Drioli E and Giorno L, *Comprehensive Membrane Science and Engineering, Vol. 1, Basic Aspects of Membrane Science and Engineering*, Kidlington, Elsevier, 91–111.
10. Ulbricht M (2006), 'Advanced functional polymer membranes', *Polymer*, **47**, 2217–2262.
11. Kickelbick G (2003), 'Concepts for the incorporation of inorganic building blocks into organic polymers on a nanoscale', *Prog Polym Sci*, **28**, 83–114.
12. Vankelecom I (2002), 'Polymeric membranes in catalytic reactors', *Chem Rev*, **102**, 3779–3810.
13. Barbieri G and Scura F (2009), 'Fundamental of chemical membrane reactors', in Drioli E and Giorno L, *Membrane Operations, Innovative Separations and Transformations*, Weinheim, Wiley-VCH, 287–308.
14. Fontananova E and Drioli E (2010), 'Catalytic membranes and membrane reactors', in Drioli E and Giorno L, *Comprehensive Membrane Science and Engineering, Vol. 3, Chemical and Biochemical Transformations in Membrane Systems*, Kidlington, Elsevier, 110–131.
15. Caro J (2008), 'Catalysis in micro-structured membrane reactors with nano-designed membranes', *Chin J Catal*, **29**, 1169–1177.
16. Ozdemir S S, Buonomenna M G and Drioli E (2006), 'Catalytic polymeric membranes: preparation and application', *Appl Catal A-Gen*, **307**, 167–183.
17. Bruggen B vander (2010), 'Pervaporation membrane reactors', in Drioli E and Giorno L, *Comprehensive Membrane Science and Engineering, Vol. 3, Basic Aspects of Membrane Reactors*, Kidlington, Elsevier, 135–162.
18. David M O, Nguyen Q T and Neel J (1992), 'Pervaporation membranes endowed with catalytic properties, based on polymer blends', *J Membr Sci*, **73**, 129–141.
19. Liu Q L, Jia P S and Chen H F (1999), 'Study on catalytic membranes of $H_3PW_{12}O_{40}$ entrapped in PVA', *J Membr Sci*, **159**, 233–241.

20. Bagnell L, Cavell K J, Hodges A M, Mau A W H and Seen A J (1993), 'The use of catalytically active pervaporation membranes in esterification reactions to simultaneously increase product yield, membrane permselectivity and flux', *J Membr Sci*, **85**, 291–299.
21. Zhu Y S and Chen H F (1998), 'Pervaporation separation and pervaporation-esterification coupling using crosslinked PVA composite catalytic membranes on porous ceramic plate', *J Memb Sci*, **138**, 123–134.
22. Figueiredo K, Salim V and Borges C (2008), 'Synthesis and characterization of a catalytic membrane for pervaporation-assisted esterification reactors', *Catal Today*, **133**, 809–814.
23. Castanheiro J E, Ramos A M, Fonseca I M and Vital J (2006), 'Esterification of acetic acid by isoamylic alcohol over catalytic membranes of poly(vinyl alcohol) containing sulfonic acid groups', *Appl Catal A-Gen*, **311**, 17–23.
24. Abetz V, Brinkmann T, Dijkstra M, Ebert K, Fritsch D, Ohlrogge K, Paul D, Peinemann K-V, Nunes S, Scharnagl N and Schossig M I (2006), 'Developments in membrane research: from material via process design to industrial application', *Adv Eng Mater*, **8**, 328–358.
25. Buonomenna M G, Choi S H and Drioli E (2010), 'Catalysis in polymeric membrane reactors: the membrane role', *Asia-Pac J Chem Eng*, **5**, 26–34.
26. Choi J S, Song I K and Lee W Y (2000), 'A composite catalytic polymer membrane reactor using heteropolyacid blended polymer membrane', *Korean J Chem Eng*, **17**, 280–283.
27. Choi J S, Song I K and Lee W Y (2001), 'Performance of shell and tube-type membrane reactors equipped with heteropolyacid-polymer composite catalytic membranes', *Catal Today*, **67**, 237–245.
28. Bengtson G, Scheel H, Theis J and Fritsch D (2002), 'Catalytic membrane reactor to simultaneously concentrate and react organics', *Chem Eng J*, **85**, 303–311.
29. Bengtson G, Oehring M and Fritsch D (2004), 'Improved dense catalytically active polymer membranes of different configuration to separate and react organics simultaneously by pervaporation', *Chem Eng Process*, **43**, 1159–1170.
30. Bengtson G, Panek D and Fritsch D (2007), 'Hydrogenation of acetophenone in a pervaporative catalytic membrane reactor with online mass spectrometric monitoring', *J Memb Sci*, **293**, 29–35.
31. Uragami T and Nishikawa M (2010), 'Syntheses of esters through poly(styrene sulfonic acid)/poly(vinyl alcohol) membrane reactor', *Asia-Pac J Chem Eng*, **5**, 3–11.
32. Wu H S and Wu Y K (2005), 'Preliminary study on the characterization and preparation of quaternary ammonium membrane', *Ind Eng Chem Res*, **44**, 1757–1763.
33. Wu H S and Li C C (2008), 'Kinetic study of phenol recovery using phase-transfer catalysis in horizontal membrane reactor', *Chem Eng J*, **144**, 502–508.
34. Wu H S and Wu Y K (2008), 'Kinetics of allylation of phenol using quaternary ammonium membranes in a membrane reactor', *J Chin Inst Chem Eng*, **39**, 29–35.
35. Shah T N and Ritchie SMC (2005), 'Esterification catalysis using functionalized membranes', *Appl Catal A-Gen*, **296**, 12–20.
36. Shah N T, Goodwin J C and Ritchie SMC (2005), 'Preparation and electrochemical characterizations of cation-exchange membranes with different functional groups', *J Membr Sci*, **251**, 81–89.

37. Randjelovic I, Bengtson G and Fritsch D (2002), 'Catalytic membranes: alkene dimerisation by means of acidic porous thin-film composite membranes', *Desalination*, **144**, 417–418.
38. Fritsch D, Randjelovic I and Keil F (2004), 'Application of a forced-flow catalytic membrane reactor for the dimerisation of isobutene', *Catal Today*, **98**, 295–308.
39. Ziegler S, Theis J and Fritsch D (2001), 'Palladium modified porous polymeric membranes and their performance in selective hydrogenation of propyne', *J Membr Sci*, **187**, 71–84.
40. Groschel L, Haidar R, Beyer A, Colfen H, Frank B and Schomacker R (2005), 'Hydrogenation of propyne in palladium-containing polyacrylic acid membranes and its characterization', *Ind Eng Chem Res*, **44**, 9064–9070.
41. Bottino A, Capannelli G, Comite A and Felice R Di (2002), 'Polymeric and ceramic membranes in three-phase catalytic membrane reactors for the hydrogenation of methylene cyclohexane', *Desalination*, **144**, 411–416.
42. Fritsch D and Bengtson G (2006), 'Developments in membrane research: from material via process design to industrial application', *Adv Eng Mater*, **8**, 386–389.
43. Fritsch D and Bengtson G (2006), 'Development of catalytically reactive porous membranes for the selective hydrogenation of sunflower oil', *Catal Today*, **118**, 121–127.
44. Bengtson G and Fritsch D. (2006), 'Catalytic membrane reactor for the selective hydrogenation of edible oil: platinum versus palladium catalyst', *Desalination*, **200**, 666–667.
45. Xu J and Bhattacharyya D (2008), 'Modeling of Fe/Pd nanoparticle-based functionalized membrane reactor for PCB dechlorination at room temperature', *J Phys Chem C*, **112**, 9133–9144.
46. Molinari R and Poerio T (2009), 'Preparation, characterisation and testing of catalytic polymeric membranes in the oxidation of benzene to phenol', *Appl Catal A-Gen*, **358**, 119–128.
47. Molinari R, Poerio T and Argurio P (2009), 'Liquid-phase oxidation of benzene to phenol using CuO catalytic polymeric membranes', *Desalination*, **241**, 22–28.
48. Molinari R, Poerio T, Granato T and Katovic A (2010), 'Fe-zeolites filled in PVDF membranes in the selective oxidation of benzene to phenol', *Microporous Mesoporous Mat*, **129**, 136–143.
49. Macanás J, Ouyang L, Bruening M L, Muñoz M, Remigy J C and Lahitte J F (2010), 'Development of polymeric hollow fiber membranes containing catalytic metal nanoparticles', *Catal Today*, **156**, 181–186.
50. Bonchio M, Carraro M, Scorrano G, Fontananova E and Drioli E (2003), 'Heterogeneous photooxidation of alcohols in water by photocatalytic membranes incorporating decatungstate', *Adv Synth Catal*, **345**, 1119–1126.
51. Bonchio M, Carraro M, Gardan M, Scorrano G, Drioli E and Fontananova E (2006), 'Hybrid photocatalytic membranes embedding decatungstate for heterogeneous photooxygenation', *Top Catal*, **40**, 133–140.
52. Fontananova E, Donato L, Drioli E, Lopez L C, Favia P and d'Agostino R (2006), 'Heterogenization of polyoxometalates on the surface of plasma-modified polymeric membranes', *Chem Mater*, **18**, 1561–1568.
53. Drioli E, Fontananova E, Bonchio M, Carraro M, Gardan M and Scorrano G (2008), 'Catalytic membranes and membrane reactors: an integrated approach to catalytic process with a high efficiency and a low environmental impact', *Chin J Catal*, **29**, 1152–1158.

54. Bellobono I R, Martini G de and Tozzi P M (2006), 'Modelling of quantum yields in photocatalytic membrane reactors immobilising titanium dioxide', *Int J Photoenergy*, **2006**, 26870.
55. Bellobono I R, Stanescu R, Costache C, Canevali C, Morazzoni F, Scotti R, Bianchi R, Mangone E S, Martini G de and Tozzi P M (2006), 'Laboratory-scale photomineralization of n-alkanes in gaseous phase by photocatalytic membranes immobilizing titanium dioxide', *Int J Photoenergy*, **2006**, 73167.
56. Bellobono I R, Rossi M, Testino A, Morazzoni F, Bianchi R, Martini G de, Tozzi P M, Stanescu R, Costache C, Bobirica L, Bonardi M L and Groppi F (2008), 'Influence of irradiance, flow rate, reactor geometry, and photo promoter concentration in mineralization kinetics of methane in air and in aqueous solutions by photocatalytic membranes immobilizing titanium dioxide', *Int J Photoenergy*, **2008**, 283741.
57. Mendes A M, Magalhaes F D and Costa C A V (2006), 'New trends on membrane science', in Conner W C and Fraissard J, *Fluid Transport in Nanoporous Materials*, NATO Science Series, Vol. 219, Dordrecht, Springer, 439–479.
58. Dioos B M L, Vankelecom I and Jacobs P (2006), 'Aspects of immobilisation of catalysts on polymeric supports', *Adv Synth Catal*, **348**, 1413–1446.
59. Gryaznov V M, Smirnov V S, Vdovin V M, Ermilova M M, Gogua L D, Pritula N A and Fedorova G K (1983) *Membrane catalyst for hydrogenation of organic-compounds and method for preparing same*, US Patent 4,394,294.
60. Carlin R T and Fuller J (1997), 'Ionic liquid-polymer gel catalytic membrane', *Chem Commun*, 1345–1346.
61. Scott K, Basov N, Jachuck R J J, Winterton N, Cooper A and Davies C (2005), 'Reactor studies of supported ionic liquids – Rhodium-catalysed hydrogenation of propene', *Chem Eng Res Des*, **83**, 1179–1185.
62. Brandao L, Fritsch D, Mendes A M and Madeira L M (2007), 'Propylene hydrogenation in a continuous polymeric catalytic membrane reactor', *Ind Eng Chem Res*, **46**, 5278–5285.
63. Brandao L, Madeira L M and Mendes A M (2007), 'Propyne hydrogenation in a continuous polymeric catalytic membrane reactor', *Chem Eng Sci*, **62**, 6768–6776.
64. Singh D, Rezac M E and Pfromm P H (2009), 'Partial hydrogenation of soybean oil with minimal trans fat production using a Pt-decorated polymeric membrane reactor', *J Am Oil Chem Soc*, **86**, 93–101.
65. Singh D, Rezac M E and Pfromm P H (2010), 'Partial hydrogenation of soybean oil using metal-decorated integral-asymmetric polymer membranes: Effects of morphology and membrane properties', *J Memb Sci*, **348**, 99–108.
66. Singh D, Pfromm P H and Rezac M E (2011), 'Overcoming mass-transfer limitations in partial hydrogenation of soybean oil using metal-decorated polymeric membranes', *AIChE J*, **57**, 2450–2457.
67. Volkov V V, Lebedeva V I, Petrova I V, Bobyl A V, Konnikov S G, Roldughin V I, Erkel J van and Tereshchenko G F (2011), 'Adlayers of palladium particles and their aggregates on porous polypropylene hollow fiber membranes as hydrogenization contactors/reactors', *Adv Colloid Interface Sci*, **164**, 144–155.
68. Espro C, Arena F, Tasselli F, Regina A, Drioli E and Parmaliana A (2006), 'Selective oxidation of propane on Nafion/PEEK-WC catalytic membranes in a multifunctional reaction system', *Catal Today*, **118**, 253–258.

69. Vankelecom I and Jacobs P (2000), 'Selective hydrocarbon oxidation using a liquid phasecatalytic membrane reactor', *Catal Today*, **56**, 131–135.
70. Parton R, Vankelecom I, Casselman M, Bezoukhanova C, Uytterhoeven J and Jacobs P (1994), 'An efficient mimic of cytochrome P-450 from a zeolite-encaged iron complex in a polymer membrane', *Nature*, **370**, 541–544.
71. Vankelecom I, Parton R, Casselman M, Uytterhoeven J and Jacobs P (1996), 'Oxidation of cyclohexane using FePcY-zeozymes occluded in polydimethylsiloxane membranes', *J Catal*, **163**, 457–464.
72. Langhendries G and Baron G V (1998), 'Mass transfer in composite polymer-zeolite catalytic membranes', *J Membr Sci*, **141**, 265–275.
73. Langhendries G, Baron G V, Vankelecom I, Parton R and Jacobs P (2000), 'Selective hydrocarbon oxidation using a liquid phase catalytic membrane reactor', *Catal Today*, **56**, 131–135.
74. Wu S, Bouchard C and Kaliaguine S (1998), 'Zeolite containing catalytic membranes as interphasecontactors', *Res Chem Intermed*, **24**, 273–289.
75. Vankelecom I, Vercruysse K, Neys P, Tas D, Janssen K, Knops-Gerrits P and Jacobs P (1998), 'Novel catalytic membranes for selective reactions', *Top Catal*, **5**, 125–132.
76. Vankelecom I and Jacobs P (2000), 'Dense organic catalytic membranes for fine chemical synthesis', *Catal Today*, **56**, 147–157.
77. Wu S, Gallot J -E, Bousmina M, Bouchard C and Kaliaguine S (2000), 'Zeolite containing catalytic membranes as interphase contactors', *Catal Today*, **56**, 113–129.
78. Buonomenna M G, Lopez L C, Barbieri G, Favia P, d'Agostino R and Drioli E (2007), 'Sodium tungstate immobilized on plasma-treated PVDF membranes: New efficient heterogeneous catalyst for oxidation of secondary amines to nitrones', *J Mol CatalA-Chem*, **273**, 32–38.
79. Vankelecom I, Tas D, Parton R, Vyver V van de and Jacobs P (1996), 'Chiral catalytic membranes', *Angw Chem Int Ed Engl*, **35**, 1346–1348.
80. Vital J, Ramos A M, Silva I F, Valente H and Castanheiro J E (2000), 'Hydration of α-pinene over zeolites and activated carbons dispersed in polymeric membranes', *Catal Today*, **56**, 167–172.
81. Vital J, Ramos A M, Silva I F and Castanheiro J E (2001), 'The effect of α-terpineol on the hydration of α-pinene over zeolites dispersed in polymeric membranes', *Catal Today*, **67**, 217–223.
82. Guerreiro L, Castanheiro J E, Fonseca I M, Martin-Aranda R M, Ramos A M and Vital J (2006), 'Transesterification of soybean oil over sulfonic acid functionalised polymeric membranes, *Catal Today*, **118**, 166–171.
83. Sundmacher K, Rihko-Struckmann L K and Galvita V (2005),'Solid electrolyte membrane reactors: Status and trends', *Catal Today*, **104**, 185–199.
84. Rihko-Struckmann L K, Munder B, Chalakov L and Sundmacher K (2010), 'Solid electrolyte membrane reactors', in Seidel-Morgenstern A, *Membrane Reactors, Distributing Reactants to Improve Selectivity and Yield*, Weinheim, Wiley-VCH Verlag, 193–233.
85. Alberti G and Casciola M (2010), 'Basic aspects in proton-conducting membranes', in Drioli E and Giorno L, *Comprehensive Membrane Science and Engineering, Vol. 2, Membrane Operations in Molecular Separations*, Kidlington, Elsevier, 433–464.

86. Itoh N, Xu WC, Hara S and Sakaki K (2000), 'Electrochemical coupling of benzene hydrogenation and water electrolysis', *Catal Today*, **56**, 307–314.
87. Kim H J, Choi S M, Nam S H, Seo M H and Kim W B (2009), 'Carbon-supported PtNi catalysts for electrooxidation of cyclohexane to benzene over polymer electrolyte fuel cells', *Catal Today*, **146**, 9–14.
88. Onder E, Koparal A S and Ogutveren U B (2009), 'Electrochemical treatment of aqueous oxalic acid solution by using solid polymer electrolyte (SPE) reactor', *Chem Eng J*, **147**, 122–129.
89. Kilic E O, Koparal A S and Ogutveren U B (2009), 'Hydrogen production by electrochemical decomposition of formic acid via solid polymer electrolyte', *Fuel Process Technol*, **90**, 158–163.
90. Prakash S, Rajesh A M and Shahi V K (2011), 'Chlorine-tolerant poly electrolyte membrane for electrochemical dye degradation', *Chem Eng J*, **168**, 108–114.
91. Li C, Qi T, Wang F, Zhang Y and Yu Z (2006), 'Variation of cell voltage with reaction time in electrochemical synthesis process of sodium dichromate', *Chem Eng Technol*, **29**, 481–486.
92. Hasnat M A, Alam M S, Mahbub-ul Karim M H, Rashed M A and Machida M (2011), 'Divergent catalytic behaviors of Pt and Pd films in the cathode of a sandwiched type membrane reactor', *Appl CatalB-Environ*, **107**, 294–301.
93. Hasnat M A, Karim M R and Machida M (2009), 'Electrocatalytic ammonia synthesis: Role of cathode materials and reactor configuration', *Catal Commun*, **10**, 1975–1979.
94. Pintauro P N, Gil M P, Warner K, List G and Neff W (2005), 'Electrochemical hydrogenation of soybean oil with hydrogen gas', *Ind Eng Chem Res*, **44**, 6188–6195.
95. Benziger J and Nehlsen J (2010), 'A polymer electrolyte hydrogen pump hydrogenation reactor', *Ind Eng Chem Res*, **49**, 11052–11060.
96. Narayanan S R, Haines B, Soler J and Valdez T I (2011), 'Electrochemical conversion of carbon dioxide to formate in alkaline polymer electrolyte membrane cells', *J Electrochem Soc*, **158**, A167–A173.
97. Jörissen J (2003), 'Electro-organic synthesis without supporting electrolyte: possibilities of solid polymer electrolyte technology', *J Appl Electrochem*, **33**, 969–977.
98. Vaghela S S, Ramachandraiah G, Ghosh P K and Vasudevan D (2002), 'Electrolytic synthesis of succinic acid in a flow reactor with solid polymer electrolyte membrane', *J Appl Electrochem*, **32**, 1189–1192.
99. Montiel V, Saez A, Exposito E, Garcia-Garcia V and Aldaz A (2010), 'Use of MEA technology in the synthesis of pharmaceutical compounds: the electrosynthesis of N-acetyl-L-cysteine', *Electrochem Commun*, **12**, 118–121.
100. Waldburger R M and Widmer F (1996), 'Membrane reactors in chemical production processes and the application to the pervaporation-assisted esterification', *Chem Eng Technol*, **19**, 117–126.
101. Zhu Y, Minet R G and Tsotsis T T (1996), 'A continuous pervaporation membrane reactor for the study of esterification reactions using a composite polymeric/ceramic membrane', *Chem Eng Sci*, **51**, 4103–4113.
102. Park B G and Tsotsis T T (2004), 'Models and experiments with pervaporation membrane reactors integrated with an adsorbent system', *Chem Eng Process*, **43**, 1171–1180.

103. Sanz M T and Gmehling J (2006), 'Esterification of acetic acid with isopropanol coupled with pervaporation. Part I: Kinetics and pervaporation studies', *Chem Eng J*, **123**, 1–8.
104. Sanz M T and Gmehling J (2006), 'Esterification of acetic acid with isopropanol coupled with pervaporation. Part II. Study of a pervaporation reactor', *Chem Eng J*, **123**, 9–14.
105. Benedict D J, Parulekar S J and Tsai S P (2003), 'Esterification of lactic acid and ethanol with/without pervaporation', *Ind Eng Chem Res*, **42**, 2282–2291.
106. Delgado P, Sanz M T, Beltran S and Nunez L A (2010), 'Ethyl lactate production via esterification of lactic acid with ethanol combined with pervaporation', *Chem Eng J*, **165**, 693–700.
107. Lauterbach S and Kreis P (2006), 'Experimental and theoretical investigation of a pervaporation membrane reactor for a heterogeneously catalysed esterification', *Desalination*, **199**, 418–420.
108. Buchaly C, Kreis P and Gorak A (2007), 'Hybrid separation processes – Combination of reactive distillation with membrane separation',*Chem Eng Proc*, **46**, 790–799.
109. Lee J K, Song I K and Lee W Y (1995), 'An experimental study on the application of polymer membranes to the catalytic decomposition of MTBE (methyl *tert*-butyl ether)', *Catal Today*, **25**, 345–349.
110. Choi J S, Song I K and Lee W Y (2000), 'Simulation and experimental study on the polymer membrane reactors for the vapor-phase MTBE (methyl *tert*-butyl ether) decomposition, *Catal Today*, **56**, 275–282.
111. Buonomenna M G and Drioli E (2008), 'Solvent free selective oxidation of benzyl alcohol to benzaldehyde using a membrane contactor unit', *Appl CatalB-Environ*, **79**, 35–42.
112. Molinari R, Poerio T and Argurio P (2006), 'One-step production of phenol by selective oxidation of benzene in a biphasic system', *Catal Today*, **118**, 52–56.
113. Molinari R, Poerio T and Argurio P (2006), 'Preparation, characterisation and reactivity of polydimethylsiloxane membranes for selective oxidation of benzene to phenol', *Desalination*, **200**, 673–675.
114. Molinari R, Caruso A and Poerio T (2009), 'Direct benzene conversion to phenol in a hybrid photocatalytic membrane reactor', *Catal Today*, **144**, 81–86.
115. Molinari R, Pirillo F, Falco M, Loddo V and Palmisano L (2004), 'Photocatalytic degradation of dyes by using a membrane reactor', *Chem Eng Proc*, **43**, 1103–1114.
116. Molinari R, Pirillo F, Loddo V and Palmisano L (2006), 'Heterogeneous photocatalytic degradation of pharmaceuticals in water by using polycrystalline TiO_2 and a nanofiltration membrane reactor', *Catal Today*, **118**, 205–213.
117. Subhani M A, Beigi M and Eilbracht P (2008), 'Polyurethane- and polystyrene-supported 2,2,6,6-tetramethylpiperidine-1-oxyl (TEMPO); facile preparation, catalytic oxidation and application in a membrane reactor', *Adv Synth Catal*, **350**, 2903–2909.
118. Wijmans J G and Baker R W (1995), 'The solution-diffusion model: a review', *J Membr Sci*, **107**, 1–21.
119. Frisch H L, Maaref S and Deng-Nemer H (1999), 'Low-temperature dehydrogenation reaction-separation membrane using zeolite 13 X polyethylacrylate', *J Membr Sci*, **154**, 33–40.

120. Castanheiro J E, Fonseca I M, Ramos A M, Oliveira R and Vital J (2005), 'Hydration of α-pinene over molybdophosphoric acid immobilized in hydrophobically modified PVA membranes', *Catal Today*, **104**, 296–304.
121. Frisch H L, Huang L and Zeng W (2000), 'Low-temperature cis to trans isomerization reaction-separation membrane using zeolite 13X polyethylacrylate', *J Membr Sci*, **170**, 35–41.
122. Yawalkar A A, Pangarkar V G and Baron G V (2001), 'Alkene epoxidation with peroxide in a catalytic membrane reactor: a theoretical study', *J Membr Sci*, **182**, 129–137.
123. Nagy E (2007), 'Mass transfer through a dense, polymeric, catalytic membrane layer with dispersed catalyst', *Ind Eng Chem Res*, **46**, 2295–2306.
124. Sousa J M, Cruz P and Mendes A M (2001), 'Modeling a catalytic polymeric non-porous membrane reactor', *J Membr Sci*, **181**, 241–252.
125. Sousa J M, Cruz P and Mendes A M (2001), 'A study on the performance of a dense polymeric catalytic membrane reactor', *Catal Today*, **67**, 281–291.
126. Sousa J M, Cruz P, Magalhães F D and Mendes A M (2002),'Modeling catalytic membrane reactors using an adaptive wavelet-based collocation method', *J Membr Sci*, **208**, 57–68.
127. Sousa J M and Mendes A M (2003), 'Modelling a dense polymeric catalytic membrane reactor with plug-flow pattern', *Catal Today*, **82**, 241–254.
128. Sousa J M and Mendes A M (2003), 'Simulation study of a dense polymeric catalytic membrane reactor with plug-flow pattern', *Chem Eng J*, **95**, 67–81.
129. Sousa J M and Mendes A M (2004), 'Simulating catalytic membrane reactors using orthogonal collocation with spatial coordinates transformation', *J Membr Sci*, **243**, 283–292.
130. Sousa J M and Mendes A M (2005), 'Modelling a catalytic membrane reactor with plug-flow pattern and a hypothetical equilibrium gas-phase reaction with $\Delta n \neq 0$', *Catal Today*, **104**, 336–343.
131. Sousa J M and Mendes A M (2006), 'Consecutive-parallel reactions in nonisothermal polymeric catalytic membrane reactors', *Ind Eng Chem Res*, **45**, 2094–2107.
132. Brandão L, Madeira L M and Mendes A M (2007), 'Mass transport on composite dense PDMS membranes with palladium nanoclusters', *J Membr Sci*, **288**, 112–122.
133. Brandão L, Fritsch D, Madeira L M and Mendes A M (2004), 'Kinetics of propylene hydrogenation on nanostructured palladium clusters', *Chem Eng J*, **103**, 89–97.
134. Brandão L, Fritsch D, Mendes A M and Madeira L M (2007), 'Propyne hydrogenation kinetics over surfactant-stabilized palladium nanoclusters', *Ind Eng Chem Res*, **46**, 377–384.

1.7 Appendix: nomenclature

1.7.1 Abbreviations

DMAc dimethylacetamide
DMF dimethylformamide

DO	dissolved oxygen
EIPS	evaporation induced phase separation
HPA	heteropolyacid
MEA	membrane electrolyte assembly
MTBE	methyl *tert*-butyl ether
NIPS	non-solvent induced phase separation
NMP	N-methyl-2-pyrrolidone
PAI	polyamideimide
PAN	polyacrylonitrile
PCMR	polymericcatalytic membrane reactor
PCB	polychlorinated biphenyl
PDMS	polydimethylsiloxane
PEBA	poly(ether block amide)
PEEK	polyetheretherketone
PEI	polyetherimide
PEM	polymer electrolyte membrane
PES	polyethersulfone or polysulfone
PI	polyimide
PIMR	polymeric inert membrane reactor
POM	polyoxomethalate
PPO	polypropyleneoxide
PSA	poly(styrene sulfonic acid)
PVA	poly(vinyl alcohol)PVDF poly(vinylidene fluoride)
PVP	poly(vinylpyrrolidone)
PW	tungstophosphoric acid
TEMPO	2,2,6,6-tetramethylpiperidine-1-oxyl
TIPS	thermally induced phase separation

2

Inorganic membrane reactors for hydrogen production: an overview with particular emphasis on dense metallic membrane materials

A. BASILE, ITM-CNR, Italy, J. TONG, Colorado School of Mines, USA and P. MILLET, University of Paris (11), France

DOI: 10.1533/9780857097330.1.42

Abstract: In this chapter, recent progress on palladium (Pd)-based membrane reactors (MRs) is outlined concentrating, in particular, on the production of pure hydrogen. Aspects of many reactions, as well as analysis of both the Pd-based and the amorphous membranes under study, and the governing equations are presented. Some critical aspects of non-Pd-based membranes are also discussed. All the preparation techniques for pure, alloyed, amorphous, non-Pd-based membranes used in MRs are briefly summarized and compared. Moreover, some problems related to the effect of contamination of the Pd-based membranes on the H_2 flux are discussed.

Key words: hydrogen, reactions, Pd, alloy, amorphous, non-Pd-based membrane reactors.

2.1 Introduction

Hydrogen is an important raw material in the chemical and petrochemical industries, mainly for its use in producing ammonia, processing petroleum, and reducing processes in metallurgy. H_2 is also widely used as an energy vector able to play a role of increasing importance in future energy systems. In fact, H_2 is becoming a key factor for new applications as a result of spectacular advances in fuel cell technology. The production of ultra-pure H_2 for use in polymer electrolyte membrane (PEM) fuel cells, for small- as well as medium-scale applications, has increased notably in the last few years. For example, fuel cells are applied in the automotive industry and in distributive power generation, as well as in small scale applications (< 250 kW), where they find application in transportation or household power supply (Palo *et al.* 2007). Unfortunately, this gas is not directly available on our planet, so it must be produced from other naturally occurring species, typically from fossil hydrocarbons and water.

World demand for hydrogen is increasing – some years ago, it was around 5 billion cubic metres per year (Lu *et al.*, 2007) – of which the greatest part is deliberately produced while the remainder is a by-product of other chemical processes (Uhrig, 2005).

Traditionally, H_2 is produced via steam reforming (SR) of hydrocarbons such as methane, naphtha oil and methanol. On an industrial scale, most H_2 is currently produced by SR of natural gas. In fact, the most important conventional technology is the SR of methane, carried out using multi-tubular fixed bed reactors, which produces approximately 90% of the total H_2 produced in the world (National Research Council and National Academy of Engineering, 2008). In small scale applications, two other main alternatives are generally considered along with SR: partial oxidation reactions, whose efficiency is significantly lower than SR, and auto-thermal reforming, where the partial oxidation and SR reactions are carried out simultaneously.

The main problem with SR, partial oxidation and auto-thermal conventional reactors is that all these reactions lead to a H_2-rich gas mixture also containing carbon oxides and other by-products as well as unreacted reagents. In other words, kinetic and/or thermodynamic constraints do not allow completion of the reaction approach. For this reason, chemical processes have traditionally been carried out in reaction units in series with one or more separation units. Meanwhile, many papers (e.g., Paiva and Malcata, 1997; Paiva *et al.,* 1999) have demonstrated that an integrated system able to perform reaction and separation simultaneously instead of sequentially is able to produce both kinetic and thermodynamic enhancements.

With respect to SR reactions, conventional systems need at least four different stages, consisting of three reactors (e.g., one SR reactor and two water gas shift (WGS) reactors) and a purification system (pressure swing adsorption or others). As an example, a simplified H_2 production process (Austin, 1986; Smith and Shantha, 2007), consisting of a reformer of natural gas followed by two shift reactors and an H_2 purifier, is shown in the upper part of Fig. 2.1. With this process, high H_2 yields are achieved, but to obtain H_2 at the desired purity expensive high temperature heat exchangers and complex energy integration among various process units are necessary, including reformer, high and low temperature reactors to carry out the WGS reaction, as well as preferential oxidation reactors or pressure swing adsorption units.

Among different technologies related to production, separation and purification of H_2, membrane technologies really seem to play a fundamental role. In particular, owing to its attractive capability of continuous operation and process miniaturization, membrane separation, an energy-saving technique without any phase transformation, can be considered a good candidate for replacing conventional systems nowadays. For example, specific thermodynamic limits can be overcome only by using innovative systems, such as

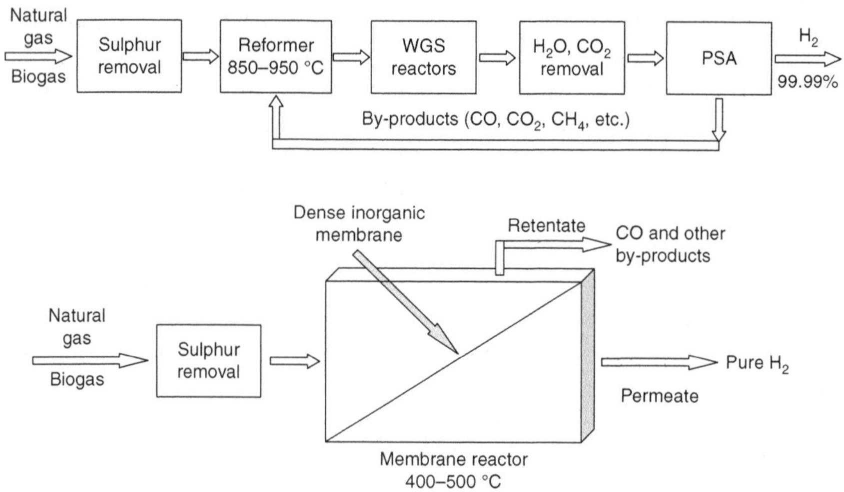

2.1 Scheme of pure hydrogen production by steam reforming of natural gas or biogas: traditional scheme and conceptual scheme of a dense inorganic membrane reactor. (PSA – pressure swing adsorption.)

MRs, engineering systems in which both reaction and separation occur in the same device. For carrying out the same SR reaction, and for obtaining an ultra-pure H_2 stream, the use of a dense Pd-based MR, in which only H_2 selectively permeates through the metal, permits the number of steps to be reduced to only one (see lower part of Fig. 2.1). With respect to a classical configuration consisting of a reactor unit in series with a separation unit, an MR represents a modern configuration in which an integrated reaction/separation unit has many potential advantages: reduced capital costs, improved yields and selectivities, and drastically reduced downstream separation costs.

In this context, in an inorganic MR, pure H_2 moves to the permeate side enabling the reactions to proceed towards completion and so making it possible to achieve: (a) conversion values higher than those obtained by traditional reactors working under the same operative conditions, or (b) the same conversion values obtained by traditional reactors but, in this case, working under milder operating conditions. To better understand this concept, the conceptual scheme of Fig. 2.2 is considered, where the conversion versus the temperature for this generic endothermic reaction is shown. Following Raich and Foley (1998), the selective and continuous removal of H_2 in Pd-based MRs is only a *necessary* condition for achieving a level of conversion higher than the thermodynamic equilibrium value for traditional reactors (TR), but it is not *sufficient*. In the case of dehydrogenation reactions, for example, the *necessary* and *sufficient* conditions have to take into account both aspects: thermodynamics and kinetics. In fact, generally,

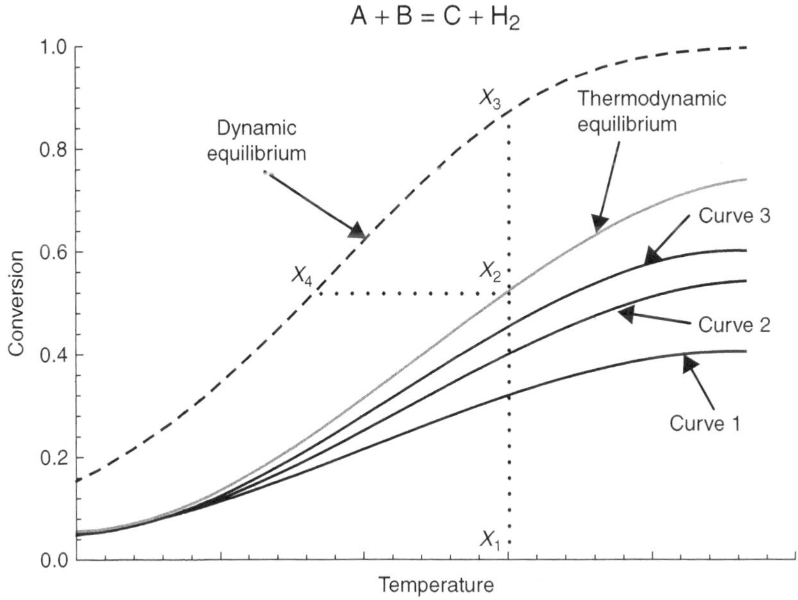

2.2 Conversion versus temperature for a generic endothermic reaction.

the rate of hydrogen permeation is not the only impediment to achieving supra-equilibrium conversions, since the kinetics of the reaction are also critical. This is clearly seen by considering a simple and generic dehydrogenation chemical reaction: A + B = C + H_2 (Fig. 2.2). The *Thermodynamic Equilibrium* curve represents the equilibrium conversion for a TR. The conversion for a TR necessarily falls under this curve (the region X_1–X_2, at a generic fixed temperature). In particular, the conversion depends on the catalyst activity used for the specific reaction considered: a very active catalyst bed gives conversion values (curve 3) closer to the thermodynamic equilibrium curve, whereas a poor active catalyst gives conversion values (curve 1) distant from the thermodynamic equilibrium curve. On the other hand, the *Dynamic Equilibrium* curve is only related to a dense metallic H_2 permselective MR, and represents the reaction not limited by any kind of chemical kinetics and thus the feed gas is assumed to be continuously at equilibrium inside the MR. As hydrogen permeates through the dense metallic layer, the feed gas composition changes due to hydrogen permeation. This means that the reaction rate is not the limiting step: all molecules in contact with the catalyst are considered to react with an infinite velocity (i.e., no kinetics resistance). The conversion for an MR necessarily falls under this curve (the region X_1–X_3, at a generic fixed temperature). Also in this case, the conversion depends on the catalyst activity used for the specific reaction: a very active catalyst bed gives conversion values possibly higher than the thermodynamic equilibrium curve (and falls in the region X_2–X_3, at a generic fixed temperature), while a poor active catalyst bed gives conversion values far from the dynamic equilibrium curve (and can fall in the region X_1–X_2, at a generic fixed temperature). It must also be observed that an MR could work in the region X_4–X_2: here the temperature is reduced (for example the one corresponding to X_4), whereas the conversion remains the same.

In the last few years, significant work has been done by scientists and chemical engineers to improve MR performance and overcome problems such as membrane durability and resistance of the membrane-module seals to high temperature and pressure. Improvements in the technology of MRs are bringing them closer to industrial utilization. Their introduction in an industrial context strongly depends on the economics of the membrane process. This chapter attests to the growing strategic role of MRs for pure H_2 production. For our purposes, mainly dense membranes will be discussed.

2.2 Development of inorganic membrane reactors (MRs)

A brief history of the study of H_2 on metallic membranes is presented elsewhere by Basile *et al.* (2008a). Nowadays, the significant progress in the field of the MRs is well reflected in the increasing number of publications. In fact, this number has grown exponentially, especially over the last ten years, as shown in Fig. 2.3. In particular, to find the data reported on this figure (related to the years *before* 2012), the Elsevier Scopus database research (http://scopees.elsevier.com), where more than 6000 journals are taken into account, was used in three different ways:

- curve 1 was obtained using the words '*membrane reactor*' in all the fields (Title, Abstract, *etc.*);
- curve 2 using '*membrane reactor*' and '*hydrogen*';
- curve 3 with both '*membrane reactor*' and '*pure hydrogen*'.

It should be said that curve 1 includes all kinds of MRs, not only the inorganic ones, whereas curve 2 restricts the field to both dense and porous inorganic membranes, and curve 3 refers only to dense inorganic membranes. It is quite evident that each curve shows an exponential increase.

Many chemical reactions have been carried out in MRs, both experimental studies and mathematical models describing the occurrence of main kinetics and fluid-dynamic phenomena. Apart from the huge number of specific papers, a general description in terms of research innovations, improvements for chemical stability, supported thin film structures, and so

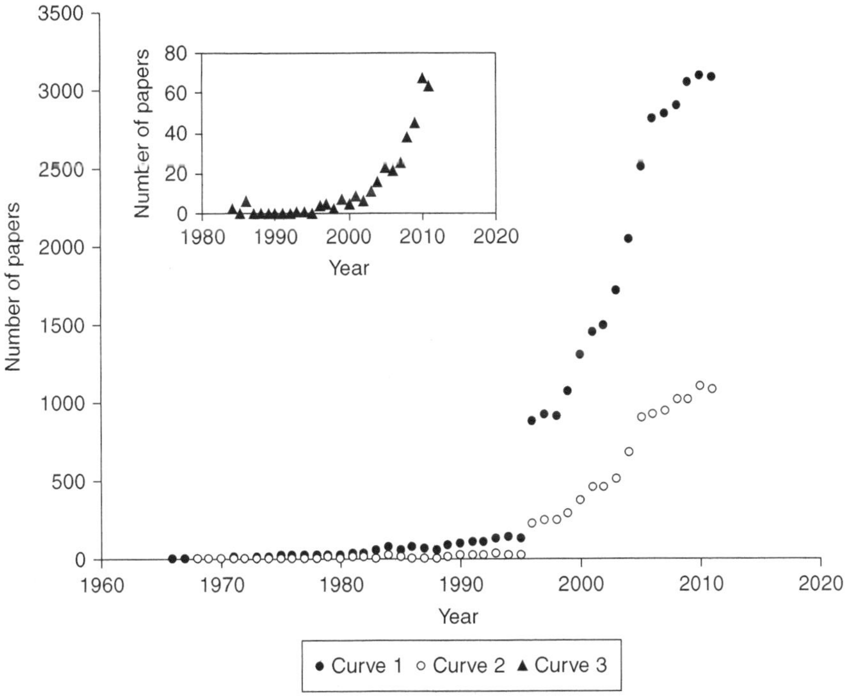

2.3 Number of publications on MRs versus time (years); curve 1 refers to only MRs; curve 2 refers to both MRs and H_2; whereas curve 3 concerns both MR and pure H_2.

on, and relevant materials, membrane preparation, transport phenomena, application case studies, modelling, etc., can be found in several significant reviews (Bouwmeester, 2003; Coronas and Santamaria, 1999; Coronas and Santamaria, 2004; Dittmeyer *et al.,* 2001; Dittmeyer *et al.,* 2004; Dixon, 1999; Dixon, 2003; Gryaznov, 1999; Gryaznov, 2000; Gryaznov *et al.,* 2005; Hoff *et al*., 2003; Hwang and Korean, 2001; Julbe *et al.,* 2001; Mc Leary *et al*., 2006; Michaels, 1968; Paglieri and Way, 2002; Paturzo *et al.,* 2002; Saracco *et al.,* 1994; Saracco and Specchia, 1994; Saracco *et al.,* 1999; Sirkar *et al.,* 1999; Specchia *et al.,* 2005; Stoukides, 2000; van de Water and Maschmeyer, 2004; Uemiya, 1999; Zaman and Chakma, 1995), books (Basile and Gallucci, 2009; Basile and Calabrò, 2010; Basile and Gallucci, 2011; Bhave, 1991; Drioli and Giorno, 2011; Hsieh, 1996; Li, 2007; Malada and Menendez, 2008; Marcano and Tsosis, 2002). In particular, the reader is also directed to the special issues recently published in international journals dedicated to MRs (e.g., Basile and Drioli, 2006; Basile and Gallucci, 2009 and 2009b; Volkov *et al.*, 2012).

The following classes of reactions in gas phase are under study with the aim of hydrogen production:

- dehydrogenation: 1,2-cyclohexanediol (Mishchenko *et al.,* 1997), 2-butanol (Keuler and Lorenzen, 2002a), ammonia (Collins and Way, 1994), butane (de Rosset and Hills, 1968), *cis*-3-hexen-1-ol (Sato *et al.,* 2007), cyclohexane (Itoh *et al.,* 2003), cyclohexene (Aguilar *et al.,* 1977), cyclohexanol (Basov *et al.,* 1993), ethane (Wang *et al.,* 2003), ethylbenzene (Abballa and Elnashaie, 1995), heptane (Zhu *et al.,* 1993), hydriodic acid (Yehenskel *et al.,* 1979), hydrogen sulphide (Edlun and Pledger, 1993), isoamylene (Orekova *et al.,* 1976), isobutane (Liang and Hughes, 2005), isopropanol (Mikhalenko *et al.,* 1986), methylcyclohexane (Ali and Baiker, 1997), *n*-hexane (Smirnov *et al.,* 1977), octane (Smirnov *et al.,* 1977), propane (Sheintuch and Dessau, 1996), β-ionol (Smirnov *et al.,* 1978), water (Omorjan *et al.,* 1999);
- dry reforming of methane (Haag *et al.,* 2007);
- oxidative dehydrogenation: ethane (Akin and Lin, 2002), isobutane (Raybold and Huff, 2000), methane (Basile *et al.,* 2001a), methanol (Eswaramooth *et al.,* 2009);
- SR: acetic acid (Basile *et al.,* 2008b), ethanol (Iulianelli *et al.,* 2010a), methane (Tong and Matsumura, 2006), methanol (Agrell *et al.,* 2002);
- WGS reaction (Liguori *et al.,* 2012).

Among all these reactions, only some of them could be considered of practical value: WGS, SR of methane, dry reforming of methane, partial oxidation of methane, methanol steam reforming, oxidative methanol steam reforming, ethanol steam reforming, steam reforming of acetic acid, glycerol steam reforming and some of the dehydrogenation reactions. A brief panoramic view of these reactions is proposed in the following sections.

2.2.1 Water gas shift (WGS) reaction

The WGS reaction, an exothermic reaction with equilibrium constant increasing with decreasing temperature, is one of the most important industrial reactions that can be used to produce hydrogen for ammonia synthesis, for adjusting the hydrogen–carbon monoxide ratio of synthesis gas, etc. The WGS reaction is catalysed by several catalysts based on Fe, Cu, Zn, Ce, Cr, Co, Ni, oxides and their combinations, as is evident in Fig. 2.4a–c. However, also noble-metal-based catalysts, particularly Pt and Au based ones, are used (Fig. 2.4a). Industrially, the most utilized catalysts are $CuZn/Al_2O_3$ and Fe_3O_4–Cr_2O_3 ranging from 200–250°C and 300–400°C, respectively. Much interest in the WGS reaction assisted by MRs has been evidenced in the literature and many studies are focused on hydrogen recovery (HR) from water by a catalytic shift reactor coupled with a membrane. For example, Seok and Wang (1990) evaluated the WGS reaction performance in MRs by using Vycor glass coated with ruthenium (III) chloride trihydrated. The

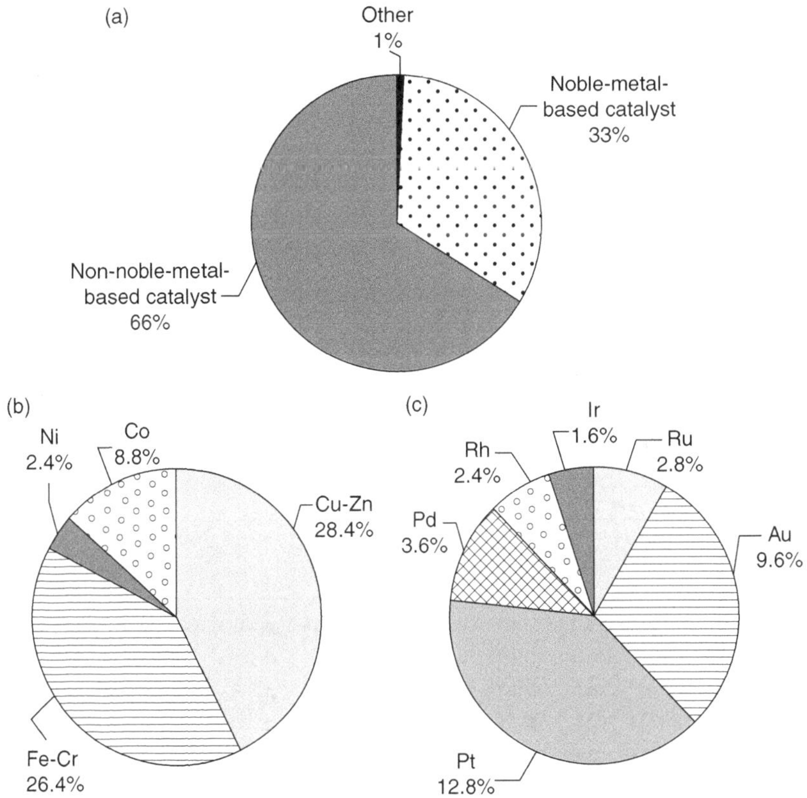

2.4 Percentage distribution of the catalysts used for carrying out the WGS reaction. a) noble and non-noble metals distribution; b) catalyst distribution based on non-noble metal catalysts; c) catalyst distribution based on noble metal catalysts.

reaction was carried out under various operating conditions of temperature, pressure and feed composition. The highest CO conversion obtained was 85% (equilibrium value 99.9%) at a relatively low temperature (160°C), while complete conversion (100%) was obtained by Kikuchi *et al.* (1989) and by Uemiya *et al.* (1991a) at 400°C, using a double tubular MR where the inner tube is represented by a thin Pd film. Recently, Gessler *et al.* (2003) compared the performance of their molecular sieve silica membranes in the WGS reaction to the Pd-composite membranes used by Basile *et al.* (1996a). Under the same experimental conditions, Gessler *et al.* (2003) obtained a CO conversion of 95% and their molecular sieve silica membranes performed better than the Pd-composite one. An excellent review on these aspects has been published recently by Lu *et al.* (2007).

Figure 2.5 presents the literature data concerning CO conversion realized in different MRs and conventional reactors, as well as the thermodynamic

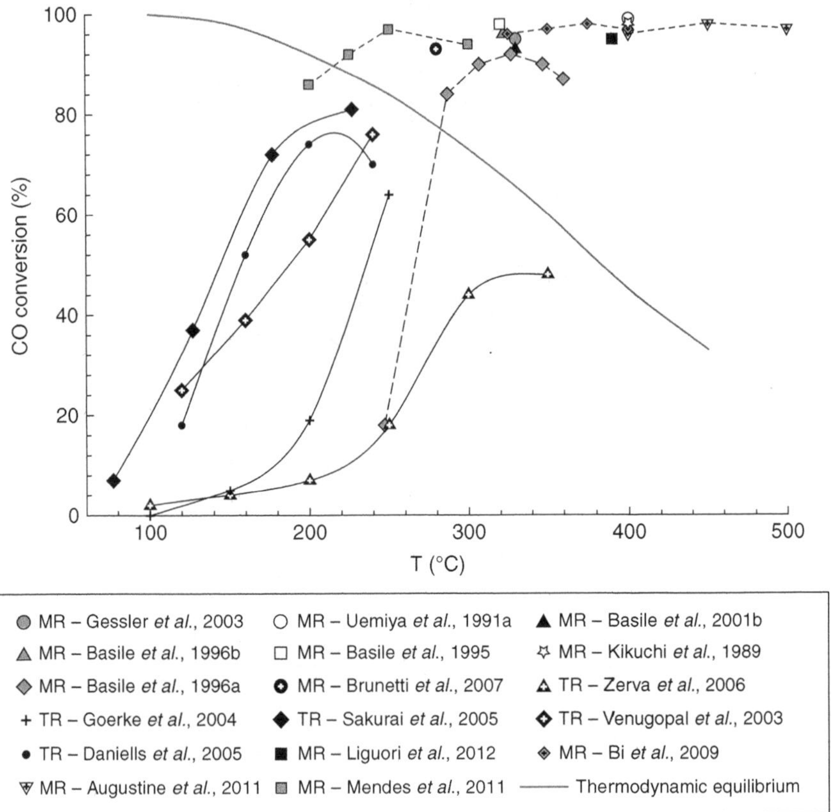

2.5 WGS reaction: CO conversion versus reaction temperature.

equilibrium conversion. Unfortunately, it is not possible to compare the results directly, owing to the different operating conditions used in each work. As a general observation, it appears quite evident that MRs show CO conversions overcoming the thermodynamic equilibrium due to the hydrogen permeation through the membrane that shifts the WGS reaction towards further products formation, allowing a higher CO conversion to be achieved.

A potential application of the WGS reaction carried out in an MR is represented by the tritium recovery process from tritiated water from breeder blanket fluids in fusion reactor systems. The hydrogen isotopes separation at low concentration in gaseous mixtures is a typical problem of the fusion reactor fuel cycle. In fact, the tritium produced in the breeder needs a proper extraction process to reach the required purity level. Yoshida *et al.* (1984) carried out experimental and theoretical studies of a catalytic reduction method which allows tritium recovery from tritiated water with a high conversion value (> 99.99%) at a relatively low temperature, while Hsu and Buxbaum (1986) studied a palladium-catalysed oxidative diffusion to extract hydrogen from metal solution at very low partial pressures. The method is useful for extracting tritium from a liquid lithium breeder blanket at low concentration. A conceptual modified version of the tritium recovery plant for a ceramic breeder in the fusion fuel cycle, working by using two membrane reaction/separation units, was studied by Violante *et al.* (1993; 1995). In these works, a catalytic ceramic MR was used to remove the hydrogen isotopes via oxidation from purge gas (He) while a Pd–Ag separator was used for tritiated water via WGS reaction. As advancement of this study, Tosti *et al.* (2003) proposed a Pd–Ag membrane reactor to recover hydrogen and its isotopes from tritiated water by using the WGS reaction. Operating with reference to a closed-loop process (studied for application in the fusion fuel cycle), the WGS reaction carried out in this MR showed the capability to achieve higher conversion than the TR, exercising with low H_2O/CO ratios, constituting a very important result in fusion nuclear applications where the tritium inventory can be reduced.

An interesting economic analysis of Pd-based MRs for the WGS reaction was made by Criscuoli *et al.* (2001), with the aim of fixed pure hydrogen production. The analysis focused on the comparison between the conventional apparatus and different MRs, in terms of both capital and operating costs. The effects of the Pd thickness and of the hydrogen permeability through the membrane on the membrane devices costs were also considered. Their conclusion was that, for a Pd thickness < 20 μm, MRs could represent a possible alternative to conventional apparatus in the specific case considered.

For the most relevant topics in WGS MR technology – catalysis and membrane science – interested readers are directed to the recent paper published by Mendes *et al.* (2010).

2.2.2 Steam reforming of methane

The reaction of SR of methane is a commercially available process for H_2 production. In the USA, for example, over 90% H_2 is nowadays produced via SR of natural gas. As already said, the current technology in the manufacture of H_2 is accomplished in several steps:

(a) steam reforming of methane
 $(CH_4 + H_2O = CO + 3H_2;\quad \Delta H^{\circ}_{298\,K} = +206\ kJ/mol)$;
(b) one or more WGS reactors;
(c) a purification reactor and/or system (preferential oxidation reactor and/or pressure swing adsorption system/palladium membrane).

The degree of H_2 purification depends on its application. For industrial H_2, pressure swing absorption systems or Pd membranes are used to produce H_2 at up to 99.999% purity; whereas for PEM or phosphoric acid fuel cells

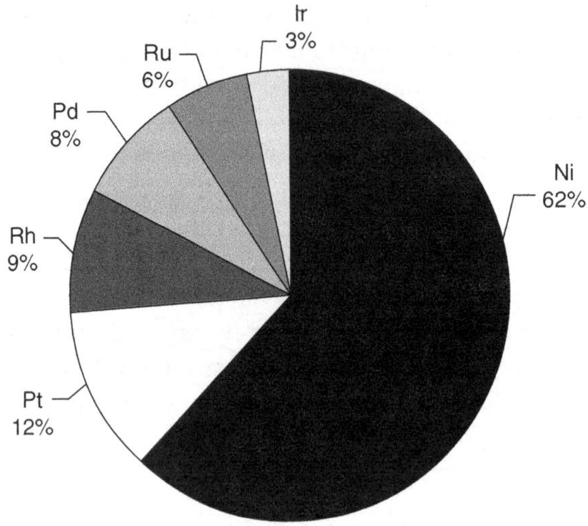

2.6 Percentage distribution of the catalysts used for carrying out the methane steam reforming reaction.

closely coupled to reformers, diluents such as CO_2 and CH_4 are tolerable (it depends on their concentration). In the specific case of PEM fuel cells, CO must be less than 10 ppm, so an additional CO removal system, such as preferential oxidation, must be used. Preferential oxidation technology is under development for use with reformers in fuel cell cogeneration systems or on board fuel cell vehicles.

In Fig. 2.6, the percentage distribution of some used catalysts for performing SR of methane, such as Ni, Ru, Rh, Pd, Ir and Pt, is depicted. It is clearly evident that Ni is the most widely used catalyst due to its lower cost. These catalysts show both high resistance towards deactivation due to poisoning and high thermal stability under harsh operating conditions, such as 800–1000°C, and 14–20 bar (Barelli *et al.,* 2008).

Using MRs it is possible to mitigate some of the drastic operating conditions in TRs. For this reason, MR steam reformers are still undergoing laboratory R&D as well. Considering, for example, the same database search cited in the Section 2.1, by using the words: 'methane steam reforming' and 'membrane', and referring to the situation until the year 2011 (*included*), the following results are obtained: number of articles published: 261; conference papers: 131; articles in press: 7; reviews: 7; patents: 3883.

In particular, the following papers, related to both experimental and simulation of methane SR carried out in dense as well as porous MRs, are particularly interesting and suggested to the reader: Agrell *et al.,* 2002; Basile *et al.,* 2005a; Basile *et al.,* 2008b; Kikuchi *et al.*, 1989; Raybold and Huff, 2000; Seok and Hwang, 1990; Tosti *et al.,* 2008a; Uemiya *et al.,* 1991a.

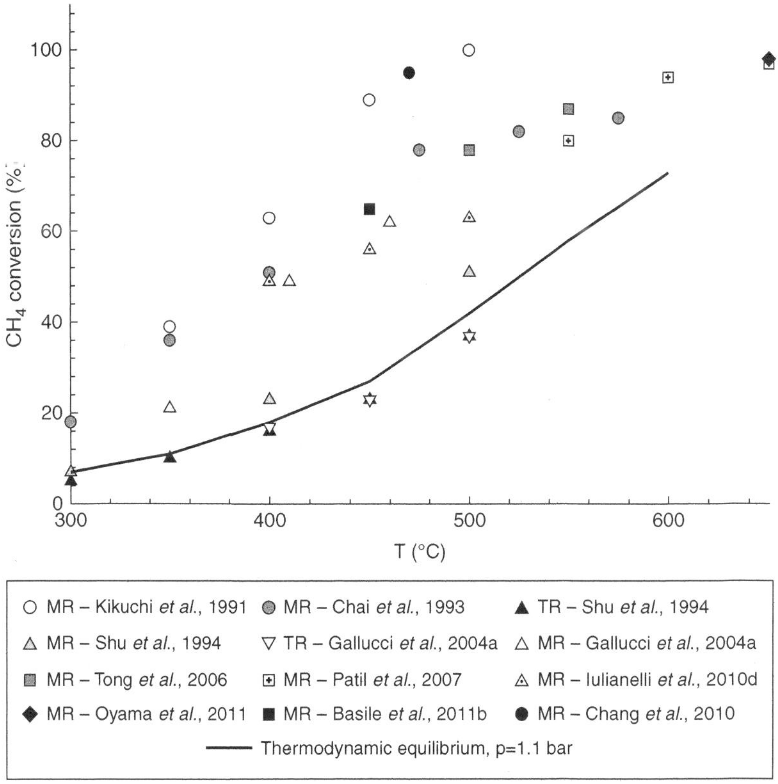

2.7 Methane steam reforming reaction: conversion versus temperature.

Two reviews on the state-of-the-art on H_2 production through SR of methane reaction carried out in MRs have been recently published by Ritter and Ebner (2007) and Barelli *et al.* (2008). Also interesting is the review published by Uemiya (2004).

Inorganic MRs are a promising technology where, as was said, the SR of methane, WGS and H_2 purification steps all take place in a single reactor. Concerning the temperature influence on the methane conversion, some of the literature data for both TRs and MRs are summarized in Fig. 2.7. In the range 300–650°C, the MRs show higher methane conversions than both the TRs and the thermodynamic equilibrium curve corresponding to the reaction pressure 1.1 bar. Moreover, TR conversions are always *below* the corresponding thermodynamic curve, whereas MR values are always *above* it. However, this latter fact should not be considered as a rigid rule. Depending on the particular experimental conditions, if H_2 removal from the reacting zone is not perfect (i.e., H_2 membrane selectivity is not infinite), the methane conversion for the MRs might also be lower than that for the TRs. In the case of porous membranes, for example, part of the reactants is lost because of their permeation before that the reaction takes place. This figure well evidences that by using MRs it is possible to get the same conversion of TRs but operating at lower temperature, with the advantage of obtaining both an energy saving and pure H_2 production from the permeate side stream.

From a modelling point of view, the SRM reaction was investigated to study the effect of different parameters on the methane conversion, such as: operating pressure, temperature, membrane thickness and membrane length. In particular, the effect of operating pressure seems to be not obvious. Concerning SRM, an increase in operating pressure (Gallucci *et al.*, 2004b):

- in a TR, always causes a decrease in methane conversion;
- in an MR, causes an increase or a decrease in methane conversion, depending on the combination of p, T, membrane thickness and reactor length.

The same considerations reported here for methane SR could be repeated for all the other SR reactions.

2.2.3 Dry reforming of methane

Another approach for hydrogen production is the dry reforming of methane, which gives a 1:1 mixture of H_2 and CO:

$$CH_4 + CO_2 = 2CO + 2H_2; \quad \Delta H^{\circ}_{298\,K} = +247 \text{ kJ/mol}$$

Although this reaction is more endothermic than methane steam reforming, an industrial application of the methane dry reforming could reduce the amount of greenhouse gases released in the atmosphere. In fact, the disadvantage owing to the reaction cost (in terms of energy input as well as the cost of materials) could be balanced by the benefit deriving from the CO_2 consumption. For this reaction, different kinds of catalyst are studied: Ni, Ru, Rh, Pd and Pt based (Chang *et al.*, 1996; Hayakawa *et al.*, 1999; Rostrup-Nielsen and Bak Hansen, 1993; Rostrup-Nielsen, 1994; Tomishige *et al.*, 1999; Wang and Lu, 1998).

An important limitation for making methane dry reforming a commercially viable reaction using TRs is due to the thermodynamic, which limits the reactants conversion. Nevertheless, methane (and carbon dioxide) conversion can be increased if both the products of the reaction (or preferentially only hydrogen) are selectively removed from the reaction side. Figure 2.8 depicts a comparison among the literature data concerning CH_4 conversion realized in different MRs and in conventional reactors (TRs). It is

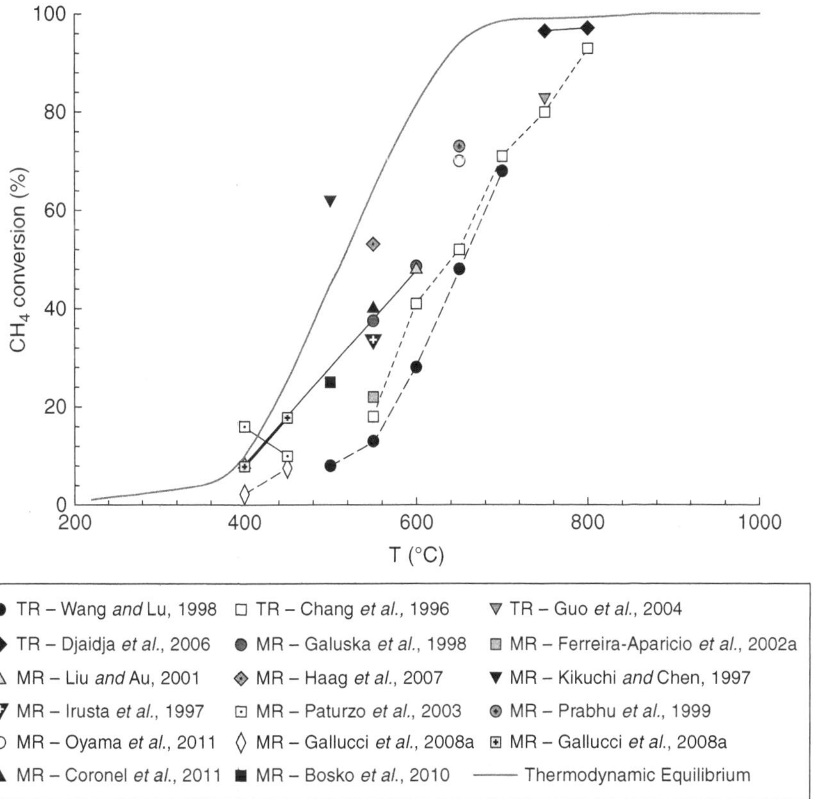

2.8 Dry reforming of methane: conversion versus reaction temperature.

evident that an MR allows higher methane conversion to be realized at lower temperature than that obtained by using a conventional reactor. However, the performance of the MR is determined by the kind of membrane used and suitable matching between the membrane properties and the operating conditions adopted. For instance, Ferreira-Aparicio *et al.* (2002a) have demonstrated that, by using a porous membrane with low ideal selectivity ($\alpha_{H_2/CH_4} = 2$), the methane conversion is not considerably enhanced. Liu and Au (2001), and Prabhu *et al.* (1999) have also shown that using zeolite and Vycor glass MRs, respectively, it was possible to obtain high CH_4 conversion only at high temperature.

However, the enhancement of methane conversion is particularly evident when dense Pd-based MRs are used. In particular, Kikuchi and Chen (1997) reached the highest methane conversion at 500°C by using a Pd-based MR, overcoming also thermodynamic equilibrium.

Also Paturzo *et al.* (2003) studied this reaction carrying it out, at first, in a TR and, successively, in a tubular MR. The tubular composite catalytic Ru-based membrane consists of a mesoporous commercial ceramic tube in which an internal two-layer deposit of Ru nanoparticles was realized. In more detail, the effect of the catalytic tubular composite Ru-based membrane, not packed with conventional catalyst, was focused on the behaviour of the reaction system in terms of reactant conversion and product selectivity.

One of the most important problems related to methane dry reforming is carbon deposition. Considering that ruthenium exhibits the highest catalytic activity towards the CO_2 reforming reaction (Sakuarai *et al.,* 2005), by using the Ru-based MR, Paturzo *et al.,* 2003 suggested a way to realize a catalytic membrane able to depress carbon formation as well as to remove hydrogen and improve conversion.

In their recent work, Gallucci *et al.* (2008a) reported that:

(a) if the dry reforming is viewed as a carbon dioxide consumption method, the best way to proceed is to use the Ni/CaAlO catalyst, the Pd/γ-Al_2O_3 catalyst or the Ni/Al_2O_3 one in the TR or in a porous MRs;
(b) *vice versa*, if the dry reforming is viewed as a hydrogen production method, the best way is to operate with (1% Rh)/TiO_2, Co/γ-Al_2O_3 or Ni/Al_2O_3 catalyst in a dense Pd-based MR.

Moreover, by using the dense Pd-based MR, carbon deposition on the catalyst is drastically reduced and a CO-free hydrogen stream is produced, which can be used in systems such as the PEM fuel cells.

2.2.4 Partial oxidation of methane

The endothermicity of both the steam reforming and dry reforming reactions requires an energy input. On the other hand, the partial oxidation of methane (POM) is a mildly exothermic reaction and could be a viable alternative reaction for the hydrogen production:

$$CH_4 + \frac{1}{2}O_2 = CO + 2H_2; \quad \Delta H^{\circ}_{298\,K} = -36\ \text{kJ/mol}$$

An important limitation to making this reaction commercially viable for TR is thermodynamics. In particular, the pressure increase gives a decrease in equilibrium methane conversion (Ostrowski *et al.,* 1998). An MR allows the possibility of overcoming these thermodynamic limitations, reaching a high methane conversion at low temperature with respect to a TR. In the literature, the POM reaction, carried out in composite and in Pd-based

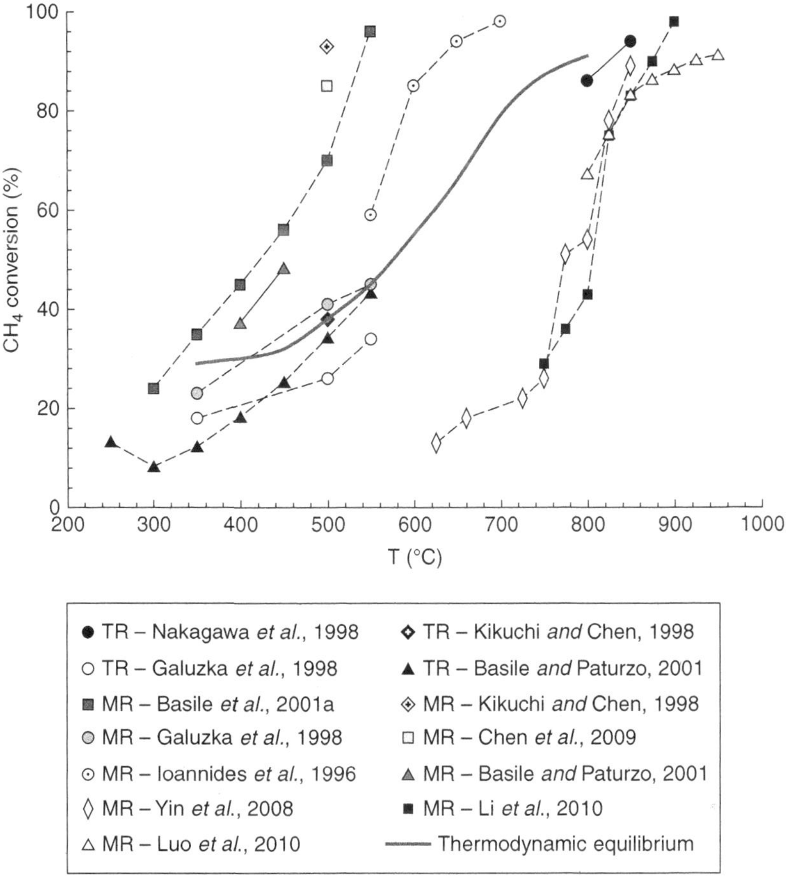

2.9 Partial oxidation of methane: conversion versus reaction temperature.

MRs, was studied from both experimental and simulation points of view. The main results, in terms of methane conversion, are compared with both those obtained using TRs and the thermodynamic equilibrium (Basile *and* Paturzo, 2001) and could be summarized as follows:

- methane conversion is remarkably higher in MRs than in the TRs, at a fixed temperature;
- the Pd-based MR shows the highest methane conversion;
- applying MRs, methane conversions exceeding equilibrium are achieved.

For details the reader should refer to Fig. 2.9.

2.2.5 Methanol steam reforming

The methanol steam reforming reaction is viewed as a very interesting and promising method for hydrogen production, useful for fuel cell applications (Agrell *et al.,* 2002; Han *et al.,* 2000; Sekizawa *et al.,* 1998; Takahashi *et al.,* 2001). In accordance with the specialized literature, the main chemical reactions taken into consideration are the following:

$$CH_3OH + H_2O = CO_2 + 3H_2 \quad \Delta H^{\circ}_{298K} = 49.7 \text{ kJ/mol}$$

$$CO + H_2O = CO_2 + H_2 \quad \Delta H^{\circ}_{298K} = -41.0 \text{ kJ/mol}$$

$$CH_3OH = 2H_2 + CO \quad \Delta H^{\circ}_{298K} = 92.0 \text{ kJ/mol}$$

The first and the third reactions are both reversible and endothermic and proceed under volume increase, suggesting that the highest methanol conversions are obtained at high temperature and low pressure. The exothermic second reaction is the well-known WGS reaction, which proceeds simultaneously with methanol SR and without any volume change. This reaction system, when carried out in a TR, leads to an H_2-containing mixture. If the H_2 produced is to be fed into a PEM fuel cell device, it would need prior purification. The purification is mainly oriented to removing CO, which poisons the anodic catalyst of the fuel cell.

In order to reduce the CO content in the gaseous mixture coming out from the reactor, the TR studies in the literature are mainly focused on catalyst optimization (Amphlett *et al.,* 1981; 1985; 1988; 1994; Dümpelmann, 2001; Jiang *et al.,* 1993a; 1993b; Peppley *et al.,* 1999a; 1999b; 2001). In Fig. 2.10 the percentage distribution of the most used catalysts for carrying out this reaction is reported. Recently, interesting work on the methanol auto-thermal reforming process has been published by Liu *et al.* (2008). Using a $ZnO–ZnCr_2O_4/CeO_2–ZrO_2$ monolithic catalyst, a complete methanol conversion has been achieved in the range of temperatures of 450–500°C (and GHSV = 14500 h^{-1}, H_2O/CH_3OH = 1.2 and O_2/CH_3OH = 0.3). The H_2 concentration and the CO_2 selectivity are around 51–52% and 86–88%, respectively, throughout the 1000 h test.

Only a few papers deal with the methanol SR reaction carried out on MRs and they mainly analyse the effects of temperature and pressure on the reaction system. Various Pd-based (Buxbaum, 1999), Pd/V/Pd, $Pd_{75}Ag_{25}$, $Pd_{60}Cu_{40}$ (Wieland *et al.,* 2002) and Pd-supported (Lin and Rei, 2000) MRs were studied in the pressure range of 1–25 atm and temperature range of 260–320°C.

Figure 2.11 shows a comparison of methanol conversion versus reaction temperature by using the literature data in both TRs and MRs. As a general

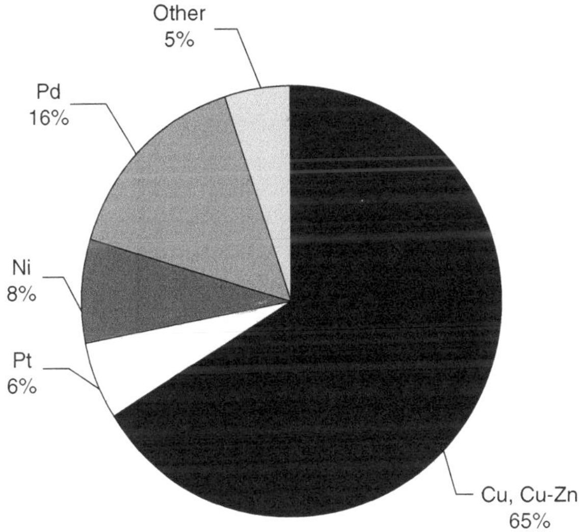

2.10 Percentage distribution of the catalysts used for carrying out the methanol steam reforming reaction.

comment, the methanol conversion is enhanced by increasing the reaction temperature. In each work, the different performance is due to the different operating conditions used by the authors, but it is also clear that MR always realizes higher CH_3OH conversion than TR, if operated under the same experimental conditions (e.g., Basile *et al.*, 2005b). Moreover, MRs allow a high-purity hydrogen stream to be obtained, avoiding or reducing further steps to separation/purification always necessary in a TR plant. As an example, recently, Basile *et al.* (2005b; 2006a) carried out the methanol SR in both composite and dense MRs. The best results were found operating at 450°C: a CO-free HR of 40%, with respect to the total hydrogen produced, was achieved in counter-current mode and high sweep gas flow rate by means of a dense Pd–Ag MR; whereas, in a recent work (Basile *et al.*, 2008c), using a new (i.e. non-commercial) CuO-Al_2O_3-ZnO-MgO catalyst, complete methanol conversion was achieved in a thickness 50 μm pinhole-free Pd–Ag MR at 300°C (and $H_2O/CH_3OH > 5/1$), while the carbon monoxide selectivity was < 1%. No carbon deposition and catalyst sintering phenomena were noticeable in the range of 250–300°C.

2.2.6 Oxidative methanol steam reforming

Only a few studies on the oxidative methanol SR for hydrogen production are present in the literature. Murcia-Mascaros *et al.* (2001), for example, proposed the oxidative methanol SR as a combination of methanol SR with methanol partial oxidation:

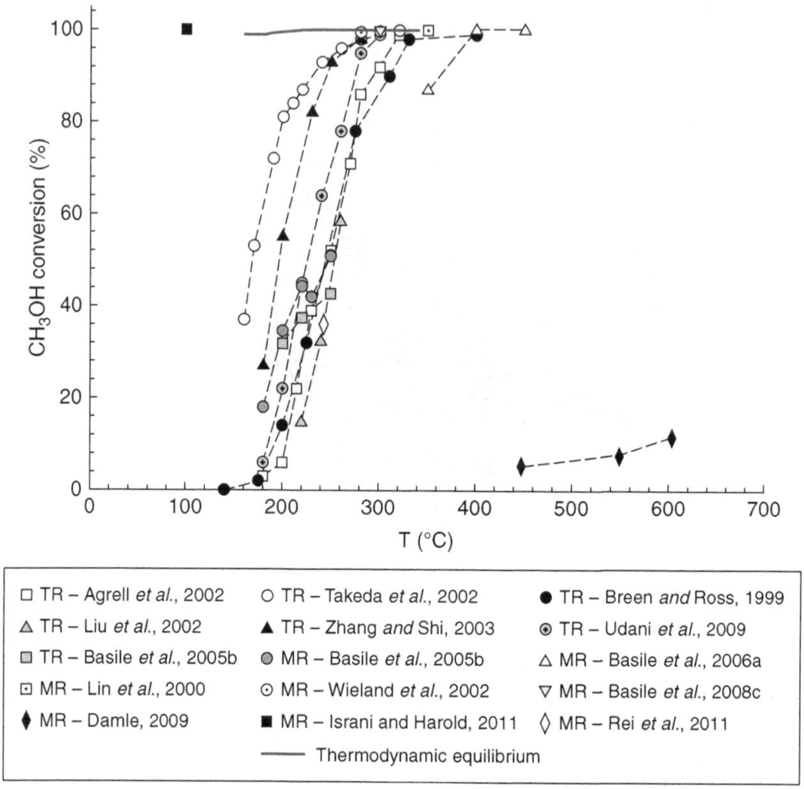

2.11 Methanol steam reforming reaction: conversion versus reaction temperature.

$$CH_3OH + H_2O = 3H_2 + CO_2; \quad \Delta H^{\circ}_{298K} = +49.4 \text{ kJ/mol}$$

$$CH_3OH + 1/2O_2 = 2H_2 + CO_2; \quad \Delta H^{\circ}_{298K} = -192.2 \text{ kJ/mol}$$

Also in this case, using TRs, the investigations are mainly related to the catalyst performances. For example, Velu *et al.* (2000) carried out this reaction system on different catalysts based on CuZnAl(Zr), obtaining 90% methanol conversion at 230°C; whereas Reitz *et al.* (2001) studied the characterization of a commercial CuO/ZnO based catalyst, finding a key influence of Cu on the methanol conversion, depending on the conversion rate of oxygen during the reaction. More recently, Lenarda *et al.* (2004) studied the same reaction system over Pd-based catalysts supported on $Zn/ZnAl_2O_4$; and they also prepared a catalyst based on Pd and Pd/Cu

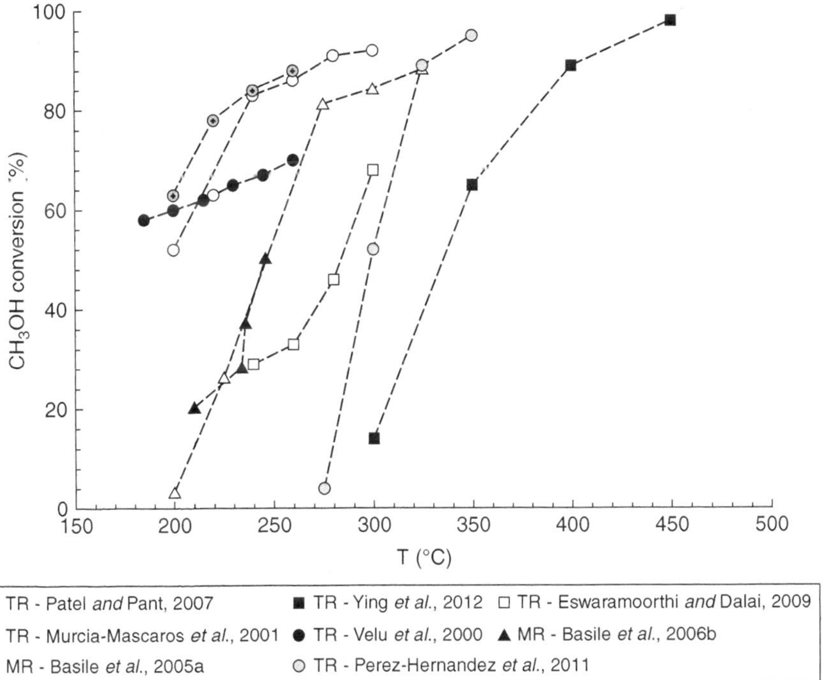

2.12 Oxidative methanol steam reforming: conversion versus reaction temperature.

supported on ordered mesoporous ceria-doped alumina for the same reaction.

Regarding MRs, Basile *et al.* (2005a), using a dense tubular Pd–Ag MR, focused attention on the influence on methanol conversion of the main operating parameters, that is, both the temperature and the O_2/CH_3OH feed molar ratio. As shown in Fig. 2.12, the highest methanol conversion of about 90% has been achieved at 260°C and $O_2/CH_3OH = 0.09$, versus a maximum value of 86% achieved at the same temperature in a TR by Patel and Pant (2007). In these experiments, all the oxygen fed into the MR was consumed and the maximum hydrogen was 28.6%, at 260°C and $O_2/CH_3OH = 0.17$, removed from the reaction zone in the permeate stream.

Oxidative methanol steam reforming was also studied in a flat Pd–Ag MR, in terms of methanol conversion as well as HR and CO selectivity (Basile *et al.,* 2006b). Also using the flat device, it was possible to produce a CO-free hydrogen stream that could be directly used as feed into a PEM fuel cell, demonstrating that it is possible to increase the methanol conversion as well as the hydrogen production and to reduce CO selectivity using a low amount of oxygen in the feed stream of the MR.

2.2.7 Ethanol steam reforming

The ethanol SR reaction is also considered as important for hydrogen production because of the CO_2 neutral nature as well as the less toxicity than methanol.

Focusing their attention on this reaction, different authors indicate that the product distribution depends on the catalyst used and, normally, it is the following: hydrogen, carbon monoxide, carbon dioxide, methane, acetaldehyde and ethylene (Breen *et al.*, 2002).

Also the catalysts used for the ethanol SR reactions are different, and the percentage distribution of the main catalysts used in TRs is illustrated in Fig. 2.13. It appears evident that non-noble-metal catalysts are the most used, probably because many researchers try to maximize hydrogen production, taking into account the economic advantage of finding catalysts cheaper than noble-metal ones.

Generally, as also reported in Fig. 2.14, the metals used for carrying out the ethanol steam reforming reaction in MRs are Rh, Ru, Pd, Pt, Ni, Co and Cu supported on oxides Al_2O_3, SiO_2, MgO and La_2O_3. For example, Liguras *et al.* (2003) reported that the catalytic activity in a TR increases following this trend: Rh >> Pt > Pd > Ru. By using 1%wt Rh/Al_2O_3 catalyst, it is possible to obtain complete ethanol conversion at 800°C, while for achieving the same conversion using a Ru/Al_2O_3 catalyst, 5%wt of metal concentration catalyst is necessary. Considering TRs again, the research in this reaction is still mainly concentrated on catalyst development, which must be able to inhibit the formation of carbon monoxide, besides high hydrogen selectivity and good coke resistance (Freni *et al.*, 2000; Vaidya and Rodrigues, 2006).

Figure 2.15 shows the most important results in terms of ethanol conversion against reaction temperature, by performing the ethanol SR reaction in both TRs and MRs. It is evident that the ethanol conversion for both MRs and TRs is favoured by a temperature increase, due to reaction endothermicity. With respect to TR technology, ethanol conversion exceeds 80% at mild temperatures (450–500°C) only by using Co-based catalysts; whereas conversions higher than 80% are achieved at T > 600°C in all the other cases. But with respect to MR technology, high conversions are also realized in the range of 300–400°C and they are not dependent on the specific catalyst used during the reaction. For instance, Gallucci *et al.* (2007a) evaluated the influence of different parameters, such as membrane type, temperature, sweep gas configuration and sweep gas flow rate in two different Pd-based MRs and in a TR, using the same commercial 5% Ru/Al_2O_3 catalyst. The best results concern the ethanol conversion in MRs, which is always higher than that obtained in the TR, in the range of 350–450°C.

One of the most important problems evidenced by ethanol steam reforming carried out in the TR at low temperature was coke formation: the carbon

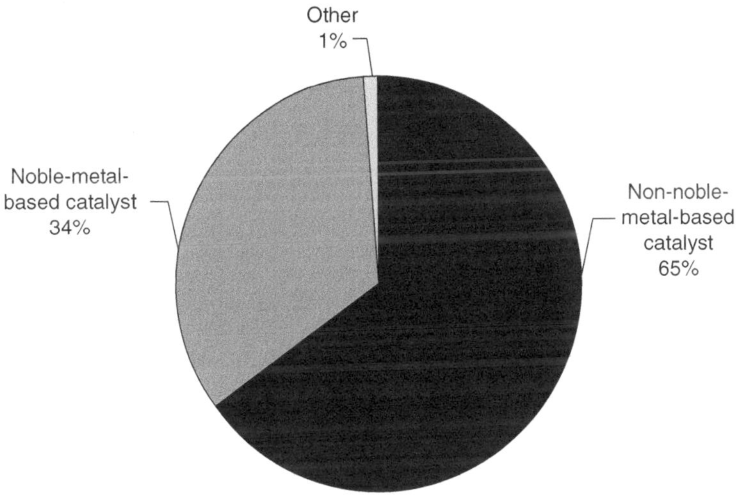

2.13 Percentage distribution of the catalysts used for carrying out the ethanol steam reforming reaction in traditional reactors.

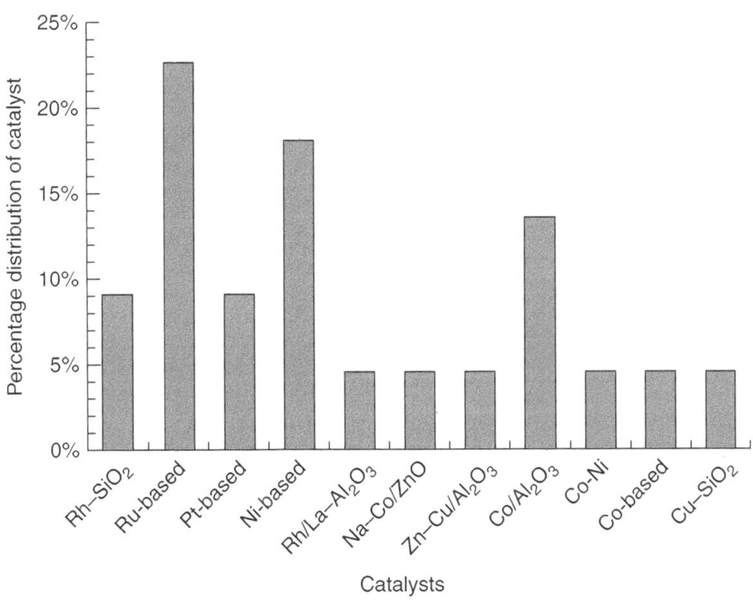

2.14 Percentage distribution of the catalysts used to carry out the ethanol steam reforming in membrane reactors.

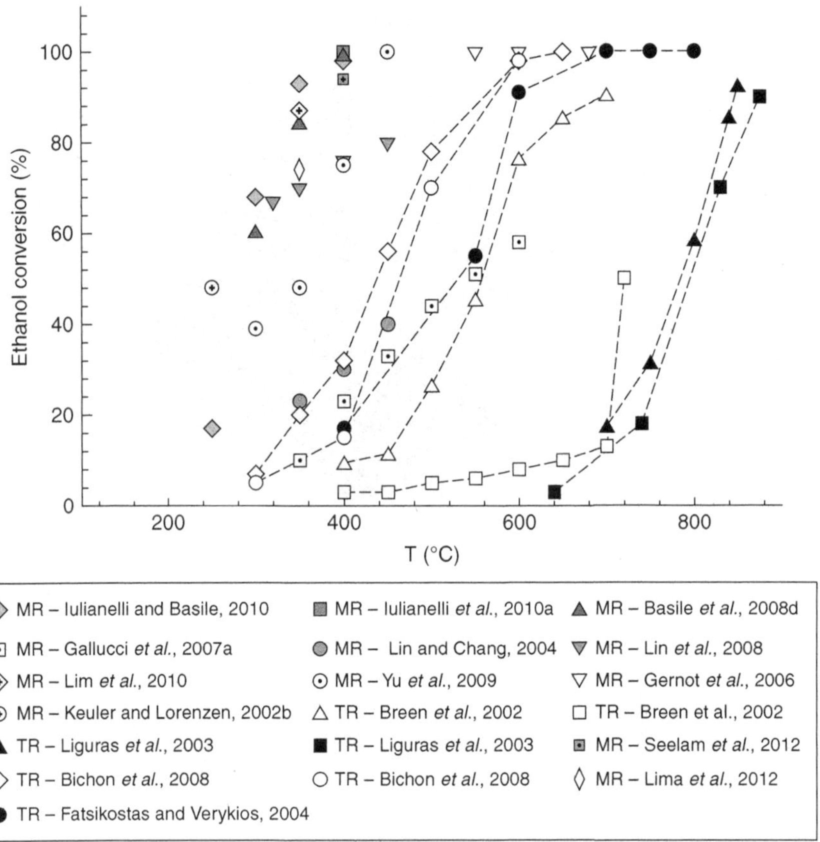

2.15 Ethanol steam reforming reaction: conversion versus reaction temperature.

deposition resulted in a catalyst deactivation during ethanol SR. However, as said above, by using a Pd–Ag MR, the same conversion of a TR was achieved but working at lower temperature of almost 400°C and, in counter-current mode and at 450°C, 40% HR of the total hydrogen produced was obtained, evidencing the advantages of the MRs with respect to the TRs.

2.2.8 Steam reforming of acetic acid

Among the various ways to produce hydrogen from biomass transformation, particularly interesting are methods in which the biomass is converted into intermediate liquid bio-fuels such as pyrolysis oil, acetic acid or ethanol. Among the others, acetic acid is renewable and can be easily obtained from biomass by fermentation. In addition, acetic acid, unlike methanol and ethanol, is non-inflammable; hence, it is a safe hydrogen carrier. Acetic acid can

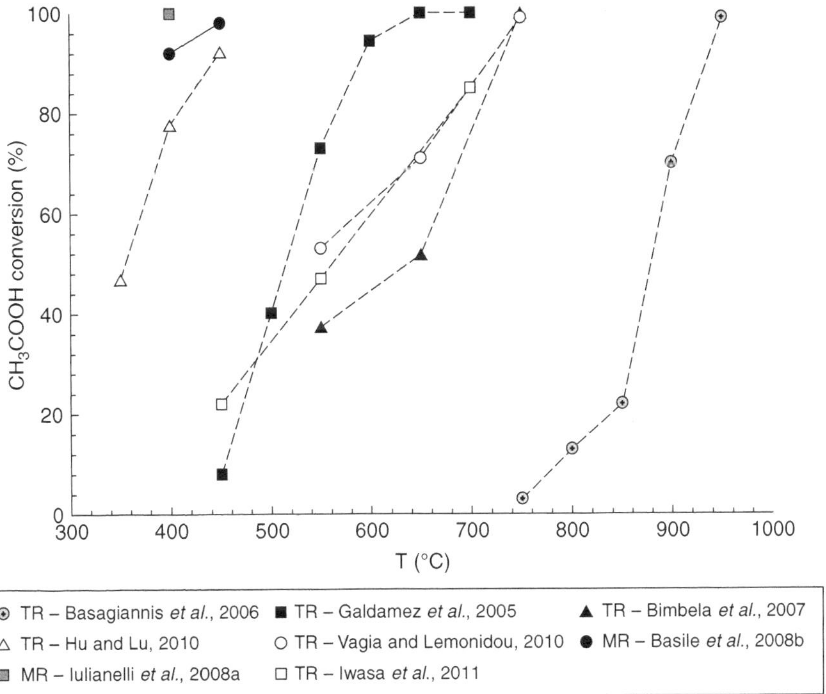

2.16 Acetic acid steam reforming: conversion versus reaction temperature.

also be easily converted to hydrogen with high-selectivity at low temperature over effective catalysts such as Ni–Co and Ru, so it may be a suitable fuel for producing hydrogen for the proton exchange membrane fuel cells.

Despite the potential importance of this reaction, there has been little investigation of catalysts in producing hydrogen from acetic acid SR and only a few papers on the subject have been published. As a comparison of the literature data, the acetic acid conversion, carried out in both TRs and MRs versus reaction temperature, is shown in Fig. 2.16. For example, Hu and Lu (2007; 2010), investigating a series of Ni–Co catalysts, completely converted acetic acid. Their main product was hydrogen: in fact, H_2 and CO_2 selectivities up to 96% and 98%, respectively, were obtained. Also Bimbela *et al.* (2007) proposed the catalytic SR of acetic acid for obtaining high hydrogen yields. They used three different Ni co-precipitated catalysts with varying Ni content (23, 28 and 33%): the 28% Ni presents the best performance. For the same reaction, Takanabe *et al.* (2004) obtained a complete conversion using Pt/ZrO_2 catalysts, whereas the hydrogen yield was close to thermodynamic equilibrium. However, the catalyst is easily deactivated by the formation of oligomers which block the active sites. Basagiannis and

Verykios (2006; 2007a; 2007b) also concentrated their work on the catalyst effects. These authors stated that the reforming reaction, and thus hydrogen production, is enhanced at high temperatures and low space velocities.

The first experimental work on the acetic acid steam reforming reaction for producing pure hydrogen was carried out in a dense Pd–Ag MR by Basile *et al.* (2008b). Two kinds of catalytic bed patterns were considered: in the first pattern, an Ni-based commercial catalyst was packed inside the membrane lumen, while in the second one both Ru-based and Ni-based commercial catalysts were packed together in the membrane lumen. Experimental tests were performed in the temperature and pressure ranges of 400–450°C and 1.5–2.5 bar, respectively. The results show that MR is able to give a rather high acetic acid conversion with a 30–35% HR. Moreover, the second catalytic bed pattern is able to both increase the hydrogen production and decrease the methane production, with respect to the first pattern. Afterwards, Iulianelli *et al.* (2008a) investigated the influence of other operating conditions, such as different flow configuration, sweep gas flow rate and reaction pressure on the Pd–Ag MR performances. As the best result of this work, complete acetic acid conversion and 70% of HR were achieved at 400°C, 1.5 bar as reaction pressure, highest sweep gas flow rate and in counter-current configuration.

2.2.9 Steam reforming of glycerol

Glycerol can be regarded as an important bio-source for producing hydrogen. Nowadays, glycerol is produced in large quantities as by-product of bio-diesel production. It is characterized by high energy density, and it is non-toxic (Xuan *et al.,* 2009).

Currently, glycerol is used in many applications, such as personal care, and for polymer and pharmaceutical use. However, growth of the bio-diesel industry has created a huge amount of glycerol which has led, as a consequence, a reduction in the glycerol market price (Adhikari *et al.,* 2007). Therefore, an alternative use for glycerol is important, and one possibility is to use it as renewable feedstock for producing hydrogen and syngas by SR reaction.

It could be useful to give an overview of the studies present in the open literature on this reaction. Currently, only a few studies are focused on glycerol SR for hydrogen production. In particular, this reaction can be carried out in either aqueous or gas phase. When the reaction is performed in aqueous phase, its low catalyst deactivation is an advantage; nevertheless, high pressures are required. On the other hand, in gas phase, the reaction can be carried out at atmospheric pressure, and, as a drawback, the catalyst is subject to a great deactivation (Hirai *et al.,* 2005).

Concerning glycerol SR reaction in TR gas phase, the researchers mainly address the effects of catalysts. For instance, Adhikari *et al.* (2008) studied nickel-based catalysts with MgO, CeO_2 and TiO_2 supports. They found that maximum hydrogen yield could be obtained at 650°C with MgO supported catalysts. Iriondo *et al.* (2008), using a modified alumina-supported Ni catalysts with Ce, Mg, Zr and La for producing hydrogen from glycerol, revealed an increase of the hydrogen selectivity. Furthermore, the authors deduced that: (a) Ce and La can increase stability of Ni; (b) Mg can enhance surface Ni concentration; and (c) Zr can improve the capacity to activate steam.

Besides Ni-based catalysts, other types of catalysts were also evaluated. For example, Ce supported Ir, Co and Ni catalysts have been studied by Zhang *et al.* (2007a), which exhibited significant activity and selectivity, since the dehydration of glycerol to ethylene or propylene did not occur, thus avoiding coke formation and catalyst deactivation. As a comparison, Hirai *et al.* (2005) developed a novel efficient catalyst for this reaction: Ru catalysts were preferred and high performance was observed for the Ru/Y_2O_3 one.

The first experimental work on the glycerol SR reaction performed in a dense Pd–Ag MR for producing pure hydrogen was carried out by Iulianelli *et al.* (2010b; 2011). In particular, the authors studied the catalyst's influence on reactor performances (glycerol conversion and pure HR), using two commercial catalysts: Co/Al_2O_3 and Ru/Al_2O_3. A glycerol conversion of 94.0% and a pure HR higher than 60.0% were obtained by using the Co/Al_2O_3 catalyst at 4.0 bar and 400°C, while 20.0% glycerol conversion and 16.0% pure HR were achieved by using Ru/Al_2O_3 catalyst at 5.0 bar. Nevertheless, the authors observed carbon formation during the reaction, affecting negatively the performance of the Pd–Ag membrane in terms of lower hydrogen permeated flux and catalyst deactivation.

In Fig. 2.17, a short summary of the most significant scientific results in terms of glycerol conversion at different temperatures, using both TRs and Pd–Ag MRs, is given. The main indication from the graph is that MRs operate at a lower temperature than TRs. This result is greatly important, because lower operating temperatures also means higher energy saving and, as a consequence, MRs could result in a cheaper solution to performing this reaction.

2.2.10 Dehydrogenation reactions

Dehydrogenation reactions are considered an important way of producing hydrogen.

It is well known that dehydrogenation reactions are endothermic and limited by thermodynamic equilibrium. Nevertheless, as already mentioned

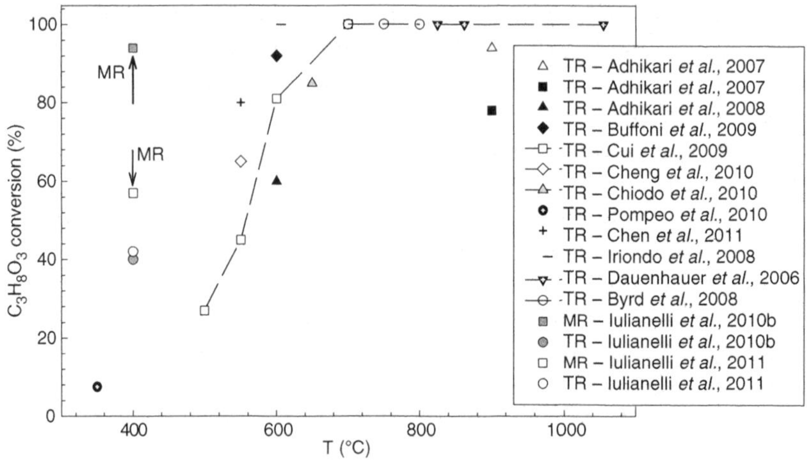

2.17 Glycerol steam reforming: conversion versus reaction temperature.

often in this chapter, thermodynamic constraints can be overcome by using dense Pd-based MRs. In particular, a consistent number of dehydrogenation reactions have been investigated using inorganic MRs (Ali *et al.,* 1994; Itoh, 1987; 1992; Matsuda *et al.,* 1993; Mondal and Ilias, 2001; Raich and Foley, 1998; She *et al.,* 2001).

For instance, Itoh *et al.* (2003) studied the dehydrogenation of cyclohexane by using Pd membrane tube packed with 0.5% Pt/Al_2O_3 catalyst. The reaction was carried out varying both reaction temperature and pressure in the ranges of 250–300°C and 1–4 bar, respectively. Moreover, the permeate pressure was controlled by vacuum pump in the range of 0.1–1 bar. The conversion at 300°C (reaction pressure 2.5 bar, permeate pressure 0.1 bar) was 97%, versus 19% equilibrium value. In a previous work, Itoh *et al.* (1988) performed the same reaction also in a Vycor glass MR with mean pore size of 40 Å, obtaining 45% of cyclohexane conversion at 200°C and 1 bar. However, carbon-based (Itoh and Haraya, 2000) and alumina-based (Tiscarno-Lechuga *et al.,* 1996) MRs have been also used for carrying out this reaction. And recently Jeong *et al.* (2003), using a zeolite MR packed with a Pt/Al_2O_3 catalyst at 200°C and ambient pressure, obtained 73% of hydrocarbon conversion with respect to 32% of equilibrium value.

Regarding isobutane dehydrogenation, Guo *et al.* (2003) have used a Pd–Ag/ceramic composite MR packed with a Cr_2O_3–Al_2O_3 pellets catalyst. Moreover, comparison with a conventional reactor working at the same MR operating conditions was given. In particular, at 450°C, the MR was able to achieve 51% isobutene conversion as opposed to 16% in the TR and 19% equilibrium.

However, Casanave *et al.* (1995) performed the same reaction in a zeolite-based MR using Pt–Sn/γ-Al_2O_3 as catalyst at 450°C. Also in this case, the MR showed better performance than the TR in terms of isobutylene yield, achieving 22.4% isobutylene yield in the MR, as a best result, with respect to 13% in the TR.

Ethylbenzene dehydrogenation is another worthy reaction for hydrogen production. For instance, Yu and Xu (2011) carried out this reaction in both a Pd-based MR and TR, obtaining 78% and 62%, respectively, of ethylbenzene conversion (at 580°C and 1.4 bar), versus 69% equilibrium conversion. On the other hand, Becker *et al.* (1993) obtained 10–15% higher ethylbenzene conversions than a TR using an alumina MR.

Other studies were focused on the use of Pd-based MRs by performing dehydrogenation reactions of methylcyclohexane (Ali and Baiker, 1996; 1997) or isobutane (Liang and Hughes, 2005) or propane (Brinker et al., 1996). In all these studies, Pd-based MRs were always able to obtain higher performances in terms of conversion and selectivities than TRs.

Table 2.1 presents a summary of the results related to the reactions cited above, in which different types of membranes have been used in MRs and, in some cases, a comparison with TRs has been also indicated. Making a qualitative analysis of the performances realized using different MRs, it is quite clear that MRs always achieve higher conversions than TRs. Moreover, dense Pd-based MRs seem to show better performance in terms of conversion and yield than other inorganic membranes.

2.3 Types of membranes

The existing membranes used to move H_2 to the permeate side in MRs are generally divided into two types: porous and dense (or non-porous) membranes. The ability of a membrane to address the reactions to proceed towards completion strongly depends on the type used. In Table 2.2 a comparison among the different types of membranes is presented. Intuitively, a porous membrane, owing to its relatively large pore size, can never perform the same separation shown by a dense membrane and, correspondingly, the porous membrane has a higher permeability than the dense one.

2.3.1 Dense inorganic membranes

Some materials, capable of being selectively permeated by only one gas, are present in nature. Generally speaking, almost all metals are permeable only to H_2, while other materials are capable of being selectively permeated only by oxygen. In particular, some dense materials (Pd and its alloys, Pt, V, Ta, Nb and SiO_2 ceramic membrane) exhibit significant H_2 permeability

Table 2.1 Dehydrogenation reactions carried out in membrane reactors

Membrane/catalyst	T (°C)	P (kPa)	Notes	References
Dehydrogenation of 1,2-cyclohexanediol $C_6H_{12}O_2 \rightarrow C_6H_6O_2 + H_2$				
Pd–Rh (7%) Foil	250	–	Y = 95.0% No phenol formation	Sarylova *et al.*, 1977
Pd–Cu (37–42%) Foil	250–400	–	At 400°C $Y_{C_6H_6O_2\,MAX}$ = 30% with Pd–Cu (42%)	Mishchenko *et al.*, 1977
Dehydrogenation of 2-butanol $C_4H_{10}O \rightarrow C_4H_8O + H_2$				
Pd–Ag (23 wt%)/Al_2O_3 Tube Cu(14.4 wt%)/SiO_2 (pellets)	190–240	101	χ = 93.0% at 240°C χ_{eq} = 80.0% at 240°C Sel $-_{C_4H_8O}$ = 96.0%	Keuler and Lorenzen, 2002a
Dehydrogenation of ammonia $NH_3 \rightarrow \frac{1}{2}N_2 + \frac{3}{2}H_2$				
Composite Pd-ceramic membrane Tube Ni/Al_2O_3	450–600	1618	At 600°C: χ_{PBMR} = 94.0% χ_{PBR} = 53.0% χ_{eq} = 58.0%	Collins and Way, 1994
Dehydrogenation of butene $C_4H_8 \rightarrow C_4H_6 + H_2$				
Ag/Cr_2O_3–Al_2O_3	538	108	$Y_{C_4H_6} = \chi_{C_4H_8}$ Sel-$_{C_4H_6} = 20 \times 0.94 = 19\%$	De Rosset and Hills, 1968
Dehydrogenation of cis-3-hexen-1-ol $C_6H_{12}O \rightarrow C_6H_{10}O + C_6H_{12}O + C_6H_{14}O + H_2$				
Pd–Ag (23%) Tube CuO-ZnO (cylinder pellet)	170–230	–	Sel-$_{1\text{-Hexanol}}$ = 17.0% Sel-*cis*$_{\text{-3-hexenal}}$ = 9.4 ÷ 65.0% Sel-$_{\text{Hexanal}}$ = 32.6 ÷ 80.4%	Sato *et al.*, 2007

Dehydrogenation of cyclohexane $C_6H_{12} \rightarrow C_6H_6 + 3H_2$				
Pd–Ag (23%) Tube	125	–	χ = 87.0%	Wood, 1968
Pd–W (5%)-Ru (1%)	300–500	100	Y = 51.4% at 492°C	Smirnov *et al.*, 1977
Pd–Ru (10%) Foil	330–575	100	Y = 91.0% at 340°C	Gryaznov *et al.*, 1977
FAU-type zeolite membrane/Al_2O_3 Tube Pt(1.0 wt%)/Al_2O_3	150–250	101	at 200°C, sweep flow = 100 $cm^3 \cdot min^{-1}$ and feed rate = 1.1 $mol \cdot h^{-1}$: χ = 72.1% χ_{eq} = 32.2%	Jeong *et al.*, 2003
Pd/Al_2O_3 Tube Pt(0.5 wt%)/Al_2O_3 (pellets)	250–300	100–400	χ = 97.0% at 300°C χ_{eq} = 18.7% HR = 93.0%	Itoh *et al.*, 2003
Dehydrogenation of cyclohexene $C_6H_{10} \rightarrow C_6H_6 + 2H_2$				
Au coated Pd–Ni	100–320	100	–	Aguilar *et al.*, 1977
Dehydrogenation of cyclohexanol $C_6H_{11}OH \rightarrow C_6H_{10}O + H_2$				
Pd–Ru (9.8%) Pd–Rh (5%), Pd–Rh (15%) Pd–Ru (4%)-Pb (1%)	150–280	100	$Y_{C6H10)}$ = 90% at 280°C Pd–Ru(9.8%) at the coupling with cyclopentadien hydrogenation	Basov *et al.*, 1993
Dehydrogenation of ethane $C_2H_6 \rightarrow C_2H_4 + H_2$				
Pd–Ag (23%)/Vycor Tube Pd/Al_2O_3 (cylindrical pellets)	387	128.7	Contact time is very important: χ(measured) = 18.4% at 1.5 $gcat \cdot s \cdot gmol^{-1}$ χ(predicted) = 26.1% at 1.5 $gcat \cdot s \cdot gmol^{-1}$	Gobina *et al.*, 1995a
Pd–Ag Tube Re/HZSM-5	500–585	100	Sel_{-C2H4} = 70.0 ÷ 80.0% at 545°C χ_{C2H6} = 9.0 ÷ 25.0% at 545°C	Wang *et al.*, 2003

(*Continued*)

Table 2.1 Continued

Membrane/catalyst	T (°C)	P (kPa)	Notes	References
Dehydrogenation of ethylbenzene $C_8H_{10} \rightarrow C_8H_8 + H_2$				
Pd Tube	700	3	With seven fluidized bed in series and 16 membrane tubes: $\chi_{C8H10} = 96.5\%$ $Y_{C8H8} = 92.4\%$	Abballa and Elnashaie, 1995
Pd–Ag/porous substrate Tube	600	–	Simulation study: $\chi = 62.1\%$ Co-current mode $\chi = 78.8\%$ Counter-current mode	Gobina *et al.*, 1995b
Pd-coated porous stainless steel Tube	620	100–250	Simulation study: $\chi = 88.8\%$ at 620°C and 250 kPa $\chi_{eq} = 71.2\%$ Sel-$_{C_8H_8} = 95.2\%$ $Y_{C_8H_8} = 83.8\%$	Hermann *et al.*, 1997
Pd/porous ceramic Tube $80\%Fe_2O_3$-$20\%K_2O$ $75\%Fe_2O_3$-$20\%K_2O$-$5\%CeO_2$ $70\%Fe_2O_3$-$20\%K_2O$-$5\%CeO_2$-$5\%Cr_2O_3$	600	200	$\chi_{PBR} = 46.5\%; Y_{C_8H_8} = 33.0\%$ $\chi_{PBMR} = 52.6\%; Y_{C_8H_8} = 41.3\%$ $\chi_{FLBR} = 34.2\%; Y_{C_8H_8} = 19.5\%$ $\chi_{FLBMR} = 67.9\%; Y_{C_8H_8} = 55.6\%$ $\chi_{PBR} = 52.4\%; Y_{C_8H_8} = 28.4\%$ $\chi_{PBMR} = 58.3\%; Y_{C_8H_8} = 39.2\%$ $\chi_{FLBR} = 37.7\%; Y_{C_8H_8} = 12.7\%$ $\chi_{FLBMR} = 73.5\%; Y_{C_8H_8} = 54.0\%$	Elnashaie *et al.*, 2001

			$\chi_{PBR} = 51.3\%; Y_{C_8H_8} = 44.3\%$ $\chi_{PBMR} = 64.8\%; Y_{C_8H_8} = 60.5\%$ $\chi_{FLBR} = 36.3\%; Y_{C_8H_8} = 23.4\%$ $\chi_{FLBMR} = 79.1\%; Y_{C_8H_8} = 75.4\%$	
Pd/ceramic Tube	580	140	$\chi = 78\%$ $\chi_{eq} = 70\%$ $\chi_{PBR} = 64\%$	Yu and Xu, 2011
Zeolite silicate supported on PSS membrane Tube Iron oxide catalyst	580–640	80	At 640°C $\chi_{MR} = 85.3\%; Y_{C_8H_8} = 74.2\%$ $\chi_{TR} = 75.8\%; Y_{C_8H_8} = 66.2\%$	Kong *et al.*, 2007
Dehydrocyclization of heptane $C_7H_{16} \rightarrow C_7H_8 + C_6H_6 + CO + CO_2 + H_2$				
Pd–W (5%)-Ru (1%)	450–590	100	$Y_{C_6H_6} = 54.7\%; Y_{C_7H_8} = 2.6\%$ at 590°C $Y_{CH_4} = 30.6\%$	Smirnov *et al.*, 1978
Pd–Rh (5%)	500–625	–	–	Gryaznov *et al.*, 1983
Dense membrane BSCFO ($Ba_{0.5}Sr_{0.5}Co_{0.8}Fe_{0.2}O_3$) Disk LiLaNiO/$\gamma$- Al_2O_3	750–950	–	$\chi = 100\%$ at 850°C $Sel_{CO} = 90.0–92.0\%$ $Sel_{H_2} = 95.0–97.0\%$	Zhu *et al*, 1993
Dehydrogenation of hydriodic acid $2HI \rightarrow H_2 + I_2$				
Pd–Ag(23%) Tube	500	–	$\chi = 4.0\%$ $\chi_{eq} = 2.0\%$	Yehenskel *et al.*, 1979

(*Continued*)

Table 2.1 Continued

Membrane/catalyst	T (°C)	P (kPa)	Notes	References
Dehydrogenation of hydrogen sulphide $H_2S \rightarrow H_2 + S$				
Composite Pt-SiO_2-V-SiO_2-Pd Disk	≥ 700	–	$\chi > 99.4\%$ $\chi_{eq} = 13.0\%$	Edlun and Pledger, 1983
Pd membrane Tube	700–1000	101	Simulation study, co-current mode $\chi \sim 82.0\%$ at 1000°C	Zaman and Chakma, 1995
Ceramic membrane Tube			High sweep rates can push the reaction to nearly 100% conversion $\chi = 39.0\%$ at 1000°C	
ZrO_2–SiO_2 composite membrane – tube	700–750	110–220	Simulation study Y = 0.35 at 750°C, 110 kPa	Fan *et al.*, 1999
Dehydrogenation of isoamylene $C_5H_{10} \rightarrow C_5H_8 + H_2$				
Pd–Ni (5.5%)	280–500	–	–	Gryaznov and Smirnov, 1974
Pd–Ni (5.9%)	275–430	–	$Y_{C_5H_8} = 30.3\%$ at 422°C	Orekova *et al.*, 1976
Pd–Ni (5%)-Ru (1%)	330–440	–	Y = 28.1% at 440°C	Smirnov *et al.*, 1977
Dehydrogenation of isobutane $C_4H_{10} \rightarrow C_4H_8 + C_4H_6 + H_2$				
SiO_2/Vycor Cr_2O_3/Al_2O_3	450–550	–	$Y_{C_4H_8} = 36.4\%$ at 450°C $Y_{TR} = 29.0\%$ $\chi_{eq} = 33.2\%$	Ioannides and Gavalas, 1993
50 wt% CrO_3/Al_2O_3 $CrO_3 + K_2CO_3/\alpha\text{-}Al_2O_3$	450–500	–	$Y_{C_4H_8} = 32.0\%$ at 450°C $Y_{TR} = 19.2\%$ $\chi_{eq} = 24.9\%$	Zhu *et al.*, 2005
Pd/Al_2O_3 CrO_3–Al_2O_3, Pt/Al_2O_3	350–450	–	$Y_{C_4H_8} = 28.5\%$ at 450°C $Y_{TR} = 6.0\%$ $\chi_{eq} = 10.5\%$	Matsuda *et al.*, 1993

Zeolite Tube Pt–Sn/γ-Al_2O_3 (cylindrical pellets)	450	–	$Y_{C_4H_8}$ = 22.4% counter-current $Y_{C_4H_8}$ = 21.7% co-current Y_{TR} = 13.5%	Casanave *et al.*, 1999
2% Ru/Pd Tube Pt(0.52%)/Al_2O_3 (pellets)	500	101	$Y_{C_4H_8}$ = 75.9% χ = 81.2% χ_{eq} = 32.4%	Sheintuch and Dessau, 1996
Zeolite composite MF1-alumina Tube Bimetallic Pt-In/SiO_2	20–450	125	At 450°C: the best selectivity: SF = 25; feed: 12.0% H_2, 15.0% C_4H_{10}, 73.0% N_2; sweep gas (N_2) = 5.4 × 10^{-5} mol·s^{-1} Counter-current mode	Ciavarella *et al.*, 2000
Zeolite membrane Cr_2O_3/Al_2O_3	440–490	101	At 490°C $Y_{C_4H_8}$,MR = 41.0% $Y_{C_4H_8,eq}$ = 28.0%	Van der Bergh *et al.*, 2011
MF1-alumina ceramic membrane Tube Cr_2O_3/Al_2O_3 (spheres)	480–540	101	WHSV = 0.5 (h^{-1}): SF = 3 counter-current: χ = 41.7% SF = 6 counter-current: χ = 49% SF = 6 co-current: χ = 48.6% WHSV = 0.8 (h^{-1}): SF = 3.5 counter-current: χ = 39.3% SF = 6 counter-current: χ = 38.5% SF = 6 co-current: χ = 46.5% WHSV = 1.6 (h^{-1}): SF = 3 co-current: χ = 38.1%	Illgen *et al.*, 2001
Pd/Ag ceramic composite membrane Tube Cr_2O_3–Al_2O_3 (pellets)	350–500	100–300	χ = 50.5% at 450°C χ_{eq} = 18.8% χ_{TR} = 15.5%	Guo *et al.*, 2003

(*Continued*)

Table 2.1 Continued

Membrane/catalyst	T (°C)	P (kPa)	Notes	References
Pd/Ag Membrane tube Pt(0.5%)/Al_2O_3 (pellets)	485–520	106–140	Sweep gas (N_2) = 1000 Nml/min $Y_{C_4H_8}$ = 20.0% at 485°C, 140kPa $Y_{C_4H_8}$ = 29.0% at 520°C, 140kPa Sel-$_{C_4H_8}$ = 92.0% at 485°C, 106kPa	Liang and Hughes, 2005
Dehydrogenation of isopropanol $C_3H_8O \rightarrow C_3H_6O + H_2$				
Pd Foil	156	4	–	Mikhlenko *et al.*, 1986
Porous Vycor glass Tube Cu (wt 34%), Cr (wt 25%)	115–150	101	χ = 23.6% at 115°C χ_{eq} = 21.4% at 115°C	Trianto *et al.*, 2001
Dehydrogenation of methylcyclohexane $C_7H_{14} \rightarrow C_7H_8 + 3H_2$				
Pd–Ag (23%) Tube Pt/Al_2O_3	320–400	1500	at 370°C: χ_{MR} = 90.0% χ_{eq} = 73.3% χ_{TR} = 65.0%	Ali and Baiker, 1997
Porous Vycor membrane Tube Pt/Al_2O_3	150–300	101	Counter-current mode χ = 65.0% at 180°C	Ferreira-Aparicio *et al.*, 2002b
Dehydrogenation of n*-hexane* $C_6H_{14} \rightarrow C_6H_6 + 3H_2$				
Pd–Ru (5%) Foil	465–575	–	–	Gryaznov *et al.*, 1977
Pd–Ru (5%)	530–575	–	$Y_{C_6H_6}$ = 50.0–58.0%	Smirnov *et al.*,1977
Pd–Ni (5.9%) Spiral tube	520	–	–	Lebedeva and Gryaznov, 1981

Dehydrogenation of octane $C_8H_{18} \rightarrow C_6H_6 + C_XH_{nx} + H_2$				
Pd–Re (5%)	450–550	–	$Y_{C_6H_6} = 16.5\%$	Smirnov *et al.*, 1977
Dehydrogenation of propane $3C_3H_8 \rightarrow C_3H_6 + H_2$				
Pd–Ag (24%) Tube Ga-ZSM-5 zeolite (pellet)	550	–	$\chi = 87.0\%$	Clayson and Howard, 1987
Pd/Al_2O_3 Tube Ga-ZSM-5 zeolite (pellet)	500	–	$\chi = 95.0\%$	Uemiya *et al.*, 1991
Pd–Ag (25%) Pd–Ru (2%) Tube Pt	550–570	101	at 550°C: $Y_{C_3H_6}$, MR = 70.0% with Pd–Ru $Y_{C_3H_6,eq} = 32.0\%$ $Y_{C_3H_6,MR} = 47.5\%$ with Pd–Ag	Sheintuch and Dessau, 1996
Pd-composite membrane Tube Pt/Al_2SiO_5	500–575	100	$Sel-_{C_3H_6} = 97.0\%$ at 500°C: $Y_{C_3H_6}$, MR = 16.0% $Y_{C_3H_6}$, TR = 12.0%	Brinker *et al.*, 1996
Silica composite membrane Pd–Ag composite Tube Chromia-alumina	500	101	$Sel-_{C_3H_6} = 85.0\%-95.0\%$ at WHSV = 0.005–0.1 h^{-1}: $\chi = 60.0\%-80.0\%$	Weyten *et al.*, 2000

(*Continued*)

Table 2.1 Continued

Membrane/catalyst	T (°C)	P (kPa)	Notes	References
Pd/Al_2O_3 composite (1) Plated Pd-stainless steel (2) Tube Chromia-alumina	560	130	$\chi_{eq} = 30.2\%$ $\chi_{TR} = 29.2\%$ $Y_{C_3H_6}$, TR = 22.4% $\chi_{MR(1)} = 34.7\%$ $Y_{C_3H_6}$, $MR_{(1)} = 27.8\%$ $\chi_{MR(2)} = 46.5\%$ $Y_{C_3H_6}$, $MR_{(2)} = 26.7\%$	Quicker, 2000
Ceramic membrane Tube Cr_2O_3/Al_2O_3 $Pt–Sn/Al_2O_3$	500–535	–	at 535°C: $\chi_{MR} = 49.0\%$ $\chi_{CFB} = 37.0\%$ $Sel\text{-}_{C_3H_6,MR} = 70.0\%$ $Sel\text{-}_{C_3H_6,CFB} = 80.0\%$ $Y_{C_3H_6,MR} = 34.0\%$, $Y_{C_3H_6,CFB} = 29.0\%$ at 500°C: $Sel\text{-}_{C_3H_6,MR} = 80.0\%$ $Sel\text{-}_{C_3H_6,CFB} = 70.0\%$ $Y_{C_3H_6,MR} = 22.0\%$, $Y_{C_3H_6,CFB} = 17.0\%$	Schäfer *et al.*, 2003
Dehydrogenation of β*-ionol* $C_{13}H_{22}O \rightarrow C_{13}H_{20}O + H_2$				
Pd–Ni (5%)	100–200	–	–	Smirnov *et al.*,1978
Dehydrogenation water $H_2O \rightarrow H_2 + \frac{1}{2}O_2$				
Pd	450–800	–	–	Compagnie des Métaux Précieux, 1976

Single membrane Double membrane	1727	–	Simulation study: χ_{single} = 25.0%, co-current mode χ_{single} = 37.0%, counter-current mode χ_{double} = 61.0%	Omorjan *et al.*, 1999
Dry reforming of methane $CH_4 + CO_2 \leftrightarrow 2CO + 2H_2$				
Pd–Ag/PSS Tube Rh/La_2O_3	450	100	χ_{CH4} = 15.0% HR = 80.0%	Bosko *et al.*, 2010
MR, dense Pd/α-Al_2O_3, Pd/γ-Al_2O_3	550	–	$\chi_{CH_4}(500°C, MR) = 37.50\%$	Smithells, 1937
Porous and dense Pd–Ag tubular membranes Ni/Al_2O_3	350–500	100–200	$\chi_{CH_4}(450°C, MR) = 26.82\%$	Gallucci *et al.*, 2008a
MR, dense Pd/Ag, Pt/CeO_2-Al_2O_3	500	100	$\chi_{CH_4}(500°C, MR) = 17.41\%$	Kikuchi, 1995
Oxidative dehydrogenation of ethane $C_2H_6 + O_2 \rightarrow 2CO + 3H_2$				
Porous membrane Tube Pt/Al_2O_3	600	–	$\chi_{C_2H_6} = 46.0\%$ Sel $-_{C_2H_4} = 96.0\%$	Champagnie *et al.*, 1990
Porous membrane Tube $MgO/LiO/Sm_2O_3$	600	–	$\chi_{C_2H_6} = 95.0\%$ Sel-$_{C_2H_4} = 53.0\%$ Sel-$_{C_2H_4,TR} = 8.1\%$ $Y_{C_2H_4} = 50.5\%$ $Y_{C_2H_4,TR} = 8.4\%$	Tonkovich *et al.*, 1996

(*Continued*)

Table 2.1 Continued

Membrane/catalyst	T (°C)	P (kPa)	Notes	References
Dense membrane Tube $Bi_{1.5}Y_{0.3}Sm_{0.2}O_3$ (BYS)	825–875	$p_{C_2H_6} = 5-15$ $p_{O_2} = 21$	$Y_{C_2H_4} = 56.0\%$ at 875°C Sel-$_{C_2H_4} = 80\%$ $Y_{C_2H_4,TR} = 11.0\%$ at 900°C Sel-$_{C_2H_4,TR} = 11.0\%$	Akin and Lin, 2002
Membrane of BSCFO ($Ba_{0.5}Sr_{0.5}Co_{0.8}Fe_{0.2}O_{3-\delta}$) Tube LiLaNiO/$\gamma$-$Al_2O_3$	800–900	101	At 900°C: Sel-$_{H_2} = 22.9\%$ Sel-$_{CO} = 32.4\%$ $\chi_{C_2H_6} = 99.1\%$	Wang *et al.*, 2002
Porous membrane Tube V_2O_5/ γ-Al_2O_3	500–680	101	$Y_{C_2H_4,FLBMR} = 36.6\%$ $Y_{C_2H_4,FLBR} = 23.3\%$ $Y_{C_2H_4,PBMR} = 34.9\%$	Ahchieva *et al.*, 2005
Oxidative dehydrogenation of isobutane $C_4H_{10} + 2O_2 \rightarrow 4CO + 5H_2$				
Pd, Pd/Ag Foils (1) Pt/α-Al_2O_3 (monoliths) (2) Rh/α-Al_2O_3 (monoliths) (3) Pt/γ-Al_2O_3 (pellets)	400–700	112	$Y_{C_4H_8} = 9.3\%$ at 668°C $Y_{TR} = 8.4\%$ at 700°C (1) $\chi = 69.0\%$ at 616°C (2) $\chi = 61.0\%$ at 628°C (3) $\chi = 55.0\%$ at 704°C	Raybold and Huff, 2000
Oxidative dehydrogenation of methane $CH_4 + 0.5O_2 \rightarrow CO + 2H_2$				
Pd Tube Ni	250–550	120–160	$\chi_{MAX} = 96.0\%$ at 550°C Sel-$_{H_2} = 88.3\%$ Sel-$_{H_2,eq} = 83.0\%$	Basile *et al.*, 2001a

Perovskite membrane Tube Ni-based catalyst	900	–	$\chi > 99.0\%$ Sel-$_{CO}$ = 94%	Luo *et al.*, 2010
Pd Tube Rh/Al_2O_3	500	160	$\chi = 85.0\%$ Sel-$_{CO}$ = trace Sel-$_{H_2}$ = 60.0%	Cheng *et al.*, 2009
Oxidative dehydrogenation of ethanol $C_2H_5OH + 0.5O_2 \rightarrow 2CO + 3H_2$				
Pd–Ag Tube Rh/Al_2O_3	450	100–300	$\chi = 100\%$ Sel-$_{H_2}$ = 34.0% $Y_{H_2} = 32.0\%$	Iulianelli *et al.*, 2010c
Oxidative dehydrogenation of methanol $CH_3OH + \frac{1}{2}O_2 \rightarrow 2H_2 + CO_2$				
Composite multi-layered membrane Tube Pt	275	$\Delta P_{Trans\text{-}Membr.}$ 0–16	$\chi_{CH_3OH} > 84.0\%$ Sel-$_{H_2}$ = 50.0 – 55.0%	Brinkmann *et al.*, 2001
Pd–Ag Tube $CuO/ZnO/Al_2O_3$ pellet	200–260	260	$\chi_{TR} = 83.0\%$ at 325°C At 260°C: $\chi_{MR} = 90.0\%$ at $O_2/CH_3OH = 0.09$ $Y_{H_2,MAX} = 28.0\%$ at $O_2/CH_3OH = 0.17$	Basile *et al.*, 2005a
Vanadium oxide/α-Al_2O_3 Tube	200–400	101	At 390°C: $\chi_{MR} = 95.0\%$	Ermilova *et al.*, 2008
Steam reforming of acetic acid $CH_3COOH + 2H_2O \rightarrow 2CO_2 + 4H_2$				
Pd–Ag Tube Two catalytic bed patterns: Ni-based commercial catalyst (1)	400–450	150–250	At 400°C: Sel-$_{H_2}$(1) = 30.0% HR(1) = 19.5% at 150 kPa HR(1) = 26.0% at 250 kPa At 450°C: Sel-$_{H_2}$(1) = 40.0%	Basile et al., 2008b

(*Continued*)

Table 2.1 Continued

Membrane/catalyst	T (°C)	P (kPa)	Notes	References
Ru(5%)-based and Ni-based commercial catalyst (2)	–	–	Sel-$_{H2}$(2) = 45.0% HR(1) = 20.0% at 150 kPa HR(1) = 32.0% at 250 kPa HR(2) = 36.0% at 250 kPa	
Pd–Ag Tube Ni/Al_2O_3	400	150–350	At SF = 29.2 Y_{H2} = 60.0% Sel-$_{H_2}$ = 70.0%	Iulianelli *et al.*, 2008a
Steam reforming of ethanol $C_2H_5OH + 3H_2O \rightarrow 2CO_2 + 6H_2$				
NiP Cu composite membrane Tube $Cu\text{-}P/SiO_2$	260–310	–	χ = 87.8%	Liu *et al.*, 2003
TiO_2–Al_2O_3 commercial membrane with a Pd–Ag deposit (MR1) ceramic membrane with a Pd–Ag deposit (MR2) Pd/Ag (MR3) Tube 5% Ru/Al_2O_3	350–600	130	MR1: Sel-$_{H_2}$ = 80.1% at 600°C Sel-$_{H_2,TR}$ = 70.3% at 600°C MR2: χ = 37.0% at 550°C χ_{TR} = 47.0% at 550°C MR3: HR = 38.7% in counter-current HR = 29.9% in co-current χ = 49.6% in counter-current χ_{TR} = 40.0%	Gallucci *et al.*, 2007a
Pd-Ag Tube 5% Ru/Al_2O_3	400–450	150–200	At 400°C, 200 kPa in co-current: $Y_{H_2,MAX}$ = 80.0%	Tosti *et al.*, 2008a

Pd–Ag Tube 15% Co/Al2O3	250–400	150	At 400°C χ = 95.0% in counter-current χ_{TR} = 84.0% Sel-$_{H_2,MR}$ = 67.0% Sel-$_{H_2,TR}$ = 64.0% Y_{H_2} = 20.0% HR = 30.0% in counter-current	Iulianelli and Basile, 2010
Pd/PSS Tube 15% Co/Al_2O_3	400	300–800	At 800 kPa χ = 100% HR = 50.0%	Basile *et al.*, 2011a
Steam reforming of methane $CH_4 + H_2O \rightarrow CO + 3H_2$				
Pd Foil	727	1963	χ = 94.0%	Nazarkina and Kirichenko, 1979
Pd Disk	700–800	–	χ_{eq} = 77.0% χ = 96.0%	Oertel *et al.*, 1987
Pd/Vycor Tube Supported Ni pellet	350–500	101–909	χ = 90.0% at 500°C	Uemiya *et al.*, 1991c
SFC-2 membrane ($Sr_lFe_lCo_{0.5}O_x$) Tube Rh	850	100	χ > 98.0% Sel-$_{CO}$ = 90.0%	Balachancran *et al.*, 1997
Pd–Ag (25%) Flat Ru(5%)/Al_2O_3	150–300	101	At 150°C: χ_{MR} = 15.4% χ_{TR} = 12.7% At 300°C: χ_{MR} = 16.5% χ_{TR} = 16.5%	Basile *et al.*, 2003

(*Continued*)

Table 2.1 Continued

Membrane/catalyst	T (°C)	P (kPa)	Notes	References
Dense Pd/CeO_2/MPSS composite membrane Tube Nickel alumina reforming catalyst	500	100–300	χ = 70.0% at 300 kPa	Tong *et al.*, 2005b
Pd-based Noble-metal catalyst	550–650	202–404	At 650°C: χ_{FLBMR} > 97.0% Sel_{-CO} < 15.0%	Patil *et al.*, 2007
Pd-Ag Tube	450	500–900	At Δp = 400 kPa and SF = 1.6 χ = 65% HR = 80.0%	Basile *et al.*, 2011b
Steam reforming of glycerol $C_3H_8O_3 + 3H_2O \rightarrow 7H_2 + 3CO_2$				
Pd–Ag Tube Co/Al_2O_3	400	100–400	At 400 kPa, χ = 94% HR = 60% Y_{H_2} = 26%	Iulianelli *et al.*, 2010b
Pd–Ag Tube Ru/Al_2O_3	400	100–500	At 500 kPa, χ = 57% HR = 60% Y_{H_2} = 28%	Iulianelli *et al.*, 2011
Steam reforming of methanol $CH_3OH + H_2O \rightarrow 3H_2 + CO_2$				
Pd–Ag Tube Cu/ZnO/Al_2O_3 pellet	175–350	101	χ_{CH_3OH} = 100.0% at 320°C	Agrell *et al.*, 2002

Pd–Ag Tube $CuO/ZnO/Al_2O_3$	200–250	125	At 250°C and H_2O/CH_3OH feed ratio 1.22: $\chi_{MR} = 51.0\%$ $\chi_{TR} = 43.0\%$ $Sel\text{-}_{H_2} = 82.8\%$ $Sel\text{-}_{CO} = 0.47\%$ $Sel\text{-}_{CO_2} = 16.7\%$ $\chi_{MR} = 76.0\%$ at H_2O/CH_3OH feed ratio 2.43 $\chi_{TR} = 90.0\%$	Basile *et al.*, 2003
TiO_2–Al_2O_3 commercial membrane with a Pd–Ag deposit (MR1) Ceramic membrane with a Pd–Ag deposit (MR2) Pd/Ag (MR3) Tube $5\%Ru/Al_2O_3$	350–600 350–600 350–450	130	MR1: $\chi = 30.0\%$ at 350°C $\chi = 100.0\%$ at 600°C MR2: $\chi = 45.0\%$ at 350°C $\chi = 65.0\%$ at 550°C MR3: $\chi = 87.0\%$ at 350°C $\chi = 100.0\%$ at 450°C $Y_{H_2} = 40\%$ in counter-current	Basile *et al.*, 2006a
Pd–Ag Foil Fe-Cr based catalyst	450	520	$\chi = 76\%$ HR = 15.0%	Damle, 2009
Pd–Ag Tube CuO-Al_2O_3-ZnO-MgO	300	300	$Y_{H_2} = 77\%$ HR = 93%	Iulianelli *et al.*, 2008b

(*Continued*)

Table 2.1 Continued

Membrane/catalyst	T (°C)	P (kPa)	Notes	References
WGS reaction $CO + H_2O \rightarrow CO_2 + H_2$				
Pd/Vycor Tube Fe_2O_3–Cr_2O_3 (pellet)	400	–	χ_{eq} = 76.0% χ = 92.0%	Kikuchi *et al.*, 1989
Porous glass Ruthenium (III) chloride trihydrate	157	–	χ_{eq} = 99.0% χ = 85.0%	Seok and Hwang, 1990
Pd on porous glass Iron-chromium oxide	400	101	χ_{eq} =75.0% χ = 98.0%	Uemiya *et al.*, 1991a
Pd on ceramic (10 μm) Tube Low-temperature shift catalyst LK-821–2	320	100	At feed flow rate of CO = 2.5×10^{-5} mol·s^{-1}: χ_{eq} = 86.0% χ = 94.0%	Basile *et al.*, 1995
Pd/γ-Al_2O_3 (0.2 μm) Tube	320	100	χ_{eq} = 99.1% χ = 99.9% Counter-current mode	Basile *et al.*, 1996a
Pd/γ-Al_2O_3 (0.1 μm) Tube	320	–	χ_{eq} ~ 93.0% χ ~ 95.0%	Basile *et al.*, 1996b
Pd (70–75 μm) Tube	320	130–150	Co-current mode With feed: CO (32%), CO_2 (12%), H_2 (4%), N_2 (52%): $\chi_{MR(Pd)}$ = 100.0% $\chi_{MR(mesoporous)}$ = 78.0% χ_{TR} = 46.3%	Criscuoli *et al.*, 2000
Pd–Ag (50–70 μm) Tube Low-temperature shift catalyst LK-821–2	230–330	100	Co-current mode Counter-current mode At W/F = 2.5×10^{-3} g_{cat}·min·mol$^{-1}_{CO}$, 330°C: χ = 96.8% χ_{eq} = 84.1% at 230°C	Basile *et al.*, 2011b

Pd–Ag (50–70 μm) Foil	325	100	At feed flow rate of CO = 1.95×10^{-5} mol·s^{-1} and feed composition: CO = 0.4, H_2O = 0.6: $\chi = 98.9\%$ $\chi_{eq} \sim 80.0\%$	Tosti *et al.*, 2000
Pd(60%)-Cu Cu–Ce (30%La)-Ox	350	–	$\chi_{eq} = 93.0\%$ $\chi \sim 94.0\%$	Flytzani-Stephanopoulos *et al.*, 2004
Pd (1) Pd (80%)-Cu (2) Tube Heterogeneous catalysts particles	900	–	$\chi_{eq} = 54.0\%$ $\chi(1) = 93.0\%$ $\chi(2) = 66.0\%$ HR(1) = 90% HR(2) = 85%	Iyoha *et al.*, 2007
Pd–Ag Tube $CuO/CeO_2/Al_2O_3$	280–320	600	At 320°C: $\chi = 90.0\%$ HR = 80%	Brunetti *et al.*, 2009

Table 2.2 Different types of membranes and their main characteristics

Membrane	φ_{pore} (nm)	Diffusion mechanism	$\alpha_{H_2/\text{other gas}}$	H_2 permeability or permeance	Reactant loss
Macroporous	> 50	Poiseuille flow (Viscous flow)	1	Very high	High
Mesoporous	2–50	Knudsen flow	$\alpha_{H_2/N_2} = 3.74$	High	Average
Microporous – also further classified into: 'Small pores': $\varphi_{pore} \sim 0.5$ nm 'Large pores': $\varphi_{pore} = 0.5–2$ nm	< 2	Activated process	High	Average	Low
Dense Pd-based	< 0.5	Fick	Infinite	Very low	–
Composite: top layer with a dense metallic-based film	–	Fick	Infinite	Average	–
Composite: top layer with a non-metallic-based film (Nanosil membrane, etc.)	–	In the case of Nanosil membranes, gas permeation occurs through 'windows'	Very high	High	Low

values. In these membranes, gases pass through the dense layer by means of a solution–diffusion mechanism. In particular, Pd possesses exclusive selectivity to H_2 ($\alpha_{H_2/\text{other gas}} = \infty$), whereas Ag to O_2($\alpha_{O_2/\text{other gas}} = \infty$). The difference between silica dense ceramic membranes and Pd-based membranes lies in the fact that H_2 diffuses through the former in a molecular form, and through Pd in the atomic one. Other metals, such as V, Ta and Nb, show a decreasing H_2 permeability with increasing temperature; moreover, because of their tendency to be fouled, these materials are not commercially used.

Generally, H_2 is taken as a reference gas because of its higher permeability with respect to all the other gases. Nevertheless, in 1994, Ohya *et al.* (1994) prepared zirconia–silica composite membranes which allowed only H_2O and HBr to permeate, but not H_2.

Apart from the solution–diffusion mechanism, dense inorganic membranes also show another transport mechanism (*electrochemical pumping*): the driving force is provided by an electrical potential gradient, imposed by means of suitable electrodes on both sides of the membrane, whereas other membranes have both electronic and ionic conductivities. In the case of perovskite, for example, oxygen ions and electrons are transported in counter-current mode, and no external electrons are needed: it is the oxygen partial pressure difference through the membrane that acts as a driving force.

A poreless ceramic membrane (denoted as *Nanosil*), with selectivity $H_2/CO_2 = 1500$, and H_2 permeance of 5.0×10^{-7} mol m^{-2} s^{-1}Pa^{-1} at 600°C, is composed of three layers:

- a support commercial macroporous α-alumina tube (with a nominal pore size of 100 nm);
- an intermediate graded layer of γ-alumina obtained by the sequential deposition and calcination of three boehmite sols of median particle sizes of 630, 200 and 40 nm; and
- a topmost layer of amorphous silica deposited by chemical vapour deposition of a silica precursor at high temperature in an inert atmosphere.

The amorphous silica may be considered poreless, in the sense that gas molecules do not permeate the material by the well-known mechanisms of Knudsen diffusion, surface diffusion, or microporous diffusion (gas translation) that apply to materials with pores. Rather, the material is better described as having solubility sites of size 0.3 nm connected by windows with permeation occurring by hopping between adjacent solubility sites (Gu and Oyama, 2007). In a more recent investigation of these windows, the same authors observed that the structure of the membranes is more open than that of the vitreous glass (Hacarlioglu *et al.*, 2008). As is well known, generally, the silica modified membranes suffer from loss of permeability (as much as 50% or greater in the first 12 h) on exposure to moisture. This has been attributed to the removal of Si–OH groups, leading to the formation of Si–O–Si bonds which closes pore channels (Iler, 1979). This phenomenon, termed *densification* (i.e., membrane coalescence at high temperature), reduces permeability and also causes embrittlement of the silica film, compromising selectivity. On the other hand, this important aspect, already studied by Oyama *et al.* in a previous paper (Prabhu and Oyama, 2000), and due to the high stability at high temperature (measured to > 300 h), does not manifest in the Nanosil-type membrane.

2.3.2 Dense Pd-based membranes

It is known that Pd is able to absorb about 600 times its volume of H_2, at room temperature (Hughes, 2001). However, there is a serious disadvantage in using a pure Pd membrane: a phase transition between the α- and β-phases occurs, especially near room temperature when exposed to H_2 (Ma, 2006). This makes the Pd material brittle and eventually leads to micro-cracks in the metal. To avoid H_2 embrittlement caused by the phase transition between α- and β-phases of Pd hydride, it is necessary to operate above both 298°C and 2 MPa, where the β-phase is not present (Shu *et al.*, 1991), or forming a Pd-rich alloy, such as Pd–Ag or Pd–Cu, in order to lower the critical temperature (Ma, 2006). Adding another metal to Pd generally improves H_2 permeability: in Gryaznov (2000), various Pd-alloys are compared (at 500°C). For a wide range of temperatures, some metals (Ti, Zr, Nb, V, and Ta) show much higher H_2 solubility than Pd. The rare earth elements, for example, lanthanum, neodymium, cerium, and radioactive elements such as thorium, show very high solubility for H_2, exceeding that of Nb, Ta and V; whereas H_2 solubility in the range 300–400°C follows the order: Ti > Zr > La > Ce > Th > Nb > V > Ta > Pd (Mundschau, 2008). A table of measured values of H_2 permeability for dense membranes based on Group IVB-VB elements is also reported by Mundschau (2008).

It should be noted that the cost of Pd-based membranes can also be reduced by developing methods for the preparation of alloy membranes with low Pd content. For example, recently Tereschenko *et al.* (2007) and Basile *et al.* (2008e) developed Ti–Ni–Pd alloy membranes with 2, 5 and 9 wt% of Pd. They found that adding Pd into Ti–Ni alloy increases the H_2 permeability of the foils. Thus, the permeability of the $Pd_{9\%}$ was about one order of magnitude higher than that of the Ti–Ni foil. Cyclohexane dehydrogenation on Pt–Re/Al_2O_3 commercial catalyst carried out in a Ti–Ni–$Pd_{9\%}$ MR showed higher conversions than those corresponding to the TR.

The mechanism of the permeation of H_2 through Pd membranes, extensively discussed in the literature is, as already said, by solution–diffusion and follows seven different activated steps (Lewis, 1967; Koros and Fleming, 1993). Depending on temperature, pressure, gas mixture composition and the thickness of the membrane, each one of these steps may control H_2 permeation through the dense film. A recent brief review of mechanisms concerning hydrogen entry into metals is recommended (Pernga and Wu, 2003). An important parameter used to quantify H_2 transfer through the Pd-based membrane is the steady-state flux of hydrogen atoms (Kikuchi, 1995). The atomic flux is described as the moles of H_2 permeating through a given area over a period of time at a fixed temperature and pressure differential, in units of $\text{mol}_{H_2} \times \text{m}^{-2} \times \text{s}^{-1}$. This H_2 flux can be expressed in generalized form by the following expression:

$$J_{H_2} = \frac{\text{Pe}_{H_2}\left(p^n_{H_2,\text{ret}} - p^n_{H_2,\text{perm}}\right)}{\delta} \qquad [2.1]$$

where J_{H_2} is the H_2 flux through the membrane (mol m^{-2} s^{-1}); Pe_{H_2} the H_2 permeability (mol m m^{-2} s^{-1} Pa$^{-0.5}$); δ the membrane thickness (m); $p^n_{H_2,\text{ret}}$ and $p^n_{H_2,\text{perm}}$ the H_2 partial pressure either on the retentate (high H_2 partial pressure) or permeate (low H_2 partial pressure) sides, respectively (Pa); n the dependence factor of the H_2 flux on the H_2 partial pressure, generally in the range $1 \geq n \geq 0.5$.

At low pressures the previous expression becomes: $n = 0.5$, and the H_2 permeation is said to follow Fick–Sieverts' law:

$$J_{H_2,\text{Sieverts}} = \frac{\text{Pe}_{H_2}\left(p^{0.5}_{H_2,\text{ret}} - p^{0.5}_{H_2,\text{perm}}\right)}{\delta} \qquad [2.2]$$

The product of the H_2 diffusion coefficient, D_{H_2} and Sieverts' constant, K_S, is sometimes referred as $\text{Pe}_{H_2,\text{Sieverts}}$, the above reported permeability coefficient of the material:

$$\text{Pe}_{H_2,\text{Sieverts}} = D_{H_2} \times K_S \qquad [2.3]$$

K_S being a constant of Sieverts' law for the material (mol m^{-3} Pa$^{-0.5}$).

Sieverts' law links the concentration of dissolved H_2 (or solubility), C_{H_2} (mol m^{-3}) at equilibrium to the square root of the partial pressure of H_2:

$$C_{H_2} = K_S \times p_{H_2}^{0.5} \qquad [2.4]$$

The last equation is related to the first gas phase studies of the solubility of H_2 in iron (and other metals) generally attributed to Sieverts (Sieverts *et al.*, 1911), who found experimentally that amounts of H_2 dissolved in metal is directly proportional to the square root of the hydrogen pressure (Equation [2.4]). Various aspects related to the degradation of mechanical properties due to the entry of hydrogen into metals or alloys can also be found in Lewis (1967) and in Paal and Menon (1988).

A relationship between the H_2 solubility in iron and the absolute temperature was also published by Armbruster (1943), represented by an equation such as:

$$C_{H_2}\,(\text{ppm}) = 42.7\, p_{H_2}^{0.5} \exp\left(\frac{-3280}{T}\right) \qquad [2.5]$$

The particular value of the exponent n (= 0.5) reported in the Equations [2.2], [2.4] and [2.5] should be observed.

Equations [2.1] and [2.2] also describe the fact that, under diffusion-controlled conditions, J_{H_2} increases with decreasing the membrane thickness (δ). Equation [2.2] is valid only for a very low value of n (i.e., 0.5), which corresponds to the α phase at relatively low pressure. At high pressures, H–H interactions within the palladium bulk are not negligible and Equation [2.2] is considered no longer valid. For instance, at 160°C, the H_2 adsorption isotherm (hydrogen loading, H/Pd, as a function of $p^{0.5}$) is linear at low pressures, but starts to curve as the miscibility gap is approached.

Recently, Li *et al.* (2008) concentrated their work on the investigation of H_2 permeation of a 2 μm thick Pd-composite membrane (substrate: α-Al_2O_3, which represented only 4% of the total resistance, and so neglected), in the range 90–500°C and p_{feed} = 1.4–3.8 bar. The pressure exponent value was evaluated in the α–β phase transition.

The term $\left(p_{H_2,\text{ret}}^{0.5} - p_{H_2,\text{perm}}^{0.5}\right)$ represents the driving force of H_2 permeation at relatively low pressures. However, the dependence factor of the H_2 flux on H_2 partial pressure sometimes deviates from 0.5. For this reason, n is often used as an indicator for the rate-controlling step of the H_2 permeation through Pd-based membranes. In fact, if the diffusion of atomic hydrogen through the dense metal layer is the rate-limiting step, then J_{H_2}

is proportional to $\left(p_{H_2,ret}^{0.5} - p_{H_2,perm}^{0.5}\right)$. In other words, Sieverts' law is valid and $n = 0.5$, that is, only where the hydrogen-to-metal atomic ratio is small (e.g., H/Pd << 1) and there is minimal interaction of the hydrogen atoms. Deviations from Sieverts' law (i.e. $n > 0.5$) have been recently reported in the literature, and are generally expected for very thin membranes where the bulk diffusion of hydrogen through the palladium layer is faster than the surface dissociation/recombination reactions at either/or both sides of the membrane (e.g., Jayaraman and Lin, 1995a; McCool and Lin, 2001; Nam *et al.,* 1999).

For thick membranes (> 1 μm), deviations from Sieverts' law can be caused by high H_2 pressure values. For thinner membranes, when higher H_2 fluxes are achieved, the surface reaction processes gain importance and become the rate-limiting steps, and the flux of H_2 is proportional to $n = 1$ power of the hydrogen partial pressure:

$$J_{H_2} = \frac{Pe_{H_2,linear}\left(p_{H_2,ret} - p_{H_2,perm}\right)}{\delta} \qquad [2.6]$$

$Pe_{H_2,linear}$ being the H_2 permeability coefficient.

However, practically diffusion and surface control interplay with each other. Unfortunately, it is not yet clear at which Pd thickness surface processes actually begin to control the H_2 permeability, and no detailed information is yet available in literature on the influence of the microstructure of differently produced Pd layers on both H_2 permeability and selectivity.

It should also be noted that in Equation [2.1], that is, the relationship between J_{H_2} and the hydrogen partial pressure difference across the membrane, is determined through pure hydrogen permeation tests. However, in practical applications, the membrane never works under this condition, being always in contact with a mixture. In this case, other phenomena arise and thus the rate-controlling step of the hydrogen flux could also depend on other factors, such as the gas phase mass-transfer transport limitations (or the concentration polarization) (for details, the reader is addressed to Hou and Hughes, 2002; Peters *et al.*, 2008; Zhang *et al.,* 2006a).

The H_2 transport through a dense palladium film is an activated process and, in many practical cases, Sieverts' law defines the pressure dependence of the H_2 permeation at constant temperature. Generally, it is implicitly assumed that n does not depend on temperature.

With regard to the influence of temperature, DH_2, KS and Pe_{H_2} are found to vary in the Arrhenius manner, at least at low hydrogen

Table 2.3 A brief list of pre-exponential factors and apparent activation energies for hydrogen permeation through Pd-based membranes, as reported in the literature

E_a (kJ/mol)	$Pe°_{H_2}$ (10^{-5} mol·m/m²· s kPa$^{0.5}$)	References
33.31	1.66	Basile *et al.*, 2005b
29.73	7.71	Basile *et al.*, 2001a
48.50	9.33	Tosti *et al.*, 1998
15.70	2.19	Koffler *et al.*, 1969
15.50	2.54	Balovnev, 1974
18.45	1.02	Itoh et al., 1992
12.48	0.38	Itoh and Xu, 1993
17.6	0.65	Iulianelli *et al.*, 2008c
8.7	0.14	Iulianelli *et al.*, 2010d
5.58	1.30	Tosti *et al.*, 2010
7.89	1.11	Borgognoni *et al.*, 2011

concentrations. So, the relationship between H_2 permeability and temperature is described as following an Arrhenius behaviour:

$$\mathrm{Pe}_{\mathrm{H}_2} = \mathrm{Pe}^{\circ}_{\mathrm{H}_2} \exp\left(\frac{-E_a}{RT}\right) \quad [2.7]$$

Consequently, when Sieverts' law is valid, the H_2 flux is given in terms of the expression of the so-called Richardson's law (Richardson, 1904), namely:

$$J_{\mathrm{H}_2} = \frac{\mathrm{Pe}^{\circ}_{\mathrm{H}_2}\left[\exp(-E_a/RT)\right]\left(p^{0.5}_{\mathrm{H}_2,\mathrm{ret}} - p^{0.5}_{\mathrm{H}_2,\mathrm{perm}}\right)}{\delta} \quad [2.8]$$

In order to determine the two parameters $\mathrm{Pe}^{\circ}_{\mathrm{H}_2}$ and E_a (and, eventually, n, if Sieverts' law is not valid), permeation tests at different pressures and temperatures are necessary. Generally, for Pd-based dense membranes, both $\mathrm{Pe}^{\circ}_{\mathrm{H}_2}$ and E_a are estimated using an Arrhenius plot of the H_2 permeance against the reciprocal temperature. A comparison of permeation parameters for various Pd–Ag dense membranes shows that typically E_a is in the range of 12.48–48.50 kJ/mol, whereas $\mathrm{Pe}^{\circ}_{\mathrm{H}_2} = 0.38 - 9.33 \times 10^{-5}$ mol m/(m^2 s kPa$^{0.5}$) (Table 2.3).

Deviation from Sieverts' law leads to values of $\mathrm{Pe}^{\circ}_{\mathrm{H}_2}$ and E_a which are significantly different (Nishimura *et al.*, 2002). In these cases, the following equations were also proposed (Murav'ev and Vandyshev, 2003):

$$J_{H_2} = \frac{Pe_{H_2}\left(p^m_{H_2,ret} - p^n_{H_2,perm}\right)}{\delta} \quad [2.9]$$

where $m \geq 0.5$ and $1 \geq n$

$$J_{H_2} = A_1\left(p^{0.5}_{H_2,ret} - p^{0.5}_{H_2,perm}\right) + B\left(p_{H_2,ret} - p_{H_2,perm}\right) \quad [2.10]$$

where the coefficients A_1 and B describe the H_2 passage through metallic membranes.

In the latter equation, the first term reflects the contribution of the flow of atomic hydrogen diffusing through the interstices of the crystal lattice and the second one characterizes the additional flow of hydrogen (presumably molecular), migrating through discontinuities of the actual metal or alloy.

Among $E_a, Pe^{\circ}_{H_2}$ and n, the most important parameter to be carefully evaluated is E_a because it deeply influences the performance of MRs. In particular, the methane SR in a Pd-based MR is mainly affected by the H_2 permeation and, especially, E_a influences the H_2 permeation rate much more than $Pe^{\circ}_{H_2}$ and n do. In designing new alloys, for example via quantum-mechanics calculations or molecular dynamics, the most important parameter to be controlled is E_a, because the H_2 permeation through the membrane is very sensitive to its value (Gallucci *et al.*, 2007b).

2.3.3 Non-Pd-based membranes, amorphous and other metals

Excessive costs limit the use of Pd membranes for wide-scale industrial use. So, there is a strong interest in developing non-Pd-based (or, at least, with a low Pd-based content) and low-cost H_2 permeation alloys (Luo *et al.*., 2006; Nishimura *et al.*., 2002). In particular, Luo *et al.* (2006) demonstrated that an Nb–Ti–Ni alloy, consisting of only the primary phase and the eutectic phase, shows a high H_2 permeability and resistance to H_2 embrittlement. Moreover, Hara *et al.* (2002) reported that Pd-coated amorphous Zr–M–Ni (M = Ti, Hf) alloy membranes were resistant enough in an H_2 atmosphere and had stable permeability only to H_2 at least in the range of 200–300°C.

Two different, excellent and extensive reviews on this argument have been recently published by Phair and Donelson (2006) and Dolan *et al.* (2006). Among other matters, Phair and Donelson (2006) also presented a chronological list of novel crystalline alloys and their H_2 permeabilities as recently reported in the specialized literature, in the period 1991–2006. The crystalline metal alloys considered show high H_2 permeability and are based on Group IV (Zr, Ti, Hf) and V (V, Nb, Ta) metals. Possible alloying elements include Ti, Co, Cr and Al in binary and ternary alloys; Fe, Mn, Mo, Cu, Ni, Ga, Ge, Sn, Si, W, La and Be are also used. Examples of non-Pd-based metal alloy are $V_{85}Ni_{15}$ (Pe_{H_2} = 3–4 × 10^{-7} mol·m^{-1}·s^{-1}·Pa$^{-0.5}$, at 400°C), $V_{85}Ni_{14.9}1Al_{0.09}$ (Pe_{H_2} = 3–4.5 × 10^{-7} mol·m^{-1}·s^{-1}·Pa$^{-0.5}$), $Nb_{28}Ti_{42}Ni_{30}$ (Pe_{H_2} = 2 × 10^{-9}–1 × 10^{-8} mol·m^{-1}·s^{-1}·Pa$^{-0.5}$), Fe_3Al (Pe_{H_2} = 60–1.1 × 10^{-10} mol·m^{-1}·s^{-1}·Pa$^{-0.5}$, at 27°C).

Amorphous metals are also under study because they often exhibit superior properties to their crystalline counterparts and are generally more easily stabilized in alloys than in pure metals. Generally, amorphous metals possess higher H_2 solubility (Gapontsev and Kondrat'ev, 2003), strength and ductility than their counterpart crystalline structure, with a more open lattice (Hara *et al.*, 2002), giving them more resilience to corrosion and failures associated with H_2 embrittlement (Yamaura *et al.*, 2003) and extreme operating conditions.

Amorphous alloys generally show higher H_2 permeability than the corresponding crystallized ones (Itoh *et al.*, 1997) and the pressure dependence of H_2 permeability follows a relationship described by a simple n^{th} power equation, as for example reported by Itoh *et al.* (1998) in their study on the H_2 solubility in amorphous $Pd_{1-x}Si_x$ (x = 0.15, 0.175, 0.2) alloys. In this work, the exponent n joins the H_2 flux through the membrane to the H_2 partial pressure and is affected by both the temperature and Si content. In the following, the equations, reported in their paper by Itoh *et al.* (1998), are briefly summarized.

The total amount of H_2 dissolved in an amorphous alloy, c_{H_2}, was described by Kirchheim (1982) and is related to the H_2 chemical potential, μ_{H_2}, via the reverse of the error function:

$$\mu_{H_2} = G^{\circ} - \sigma \times erf^{-1}\left(1 - 2 \times c_{H_2}\right) \quad [2.11]$$

where σ is the width of the Gaussian distribution, G° the mean free energy related to the H_2 dissolution, at the reference pressure (p°).

The chemical potential is estimated using the following expression:

$$\mu_{H_2} = 1/2\, RT \ln\left(\frac{p_{H_2}}{p^{\circ}}\right) \quad [2.12]$$

where p_{H_2} is the hydrogen partial pressure in the gas phase, and p° the reference pressure.

Combining Equations [2.11] and [2.12], the relationship between c_{H_2} and p_{H_2} is obtained:

$$c_{H_2} = 1/2\left\{1 - erf\left(\frac{G^\circ - RT\ln\sqrt{p_{H_2}/p^\circ}}{\sigma}\right)\right\} \qquad [2.13]$$

In this case, the hydrogen permeation flux through the amorphous membrane follows Fick's law and is expressed by:

$$J_{H_2} = \frac{D_{H_2}A\left(c_{H_2,\text{ret}} - c_{H_2,\text{perm}}\right)}{\delta} \qquad [2.14]$$

where A is the membrane superficial area, $c_{H_2,\text{ret}}$ and $c_{H_2,\text{perm}}$ the H_2 concentration at the membrane sides 'retentate' and 'permeate', respectively.

Generally, Equation [2.11] is approximated by this simple general expression:

$$\mu = B + C\ln\left(2c_{H_2}\right) \qquad [2.15]$$

where B and C are constants to be determined.

In this case, the $c_{H_2} - p_{H_2}$ relation becomes:

$$c_{H_2} = k_a\left(\frac{p_{H_2}}{p^\circ}\right)^{n_a} \qquad [2.16]$$

where k_a and n_a are constants related to the amorphous state of the membrane.

The last equation reflects the fact that the hydrogen content in the amorphous membrane is, similarly to the crystalline state (in which $n = 0.5$), proportional to the nth power of the H_2 partial pressure. Nevertheless, differently from the crystalline state, c_{H_2} represents the ratio between the number of hydrogen atoms versus the number of available interstices (which is unknown in these kinds of membranes).

Finally, the basic equation for the H_2 permeation through an amorphous membrane is given by:

$$J_{H_2} = \frac{D_{H_2} A K_a}{\delta}\left[\left(\frac{p_{H_2}}{p^\circ}\right)_{ret}^{n_a} - \left(\frac{p_{H_2}}{p^\circ}\right)_{perm}^{n_a}\right] \quad [2.17]$$

Considering the reference pressure $p^\circ = 1$ atm, Equation [2.17] becomes:

$$J_{H_2} = \frac{D_{H_2} A k_a}{\delta}\left[\left(p_{H_2}\right)_{ret}^{n_a} - \left(p_{H_2}\right)_{perm}^{n_a}\right]$$

In other words, an nth power equation with respect to H_2 partial pressure is able to describe the H_2 flux through amorphous membranes; whereas, for the crystallized ones, the 0.5-power equation must be applied.

As already said, the nth power equation was used, for example, by Itoh *et al.* (1998) in determining the H_2 permeability in amorphous alloys, at elevated temperatures up to 390°C. It should also be added that the main conclusions of their work were: (a) H_2 permeability strongly depends on the Si content and increases with it; (b) the amorphous alloy shows higher hydrogen permeability than the crystallized one and shows resistance to hydrogen embrittlement and (c) the amorphous membrane must be used below its crystallization temperature, due to the strong reduction of H_2 permeability in the crystallization state.

2.3.4 Composite membranes

There are two types of membranes available for ultra-pure H_2 production: composite and self-supported thin ones. In both cases, the major aim of the study concerning Pd-based membranes is the reduction of their thickness. The H_2 flux increases as the membrane gets thinner, therefore the overall costs decrease because of both the reduced operating and material costs. In accordance with the applications where ultra-pure H_2 production is required, commercial self-supporting dense Pd-based membranes with wall thicknesses larger than 100 μm keep a sufficient mechanical strength and assure a defect-free surface and complete H_2 selectivity. However, these membranes are too thick to obtain a satisfactory H_2 flux. In addition to the low H_2 *permeance* (as customarily used for composite membranes and defined as the ratio of Pe_{H_2}/δ), thick Pd-based membranes are too expensive. Indeed, the price of palladium does not permit the production of low-cost permeators for a wide use in medium–large scale applications. For practical use, it is necessary to reduce the thickness of the palladium layer and also to find selective Pd-based membranes with low Pd content.

In order to meet the dual challenge of both high H_2 selectivity and permeability, a compromise could be reached: one or more thin dense layers of highly selective membrane material are deposited on supports having high permeability. These materials are called composite membranes.

Porous membranes (low H_2 selectivity) are used as substrate onto which a very thin but continuous film of a selective metal, generally Pd-based, is deposited. Dense composite membranes are also fabricated by sintering together powders of highly H_2 permeable metals (Pd, Pt, Nb, Ta, Ti, V, Zr and their alloys) with powders of a second metal or alloy which is non-permeable to H_2 (Mundschau, 2005), having the function of providing the mechanical support. Some fundamental science related to the H_2 separation using dense composite membranes is reported by Mundschau (2008).

In the field of surface-engineering techniques, the fabrication of thin films has evolved greatly in recent years. Unfortunately, no method is exhaustive, and for every choice various compromises are necessary. In fact, each method has advantages as well as disadvantages. An exhaustive list of the various methods, and relative discussion, can be found, for example in Basile *et al.* (2008a).

Composite Pd-based membranes add to the problems of Pd-based membranes (embrittlement and its sensitivity to poisoning) (Armor, 1992). For example, support–metallic adherence is one of the most important problems to be better understood and studied. Especially in Pd-ceramic composite membranes, the Pd–Ag layer thickness plays a key role in terms of physical and chemical stability as well as of membrane selectivity. Finally, a comparison of some Pd and Pd-based composite membranes prepared using various deposition techniques is shown in Table 2.4. In conclusion, composite metal membranes seem also to be a good choice to reduce Pd-based thickness, and especially porous stainless steel is considered as a valid support because of its mechanical durability, its thermal expansion coefficient being close to that of palladium, and the ease of gas sealing. Unfortunately, this support forms an alloy with the palladium at relatively high temperatures, leading to the reduction of H_2 permeability. To prevent the inter-metal diffusion, materials such as porous silica, tungsten, tantalum oxide, magnesia and alumina have been used as a diffusion barrier. For these composite membranes another important problem should also be considered: the lack of long term durability tests.

2.4 Preparation of dense metallic membranes

Based on the different compositions, structures and configurations, the dense metallic membranes can be prepared using various methods. As described earlier, the dense metallic hydrogen separation membrane can be

Table 2.4 Permeation data of various inorganic membranes

Membrane type	T (K)	Δp (kPa)	Thickness (μm)	H_2 flux (mol/m^{2}s)	H_2 permeance (mol/m^2 s Pa) or (mol/m^2 s Pa$^{0.5}$)	Pe_{H_2} (mol m/m^2 s Pa) or (mol m/m^2 s Pa$^{0.5}$)	α_{H_2/N_2} (kJ/mol)	α_{H_2/N_2} (—)	Preparation method	Reference
Pd/PSS-YSZ	673	–	7–10	2.5×10^{-2}	–	4.7×10^{-9}	8.0	800–900	ELP	Huang and Dittmeyer, 2007
Pd/Al_2O_3	473	10	15	2.2×10^{-1}	$2.2 \times 10^{-5*}$	$3.3 \times 10^{-10*}$	–	7	ELP	Altinisik *et al.*, 2005
Pd	473	51	95	2.7×10^{-2}	$5.2 \times 10^{-7*}$	$5.0 \times 10^{-11*}$	23.2	–	Oxidation/ rolling	Zhang *et al.*, 2006a
Pd/glass	623–773	400	2	–	1.7×10^{-6}	$3.4 \times 10^{-12*}$	12.2	1140–12900	ELP	Wang *et al.*, 2004a
Pd/Al_2O_3	723	–	4.8	–	3.0×10^{-6}	$1.4 \times 10^{-11*}$	–	60	ELP	Van Dyk *et al.*, 2003
Pd/Al_2O_3	573	30	2–4	$1.0–2.0 \times 10^{-1}$	$6.7 \times 10^{-6*}$	$1.3–2.7 \times 10^{-11*}$	–	5000	CVD	Itoh *et al.*, 2005
Pd/Al_2O_3	801	–	2–3	–	1.2×10^{-6}	$3.5 \times 10^{-12*}$	–	< 18	ELP	Kleinert *et al.*., 2005
Pd/Al_2O_3	673	100	5	1.6×10^{-1}	$1.6 \times 10^{-6*}$	$7.8 \times 10^{-12*}$	–	100—200	ELP	Liang and Hughes 2003
Pd/BaZrO3	873	–	41	–	1.2×10^{-9}	–	–	5.7	CVD	Okada *et al.*, 2007
Pd/MPSS	773	100	6	3.0×10^{-1}	$3.0 \times 10^{-6*}$	$1.8 \times 10^{-11*}$	16.7	–	ELP	Tong *et al.*, 2006a
Pd/PNS	773	358	–	8.3×10^{-2}	–	–	–	3.7	MS	Ryi *et al.*, 2006
Pd/PSS	793	150	10	1.8×10^{-1}	$1.2 \times 10^{-6*}$	$1.2 \times 10^{-11*}$	–	–	ELP	Basile *et al.*, 2008e
Pd/TiO_2	773	45	0.3–0.4	$2.8 \times 10^{-1*}$	$6.3 \times 10^{-6*}$	$1.9–2.5 \times 10^{-12*}$	–	1140	ELP–UV	Wu *et al.*, 2000
Pd/ZrO_2/PSS	773	100	10	8.3×10^{-2}	$8.3 \times 10^{-7*}$	$8.3 \times 10^{-12*}$	7.1	–	ELP	Wang *et al.*, 2004
Pd/αAl_2O_3	573	290	1	2.1	$3.8 \times 10^{-3*}$	$3.8 \times 10^{-9*}$	–	10^{-6}	MCVD	Sato *et al.*, 2005
Pd/αAl_2O_3	643	400	1	4.0×10^{-1}	–	–	–	3000–8000	ELP	Nair and Harold, 2007a
$Pd_{84}–Cu_{16}$/ ZrO_2–PSS	753	250	5	6.0×10^{-1}	$5.3 \times 10^{-4*}$	$2.6 \times 10^{-9*}$	14.5	∞	ELP	Gao *et al.*, 2005
$Pd_{90}–Ag_{10}/\alpha Al_2O_3$	473–616	80–250	20	1.4×10^{-1}	$1.3 \times 10^{-6*}$	$2.5 \times 10^{-11*}$	–	30–178	ELP	Huang *et al.*, 2003

Pd–Ag/Al_2O_3	–	142	10	1.0×10^{-1}	$1.0 \times 10^{-6*}$	$1.0 \times 10^{-11*}$	–	1500	ELP	Liang and Hughes 2005
Pd–Ag/Ni	673	–	1.2–3	–	2.0×10^{-5}	–	–	–	MEMS	Zhang *et al.*, 2007b
Pd–Ag	673–773	162	50	–	–	–	33.3	∞	LT	Basile *et al.*, 2005a
Pd–Ag/PSS	673–773	100	2–3	3.0×10^{-1}	$3.0 \times 10^{-6*}$	$6.0 \times 10^{-12*}$	25.7	–	ELP	Tong *et al.*, 2005a
Pd–Ag/αFe_2O_3/PSS	773	100–500	16–20	–	4.9×10^{-4}	$8.9 \times 10^{-9*}$	–	3500–5000	SELP	Yepes *et al.*, 2006
Pd–Ag/α–Al_2O_3	823	400	11	7.0×10^{-2}	5.0×10^{-4}	–	21.0	~1000	ELP	Nair *et al.*, 2007b
Pd–Cu	725	–	0.75	1.6	–	–	–	500 $_{(H2/He)}$	DST	Hoang *et al.*, 2004
Pd–Cu/αAl_2O_3	723	345	11	8.0×10^{-1}	$2.3 \times 10^{-6*}$	$2.6 \times 10^{-11*}$	–	1150	ELP	Roa *et al.*, 2003
Pd–Ni	673	20	2.5	3.1×10^{-1}	$1.2 \times 10^{-5*}$	$2.9 \times 10^{-11*}$	–	–	DC SP	Zhang *et al.*, 2006b
Pd–Ni/SS	623–823	68	0.8	$1.4 \times 10^{-1} – 1.3$	$2.0 – 1.9 \times 10^{-5*}$	$1.6 – 1.5 \times 10^{-11*}$	–	300–4700	VED	Nam *et al.*, 1999
Pd–PSS	593–773	–	20	–	–	–	12.5	–	ELP	Aspen system, 1999
Pd–Ru/SS	645	100	1.5	4.9×10^{-2}	$4.9 \times 10^{-7*}$	$7.4 \times 10^{-13*}$	–	∞	MCVD	Yan *et al.*, 1994
Ti–Ni–Pd	723	300	45	$\sim 3.3 \times 10^{-3}$	–	1.7×10^{-10}	42.2	∞	Cold rolling	Basile *et al.*, 2008e

*Calculated from reference data.

categorized as pure metal, crystalline alloy, amorphous alloy, etc. Therefore, in this section, the preparation techniques of non-Pd pure-metal membranes will be briefly introduced first. After that, the techniques for the preparation of Pd and Pd-based crystalline alloy membranes will be reviewed. Finally, the preparation techniques of the recently developed amorphous alloy membranes will be briefly summarized.

2.4.1 Preparation of pure non-Pd metal membranes

A pure-metal H_2 permeation membrane, in the present discussion, refers to any crystalline, single element metal which exists in a membrane configuration and has H_2 permeability. Except for the well-known H_2 permeation metal Pd, several other metals, including Nb, Ta, V, Fe, Cu, Ni, Pt, Hf, Ti and Zr, have been studied for H_2 permeation (Adhikari and Fernando, 2006; Phair and Donelson, 2006). It is recognized that both high H_2 solubilities and low activation energies of H_2 diffusion are desirable, since they lead to higher H_2 permeability. Nb, Ta and V are all more permeable than Pd and have lower energies for bulk diffusion of H_2. However, their enthalpies for hydride formation are negative, indicating they are more susceptible to embrittlement than Pd. Thus, Zr, Ti, Nb, Ta and V exposed to high partial pressures of H_2 are likely to form hydrides, which lead to degradation of their mechanical properties. Group IV or V metals, such as Nb, Ta and V, have another drawback, in that their catalysis of the dissociation and re-association of H_2 is too slow for high flux rates. Moreover, they can form a tightly held oxide on the surface, which may impede the dissociation of H_2 molecules and the subsequent dissolution and absorption of H atoms within the metal. This drastically limits their utilization as H_2 separation membranes unless the surface resistance is somehow removed or modified.

It is obvious that the pure-metal membranes can be easily prepared by conventional metallurgical processes in the configurations of tube and disk, which have a thickness greater than 25 μm and can be used as unsupported H_2 permeation membranes. Because of the two obvious weaknesses of embrittlement and slow surface kinetic, the preparation of these metal-based H_2 permeation membranes mainly focuses on preparation of alloys in order to improve the embrittlement and modification of the surface for improving the surface kinetic. The preparation of an extra modified layer on these pure non-Pd metal membranes is the subject of the current discussion.

Makrides *et al.* (1967) patented a H_2-extraction membrane where refractory metals, such as Ta, Nb and V, were coated with Pd to facilitate H_2 ingress and egress and to prevent oxidation of the refractory metal surfaces. They started with commercially available foils. After being etched electrolytically in hydrofluoric acid and washed with acetone, the wet foils were placed in a vacuum chamber where they were further dried by evacuation. After that, the Pd thin films were sputtered onto both sides in argon atmosphere. Especially, the vanadium membrane was coated by omitting the etching step and simply heating it in the vacuum chamber to about 1000°C to get rid of surface oxide. The resulting membranes showed greater hydrogen transport than with Pd or its alloys, were strong, produced 100% pure hydrogen, and had a high tolerance to pinhole defects. Although the coat was not continuous, selectivity remained 100% because the non-porous substrate blocked non-hydrogen gas transport that penetrated defects in the coat. Makrides *et al.*'s membranes achieved limited application though, because the coating process and Pd vapour deposition at high vacuum was expensive and deficient. Vapour deposition does not coat inside tubes, the preferred extractor configuration, and generally retains significant surface transport resistance. Hydrogen transport, while higher than with Pd, was one-tenth that predicted from the permeabilities of the Ta, Nb and V. Makrides *et al.*'s vapour-coated membranes found use only in the nuclear industry where the high cost is offset.

Buxbaum *et al.* (1992; 1993; 1996) started with commercial-grade tantalum heat exchanger tubes in four wall thicknesses (0.054, 0.037, 0.027, and 0.007 cm) and with a niobium tube. The surfaces of these tubes arrive from the manufacturer coated with a layer of oxide and often with a layer of oil. Both of these interfere with the coating process and must be removed in order to make an acceptable membrane. Gross oxide and oil was removed by abrasion and detergent. After further electro-polishing and hydriding, the Pd plating was applied. The electroless Pd bath solution used was $PdCl_2$, 2 g/L; HCl (38%), 4 mL/L; NH_4OH (28%), 160 mL/L; $NaH_2PO_2{\cdot}H_2O$, 10 g/L. The temperature and the pH value were 50°C and 9.8, respectively. For several of the coated tubes, they used hydrazine as the reducing agent instead of hypophosphate. Hypophosphate was found to deposit Pd-5% phosphorus and superior performance with a purer Pd coat produced by reducing with hydrazine. Using the same technique, the vanadium membrane was also coated with Pd protective layer. The H_2 permeation fluxes were as high as 1.47 kmol/($m^2{\cdot}s{\cdot}Pa^{1/2}$) at 420°C. The main transport resistance is in the refractory metal substrate. Durability tests showed a 15% reduction in flux for 31 days of continuous membrane operation. Assuming durability is maintained for at least one year, this price and flux should allow competitive application for H_2 recovery in petrochemical plants and for MRs.

2.4.2 Preparation of Pd and Pd-based alloy membranes

The thick-wall unsupported Pd and Pd-based alloys can also be prepared by conventional metallurgical processes. Self-supporting dense Pd-based membranes possess wall thicknesses greater than 50–100 μm to keep a sufficient mechanical strength. These membranes are too thick to obtain a satisfactory hydrogen flux. Besides the low permeance, thick Pd membranes are too expensive for economical use, in particular since the price of Pd has tremendously increased over the last few years. For practical use it is necessary to reduce the thickness of the Pd or Pd-based alloy layer. Therefore, the composite membrane configuration of ultrathin Pd or Pd-based alloy films deposited on porous supports is a promising application of the Pd-based hydrogen separation membrane. Therefore, we will focus on the preparation of this kind of composite Pd-based membrane.

2.5 Preparation of Pd-composite membranes

Pure Pd-composite membranes are the most popular H_2 permeation membranes, whose preparation methods have been extensively studied. In what follows, some main and novel methods are summarized.

2.5.1 Electroplating techniques

Electroplating techniques are extensively used for the deposition of pure-metal and alloy thin films. In the electroplating method, reduction takes place by supplying current externally and the sites for the anodic and cathodic reactions are separate. Therefore, substrates have to be conductive. Electroplated Pd films have a fine structure and have valuable physical properties, such as hardness, high reflectivity, etc. A great advantage of the electroplating deposition is that the thickness of the layer can be controlled to a fraction of a micron. However, the electroplating has to be carried out on conductive substrates and the environmental pollution is serious due to the special electroplating bath.

Chen *et al.* (2008) prepared a Pd membrane on an AISI 316L porous stainless steel substrate by a modified electroplating process with a proper control of rotation speed of the substrate. The use of low-current density with high-rotation speed enabled them to avoid H_2 absorption during electroplating and to prepare smooth thin surfaced membrane without any defects. The resultant Pd membrane showed high H_2 permeation flux and excellent selectivity (H_2/He > 100 000), and was used in a membrane-assisted steam reformer of methanol for the production of high-purity H_2. It was also found that the membrane was resistant to H_2 embrittlement under the phase-transition temperature of 280°C, when operated in the temperature range of 250–350°C under the H_2 pressure of 0.9 MPa.

2.5.2 Electroless plating

Electroless plating, in particular, is an autocatalytic process which allows obtaining thin films with good adhesion (Basile and Gallucci, 2011c). The Pd deposition occurs owing to an autocatalytic reduction of Pd ions to Pd metal in an electroless plating bath of a Pd salt and a chemical reducing agent without the use of external electric power. As a consequence, the deposition process can occur on both conductive and non-conductive substrates, such as stainless steel, glass, ceramics and plastics. Furthermore, it does not require expensive set up and is relatively easy to scale up from laboratory to industrial level. Because of the above characteristics, electroless plating is nowadays a widespread industrial practice and the most common method for the preparation of Pd-composite membranes for H_2 permeation.

The Pd electroless plating must be carried out on different kinds of porous substrates with different pore size and roughness. In order to get thin and dense Pd films, after proper cleaning, the porous substrates normally need to be modified and activated. This modification can not only smooth the surface and decrease the surface open pore size, but can also introduce a buffer layer to avoid the diffusional reaction between substrate and Pd membrane. Furthermore, the modification can simultaneously introduce well-dispersed Pd seeds used for electroless plating catalysts. After that, Pd membrane growth can be carried out through the autocatalytic reaction in the electroless plating bath of a Pd salt, a complexing agent, a reducing agent and a stabilizing agent. Therefore, the preparation of Pd membranes on porous substrates by electroless plating can be divided into three steps: (a) clean substrate; (b) modify and activate the substrate surface; and (c) electroplate Pd membranes.

The porous substrates have to be cleaned to remove contaminants such as grease oil, dirt and corrosion products. Substrate cleaning must be carried out very carefully, since impurities can inhibit autocatalytic sites formation in the subsequent activation, which can adversely affect the formation of a pinhole-free and adherent membrane on the substrate. The ceramic substrates are usually cleaned by the following procedures: ultrasonic clean in an organic media, basic cleaning, acid cleaning and final rinsing in deionized water. Regarding metallic supports, cleaning can be performed in an ultrasonic bath with an alkaline solution at 60°C, followed by rinsing in tap water and in deionized water to remove all the alkaline solution and, finally, by cleansing in an organic solvent, such as isopropanol. At the end of the cleaning procedure, the supports are dried overnight in an oven at a temperature of about 120°C.

The traditional activation of the substrates is usually performed by dipping the substrates first into a stannous chloride solution (sensitization) and, thereafter, in a Pd chloride solution (activation), several times. A typical composition of the sensitization and activation solutions is shown in Table 2.5 (Basile and Gallucci, 2011c). It can be noted that activation is performed at acidic pH in order to maintain a good salt solubility. The overall reaction is:

$$SnCl_2 + PdCl_2 = SnCl_4 + Pd \quad [2.18]$$

Table 2.5 Typical composition of solutions used for sensitization/activation of substrates, at room temperature

	Chemical	Concentration
Sensitization	$SnCl_2 \cdot 2H_2O$	$0.1\ g \cdot L^{-1}$
	HCl (37%)	$1\ mL \cdot L^{-1}$
Activation	$PdCl_2$ 99.9% (metal basis)	$0.1\ g \cdot L^{-1}$
	HCl (37%)	$1\ mL \cdot L^{-1}$

The sensitizing/activation cycle is performed at room temperature and the number of dipping cycles depends on the substrate material, its pore size distribution, and surface morphology. Stannous solution, in fact, enters the pores of the support and reduces Pd ions, thus creating autocatalytic sites where Pd plating can occur. Autocatalytic sites must be uniformly distributed on the surface of the support and should be sufficient in number to allow the subsequent Pd deposition. Typically the sensitization/activation cycle is repeated from two up to ten times. Rinsing with deionized water between sensitization and activation is carried out in order to avoid deposition/adsorption of the products of stannous ions hydrolysis such as $SnOH_{1.5}Cl_{0.5}$ and other hydroxyl chlorides on the surface of the substrate. An excess of stannous ions, in fact, can result in the formation of a loose Pd layer, while a deficiency of stannous ions can lead to a non-uniform seeding of Pd nuclei. In some works, rinsing with a 0.01 NHCl solution after the activation step is also performed to prevent the hydrolysis of Pd ions.

The surface modification can be done separately or combined with the activation process and has been extensively studied for the preparation of Pd and Pd-based alloy membrane on porous substrates. Except for obtaining a smooth surface, the modification can simultaneously introduce a buffer layer to avoid diffusion reaction between Pd/Pd-based alloy and substrate. Furthermore, the modification process can introduce a well-dispersed Pd-seed for the catalyst of the electroless plating Pd.

Li *et al.* (1996) prepared thin Pd membrane on porous α-Al_2O_3 disc substrates with an average pore size of 0.2 μm. A multilayer of γ-Al_2O_3 with an average pore size of about 5 nm and thickness of 3–5 μm was deposited on the α-Al_2O_3 substrate by a sol–gel technique. After calcination, the modified substrate was subjected to activation and Pd deposition. Furthermore, the same porous α-Al_2O_3 was modified by another method, which combined the surface modification and activation into one step. The $PdCl_2$ solution was added into the γ-AlOOH sol using the technique of MTPS (modification technique of particle surface), which permits the Pd ions to be adsorbed on the surface of AlOOH nanoparticles. The $PdCl_2$ improved γ-AlOOH sol was coated on the surface of the porous α-alumina disc substrate by a slip-casting process. After calcination at 550°C in air, the Pd ions in the γ-Al_2O_3 top layer were reduced to metallic Pd particles in H_2 at 400°C for 2 h. The derived γ-Al_2O_3 layer with Pd nuclei was plated in an electroless plating Pd bath directly, without further activation process. Certainly, the process combined the smoothing surface and catalysing surface into one step.

Using a shot peening treatment, Jema *et al.* (1996) modified a porous 316L stainless substrate with a thickness of 1 mm and a nominal particle retention size of 0.5 μm. Iron shot with an average diameter of less than 125 μm was used to hit the substrate. For comparison, iron shot with diameter larger than 125 μm was also tested. The shot was released under a peening pressure of 5.1×10^5 Pa from an ordinary sand-blasting machine. The SEM (scanning electron microscopy) characterization indicates that the original porous substrate has surface pore openings as large as 5 μm, which were significantly reduced after the shot peening. The resultant pore size was uniform and was estimated to be around 1 μm. Furthermore, the surface roughness decreased relatively after shot peening treatment. A thin defect-free Pd membrane with thickness of 6 μm was deposited on the modified porous stainless steel substrate by following electroless plating process.

Using ZrO_2 nanoparticles, Wang *et al.* (2004b) modified a porous stainless steel tube with a major surface pore size of 50–100 μm and pore population of 20 mm^{-2}. In their experiment, colloidal zirconium oxide (NSZ-30 A) was used and the pH value of the colloidal was controlled by the addition of ammonia. The colloidal was sucked through the porous tube in an ultrasonic bath for 1 h at room temperature by evacuation from inside the tube for modification of the pores with ZrO_2 particles. The excessive ZrO_2 adhering to the substrate surface was removed with rinsing water. After that, the tube was heated at 300°C for 2 h. The same procedure was repeated several times in order to get better modification results. After the deposition of ZrO_2 on the porous tube, the permeation rate of He decreased considerably, indicating that the oxide particles had been successfully introduced into the large surface pores. A Pd membrane of ~10 μm thickness was prepared on this ZrO_2-modified porous stainless steel tube by electroless plating using a commercial electroless plating bath. The H_2 permeation flux was found to be lower than that for a Pd membrane prepared on a porous glass. The lower activation energy shows that ZrO_2 particles are resistance of H_2 diffusion inside the pores.

Using SiO_2 colloidal suspensions, Su *et al.* (2005) modified the porous stainless steel tube with a nominal particle retention size of 0.2 μm. The SiO_2 colloids with particle sizes 70–100 nm and 20–50 nm were coated on the substrate in a clean room at 22°C and 50% relative humidity. Each of the two suspensions was prepared by mixing Snowtex and ethanol at a weight ratio of 1:5 and was coated on the outer surface of the substrate four times at a dipping withdrawal speed of 1 mm/s. The modified substrate was dried at 300°C. The original porous substrate had opening pores of 10–20 μm on the surface and became narrow inside with an average pore size of about 2.7 μm. After modification by SiO_2 colloid, the substrate was filled with SiO_2 and the surface roughness was greatly improved. The pore size of the upper SiO_2 layer was about several tens of nanometres. The SiO_2 particles entered into the substrate pores are not deeper than 50 μm below the surface.

Using a cerium hydroxide suspension, Tong *et al.* (2004; 2005b) modified the similar porous stainless tube in their previous work (Wang *et al.*, 2004b). The cerium hydroxide suspension was prepared from cerium nitrate aqueous solution (0.03 M) by the addition of sodium carbonate aqueous solution (0.05 M) with strong stirring at 50°C. The pH value of the resultant suspension was ~8. The solid was separated by vacuum filtering and washed using distilled water, then was dispersed into distilled water with ultrasonic bath treatment. The average particle size of cerium hydroxide suspension was 1–4 μm. The cerium hydroxide particles were filled into the pores of the substrate for 2 h at room temperature with an ultrasonic bath and vacuum suction, and then followed by only vacuum suction for 1 h. The excessive cerium hydroxide particles adhering to the substrate surface and in the surface pore entrance were removed by strong distilled water rinsing. So, the cerium hydroxide particles were mainly filled into the inside pores of the substrate. By omitting the calcination step, Pd seeds for electroless plating were *in situ* planted not only on the substrate wall but also on the cerium hydroxide particles using the commercial activating solutions. After that, a mild electroless plating was performed in a commercial Pd solution with pH 6–8 at 30–45°C for retention of the cerium hydroxide particles. The formal plating was performed in the same Pd solution with pH 5–6 at 50–60°C for 7 h. The as-resulted Pd-composite membrane tube had a thickness of 13 μm. An H_2 permeation flux as high as 0.275 mol/(m^2·s) was obtained at 550°C and 200 kPa. In their continuing work, Tong *et al.* (2005c) improved their method by using Pd-containing aluminium hydroxide sol to modify the porous stainless steel substrates. As shown in Fig. 2.18, the improved electroless plating method is based on the multi-dimensional plating mechanism. Normal electroless plating is based on the pore-covering plating mechanism (Figure 2.18c) whose process is composed of Pd-seed planting and Pd plating. The Pd seeds can be planted only on the substrate wall and the amount of Pd seeds is very small. Spherical growth of Pd metal occurs around the seeds on the wall. The pore size gradually decreases with the growth of the Pd layer in the pore, and finally, the pore is choked with the Pd metal. As a result, a thick Pd layer is required to cover the large pore, due to the large pore size of the porous stainless steel substrate. Thus, the Pd layer on the outer surface of the substrate wall becomes thick, while the thickness is small in some parts near the middle of substrate pores. The procedure causes a low hydrogen flux and formation of pinholes in the middle of the substrate pores. In the case of the multi-dimensional plating mechanism (Fig. 2.18a), the pores of the substrate are filled with a dense aluminium hydroxide gel. The water flux under the pressure difference of c. 100 kPa is negligible, showing that the tube filter is almost pore-free. The Pd seeds are planted not only on the support wall, but also on the aluminium hydroxide gel inside the pores. Hence, the growth of Pd metal occurs on the aluminium hydroxide inside the pores as well as on the support surface. This results in formation of a very thin and defect-free Pd layer on the pore-free substrate. The volume of the aluminium hydroxide layer will decrease greatly by dehydration with the thermal treatment up to 500°C, and the original big substrate pores beneath the Pd layer are recovered. The improved multi-dimensional plating mechanism (Fig. 2.18b) is the same as the multi-dimensional plating mechanism in Fig. 2.18a, except that Pd/aluminium hydroxide gel is used instead of aluminium hydroxide gel as a filling material. Since the amount of the Pd seeds inside the substrate is large, due to the introduction of Pd species into the filling material, the Pd plating rate inside the substrate pores will be larger than that without the Pd seeds in the filling material. This can produce a thin uniform membrane with high stability. The formation of Pd metal inside the substrate pores by the reduction of Pd species on the filling material will increase the Pd effective area inside the membrane. A pinhole-free and stable Pd membrane with thickness of 6 μm was prepared and the H_2 permeation flux is as high as 0.302 mol/(m^2·s) with the pressure difference of 100 kPa at 500°C. The flux relates linearly to the pressure difference, suggesting that the hydrogen permeation is controlled by the surface reactions.

Iron oxide and YSZ layers were introduced on the top of porous stainless steel substrates by Zhang *et al.* (2009). For iron oxides, the substrates were oxidized in stagnant air at 600 or 800°C for 8 h with a 5°C/min heating rate, during which the substrate was oxidized to form a thin oxide layer, and then cooled down to room temperature by natural cooling. This method is advantageous for its convenience and simplicity. For the YSZ layer, stable sol was prepared first. Zirconylchloride octahydrate ($ZrOCl_2·8H_2O$) and hexamethylenetetramine ($C_6H_{12}N_4$) were dissolved in deionized water to obtain 0.5 M solutions at room temperature. The hexamethylenetetramine solution was slowly added to the $ZrOCl_2·8H_2O$ solution while stirring with a molar ratio of zirconyl chloride to hexamethylenetetramine (designated as 'Zr/N') of 1.25–2.5, resulting in a zirconia sol at pH 3–4. Then yttrium nitrate solution (0.07 M) was added to zirconia sol in proportion to make 8 mol% YSZ. Finally, appropriate amounts of 5 wt% polyvinyl alcohol (PVA) solution was added to obtain the final YSZ sol. The supported YSZ layers were prepared by dip-coating the sol on substrates. They were then calcined for 3 h at 450°C. After activation using commercial activation solution, the electroless

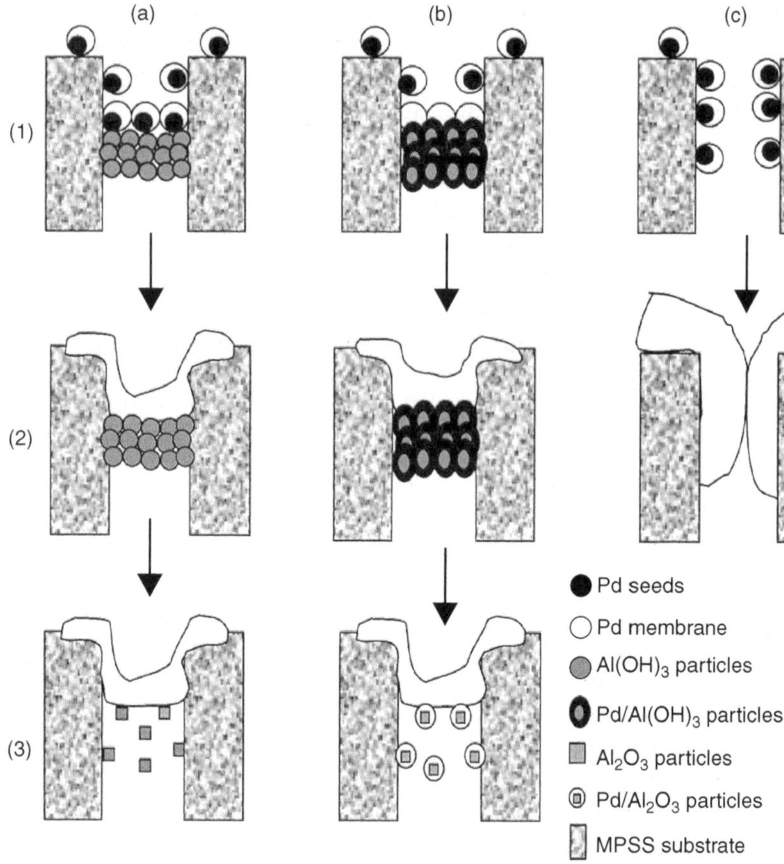

2.18 Different preparation mechanisms of Pd/MPSS membranes. (a) Multi-dimensional plating mechanism: (1) $Al(OH)_3$ particles filling and Pd seeds planting; (2) thin Pd membrane plating; (3) recovering and activating of substrate pores. (b) Improved multi-dimensional plating mechanism: (1) Pd/$Al(OH)_3$ particles filling and Pd seeds planting; (2) thin Pd membrane plating; (3) recovery of substrate pores and activating of Pd/Al_2O_3 inside pores. (c) Normal pore-coverplating mechanism: (1) Pd seeds planting; (2) thick Pd membrane plating. (Tong *et al.*, 2005c).

plating was done using the traditional electroless plating bath. High temperature permeation and 100 h stability tests showed that both intermediate layers were effective as the diffusion barrier for Pd membranes on porous substrates in the temperature range of 500–600°C. At temperatures above 600°C, only the YSZ intermediate layer was effective in preventing intermetallic diffusion and gave a stable Pd membrane.

The electroless plating process is normally carried out in a traditional bath consisting of a palladium ion source ($PdCl_2$,$Pd(NH_3)_4Cl_2$, $Pd(NH_3)_4(NO_3)_2$),

Table 2.6 Typical compositions of palladium plating baths

	Component	Value
Pd ion source	$Pd(NH_3)4Cl_2$ (g·L^{-1})	4–5
Complexant	Na_2ETDA (g·L^{-1})	40–80
Reducing agent	N_2H_4 1 M (ml ·l^{-1})	6–10
Buffer, stabiliser	$NH_4(OH)$ 28% (mL·L^{-1})	200–650
pH		10–12
T (K)		313–333

a complexant (ethylene di-amine tetra acetic acid, EDTA; ethylenediamine, EDA), a reducing agent (hydrazine, sodium hypophosphite), and a pH controller (ammonia) (Basile *and* Gallucci, 2011c). Hydrazine is the most common and suitable reducing agent for electroless plating of Pd. Hypophosphite-based baths with the use of EDA as complexant were also tried in the past, since they have a better efficiency than hydrazine-based baths; however, H_2 evolution during the plating process is absorbed in the Pd layer and causes the formation of cracks and delamination. A typical composition of a hydrazine-based bath is shown in Table 2.6 (Basile *and* Gallucci, 2011c). Deposition is carried out at basic pH (10–12) and at a controlled temperature ranging between 40°C and 60°C. In more detail, the electroless technique is based on the following redox reactions, which occur simultaneously in the solution:

$$\text{Anodic reaction: } N_2H_4 + 4OH^- \rightarrow N_2 + 4H_2O + 4e^- \qquad E^\circ = 1.12\,V \tag{2.19}$$

$$\text{Cathodic reaction: } 2Pd\left(NH_a\right)_4^{2+} + 4e^- \rightarrow 2Pd^\circ + 8NH_a \qquad E^\circ = 0.95\,V \tag{2.20}$$

$$\text{Overall reaction: } 2Pd\left(NH_a\right)_4^{2+}\ N_2H_4 + 4OH^- \rightarrow 2Pd^\circ + N_2 + 8NH_a + 4H_2O \qquad E^\circ = 2.07\,V \tag{2.21}$$

The reaction occurs at the surface of the substrate, preferentially at the Pd seeds. Hydrazine reacts with hydroxide ions, forming nitrogen and water and with the release of electrons which are used to reduce the Pd^{2+} complex into Pd metal. Nitrogen gas is released as bubbles during the process. The rate of palladium deposition and hydrazine consumption are given, respectively, by the following kinetic equations:

$$-\gamma_{\mathrm{Pd}} = \frac{\mathrm{d}[\mathrm{Pd}]}{\mathrm{d}t} = k_1 [\mathrm{Pd}]^{\alpha} [\mathrm{N_2H_4}]^{\beta} \qquad [2.22]$$

$$-\gamma_{\mathrm{N_2H_4}} = \frac{\mathrm{d}[\mathrm{N_2H_4}]}{\mathrm{dt}} = \frac{1}{2}\frac{\mathrm{d}[\mathrm{Pd}]}{\mathrm{d}t} + k_2 [\mathrm{N_2H_4}]^{4} \qquad [2.23]$$

where γ_{Pd} and $\gamma_{\mathrm{N_2H_4}}$ are the rates of Pd and hydrazine consumption; [Pd] and [N_2H_4] the concentration of Pd and hydrazine in the plating bath, k_1 and k_2 the rate constants, and a and b constant values (usually < 1). Constant values can vary accordingly to the plating conditions, such as temperature, concentration of reagents, and plating surface/volume of plating solution ratio. It is clear that hydrazine is consumed both by plating and by decomposition. It has been observed experimentally that hydrazine decomposes rapidly in the presence of Pd metal; this reaction is responsible for the low efficiency, typically less than 20%, of hydrazine-based plating baths. Plating efficiency, defined as the ratio between deposited Pd and the initial amount of Pd in the plating bath, can be increased by increasing Pd concentration in the bath. However, an excess of hydrazine can result in bulk precipitation and cause non-uniform coating. From a practical point of view, hydrazine can be added several times during the plating period in order to maintain a high concentration in the bath and increase plating rate. Electroless plating should be performed in controlled conditions, in order to obtain the formation of pinhole-free, adherent palladium film. It should be noted that the membrane support is inserted in a reactor and maintained in rotation at constant velocity by a variable speed motor to allow the removal from the reaction zone of nitrogen produced by the electroless reaction. The reactor is put inside a thermostatic bath to keep the temperature at a constant value. Hydrazine can be periodically added from the tube inserted in the middle of the reactor, while nitrogen can be evacuated from the reactor through the tube on the right side.

2.5.3 Chemical vapour deposition (CVD)

Chemical vapour deposition (CVD) involves the dissociation and/or chemical reactions of gaseous reactants in an activated (heat, light, plasma) environment, followed by the formation of a stable solid product (Choy, 2003). The deposition involves homogeneous gas-phase reactions, which occur in the gas phase, and/or heterogeneous chemical reactions which occur on/near the vicinity of a heated surface leading to the formation of powders or films, respectively. Figure 2.19 shows a schematic diagram of CVD of coatings. Although CVD is a complex chemical system, it has the following distinctive advantages:

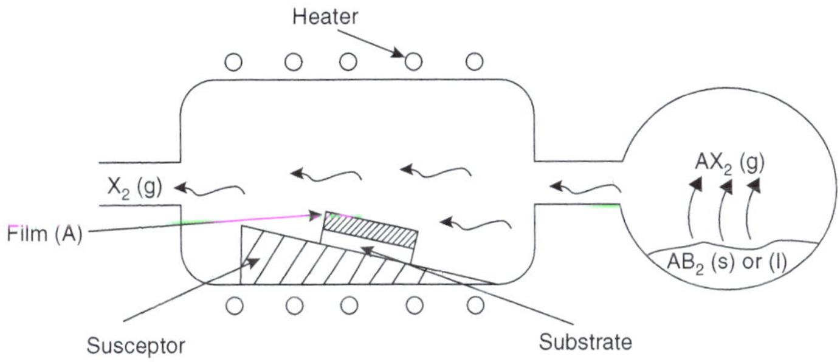

2.19 Schematic of CVD deposition (Choy, 2003). AB_2: solid or liquid precursor; AX_2: gas precursor; X_2 (g): gas product; and A: deposited thin metallic film.

- the capability to produce highly dense and pure materials;
- produces uniform films with good reproducibility and adhesion at reasonably high deposition rates;
- can be used to coat complex shaped components uniformly and deposit films with good conformal coverage;
- the ability to control crystal structure, surface morphology and orientation of film by controlling process parameters;
- deposition rate can be adjusted readily;
- relatively low deposition temperatures.

However, the disadvantages of such high equipment cost and high operation vacuum affect its extensive practical application in large scale for the preparation of Pd membrane.

Yan *et al.* (1994) prepared thin Pd membranes using the metal–organic chemical vapour deposition (MOCVD) method in the macropores of an α-alumina support tube. The best Pd membrane was obtained under a pressure of 100–120 Pa inside the reactor and a heating rate of 10°C/min, at 300°C. The H_2 permeability was equivalent to that of the membrane prepared by Uemiya *et al.* (1991d), and the selectivity was higher than 1000 at a permeation temperature of 100–300°C. The H_2 permeability was proportional to the first order of the H_2 partial pressure, suggesting that the diffusion of dissolved H_2 was not rate-determining. H_2 embrittlement was restrained at a temperature as low as 100°C, and the membrane was resistant to abrasion in spite of its thinness.

Xomeritakis and Lin (1996) prepared a thin, gas-tight Pd membrane using the counter-diffusion CVD process employing $PdC1_2$ vapour and H_2 as Pd precursors. A disk-shaped, two-layer porous ceramic membrane consisting of a fine pore γ-$A1_2O_3$ top layer and a coarse-pore α-$A1_2O_3$ substrate was used as Pd membrane support. A 0.5–1.0 μm thick Pd membrane was deposited in the γ-Al_2O_3 top layer very close to its surface, as verified by

XRD (X-ray diffraction) and SEM with a backscattered electron detector. The most important parameters affecting the CVD process were reaction temperature, reactant concentrations, and top layer quality. Deposition of Pd in the γ-Al_2O_3 top layer resulted in a 100- to 1000-fold reduction in He permeance of the porous substrate. The H_2 permeation flux of these membranes was in the range 0.5–1.0 × 10^{-6} mol m^{-2} s^{-1} Pa^{-1} at 350–450°C. The H_2 permeation data suggest that surface reaction steps are rate-limiting for H_2 transport through such thin membranes in the temperature range studied.

Itoh *et al.* (2005) prepared a thin and firmly deposited Pd membrane applicable to surface catalysis. The technique and equipment developed in this study is based on CVD under a forced flow, where due to a pressure difference applied between the outside and the inside of the support tube the chemical vapours enter the porous layer of the support where they decompose. $(CH_3COO)_2Pd$ was used as a Pd source. The tubular support made from α-alumina powder was porous and have an average pore diameter of 0.15 mm. The forced flow CVD was carried out by heating according to a temperature program under regulated vacuum pressure. The palladium membrane thus obtained was as thin as 2–4 μm and had a H_2/N_2 selectivity > 5000.

2.5.4 Sputter deposition

Sputter deposition is a physical vapour deposition (PVD) process involving the removal of atoms from a solid target by bombarding the target with positive ions (Basile *and* Gallucci, 2011c). Since the sputtering mechanism has a mechanical nature, refractory materials can be easily deposited at temperatures well below their melting point. In addition, it is very useful for the deposition of alloys and compounds, because the resulting film composition, generally, matches that of the source material. Finally, the film structure can be tailored using the proper deposition parameters. In the sputtering process, an energetic particle bombards a target material with sufficient energy to produce the ejection of one or more atoms from the atomic surface layers. The ejected species, mostly neutral atoms, are then transported in the form of vapour to the substrate, where they condense forming the deposited film.

Jayaraman *et al.* (1995b) deposited ultrathin Pd films (< 500 nm) on porous ceramic substrates using the sputter deposition technique. The following two parameters were found to be most critical to the synthesis of the gas-tight metal–ceramic composite: substrate type (surface roughness) and deposition temperature. Fairly gas-tight Pd films with good adhesion could be coated on sol–gel derived fine pore γ-alumina substrates but not on coarse α-alumina substrates. Poor adhesion between the coated film and the

γ-alumina substrate was observed for films coated at room temperature, to a thickness of 300 nm or larger. Both coating temperature (80–600°C) and substrate type affected the grain size, nitrogen gas-tightness and the adhesion of the deposited Pd films. Characterization results show that 400°C was the optimum coating temperature. XRD and SEM showed that the films were fairly crystalline, with a uniform and smooth surface morphology.

Checchetto *et al.* (2004) deposited 5 μm Pd thin films by using radio frequency (RF) magnetron sputtering on porous stainless steel discs (0.5 μm rejection grade) as membrane for hydrogen filtering. In order to fill the steel pores and prepare a flat surface for the Pd coating, a polymeric layer made of commercial polycarbonate was deposited on the steel surface by the spin-coating technique. The polycarbonate layer showed a relatively smooth surface morphology and strong adhesion to the porous substrate. Secondary electron microscopy analysis of the fractured sample showed, in fact, branches of polymeric material which departed from the PC surface and entered into the substrate pores to a length of 10 μm, ensuring the necessary anchorage. Pd sputter deposition produced rough and pinhole-free Pd coating adhering well to the PC buffer layer. Gas selectivity tests by permeation analysis evidenced an H_2 permeance of 5×10^{-7} mol/(m^2·s·Pa) of the as-prepared Pd–PC composite membrane and high H_2/N_2 selectivity.

2.6 Preparation of Pd–Ag alloy membranes

Alloying Pd with Ag to form Pd–Ag alloy H_2 permeation membrane can not only improve the embrittlement but also improve the H_2 permeability. Simultaneously, less use of Pd greatly decreases the membrane cost. Therefore, the Pd–Ag membranes were extensively studied for preparation and applications. The techniques used for the pure Pd membrane preparation were also tried for Pd–Ag alloy membrane preparation. As mentioned before, the preparation of Pd–Ag membranes also focused on the supported composite Pd–Ag configuration.

2.6.1 Cold rolling and diffusion welding

Cold rolling and diffusion welding have been extensively used to prepare thin-walled (~25 μm) Pd–Ag alloy tubes (Gallucci *et al.*, 2008b; Tosti and Bettinali, 2004; Tosti *et al.*, 2008b). Cold rolling is a practical metal working applied to reduce the thickness of metal sheets: several kinds of rolling mills are used, according to the production requirements (hardness of the material, rolling speed, minimum thickness, etc.). Specifically, the minimum thickness which can be obtained for a given material is determined by the diameter of the milling rolls: the smaller the diameter the thinner

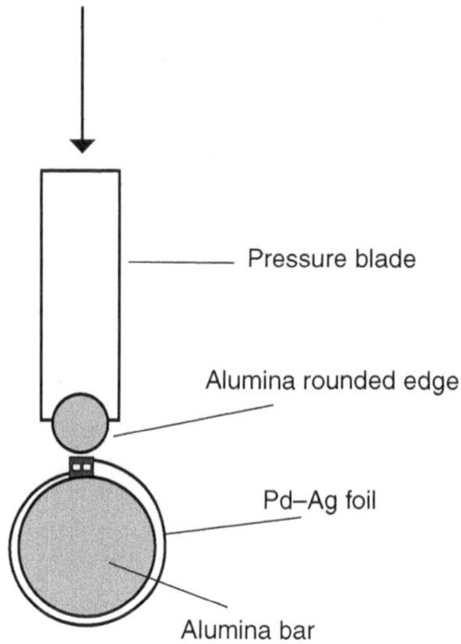

2.20 Scheme of the diffusion welding device (Basile and Gallucci, 2011c).

the worked metal sheet. However, the use of small diameter rolls involves their bending under operation: such behaviour produces curved metal foils. Consequently, in order to produce both very thin and straight foils, a four-high rolling mill was used: in this device two larger support rolls constrained the working rolls of small diameter by avoiding their bending (Fig. 2.20). After cold rolling, the Pd–Ag foils were joined to form the permeator tubes by diffusion welding. In general, diffusion welding is a technique capable of joining most metals and some non-metals: usually, the load applied provokes no macroscopic deformation of the material while the bonding temperature is 50–75% of the metal melting point.

The main characteristic of thin wall Pd–Ag tubes is their complete H_2 selectivity, which permits the production of ultra-pure hydrogen. The hydrogen permeability and the chemical and physical stability have been verified in long term tests (Tosti and Bettinali, 2004; Tosti *et al.*, 2008b). Under operating conditions of 300–350°C and differential transmembrane pressure of 200 kPa, about 3 $Nm^3\ m^{-2}\ h^{-1}$ H_2 permeation rates have been measured. Membrane reactors for dehydrogenation reactions are important applications of the thin wall Pd–Ag membrane tubes. Especially, the Pd-based membrane reactors combine a fixed bed catalytic reactor with a permselective membrane: the H_2 removal through the membrane promotes the reaction conversion beyond the thermodynamic equilibrium (shift effect). Ample experimental work has been carried out by membrane reactors using thin wall permeators: these tests studied the production of ultra-pure

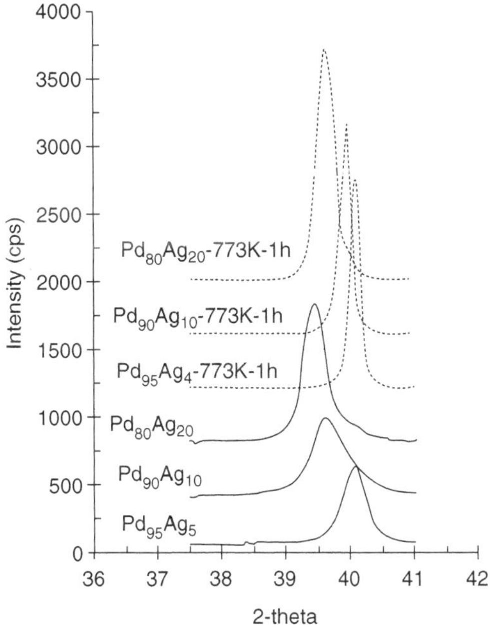

2.21 XRD patterns (111-face) of Pd–Ag alloys with different compositions (Tong *et al.*, 2006b).

hydrogen via steamer forming of ethanol (Gallucci *et al.*, 2008b) and acetic acid (Basile *et al.*, 2008b), methane dry reforming (Gallucci *et al.*, 2008c) and methanol steam reforming (Basile *et al.*, 2008c). The capability of the thin wall membranes to produce ultra-pure hydrogen was demonstrated: moreover, these membrane reactors permitted reaction conversions to be attained and hydrogen yields higher than traditional reformers produce.

2.6.2 Electroplating

Electroplating was used to simultaneously deposit thin Pd–Ag alloy membranes on commercial asymmetric porous stainless steel (APSS) tubes (Tong *et al.*, 2005a; Tong *et al.*, 2006b). Pd–Ag alloy was electroplated in a bath containing palladium and silver ammine complexes with nitrate and sulfonate anions at room temperature. The XRD patterns proved that the Pd–Ag alloy structure was formed at room temperature (Fig. 2.21). The permeation behaviour was investigated in detail, revealing that the hydrogen permeation flux, which was as high as 0.28 mol/(m^2·s), and the infinite hydrogen selectivity versus argon, were achieved at a temperature of 500°C with a pressure difference of 100 kPa. The good membrane stability was proven by the temperature-changing cycles and the gas-exchanging cycles.

Electron probe microanalysis indicated that the Pd–Ag alloy composite membranes had a stable structure and a homogeneous Pd–Ag alloy layer. In addition, the membrane reactor for methane SR was constructed and the reaction performance was tested. The methane conversion, which was as high as 80%, was achieved at a lower temperature of 500°C and a pressure of 500 kPa.

2.6.3 Radio frequency (RF) magnetron sputtering deposition technique

RF magnetron sputtering deposition technique was used by Jayaraman *et al.* (1995a) to deposit ultrathin Pd–Ag films on porous ceramic substrates. Pd–Ag films with a thickness ranging from 250 nm to 500 nm are coated on the surface of 3 nm pore sol–gel derived γ-alumina substrate. The coated Pd–Ag membranes exhibited the same composition and phase structure as those of the Pd–Ag foil used as the target in sputter deposition. The H_2/N_2 separation factor of the ultrathin Pd–Ag membrane is 5.7 at 250°C and increases with increasing temperature. Under proper preparation conditions, use of a pinhole-free γ-alumina substrate is the key to ensuring the gas-tightness and high-selectivity of the coated Pd–Ag membranes.

Using the simultaneous sputtering deposition technique, Tong *et al.* (2005d) synthesized submicron Pd–Ag alloy films with 23 wt% of silver (Pd–Ag) from pure targets of Pd and Ag. Full characterization of the deposited films was performed using X-ray photoelectron spectroscopy, high-resolution SEM, high-resolution transmission electron microscopy and XRD. The analytical results revealed that the deposited Pd–Ag alloy had an Ag content of about 21–22 wt%, very close to and within measurement error of the expected Ag content of 23 wt%. The as-deposited Pd–Ag alloy had a fine microstructure. The characterized Pd–Ag alloy films were then deposited on a supporting microsieve to form Pd–Ag membranes for H_2 separation. The submicron thick Pd–Ag membranes obtained high separation fluxes up to 4 mol/(m^2·s) with an H_2/He selectivity higher than 1500.

2.6.4 Electroless plating

Electroless plating was used by Shu *et al.* (1993) to deposit simultaneously Pd–Ag membrane on porous stainless steel substrate. This work investigated the co-deposition behaviour of Pd and Ag on porous stainless steel by means of electroless plating. Although Pd and Ag can be deposited independently to a considerable thickness, simultaneous deposition of palladium and silver was passivated by the preferential deposition of Ag in an electroless bath containing EDTA as a complexing agent. After effective activation of the substrate by Pd pre-deposition, Pd and Ag could be successfully co-deposited. A combination of XRD and EDX analyses revealed that co-deposited Pd and Ag were in separate phases, and that small amounts of an amorphous Pd rich phase were also present in the deposited material. Thus a code position mechanism of Pd and Ag was proposed. Thermal treatment of as-deposited Pd–Ag films in H_2 atmosphere provided composite Pd–Ag alloy membranes on the porous stainless steel substrate.

Cheng and Yeung (1999) also co-deposited Pd–Ag membranes using electroless plating in a hydrazine-based plating bath containing both palladium and silver precursors. The electroless plating kinetics had been determined for co-deposition of Pd and Ag from the mixed plating bath. A mathematical model was developed to predict the film thickness, plating rate and composition profile as a function of plating parameters (i.e., reactant concentrations and time). The evolution of the microstructure and composition of the film during electroless plating was also investigated. A hydrogen permeation study demonstrated that the Pd–Ag alloy membrane had superior performance compared with a pure palladium membrane of similar thickness. In addition, the formation of a single-phase alloy resulted in substantial enhancement in the hydrogen permeation rate.

2.6.5 Spray pyrolysis

Spray pyrolysis was used by Li *et al.* (1993) to prepare a Pd–Ag alloy membrane on the outer surface of a porous alumina hollow-fibre substrate using spray pyrolysis of a Pd $(NO_3)_2$ and $AgNO_3$ solution in an H_2–O_2 flame. The mass fraction of silver in the membrane obtained at a substrate-surface temperature of 967–1067°C was as low as 0.04, while its fraction in the total metal mass dissolved in the spray solution was 0.1–0.4. This is explained by the higher partial pressure of Ag compared to that of Pd. An additional spray pyrolysis with a silver nitrate solution provided a proper silver content in the alloy membrane. The thickness of the alloy membrane was 1.5–2.0 μm, and the separation factor of hydrogen to nitrogen was about 24 at 500°C. The membrane surface was observed by high-resolution SEM, and the evolution of the morphology was discussed from the view of the deposition mechanism.

2.7 Preparation of Pd–Cu alloy composite membranes

Pd–Cu alloy composite membranes are not susceptible to the mechanical, embrittlement, and poisoning problems that have prevented widespread industrial use of Pd for high temperature H_2 separation. In addition, the H_2 permeability was confirmed to be higher than pure Pd membrane and the cost is much lower because of the large amount of Cu in the Pd–Cu alloy membranes. Therefore, the preparation of Pd–Cu alloy composite membrane was also studied extensively.

2.7.1 Electroplating method

The electroplating method was used by Gao *et al.* (2009) to deposit Pd–Cu alloy films on 316L stainless steel. The electroplating condition was as follows: $PdCl_2$: 15 g/L, $NH_3{\cdot}H_2O$ (28%): 15 mL/L, NH_4Cl: 50 g/L, $(NH_4)_2PO_4$: 50 g/L, EDTA·Cu: 0.6–3 g/L, EDTA·$2NH_4$: 80–150 g/L, current density: 1 A/dm^2, temperature: 40–50°C, pH 7–8, time: 3–5 min. The films showed very good morphology and could be used as H_2 separation membranes. In their study, the corrosion was tested. In boiling mixture of 90% acetic acid + 10% formic acid + 400 ppm Br under stirring (625 r/min), the Pd–Cu films showed better corrosion resistance than Pd film. The Pd–5.66%Cu films showed the lowest corrosion rate almost three orders of magnitude lower than that of 316L matrix. The increased corrosion resistance of Pd–Cu films was attributed to the improved passivity, better barrier effect, increased surface hardness and the effect of Cu to resist pitting.

Moreover, thin (less than 2 μm thickness) and pinhole-free Pd–Cu alloy composite membranes with a diffusion barrier have been fabricated on mesoporous stainless steel supports (MSSS) by vacuum electro-deposition (Nam and Lee, 2001). The deposition film was fabricated by multilayer coating and diffusion treatment and the formation of Pd–Cu alloys was achieved by annealing the as-deposited membranes at 450°C in nitrogen atmosphere. To improve the structural stability of Pd alloy/Ni–MSSS composite membranes, a thin intermediate layer of silica by sol–gel method was introduced as a diffusion barrier between Pd–Cu active layer and a modified MSSS substrate. The composition and phase structures of the alloy film were studied by energy dispersive analysis (EDS) and XRD; the typical Pd–Cu plating had a composition of 63%Pd and 37%Cu and the atomic inter-diffusion of Pd and Cu resulted in Pd–Cu alloys in an fcc structure. The electron probe microanalyser (EPMA) profiling analysis indicated that the improved membranes were structurally stable. The Pd–Cu alloy composite membrane obtained in this study yielded excellent separation performance with H_2 permeance of 2.5×10^{-2} $cm^3/(cm^2{\cdot}cmHg{\cdot}s)$ and H_2/N_2 selectivity above 70000 at 450°C.

2.7.2 Electroless plating

Electroless plating was used to prepare Pd–Cu composite membranes by successive deposition of Pd and then Cu onto various tubular porous ceramic substrates (Roa *et al.*, 2002). The symmetric 0.2 μm cut-off α-alumina tubes and asymmetric 0.05 and 0.02 μm cut-off zirconia coated α-alumina tubes were used as substrates. A seeding procedure was carried out prior to the Pd plating in order to ensure adhesion between the metallic film and the ceramic surface. That step involved impregnation of the ceramic support using an organic Pd salt solution, followed by calcination and reduction in flowing H_2. Common Pd and Cu electroless plating baths were then used in combination with osmotic pressure gradients to deposit films ranging from 1 μm to 25 μm in thickness. The osmotic pressure, generated by circulating concentrated sucrose solutions on the outside of the tubes, ensured reduced porosity and promoted surface homogeneity and densification of the plated Pd film. Composite membranes made by this procedure were leak-tested by pressurizing the composite membranes with N_2 and soaking them in a solution of isopropyl alcohol/water to detect the amount and size of bubbles. Membranes showing low N_2 leakage rate, less than 5×10^{-5} mol/(m^2·s), were tested at high temperature. The highest H_2 permeability was observed at an alloy composition of 60 wt% Pd at a constant temperature of 350°C. A typical H_2 flux for this membrane was 0.81 mol/(m^2·s) at 350°C and 255 kPa H_2 partial pressure.

Dual sputtering deposition technique was used to prepare submicron thin Pd–Cu alloy films, which allowed a high composition control of the layer (Hoang *et al.*, 2004). The composition, surface morphology and phase structure of the sputtered layers were investigated by EDS, X-ray, XPS, SEM, TEM and XRD. For example, the XRD data proved that the Pd–Cu layers were an alloy of Pd and Cu. Subsequently, the characterized Pd–Cu alloy layers were deposited on a silicon support structure to create a 750 nm thin Pd–Cu membrane for H_2 separation. The reported membrane obtained a high H_2 flux of 1.6 mol/(m^2·s) at a temperature of 452°C, while the selectivity was at least 500 for H_2/He.

2.8 Preparation of Pd–Au membranes

The early research into the permeation behaviour of Pd–Au alloys was done using cold-worked foils. While this fabrication technique produces homogeneous films with extremely precisely controlled compositions, it requires substantial capital investment, the foils produced are generally limited to greater than 25 μm in thickness and the technique cannot be modified to produce supported foils. Other techniques, such as chemical or physical vapour deposition, have been used successfully to produce high-quality alloy films, but require costly vacuum equipment and targets. A third option is electroplating or electroless plating. Films produced by plating processes can be markedly thinner than cold-worked films, resulting in higher flux membranes with substantially reduced precious metal costs. Palladium can

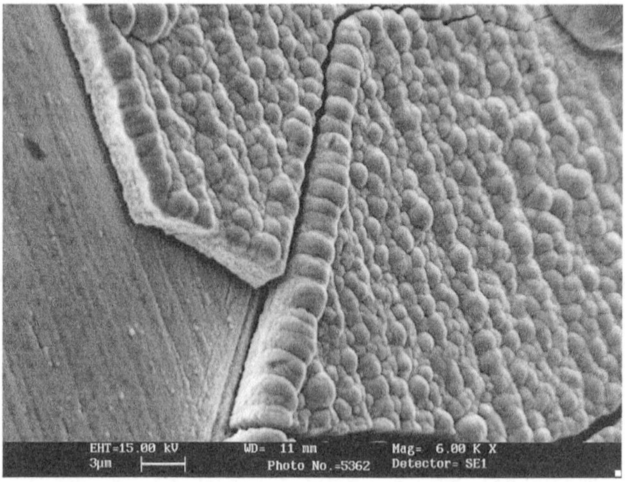

2.22 SEM images of the surface of Pd–Au alloys (75.8% Pd in the bulk) taken after H_2 absorption/desorption experiments (Lukaszewski and Czerwinski, 2003).

be successfully electroless plated on any appropriately activated surface, alloys can be produced either by co-plating or sequential plating, and equipment costs are very small. Sequential plating obviates the issue of controlling composition intrinsic to co-plating, but creates a material with a layered structure rather than a homogeneous composition.

Woods (1969) co-deposited Pd–Au alloy film as electrode on tantalum foil by the electroplating technique at room temperature. The electroplating bath was made up from 2% solutions of $PdCl_2$ and AuCl in 1 M HCl in varying proportions. XRD patterns indicated that Pd and Au formed homogenous solid solution. Using the similar process, Lukaszewski and Czerwiński (2003) co-deposited Pd–Au thin film by electroplating technique on gold wire from a bath containing $PdCl_2$ and $HAuCl_4$ in 1 M HCl. The deposition potential was always higher than the potential of hydrogen sorption in order to avoid hydrogen insertion into the alloy being formed. The composition of the alloys can be altered by employing different electrolyte compositions and deposition potentials. The thickness of the obtained Pd–Au film was in the range of 1.5–3.0 µm. The roughness of the freshly deposited alloys was in the range of 100–500. As shown in Fig. 2.22, the Pd–Au film were quite dense although a number of large cracks due to the desorption of oxygen formed during electroplating.

Su *et al.* (2004) also electroplated Pd–Au alloy in an ambient temperature ionic liquid electrolyte, 1-ethyl-3-methylimidazolium chloride tetrafluoroborate system (EMI-Cl-BF4), containing Pd (II) and gold (I). Compact alloy deposits were obtained with galvanostatic electrolysis. The Pd content in the deposited Pd–Au alloy increased with both increasing current density and Pd mole fraction in the plating bath. For deposits obtained under mass-transport-limited current, the Pd content in the deposited alloys approximated the Pd mole fraction in the plating bath. The deposition temperature affected the homogeneity of the deposited Pd–Au solid solution. Data collected from XRD experiments showed that Pd–Au alloys deposited at temperatures higher than 80°C were homogeneous Pd–Au solid solutions. The scanning electron micrographs of the deposits revealed nodular particles. The particle size decreased with increasing deposition current.

Although the electroplating technique has been used to deposit Pd–Au films, it suffered from the formation of micro-cracks and has not been extensively used for Pd–Au alloy membrane preparation for hydrogen separation.

Okazaki *et al.* (2008) fabricated defect-free Pd–Au alloy membrane (4–5 μm thickness) on porous α-Al_2O_3 tube by the sequential electroless plating method. The Pd layer was electroless plated on the porous substrate using the conventional bath consisting of hydrazine, palladium chloride, ammonia and Na_2EDTA, etc., at 45°C for 24 h. The Au layer was electroless plated on the top of Pd layer using the bath consisting of $K[Au(CN)_4]$, KOH and hydrazine, etc., at 90°C for 6 h. By adjusting the concentration of $K[Au(CN)_4]$ in the Au plating solution, the thickness of the gold layer could be controlled. The dual layer of Pd and Au was heated at 600°C for 12 h under nitrogen atmosphere, and then hydrogen atmosphere. After that, the temperature was increased to 750°C and kept for 24 h to finish the alloying process.

Furthermore, Gade *et al.* (2009) developed a sequential electroplating and electroless plating procedure to prepare freestanding planar Pd–Au membranes with Au contents ranging from 0 to 20 wt%, consisting of Au layers on both sides of a pure Pd core. High temperature X-ray diffraction (HTXRD) results indicated that the pre-treatment at 750°C with hydrogen was also needed to form Pd–Au alloy structure and achieve reasonably high hydrogen permeability.

2.9 Preparation of amorphous alloy membranes

Amorphous alloy membranes are at present being developed to tackle the problems of H_2 embrittlement, sintering and cost that occur with crystalline alloy membranes, while still providing H_2 permeance comparable to Pd. Fabricated from non-Pd elements, these membranes are low price and inherently resistant to embrittlement because of their amorphous structure. Much development is still required, however, to improve stability and achieve the performance of existing Pd-based crystalline alloy

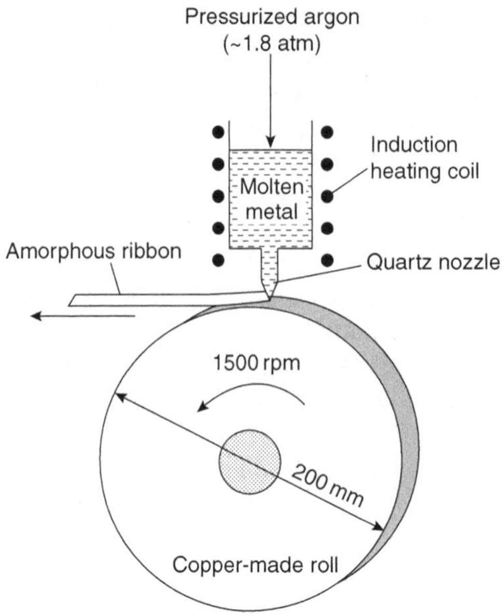

2.23 A single-roller melt spinning apparatus for preparing amorphous alloy ribbon (Itoh *et al.*, 1997).

membranes under the conditions above specified (Dolan *et al.*, 2006). In this part, we just briefly discuss the preparation of some typical amorphous membranes.

2.9.1 Single-roller melt spinning technique

A single-roller melt spinning technique was used to prepare the amorphous $Zr_{36}Ni_{64}$ alloy membranes (Hara *et al.*, 2000). The eutectic $Zr_{36}Ni_{64}$ alloy was prepared by arc melting under a protective Ar atmosphere. As schematically shown in Fig. 2.23 (Itoh *et al.*, 1997), the roller made of copper (20 mm wide, and 200 mm in diameter) was regulated to rotate at 1500 rpm. The alloy sample was put into a quartz tube, which had a thin slit (6 × 0.5 mm), heated to melt by an induction coil connected to a high frequency generator, and then sprayed with 1.8 atm pressurized Ar. The ribbon samples thus prepared had a mean thickness of about 30 μm and c. 5 mm in width. The wider ribbon samples can be easily made by making use of a wide-slit nozzle. The membrane obtained was too tough to yield to bending, which is a characteristic of amorphous alloys. The H_2 permeation rate was more than $1\ \text{cm}^3{}_{H_2}\ (\text{STP})/\left(\text{cm}^2 \cdot \text{min}\right)$. Using the same method, the amorphous membranes of Ni–Nb–Zr (Yamaura *et al.*, 2003), Zr–M–Ni

(M = Ti, Hf) (Hara *et al.*, 2002), $Zr_{60}Al_{15}Co_{2.5}Ni_{7.5}Cu_1$ (Shimpo *et al.*, 2004) were also prepared.

2.9.2 Cold rolling

Cold rolling was also used to prepare amorphous alloy membranes for H_2 permeation (Tang *et al.*, 2007). Maximum cold rolling reduction rates of as-cast $Nb_{40}Ti_{30-x}Zr_xNi_{30}$ alloys were measured to evaluate ductility by measuring the change in the thickness. Reduction rates of 50% or higher were obtained for the alloys containing the Zr content of 12 mol% or less, but lower values were obtained for the alloys substituted with more Zr. Changes of H_2 permeability and microstructure of the $Nb_{40}Ti_{18}Zr_{12}Ni_{30}$ alloy caused by cold rolling and subsequent vacuum annealing were investigated. Although the eutectic phase disappeared and was replaced by a small spherical (Nb, Ti, Zr) phase embedded in the (Ti, Zr)Ni matrix after rolling and subsequent annealing, these alloys showed good resistance to hydrogen embrittlement at 250°C or more.

2.9.3 Magnetron sputtering deposition

Magnetron sputtering deposition was used to prepare $Nb_{40}Ti_{30}Ni_{30}$ amorphous membrane for H_2 permeation (Xiong *et al.*, 2010). $Nb_{40}Ti_{30}Ni_{30}$ membranes were first fabricated on Mo substrate by magnetron sputtering method under pure Ar atmosphere at 0.6 Pa for 60 min, and the base temperature was maintained at 25, 300 and 500°C, respectively. The membranes deposited at base temperature of 25°C were then annealed at 700°C for 30 min in vacuum. Microstructure and mechanical properties of $Nb_{40}Ti_{30}Ni_{30}$ membranes fabricated by magnetron sputtering were investigated. Deposited and annealed $Nb_{40}Ti_{30}Ni_{30}$ membranes consisted of amorphous and crystalline phases, respectively. Higher base temperature was shown to increase the hardness and elastic modulus of the $Nb_{40}Ti_{30}Ni_{30}$ membrane. Furthermore, a porous nickel substrate was successfully prepared by uniaxial compression of nickel powders. After that, the Pd/$Nb_{40}Ti_{30}Ni_{30}$/Pd/porous nickel support composite membranes were then fabricated using a multilayer magnetron sputtering method. The H_2 permeability of the composite membranes with amorphous and crystallized $Nb_{40}Ti_{30}Ni_{30}$ metal layer was measured and compared with that of self-supported $Nb_{40}Ti_{30}Ni_{30}$ and Pd-alloys. Solid-state diffusion was shown to be the rate-controlling factor when the thickness of the $Nb_{40}Ti_{30}Ni_{30}$ layer was about 12 μm or greater, while other factors were in effect for thinner layers (such as 6 μm). The Pd/$Nb_{40}Ti_{30}Ni_{30}$/Pd/porous nickel support composite membrane exhibited excellent permeation capability and satisfactory mechanical properties.

2.9.4 Electroless plating technique

Electroless plating technique was used to prepare Ni–P, Ni–Ru–P, Ni–B amorphous membranes for H_2 permeation (Liu *et al.*, 1997; Wu *et al.*, 1998; Xue and Deng, 2001). The porous ceramic tubes were used as substrate. After the trapped air in substrate pores was removed by vacuum and water pumping, electroless plating was carried out. The plating bath consisted of 30 g/L $NiCl_2 \cdot 6H_2O$, 60 mL/L ethylenediamine, 1.5 g/L KBH_4, 20 g/L KOH and 0.6 mg/L $CdSO_4$. The solution temperature was 90°C. With the vacuum pump working, the activated tube was taken out of water and quickly dipped into the vigorously stirred plating solution and remained there for 40 min. A desired Ni–B membrane was obtained in this way.

2.10 Degradation of dense metallic membranes

The degradation of dense metallic membranes usually arises because the H_2 permeability, the H_2/other gases selectivity, and the membrane mechanical strength gradually deteriorate with increasing operation period, because of the H_2 transportation at different temperatures. This degradation, when the membrane is combined with chemical reactions, will be briefly discussed in this section. In particular, the degradations of embrittlement (pure metal), oxidation (some pure metal and amorphous alloy membrane), polarization (Pd-based alloy and amorphous membranes), interaction with support (supported membranes) and dense metallic MRs (interaction with catalysts, coke, sulphur, morphology change) will be summarized briefly.

2.10.1 Degradation of membranes themselves

H_2 embrittlement is a consequence of H_2/metal interactions. It has been established that severe mechanical degradation occurs and is manifested in a decrease of fracture resistance. From a materials-classification standpoint, one may distinguish between microstructurally stable materials, in which the metal and solute H_2 interact, and those which require attention to phase stability. Stress-induced hydride formation and cleavage mechanism is the main H_2 embrittlement mechanism for pure-metal H_2 permeation membranes. The H_2 permeation metals Pd, IVB (Ti, Zr, Hf) and VB (V, Nb, Ta) can all easily form hydride when exposed to H_2 at relatively low temperatures (Buxbaum and Markerb, 1993; Buxbaum and Kinney, 1996). Alloying with other metals can stabilize the structure and improve the embrittlement effect (Yamaura *et al.*, 2004; Yukawa *et al.*, 2008).

An oxide layer is commonly formed on the surface of the pure metals of IVB (Ti, Zr, Hf) and VB (V, Nb, Ta) due to being thermodynamically favourable to forming respective oxides. Therefore, the pure metallic H_2 permeation membranes and the amorphous alloy membranes consisting of these metal elements are easily oxidized on the surface. The oxide layer can decrease the H_2 dissociate adsorption rate and associate desorption rate greatly, which will furthermore slow down the H_2 permeation (Yukawa *et al.*, 2008). In order to avoid oxidation, the common improvement method is to deposit Pd or Pd alloy layer on both surfaces of these kinds of membranes (Busnyuk *et al.*, 2010).

Interaction with substrate is a degradation that frequently happens for the supported composite H_2 permeation membranes, because of morphology change or chemical reaction. Pd membrane with a thickness of 22 μm, supported on porous stainless steel substrate, after high-pressure (100–2800 kPa) was found to be greatly decreased in H_2/He selectivity (from infinite to 12) and permeance (decreased 35%). SEM analysis revealed that extremes in the Pd film thickness, ranging from about 10–50 μm, with Pd 'fingers' extending into the pore structure anchoring the Pd layer to the substrate. Although surface characterization could not pinpoint the source of the degradation, intermetallic diffusion is considered to be a possibility (Rothenberger *et al.*, 2004).

Preferential segregation of certain elements from dense alloy membranes can also result in degradation of the performance of H_2 permeation membranes. For example, Pd–Ag films (~2.4% Ag, 20–26 μm thick) were deposited by sequential electroless plating onto porous tubular stainless steel substrates with Al_2O_3 oxide layers to modify the substrate pore size and to prevent intermetallic diffusion of the stainless steel components into the Pd–Ag layer (Bosko *et al.*, 2011). Composite membranes annealed at temperatures of 500–600°C were characterized for film structure (XRD), morphology (SEM), bulk and surface component distribution (EDS, XPS), and H_2 permeance. Composition measurements within the Pd–Ag layer revealed preferential segregation of the Ag component to the top surface. This result is consistent with the lower surface free energy of Ag.

2.10.2 Degradation of dense metallic MRs

Coke is another problem associated with metal when an initially defect-free Pd-based composite membrane is used in high temperature catalytic applications. The further diffusion of these deposited carbonaceous impurities into the bulk phase of the membrane can lead to defects in the membrane (Lu *et al.*, 2007). Lin (2001) has conducted some systematic investigations into this. Figure 2.24 shows the permeance and separation results of a thin Pd/Ag membrane prepared by sputter deposition before and after being exposed to a graphite ring (surrounding the membrane disk) at 600°C

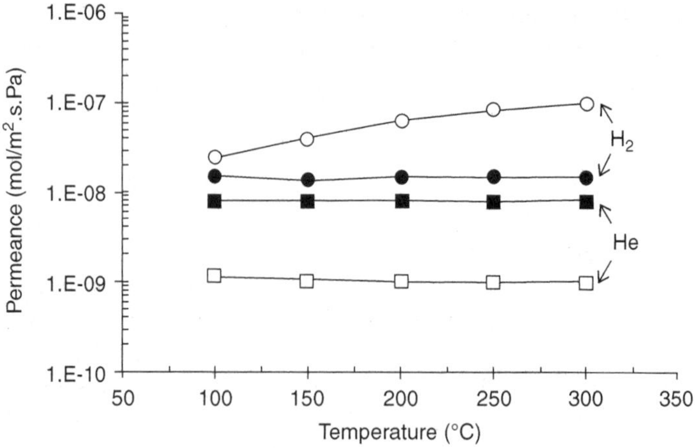

2.24 Hydrogen and helium permeance (with the feed of 1:1 H_2 and He mixture) through a 200 nm thick Pd–Ag membrane before and after being exposed to a carbon source at 600°C (Lin, 2001). Open symbols – before carbon poisoning; closed symbols – after carbon poisoning.

overnight. XRD analysis shows expansion of the Pd–Ag lattice, indicating carbon diffusion into the lattice after exposing the Pd–Ag membrane to the carbon-containing source. The increase in He permeance after poisoning indicates a change to the Pd–Ag membrane microstructure after the expansion of Pd–Ag lattice, creating defects or enlarging the grain-boundary. The incorporation of carbon in the Pd–Ag lattice could reduce hydrogen solubility, decreasing the hydrogen permeability of the membrane. Re-exposure of the poisoned Pd–Ag to H_2 atmosphere could remove the poisoning agent, but cannot restore the mechanical integrity of Pd–Ag membrane that was destroyed by the poisoning.

2.10.3 The effect of contamination of Pd-based membranes on H_2 flux

Chemical stability of the surfaces of membranes used in MRs is another important issue. Several researchers have demonstrated the negative influence of various chemical species, among which are O_2, H_2S, CO_2 and H_2O, on the performances of MRs (Gabitto and Tsouris, 2009). Oxygen exposure is able to influence the membrane surface morphology greatly, even if Pd-oxide formation is not observed: its influence is related to the increase in the roughness of the membrane surface (and thus of the membrane area), causing problems to the integrity of the thin metallic film. On the other hand, the presence of CO_2 and H_2O is able to block the hydrogen adsorption sites causing a reduction in performance, whereas the presence of sulphur compounds, even in small amounts, deactivates the membrane surface compromising its catalytic performance.

It must be also added that surface contamination by Hg vapour, H_2S, SO_2, thiophene, arsenic, unsaturated hydrocarbons or chlorine carbon from organic materials and compounds containing chemical elements, such as S, As, I, Br or zinc, is also responsible for deviations from Sieverts' law. In particular, CO and possibly also CO_2, owing to their strong adsorption on the Pd surface, are the major compounds in the methane SR that might poison the Pd membrane surface. In these cases, membranes show a permeation controlled by the rate of adsorption on heavily contaminated surfaces: the rate-limiting step is not yet bulk H_2 diffusion, whereas H_2 dissociative adsorption plays an important role. For thick films, deviations from Sieverts' law can also be due to a decrease in the surface reaction rate after adsorption of contaminants, such as C, CO, CO_2, hydrocarbons, sulphur or chlorine, on the Pd surface. In these cases, for Pd-based membranes, an air oxidation at 300–450°C is necessary for removing adsorbed contaminants (Basile *et al.* 2007; Keuler *and* Lorenzen 2002c). In particular, the influence of CO is slightly more complex: CO decreases H_2 permeation through a Pd-based membrane because of a blockage of the surface: the lower the temperature the stronger the effect. For example, Amandusson *et al.* (2001) observed no permeation of H_2 through the membrane at 100°C; on the other hand, at 250°C a decrease in the H_2 flux of only 10% was observed; while at above 350°C, CO showed no influence on the H_2 flux. Many explanations were proposed for the observed deviation: contamination of the membrane surface, dependence of the H_2 diffusion coefficient on the dissolved H_2, H_2 leakages through cracks or defects, and so on. To the best of our knowledge, up to now no definite evidence one way or the other has yet been presented. In other cases, in an H_2 rich environment of methane SR at 500°C, the effect of CO and CO_2 on the composite Pd membrane was negligible, and also Pd–Au alloy membranes show sulphur resistance (Hughes, 2001).

In this last part of the chapter, the effect of H_2S on the performance of Pd-based membranes is considered because, as has been said, even a small quantity of this gas (in the syngas, for example, as a by-product of the coal gasification process) can degrade the performance of Pd-based membranes (Chen and Ma, 2010). Sulphur poisoning of Pd-based membranes is generally discussed in terms of two different processes (Gabitto and Tsouris, 2009): (a) the sulphur deactivates the membrane surface by adsorbing on surface sites and so reducing their catalytic activity; and (b) the sulphur chemically reacts with the membrane producing corrosion products on the surface that block the H_2 flux.

An example of the second process is furnished by Kulprathipanja *et al.* (2005), who hypothesize that the reduction of the performance was due to 'site-blocking' with complete inhibition at H_2S concentrations > 100 ppm. In a recent experiment, carried out by Chen and Ma (2010), a Pd membrane, exposed at 400°C to a 55 ppm H_2S/H_2 mixture, resulted in a two-stage H_2 permeance decline, owing to: (a) surface blocking, because of the dissociative adsorption of H_2S; and (b) black sulphidation of Pd with the formation of the Pd_4S compound, which caused irreversible permeance loss owing to the formation of pinholes which gave rise to membrane failure.

Generally, the poison process starts with H_2S adsorption on the Pd membrane surface, followed by H_2S dissociation into adsorbed HS and sulphur, which is then followed by the formation of Pd_4S (or Pt_4S) compounds, that is, one sulphur atom per four palladium (or platinum) atoms.

What is important to note is that, for example, Pd/Au alloy membranes (13 wt% Au) do not show significant structural change due to the bulk sulphide formation (Chen and Ma, 2010), and also that Pd/Cu alloy membranes show very high sulphur resistance (Pomerantz and Ma, 2009).

Finally, the reader interested in the prediction of hydrogen flux through sulphur tolerant binary alloy membranes should refer to Kamakoti *et al.* (2005).

2.11 Conclusions and future trends

The exponential trend of significant progress in the field of the inorganic MRs is well reflected in the increasing number of publications and patents, evidencing that the performance of the Pd-based membrane used in MRs continues to improve due to the strong and collective efforts of the scientific community. For example, hydrogen permeance has increased by one order of magnitude with respect to the first experience in the 1960s.

Dense inorganic membranes used in MRs are selective towards only one of the molecules hydrogen and oxygen. The aim of this chapter is to attest to the growing strategic role of MRs for pure H_2 production through certain chemical reactions in the phase gas.

As compared to a classical configuration consisting of a reactor unit in series with a separation unit, an MR is a modern configuration with an integrated reaction/separation unit which brings many potential advantages: reduced capital costs, improved yields and selectivities, and drastically reduced downstream separation costs. In a dense inorganic MR, pure H_2 moves to the permeate side, enabling the reactions to proceed towards completion and so making it possible to achieve: (a) conversion values higher than those obtained by TRs working under the same operational conditions; or (b) the same conversion values as obtained by traditional reactors but under milder operational conditions.

In this chapter, as stated, efforts have been concentrated solely on hydrogen production and so all classes of reactions in the gas phase still under study have been proposed to the reader: dehydrogenation reactions, oxidative dehydrogenation reactions and steam reforming reactions. For each of these reactions, a general overview has been presented in which pressure, temperature, type of catalyst and membrane used, and the best results in terms of conversions, yields and selectivity have been summarized. Since only some of them can be considered of practical value (WGS, SR of methane, dry reforming of methane, POM, methanol SR, oxidative methanol SR, ethanol steam reforming, SR of acetic acid, glycerol SR and some of the dehydrogenation reactions), a brief panoramic overview, also in terms of the various types of catalysts used, of these reactions more in-depth has also been proposed. From comparison in terms of reaction conversion and yield, a qualitative analysis shows that the performance of inorganic MRs is always higher than the traditional/conventional ones. Moreover, in particular, it also shows that dense Pd-based MRs perform better than porous inorganic membranes.

It was also clarified that *Thermodynamic Equilibrium* conversion is only related to TRs (closed systems), and has nothing to do with MRs (open systems). The conversion limit of MRs has been indicated as a *Dynamic Equilibrium* curve, which represents the reaction not limited by any kind of chemical kinetics (no kinetic resistance, that is, all molecules in contact with the catalyst are considered to react with an infinite velocity) and thus the feed gas is assumed to be continuously at equilibrium inside the MR.

An important aspect related to inorganic MRs, always to be taken into consideration, is the kind of membrane used. Many metals are permeable only to H_2, others to O_2. In particular, some dense materials (Pd and its alloys, Pt, V, Ta, Nb and SiO_2 ceramic membrane) exhibit significant H_2 permeability values and very high H_2 selectivities in accordance with a solution–diffusion mechanism. In particular, dense Pd membrane possesses exclusive selectivity to $H_2(\alpha_{H_2/\text{other gas}} = \infty)$, whereas porous materials show a lower selectivity (but a higher H_2 flux). Equations and the various laws (Fick, Sieverts, Armbruster, Arrhenius, Richardson and Murav'ev-Vandyshev) governing the hydrogen solution and flux in/through dense Pd-based membranes are well known today and are discussed in this chapter.

Another important class of membranes used in MRs is the so-called *non-Pd-based membrane*. The interest in developing these membranes (or, at least, with a low Pd-based content) is mainly due to the high cost of Pd material, which strongly limits its use in wide-scale industrial use. Moreover, these membranes also show very high values of H_2 permeability and resistance to H_2 embrittlement (for example, an Nb–Ti–Ni alloy). Other non-Pd-based membranes, such as Pd-coated amorphous Zr–M–Ni (M = Ti, Hf) alloy membranes, are resistant enough in an H_2 atmosphere and show stable permeability only to H_2 at least in the range of 200–300°C.

Amorphous metals are also under study because of their superior properties with respect to their crystalline counterparts. In fact, amorphous metals possess higher H_2 solubility, strength and ductility and are more resilient to corrosion and failures associated with H_2 embrittlement, and are able to operate in more extreme operating conditions than their crystalline structure counterparts. Owing to the increasing importance of amorphous alloys, the equations governing hydrogen dissolution in, and the hydrogen transport through, these membranes have been also briefly summarized.

Because of the cost of palladium, the major aim of the study concerning Pd-based membranes is the reduction of the thickness of the two types of membranes currently available for pure H_2 production (composite and self-supported thin).When ultra-pure H_2 production is required, commercial self-supporting dense Pd-based membranes with wall thicknesses larger than 100 μm maintain sufficient mechanical strength, and assure a defect-free surface and complete H_2 selectivity. Unfortunately, however, these membranes are too thick to obtain a satisfactory H_2 flux. For a practical use, it is necessary to reduce the thickness of the palladium layer or/and to find selective Pd-based membranes with low Pd content. A compromise is reached using composite membranes in which one or more thin dense layers of highly selective membrane material are deposited on porous supports (i.e., with high H_2 permeability).

In the field of surface-engineering techniques, the fabrication of thin film has evolved greatly in recent years. Unfortunately, there is no method that is exhaustive (each method has advantages and disadvantages) and various compromises are necessary for every choice. For this reason, the rest of the chapters show, extensively and in detail, the various aspects and techniques related to the preparation of the various membranes considered.

The last part of the chapter is dedicated to the degradation of dense metallic (pure, alloy, amorphous) membranes. The H_2 permeability, the H_2/other gases selectivity, and the membrane mechanical strength gradually degrade with increasing operation because of the H_2 transportation at different temperatures. This degradation, when the membrane is combined with chemical reactions, is briefly discussed with particular attention to the degradation due to embrittlement (pure metal), oxidation (some pure metal and amorphous alloy membrane), the polarization effect (Pd-based alloy and amorphous membranes), the interaction with support (supported membranes) and the degradation of dense metallic MRs (interaction with catalysts, coke, sulphur and morphology change).

To summarize, in the last few years, significant work has been done by scientists and chemical engineers to improve inorganic MRs performance and overcome such problems as membrane durability and resistance of the membrane-module seals to high temperature and pressure. Improvements in the technology of inorganic MRs are bringing them closer to industrial utilization. Their introduction in an industrial context strongly depends on the economics of the membrane process, as is fully discussed in many chapters of this handbook.

Finally, it should also be said that most of the information presented in this chapter is also intended as an anticipation of some important concepts presented and developed in the chapters that follow.

2.12 Acknowledgements

The author is grateful to Dr Simona Liguori for her help in preparing some of the figures and tables.

2.13 References

Abballa B.K. and Elnashaie S.S.E.H. (1995), *J. Membrane Sci.*, **101**, 31.
Adhikari S., Fernando S., Gwaltney S.R., To S.D.F., Bricka R.M., Steele P.H. and Haryanto A. (2007), *Int. J. Hydrogen En.*, **32**, 2875.
Adhikari S., Fernando S.D. and Haryanto A. (2008), *Renew. Energy*, **33**, 1097.
Adhikari S. and Fernando S. (2006), *Ind. Eng. Chem. Res.*, **45**, 875.
Agrell J., Birgersson H. and Boutonnet M. (2002), *J. Power Sou.*, **106**(1–2), 249.
Aguilar G., Gryaznov V.M., Pavlova L.F. and Yagodovskii V.D. (1977), *React. Kinet. Catal. Lett.*, **7**, 181.
Ahchieva D., Peglow M., Heinrich S., Mörl L., Wolff T. and Klose F. (2005), *Applied Catal. A: General*, **296**, 176.
Akin F.T. and Lin Y.S. (2002), *J. Membrane Sci.*, **209**, 457.
Ali J.K., Newson E.J. and Rippin D.W.T. (1994), *Chem. Eng. Sci.*, **49**, 2129.
Ali J.K. and Baiker A. (1996), *Appl. Catal. A: Gen.*, **140**, 99.
Ali J.K. and Baiker A. (1997), *Applied Catal. A: General*, **155**, 41.
Altinisik O., Dogan M. and Dogu G. (2005), *Catal. Today*, **105**, 641.
Amandusson H., Ekedahl L.G. and Dannetun H. (2001), *J. Membrane Sci.*, **193**, 35.
Amphlett J.C., Evans M.J., Jones R.A., Mann R.F. and Weir R.D. (1981), *Can. J. Chem. Eng.*, **59**, 720.
Amphlett J.C., Evans M.J., Mann R.F. and Weir R.D. (1985), *Can. J. Chem. Eng.*, **63**, 605.
Amphlett J.C., Mann R.F. and Weir R.D. (1988), *Can. J. Chem. Eng.*, **66**, 950.
Amphlett J.C., Mann R.F. and Peppley B.A. (1994), *Stud. Surf. Sci. Catal.*, **81**, 409.
Armbruster M.H. (1943), *J. Am. Chem. Soc.*, **65**, 1043.
Armor J.N. (1992), *Chem-Tech*, **22**, 557.
Aspen Systems, (1999), Proc. of the USDOE CARAT Program Review, Troy (Michigan), Sept. 23.
Augustine A.S., Ma Y.H. and Kazantzis N.K. (2011), *Int. J. Hydrogen En.*, **36**, 5350.
Austin G.T. (1986), *Shreve's chemical process industries*, (McGraw Hill, New York).
Balachandran U., Dusek J.T., Maiya P.S., Ma B., Mieville R.L., Kleefisch M.S. and Udovich C.A. (1997), *Catal. Today*, **36**, 265.
Balovnev Y.A. (1974), *Russ. J. Phys. Chem.*, **48**(3), 719.
Barelli L., Bidini G., Gallorini F. and Servili S. (2008), *Energy*, **33**, 554.

Basagiannis A.C. and Verykios X.E. (2006), *Appl. Catal. A: Gen.*, **308**, 182.
Basagiannis A.C. and Verykios X.E. (2007a), *Catal. Today*, **127**, 256.
Basagiannis A.C. and Verykios X.E. (2007b), *Int. J. Hydrogen En.*, **32**, 3343.
Basile A. and Calabrò V. (Eds), (2010), *Sustainable Hydrogen and Energy Production from Renewable Sources – Catalytic and Bio-catalytic Membrane Reactors – Fuel Cells*, University of Calabria, Centro Editoriale e Librario press – Rende (Italy), ISBN: 978-88-7458-118-4.
Basile A. and Drioli E. (2006), *Catal. Tod.*, **118**(1–2), 1–265.
Basile A. and Gallucci F. (2009a), I Special Issue on Membrane Reactors, *Asia-Pac. J Chem. Eng.*, **4**(3), 237–386
Basile A. and Gallucci F. (2009b), II Special Issue on Membrane Reactors, *Asia-Pac. J. Chem. Eng.*, **5**(1), 1–255.
Basile A. and Gallucci F. (Eds), (2009), *Simulation of Membrane Reactors*, Nova Science Pub., New York, ISBN: 978-1-60692-425-9.
Basile A. and Gallucci F. (Eds), (2011c), 'Innovations in catalytic inorganic membrane reactors', *Membranes for Membrane Reactors: Preparation, Optimization and Selection*, John Wiley & Sons, Chichester, UK, ISBN 978-0-470-74652-3.
Basile A. and Paturzo L. (2001), *Catal. Today*, **67**, 55.
Basile A., Violante V., Santella F. and Drioli E. (1995), *Catal. Today*, **25**(3–4), 321.
Basile A., Criscuoli A., Santella F. and Drioli E. (1996a), *Gas Sep. Purif.*, **10**, 243.
Basile A., Drioli E., Santella F., Violante V. and Capannelli G. (1996b), *Gas Sep. Purif.*, **10**(1), 53.
Basile A., Paturzo L. and Laganà F. (2001a), *Catal. Today*, **67**, 65.
Basile A., Chiappetta G., Tosti S. and Violante V. (2001b), *Sep. Purif. Technol.*, **25**, 549.
Basile A., Paturzo L. and A. Vazzana, (2003), *Chem. Eng. J.*, **93**, 31.
Basile A., Gallucci F. and Paturzo L. (2005a), *Catal. Today*, **104**, 251.
Basile A., Gallucci F. and Paturzo L. (2005b), *Catal. Today*, **104**, 244.
Basile A., Tosti S., Capannelli G., Vitulli G., Iulianelli A., Gallucci F. and Drioli E, (2006a), *Catal. Today*, **118**, 237.
Basile A., Tereschenko G.F., Orekhova N.V., Ermilova M.M., Gallucci F. and Iulianelli A. (2006b), *Int. J. Hydrogen En.*, **31**, 1615.
Basile A., Gallucci F. and Iulianelli A. (2007), *Membranes*, **2**(34), 3.
Basile A., Gallucci F. and Tosti S. (2008a), 'Synthesis, characterization and applications of palladium membranes', in R. Malada and M. Menendez (Eds), *Inorganic Membranes: Synthesis, Characterization and Applications, Membrane Science and Technology*, Elsevier, Amsterdam, ch. 8, 255–323.
Basile A., Gallucci F., Iulianelli A., Borgognoni F. and Tosti S. (2008b), *J. Membrane Sci.*, **311**, 46.
Basile A., Parmaliana A., Tosti S., Iulianelli A., Gallucci F., Espro C. and Spooren J. (2008c), *Catal. Today*, **1**, 17–22.
Basile A., Gallucci F., Iulianelli A. and Tosti S. (2008d), *Fuel Cells*, **1**, 62.
Basile A., Gallucci F., Iulianelli A., Tereschenko G.F., Ermilova M.M. and Orekhova N.V. (2008e), *J. Membrane Sci.*, **310**, 44.
Basile A., Pinacci P., Iulianelli A., Broglia M., Drago F., Liguori S., Longo T. and Calabrò V. (2011a), *Int. J. Hydrogen En.*, **36**, 2029.
Basile A., Campanari S., Manzolini G., Iulianelli A., Longo T., Liguori S., De Falco M. and Piemonte V. (2011b), *Int. J. Hydrogen En.*, **36**, 1531.

Basov N.L., Gryaznov V.M. and Ermilova M.M. (1993), *Russ. J. Phys. Chem.*, **67**, 1967.
Becker Y.L., Dixon A.G., Moser W.R. and Ma Y.H. (1993), *J. Membrane Sci.*, **77**, 233.
Bhave R.R. (1991), *Inorganic membranes: Synthesis, Characteristics and Applications*, (Van Nostrand Reinhold, New York).
Bi Y., Xu H., Li W. and A. Goldbach, (2009), *Int. J. Hydrogen En.*, **34**, 2965.
Bichon P., Haugom G., Venvik H.J., Holmen A. and Blekkan E.A. (2008), *Top. Catalysis*, **49**, 38.
Bimbela F., Oliva M., Ruiz J., Garcıa L. and Arauzo J. (2007), *J. Anal. Appl. Pyrolysis*, **79**, 112.
Borgognoni F., Tosti S., Vadrucci M. and Santucci A. (2011), *Int. J. Hydrogen En.*, **36**, 7550.
Bosko M.L., Múnera J.F., Lombardo E.A. and Cornaglia L.M. (2010), *J. Membrane Sci.*, **364**, 17.
Bosko M.L., Miller J.B., Lombardo E.A., Gellman A.J. and Cornaglia L.M. (2011), *J. Membrane Sci.*, **369**, 267.
Bouwmeester H.J.M. (2003), *Catal. Today*, **82**, 141.
Breen J.P. and Ross J.R.H. (1999), *Catal Today*, **51**, 521.
Breen J.P., Burch R. and Coleman H.M. (2002), *Appl. Catal. B: Env.*, **39**, 65.
Brinker C.J., Collins J.P., Schwartz R., Sehgal R., Ward T., Hagen G. and Udovich C. (1996), *Ind. Eng. Chem. Res.*, **35**, 4398.
Brinkmann T., Perera S.P. and Thomas W.J. (2001), *Chem. Eng. Sci.*, **56**, 2047.
Brunetti A., Barbieri G., Drioli E., Lee K.H., Sea B. and Lee D.W. (2007), *Chem. Eng. Prog.*, **46**, 119.
Brunetti A., Barbieri G. and Drioli E. (2009), *Chem. Eng. Sci.*, **64**, 3448.
Buffoni I.N., Pompeo F., Santori G.F. and Nichio N.N. (2009), *Catal. Comm.*, **10**, 1656.
Busnyuk A.O., Notkin M.E., Grigoriadi I.P., Alimov V.N. and Lifshitz, A.I. (2010), *Tech. Phys.*, **55**, 117.
Buxbaum R.E. (1999), *Sep. Sci. Technol.*, **34**, 2113.
Buxbaum R.E. and Hsu P.C. (1992), US Patent 5,149,420, Sept. 22.
Buxbaum R.E. and Markerb T.L. (1993), *J. Membrane Sci.*, **85**, 29.
Buxbaum R.E. and Kinney A.B. (1996), *Ind. Eng. Chem. Res.*, **35**, 530.
Byrd A.J., Pant K.K. and Gupta R.B. (2008), *Fuel*, **87**, 2956.
Campanari S., Macchi E. and Manzolini G. (2008), *Int. J. Hydrogen En.*, **33**,1361.
Casanave D., Giroir-Fendler A., Sanchez J., Loutaty R. and Dalmon J.A. (1995), *Catal. Today*, **25**, 309.
Casanave D., Ciavarella P., Fiaty K. and Dalmon J.A. (1999), *Chem. Eng. Sci.*, **54**, 2807.
Chai M., Machida M., Educhi K. and Arai H. (1993), *Chem. Lett.*, **41**, 425.
Champagnie A.M., Tsotsis T.T., Minet R.G. and Webster I.A. (1990), *Chem. Eng. Sci.*, **45**, 2423.
Chang H.F., Pai W.J., Chen Y.J. and Lin W.H. (2010), *Int. J Hydrogen En.*, **35**, 12986.
Chang J.S., Park S.E. and Chon H. (1996), *Appl. Catal. A: Gen.*, **145**, 111.
Checchetto R., Bazzanella N., Patton B. and Miotello A. (2004), *Surf. Coat. Technol.*, **177–178**, 73.
Chen S.C., Tu G.C., Hung C.C.Y., Huang C.A. and Rei M.H. (2008), *J. Membrane Sci.*, **314**, 5.

Chen C.-H. and Ma Y.H. (2010), *J. Membrane Sci.*, **362**, 525–544.
Chen H., Ding Y., Cong N.T., Dou B., Dupont V., Ghadiri M. and Williams P.T. (2011), *Ren. Energy*, **36**, 779.
Cheng C.K., Foo S.Y. and Adesina A.A. (2010), *Catal. Comm.*, **12**, 292.
Cheng Y.S. and Yeung K.L. (1999), *J. Membrane Sci.*, **158**, 127.
Cheng Y.S., Pena M.A. and Yeung K.L. (2009), *J. Taiwan Inst. Chem. Eng.*, **40**, 281.
Chiodo V., Freni S., Galvagno A., Mondello N. and Frusteri F. (2010), *Appl. Catal. A: Gen.*, **381**, 1.
Choy K.L. (2003), *Prog. Mat. Sci.*, **48**, 57.
Ciavarella P., Moueddeb H., Miachon S., Fiaty K. and Dalmon J.A. (2000), *Catal. Today*, **56**, 253.
Clayson D.M. and Howard P. (1987), GB Patent No. 2,190,397.
Collins J.P. and Way J.D. (1994), *J. Membrane Sci.*, **96**, 259.
Compagnie des Métaux Précieux, (1976), French Patent No. 2,302,273.
Coronas J. and Santamaria J. (1999), *Catal. Today*, **51**, 377.
Coronas J. and Santamaria J. (2004), *Topics in Catal.*, **29**, 29.
Coronel L., Munera J.F., Lombardo E.A. and Cornaglia L.M. (2011), *Appl. Catal. A: Gen.*, **400**, 185.
Criscuoli A., Basile A. and Drioli E. (2000), *Catal. Today*, **56**, 53.
Criscuoli A., Basile A., Drioli E. and Loiacono O. (2001), *J. Membrane Sci.*, **181**, 21.
Cui Y., Galvita V., Rihko-Struckmann L., Lorenz H. and Sundmacher K. (2009), *Appl. Catal. B: Env.*, **90**, 29.
Damle A.S. (2009), *J. Power Sou.*, **186**, 167.
Daniells S.T., Makkee M. and Moulijn J.A. (2005), *Catal. Lett.*, **100**, 39.
Dauenhauer P.J., Salge J.R. and Schmidt L.D. (2006), *J. Catal.*, **244**, 238.
De Rosset A.I. and Hills C. (1968), US Patent, No. 3,375,288.
Dittmeyer R., Hollein V. and Daub K. (2001), *J. Molecular Catal. A: Chemical*, **173**, 135.
Dittmeyer R., Svajda K. and Reif M. (2004), *Topics in Catal.*, **29**, 3.
Dixon A.G. (1999), 'Innovations in catalytic inorganic membrane reactors', in J.J. Spivey (Ed.), *Specialist Periodical Reports: Catalysis*, Royal Society of Chemistry, London, Vol. 14, 40–92.
Dixon A.G. (2003), *Int. J. Chem. Reactor Eng.*, **1**, 1.
Djaidja A., Libs S., Kiennemann A. and Barama A. (2006), *Catal. Today*, **113**, 194.
Dolan M.D., Dave N.C., Ilyushechkin A.Y., Morpeth L.D. and McLennan K.G. (2006), *J. Membrane Sci.*, **285**, 30.
Drioli E. and Giorno L. (Eds), (2011), Catalytic membranes and catalytic membrane reactors, Elsevier, Vol. 4. ISBN: 9780080932507.
Dümpelmann R. (2001), Ph.D. Thesis, cited on M. Schuessler, O. Lamla, T. Stefanovsky, C. Klein and D. Zur Megede, *Chem. Eng. Technol.*, **24**(11), 1141.
Edlun D.J. and Pledger W.A. (1993), *J. Membrane Sci.*, **77**, 255.
Elnashaie S.S.E.H., Abdallah B.K., Elshishini S.S., Alkhowaiter S., Noureldeen M.B. and Alsoudani T. (2001), *Catal. Today*, **64**, 151.
Ermilova M.M., Orekhova N.V., Tereshchenko G.F., Malygin A.A., Malkov A.A., Basile A. and Gallucci F. (2008), *J. Membrane Sci.*, **317**, 88.
Eswaramoorthi I. and Dalai A.K. (2009), *Int. J. Hydrogen En.*, **34**, 2580.
Fan J., Ohya H., Ohashi H., Aihara M., Takeuchi T., Negishi Y. and Semenova S.I. (1999), *J. Membrane Sci.*, **162**, 125.

Fatsikostas A.N. and Verykios X.E. (2004), *J. Catalysis*, **225**, 439.
Ferreira-Aparicio P., Rodrìguez-Ramos I. and Guerrero-Ruiz A. (2002a), *Appl. Catal. A: Gen.*, **237**, 239.
Ferreira-Aparicio P., Rodriguez-Ramos I. and Guerriero-Ruiz A. (2002b), *J. Catal.*, **212**, 182.
Flytzani-Stephanopoulos M., Qi X. and Kronewitter S. (2004), Final report to DOE, Grant # DEFG2600-NT40819, 1–38.
Freni S., Mondello N., Cavallaro S., Cacciola G., Pardon V.N. and Sobyanin V.A. (2000), *React. Kinet Catal. Lett.*, **71**, 143.
Gabitto J.F. and Tsouris C. (2009), *Int. Rev. Chem. Eng.*, **1**(5), 394–411.
Gade S.K., Payzant E.A., Park H.J., Thoen P.M. and Way J.D. (2009), *J. Membrane Sci.*, **340**, 227.
Galdamez J.R., Garcia L. and Bilbao R. (2005), *Energ. Fuel.*, **19**, 1133.
Gallucci F., Paturzo L., Famà A. and Basile A. (2004a), *Ind. Eng. Chem. Res.*, **43**, 928.
Gallucci F., Paturzo L. and Basile A. (2004b), *Int. J. Hydrogen En.*, **29**, 611.
Gallucci F., Basile A., Tosti S., Iulianelli A. and E. Drioli, (2007a), *Int. J. Hydrogen En.*, **32**, 1201.
Gallucci F., De Falco M., Tosti S., Marrelli L. and Basile A. (2007b), *Int. J. Hydrogen En.*, **32**, 4052.
Gallucci F., Tosti S. and A. Basile, (2008a), *J. Membrane Sci.*, **317**, 96.
Gallucci F., De Falco M., Tosti S., Marrelli L. and Basile, A. (2008b), *Int. J. Hydrogen En.*, **33**, 6165.
Gallucci F., Tosti S. and Basile A. (2008c), *J. Membrane Sci.*, **317**, 96.
Galuszka J., Pandey R.N. and Ahmed S. (1998), *Catal. Today*, **46**, 83.
Gao H., Lin J.Y.S., Li Y. and Zhang B. (2005), *J. Membrane Sci.*, **265**, 142.
Gao X., Tang J., Zuo Y., Tang Y. and Xiong J. (2009), *Corrosion Sci.*, **51**, 1822.
Gapontsev A.V. and Kondrat'ev V.V. (2003), *Phys.-Usp.*, **46**(10), 1077.
Gernot E., Aupretre F., Deschamps A., Epron F., Merecot P., Duprez D. and Etievant C. (2006), 'Production of hydrogen from bioethanol in catalytic membrane reactor', Proc. of the 16th Confèrence Mondiale del Hydrogène Energie (WHEC16), 13–16 June, Lyon (France), pp. 1–7.
Gessler S., Jordan L., da Costa J.C.D. and Lu G.Q.M. (2003), *Sep. Purif. Technol.*, **32**, 255.
Gobina E., Hou K. and Hughes R. (1995a), *Catal. Today*, **25**, 365.
Gobina E., Hou K. and Hughes R. (1995b), *J. Membrane Sci.*, **105**, 163.
Goerke O., Pfeifer P. and Schubert K. (2004), *Appl. Catal., A: Gen.*, **263**, 11.
Gryaznov V.M. (1999), *Catal. Today*, **51**, 391.
Gryaznov V.M. (2000), *Sep. Purif. Methods*, **29**(2), 171.
Gryaznov V.M. and Smirnov V.S. (1974), *Russ. Chem. Rev.*, **43**, 821.
Gryaznov V.M., Polyakova V.P., Savitsky E.M. and Khrapova E.V. (1977), US Patent No. 4,026,958.
Gryaznov V.M., Ermilova M.M., Morozova L.S. and Orekhova N.V. (1983), *J. Less Common Metals*, **89**, 529.
Gryaznov V.M., Ermilova M.M., Orekhova N.V. and Tereschenko G.F. (2005), Reactors with Metal and Metal-Containing Membranes, in *Structured catalysts and* reactors, A. Cybulski and J.A. Moulijn (Eds.), (Taylor & Francis, London) p. 579.

Gu Y. and Oyama S.T. (2007), *Adv. Mater.*, **19**, 1636.
Guo J., Lou H., Zhao H., Chai D. and Zheng X. (2004), *Appl. Catal. A: Gen.*, **273**, 75.
Guo Y., Lu G., Wang Y. and R. Wang, (2003), *Sep. Pur. Sci.*, **32**(1–3), 271.
Haag S., Burgard M. and Ernst B. (2007), *J. Catal.*, **252**, 190.
Hacarlioglu P., Lee D., Gibbs G.V. and Oyama S.T. (2008), *J. Membrane Sci.*, **313**, 277.
Han J., Kim I.S. and Choi K.S. (2000), *J. Power Sou.*, **86**, 223.
Hara S., Sakaki K., Itoh N., Kimura H.-M., Asami K. and Inoue A. (2000), *J. Membrane Sci.*, **164**, 289.
Hara S., Hatakeyama N., Itoh N., Kimura H.M. and Inoue A. (2002), *Desalination*, **144**(1–3), 115.
Hayakawa T., Suzuki S., Nakamura J., Uchijima T., Hamakawa S., Suzuki K., Shishido T. and Takehira K. (1999), *Appl. Catal. A: Gen.*, **183**, 273.
Hermann C., Quicker P. and Dittmeyer R. (1997), *J. Membrane Sci.*, **136**, 161.
Hirai T., Ikenaga N., Miyake T. and Suzuki T. (2005), *Energ. Fuel.*, **19**, 1761.
Hoang H.T., Tong H.D., Gielens F.C., Jansen H.V. and Elwenspoek M.C. (2004), *Mater. Letters*, **58**, 525.
Hoff K.A., Poplsteinova J., Jakobsen H.A., Falk-Pedersen O., Juliussen O. and Svendsen H.F. (2003), *Int. J. Chem. Reactor Eng.*, **1**, A9.
Hou K. and Hughes R. (2002), *J. Membrane Sci.*, **206**, 119.
Hsieh H.P. (1996), *Inorganic Membranes for Separation and Reaction,* Elsevier, Amsterdam.
Hsu C. and Buxbaum R.E. (1986), *J. Nucl. Mater.*, **141**, 238.
Hu X. and Lu G. (2007), *J. Mol. Catal. A: Chem.*, **261**, 43.
Hu X. and Lu G. (2010), *Catal. Comm.*, **12**, 50.
Huang T.C., Wei M.C. and Chen H.I. (2003), *Sep. Pur. Techn.*, **32**, 239.
Huang Y. and Dittmeyer R. (2007), *J. Membrane Sci.*, **302**, 160.
Hughes R. (2001), *Membrane Tech.*, **131**, 9.
Hwang S.T. and Korean J. (2001), *Chem. Eng.*, **18**(6), 775.
Iler R.K. (1979), *The Chemistry of Silica*, (Wiley, New York).
Illgen U., Schäfer R., Noack M., Kölsch P., Kühnle A. and Caro J. (2001), *J. Catal. Comm.*, **2**, 339.
Ioannides T. and Gavalas G.R. (1993), *J. Membrane Sci.*, **77**, 207.
Ioannides T. and Verykios X.E. (1996), *Catal. Lett.*, **36**, 165.
Iriondo A., Barrio V.L., Cambra J.F., Arias P.L., Güemez M.B., Navarro R.M., Sánchez-Sánchez M.C. and Fierro J.L.G. (2008), *Top. Catal.*, **49**, 46.
Irusta S., Munera J., Carrara C., Lombardo E.A. and Cornaglia L.M. (1997), *Appl. Catal. A Gen.*, **287**, 147.
Israni S.H. and Harold M.P. (2011), *J. Membrane Sci.*, **369**, 375.
Itoh N. (1987), *AIChE J.*, **33**, 1576.
Itoh N., Shindo Y., Haraya H. and Hakuta, T. (1988), *J. Chem. Eng. Jpn.*, **21**, 399.
Itoh N., Xu W.C. and Haraya K. (1992), *J. Membrane Sci.*, **66**, 149.
Itoh N. and Xu W.C. (1993), *Appl. Catal. A*, **107**, 83.
Itoh N., Xu W.C., Hara S., Kimura H.M. and Masumoto T. (1997), *J. Membrane Sci.*, **126**(1), 41.
Itoh N., Xu W.C., Hara S., Kimura H.M. and Masumoto T. (1998), *J. Membrane Sci.*, **139**, 29.
Itoh N. and Haraya, K. (2000), *Catal. Today*, **56**, 103.

Itoh N., Tamura E., Hara S., Takahashi T., Shono A., Sato K. and Namba T. (2003), *Catal. Today*, **82**, 119.
Itoh N., Akiha T. and Sato T. (2005), *Catal. Today*, **104**, 231.
Iulianelli A. and Basile A. (2010), *Int. J. Hydrogen En.*, **35**, 3170.
Iulianelli A., Longo T. and Basile A. (2008a), *Int. J. Hydrogen En.*, **33**, 4091.
Iulianelli A., Longo T. and Basile A. (2008b), *Int. J. Hydrogen En.*, **33**, 5583.
Iulianelli A., Longo T. and Basile A. (2008c), *J. Membrane Sci.*, **323**, 235.
Iulianelli A., Liguori S., Longo T., Tosti S., Pinacci P. and Basile A. (2010a), *Int. J. Hydrogen En.*, **35**, 3159.
Iulianelli A., Longo T., Liguori S. and Basile A. (2010b), *Asia-Pac. J. Chem. Eng.*, **5**, 138.
Iulianelli A., Liguori S., Calabrò V., Pinacci P. and Basile A. (2010c), *Int. J. Hydrogen En.,* **35**, 12626.
Iulianelli A., Manzolini G., De Falco M., Campanari S., Longo T., Liguori S. and Basile A. (2010d), *Int. J. Hydrogen En.*, **35**, 11514.
Iulianelli A., Seelam P.K., Liguori S., Longo T., Keiski R., Calabrò V. and Basile A. (2011), *Int. J. Hydrogen En.*, **36**, 3827.
Iwasa N., Yamane T. and Arai M. (2011), *Int. J. Hydrogen En.*, **36**, 5904.
Iyoha O., Enick R., Killmeyer R., Howard B., Morreale B. and Ciocco M. (2007), *J. Membrane Sci.*, **298**, 14.
Jayaraman V. and Lin Y.S. (1995a), *J. Membrane Sci.*, **104**, 251.
Jayaraman V., Lin Y.S., Pakala M. and Lin R.Y. (1995b), *J. Membrane Sci.*, **99**, 89.
Jema N., Shu J., Kaliaguine S. and Grandjean B.P.A. (1996), *Ind. Eng. Chem. Res.*, **35**, 973.
Jeong B.H., Sotowa K.I. and Kusakabe K. (2003), *J. Membrane Sci.*, **224**, 151.
Jiang C.J., Trimm D.L., Wainwright M.S. and Cant N.W. (1993a), *Appl. Catal. A: Gen.*, **93**(2), 245.
Jiang C.J., Trimm D.L., Wainwright M.S. and Cant N.W. (1993b), *Appl. Catal. A: Gen.*, **97**(2), 145.
Julbe A., Farrusseng D. and Guizard C. (2001), *J. Membrane Sci.*, **181**, 3.
Kamakoti P., Morreale B.D., Ciocco M.V., Howard B.H., Killmeyer R.P. and Cugini A.V. (2005), *Science*, **307**, 569.
Keuler J.N. and Lorenzen L. (2002a), *J. Membrane Sci.*, **202**, 17.
Keuler J.N. and Lorenzen L. (2002b), *Ind. Eng. Chem. Res.*, **41**, 1960.
Keuler J.N. and Lorenzen L. (2002c), *J. Membrane Sci.*, **195**, 203.
Kikuchi E., Uemiya S., Sato N., Inoue H., Ando H. and Matsuda T. (1989), *Chem. Lett.*, **489**.
Kikuchi E., Uemiya S. and Matsuda T. (1991), *Nat. Gas Conv.*, **509**.
Kirkchheim R. (1982), *Acta Metall.*, **30**, 1069.
Kleinert A., Grubert G., Pan X., Hamel C., Morgenstern A.S. and Caro J. (2005), *Catal. Today*, **104**, 267.
Koffler S.A., Hudson J.B. and Ansell G.S. (1996), *Trans. AIME*, **245**, 1735.
Kong C., Lu J., Yang J. and Wang J. (2007), *J. Membrane Sci.*, **306**, 29.
Koros W.J. and Fleming G.K. (1993), *J. Membrane Sci.*, **83**(1), 1.
Kulprathipanja A., Alptekin,G.O., Falconer J.L., and Way, J.D. (2005), *J. Membrane Sci.*, **254**(1–2), 49–62.
Lebedeva V.I. and Gryaznov V.M. (1981), *Bull. Acad. Sci. USSR Chem. Ser.*, **3**, 452.
Lenarda M., Storaro L., Ganzerla R., Casagrande M., Marchiori M., Capannelli G., Uliana C., Ferrari F. and Frattini R. (2004), Proc. XIV Congresso Nazionale

di Catalisi GIC2004: *Le Scienze Molecolari al Servizio dell'Uomo*, Lerici (La Spezia, Italy), Book of Abstracts, **2**, 143.

Lewis F.A. (1967), *The Palladium Hydrogen System*, Academic Press, London.

Li K. (2007), *Ceramic Membranes for Separation and* Reaction, J. Wiley & Sons, England.

Li Z.Y., Maeda H., Kusakabe K., Morooka S., Anzaib H. and Akiyama S. (1993), *J. Membrane Sci.*, **78**, 247.

Li A., Xiong G., Gu J. and Zheng L. (1996), *J. Membrane Sci.*, **110**, 257.

Li Q., Zhu X., He Y. and Yang W. (2010), *Cat. Today*, **149**, 185.

Li H., Xu H. and Li W. (2008), *J. Membrane Sci.*, **324**, 44–49.

Liang W. and Hughes R. (2003), *Chem. Eng. J.*, **112**, 81.

Liang W. and Hughes R. (2005), *Catal. Today*, **104**, 238.

Liguori S., Pinacci P., Seelam P.K., Keiski R., Drago F., Calabrò V., Basile A. and Iulianelli A. (2012), *Catal. Today*, doi:10.1016/j.cattod.2012.02.005.

Liguras D.K., Kondarides D.I. and Verykios X.E. (2003), *Appl. Catal.,* **43**, 345.

Lim H., Gu Y. and Oyama S.T, (2010), *J. Membrane Sci.*, **351**, 149.

Lim H., Gu Y. and Oyama S.T. (2012), *J. Membrane Sci.*, **396**, 119.

Lin Y.S. (2001), *Sep. Purif. Technol.*, **25**, 39.

Lin Y.M. and Rei M.H. (2000), *Int. J. Hydrogen En.*, **25**, 211.

Lin W.H. and Chang H.F. (2004), *Cat. Today*, **97**, 181.

Lin W.H., Hsiao C.S. and Chang H.F. (2008), *J. Membrane Sci.*, **322**, 360.

Liu B.S. and Au C.T. (2001), *Catal. Lett.*, **77**, 67.

Liu B., Li H., Cao Y., Deng J.F., Sheng C. and Zhou S. (1997), *J. Membrane Sci.*, **135**, 33.

Liu B.S., Lian P.Y. and Zhao X.H. (2003), *Sep. Purif. Tech.*, **32**, 281.

Liu N., Yuan Z., Wang S., Zhang C., Wang S. and Li D. (2008), *Int. J.* Hydrogen E., **33**, 1643.

Liu Y., Hayakawa T., Suzuki K., Hamakawa S., Tsunoda T., Ishii T. and Kumagai M. (2002), *Appl. Catal. A: Gen.*, **223**, 137.

Lu G.Q., da Costa J.C.D., Duke M., Giessler S., Socolow R., Williams R.H. and Kreutz T. (2007), *J. Colloid Interface Sci.*, **314**, 589.

Lukaszewski M. and Czerwinski A. (2003), *Electrochim. Acta*, **48**, 2435.

Luo H., Wei Y., Jiang H., Yuan W., Lv Y., Caro J. and Wang H. (2010), *J. Membrane Sci.*, **350**, 154.

Luo W., Ishikawa K. and Aoki K. (2006), *J. Alloy. Compd.*, **407**, 115.

Ma Y.H. (2006), Proc. of the 9th Int. conf. on Inorganic Membranes, Lillhammer-Norway, June 25–29.

Makrides A.C., Wright M.A. and Jewett D.N. (1967), US Patent 3,350,846, Nov. 07.

Malada R. and Menendez M. (2008), *Inorganic membranes: synthesis, characterization and applications*, Membrane Science and Technology, (Elsevier, Amsterdam).

Manzolini G., Macchi E., Dijkstra J.W. and Jansen D. (2006) Proc. of the ASME TURBO EXPO 2006, Barcelona (Spain), May 9–12.

Marcano J.G.S. and T.T. Tsosis (2002), *Catalytic membranes and membrane reactors*, (Wiley VCH, Weinheim).

Matsuda T., Koike I., Kubo N. and Kikuchi E. (1993), *Appl. Catal. A: General*, **96**, 3.

Matzakos A. (2006), NHA Annual Hydrogen Conference, Long Beach, CA, March 15, cited on (May, 2006).

Mc Leary E.E., Jansen J.C. and Kapteijn F. (2006), *Microp. Mesop. Mat.*, **90**, 198.

McCool B.A. and Lin Y.S. (2001), *J. Mater. Sci.*, **36**, 3221.

Mendes D., Mendes A., Madeira L.M., Iulianelli A., Sousa J.M. and Basile A. (2010), *Asia-Pac. J. Chem. Eng.*, **5**, 111.
Mendes D., Sa S., Tosti S., Sousa J.M., Madeira L.M. and Mendes A. (2011), *Chem. Eng. Sci.*, **66**, 2356.
Miachon S. and Dalmon J.A. (2004), *Topics in Catal.*, **29**(1–2), 59.
Michaels A.S. (1968), *Chem. Eng. Prog.*, **64**, 31.
Mikhalenko N.N., Khrapova E.V. and Gryaznov V.G. (1986), *Kinet. Catal.*, **26**, 125.
Mishchenko A.P., Sarylova M.E., Gryaznov V.M., Smirnov V.S., Roshan N.R. and Polyakova V.P. (1997), *Bull. Acad. Sci. USSR Chem. Ser.*, **7**, 1490.
Mondal A.M. and Ilias S. (2001), *Sep. Sci. Technol.*, **36**, 1101.
Mundschau M. (2005), US Patent, N. 6,899,744 B2.
Mundschau M. (2008), Hydrogen Separation Using Dense Composite Membranes, Part 1 Fundamentals, in: *Inorganic Membranes for Energy and Fuel Applications*, (A. Bose Ed., Springer, Berlin).
Murav'ev L.L. and Vandyshev A.B. (2003), *Chem. & Petroleum Eng.*, **39**(9–10), 571.
Murcia-Mascaros S., Navarro R.M., Gomez-Sainero L., Costantino U., Nocchetti M. and Fierro J.L.G. (2001), *J. Catal.*, **198**(2), 338.
Nair B.K.R. and Harold M.P. (2007a), *J. Membrane Sci.*, **290**, 182.
Nair B.K.R., Choi J. and Harold M.P. (2007b), *J. Membrane Sci.*, **288**, 67.
Nakagawa K., Ikenaga N., Suzuki T., Kobayashi T. and Haruta M. (1998), *Appl. Catal. A: Gen.*, **169**, 281.
Nam S.E. and Lee K.H. (2001), *J. Membrane Sci.*, **192**, 177.
Nam S.E., Lee S.H. and Lee K.H. (1999), *J. Membrane Sci.*, **153**, 163.
National Research Council and National Academy of Engineering, (2008), *The hydrogen economy: opportunities, costs, barriers, and R&D needs*, National Academy Press, Washington, DC, 2004, cited on: F. Joseck, M. Wang and Y. Wu, Int. J. Hydrogen En., **33**, 1445.
Nazarkina E.B. and Kirichenko N.A. (1979), *Khim. Technol. Topl. Masel.*, **3**, 5.
Nishimura C., Komaki M., Hwang S. and Amano M. (2002), *J. Alloy. Compd.*, **330–332**, 902.
Oertel M., Schmitz J., Weirich W., Jendryssek-Neumann D. and Schulten R. (1987), *Chem. Eng. Techn.*, **10**, 248.
Ohya H., Nakajima H., Togami N., Aihara M. and Negishi Y. (1994), *J. Membrane Sci.*, **97**, 91.
Okada S., Mineshige A., Kikuchi T., Kobune M. and Yazawa T. (2007), *Thin Solid Films*, **515**, 7342.
Okazaki J., Tanaka D.A.P., Tanco M.A.L., Wakui Y. and Ikeda T. (2008), *Mater. Trans.*, **49**, 449.
Omorjan R.P., Paunovic R.N. and Tekic M.N. (1999), *J. Membrane Sci.*, **154**, 273.
Orekova N.V., Ermilova M.M., Smirnov V.S. and Gryaznov V.M. (1976), *Bull. Acad. Sci. USSR Chem. Ser.*, **11**, 2421.
Ostrowski T., Girior-Fendler A., Mirodatos C. and Mleczko L. (1998), *Catal. Today*, **40**, 181.
Oyama S.T., Hacarlioglu P., Gu Y. and Lee D. (2011), *Int. J. Hydrogen En.*, doi:10.1016/j.ijhydene.2011.09.149.
Paal Z. and Menon P.G. (Eds.), (1988), *Hydrogen effects in catalysis – Fundamental and practical applications*, (Marcell Dekker Inc., New York).
Paglieri S.N. and Way J.D. (2002), *Sep. Purif. Methods*, **31**(1), 1.

Paiva A.L. and Malcata F.X. (1997), *Chem. Eng. Sci.*, **52**, 449.
Paiva A.L., van Rossum D. and Malcata F.X. (1999), *Chem. Eng. Sci.*, **54**, 1825.
Palo D.R., Dagle R.A. and Holladay J.D. (2007), *Chem. Rev.*, **107**, 3992.
Patel S. and Pant K.K. (2007), *Fuel Proc. Tech.*, **88**, 825.
Patil C.S., Annaland M.V.S. and Kuipers J.A.M. (2007), *Chem. Eng. Sci.*, **62**(11), 2989.
Paturzo L., Basile A. and Drioli E. (2002), *Rev. Chem. Eng.*, **18**(6), 511.
Paturzo L., Gallucci F., Basile A., Vitulli G. and Pertici P. (2003), *Catal. Today*, **82**, 57.
Peppley B.A. (2001), Ph.D. Thesis, cited on Schuessler, O. Lamla, T. Stefanovsky, C. Klein and D. zur Megede, *Chem. Eng. Technol.*, **24**(11), 1141.
Peppley B.A., Amphlett J.C., Kearns L.M. and Mann R.F. (1999a), *App. Catal. A: Gen.*, **179**(1–2), 21.
Peppley B.A., Amphlett J.C., Kearns L.M. and Mann R.F. (1999b), *Appl. Catal. A: Gen.*, **179**(1–2), 31.
Perez-Hernandez R., Gutierrez-Martınez A., Palacios P., Vega-Hernandez M. and Rodrıguez-Lugo V. (2011), *Int. J. Hydrogen En.*, **63**, 6601.
Pernga T.P. and Wu J.K. (2003), *Mat. Lett.*, **57**, 3437.
Peters T.A., Stange M., Klette H. and Bredesen R. (2008), *J. Membrane Sci.*, **316**, 119.
Phair J.W. and Donelson R. (2006), *Ind. Eng. Chem. Res.*, **45**, 5657.
Pomerantz N. and Ma Y.H. (2009), *Ind. Eng. Chem. Res.*, **48**, 4030.
Pompeo F., Santori G. and Nichio N.N. (2010), *Int. J. Hydrogen En.*, **35**, 8912.
Prabhu A.K. and Oyama S.T. (2000), *J. Membrane Sci.*, **176**, 233.
Prabhu A.K., Radhakrishnan R. and Oyama S.T. (1999), *Appl. Catal. A Gen.*, **183**, 241.
Quicker P. (2000), Doctoral Thesis, University of Erlangen.
Raich B.A. and Foley C.H. (1998), *Ind. Eng. Chem. Res.*, **37**, 3888.
Raybold T.M. and Huff M.C. (2000), *Catal. Today*, **56**, 35.
Rei M.H., Yeh G.T., Tsai Y.H., Kao Y.L. and Shiau L.D. (2011), *J. Membrane Sci.*, **369**, 299.
Reitz T.L., Lee P.L., Czaplewski K.F., Lang J.C., Popp K.E. and Kung H.H. (2001), *J. Catal.*, **199**(2),193.
Richardson O, (1904), Philos. Mag., 7, 266, cited by Forcely K.S., Ross D.K. and Simpson J.C.B. (1988), *J. Nuclear Mat.*, **160**, 117–124.
Ritter J.A. and Ebner A.D. (2007), *Sep. Sci. Techn.*, **42**(6), 1123.
Roa F., Way J.D., McCormik R.L. and Paglieri S.N. (2003), *Chem. Eng. J.*, **93**, 11.
Roa F., Block M.J. and Way J.D. (2002), *Desalination*, **147**, 411.
Rostrup-Nielsen J.R. and Bak Hansen J.H. (1993), *J. Catal.*, **144**, 38.
Rostrup-Nielsen J.R. (1994), *Natural Gas Conversion II*, (Elsevier, Amsterdam).
Rothenberger K.S., Cugini A.V., Howard B.H., Killmeyer R.P., Ciocco M.V., Morreale B.D., Enick R.M., Bustamante F., Mardilovich I.P. and Ma Y.H. (2004), *J. Membrane Sci.*, **244**, 55.
Ryi S.K., Park J.S., Kim S.H., Cho S.H., Park J.S. and Kim D.W. (2006), *J. Membrane Sci.*, **279**, 439.
Sakurai H., Akita T., Tsubota S., Ciuchi M. and Haruta M. (2005), *Appl. Catal. A: Gen.*, **291**, 179.
Saracco G. and Specchia V. (1994), *Catal. Rev. Sci. Eng.*, **36**, 305.
Saracco G., Versteeg G.F. and van Swaij W.P.M. (1994), *J. Membrane Sci.*, **95**, 105.

Saracco G., Neomagus H.W.J.P., Versteeg G.F. and van Swaij W.P.M. (1999), *Chem. Eng. Sci.*, **54**, 1997.
Sarylova M.E., Mishchenko A.P., Gryaznov V.M. and Smirnov V.S. (1977), *Bull. Acad. Sci. USSR Chem. Ser.*, **2**, 389.
Sato K., Hanaoka T., Niwa S., Stefan C., Namba T. and Mizukami F. (2005), *Catal. Today*, **104**, 260.
Sato T., Yokoyama H., Miki H. and Itoh N. (2007), *J. Membrane Sci.*, **289**, 97.
Schäfer R., Noack M., Kölsch P., Stöhr M. and Caro J. (2003), *Catal. Today*, **82**, 15.
Seelam P.K., Liguori S., Iulianelli A., Pinacci P., Drago F., Calabrò V., Huuhtanen M., Keiski R., Piemonte V., Tosti S., De Falco M. and Basile A. (2012), in press to *Cat. Today*, 10.1016/j.cattod.2012.01.00
Sekizawa K., Yano S., Educhi K. and Arai H. (1998), *Appl. Catal. A: Gen.*, **169**, 291.
Seok D.R. and Hwang S.T. (1990) in: M. Misono, Y. Morooka and S. Kimura, *Future Opportunities in Catalytic and Separation Technology* (Elsevier, Amsterdam) 248.
She Y., Han J. and Ma Y.H. (2001), *Catal. Today*, **67**, 43.
Sheintuch M. and Dessau R.M. (1996), *Chem. Eng. Sci.*, **51**(4), 535.
Shimpo Y., Yamaura S., Okouchi H., Nishida M., Kajita O., Kimura H. and Inoue A. (2004), *J. Alloy. Compd.*, **372**, 197.
Shu J., Grandjean B.P.A., Vanneste A. and S. Kaliaguine, (1991), *Can. J. Chem. Eng.*, **89**, 1036.
Shu J., Grandjean B.P.A., Ghali E. and Kaliaguine S. (1993), *J. Membrane Sci.*, **77**, 181.
Shu J., Grandjean B.P.A. and Kaliaguine S. (1994), *Appl. Catal. A: Gen.*, **119**, 305.
Sieverts A., Krumbhaar W. and Jurisch E. (1911), *Z. Phys. Chem.*, **77**, 591–613.
Sirkar K.K., Shanbhag P.V. and Kovvali A. S. (1999), *Ind. Eng. Chem. Res.*, **38**, 3715.
Smirnov V.S., Gryaznov V.M., Lebedeva V.I., Mishchenko A.P., Polyakova V.P. and Savitsky E.M. (1977), US Patent, No. 4,064,188.
Smirnov V.S., Gryaznov V.M., Miropolskay M.A., Samokhvalov G.I., Kozlovskaya N.K., Ermilova M.M. and Mishchenko A.P. (1978), Soviet Patent, No. 437,743.
Smith B. and Shantha M.S. (2007), *Int. J. Chem. Reactor Eng.*, **5**(A84), 1.
Smithells C.J. (1937) *Gases and Metals*, Wiley, New York.
Specchia S., Fino D., Saracco G. and Specchia V. (2005), 'Inorganic membrane reactors', in A. Cybulski and J.A. Moulijn (Eds.), *Structured Catalysts and Reactors*, Taylor & Francis, London, p. 615.
Stoukides M. (2000), *Catal. Rev. Sci. Eng.*, **42**, 1.
Su C., Jin T., Kuraoka K., Matsumura Y. and Yazawa T. (2005), *Ind. Eng. Chem. Res.*, **44**, 3053.
Su F.Y., Huang J.F. and Sun I.W. (2004), *J. Electrochem. Soc.*, **151**(12), C811.
Takahashi T., Inoue M. and Kai T. (2001), *Appl. Catal. A: Gen.*, **218**, 189.
Takanabe K., Aika K., Seshanb K. and Lefferts L. (2004), *J. Catal.*, **227**, 101.
Takeda K., Baba A., Hishinuma Y. and Chikahisa T. (2002), *JSAE Rev.*, **23**, 183.
Tang H., Ishikawa K. and Aoki K. (2007), *Mater. Trans.*, **48**, 2454.
Tereschenko G.F., Ermilova M.M., Mordovin V.P., Orekhova N.V., Gryaznov V.M., Iulianelli A., Gallucci F. and Basile A. (2007), *Int. J. Hydrogen En.*, **32**, 4016.
Tiscarno-Lechuga F., Hill C.G. and Anderson M.A. (1996), *J. Membrane Sci.*, **118**, 85.
Tomishige K., Chen Y.G. and Fujimoto K. (1999), *J. Catal.*, **181**, 91.

Tong J. and Matsumura Y. (2006), *Catal. Today*, **111**(3–4), 147.

Tong J., Xu H., Wang D. and Matsumura Y. (2004), *J. Japan Petroleum Inst.*, **47**, 64.

Tong J., Shirai R., Kashima Y. and Matsumura Y. (2005a), *J. Membrane Sci.*, **260**, 84.

Tong J., Matsumura Y., Suda H. and Haraya K. (2005b), *Sep. Purif. Tech.*, **46**, 1.

Tong J., Suda H., Haraya K. and Matsumura Y. (2005c), *J. Membrane Sci.*, **260**, 10.

Tong H.D., van den Berg A.H.J., Gardeniers J.G.E., Jansen H.V., Gielens F.C. and Elwenspoek M.C. (2005d), *Thin Solid Films*, **479**, 89.

Tong J., Suda H., Haraya K. and Matsumura Y. (2006a), *J. Membrane Sci.*, **260**, 10.

Tong J., Su L., Kashima Y., Shirai R., Suda H. and Matsumura Y. (2006b), *Ind. Eng. Chem. Res.*, **45**, 648.

Tonkovich A.L.Y., Zilka J.L., Jimenez D.M., Roberts G.L. and Cox J.L. (1996), *Chem. Eng. Sci.*, 51, 789.

Tosti S. and Bettinali L. (2004), *J. Membrane Sci.*, **39**, 3041.

Tosti S., Bettinali L., Violante V., Basile A., Drioli E., Chiappetta G., Criscuoli A. and Rizzello C. (1998), Proc. of the 20th Symposium on Fusion Technology, 1033–1036, Marseille (France), Sept. 7–11.

Tosti S., Violante V., Basile A., Chiappetta G., Castelli S., De Francesco M., Scaglione S. and Sarto F. (2000), *Fusion Eng. Des.*, **49–50**, 953.

Tosti S., Bettinali L., Castelli S., Sarto F., Scaglione S. and Violante V. (2002), *J. Membrane Sci.*, **196**, 241.

Tosti S., Basile A., Chiappetta G., Rizzello C. and Violante V. (2003), *Chem. Eng. J.*, **93**, 23.

Tosti S., Basile A., Bettinali L., Borgognoni F., Chiaravalloti F. and Gallucci F. (2006), *J. Membrane Sci.*, **284**, 393.

Tosti S., Basile A., Borgognoni F., Capaldo V., Cordiner S., Di Cave S., Gallucci F., Rizzello C., Cantucci A. and Traversa E. (2008a), *J. Membrane Sci.*, **308**, 250.

Tosti S., Basile A., Bettinali L., Borgognoni F., Gallucci F. and Rizzello C. (2008b), *Int. J. Hydrogen En.*, **33**, 5098.

Tosti S., Borgognoni F. and Santucci A. (2010), *Int. J. Hydrogen En.*, **35**(15), 7796.

Trianto A., Quiao W.L. and Kokugan T. (2001), *J. Chem. Eng. Jpn*, **34**(8), 1065.

Udani P.P.C., Gunawardana P.V.D.S., Lee H.C. and Kim D.H. (2009), *Int. J. Hydrigen En.*, **34**, 7648.

Uemiya S. (1999), *Sep. Purif. Methods*, **28**(1), 51.

Uemiya S. (2004), *Top. Catal.*, **29**(1–2), 79.

Uemiya S., Sato N., Ando H. and Kikuchi E. (1991a), *Ind. Eng. Chem. Res.*, **30**, 585.

Uemiya S., Matsuda T. and Kikuchi E. (1991b), *Appl. Catal.*, **76**, 171.

Uemiya S., Sato N., Ando H. and Matsuda T. (1991c), *Appl. Catal.*, **67**(2), 223.

Uemiya S., Sato N., Ando H., Kude Y., Matsuda T. and Kikuchi E. (1991d), *J. Membrane Sci.*, **56**, 303.

Uhrig R.E. (2005), Proc. of the Int. Hydrogen En. Congress and Exhibition, IHEC 2005, Istanbul (Turkey), July 13–15.

Utaka T., Okanishi T., Takeguchi T., Kikuchi R. and Eguchi K. (2003), *Appl. Catal. A.*, **245**, 343.

Vagia E. and Lemonidou A.A. (2010), *J. Catal.*, **269**, 388.

Vaidya P.D. and Rodrigues A.E. (2006), *Chem. Eng. J.*, **117**, 39.

van de Water L.G.A. and Maschmeyer T. (2004), *Top. Catal.*, **29**, 67.

van den Bergh J., Gücüyener C., Gascon J. and Kapteijn F. (2011), *Chem. Eng. J.*, **166**, 368.

van Dyk L., Miachon S., Lorenzen L., Torres M., Fiaty K. and Dalmon J.A. (2003), *Catal. Today*, **82**, 167.
Velu S., Suzuki K., Kapoor M.P., Ohashi F. and Osaki T. (2000), *Appl. Catal. A: Gen.*, **194**(2), 373.
Venugopal A. and Scurrell M.S. (2003), *Appl. Catal., A: Gen.*, **245**, 137.
Violante V., Drioli E. and Basile A. (1993), *Fusion Eng. Des.*, **22**, 257.
Violante V., Drioli E. and Basile A. (1995), *J. Membrane Sci.*, **104**, 11.
Volkov V., Basile A., Orekhova N. and Sanchez-Marcano J. (2012), Catalysis in membrane reactors, *Catal. Tod.*, **193**(1) 1–235.
Wang D., Flanagan T.B. and Shanahan K.L. (2004a), *J. Alloy. Compd.*, **372**, 158.
Wang D., Tong J., Xu H. and Matsamura Y. (2004b), *Catal. Today*, **93–95**, 689.
Wang H., Cong Y. and Yang W. (2002), *J. Membrane Sci.*, **209**, 143.
Wang L., Murata K. and Inaba M. (2003), *Catal. Today*, **82**, 99.
Wang S. and Lu G.Q. (1998), *Appl. Catal. A: Gen.*, **169**, 271.
Weyten H., Luyten J., Keizer K., Willems L. and Leysen R. (2000), *Catal. Today*, **56**, 3.
Wieland S., Melin T. and Lamm A. (2002), *Chem. Eng. Sci.*, **57**(9), 1571.
Wood B.J. (1968), *J. Catal.*, **11**, 30.
Woods R. (1969), *Electrochim. Acta*, **14**, 632.
Wu G.H., Liu B.S. and Deng J.F. (1998), *Chem. Lett.*, **9**, 889.
Wu L.Q., Xu N. and Shi J. (2000), *Ind. Eng. Chem. Res.*, **39**, 342.
Xiong L., Liu S. and Rong L. (2010), *Int. J. Hydrogen En.*, **35**, 1643.
Xomeritakis G. and Lin Y.S. (1996), *J. Membrane Sci.*, **120**, 261.
Xuan J., Leung M.K.H., Leung D.Y.C. and Ni M. (2009), *Ren. Sust. En. Rev.*, **13**, 1301.
Xue D. and Deng J.F. (2001), *Mater. Lett.*, **47**, 271.
Yamaura S.I., Shimpo Y., Okouchi H., Nishida M., Kajita O., Kimura H. and Inoue A. (2003), *Mater. Trans.*, **44**(9), 1885.
Yamaura S., Shimpo Y., Okouchi H., Nishida M., Kajita O. and Inoue A. (2004), *Mater. Trans.*, **45**, 330.
Yan S., Maeda H., Kusakabe K. and Morooka S. (1994), *Ind. Eng. Chem. Res.*, **33**, 616.
Yehenskel J., Leger D. and Courvoisier F. (1979), *Adv. Hydrogen En. Progr.*, **2**(1), 569, cited on (Gryaznov, 1999).
Yepes D., Cornaglia L.M., Irusta S. and Lombardo E.A. (2006), *J. Membrane Sci.*, **274**, 92.
Yin X., Honga L., Liu Z. (2008), *J. Membrane Sci.*, **311**, 89.
Ying L.A., Liu J., Mo L., Lou H. and Zheng X. (2012), *Int. J. Hydrogen En.*, **37**, 1002.
Yoshida H., Takeshita H., Konishi S., Ohno H., Kurasawa T. and H. Watanabe, (1984), *Nucl. Tech. Fusion,* **5**, 178.
Yu C. and Xu H. (2011), *Sep. Pur. Tech.*, **78**, 249.
Yu C.Y., Lee D.W., Park S.J., Lee K.Y. and Lee K.H. (2009), *Int. J. Hydrogen En.*, **34**, 2947.
Yukawa H., Nambu T., Matsumoto Y., Watanabe N., Zhang G. and Morinaga M. (2008), *Mater. Trans.*, **49**, 2202.
Zaman J. and Chakma A. (1995), *Int. J. Hydrogen En.*, **20**(1), 21.
Zerva C. and Philippopoulos C.J. (2006), *Appl. Catal., B: Env.*, **67**, 105.
Zhang B.C., Tang X.L., Li Y., Xu Y.D. and Shen W.J. (2007a), *Int. J. Hydrogen En.*, **32**, 2367.

Zhang J., Liu D., He M., Xu H. and Li W. (2006a), *J. Membarne Sci.*, **274**, 83.
Zhang X. and Shi P. (2003), *J. Molecular Catal A: Chem.*, **194**, 99.
Zhang Y., Gwak J., Murakoshi Y., Ikeara T., Maeda R. and Nishimura C. (2006b), *J. Membrane Sci.*, **277**, 203.
Zhang Y., Lu J., Maeda R. and Nishimura C. (2007b), *J. Alloy. Compd.*, **446–447**, 579.
Zhang K., Gao H., Rui Z., Liu P., Li Y. and Lin Y.S. (2009), *Ind. Eng. Chem. Res.*, **48**, 1880.
Zhu W., Han W., Xiong G. and Yang W. (2005), *Catal. Today*, **104**, 149.
Zhu Y., Minet R.G. and Tsotsis T.T. (1993), *Catal. Lett.*, **18**, 49.

2.14 Appendix: nomenclature

2.14.1 Notation

A	membrane superficial area
A_1	coefficient that describes the passage of hydrogen through metallic membranes
B	coefficient that describes the passage of hydrogen through metallic membranes
$c_{H_2,ret}$	H_2 concentration at the membrane side 'permeate'
$c_{H_2,perm}$	H_2 concentration at the membrane side 'retentate'
C_{H_2}	concentration of H_2 dissolved in the metallic membrane surface
Δ_{H_2}	diffusion coefficient
n	dependence factor of the H_2 flux on the H_2 partial pressure
n_a	constant related to the amorphous state of the membrane
G°	mean free energy related to the H_2 dissolution at 1 atm
J_{H_2}	hydrogen flux through the membrane
$J_{H_2,Sieverts-Fick}$	hydrogen flux through the membrane according to Sieverts–Fick law
k_a	constant related to the amorphous state of the membrane
k_1	rate constant
k_2	rate constant
K_S	constant of Sieverts' law
E_a	apparent activation energy
$Pe^\circ_{H_2}$	pre-exponential factor
Pe_{H_2}	H_2 permeability
p°	reference pressure
p_{H_2}	hydrogen partial pressure in the gas phase
$p^n_{H_2,ret}$	H_2 partial pressure either on the retentate (high H_2 partial pressure) side
$p^n_{H_2,perm}$	H_2 partial pressure either on the permeate (low H_2 partial pressure) side

R	universal gas constant
r_{pd}	rate of Pd consumption
$r_{N_2H_4}$	rate of hydrazine consumption
T	absolute temperature
t	time
Sel	selectivity
Y	yield
α	ideal selectivity
δ	membrane thickness
σ	width of the Gaussian distribution
μ_{H_2}	chemical potential
χ	conversion
χ_{eq}	equilibrium conversion

2.14.2 Abbreviations

APSS	asymmetric porous stainless steel
$C_6H_{12}N_4$	hexamethylenetetramine
CFB	circulating fluidized bed
CVD	chemical vapour deposition
EDA	ethylenediamine
EDS	dispersive electronic analysis
EDTA	ethylene di-amine tetra acetic acid
EMI-Cl-BF4	1-ethyl-3-methylimidazolium chloride tetrafluoroborate system
EPMA	electron probe microanalyser
ESR	ethanol steam reforming
FLBR	fluidized bed reactor
FLBMR	fluidized bed membrane reactor
HR	hydrogen recovery
HTXRD	high temperature X-ray diffraction
MOCVD	metal organic chemical vapour deposition
MR	membrane reactor
MSSS	mesoporous stainless steel substrate
MTPS	Modification technique of particle surface
PBMR	packed bed membrane reactor
PBR	packed bed reactor
PEM	polymer electrolyte membrane
POM	partial oxidation of methane
PVA	polyvinyl alcohol
PVD	physical vapour deposition
RF	radio frequency

SEM	scanning electron microscopy
SF	sweep factor
SR	steam reforming
SMR	steam methane reforming
TEM	transmission electron microscopy
TR	traditional reactor
WGS	water gas shift
XRD	X-ray diffractometer
XPS	X-ray photoelectron spectroscopy
YSZ	yttria-stabilized zirconia

3

Palladium-based composite membranes for hydrogen separation in membrane reactors

P. PINACCI, Research on the Energetic System (RSE) S.p.A., Italy and A. BASILE, ITM-CNR, Italy

DOI: 10.1533/9780857097330.1.149

Abstract: This chapter presents methods of preparing composite membranes for hydrogen separation. Particular emphasis is given to palladium and palladium alloy composite membranes prepared via electroless plating and magnetron sputtering. The chemical and physical stability of these membranes, in relation to specific applications, is discussed. This chapter also assesses the performance of current membrane reactors and discusses significant developments, including current and planned prototype membranes.

Key words: palladium composite membranes, electroless plating, magnetron sputtering, hydrogen separation.

3.1 Introduction

When dealing with composite membranes, it is important to work on the premise that the substrate used for depositing the thin layers of metals is, in fact, a porous membrane. The International Union of Pure and Applied Chemistry (IUPAC) directives classify porous membranes, according to their pore diameter, as macroporous, mesoporous and microporous membranes. Examples are shown in Table 3.1 with reference to hydrogen permeation (and separation). Porous membranes with an average pore diameter (ϕ_{pore}) greater than 50 nm are classified as macroporous, those with $2 < \phi_{pore} < 50$ nm as mesoporous, and when $\phi_{pore} < 2$ nm, the membrane is classified as microporous. Dense Pd-based membranes are also shown in Table 3.1 for comparison. The effect of pore diameter on the transport mechanism through the membrane, H_2 selectivity versus other gases, H_2 permeance and the reactant loss can be clearly seen. Macroporous membranes have no selectivity, meaning that the selectivity of the gas in question, for example hydrogen, versus another gas is equal to 1, for example, the permeance of hydrogen and the permeance of the other gas are the same. In macroporous membranes the greater part of the reactant leaves the shell side without any reaction (this is called *reactant loss problem*). In comparison, an increase in both selectivity and permeance, and a decrease in reactant loss, are observed in both mesoporous and microporous membranes. Dense

Table 3.1 Main characteristics of porous and dense membranes

Membrane	ϕ_{pore} (nm)	Diffusion mechanism	$\alpha_{H_2/other\ gas}$*	H_2 permeance	Reactant loss
Macroporous	> 50	Poiseuille (viscous flow)	1	Very high	High
Mesoporous	2–50	Knudsen	$\alpha_{H_2/N_2} = 3.74$	High	Average
Microporous	< 2	Activated process	High	Average	Low
Dense Pd	–	Fick	Infinite	Low	–

*The selectivity, $\alpha_{H_2/other\ gas}$, is defined as the ratio between the permeance of pure H_2 versus the permeance of the other (pure) gas.

palladium membranes, however, show an infinite selectivity of hydrogen versus all other gases, a lower level of H_2 permeance and zero reactant loss. If a sweep gas is fed into the permeate side of porous membranes, the back-permeation of the inert sweep gas, from the permeate side to the reaction side, must also be taken in consideration. This is obviously not an issue when using Pd-based membranes.

H_2 flux in Pd-based membranes is highly limited by the membrane thickness. As a consequence, therefore, they must be designed taking into account both mechanical strengths and membrane durability. Commercial dense palladium membranes with a wall thickness of around 100–50 μm show two main drawbacks: (i) their hydrogen fluxes are relatively very low, and (ii) their cost – palladium costs around 15 €/g (cost based on prices at November 2011) making these membranes very expensive. The major aim of any study concerning Pd-based composite membranes is, therefore, to reduce palladium thickness without compromising the mechanical stability of the membrane. The flux increases as the Pd-layer gets thinner, meaning that the overall cost of the composite membrane decreases because of the reduced amount of palladium used.

To meet the dual challenge of achieving high selectivity and high fluxes the following compromise could be considered: one or more thin layers of a highly selective membrane material are deposited on supports, which have little or no resistance to fluxes. For example, porous membranes could be used as substrates onto which a very thin, but continuous, film of a selective metal, such as palladium or palladium alloy, can be deposited. The result is a palladium-based composite membrane, characterized by a highly selective metallic layer, deposited on a substrate with a high porosity. Composite Pd-based membranes, in particular, can be obtained by depositing a very thin Pd-based layer onto the porous support, achieving an increased mechanical resistance at a reduced cost.

These thinner composite membranes offer several clear advantages over the thick-walled palladium-based ones including higher hydrogen fluxes, very high H_2 selectivity and reduced palladium costs.

3.2 Development of composite membranes

The development of composite membranes for hydrogen separation in membrane reactors has received a great deal of attention, ever since the early years when research was focused on specific applications such as methane steam reforming and dehydrogenation reactions (Armor, 1995; Kikuchi, 1995; Saracco and Specchia, 1994). Interest in composite membrane development has surged during the last decade, however, in correspondence to the growing interest of industries in hydrogen economy and carbon capture in power generation plants. The CO_2 Capture Project (CCP), promoted by seven major energy companies, is a significant example (Miracca, 2009). The more relevant research projects, both in the USA (Department of Energy (DOE)/National Energy Technology Laboratory (NETL, 2010)) and Europe (Kougionas, 2010) are now co/financed as carbon-capture R&D programs.

Most of these projects deal with the development of composite membranes consisting of a dense layer, in which hydrogen separation occurs, deposited on cheaper porous support materials using various methods.

Candidate materials are usually very expensive and their costs fluctuate greatly. For example, the cost of palladium increased from about 10 to 18 €/g in 2010. Forecasts for the near future indicate further increases, due to the growing demand from emerging economies (e.g., China and India) for large-scale applications, such as catalytic silencers, in the automotive industry. Similarly the costs of rare-earth elements such as Lanthanum and Cerium have increased by about 420% during 2010.

In general, regardless of the type of membrane and mechanism of separation, hydrogen flux can be considered as inversely proportional to thickness. A reduction in membrane thickness of a factor ‘m’, therefore, could determine a corresponding saving of a factor ‘m^2’ of metal quantity necessary to achieve the same performances, in terms of flux. Moreover, membrane-reactor technology could realize further advantages, including the minimization of the equipment’s process intensification, which can vary depending on the specific application.

Figure 3.1 shows an example of a parametric study performed by Haldor Topsoe in the framework of the CCP project, indicating that the cost of CO_2 capture in integrated gasification combined cycle (IGCC) can be significantly reduced by approximately 30% using thin composite Pd membranes (less than 10 μm thick) for the water-gas shift (WGS) reaction of the syngas in a membrane reactor (Middleton *et al.*, 2002).

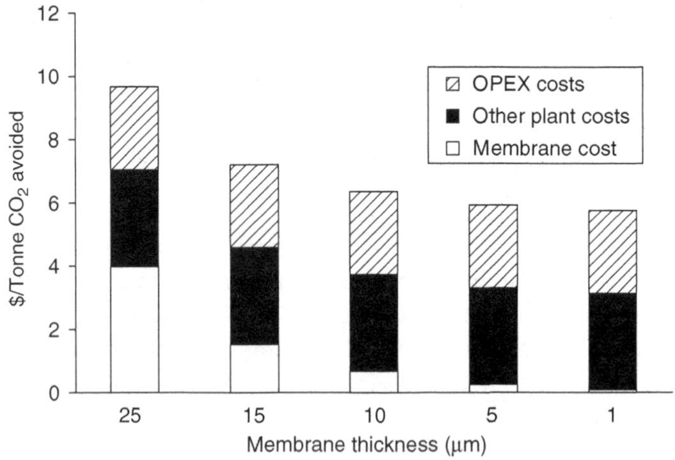

3.1 Projected cost of CO_2 capture in coal gasification plants (IGCC) by using composite Pd membranes for the WGS of the syngas in a membrane reactor. (After Middleton *et al.*, 2002.)

Table 3.2 Main typologies of membranes for hydrogen separation

Membrane type	Temperature (°C)	
Mixed conductors (ceramics, dual phase ceramics, cermets)	750–950	↓ H_2 permeability increase
Pd and Pd alloys (Pd–Ag, Pd–Cu, Pd–Au)	300–600	
Composite metallics (sandwich structure FCC/BCC/FCC)	300–400	

The development of composite membranes, however, raises challenging issues of its own, such as minimizing defects in the deposited layer and maintaining chemical and mechanical stability in the operating conditions. Chemical stability is necessary to avoid an inter-diffusion of elements between the deposited layer and the support, and to minimize surface fouling and poisoning. Mechanical stability is also necessary to minimize the stresses induced by the different thermal expansion coefficients of the deposited layer and the support.

Finally, while evaluating the potential benefits of composite membranes, their target performance and stability should also be evaluated on the basis of the specific application. For example, the H_2 purity for fuel cells should exceed 99.999% whereas, for other applications such as internal combustion engines and/or gas turbines, less pure hydrogen of approximately 93.5% is acceptable (Kreutz *et al.*, 2002). The main typology of membranes for hydrogen separation is shown in Table 3.2

Membranes for hydrogen separation can be conveniently classified into three groups, which are suitable for applications in different ranges of temperatures, and which exhibit increasing hydrogen permeability.

3.2.1 Mixed conducting materials

These are ceramic materials (such as perovskites or fluorites) in which both electronic and proton conduction occurs. Hydrogen flux $\left(J_{H_2}\right)$, at a given temperature (T), is proportional to the natural logarithm of the pressure gradient across the membrane, based on Nernst potential (Bredesen *et al.*, 2004):

$$J_{H_2} = -\frac{RT}{16F^2\delta}\int_{\ln pH_2'}^{\ln pH_2''} \frac{\sigma_i \sigma_e}{\sigma_i + \sigma_e} d\ln pH_2 \qquad [3.1]$$

where R is the ideal gas constant, F the Faraday constant, δ the membrane thickness and σ_i and σ_e represent ionic and electronic conductivity of the material, respectively.

A key issue for obtaining a significant hydrogen flux is to increase σ_e, which is usually much lower than σ_i. Proton-conducting oxides exhibit low permeation fluxes and require the application of an external electrical field to drive the separation (Serra *et al.*, 2007). Hydrogen flux can be increased, either by adding a pure electronic conducting material, and thereby obtaining a dual phase ceramic, or by dispersing a pure metal into the ceramic matrix, which would result in a cermet material, such as Pd-yttria-stabilized zirconia (YSZ) (Balachandrian *et al.*, 2006). These mixed conducting materials are in the early stages of development, but have already proved suitable for high temperature applications (700–900°C) such as methane steam reforming (Vigeland and Aasen, 2006). An extensive review of these concepts can be found in Chapter 7.

3.2.2 Palladium and palladium alloy membranes

Hydrogen permeation through palladium membranes occurs through a solution–diffusion mechanism. The main steps involved in hydrogen transport across the membrane, under a partial pressure gradient, are as follows (Uemiya, 1999):

- diffusion of molecular hydrogen to the membrane surface (feed side); hydrogen chemi-adsorption and dissociation in atomic form, proton and electron, on the membrane surface;

- proton solubilization and diffusion through the metal lattice;
- proton and electron recombination in hydrogen molecules and desorption on the opposite membrane surface (permeate side);
- molecular hydrogen diffusion away from the surface.

Hydrogen flux through the membrane (J_{H_2}) can be shown using the following equation (Uemiya, 1999):

$$J_{H_2} = \frac{\text{Pe}\left(p_f^n - p_p^n\right)}{\delta} \quad [3.2]$$

where p_f and p_p represent hydrogen partial pressure on the feed and on the permeate side, respectively, δ represents the palladium layer thickness and Pe the permeability that is generally expressed by an Arrhenius-type equation:

$$\text{Pe} = \text{Pe}_0 \exp\left(-\frac{E_a}{RT}\right) \quad [3.3]$$

Exponent n in Equation [3.2] is related to the transport step controlling hydrogen flux and can vary between 0.5 and 1. In membranes with a high thickness (more than 30 μm), transport is controlled by solution–diffusion through the lattice and $n = 0.5$; whereas, in membranes with lower thicknesses, diffusion from the bulk to the membrane surface and chemi-adsorption tend to prevail, and n can increase up to 1 (Ward and Dao, 1999). The n-value is also influenced by temperature (Zhang *et al.*, 2006) and pressure (Morreale *et al.*, 2003; Li *et al.*, 2008).

Deviations from the Sieverts' law behaviour have been reported by various authors and are attributed to factors such as surface processes, surface poisoning and grain boundaries (Guazzone *et al.*, 2006). An extensive review of these concepts can be found in Chapter 2.

3.2.3 Composite (sandwich) metallic membranes

Body centred cubic (BCC) metals, such as Ta, Nb, V and Zr, exhibit a hydrogen solubility higher than *f*ace *c*entred *c*ubic (FCC) metals, such as palladium, up to a factor of 10–100 (Dolan, 2010). Hydrogen permeation through these metals occurs by a solution–diffusion mechanism, similar to the one described above for palladium. These metals, however, exhibit a tendency towards embrittlement. Moreover, they are very reactive and can form carbides, nitrites and oxides when used in operating environments. In order to avoid this poisoning, BCC metals can be coated on both sides by a thin film (less than 1 μm) of a noble metal, which acts as a catalyst and can prevent poisoning effects and favour hydrogen adsorption and desorption from the metal surface. In this way a typical sandwich structure is obtained (Mundschau *et al.*, 2006). Due to the above mentioned problems, these membranes are suitable for low temperature applications (300–400°C), such as the WGS of syngas produced in coal gasification plants (Evenson, 2011).

This chapter will focus on composite membranes for hydrogen separation and, specifically, on Pd and Pd-alloy composite membranes. Preparation methods will be described, and chemical and physical stability in relation to specific applications will be discussed. Performances in membrane reactors and significant developments up to and including the prototype scale will be also presented.

3.3 Palladium and palladium-alloy composite membranes for hydrogen separation

In the last 20 years, composite palladium and palladium alloy membranes have been extensively studied for hydrogen separation applications in membrane reactors (Bredesen *et al.*, 2004; Middleton *et al.*, 2005; Chiesa *et al.*, 2006; Dijkstra *et al.*, 2006). Research efforts have been targeted at obtaining thin deposits of cheaper support materials, such as ceramics and metals, using several techniques (Uemiya, 1999), including sputtering (Mc Cool *et al.*, 1999; Zhao *et al.*, 2000), chemical vapour deposition (Morooka *et al.*, 1995; Itoh *et al.*, 2005), electroless plating (Li *et al.*, 1999; Cheng and Yeung, 2001 and electroplating (Nam and Lee, 2001). Electroless plating and magnetron sputtering have progressively grown in importance, however, and, as a result, manufacturing processes were scaled up and, currently, several projects for prototype applications are running or have been scheduled (DOE/NETL, 2010; Kougionas, 2010).

This chapter, therefore, will mainly focus on the preparation of membranes using these two methods. Critical issues related to their specific applications will be discussed, and an insight concerning on-going projects, and current and future developments will be given.

3.3.1 Preparation via electroless plating

Electroless plating is an autocatalytic process in which deposition occurs by the reaction in solution with a metal salt and a chemical reducing agent, without the supply of any electric power. As a consequence, deposition occurs both on conductive and non-conductive materials, such as ceramics and plastics; moreover, it does not require expensive equipment and is relatively easy to scale up from laboratory to industrial scale. Electroless plating is nowadays a common industrial practice: it is used in engineering coating applications where wear resistance, hardness and corrosion protection is required (Mallory and Hajdu, 1990).

Generally the electroless plating bath consists of a metallic salt, a complexing agent, a reducing agent and a stabilizing agent. Electroless plating, in fact, involves the reduction of the metallic salt on a catalytic surface. However, common substrates such as metals, ceramics and glass need very long induction times to start the nucleation process; therefore, an activation of the support is necessary to initiate the autocatalytic deposition process.

Starting from the 1990s, dense Pd and Pd-alloy membranes have been obtained on porous supports such as glass (Uemiya *et al.*, 1991a), ceramics (Uemiya *et al.*, 1991b), (Collins and Way, 1993) and sintered metal alloys (Shu *et al.*, 1993).

Pure palladium membranes present the problem of hydrogen embrittlement caused by the transition between α and β phases of palladium hydride, which occurs below the critical temperature of about 300°C (Uemiya, 1999). Phase transition and hydrogen embrittlement can be avoided by the addition to palladium of alloying elements such as Ag, Au, Cu, Y and Ce. Moreover, the addition of such elements in a proper amount can increase hydrogen permeability as shown by Knapton (1977) and Gryaznov (2000). Finally the addition of Cu and Au can increase resistance to sulphur poisoning (McKinley, 1969a), as will be discussed in the next paragraph.

A detailed description of the preparation procedure of Pd-alloy membranes is reported by Broglia *et al.* (2011). The preparation procedure involves the following main subsequent steps (Fig. 3.2)

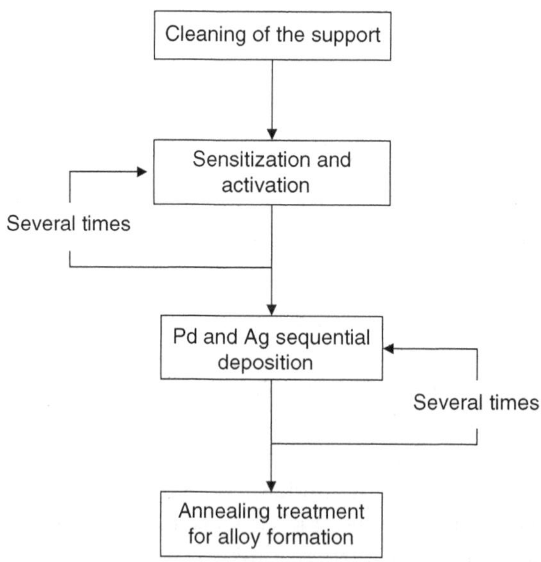

3.2 Steps for preparation of Pd–Ag membranes via electroless plating.

Table 3.3 Typical composition of solutions used for sensitization/activation of metallic supports for electroless deposition. After Ma *et al.* (2004)

	Chemical	Concentration
Sensitization	$SnCl_2 \cdot 2H_2O$	0.1 g/L
	HCl (37%)	1 mL/L
Activation	$PdCl_2$ 99.9% (metal basis)	0.1 g/L
	HCl (37%)	1 mL/L

Cleaning of the support

In order to remove contaminants, such as grease oil, dirt and corrosion products, from the surface of the support. Cleaning should be carried out very carefully since impurities can inhibit autocatalytic sites formation in the subsequent activation phase.

Activation of the support

Activation is usually performed at room temperature by dipping the support first into a stannous chloride solution (sensitization) and then in a palladium chloride solution (activation), several times. A typical composition of the sensitization and activation solutions is shown in Table 3.3. Activation is performed at acidic pH in order to maintain a good salt solubility. The overall reaction is:

$$SnCl_2 + PdCl_2 \rightarrow SnCl_4 + Pd \qquad [3.4]$$

The number of dipping cycles depends on the support material, its pore size distribution and surface morphology. Typically, the sensitization/activation cycle is repeated from two to ten times (Ma *et al.*, 2000; Bottino *et al.*, 2006).

Metal plating

Hydrazine is the most common and suitable reducing agent for electroless plating of palladium. A typical composition of a hydrazine-based bath is shown in Table 3.4.

The plating bath consists of a palladium ion source $[Pd(NH_3)_4Cl_2]$, a complexant (ethylene-di-amine tetra acetic acid (EDTA)) and a pH controller (ammonia). Deposition is carried out at basic pH (10–12) and at a controlled temperature ranging between 40 and 60°C. In more detail, the electroless technique is based on the following redox reactions, which occur simultaneously in the solution (Bottino *et al.*, 2006):

Table 3.4 Typical composition of palladium electroless plating baths reported in literature

	Component	Value
Pd ion source	$Pd(NH_3)_4Cl_2$	4–5 (g/L)
Complexant	Na_2EDTA	40–80 (g/L)
Reducing agent	N_2H_4 1 M	6–10 (mL/L)
Buffer agent	NH_4 (OH) 28%	200–650 (mL/L)
pH	–	10–12
T(°C)	–	40–60

$$\text{Anodic: } N_2H_4 + 4OH^- \rightarrow N_2 + 4H_2O + 4e^- \quad E^\circ = 1.12\text{ V} \qquad [3.5]$$

$$\text{Cathodic: } 2Pd[NH_3]_4^{2+} + 4e^- \rightarrow 2Pd^0 + 8NH_3 \quad E^\circ = 0.95\text{ V} \qquad [3.6]$$

The overall reaction is:

$$2Pd[NH_3]_4^{2+} + N_2H_4 + 4OH^- \rightarrow 2Pd^0 + 8NH_3 + N_2 + 4H_2O \quad E^\circ = 2.07\text{ V} \qquad [3.7]$$

Deposition should be performed in controlled conditions to obtain the formation of pinhole-free, adherent palladium film; a typical laboratory apparatus is shown in Fig. 3.3. The membrane support is inserted in a reactor and maintained in rotation at constant velocity by a variable-speed motor to allow the removal from the reaction zone of nitrogen produced by hydrazine reduction (Equation [3.5]). The reactor is immersed in a thermostatic bath to keep the temperature at a constant value. Hydrazine can be periodically added from the tube inserted in the middle of the reactor, while nitrogen can be evacuated from the reactor through the tube on the right side.

In case an alloy is formed, such as palladium-silver or palladium-copper, either metal co-deposition (Cheng and Yeung, 1999; Huang *et al.*, 2003) or sequential deposition can be made (Kikuchi and Uemiya, 1991; Ma *et al.*, 2004); in both cases, however, an annealing treatment should be made to obtain alloying by intermetallic diffusion (Uemiya *et al.*, 1991b; Roa *et al.*, 2003).

The morphology of deposited Pd and Pd alloys depends both on the porous support and on the plating bath characteristics, as discussed in detail by Broglia *et al.* (2011). The initial growth of a film, however, is strongly influenced by the support material, surface roughness and surface contaminants, while subsequent stages of growth are more dependent on bath physical–chemical parameters.

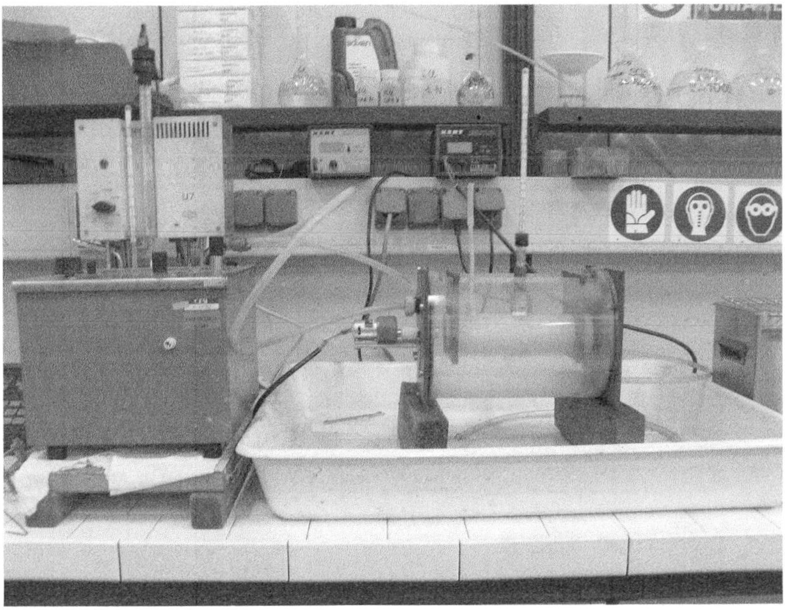

3.3 Laboratory cell used for electroless deposition of Pd membranes at RSE laboratories.

Porous ceramics substrates such as α-alumina have always aroused great attention because of the relatively smooth surface and narrow pore distribution (Uemiya *et al.*, 1991b). Figure 3.4 shows the scanning electron microscopy (SEM) images of a Pd membrane obtained by electroless plating on the external surface of an α-alumina support with a 200 nm pore size. Palladium penetrates deeply into the pores of the support, providing good anchoring for the deposited film. The above characteristics allow us to obtain a dense composite Pd-alloy membrane with a thickness as low as 5 μm (Dijkstra *et al.*, 2006).

Ceramic substrates, however, present the problem of leak-tight seals at the ceramic–metal interface, as well as an intrinsic fragility which, in perspective, makes it difficult to assemble them into a module, affecting their potential industrial application as well. Because of these problems, research has progressively addressed considering macroporous metallic alloys as candidate materials for supports, such as austenitic stainless steels and Ni–Cr super-alloys (Inconel and Hastelloy) (Ma, 2006). These materials have higher robustness and mechanical stability, and can be easily welded and assembled in a module. Commercially available macroporous metals, however, present high roughness and large pore dimensions (in the micron range); this causes high thickness of the palladium deposit (up to 20–30 μm) in order to obtain dense membranes (Mardilovich *et al.*, 2002).

3.4 SEM images of Pd deposited on a ceramic support via electroless plating: (a) surface and (b) cross section. (Courtesy of RSE laboratories.)

An SEM image of the surface and a section of a Pd membrane deposited on a macroporous AISI 316L support is shown in Fig. 3.5. Single crystals of palladium have agglomerated to form clusters of a few microns in size, obtaining the characteristic 'cauliflower type' structure (Fig. 3.5a); the palladium layer uniformly covers the surface and is well anchored to the support; the thickness is about 25 μm (Fig. 3.5b).

High operating temperatures and long exposure times can determine the intermetallic diffusion of metal elements (such as iron, chromium, nickel) from the support into the palladium layer, affecting H_2 permeance of the

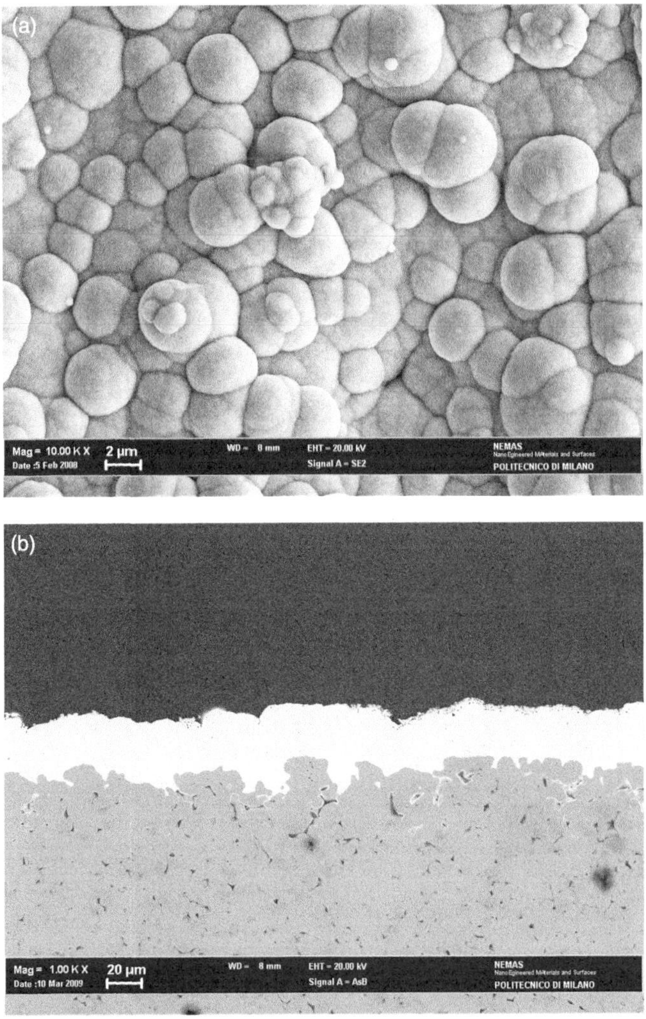

3.5 SEM images of Pd deposited on a metallic support via electroless plating: (a) surface and (b) cross section. (Courtesy of RSE laboratories.)

membrane. Specifically, the activation energy for hydrogen permeation through Pd–Fe alloys is much higher than through pure Pd and Pd alloys (Flanagan *et al.*, 1977), so leading to a sharp decrease of hydrogen permeance. In more detail, the diffusion rate of metal elements at the interface between two metallic layers becomes significantly close or above their Tamman temperature. The Tamman temperature (T_{Ta}) of a metal is defined as the temperature above which a significant increase of atom vibration in the lattice occurs, thus bringing about an increase in its mobility and diffusion. T_{Ta} is conventionally assumed as one-half of its melting temperature (in K). In the case of palladium and stainless steel, Tamman temperature is 640°C and 550–560°C, respectively (Ma *et al.*, 2000). This suggests that Fe diffuses into palladium more rapidly than vice versa, as reported by Liu *et al.* (2010).

Hydrogen flux decline, however, has been observed at temperatures as low as 300–50°C (Rothenberger *et al.*, 2004), possibly due to the fact that, in addition to diffusion through the lattice, other processes, such as iron diffusion through palladium grain boundaries and palladium clusters, can occur. For practical purposes, therefore, operation of Pd and Pd alloys above 300°C requires the introduction of a suitable barrier layer which could mitigate iron diffusion into the palladium layer. Several types of barrier layers have been considered, including titanium nitride (Shu *et al.*, 1996), mixed oxides grown by oxidation of the support (Ma *et al.*, 2004), a bi-metallic deposited multilayer (Ayturk *et al.*, 2006), an electro-deposited chromium oxide (Samingptai *et al.*, 2010), ceramic materials such as zirconia oxide (Gao *et al.*, 2006), zirconia-yttria (Singh *et al.*, 2006), cerium oxide (Tong *et al.*, 2006) and alumina(Tong *et al.*, 2005a; Ma and Guazzone, 2006; Broglia *et al.*, 2009).

Grown-on oxides have proven to be effective in avoiding intermetallic diffusion up to 400°C throughout a time period of thousands of hours (Ma *et al.*, 2000; Pinacci *et al.*, 2011). Oxidation of macroporous AISI 316L for temperatures around 500°C, in fact, brings about the formation of a protective thin layer of chromium oxide. For temperatures above 400°C, however, a rapid H_2 flux decline can be observed (Ayturk *et al.*, 2006). The more promising approach to obtaining an effective barrier up to 600–650°C, appears to be to deposit a ceramic material. As shown in Fig. 3.6, ceramic materials, such as γ-alumina, could allow smooth surfaces with narrow pore size, resulting in the preparation of a dense membrane with lower thickness (10 μm, approximately). Due to the morphological and topographic characteristic of the support, a multi-graded layer has been often deposited in order to reduce surface rugosity (Ma and Guazzone, 2006).

The preparation procedure, as described by Azzurri *et al.* (2011), can follow the following subsequent steps: (1) oxidation of the porous metal; (2) dip-coating deposition of a ceramic barrier layer; (3) dip-coating deposition of an activation layer, containing $PdCl_2$; and (4) Pd deposition by electroless plating.

The challenge, for these membranes, is related to stability under thermal cycling, due to a mismatch in thermal expansion coefficients between the ceramic and the palladium, particularly when hydrogenated (Ayturk *et al.*, 2006). At high temperature, total stress can be considered as the sum of thermal stress and H_2 stress due to H_2 dissolution within the Pd lattice. As shown in Table 3.5, the thermal expansion coefficient of stainless steel is about three times higher than that of alumina. It is also interesting to note that Hastelloy C and zirconia oxide exhibit very close values. Pall

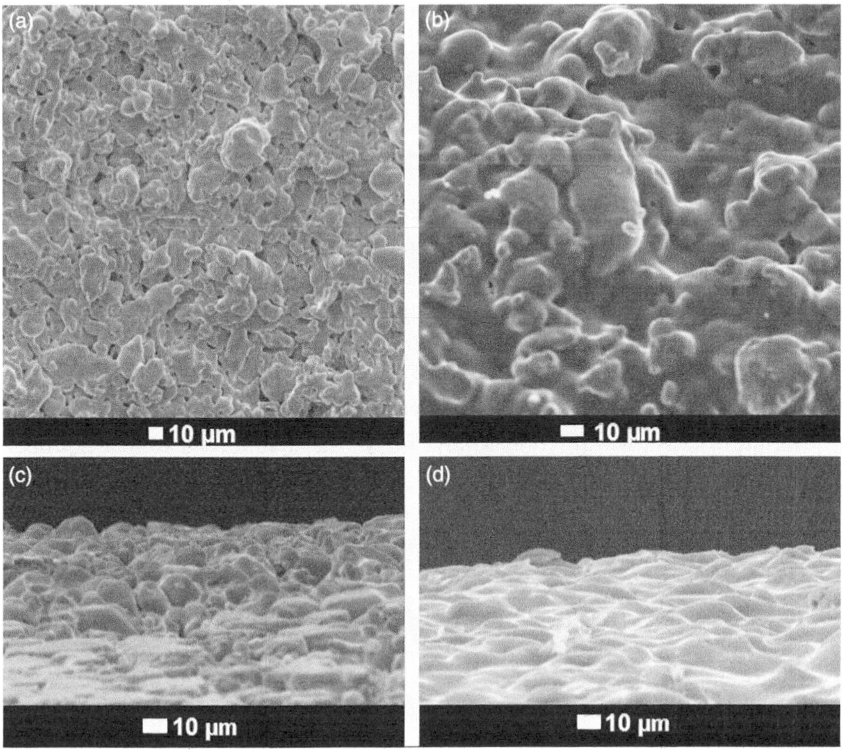

3.6 Top and perspective view of the surface of a macroporous support of AISI 316L: (a) before and (b) after γ-Al_2O_3 deposition. (Reprinted from Broglia *et al.*, 2009. Copyright (2009) with permission from Elsevier.)

Table 3.5 Thermal expansion coefficients for various materials. After Ayturk *et al.* (2006)

Metal	Thermal expansion coefficient $(L-L_0)/L_0$ (°K^{-1})
AISI 316 L	1.73×10^{-5}
Hastelloy	1.20×10^{-5}
Palladium	1.17×10^{-5}
Silver	1.95×10^{-5}
Nickel	1.30×10^{-5}
Iron	1.20×10^{-5}
ZrO_2	1.0×10^{-5}
γ-Alumina	0.65×10^{-5}

L_0 = initial length; L = final length after temperature change.

Corporation is currently developing Pd-alloy membranes on AISI 310SC steel coated with a multi-graded ZrO_2 barrier for use up to 550°C and 20 bar. Pd–Au membranes, with thickness varying in the 3–9 μm range and alloy composition in the range 0–30% have been synthesized. Membranes have been scaled up to 50 cm and exhibit very high permeation; however, till now they have been tested only for a few hundred hours up to 400°C, in simulated methanol reformate streams (Damle, 2011).

3.3.2 Preparation via magnetron sputtering

The magnetron sputtering technique allows the preparation of thin homogenous film using multi-component targets with good thickness and composition control (Mc Cool *et al.*, 1999; Zhao *et al.*, 2000).

The sputtering deposition process makes use of plasma formed very close to the target surface as a source of energetic bombarding ions which produce the ejection of one of more atoms from the surface layers. The ejected species are then transported in the form of vapour to the substrate where they condense to form the deposited film. In the magnetron sputtering process, a magnetic field is placed into the plasma region close to the cathode in order to increase deposition rates (Cecchetto *et al.*, 2011).

In more detail, composite Pd-alloy membranes are prepared by a two-steps process (Bredesen and Klette, 2000). Alloy is first deposited by DC magnetron sputtering on a substrate with a polished surface, such as a silicon mono-crystalline wafer. The film is then removed from the wafer. This film can either be used as a self-supported membrane or integrated with various supports, depending on the specific application (Mejdell *et al.*, 2009). Magnetron sputtering appears to be well suited for the preparation of Pd alloy such as Pd–Ag, since it allows homogeneous alloys to be obtained without any annealing post-treatment, which has proved to be critical for membranes prepared via electroless plating.

By using the two step procedure described above, Sintef (Norway) has produced Pd–23%Ag membranes, 2–5 μm thick, on a tubular macroporous AISI 316 L support 1/2 inch outer diameter (OD), supplied by Pall Corporation (Peters *et al.*, 2009b). Sealing was obtained by clamping the film to the dense stainless steel tube ends (Klette and Bredesen, 2008). In this way membranes with a length of up to 50 cm were prepared and tested.

Membrane development have been focused on applications in pre-combustion power generation cycles with CO_2 capture and, specifically, on WGS of syngas mixtures. In WGS conditions (57.5% H_2, 18.7% CO_2, 3.8% CO, 1.2% CH_4 and 18.7% steam), a H_2 permeance of 1.1×10^{-3} mol m^{-2} s^{-1} $Pa^{-0.5}$ was obtained at 400°C and 26 bar feed pressure. The influences on H_2 of dilution, concentration polarization and surface effects due to reversible absorption of CO and CO_2 have been also investigated (Peters *et al.*, 2008).

The long-term stability of these membranes has been assessed in H_2/N_2 and WGS mixtures for over a year. Stable membrane operation has been obtained while operating at 400°C; however, increasing the temperature

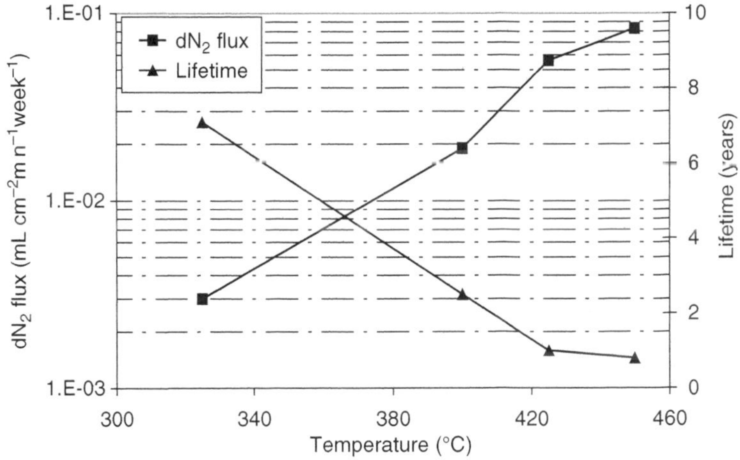

3.7 Pd–Ag membranes, 2.6 μm thick, obtained by magneton sputtering: increase in N_2 leakage and lifetime estimation as a function of temperature. (After Peters *et al.*, 2009b.)

to 450°C resulted in a sharp increase of N_2 flux, due to the formation of pinholes in the membranes (Peters *et al.*, 2009a). Detailed microstructural analysis, both at the surface and in the bulk, suggested that, at elevated temperature, inhomogeneous stress distribution in the film, possibly originating from hydrogen dissolution and mismatch in thermal expansion coefficients, could cause microstructural changes and pinhole formation. Based on the above results, the lifetime of such membranes as a function of temperature has been estimated, by assuming that they are maintained in operation until H_2/N_2 permselectivity drops below 400 (Peters *et al.*, 2009b). As shown in Fig. 3.7, by increasing temperature from 325°C to 400°C, the expected lifetime of membranes decreases from 7.1 to 2. 5 years and, at 450°C, drops to less than a year.

In summary, the above membranes, although not yet tested in a membrane reactor configuration, appear as an interesting candidate for H_2 separation of a syngas mixtures below 400°C.

Due to its intrinsic characteristics, magnetron sputtering is nowadays receiving a lot of attention for the preparation of sulphur-resistant membranes for hydrogen separation. Hydrogen production from coal, for example, from syngas in an IGCC plant, is challenging for many Pd-based membranes due to their limited stability towards sulphur. As shown in Fig. 3.8, even a few ppmv of sulphur in a gas mixture can reduce hydrogen flux drastically, due to strong surface adsorption, and may further lead to degradation of the material caused by formation of bulk Pd_4S according to the reaction.

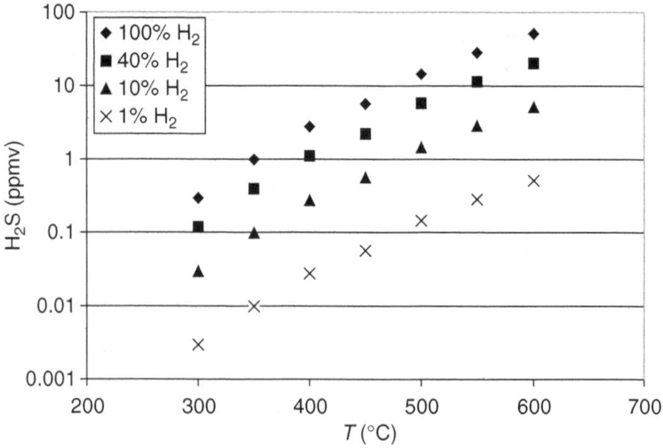

3.8 Calculated concentrations of H_2S in gas mixtures required to transform Pd in Pd_4S, as a function of temperature. (After Mundschau *et al*, 2006.)

Figure 3.9 shows the SEM image of a Pd membrane exposed to H_2S accidentally released from seals during steam reforming of an ethanol mixture at 400°C (Basile *et al*., 2011). The accidental release caused the formation of pinholes and a severe decline of separation properties of the membrane. The presence of H_2S, in a few ppmv, can be also detrimental for Pd–Ag alloys, while FCC Pd–Cu and Pd–Au show some resistance, as first evidenced by McKinley (1969a, 1969b), and more recently confirmed by Kulprathipanja *et al.* (2005), and by Chen and Ma (2010). An extensive and interesting review on sulphur poisoning of hydrogen-permeable metal membranes made of palladium and its alloys was recently published by Gabitto and Tsouris (2009). The authors concluded that poisoning of metal surfaces by sulphur compounds is a very serious problem affecting several metals (Cu, Ni, Pd, Pt and so on) and their alloys, widely used in many industrial processes. In their opinion, the best combination of permeation flux and sulphur resistance is presented by the $Pd_{60}Cu_{40}$ alloy.

In more detail, a lot of effort is currently being devoted to development of sulphur-resistant alloys which can operate in flue gas containing up to 20 ppm of H_2S (NETL, 2008). Recently, atomistic modelling of membrane materials based on density functional theory (DFT) has provided guidelines for the identification of alloys which can exhibit improved resistance to sulphur and, at the same time, higher H_2 fluxes (Kamakoti and Sholl, 2006). Starting from FCC Pd–Cu such as $Pd_{60}Cu_{40}$ and Pd–Ag alloys such as $Pd_{77}Ag_{23}$, the focus is primarily on the development of ternary alloys such as Pd–Cu–TM an Pd–Ag–TM, where TM stands for a transition metal such as Nb, Ta, Y, Ru or Mo. Based on this approach, several projects in which

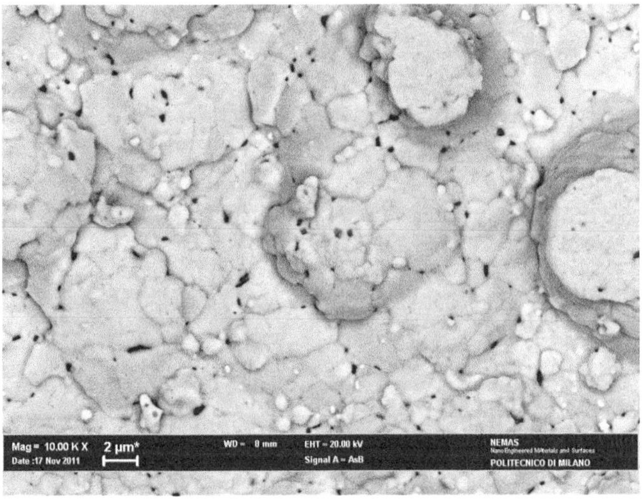

3.9 SEM image of the surface of a Pd membrane poisoned with sulphur.

ternary alloys are manufactured by magnetron sputtering are now running, in early stage, both in USA (Coulter, 2009; Henkel, 2010) and in Europe (Bredesen *et al.*, 2011).

3.3.3 Pore-filled composite membranes

Another interesting and quite recent way for improving membrane performance (mainly stability, but also increasing hydrogen permeating flux and avoiding the contact between catalyst and Pd-layer which is responsible for the formation of pine-hole) takes into consideration the control of Pd-particles deposition by confining them to the nanometre space of the (generally mesoporous) support.

This interesting alternative to the conventional composite membranes described above is also called 'pore-filled' type membrane, because the metal (generally palladium) fills the pores of the support.

An excellent example of such a composite membrane type is represented by the work recently published by Tanaka *et al.* (2008). The starting material used for preparing the novel pore-filled palladium membrane was the YSZ supported on macroporous α-alumina. At this step, palladium particles could be successively deposited as a top layer of the YSZ–γ-Al_2O_3 membrane (Fig. 3.10) or as nano-size pores of ceramic γ-Al_2O_3 (Fig. 3.11d).

In the configuration 'pore-fill membranes', Pd is not directly exposed on the surface and thus, for example, interaction with the catalyst is avoided (and so hot-spots problems are also reduced).

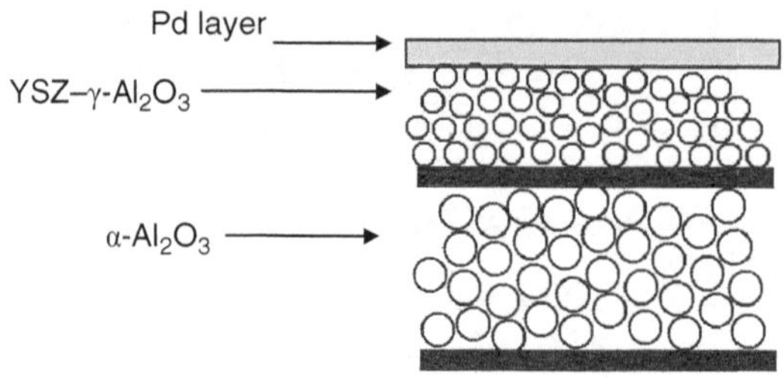

3.10 Pd membrane on the top of a YSZ–γ-Al_2O_3 intermediate layer. (After Tanaka *et al.*, 2008.)

Since the thermal expansion coefficient value of zirconia is between those of palladium and alumina, this configuration suppresses the thermal stress associated with the α–β phase transition of palladium.

Moreover, since also the mismatch of palladium and alumina thermal expansion coefficients is reduced, high temperature delamination and defects formation in the membrane are also reduced. It is also possible to control the palladium thickness (to about 5 μm) by increasing the hydrogen flux and so minimizing the cost of the composite membrane.

In detail, the process of fabrication of 'pore-filled' type Pd–YSZ–β-Al_2O_3 composite membrane supported on α-Al_2O_3 consists of four different steps (Fig. 3.11):

1. Coating of mesoporous YSZ–γ-Al_2O_3 layer (3.2 μm thickness) on the macroporous α-alumina tube (Fig. 3.11a).
2. Activation of YSZ–γ-Al_2O_3 layer by seeding with Pd nuclei (Fig. 3.11b).
3. Coating of an additional YSZ–γ-Al_2O_3 layer as the protection barrier (Fig. 3.11c).
4. Vacuum-assisted electroless plating of Pd at the activated layer (Fig. 3.11d).
5. Calcination under H_2 atmosphere (3 h at 550°C).

Palladium fills the pore networks of the YSZ–γ-Al_2O_3 layer to a thickness of about 5 μm, which is less than 1/5 of that used for a conventional Pd membrane having similar hydrogen permeation.

The Pd–YSZ–γ-Al_2O_3 showed improved thermal stability in long-term operation at $T > 500$°C and, in particular, pure gas permeation experimental results showed a separation factor of H_2/N_2 above 350 at 500°C.

Future work will address the improvement of permeation performance by: (a) fabrication of Pd–Ag alloy pore-filled membranes; (b) modification of the YSZ–γ-Al_2O_3 composition; and (c) plating conditions.

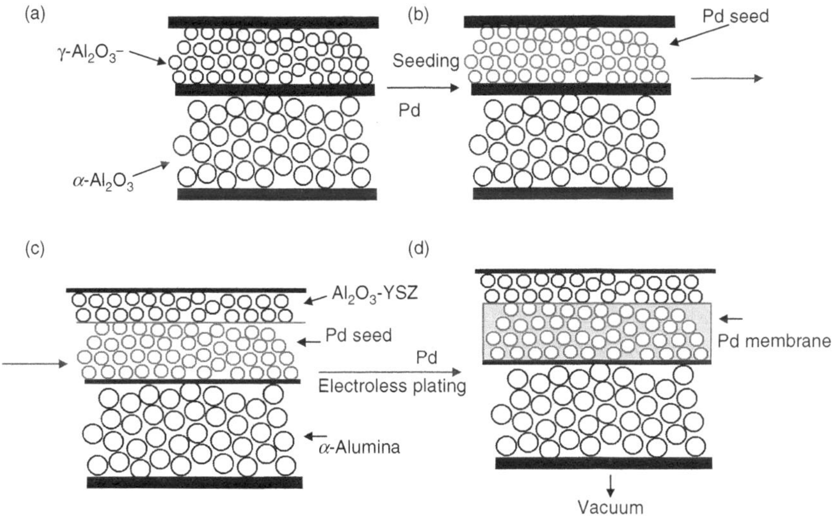

3.11 Process of fabrication of Pd 'pore-filled membranes'. (After Tanaka *et al.*, 2008.)

3.3.4 Palladium-loaded polymeric membranes

The use of a polymer for supporting the preparation of composite membranes with thin selective palladium layers was first established by Gryaznov *et al.* (1969).

As a polymer support, poly(dimethylsiloxane) (PDMS), considered the most highly permeable rubbery polymer, was used. The composite membranes showed 89% conversion of cyclopentadiene at 151°C and 93% selectivity towards cyclopentane. Apart from this high selectivity, another important aspect is related to the preparation of the catalytic membrane, for which the amount of Pd used was 100 times lower as compared with the case of the Pd foil. After this first study, the importance of these catalytic membranes was clear to Gryaznov and co-workers because of the advantages in the potential reduction of both cost and consumption of expensive noble metals. And so began the development of catalytic membranes based on polymer–metal composites. In a successive study, the polymer matrix used was PDMS and other highly permeable poly(organosiloxanes) containing, after preliminary immobilization on silica gel, palladium nanoparticles (Gryaznov *et al.*, 1983).

The reaction carried out was the hydrogenation of cyclopentadiene, for which the maximum conversion of 98.9% was attained, whereas the selectivity towards cyclopentene was 79%.

Follow-up studies in this direction were performed in the two following directions:

(i) dense catalytic membranes; and
(ii) porous catalytic membranes.

Recently, interesting experimental results have been published by various authors and discussed in comprehensive and excellent reviews on membrane catalysis by Vankelecom (2002), Dioos *et al.* (2006), Ozdemir *et al.* (2006) and Buonomenna *et al.* (2011).

A comprehensive survey devoted to the preparation of Pd-loaded on polymeric catalytic membranes for hydrogenation reactions has been published recently by Volkov *et al.* (2011).

3.4 Performances in membrane reactors

In the early 1990s, membranes prepared by electroless plating were used to prove the feasibility of the concept of the membrane reactor in several reactions. Kikuchi (1995) showed that, by using composite Pd and Pd–Ag membranes 5–22.5 μm thick deposited on α-alumina graded supports, methane conversion in steam reforming ranged from 65% to 80% while operating at 500°C, with a steam-to-carbon-ratio of 3 at 1 bar, for example well above the equilibrium thermodynamic values (about 45%). Similar results were found by Shu (1994) with a 10.3 μm thick Pd–Ag composite membrane on a porous stainless steel support. Other reactions, such as WGS (Uemiya *et al.*, 1991c), dehydrogenation of iso-butane (Matsuda *et al.*, 1993) and dehydrocyclization of propane to aromatics (Uemiya *et al.*, 1991d), were also investigated.

Research has, thereafter, focused on membrane development and progressively has addressed problems related to their scale-up and related to specific applications.

Among the more interesting on-going developments, Pd membranes synthesized at Dalian Laboratories (China) should be mentioned. Pd membranes are obtained by electroless plating on graded alumina tubes with 0.1 μm pore size, manufactured by ECN (The Netherland). Prior to deposition of the Pd-layer by electroless plating, the pores of the support tubes were pre-filled with an inorganic gel to obtain a smooth surface and to prevent Pd from entering the pore structures. These pores are reopened after completion of the Pd deposition by decomposing the gel at 500°C in H_2 (Hou *et al.*, 2005). Membranes from 2 to 10 μm thick, about 10 cm long and 10 mm OD, have been first tested in gas mixtures simulating both WGS and reformate composition (Jiang *et al.*, 2008). Membranes have been then scaled up: three Pd membranes, 44 cm long, have been tested in a parallel configuration in a process-development unit built at ECN (Dijkstra *et al.*, 2009)

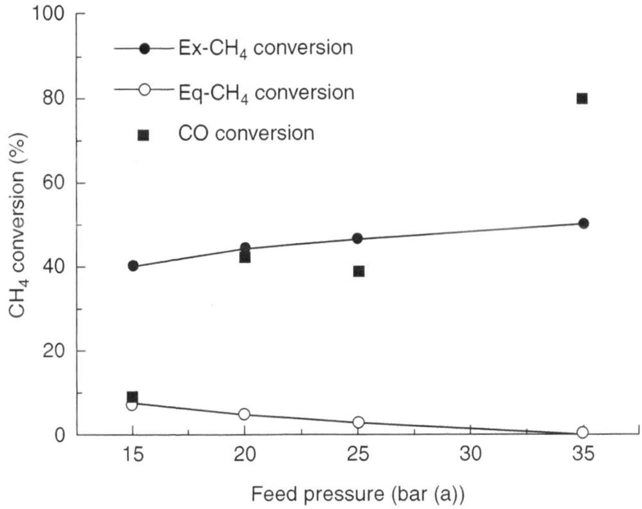

3.12 Feed pressure dependence of CH_4 conversion in a bench-scale tubular membrane reformer. Ex stands for experimental and Eq for equilibrium. (After Li *et al.*, 2011.)

with a stream simulating a shifted gas mixture of a reformate gas (Li *et al.*, 2010); membranes were fitted in the test section with proprietary graphite compression seals (Rusting *et al.*, 2001). High H_2 permeance, for example, 8.0 $\times 10^{-7}$ mol/m^2 s Pa, was measured at 400°C at 15 bar; however, at the end of the tests (23 days) the ideal selectivity of H_2/Ne decreased from 4020 to 330, possibly due to leakages through the membrane seals. Eight composite membranes of the same type, 10.9–13.8 μm thick, were used to carry out the methane steam reforming reaction at 550°C with a pressure up to 35 bar (Li *et al.*, 2011). The experiment was carried out with a feed gas simulating both a classical reforming mixture (steam/CH_4 ratio = 3) and a pre-reformed mixture in a natural gas combined cycle (NGCC) power plan with a steam/CH_4 ratio of 3.2 (0.2% CO, 3.5% CO_2, 14.6% H_2, 19.4% CH_4 and 62.2% H_2O). As shown in Fig. 3.12, despite unfavourable thermodynamics, CH_4 conversion increases with increasing feed pressure and remains well above the corresponding thermodynamic equilibrium values. The maximum CH_4 conversion and H_2 production rate were 73.4% and 1.3 Nm^3/h, respectively, at a feed/permeate pressure of 35/5 bar. Stable performance was found both for the membranes and the membrane reactor for a period of about 30 days. For comparison, performances obtained by other researchers with Pd membranes deposited by electroless plating, on various supports, are shown in Table 3.6.

The WGS reaction has been extensively studied, as reported by Basile *et al.* (2010) in an exhaustive literature review; experiment has been initially focused on WGS at low temperatures, for example, 250–350°C *et al.*, 1996),

Table 3.6 Results of steam reforming tests in membrane reactors obtained in laboratory and bench-scale tests

Membrane type	Thickness (μm)	Operating conditions			Catalyst	CH_4 conversion (%)	Reference
		T (°C)	*P* (bar)	S/C			
Pd/Al_2O_3	8	500	1	3	Ni	83.4	Kikuchi *et al.* (2000)
Pd/Al_2O_3	4	550	9	3	Ni-La/ Mg–Al	86	Chen *et al.* (2008)
Pd/PSS	6	500	3	3	Ni/Al_2O_3	65	Tong *et al.* (2005b)
Pd/PSS	10.3	500	1.36	3	Ni/Al_2O_3	63	Shu *et al.* (1994)
Pd/Al_2O_3	10.9–13.8	550	35	3.2	Ni-based	73.4	Li *et al.* (2011)

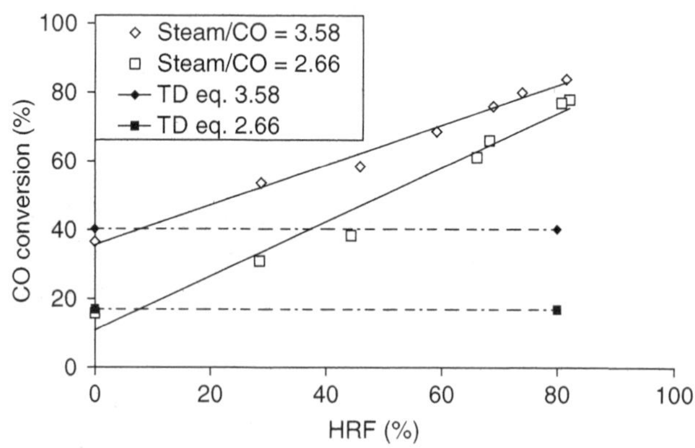

3.13 CO conversion as a function of hydrogen recovery factor (HRF) in WGS tests with a Pd membrane reactor: comparison with thermodynamic (TD) equilibrium values. (After Pinacci *et al.*, 2010.)

(Basile *et al.*, 1996; Arstad *et al.*, 2006; Pinacci *et al.*, 2007) then, more recently, on higher temperatures, 400–450°C, once the stability of Pd membranes has been sufficiently proved.

As an example, Fig. 3.13 shows the results of WGS tests carried out at 410°C in a Pd membrane reactor with gas mixtures simulating a syngas produced in IGCC plants (Pinacci *et al.*, 2010). A composite palladium-porous stainless steel membrane, 29 μm thick, obtained by electroless plating on a porous stainless steel support, has been used for this purpose. The reactor was fed with a shift gas mixture with a 7.6% CO concentration and H_2O/CO ratio of 2.7 and 3.6, respectively. CO conversions up to the 85.0% and 78%, respectively, have been reached, while operating at a feed pressure of 6 bar. These values can be compared with the corresponding conversions obtained with a conventional shift reactor, 36.6% and 15.8%, respectively, and are well above maximum thermodynamic equilibrium values (dotted line in the figure).

WGS tests have been also performed with gas mixtures simulating a syngas produced by autothermal reforming of natural gas (Bi *et al.*, 2009). The performances of two different catalysts (Fe–Cr vs. $Pt/Ce_{0.6}Zr_{0.4}O_2$) have been evaluated in a wide range of operating conditions, in a membrane reactor equipped with a 1.4 μm thick Pd membrane on a ceramic support. CO conversion remained above thermodynamic equilibrium up to feed space velocities of 9100 L kg^{-1} h^{-1} at 350°C, pressure 12 bar and steam-to-carbon ratio S/C = 3.

It is interesting to note that, in both the above studies, attention has been focused on specific phenomena which can decrease H_2 flux through the membrane, such as concentration polarization in the boundary layer and competitive adsorption of other gases (namely CO and CO_2) on the membrane surface; these aspects are currently objects of specific studies (Li *et al.*, 2000; Unemoto *et al.*, 2007).

More recently, extensive WGS tests were conducted by Augustine *et al.* (2011) with Pd membranes, 10 μm thick, prepared at the Worchester Polytechnic Institute by electroless plating on porous Inconel supports modified by an alumina (Ma and Guazzone, 2006) and a Pd/Ag anti-diffusion barrier (Ayturk *et al.*, 2006). Experiments were conducted in the 350–500°C temperature range by increasing feed pressure up to 14.4 bar; both a CO/steam feed and a syngas feed were used and the effects of steam/CO ratio and of space velocity on the reactor performance were studied. As shown in Fig. 3.14, CO conversions as high as 98% were achieved by operating at a feed pressure of 14 bar, while simultaneously recovering up to 88% of the H_2, at a temperature of 450°C. However membranes operated in the 350–500°C, up to 1000 h, showed a decline in selectivity due the formation of pinholes according to mechanism described by Guazzone and Ma (2008).

Pd and Pd-alloy membranes developed at Worchester Polytechnic Institute have been the object of a technology spin-off to CRI/Criterion, a company controlled by Shell. CRI has produced Pd and Pd–Ag membranes, 5–12 μm thick, on a porous Inconel support with a bi-metallic anti-diffusion barrier layer, as much as two inches OD and up to 48 inches long (Engwall *et al.*, 2009). These membranes have been tested in separations tests at temperatures of 300–500°C and differential pressures of 26–42 bar, for periods exceeding 2000 h, obtaining stable hydrogen flux and selectivity; hydrogen permeance in the range of 50–70 Nm^3/m^2 h $bar^{0.5}$ and hydrogen purity exceeding 99% have been reported. Steam reforming experiments with a 1 inch OD, 6 inches long Pd membrane have been run at 500°C and 30 bar in an electrically heated shell and tube reactor, with a commercial catalyst packed in the annular chamber between membrane and reactor. Membrane performances were quite stable for a period of 51 days, and

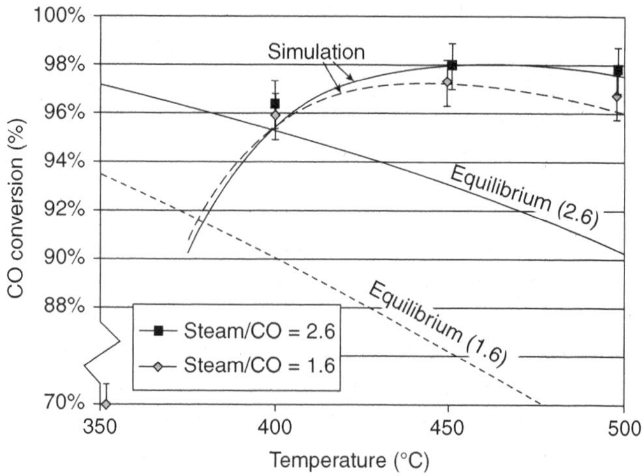

3.14 CO conversion as a function of temperature in WGS tests with a Pd membrane reactor at 14.4 bar. (Reprinted from Augustine *et al.*, 2011. Copyright (2011) with permission from Elsevier.)

resulted in conversion of about 94% and hydrogen purity exceeding 98% (Engwall *et al.*, 2007).

3.5 Conclusions and future trends

Development of palladium composite membranes began in the early 1990s. Since then several types of membranes, obtained by various deposition processes, have been successfully tested under simulated power plant conditions in laboratory-scale experiments; reactions in membrane reactor configuration have also been studied. Among the deposition processes, electroless plating and magnetron sputtering have progressively grown in importance; nowadays, several projects for field-testing of prototype membranes are running or have been planned. Some important R&D issues, however, are still open, including materials' long-term stability in the presence of a catalyst, poisoning by CO and/or sulphur, cost reduction by developing new alloys. Furthermore, mass manufacture of membranes and efficient module design, aimed to keep costs as low as possible, are issues on which R&D is still needed before commencing the technology demonstration phase.

3.6 Acknowledgements

The authors acknowledge the Electrochemistry and Membrane Laboratory of RSE for permission to reproduce membrane micrographs and photographs of equipment.

3.7 References

Armor J. N., Membrane catalysis: where is it, what needs to be done? *Catal. Today*, **25** (1995) 199–207.

Arstad B., Venvik H., Klette H., Walmsley J. C., Tucho W. M., Holmestad R., Holmen H. and Bredesen R., Studies of self-supported 1.6 μm Pd/23 wt.% Ag membranes during and after hydrogen production in a catalytic membrane reactor, *Catal. Today*, **118** (2006) 63–72.

Augustine A., Ma Y. H. and Kazantzis N. K., High pressure palladium membrane reactor for the high temperature water-gas shift reactor, *Int. J. Hydrogen En.*, **36** (2011) 5350–5360.

Ayturk M. E., Mardilovich I. P., Engwall E. E. and Ma Y. H., Synthesis of composite Pd-porous stainless steels (PSS) membranes with a Pd-Ag intermetallic diffusion barrier, *J. Membrane Sci.*, **285** (2006) 385–394.

Azzurri F., Bottino A., Broglia M., Capannelli G., Comite A., Drago F., Pinacci P. and Scrignari M., Preparation of palladium membranes with ceramic barrier layer, *Proc. of ICCMR 11*, St Petersburgh (Ru), June 20–24, 2011.

Balachandrian U., Lee T. H., Chen L., Song S. J. and Dorris S. E., Dense Ceramic-Metal Composite Membranes for Hydrogen Separation, *Proc. 9th Int. Conf. on Inorganic membranes*, Lillehammer (Norway), June 25–29, 2006.

Basile A., Criscuoli A., Santella A. and Drioli E., Membrane reactor for water gas shift reaction, *Gas Sep. Pur. Tech.*, **10** (1996) 243–254.

Basile A., Pinacci P., Tosti S., De Falco M., Evangelisti C., Longo T., Liguori S. and Iulianelli A., Water gas shift reaction in Pd-based Membrane Reactors, *Advances Sci. Technol.*, **72** (2010) 99–104.

Basile A., Pinacci P., Iulianelli A., Broglia M., Drago F., Liguori S., Longo T. and Calabrò V., Ethanol steam reforming reaction in a porous stainless steel supported palladium membrane reactor, *Int. J. Hydrogen En.*, **36** (2011) 2029–2037.

Bi Y., Xu H., Li W. and Goldbach A., Water-gas shift reaction in a Pd membrane reactor over $Pt/Ce_{0.6}Zr_{0.4}O_2$ catalyst, *Int. J. Hydrogen En.*, **34** (2009) 2965–2971.

Bottino A., Capannelli G., Comite A., Di Felice R., Broglia M. and Pinacci P., Pd and Pd/Ag composite membranes prepared by the electroless plating technique: preparation, characterization and interpretation of the H_2 transport mechanism, *Proc. of ICIM 9*, Lillehammer (NO), June 25–29, 2006.

Broglia M., Pinacci P., Radaelli M., Bottino A., Capannelli G., Comite A., Vanacore G. and Zani M., Synthesis and characterization of Pd membranes on alumina modified stainless steel supports, *Desalination*, **245** (2009) 508–515.

Broglia M., Pinacci P. and Basile A., Membranes prepared via Electroless Plating, in A. Basile and F. Gallucci (Eds), *Membranes for Membrane Reactors: Preparation, Optimization and Selection*, John Wiley & Sons, Chichester, UK, (2011) 289–312.

Bredesen R. and Klette H., *Methods of manufacturing thin metals membranes*, (2000), US Patent 6,086,729.

Bredesen R., Jordal K. and Bolland O., High temperature membranes in power generation with CO_2 capture, *Chem. Eng. and Process.*, **43** (2004) 1129–1158.

Bredesen R., Peters T. A., Kaleta T., Stange M. and Lovvik O. M., Performance of Pd-alloy membranes in water gas shift mixtures containing sulphur, *Proc. of ICCMR 10*, St. Petersburgh (Ru), June 20–24, 2011.

Buonomenna M. G., Choi S.-H., Galiano F. and Drioli E., Membranes Prepared via phase inversion, in A. Basile, F. Gallucci (Eds), *Membranes for Membrane Reactors: Preparation, Optimization and Selection*, John Wiley & Sons, Chichester, UK, (2011) 475–489.

Cecchetto R., Brusa R. S., Miotello A. and Basile A., 'PVD Techniques for Metallic Membrane Reactors', in A. Basile and F. Gallucci (Eds), *Membranes for Membrane Reactors: Preparation, Optimization and Selection*, John Wiley & Sons, Chichester, UK, (2011) 289–312.

Chen C. H., Ma Y. H., The effect of H_2S on the performance of Pd and Pd/Au composite membrane, *J. Membrane Sci.*, **362** (2010) 535–544.

Chen Y. Z., Wang Y. Z., Xu H. Y. and Xiong G. X., Hydrogen production capacity of membrane reformer for methane steam reforming near practical working conditions, *J. Membrane Sci.*, **322** (2008) 453–459.

Cheng Y. S. and Yeung K. L., Palladium-silver composite membranes by electroless plating technique, *J. Membrane Sci.*, **158** (1999) 127–141.

Cheng Y. S. and Yeung K. L., Effects of electroless plating chemistry on the synthesis of palladium membranes, *J. Membrane Sci.*, **182** (2001) 195–203.

Chiesa P., Manzolini G., Vigano F., Savoldelli P., Pinacci P. and Mazzocchi L., CO_2 capture in IGCC power plants by means of Pd-Ag membranes, *Proc. of GHGT8*, Trondheim (NO), June 18–23, 2006.

Collins J. P. and Way D., Preparation and characterization of composite palladium-ceramic membrane, *Ind. Eng. Chem. Res.*, **32** (1993) 3006–3013.

Coulter K., High permeability ternary palladium alloy membranes with improved sulphur and halide tolerance, *Proc. of DOE Hydrogen Program Annual Merit Review*, Arlington (USA) May 22, 2009.

Damle A., High performance, durable palladium ally membranes for hydrogen separation and purification, *Proc. of DOE Merit*, Washington (USA), May 9–13, 2011.

Dijkstra J. W., van Delft Y. C., Jansen D. and Pex P., Development of hydrogen membrane reactor for power production with pre-combustion decarbonisation, *Proc. of GHGT 8*, Trondheim (NO), June 18–23, 2006.

Dijkstra J. W., Jansen D., Van den Brink R. W. and Peters T. A., Development of Hydrogen Membrane Reactors for CO_2 Capture, in Eide L. I. (Ed.), *Carbon Dioxide Capture for Storage in Deep Geological Formations – Results from the CO_2 Capture project*, CPL Press, 3 (2009) 121–134.

Dioos B. M. L., Vankelecom I. F. J. and Jacobs P. A., Aspects of immobilisation of catalysts on polymeric supports, *Adv. Synth. Catal.*, **348** (2006) 1413.

DOE/NETL advanced carbon dioxide capture R&D program: Technology update, Sept. 2010.

Dolan M. D., Non-Pd BCC alloy membranes for industrial hydrogen separation, *J. Membrane Sci.*, **362** (2010) 12–28.

Engwall E., Saukaitis J., Komplin G., Joshi M., Del Paggio A. and Nijmeijer A., Reducing CO_2 footprint with hydrogen separation membranes, *Proc. of Euromembrane 2009*, Montpellier (Fr), Sept. 6–10, 2009.

Engwall E., Saukaitis J., Joshi M., Matzakos A. and Del Paggio A., Hydrogen production with a membrane steam reformer, *Proc. of AIChE national meeting*, Nov. 2007.

Evenson C., Scale-up of hydrogen transport membranes for IGCC and Future Gen plants, *Proc. of DOE Merit*, Washington (USA), May 9–13, 2011

Flanagan, T. B., Gross G. and Clewley J. D., Absorption and diffusion of hydrogen in palladium/iron alloys, *Proc. of the 2nd international congress on Hydrogen in metals*, Paris (Fr), June 6–10, 1977.

Gao H., Li Y., Lin J. Y. S. and Zhang B., Characterization of zirconia modified porous stainless steel supports for Pd membranes, *J. Porous Mater.*, **13** (2006) 419.

Gabitto J. F. and Tsouris C., Sulfur poisoning of metal membranes for hydrogen separation, *Int. Review of Chemical Eng. (IRECHE)*, **1**(5) (2009) 394–411.

Gryaznov V.M., Smirnov V.S., Vdovin V.M., Ermilova M.M., Gogua L.D., Pritula N.A. and Litvinov I.A., (1969), *Method of preparing a hydrogen-permeable membrane catalyst on a base of palladium or its alloys for the hydrogenation of unsaturated organic compounds*, US Patent 4,132,668.

Gryaznov V. M., Smirnov V. S., Vdovin V. M., Ermilova M. M., Gogua L. D., Pritula N. A. and Fedorova G. K., (1983), *Membrane catalyst for hydrogenation of organic compounds and method for preparing same*, US Patent 4,394,294.

Gryaznov V. M., Metal containing membranes for the production of ultrapure hydrogen and the recovery of hydrogen isotopes, *Sep. & Pur. Meth.*, **29** (2000) 171–187.

Guazzone F., Engwall E. E. and Ma Y. H., Effect of surface activity, defects and mass transfer on hydrogen permeance and *n*-value in composite palladium-porous stainless steel membranes, *Catal. Today*, **118** (2006) 24–31.

Guazzone F. and Ma Y.H., Leak growth mechanism in composite palladium membranes prepared via electroless deposition method, *AICHE J.*, **54** (2008) 487–494.

Henkel D., Combinatorial design of Pd ternary alloys for sulphur/carbon tolerant hydrogen separation, *Proc. of 2010 NETL CO_2 Capture Technology Meeting*, Pittsburgh (USA), Sept. 10, 2010.

Hou S. F., Jiang K., Li W. Z., Xu H. Y. and Yuan L. X. (2005), *A metal palladium composite membrane or alloy palladium composite membrane and their preparation methods*, WO Patent 2005/065,806.

Huang T. C., Wei M. C. and Chen H. Y., Preparation of hydrogen-permselective palladium-silver alloy composite membranes by elctroless co-deposition, *Sep & Pur. Techol.*, **32** (2003) 239–245.

Itoh N., Akiha T. and Sato T., Preparation of thin palladium composite membrane tube by a CVD technique and its hydrogen permselectivity, *Catal. Today*, **104** (2005) 231.

Jiang K., Goldbach A., Bao C., Yu C.Y. and Xu H.Y., *Proc. of 10th Int. Conf. on Inorganic Membranes ICIM 10*, Tokyo (Jp), (2008) paper 1a07.

Kamakoti P. and Sholl D.S., Towards first principles-based identification of ternary alloys for hydrogen purification membranes, *J. Membrane Sci.*, **279** (2006) 94.

Kikuchi E. and Uemiya S., Preparation of supported thin palladium-silver alloy membranes and their characteristics for hydrogen separation, *Gas Sep. & Purif.*, **5** (1991) 261–266.

Kikuchi E., Palladium/ceramic membranes for selective hydrogen permeation and their application to membrane reactors, *Catal. Today*, **25** (1995) 333–337.

Kikuchi E., Nemoto Y., Kajivara M., Uemiya S. and Kojima T., Steam reforming of methane in membrane reactors: comparison of electroless plating and CVD membranes and catalyst packaging modes, *Catal. Today*, **56** (2000) 75–81.

Klette H., Bredesen R. and Slotfeldt-Ellingsen D., 2008, *Leak-proof membrane element and method of manufacturing such an element*, International Patent application WOPatent2008/039,080.

Knapton A. G., Palladium alloys for hydrogen diffusion membranes, *Platinum Metals Rev.*, **21** (1977) 44–50.

Kougionas V., The SET PLAN CCS- research perspective, *Proc. of Fenco Conference*, Bruxelles (Be), Nov. 3–4, 2010.

Kreutz T. G., Williams R. H., Socolow R. H., Chiesa P. and Lozza G., Production of hydrogen and electricity from coal with CO_2 capture, *Proc. of GHGT-6*, Kyoto (Jp), October 1-4, 2002.

Kulprethipanja A., Alptekin G. O., Falconer J. L. and Way J.D., Pd and Pd-Cu membranes: inhibition of H_2 permeation by H_2S, *J. Membrane Sci.*, **254** (2005) 49–62.

Li A., Liang W. and Hughes R., Repair of a Pd/alpha-Al_2O_3 composite membrane containing defects, *Sep. & Purif. Technol.*, **15** (1999) 113–119.

Li A., Liang W. and Hughes R., The effect of carbon monoxide and steam on the hydrogen permeability of a Pd/stainless steel membrane, *J. Membrane Sci.*, **165** (2000) 135–141.

Li H., Xu H. and Li W., Study of *n* value and α/β palladium hydride phase transition within the ultra-thin palladium composite membrane, *J. Membrane Sci.*, **324** (2008) 44–49.

Li H., Dijistra J. W., Pieterse J.A.Z., Boon J., van den Brink R.W. and Jansen D., Towards full scale demonstration of hydrogen-selective membranes for CO_2 capture: Inhibition effect of WGS components on the H_2 permeation through three Pd membranes of 44 cm long, *J. Membrane Sci.*, **363** (2010) 204–211.

Li H., Pieterse J. J.A.Z, Dijistra W., Haije W.G., Xu H.Y., Bao C., van den Brink R.W. and Jansen D., Performance tests of a bench-scale multi-tubular membrane reformed, *J. Membrane Sci.*, **373** (2011) 43–52.

Liu Y., Wang J., Du Y., Zhang L. and Liang D., Mobilities and diffusivities in fcc Fe-X (X = Ag, Au, Cu, Pd and Pt) alloys, *CALPHAD: Computer Coupling of Phase Diagrams and Thermochemistry*, **34** (2010) 253–262.

Ma Y. H., Mardilovich P.P. and She Y. (2000), *Hydrogen gas-extraction module and method of fabrication*, US Patent 6,152,987.

Ma Y. H., Akis B. C., Ayturk M. E., Guazzone F., Engwall E. E. and Mardilovich I. P., Characterization of intermetallic diffusion barrier and alloy formation for Pd/Cu and Pd/Ag porous stainless steel composite membranes, *Ind. Eng. Chem. Res.*, **43** (2004) 2936–2945.

Ma Y. H., Pd-based hydrogen separation membranes – status and prospective, *Proc. of 9th Int. Conf. on Inorganic Membranes*, Lillehammer (No), June 25–29, 2006.

Ma Y.H. and Guazzone F. (2006), *Composite gas separation modules having a layer of particles with a uniform binder distribution*, US Patent 7,172,644.

Mallory G. O. and Hajdu J. B., *Electroless plating – fundamentals and applications*, William Andrew Publishing/Noyes, New York, 1990.

Mardilovich I. P., Engwall E. and Ma Y. H., Dependence of hydrogen flux on the pore size and plating surface topology of asymmetric Pd-porous stainless steel membranes, *Desalination*, **144** (2002) 85.

Matsuda T., Koike I., Kubo N. and Kikuchi E., Dehydrogenation of isobutene to isobutene in a palladium membrane reactor, *Appl. Catal.*, **96** (1993) 3.

Mc Cool B., Xomeritakis G. and Lin Y. S., Composition control and hydrogen permeation characteristics of sputter deposited palladium silver membranes, *J. Membrane Sci.*, **161** (1999) 67–76.

McKinley D.L., (1969a), *Metal alloy for hydrogen separation and purification*, US Patent 3,350,845.

McKinley D.L., (1969b), *Method for hydrogen separation and purification*, US Patent 3,439,474.

Mejdell A.L., Peters T., Stange M., Venvik H.J. and Bredesen R., Performance and application of thin Pd-alloy hydrogen separation membranes in different configurations, *J. Taiwan Institute of Chem. Eng.*, **40** (2009) 253–259.

Middleton P., Solgaard-Andersen H. and Rostup Nielsen H.T., Hydrogen production with CO_2 capture using membrane reactors, *Proc. of GHGT-6, Kyoto (Jp)*, October 1–4, 2002.

Middleton A., Hurst P., Walker P. and Grace P., Pre-combustion decarbonisation hydrogen membrane study, in D. C. Thomas, S. M. Benson *Carbon Dioxide Capture for Storage in Deep Geological Formations*, Vol. 1, Elsevier, London, (2005) 409–425.

Miracca I., Introduction to the CCP2 capture technology portfolio, in Eide L.I. (Ed), *Carbon Dioxide Capture for Storage in Deep Geological Formations – Results from the CO_2 Capture project*, CPL Press, Newbury, Berkshire, UK, 3 (2009) 135–156.

Morooka S., Yan S., Aoki K., Yokoyama S. and Kusakabe K., Palladium membrane formed in macropores of support tube by chemical vapour deposition with crossflow through a porous wall, *Sep. Sci. Technol.*, **30** (1995) 2877–2889.

Morreale B. D., Ciocco M. V., Enick R. M., Morsi B. I., Howard B. H., Cugini A. V. and Rothenberger K. S., The permeability of hydrogen in bulk palladium at elevated temperatures and pressures, *J. Membrane Sci.*, **212** (2003) 87–97.

Mundschau M. V., Xie X., Everson IV C. R. and Sammels A. F., Dense inorganic membranes for production of hydrogen from and coal with carbon dioxide sequestration, *Catal. Today*, **118** (2006) 12–23.

Nam S. E. and Lee K. H., Hydrogen separation by palladium alloy composites membranes: introduction of a diffusion barrier, *J. Membrane Sci.*, **192** (2001) 1977.

NETL test protocol, testing of hydrogen separation membranes, DOE/NETL – 2008/1335, 2008.

Ozdemir S. S., Buonomenna M. G. and Drioli E., Catalytic polymeric membranes: preparation and application, *Appl. Catal. A: General*, **307** (2006), 167.

Peters T. A., Stange M., Klette H. and Bredesen R., High pressure performance of thin Pd-23%Ag/stainless steel composite membranes in water gas shift mixture; influence of dilution, mass transfer and surface effects on hydrogen flux, *J. Membrane Sci.*, **316** (2008) 119–127.

Peters T. A., Tucho W. M., Ramachandran A., Stange M., Walmsley J. C., Holmestad R., Borg A. and Bredesen R., Thin Pd-23%Ag/stailess steel composite membranes: Long-term stability, life-time estimation and post-process characterization, *J. Membrane Sci.*, **326** (2009a) 572–581.

Peters T., Stange M. and Bredesen R., Development of Thin Pd-23% Ag/Stainless Steel Composite Membranes for Application in Water Gas Shift Membrane Reactors, in Eide L.I. (Ed), *Carbon Dioxide Capture for Storage in Deep Geological Formations – Results from the CO_2 Capture project*, CPL Press, 3 (2009b) 135–156.

Pinacci P., Broglia M., Radaelli M., Bottino A., Capannelli G. and Comite A., Development of Palladium Alloy Membranes for Hydrogen Separation from Synthesis Gas in Membrane Reactors, *Proc. of 3rd Int. Conf. on Clean Coal Technologies for our Future*, Cagliari (I), May 15–17, 2007.

Pinacci P., Broglia M., Valli C., Capannelli G. and Comite A., Evaluation of the water gas shift reaction in a palladium membrane reactor, *Catal. Today*, **156** (2010) 165–172.

Pinacci P., Broglia M. and Drago F., Development of Pd–composite membranes for hydrogen production in membrane reactors, *Proc. of EFC 2011*, Rome (I), December 14–16, 2011.

Roa F., Way J. D., McCormick R. L. and Paglieri S., Preparation and characterization of Pd-Cu composite membranes for hydrogen separation, *Chem. Eng. J.*, **93** (2003) 11–22.

Rothenberger K. S., Cugini A. V., Howard B. H., Killemeyer R. P., Ciocco M. V., Morreale B. D., Enick R. M., Bustamante F., Mardilovich I. P. and Ma Y. H., High pressure hydrogen permeance of porous stainless steel coated with a thin palladium film via electroless plating, *J. Membrane Sci.*, **244** (2004) 55–68.

Rusting F. T., Jong G. D., Pex P. P. A. C. and Peters J. A. J., 2001. *Sealing socket and method for arranging a sealing socket to a tube*, International Patent WOPatent01/63162A1.

Samingptai S., Tantayanon S. and Ma Y. H., Chromium oxide intermetallic diffusion barrier for palladium membranes supported on stainless steel, *J. Membrane Sci.*, **347** (2010) 8–16.

Saracco G. and Specchia V., Catalytic inorganic membrane reactors: present experience and future opprtunities, *Catal. Rev. Sci. Eng.*, **36** (1994) 305–384.

Serra J. M., Büchler O., Meulenberg W. A. and Buchkremer H. P., Thin $BaCe_{0.8}Gd_{0.2}O_3$-δ protonic electrolytes on porous $Ce_{0.8}Gd_{0.2}O_{1.9}$-Ni substrates, *J. Electrochem. Soc.*, **154**(3) (2007) B334–B340.

Shu J., Grandjean B. P. A., Ghali E. and Kaliaguine S., Simultaneous deposition of Pd and Ag on porous stainless steel by electroless plating, *J. Membrane Sci.*, **77** (1993) 181–195.

Shu J., Grandjean B.P. and Kaliaguine S., Methane steam reforming in asymmetric Pd and Pd-Ag /porous SS membrane reactors, *Appl. Cat. A: Gen.*, **119** (1994) 305–325.

Shu J., Adnot A., Grandjean B. P. A. and Kaliaguine S., Structurally stable composite Pd-Ag alloy membranes: Introduction of a diffusion barrier, *Thin Solid Films*, **286** (1996): 72–79.

Singh R. P., Powers T., Rekciz K., Opkins S. and Love C., Supports for high temperature gas separation membranes, *Proc. of ICIM9*, Lillehammer (No), June 25–29, 2006.

Tanaka D. A. P., Llosa Tanco M. A., Okazaki J., Wakui Y., Mizukami F. and Suzuki T. M., Preparation of "pore-fill" type Pd–YSZ–γ-Al_2O_3 composite membrane supported on α-Al_2O_3 tube for hydrogen separation, *J. Membrane Sci.*, **320** (2008) 436–441.

Tong J., Suda H., Haraya K. and Matsumura Y., A novel method for the preparation of thin dense Pd membrane on macroporous stainless steel tube filter, *J. Membrane Sci.*, **260** (2005a) 10.

Tong J. H., Matsumura Y., Suda H. and Haraya K., Experimental study of steam reforming of methane in a thin (6 μm) Pd-based membrane reactor, *Ind. Eng. Chem. Res.*, **44** (2005b) 1454–1465.

Tong J., Su C., Kuraoka K., Suda H. and Matsumura Y., Preparation of thin Pd membrane on CeO_2-modified porous metal by a combined method of electroless plating and chemical vapour deposition, *J. Membrane Sci.*, **269** (2006) 101.

Uemiya S., Sato N., Ando H., Kude Y., Matsuda T. and Kikuchi N., Separation of Hydrogen through Palladium thin film supported on a porous glass tube, *J. Membrane Sci.*, **56** (1991a) 303–313.

Uemiya S., Matsuda T. and N. Kikuchi., Hydrogen permeable palladium silver alloy membrane supported on porous ceramic, *J. Membrane Sci.*, **56** (1991b) 315–325.

Uemiya S., Sato N., Ando H. and Kikuchi. E., The water gas shift reaction assisted by a palladium membrane reactor, *Ind. Eng. Chem. Res.*, **30** (1991c) 585–589.

Uemiya S., Koike I., and Kikuchi E., Promotion of the conversion of propane to aromatics by use of a palladium membrane, *Appl. Catal.*, **76** (1991d), 171–181.

Uemiya S., State-of-the-art of supported metal membranes for gas separation, *Sep. & Purif. Meth.*, **28**(1) (1999), 51–85.

Unemoto A., Kaimai A., Sato K., Otake T., Yashiro K., Mizusaki J., Kawada T., Tsuneki T., Shirasaki Y. and Yasuda I., The effect of co-existing gases from the process of steam reforming reaction on hydrogen permeability of palladium alloy membrane at high temperatures, *Int. J. Hydrogen En.*, **32** (2007) 2881–2887.

Vankelecom I. F. J., Polymeric membranes in catalytic reactors, *Chem. Rev.*, **102** (2002), 3779.

Vigeland B. and Aasen N., Development of an hydrogen membrane for HMR process concept, *Proc. of the GHGT8*, Trondheim (NO) June 19–22, 2006.

Volkov V. V., Petrova I. V., Lebedeva V. I., Roldughin V. I. and Tereshchenko G.F. (2011), Palladium-loaded polymeric membranes for hydrogenation in catalytic membrane reactors, in A. Basile and F. Gallucci (Eds), *Membranes for Membrane Reactors: Preparation, Optimization and Selection*, John Wiley & Sons Ltd, Chichester, UK, 531–548.

Ward T. L. and Dao T., Model of hydrogen permeation behavior in palladium membrane, *J. Membrane Sci.*, **153** (1999) 221–231.

Zhang Y., Gwak. J., Murakoshi Y., Ikehara T., Maeda R. and Nishimura C., Hydrogen permeation characteristics of thin Pd membrane prepared by microfabrication technology, *J. Membrane Sci.*, **277** (2006) 203.

Zhao H. B., Xyong G. X. and Baron G. V., Preparation and characterization of palladium-based composite membranes by electroless plating and magnetron sputtering, *Catal. Today*, **56** (2000) 89–96.

3.8 Appendix: nomenclature

3.8.1 Notation

E_a apparent activation energy
F Faraday constant
J_{H_2} hydrogen flux through the membrane
m factor of membrane thickness reduction
n dependence factor of the hydrogen flux on the hydrogen partial pressure
Pe permeability
Pe_0 pre-exponential factor

p_{H_2}	hydrogen partial pressure
p_f	hydrogen partial pressure on the feed side
p_p	hydrogen partial pressure on the permeate side
R	ideal gas constant
T	temperature
ϕ_{pore}	pore diameter
δ	membrane thickness
σ_i	ionic conductivity of material
σ_e	electronic conductivity of material

3.8.2 Abbreviations

AISI 316L	stainless steel (Cr/Ni = 18/10 %)
AISI 310SC	stainless steel (Cr/Ni = 25/20 %)
BCC	body centred cubic
CCP	CO_2 capture project
DFT	density functional theory
DOE	Department of Energy
ECN	Energy Research Centre of Netherlands
FCC	face centred cubic
IGCC	integrated gasification combined cycle
NETL	National Energy Technology Laboratory
NGCC	natural gas combined cycle
OD	outer diameter
PDMS	polydimethylsiloxane
R&D	research and development
SEM	scanning electron microscopy
TM	transition metal
WGS	water-gas shift
YSZ	yttria-stabilized zirconia

4

Alternatives to palladium in membranes for hydrogen separation: nickel, niobium and vanadium alloys, ceramic supports for metal alloys and porous glass membranes

A. SANTUCCI and S. TOSTI, ENEA, Italy
and A. BASILE, ITM-CNR, Italy

DOI: 10.1533/9780857097330.1.183

Abstract: Palladium-based membranes have been widely studied in terms of their use in hydrogen separation processes. The high permeability and selectivity of these membranes mean that they are suitable for a variety of applications, although these are limited by the high cost of the precious metal. In order to reduce the cost, several researchers have considered the use of thin metal film membranes, as well as low-cost metals and their alloys, as an alternative to Pd. This chapter focuses on the development of membranes based on metals, such as Ni, Nb, V and Ti, which are promising substitutes for Pd alloys. The main issues related to the synthesis of these membranes and the effect of alloying on their chemical and physical properties are described. Properties relating to hydrogen solubility and permeability, as well as embrittlement under hydrogenation cycling, are also reported. Finally, this chapter discusses ceramic and glass porous membranes. Ceramic porous membranes are examined in terms of their applications as support for new metal alloys, while the use of glass porous membranes in gas separation and in membrane reactors is treated in detail.

Key words: hydrogen separation, metal membranes, metal alternatives to Pd alloys.

4.1 Introduction

The application of membrane technology in the field of hydrogen purification and separation processes has been intensively studied over the last three decades, due to the development of a hydrogen energy system.[1,2] In particular, palladium-based membranes have attracted considerable interest because of their high permeability and selectivity.[3–5] However, the high cost and limited availability of Pd have driven research trends towards the investigation of alternative materials and new techniques for reducing the use of Pd. This chapter provides an overview of recent activities related to the development of new materials (porous ceramic, porous glass and, particularly, dense metal) and techniques for preparing hydrogen separation membranes.

Gas separation membranes can be divided into two groups: porous and non-porous (dense) membranes. The majority of porous membranes are made up of metal oxides: these porous ceramic membranes were among the first to be applied to separation processes when the pure isotope 235 of uranium was first separated from natural uranium in the mid-1940s.[6,7]

In porous membranes, gas separation can occur when the pore size is decreased to the same level as the mean free path of a gas, which is defined as the average distance between two successive collisions of a gas molecule. In this case, the permeation flux is inversely proportional to the square root of the molecular weight of a gas molecule (Knudsen diffusion).[8–11] The best oxides to use are alumina (Al_2O_3), zirconia (ZrO_2), titania (TiO_2) and silica (SiO_2), although their mixtures are also frequently used. In order to obtain the desired permeability and selectivity, a porous ceramic membrane is constructed using a supporting layer with a very open pore structure and a thin top layer with smaller pores. However, these membranes can also be impregnated with Pd and other metals, in order to improve their performance in terms of hydrogen permeability and selectivity. Due to their good chemical and thermal stability, porous ceramic membranes are mainly used in applications where high temperatures or aggressive chemicals are present.[9,10]

Porous glass membranes (PGMs) are used for many applications, especially in membrane separation processes such as micro- and ultra-filtration,[12] gas separation,[13] demulsification medium[14] and membrane emulsification,[15] gas–liquid contacting process[16] and gas dispersion process.[17]

Porous metal membranes are applied as supports to Pd-based layers of Pd-composite membranes, and the pore size of commercial disks or tubes made of stainless steel can range from 0.1 μm to 100 μm. These metal supports are less fragile than their ceramic counterparts and can be sealed to the membrane modules more easily.[18]

As described in detail elsewhere in this book, the hydrogen molecule can dissociate into atoms on the surface of a dense metal membrane and then diffuse through the membrane lattice. In practice, a dense metal membrane is known to be permeable only to hydrogen; a palladium alloy membrane typically separates hydrogen with infinite selectivity.[19] Cost reduction of commercial Pd alloy membranes can be achieved by producing thin Pd films, or by using an alternative metal that is cheaper than Pd. The thickness of the palladium film can be reduced to 50 μm by cold-rolling,[20] to 2–3 μm by electroless plating,[21] and to about 1 μm by chemical vapour deposition (CVD).[22] Several metals (i.e. those belonging to groups IV and V) which show high hydrogen permeability have been studied as cheaper potential alternatives to Pd.

4.2 Materials

To establish the suitability of a membrane material for hydrogen separation, it is important to consider that around 90% of the current global H_2 production is obtained by gas synthesis through steam reforming of methane or other hydrocarbons and alcohols. Therefore, a suitable H_2-separation membrane should fulfil the requirements to reliably operate in a reforming reactions environment and, at the same time, to ensure high performance in terms of permeability and selectivity.

Hydrogen permeation can be described by the following general equation:

$$J = \frac{P\left(p_h^n - p_l^n\right)}{L} \qquad [4.1]$$

In this expression, J is the hydrogen flux, P the permeability, L the thickness, p_h and p_l the partial pressures of hydrogen on the high pressure (feed) side and the low pressure (permeate) side respectively, and n the pressure exponent. The permeability cannot be given for composite membranes; in this case, the ratio of hydrogen flux to transmembrane differential pressure $J/\left(p_h^n - p_l^n\right)$, defined as 'permeance', is introduced. In practice, for a membrane made of a single material, the permeance is obtained by dividing the permeability by the thickness, while the permeance of a composite membrane depends on the permeability and thickness of its different layers.

Selectivity is the term used when measuring the difference in permeability of different components. In other words, it is a measure of the effectiveness of the membrane separation. The selectivity factor $\alpha_{A/B}$ of two components A and B in a mixture is defined as:

$$\alpha_{A/B} = \frac{y_A / y_B}{x_A / x_B} \qquad [4.2]$$

where y_A and y_B are the fractions of components A and B in the permeate, and x_A and x_B are the fractions of the components A and B in the feed (A and B are usually chosen in such a way that the selectivity factor is greater than the unity). If the selectivity factor is equal to one, there is no separation.

Yun *et al.*[18] reported a comparison in terms of selectivity/permeance ratio of Pd-based and polymeric membranes for hydrogen permeators, as shown in Fig. 4.1. The polymeric membranes exhibit low hydrogen permeance with moderate H_2/N_2 selectivity, and the selectivity/permeance upper limit of the polymeric membranes is given by the Robeson plot. From this analysis, it is

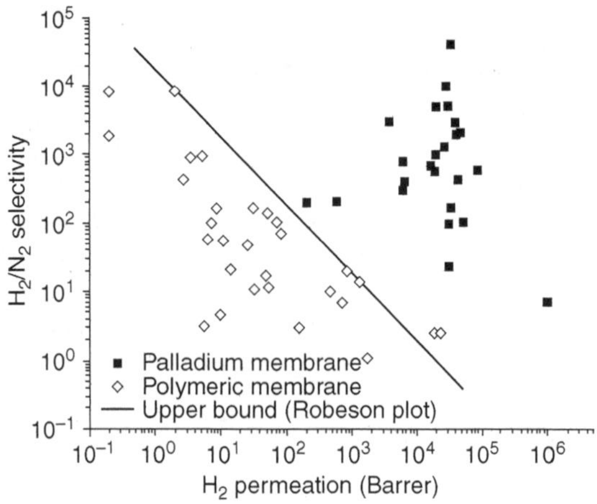

4.1 H_2/N_2 selectivity as a function of H_2 permeation of different membrane categories. (Reprinted from Reference 18.)

evident that the Pd-based membranes give a better performance. The chemical stability of a material or a membrane can be defined as its inertness against corrosive attack by aggressive liquids, while the thermal stability represents the ability of a material to resist changes in physical shape or size as its temperature changes.

As established by the US DOE,[23] the most important characteristics for an H_2-separation membrane are as follows: an operating temperature of 250–500°C, a flux of 150 $cm^3\ cm^{-2}\ min^{-1}$ with $\Delta P(H_2)$ of about 7 bar (100 psi), a cost of about \$1000 m^{-2} and a durability of five years. Materials with a strongly negative free energy and total energy of formation, such as metal oxides, are expected to be chemically very stable. Yttria, thoria, alumina, titania and zirconia are recognized for their potential chemical stability.[8] These ceramic materials are synthesized in the form of porous membranes and usually exhibit high permeability and low selectivity. On the other hand, dense metallic Pd-based membranes provide very high performance in terms of selectivity but, beside their high cost, present disadvantages related to thermal instability and brittleness resulting from the strong interaction of the hydrogen with the metal. Table 4.1 provides the main features of the different membrane classes.[24,25]

4.2.1 Ceramic porous membranes

The most useful characteristics of porous ceramic membranes are their chemical and thermal stability and their high mechanical strength.[7,26] These membranes are used for a variety of commercial applications, and their use

Table 4.1 Comparison of membrane classes

	Metallic Pd, Ta, V, Nb, and alloy	Ceramic SiO_2, Al_2O_3, TiO_2, and zeolites		Carbon, porous carbon, nanotubes	Polymers polyesters, ethers, etc.	
	Dense	Dense	Porous	Porous	Dense	Porous
T range (°C)	300–600	600–900	200–600	500–900	< 100	
Selectivity	>1000	>1000	5–139	4–20	Low	
Flux (m^3 m^{-2} s^{-1})	60–300	6–80	60–300	10–200	Low	
Mechanical issues	Phase transition	Brittle		Very brittle	Swelling and compaction	
Chemical stability	Poisoned by H_2S, HCl, SO_x	Potential degradation with H_2O, H_2S or CO_2		Oxidizing and susceptible to organic vapours	Degradation in H_2S, HCl, CO_2, SO_x	
Transport mechanism	SolD	SolD	MS	SolD/MS	SolD	MS

Notes: SolD, solution diffusion; MS, molecular sieving.
Source: Reprinted from Reference 25.

in liquid filtration techniques such as ultra-filtration and microfiltration helps to keep costs low.

In order to increase their selectivity, ceramic porous membranes are usually asymmetric. They usually have a multilayered structure, consisting of a support with large pores and a top made of one or more thin layers with small pores. The support could be produced by slip casting or extrusion, while the intermediate and top layers can be produced using one of several techniques.

4.2.2 Porous glass membranes

The main applications of PGMs are in gas dispersion process and membrane emulsification, both of which have potential applications in the food, pharmaceutical, chemical and cosmetic industries.

The main advantages of PGMs over both porous polymeric and ceramic membranes are:

- uniform pore size,
- a wide range (4 nm–50 μm) of available mean pore diameters,
- interconnected porous cylindrical pore geometry,
- possibility of surface modification.

It should be noted that no other type of porous membranes has all of these unique properties.

Table 4.2 Interaction properties of H_2 for pure metals

Packing	Metal	Hydride composition	H solubility (H/M at 27°C)	Hydride ΔH formation (kJ mol^{-1})	H_2 permeability at 500°C (mol m^{-1} s^{-1} Pa$^{-0.5}$)
fcc	Ni	Ni_2H	≈ 7.6 × 10^{-5}	−6	7.8 × 10^{-11}
	Cu		≈ 8 × 10^{-7}		4.9 × 10^{-12}
	Pd	PdH	0.03	+ 20	1.9 × 10^{-8}
	Pt	PtH	≈ 1 × 10^{-5}	+26	2 × 10^{-12}
bcc	V	VH_2	0.05	−54	1.9 × 10^{-7}
	Fe	FeH	≈ 3 × 10^{-8}	+14	1.8 × 10^{-10}
	Nb	NbH_2	0.05	−60	1.6 × 10^{-6}
	Ta	Ta_2H	0.20	−78	1.3 × 10^{-7}

Table 4.3 Cost of metals alternative to Pd

Material	$ kg^{-1}	Notes
Steel	0.41*	Hot rolled bar
18-8 stainless steel	0.59*	Scrap
Nickel	4.63*	
Niobium	6.61†	
Palladium	9323.70*	
Tantalum	74.96†	
Vanadium	12.96‡	

Notes: *Annual average price (1998).
†Year end concentrate pentoxide price (1998).
‡Annual average pentoxide price (1998).
Source: Reprinted from Reference 27.

4.2.3 Metal membranes

The hydrogen permeability of dense metallic membranes is strictly related to their lattice structure, as well to the presence of lattice defects, and to their reactivity with H_2 or other feed stream gases.

As shown in Table 4.2, body-centred cubic (bcc) forms of Nb, V and Ta commonly exhibit exceptionally high H_2 permeability values. Face-centred cubic (fcc) metals such as Ni and Pd also demonstrate good H_2 permeability.

In practice, V group metals (i.e. Nb, Ta and V) present higher permeability than Pd, while other metals such as Ni and Fe have permeability values that are lower than those for Pd, but are still interesting in terms of their practical applications due to their low cost.

Table 4.3 shows the cost of some of these alternative metals[27] and the graph of Fig. 4.2 reports their H_2 permeability values.[28] From this figure, it is evident that the refractory metals exhibit a negative temperature coefficient

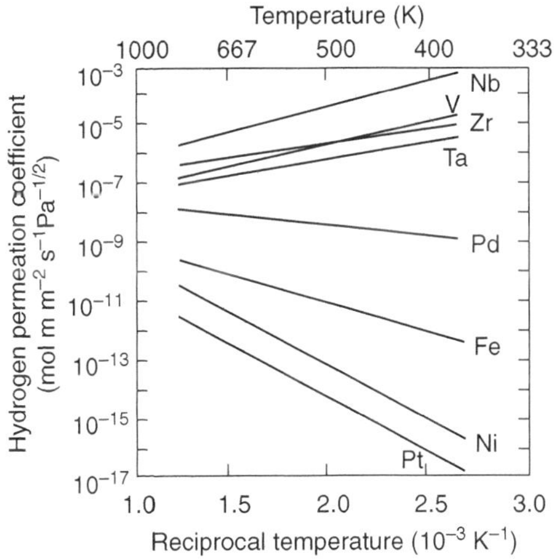

4.2 Hydrogen permeability (or permeation coefficient) of various metals. (Reprinted from Reference 28.)

as well as high permeability values. In fact, the permeability of these metals decreases with temperature as a consequence of the prevalent effect of solubility on the hydrogen mass-transfer process.

To explain this behaviour, it is worth noting that for the metals, the hydrogen permeability coefficient is given by the product of the diffusion and the solubility coefficients:

$$P = D \times S$$

where P is the permeability coefficient (mol m^{-1} s^{-1} Pa$^{-0.5}$), D the diffusion coefficient (m^2 s^{-1}) and S the solubility coefficient (mol m^{-3} Pa$^{-0.5}$). Both diffusion and solubility are activated processes, and the dependence of their coefficients on temperature is governed by an Arrhenius-type law:

$$D = D_0 \exp(-E_D / RT)$$

$$S = S_0 \exp(-E_S / RT)$$

where D_0 and S_0 are the pre-exponential coefficients and E_D and E_S are the apparent activation energies of diffusivity and solubility, respectively.

Accordingly, the permeability against the temperature can be written:

$$P = P_0 \exp(-E_P / RT)$$

where P_0 is the pre-exponential coefficient and E_P is the apparent activation energy of permeability. This results in the following equations:

Table 4.4 Solubilities and diffusivities values of the pre-exponential coefficient and activation energies for Ni, V, Ta, Nb

	D_0 ($m^2 s^{-1}$)	E_D (J mol^{-1})	S_o (mol m^{-3} $Pa^{-0.5}$)	E_s (J mol^{-1})	Reference
Ni	6.73×10^{-7}	−39700	5.49×10^{-1}	−15800	29
V	2.90×10^{-8}	−4155	1.38×10^{-1}	29000	30
Ta	4.40×10^{-8}	−13462	1.32×10^{-1}	33655	30
Nb	5.00×10^{-8}	−10221	1.26×10^{-1}	35234	30

$$P_0 = D_0 \times S_0$$

$$E_{\mathrm{P}} = E_D + E_S$$

In Table 4.4, the values of the pre-exponential coefficients and the apparent activation energies of diffusion and solubility for some of the refractory metals are reported and compared with those for Ni. It is evident that, for the refractory metals, the solubility activation energy is strongly positive, meaning that the permeability activation energy E_{P} (= $E_D + E_S$) is also positive. This explains both the negative temperature coefficient and the embrittlement of the refractory metals. In fact, the high hydrogen solubility of these metals is responsible for their high levels of embrittlement under hydrogenation, which could result in membrane failure.

Therefore, in terms of the practical use of these alternative metals, two main drawbacks have to be faced: embrittlement and surface reactivity with gases. Firstly, the alloying of metals has been studied in order to reduce the high hydrogen solubility that is a characteristic of hydride formation. The formation of stable hydrides increases the risk of hydrogen embrittlement resulting from changes in the chemical structure and lattice constant, which introduce stresses during the hydrogenation cycles. Secondly, the metals proposed as alternatives to Pd exhibit a strong surface resistance to hydrogen transport as a consequence of their high reactivity with gases. In fact, chemical compounds such as oxides and nitrides could be formed on their surfaces under the typical operating conditions of membrane applications, thus reducing hydrogen permeability. In order to avoid the formation of these surface compounds, membrane surfaces are coated with thin Pd-based films using several different techniques.

4.3 Membrane synthesis and characterization

This section describes the procedures for synthesizing membranes using new materials, with emphasis on the metals cited above. Examples of porous ceramic membranes of interest as supports for composite ceramic/

metal membranes are also provided. In recent years, wide research has been carried out on metals and metal alloys which could be used as alternatives to Pd, with the main focus on achieving high hydrogen permeability and durability. These properties are strongly affected by the alloy composition and preparation procedures.

4.3.1 Ceramic (porous) membranes

Ceramic porous membranes can be used as supports for composite membranes using metal active layers. The following sections describe a few examples of multilayer ceramic membranes obtained via sol–gel and CVD techniques.

Sol–gel technique

The sol–gel technique is a very common method for the fabrication of materials, such as metal oxides, and several procedures are described in the literature. The process typically starts from a colloidal solution (*sol*) that acts as the precursor for the *gel* formation. The most common precursors are metal alkoxides and metal salts such as chlorides, nitrates and acetates. The precursor works in hydrolysed form, while a further condensation completes the process. Figure 4.3 provides a schematic view of the sol–gel method.

Using metal alkoxide precursors, composite porous ceramic membranes have been prepared by coating commercial alumina tubes with glass layers of ZrO_2–Y_2O_3–SiO_2. The membranes with an estimated pore size in the top layer of 2–3 nm and 90% of ZrO_2 content demonstrated hydrogen permeance higher than 10^{-6} mol m^{-2} s^{-1} Pa^{-1}, and have been tested for separating hydrogen from water and HBr in thermochemical water decomposition processes.[29]

Van Gestel *et al.* focused their attention on ceramic materials with very high stability such as ZrO_2, Y_2O_3-stabilized ZrO_2 and TiO_2-doped ZrO_2.[30] These researchers synthesized various silica and non-silica sol–gel thin-film membranes which were tested for H_2/CO_2 separation. Table 4.5 summarizes

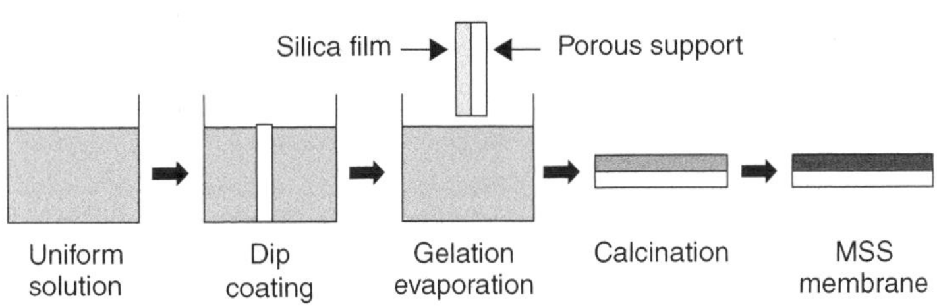

4.3 Schematic process of sol–gel method. (Printed from Reference 1.)

Table 4.5 Gas permeation of membrane with SiO_2, ZrO, $8Y_2O_3$-ZrO_2 and TiO_2-ZrO_2 top layers

Membrane	Temperature (°C)	ΔP (bar)	He (L h^{-1} m^{-2} bar^{-1})	H_2 (L h^{-1} m^{-2} bar^{-1})	CO_2 (L h^{-1} m^{-2} bar^{-1})	N_2 (L h^{-1} m^{-2} bar^{-1})	H_2/CO_2
SiO_2-S1	500	4	716	382	16.7	0	22.87
SiO_2-S2	500	4	687	168	0	0	100%
SiO_2-S3	500	4	630	138	0	0	100%
SiO_2-S4	500	2.5	817	216	11.5	0	18.78
SiO_2-S5	500	2.5	927	225	18.3	0	12.30
SiO_2-S6	800	2.5	68.8	0	0	0	
SiO_2-S6	800	4	74.0	0	0	0	
ZrO_2-S1	400	2.5	0	0	0	0	
ZrO_2-S1	500	2.5	0	4.7	0	0	
ZrO_2-S1	600	2.5	71.1	81.0	14.5	16.8	
ZrO_2-S1	400	4	7.8	4.7	0	0	
ZrO_2-S1	500	4	12.4	14.3	0	0	
ZrO_2-S1	600	4	66.4	83.6	18.6	29.0	
ZrO_2-S2	400	2.5	11.4	17.6	0	0	
ZrO_2-S2	400	4	22.4	38.2	4.7	0	
ZrO_2-S3	400	2.5	9.1	12.4	0	0	
ZrO_2-S3	400	4	21.9	28.2	4.7	0	
$8Y_2O_3$-ZrO_2-S1	400	2.5	3.3	5.2	0	0	
$8Y_2O_3$-ZrO_2-S1	500	2.5	0	3.8	0	0	
$8Y_2O_3$-ZrO_2-S1	600	2.5	8.4	8.4	0	0	
$8Y_2O_3$-ZrO_2-S1	400	4	11.9	15.3	0	0	
$8Y_2O_3$-ZrO_2-S1	500	4	7.2	9.1	0	0	
$8Y_2O_3$-ZrO_2-S1	600	4	13.4	8.6	0	0	
$8Y_2O_3$-ZrO_2-S2	400	2.5	0	0	0	0	
$8Y_2O_3$-ZrO_2-S2	500	2.5	0	0	0	0	
$8Y_2O_3$-ZrO_2-S2	600	2.5	6.1	7.6	0	0	
$8Y_2O_3$-ZrO_2-S2	400	4	2.4	4.8	0	0	
$8Y_2O_3$-ZrO_2-S2	500	4	1.4	1.4	0	0	
$8Y_2O_3$-ZrO_2-S6	600	4	11.9	14.3	0	0	
$8Y_2O_3$-ZrO_2-S3	500	4	23.9	—	—	—	
$8Y_2O_3$-ZrO_2-S4	500	4	4.7	—	—	—	
$8Y_2O_3$-ZrO_2-S1	600	4	0	—	—	—	
20 % SiO_2-ZrO_2	500	4	12	—	—	—	
TiO_2-ZrO_2-S1	400	2.5	0	0	0	0	
TiO_2-ZrO_2-S1	500	2.5	0	0	0	0	
TiO_2-ZrO_2-S1	600	2.5	0	0	0	0	
TiO_2-ZrO_2-S1	400	4	0	0	0	0	
TiO_2-ZrO_2-S1	500	4	0	0	0	0	
TiO_2-ZrO_2-S1	600	4	0	0	0	0	
TiO_2-ZrO_2-S2	400	2.5	2.3	4.6	0	0	
TiO_2-ZrO_2-S2	500	2.5	9.9	15.3	—	—	
TiO_2-ZrO_2-S2	600	2.5	16.8	22.9	2.3	—	
TiO_2-ZrO_2-S2	400	4	7.2	9.5	0	0	
TiO_2-ZrO_2-S2	500	4	13.4	19.1	1.6	3.8	
TiO_2-ZrO_2-S2	600	4	25.8	35.3	5.7	—	
ZrO_2*	500	4	31.5	—	—	7.2	
$8Y_2O_3$-ZrO_2*	600	4	95.4	—	—	—	

Notes: S1, S2, S3, etc.: Sample 1, Sample 2, Sample 3, etc.
—: no data.
The value of the ratio H_2/CO_2 of 100% corresponds to infinite selectivity to hydrogen (i.e. CO_2 permeation flux equal to zero).
*Sample with $8Y_2O_3$-ZrO_2 mesoporous sublayer.
Source: Reprinted from Reference 30.

the results of the He, H_2, CO_2 and N_2 permeation tests of membranes with top layers of silica, zirconia and doped zirconia, using mesoporous γ-Al_2O_3 as sub-layers. Generally, the hydrogen permeability of these membranes is very low, while the permeability of gases with larger molecules (e.g. CO_2 and N_2) is in some cases reduced to zero.

CVD (chemical vapour deposition)

CVD is an attractive technique for thin layer deposition. To perform CVD, a precursor gas (often diluted in carrier gases) is fed into a reaction chamber, where it reacts or decomposes to form a solid phase which is deposited onto the substrate. This technique was used by Gu *et al.* for the synthesis of ceramic composite membranes with excellent thermal stability.[31] Commercial alumina supports of pore size 5 nm were coated with layers of titania and silica to prepare composite ceramic membranes. Titanium isopropoxide (TIP) and tetraethylorthosilicate (TEOS) were used as precursors of Ti and Si for CVD at 500–700°C, which provided top layers of 10–20 nm. The membrane prepared at 600°C using a molar ratio of TIP/TEOS = 0.10 exhibited the best performance: permeation tests at 600°C showed a hydrogen permeance of 2.3×10^{-7} mol m^{-2} s^{-1} Pa^{-1} and selectivities of H_2/CH_4 = 37 and H_2/CO_2 = 57.

4.3.2 Porous glass membranes

Depending on the method used, commercially available PGMs can be classified as:

- sintered porous glass, made by sintering finely ground glass particles fused together at high temperature,
- porous glass, prepared using phase separation.

Commercial PGMs prepared using the phase separation method include porous Vycor glass and Shirasu porous glass (SPG). The main difference between the two is the chemical composition of the membranes, which greatly influences their physico-chemical properties.

SPG membranes were firstly developed by Nakashima and Kuroki,[32] starting from primary glasses in the Na_2O–B_2O_3–SiO_2–Al_2O_3–CaO system. After acid leaching, the primary glass is formed into a tube. Heat treatment, at 923–1023 K from several hours to tens of hours, causes homogenous primary glass to be separated into two glass phases: an acid-soluble phase (Na_2O–CaO–MgO–B_2O_3) and an acid-insoluble phase (Al_2O_3–SiO_2). The primary glass in this case was made of 'Shirasu' (calcium carbonate and boric acid), which is a volcanic ash found in the southern area of Kyushu (Japan). Shirasu is a source of SiO_2 and Al_2O_3 and also contains Na_2O, K_2O,

CaO, MgO and F_2O_3 as impurities.[16] The phase-separated glass tube is subsequently leached out with a hydrochloric acid solution, resulting in an SPG membrane. The mean pore diameter of the resultant membrane depends on the heat treatment conditions (i.e. the temperature and length of the treatment) and/or on the chemical composition of the primary glass. SPG membranes formed using this method have interconnected uniform pores in the range 50 nm–20 μm and also show a high resistance to water, due to the relatively high concentration of Al_2O_3 (around 10–15% wt), enabling their application in the liquid phase.

The first membranes were prepared in the form of a symmetric structure. However, a disadvantage of these early SPG membranes was their relatively low dispersed-phase flux. This is due to the high hydrodynamic resistance of the membranes, which reduces their permeability to gases and liquids and consequently limits their application in membrane separation, membrane emulsification and gas dispersion processes. To increase the dispersed-phase flux through these membranes, the membrane thickness must be reduced as much as possible. One way in which this has been achieved is by making the cross-section of the membrane structure asymmetric, a technique recently developed by Kukizaki *et al.*[33] The new tubular SPG membrane has two types of porous microstructures in a cross-section: a thin porous glass layer with a smaller pore size (an inner skin layer), and a porous glass support layer with a larger pore size below the skin layer (a support layer).

The procedure used by Kukizaki *et al.*[33] for preparing asymmetric SPG membranes is reported in the following. The starting material is a primary glass in the Na_2O–CaO–Al_2O_3–B_2O_3–SiO_2–ZrO_2–SiO_2 system, which is separated, under controlled temperature and chemical composition, into two different phases – acid-soluble (Na_2O–CaO–B_2O_3) and acid-insoluble (Al_2O_3–ZrO_2–SiO_2). The kinetics of phase separation is related to the chemical composition of the primary glass: the growth rate of the phase separation decreases as the Al_2O_3 content increases, whereas it can be increased by increasing the B_2O_3 content. Therefore, the Al_2O_3-rich primary glass was used for the inner skin layer with a smaller pore size, whereas the B_2O_3-rich primary glass was used for the support layer with a larger pore size. In order to form a tube, the Al_2O_3-rich and B_2O_3-rich primary glasses were laminated in two layers. To separate the phases, the final two-layered primary glass tube was heat-treated and then leached out with an acid. The result of this procedure was the fabrication of an asymmetric SPG membrane.

In the 1940s, Corning Glass Works (USA) developed porous Vycor glass membranes[34] as a precursor to a high siliceous glass (Vycor glass). These membranes are prepared by using the well-known process of phase inversion separation of Na_2O–B_2O_3–SiO_2 type glass and subsequently leaching the glass with a mineral acid solution. The average chemical composition of a typical primary glass (generally indicated as 'Vycor' glass composition) is: Na_2O 7%, B_2O_3 23%, SiO_2 70%. At around 873 K, the primary glass separates into two different phases, one of which is soluble in acid solutions (Na_2O–B_2O_3-rich) and the other insoluble (SiO_2-rich). The two phases are then leached out with mineral acids. Leaching of the phase-separated glass in an acid solution removes the acid-soluble phase, leaving behind a porous glass consisting of the insoluble phase.[35] The membrane is then complete, and is notable in that it will have a very high silica content (around 96%).

Disadvantages of porous Vycor glass membranes include their chemical instability, the limited range of available pore sizes (4–150 nm) and weak resistance to water and alkali even at room temperature. If the rate of cooling from the melt is increased or the annealing treatment is prolonged, the size of the domains of the two phases increases and the pore size of the leached glass increases too. Under controlled conditions, final pore diameters of 3–5 nm can be achieved. This kind of porous glass has some industrial applications in membranes for nanofiltration; however, the very low selectivity of these membranes makes them inadequate for gas separations, for which a pore diameter of less than 0.8 nm is required.

Hammel[36] developed microporous hollow glass fibres (O.D. 50–100 μm and wall thickness 5–25 μm). However, these membranes proved to be fragile and difficult to handle. For this reason, Wang and Gavalas[35] prepared composite supports consisting of a thin layer of mesoporous glass film coated onto symmetric macroporous alumina. The coating was created using a suspension of sodium borosilicate particles sintered to a dense layer and leached in acid. The final pore diameter was in the range 1–4 nm and the pore structure of the glass layer was generated by phase separation of a sodium borosilicate glass, with an initial composition similar to that used in the Vycor process. The glass–alumina composites demonstrated a permeance 10–100 times higher than that of porous Vycor tubing and can be considered suitable as supports for microporous or dense membranes.

In order to improve the chemical stability of PGMs, another method has been developed by Yazawa and Tanaka.[37] The primary glass was based on the $RO–ZrO_2–B_2O_3–SiO_2$ system (R = Mg, Ca, Sr, Ba and Zn). As in the method developed by Kukizaki, the primary glass is separated into acid-soluble B_2O_3-rich and acid-insoluble SiO_2-rich glass phases, under controlled temperature and chemical composition. These membranes contain up to 15 wt% ZrO_2 and show uniform pore size in the range 17 nm–1.5 μm. Using this method, the alkali resistance of the PGMs is greatly improved in comparison to the porous Vycor glass membranes.

An alternative way to prepare PGMs is the sol–gel procedure. This is a process in which a liquid solution gels (in crystalline or non-crystalline form) into ceramic oxides. In the final membrane, interconnected pores with a pore diameter of less than 10 nm are present. Porosity is also present at relatively high temperatures. The main advantages of this method, with respect to the conventional dry melting one, are:

- PGMs are not restricted by the chemical composition of the primary product;
- high temperatures are not necessary.

This method allows for the preparation of a large combination of glass compositions, depending on the ranges of porosity, pore size and size distribution. However, the limitation of the available pore size and difficulty in scaling up must be cited as a disadvantage of this method.

4.3.3 Alternatives to Pd alloys

As discussed before, several metals other than Pd exhibit excellent characteristics in terms of hydrogen permeability. Alloys are commonly used to improve the characteristics of pure metals. Some of the advantages of a metal alloy compared to pure metal are the following:

- improvement of the physical properties (strength, durability, degradation resistance),
- reduction in surface susceptibility to gaseous impurities (e.g. H_2S, CO, H_2O),
- increased resistance to H_2 embrittlement.

The metal alloys studied for hydrogen separation usually have either a binary or ternary composition. In particular, alloys from the highly permeable groups IV and V are used to reduce the susceptibility to hydride formation and increase the resistance to H_2 embrittlement, while alloying with Cu, Ni, Ag or Fe usually reduces surface susceptibility and subsequent surface contamination.[19] Moreover, due to the compositional flexibility of metal alloys (and amorphous metals), it is possible to increase the catalytic surface activity in order to enhance H_2-surface interactions. Because of their relatively high H_2 permeability and low cost, most of the alloys considered for H_2-separation are Nb-, V- and Ni-based; Table 4.6 summarizes the permeability data of these alloys.

Ni-based membranes

The nickel–hydrogen system has been widely studied due to its important applications in chemical catalysis and in hydrogen storage. The difference

Table 4.6 H_2 permeability data for some recent alloys reported in literature

Alloy	H_2 permeabilities (mol m^{-1} s^{-1} Pa$^{-0.5}$)	Temperature (°C)
VCr_4Ti_4	1×10^{-5} to 1.3×10^{-8}	500–650
Ni_3Al -6Fe	4×10^{-12}	375
Ni_3Al-Zr	1×10^{-12}	375
$V_{99.98}Al_{0.02}$	$0.7–1.8 \times 10^{-9}$	250–400
$V_{97.1}Al_{2.9}$	$0.7–1.8 \times 10^{-9}$	250–400
$V_{90.2}Al_{9.8}$	$2–3 \times 10^{-9}$	250–400
$V_{81.3}Al_{18.7}$	$3.7–6 \times 10^{-8}$	250–400
$V_{71.8}Al_{28.2}$	$0.7–1.8 \times 10^{-9}$	250–400
$V_{90}Al_{10}$	$1.3–2 \times 10^{-7}$	250–400
$V_{70}Al_{30}$	$0.7–1.8 \times 10^{-9}$	250–400
$V_{85}Ni_{14.91}Al_{0.09}$	$3–4.5 \times 10^{-7}$	250–400
$V_{85}Ni_{14.1}Al_{0.9}$	$3–4.5 \times 10^{-7}$	250–400
$V_{85}Ni_{12.4}Al_{2.6}$	$4–6 \times 10^{-7}$	250–400
$V_{85}Ni_{10.5}Al_{4.5}$	$5–7 \times 10^{-7}$	250–400
$Nb_{10}Zr_{45}Ni_{45}$	$\approx 2.5 \times 10^{-8}$	350
$Nb_{95}Zr_5$	$\approx 1.3 \times 10^{-7}$	300
$Nb_{95}Mo_5$	$\approx 1.3 \times 10^{-7}$	300
$Nb_{95}Ru_5$	$\approx 1.3 \times 10^{-7}$	300
$Nb_{95}Pd_5$	$\approx 1.3 \times 10^{-7}$	300
Fe_3Al	$0.6–1.01 \times 10^{-10}$	25
$Nb_{29}Ti_{31}Ni_{40}$	$1.5–7 \times 10^{-9}$	250–400
$Nb_{17}Ti_{42}Ni_{41}$	$1.1–6 \times 10^{-9}$	250–400
$Nb_{10}Ti_{50}Ni_{40}$	$0.55–4.5 \times 10^{-9}$	250–400
$Nb_{39}Ti_{31}Ni_{30}$	$0.3–2 \times 10^{-8}$	250–400
$Nb_{28}Ti_{42}Ni_{30}$	$0.3–2 \times 10^{-8}$	250–400
$Nb_{21}Ti_{50}Ni_{29}$	$0.3–1 \times 10^{-8}$	250–400
$V_{90}Ti_{10}$	2.7×10^{-7}	400
$V_{85}Ti_{15}$	3.6×10^{-7}	435
$V_{85}Ni_{15}$	3×10^{-8}	400
$V_{90}Co_{10}$	1.2×10^{-7}	400
$V_{85}Al_{15}$	6×10^{-8}	435
α-$Zr_{36}Ni_{64}$	1.2×10^{-7}	350
$(Zr_{36}Ni_{64})_{1-\alpha}(Ti_{39}Ni_{61})_{\alpha}$	$0.1–3.5 \times 10^{-9}$	200–400
$(Zr_{36}Ni_{64})_{1-\alpha}(Ti_{36}Ni_{64})_{\alpha}$	$0.15–3.5 \times 10^{-9}$	200–400
$Zr_{36-x}Hf_xNi_{64}$	$0.6–3 \times 10^{-9}$	200–400
$Ni_{65}Nb_{25}Zr_{10}$	$\approx 5 \times 10^{-9}$	400
$Ni_{45}Nb_{45}Zr_{10}$	$\approx 3 \times 10^{-9}$	400
$Ni_{50}Nb_{50}$	$\approx 2 \times 10^{-9}$	400

Source: Reprinted from Reference 25.

in lattice size between the α and β hydride is slightly larger than that of the Pd–Ag alloy, thus generate wider expansion/contraction under the hydrogenation/dehydrogenation cycling, see Fig. 4.4.[18]

As mentioned above, because the non-noble metals are easily oxidized, the formation of an oxide layer disturbs the dissociation of the hydrogen molecule on the metal surface, thus reducing their permeability. Therefore,

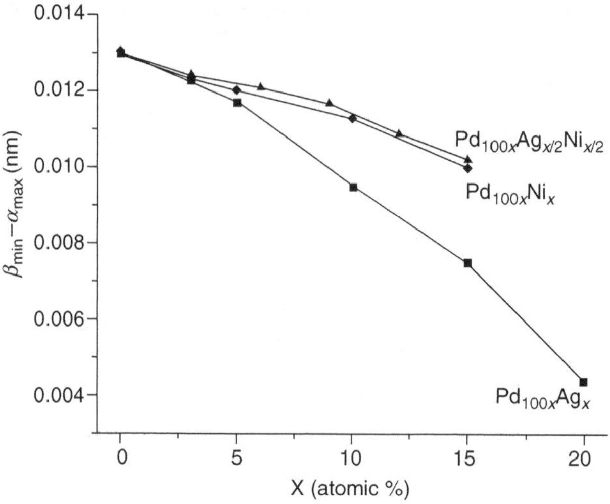

4.4 Correlation between the lattice parameter differences of palladium and palladium alloys and composition (room temperature). (Reprinted from Reference 18.)

membranes made of non-noble metals need to undergo surface treatment in order to reduce their reactivity with gases. Several techniques are applied in order to cover non-noble metal membranes with thin layers of Pd or Pd alloys. For example, a dense permeator tube consisting of Ni covered with Pd–Ag has been prepared using a diffusion welding and cold-rolling procedure.[27] Diffusion welding is a technique commonly used for joining metals and some non-metals: the parts to be joined are pressed during a thermal treatment carried out at a high temperature (50–75% of the metal melting point) at which the diffusion of the atoms is very high.[38] In this application, a Ni foil of thickness 500 μm was covered by two Pd–Ag foils of thickness 28 μm via diffusion welding at 900–1000°C under inert gas atmosphere or vacuum, to form a thick planar composite membrane (Pd–Ag/Ni/Pd–Ag). Afterwards, the composite membrane was cold-rolled down to a thickness of 141 μm (about 127 μm of Ni bulk with two Pd–Ag layers of 7 μm). Figure 4.5 shows the cross-section of this membrane. The composite metal membrane was then used for preparing a permeator tube which exhibited permeance values from 1×10^{-8} to 1×10^{-9} mol m^{-2} s^{-1} Pa$^{-0.5}$ in the temperature range 250–400°C.

Ryi *et al.*[39] successfully fabricated nickel porous membranes using uniaxial pressing. They also investigated the effect of thickness and fabrication pressure on hydrogen permeance and H_2/N_2 selectivity at room temperature. The permeance value of the Ni porous disk measures between 6×10^{-7} and 9×10^{-7} mol m^{-2} s^{-1} Pa^{-1}, while the selectivity is very poor (much lower

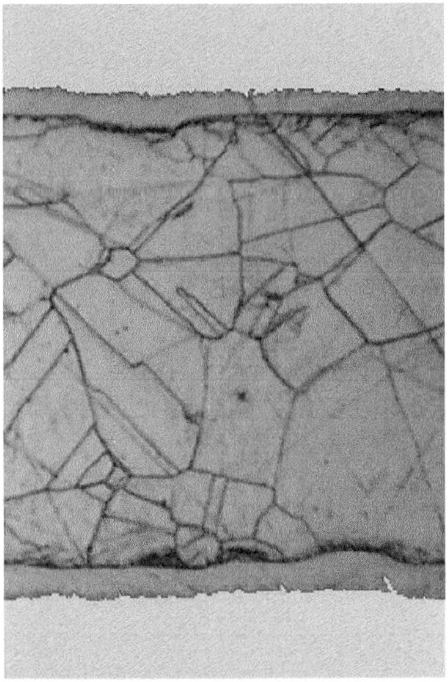

4.5 Cross-section of the laminated nickel membrane. (Reprinted from Reference 27.)

than the Pd dense membrane). This work demonstrated that a Ni porous membrane represents a good support either for dense or porous membrane. In fact, coating the Ni porous membrane with a Pd thin-film would not compromise the permeance and would definitely increase the selectivity.

Xiong *et al.*[40] studied a multilayered metal membrane (Pd/$Nb_{40}Ti_{30}Ni_{30}$/Pd/porous nickel support). The Ni porous support was obtained by uniaxial pressing of powders, while several metal layers were deposited using magnetron sputtering. The tests demonstrated the complete selectivity of the composite membranes with a $Nb_{40}Ti_{30}Ni_{30}$ layer > 3 μm thick. The results of the permeation experiments are reported in Fig. 4.6, which shows hydrogen flux against the inverse of the temperature. The composite membranes both in amorphous and crystallized form with metal layers of 6 and 12 μm are compared with the dense Pd membrane (0.5 mm), the self-supported $Nb_{40}Ti_{30}Ni_{30}$ membranes (0.5 mm) and the porous nickel support. It is evident that the 6 μm composite membranes exhibit very high permeation values.

In order to further reduce the use of Pd, Paglieri *et al.*[41] have investigated amorphous alloys based on Ni–Nb–Zr. In particular, they have systematically studied the effects of Ta addition on the performance of a series of Ni–Nb–Zr amorphous alloys with various Zr content. A series of amorphous alloy membranes consisting of $Ni_{60}Nb_{20}Zr_{20}$, $(Ni_{0.6}Nb_{0.4})_{100-x}Zr_x$ and $(Ni_{0.6}Nb_{0.3}Ta_{0.1})_{100-x}Zr_x$ (where $x = 0, 10, 20$ and 30) were prepared using melt

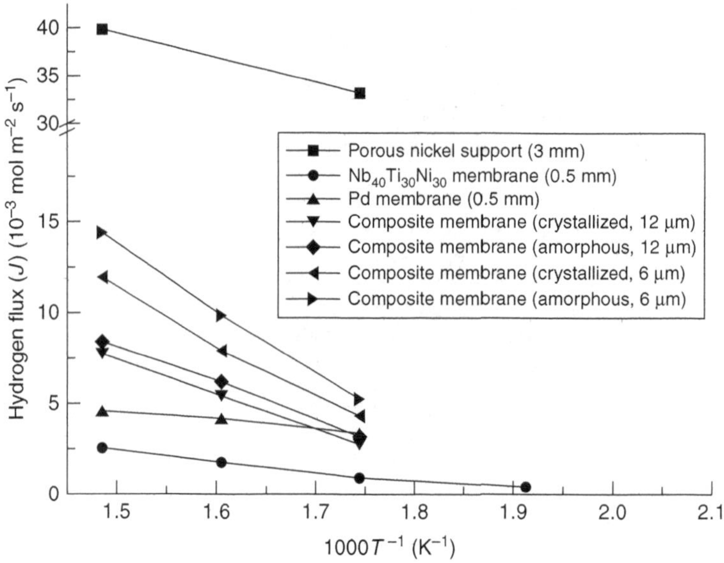

4.6 The effect of temperature on hydrogen flux in the case of upstream pressure at 0.20 MPa for different permeation membranes. The key shows membrane thicknesses of self-supported $Nb_{40}Ti_{30}Ni_{30}$ alloy, pure Pd and porous nickel support, respectively. The parameters in parentheses for the composite membranes are the phase structure and thickness of the $Nb_{40}Ti_{30}Ni_{30}$ layer. (Reprinted from Reference 40.)

spinning, by coating the foil surface with a thin (500 nm) layer of Pd through physical vapour deposition (PVD). Ta was then added in order to improve the stability of the membrane, providing an added advantage in that it has a very high melting point and is very permeable to hydrogen. Ta incorporation into the Ni–Nb–Zr amorphous alloys should therefore increase the stability without drastically reducing the hydrogen permeability. This effect is shown in Fig. 4.7, where the highest hydrogen permeability measured for each alloy is reported versus the Zr concentration. The permeability increases with the Zr content, while the partial substitution of Nb with Ta results in slightly lower permeability. It is especially noticeable that alloys with a Zr content lower than 10 atomic% exhibit very low permeability. However, in membranes with a higher Zr content, the hydrogen flux rapidly decreases with time at 400°C and 450°C. This phenomenon has been explained as a consequence of the metallic interdiffusion between the Pd surface layer and the membrane alloy, with possible change to the bulk surface structure. The substitution of Nb with Ta decreases the susceptibility of the membrane to hydrogen embrittlement.

The tendency of amorphous alloy membranes to crystallize limits their applications, as this reduces their hydrogen permeability and mechanical

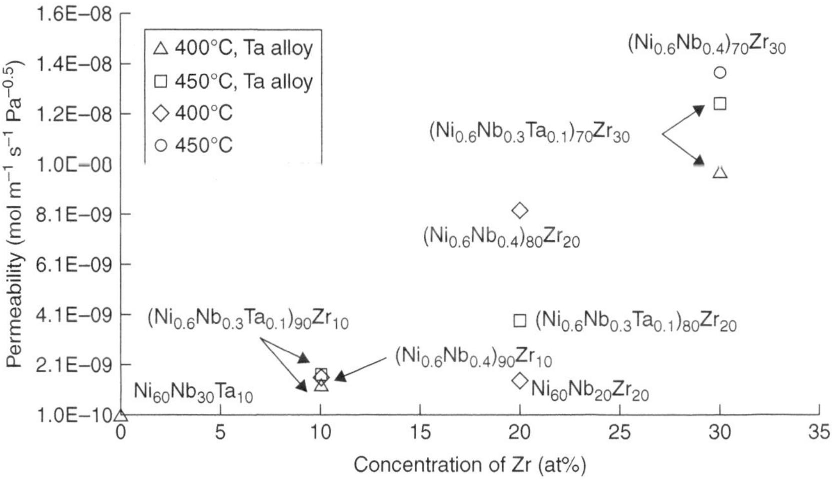

4.7 Relationship between Zr content and the hydrogen permeability of amorphous alloys with or without Ta at 400°C and 450°C. (Reprinted from Reference 41.)

strength. Dolan *et al.*[42] studied a Ni-based alloy with Nb and Zr and found that the permeability was proportional to the Zr content, while the thermal stability was proportional to the Nb content. A good compromise in terms of permeability and thermal stability is found at 400°C by the composition $Ni_{60}Nb_{40-X}Zr_X$ with $10 < X < 20$.

Nb-based membranes

An experimental work aimed at studying the effect of alloying Nb in terms of hydrogen solubility and the consequent sensitivity to embrittlement has been carried out by Japanese researchers. Addition of Ru and W to Nb decreases the hydrogen solubility and reduces embrittlement, while the hydrogen permeability is very high for Ru or W content of 5%, as shown in the Fig. 4.8.[43]

Similar improvements in the resistance to embrittlement and high values of hydrogen permeability have been reported for the Nb–W–Mo alloy membranes whose permeation tests are reported in Fig. 4.9.[44] Finally, an alloy of V with W 5 mol% has also been studied. This alloy exhibited better mechanical properties (strength, ductility) than the Nb-based alloy, as well as excellent hydrogen permeability without hydrogen embrittlement.[45] Figure 4.10 provides the results of the permeation test of this V–W alloy.

Other studies on Nb-based membranes have been developed and proposed for the separation of hydrogen isotopes (deuterium and tritium) from He in the fuel cycle of fusion reactors.[46] As mentioned in previous sections,

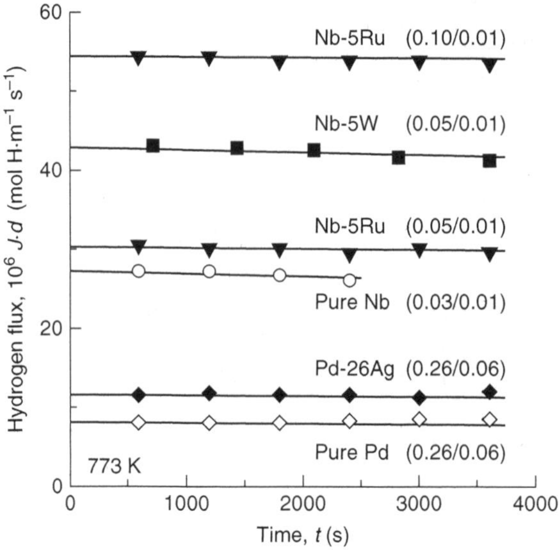

4.8 Changes in the normalized hydrogen flux during measurement at 500°C (773 K). The inlet and outlet hydrogen pressures for each measurement are indicated in parentheses in the figure as (inlet/outlet) in MPa. (Reprinted from Reference 43.)

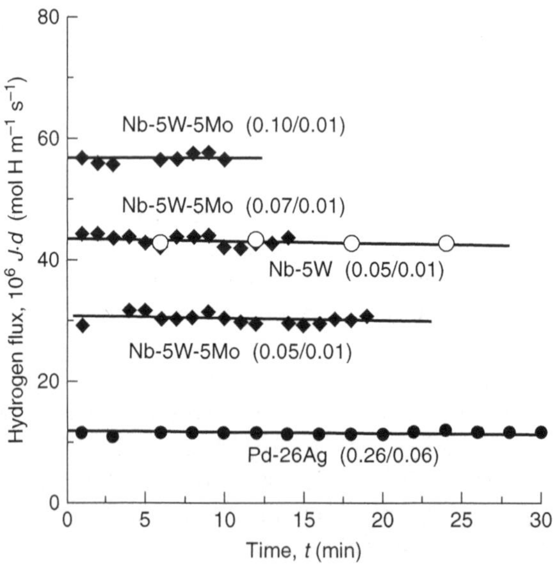

4.9 Changes in the normalized hydrogen flux during the measurements at 500°C (773 K). The inlet and outlet hydrogen pressures for each measurement are indicated in parentheses in the figure as (inlet/outlet) in MPa. Reprinted from Reference 44.

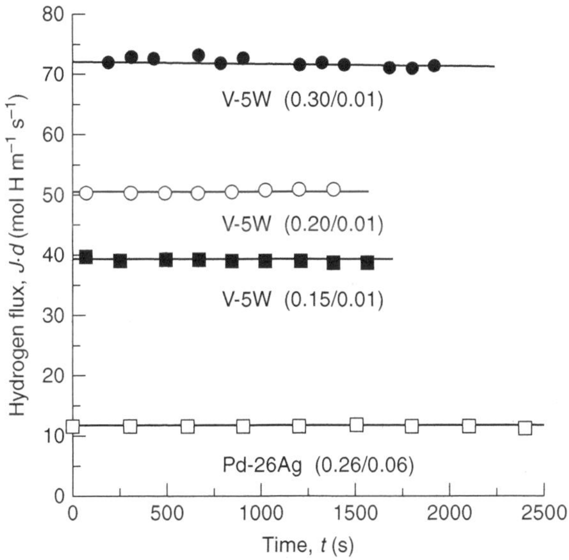

4.10 Changes in the normalized hydrogen flux during measurements at 773 K. The inlet and outlet hydrogen pressures for each measurement are indicated in parentheses in the figure as (inlet/outlet) in MPa. (Reprinted from Reference 45.)

when these metal membranes are used to separate hydrogen from the gas mixtures, two problems have to be faced. These are the formation of a superficial non-metallic film (mainly made up of oxides) that blocks the dissociative adsorption of molecular hydrogen, and the corrosive nature of some chemically active gases at the required operation temperature (> 300°C). As a solution to these problems, Alimov *et al.*[47] covered the Nb surface with a layer of Pd. A 1 cm (diameter) disk of pure Nb (99.9%) of 0.01 cm thickness and was plated with a 2 μm layer of Pd by plasma deposition: the membrane disc is shown in Fig. 4.11b. These studies demonstrated that hydrogen permeation through the composite $Pd_{2\,\mu m}$–$Nb_{100\,\mu m}$–$Pd_{2\,\mu m}$ membrane exceeds the permeation through pure Pd membranes of the same thickness. However, after heating to 500–600°C, the permeation of the membrane decreases due to interdiffusion between Pd and Nb, which results in the appearance of Nb on the membrane surface.

The diffusion welding and cold-rolling procedure previously described for the synthesis of the Ni-based membrane has been also applied to the production of a Pd–Ag/Nb/Pd–Ag composite membrane.[27] In this application, two Pd–Ag foils with a thickness of 25 μm were used to cover a Nb plate with a thickness of 1 mm. This composite membrane was rolled down to a thickness of 128 μm (of which about 122 μm was Nb bulk, between two very thin Pd–Ag layers of 3 μm). The permeation tests carried out at 180°C

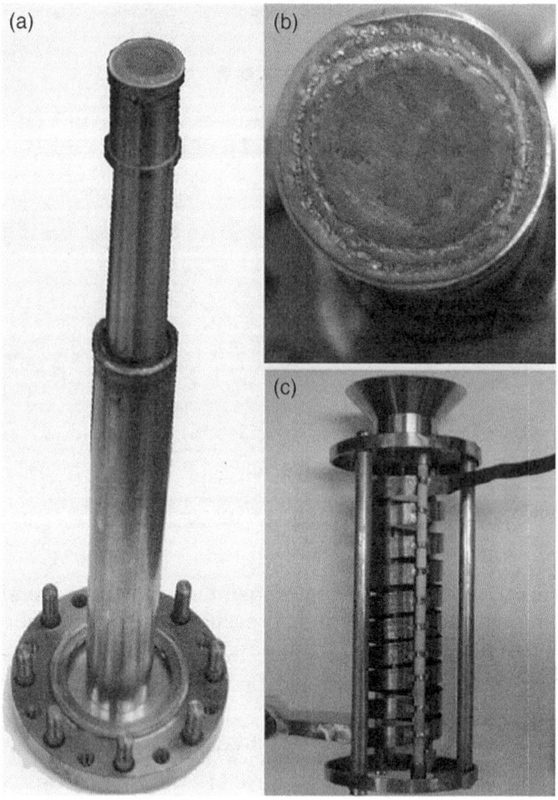

4.11 (a) Membrane unit, (b) membrane and (c) membrane heater. (Reprinted from Reference 47.)

and 200 kPa showed quick embrittlement of the material, with failure of the membrane occurring as a consequence of high hydrogen solubilization, as shown in Fig. 4.12.

The cold-rolling technique has been applied to synthesize new types of dense metal membranes made of Nb-based alloys. In particular, Luo *et al.* verified that an Nb–Ti–Ni alloy consisting only of the primary phase and the eutectic phase exhibited high hydrogen permeability and good resistance to hydrogen embrittlement.[48]

V-based membranes

The improvement of the resistance to embrittlement of V has been studied for the ternary alloy of V and Ni with a third metal (M) according to the formula $V_{85}Ni_{10}M_5$, where M is Ti, Si, Mn, Fe, Co, Ni, Cu, Pd, Ag or Al.[49] The hydrogen solubility of V-based alloys is reported in Fig. 4.13.[49] It can

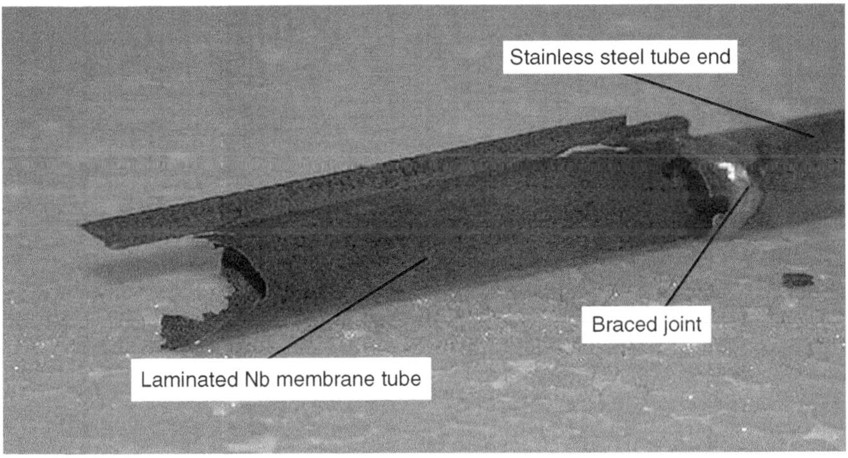

4.12 The laminated Pd–Ag/Nb/Pd–Ag membrane tube after hydrogenation. (Reprinted from Reference 27.)

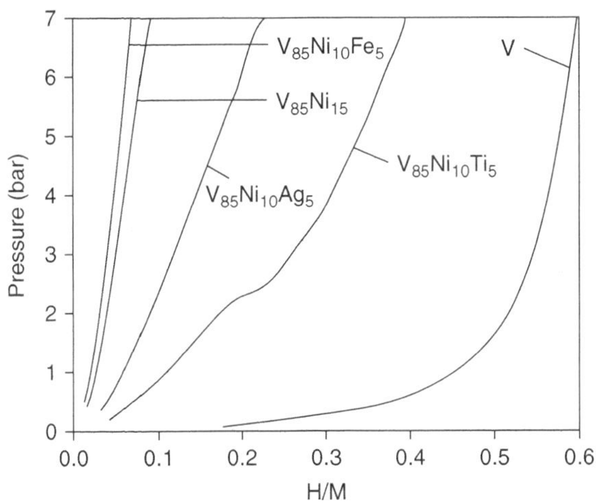

4.13 Measured hydrogen solubility (expressed as H/M) of V and several V-based alloys at 400°C. (Reprinted from Reference 49.)

be seen that Ni significantly reduces the solubility, while Ti increases it. The results of the permeation tests on V alloy membranes covered with 500 nm of Pd are reported in Fig. 4.14. Especially, the multiphase $V_{85}Ni_{10}Ti_5$ alloy exhibited the higher hydrogen permeability (9.3×10^{-8} mol m^{-1} s^{-1} $Pa^{-0.5}$ at 400 °C).

Other studies of vanadium alloys have been carried out on membranes consisting of V–15%Ni covered by thin layers of Pd.[50–52] Ozaki *et al.*[52] focused their work on the V–Ni–Al alloy which exhibited hydrogen permeability

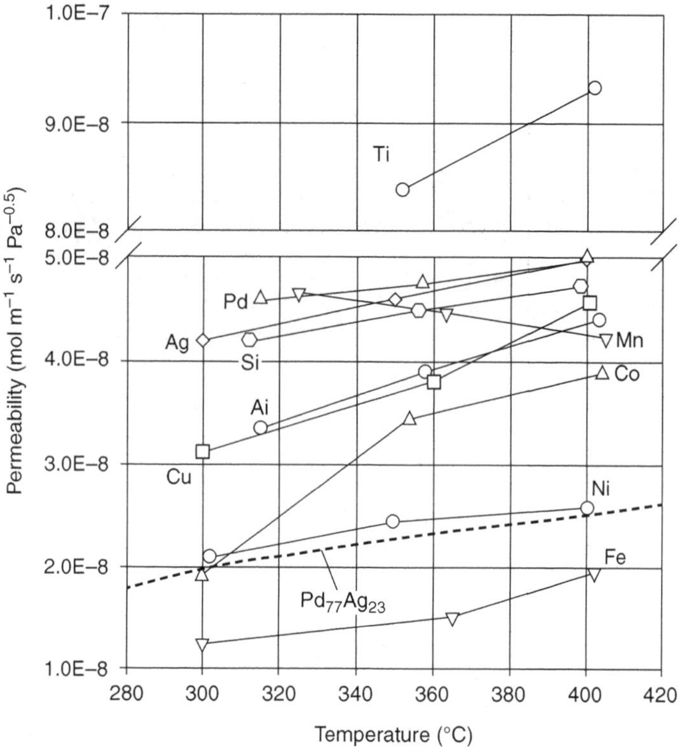

4.14 Permeability of $V_{85}Ni_{10}M_5$ alloys between 300°C and 400°C. (Reprinted from Reference 49.)

values of 6.29×10^{-8} mol m^{-1} s^{-1} Pa$^{-0.5}$ at 350°C (twice the value of that exhibited by the V–15%Ni alloy).

4.3.4 Ceramic/metal membranes

In order to couple the high chemical and thermal stability of ceramic materials with the high performance of metallic membranes, recent research has been focused on the properties of membranes made of Pd/alumina and Pd/titania.

Several methods (wetness-impregnation, electroless plating and sol–gel) can be used to dope ceramic alumina membranes with Pd. Ahmad *et al.* performed a sol–gel synthesis of palladium–alumina dried gel at 700 °C.[53] They used Autosorb, FTIR and XRD analysis to demonstrate the stability of a palladium–alumina membrane structure with a pore size of 10 nm. However, although this research suggests an application in the H_2-separation process, results concerning selectivity and permeability are not provided. Using a similar synthesis procedure, the same research group prepared TiO_2 membranes impregnated with palladium by the sol–gel method.[54] The Pd–TiO_2 membranes show gas permeability slightly higher than the TiO_2 membranes; in particular, the H_2 permeability is increased by a factor of 2–3 due to the affinity between the Pd element on the membrane surface and H_2.

Metal-dispersed alumina membranes have also been prepared by sol–gel.[55] A porous planar substrate made of Al_2O_3 (76%) and SiO_2 (23%) with a pore size of 500 nm and a porosity of 45% were dispersed via a sol–gel procedure using an alkoxide-derived sol and salts solutions of several metals: Ru, Rh, Pd, Pt and Ni. These tests did not indicate any difference in permeability for gases other than hydrogen, while the permeation rate of hydrogen and the separation factor H_2/N_2 were higher than those expected from the Knudsen diffusion mechanism, thus demonstrating that the metal particles dispersed in the alumina support were effective in promoting hydrogen permeation.

Composite ceramic/metal membranes Al_2O_3/Ni and Al_2O_3/Co have been made from γ-alumina powders produced via sol–gel.[56] The alumina and metal powders were synthesized through a rolling mechanical alloying process followed by hot press sintering. The permeation tests carried out from ambient temperature to 200°C under increasing pressure in the range 0.1–2.0 bar showed that the added metal powder promoted hydrogen permeation.

Despite its larger lattice distortion when hydrogenated, nickel has been investigated as an alternative to Pd because of its low cost. With this in mind, Ernst *et al.* set out to characterize the hydrogen permeation properties (permselectivity and stability at elevated temperatures) of a nickel/ceramic composite membrane prepared by nickel electroless plating on an asymmetric alumina substrate.[57] Two different membranes were prepared: Ni–Pd/ceramic and Ni/ceramic. The structure of the deposited films was characterized by X-ray diffraction, the surface morphology and the thickness were determined by scanning electron microscope, and the specific surface area was established using a porosimeter. The results of permeation tests showed that the γ-alumina mesoporous support presented no hydrogen selectivity, while, after the deposition of the thin (1–1.5 μm) Ni film, the Ni/ceramic membrane exhibited a good separation factor $\left(\alpha_{H_2/N_2} = 28\right)$ using argon as the sweep gas. Good adhesion of the nickel film and high hydrogen permeability were also observed (H_2 permeation flux: 2.5×10^{-3} mol m^{-2} s^{-1} at 600°C).

Composite membranes using glass supports have also been studied. Porous membranes made of Vycor glass have been used as supports for Pd-composite membranes (both Pd–Ag and Pd–Cu). The main drawback of these composite membranes is that they were characterized by low selectivity beside an important fragility.[18,58,59]

4.4 Applications

As the applications of ceramic porous membranes are covered in other chapters, this section will consider the applications of PGMs and membranes using new metals as alternatives to palladium.

Over the last 40 years, a significant amount of research has been devoted to the use of PGMs for gas separation.[13] Most of this research was carried out on commercial PGMs (e.g. Vycor).[60] However, gas separation membranes should have smaller pores than the Vycor glass membranes, in order to work efficiently. The preparation of such membranes requires optimal phase separation and leaching conditions, as well as the possibility of post-synthetic modification.[61]

Both mesoporous and microporous glass membranes have recently been prepared by Markovic *et al.*[61] Three different mesoporous glass membranes were prepared using phase separation, starting from an initial glass consisting of 70% SiO_2, 23% B_2O_3 and 7% Na_2O and with a pore diameter in the range 2.3–4.2 nm. Details can be found in Reference 61. The permeability of these membranes to different gases (He, N_2, Ar, CO_2 and C_3H_8), and their adsorption equilibrium properties, were also determined. The measured permeability values showed that the main methods of gas phase transport used are Knudsen diffusion and viscous flow mechanisms, whereas if an adsorbed phase is involved, surface diffusion is most common. In an attempt to improve the gas separation selectivity without reducing the permeance, one of the membranes (the one having pore diameter 2.3 nm) was modified with hexamethyldisilazane. This did not reduce the pore size, but lowered the permeability of the modified membrane in comparison with the unmodified membrane. In practice, the gas transport was not changed, whereas the selective surface flow changed significantly. Concerning the other two membranes, in particular the one with pore diameter 4.2 nm, it showed the largest permeation fluxes, while the selectivity for different pairs of gases at various membranes reached, as a best value, the Knudsen ratio at higher temperatures. It can therefore be stated that the range of pore sizes in all three of these membranes is still too large for achieving significant separation factors beyond the Knudsen selectivities, even when modified as described.

In order to further improve the membrane performance, subsequent researchers, starting from the same basic material, synthesized a microporous glass membrane with a reduced pore diameter of 1.4 nm by optimizing the cooling process in order to provide a very low degree of phase separation. In a successive work, Markovic *et al.*[62] characterized the microporous membrane by activated diffusion with significant selectivity but relatively low permeability. For example, the ideal selectivity of pair CO_2/N_2 was around 15 at 293 K, significantly higher than the corresponding Knudsen value (0.80).

Concerning the application of PGMs in membrane reactors, only few studies are present in the specialized literature. Their main findings are presented in the following.

Uemiya *et al.*[63] were the first to carry out steam reforming of methane in a membrane reactor consisting of a thin palladium film supported on a porous Vycor glass cylinder. They found that under the same conditions the two types of membrane used (Pd-composite and porous Vycor glass membrane) exhibited different effects on the shift of equilibrium. In the next year, Song and Hwang[64] investigated the feasibility of using a porous Vycor glass membrane reactor to shift the equilibrium of formaldehyde production from methanol. Both the experimental and simulated results of this work pointed to the necessity of developing an improved permselective inorganic membrane material in order to make the membrane reactor as an attractive alternative in many industrial applications. Some years later, Kokugan *et al.*[65] used three different kinds of membrane reactor to carry out the dehydrogenation of pure cyclohexane. The reactors used were porous Vycor glass, ceramic sol–gel and Pd–Ag membrane reactors. The experimental results showed that the higher conversions exceeding the equilibrium value (6.4%), were up to 27% (Pd–Ag) and 11% (porous Vycor glass membranes). Porous Vycor glass membrane reactors were also used by Trianto and Kokugan[66] for carrying out their experimental work on the effects of the amount of catalyst used to achieve a high purity organic product.

Recently, some aspects of modelling and simulation have also been considered. Kumar *et al.*[67] simulated a model related to a porous Vycor glass membrane reactor packed with an alumina-supported Rh catalyst for studying the production of syngas by dry reforming. In this study, the authors demonstrated various effects on methane conversion, yields and ratio of H_2 and CO: temperature, sweep gas flow rate, dilution ratio and feed ratio.

Although the development of new metals as alternatives to Pd in hydrogen separation membranes is very promising, many aspects concerning the stability and reliability of these innovative membranes have yet to be defined. The main problems to be solved are embrittlement under hydrogenation cycles and permeance/selectivity performances appropriate to the hydrogen separation processes of practical interest. Therefore, only a few application studies are reported here.

Due to its high thermal and chemical stability, three alloys of Ti, Ni and Pd ($Ti_{49.50}Ni_{49.59}Pd_{1.01}$, $Ti_{50.02}Ni_{47.42}Pd_{2.56}$, $Ti_{50.864}Ni_{44.961}Pd_{4.175}$, respectively) have been investigated by Tereshenko *et al.*[68] The three alloys, with the Ti/(Ni + Pd) atomic ratio 1:1, have been produced from 99.9% pure starting materials by arc melting. After several cycles of cold-rolling and annealing at about 700–1000°C, foils with a thickness of 40 μm were obtained. Characterization of these membranes was carried out in both permeation and reaction tests where a flat membrane reactor (MR) was used for the cyclohexane dehydrogenation reactions. The performance of the MR (obtained by the combination of the membrane with the highest Pd content and 2.8 g of Pt–Re/Al_2O_3 commercial catalyst) were measured and compared with the performance of a traditional reactor (TR). According to the shift effect provided by the membrane, the results demonstrate that the MR provides a higher level of cyclohexane conversion with respect to the traditional reaction (i.e. at 327°C the reaction conversion is 100% with the MR and only 55% with the TR).

Another study looked at the application of group VB metals to the thermochemical iodine–sulphur (IS) process for separating hydrogen from hydroiodic acid through the decomposition reaction:[69]

$$HI(g) \Leftrightarrow \frac{1}{2}H_2(g) + \frac{1}{2}I_2(g) \quad \left(\Delta H = 6\ \text{kJ mol}^{-1}\right)$$

This is an equilibrium reaction, whose conversion can be promoted in an MR through the shift effect of the membrane. The use of membrane tubes made of Nb and Ta with a wall thickness of 250 μm has been considered as an alternative to Pd alloys, which cannot withstand the corrosive environment of the IS process. The model study[69] demonstrates that, when operating at 427–527°C, the HI conversion is very close to 100% for Nb membranes, while it is in the range 75–85% for Ta membranes, as shown in Fig. 4.15.

Because of their resistance to the corrosive HI acid environment, composite ceramic membranes have also been proposed for use in hydrogen recovery from hydroiodic acid decomposition.[70] These composite membranes

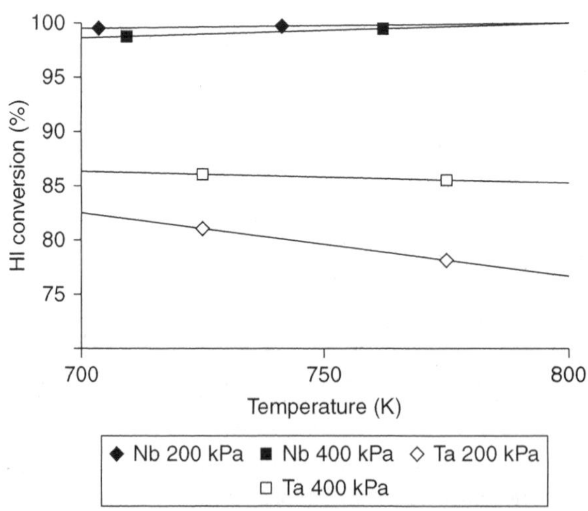

4.15 HI reaction conversion versus temperature for the studied Nb and Ta membrane reactors. (Printed from Reference 69.)

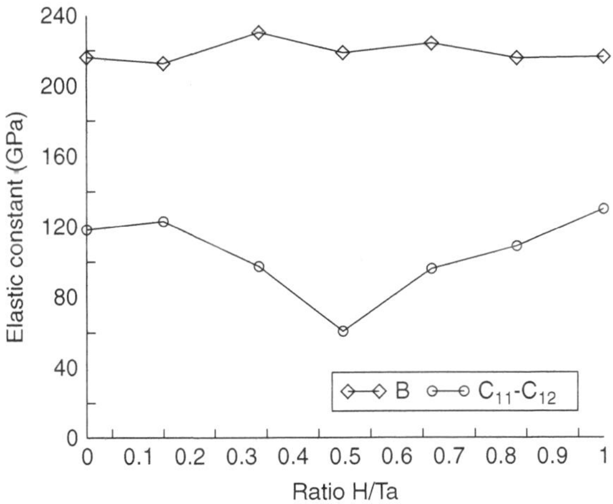

4.16 Microscopic elastic constants B (bulk modulus) and C_{11}-C_{12} (one of the two shear moduli) with respect to *x* (the ratio H/Ta). (Printed from Reference 71.)

consisted of a substrate made of γ-alumina with a pore size of 4 nm, covered with silica. A HI reaction conversion of 76.8% at 450°C has been measured for the silica MR, despite an equilibrium value of 25% at 450°C.

A model study of hydrogen interaction with the lattice of Ta has been carried out by Grena *et al.* through the Density Functional Theory (DFT), which allows us to describe the modifications of the Ta lattice induced by the presence of interstitial hydrogen.[71] In particular, the DFT evaluated the elastic constants, which play an important role in the embrittlement of the hydrogenated material. The elastic constants evaluated in this work are shown in Fig. 4.16. The main result of the study was to show that the bulk modulus is unaffected by the presence of hydrogen, while the shear modulus presents a minimum (at H/M = 0.5, that is, moles of hydrogen per mole of metal) that could induce brittle behaviour.

4.5 Conclusions

Cost reduction is a key issue to be addressed in promoting the widespread use of metal membranes for hydrogen separation. In fact, despite the proven applicability of Pd-based membranes in several hydrogen production processes, the high cost of the Pd limits the use of these membranes to niche applications, typically small scale applications where highly pure hydrogen is required.

In order to reduce the cost of Pd membranes, membranes using thin metal films have been proposed, while, in a different approach, low-cost metals and their alloys have been studied as alternatives to Pd. In addition to this, recent research has focused on pure and alloyed metals such as Ni, Nb, V and Ti.

Ni is a low-cost metal with lower permeability than Pd; it is used for preparing both porous and dense metal membranes. Among its alloys, Ni–Nb–Zr has been investigated, and it has been demonstrated that the Nb in this alloy reduces the embrittlement while the Zr increases the permeability.

Nb exhibits a very high hydrogen permeability which is linked to high hydrogen solubility and embrittlement. Alloying with Ru, W and Va has been verified as effective for reducing embrittlement. Nb–Ni–Ti has also been studied as an alloy with good permeability and stability.

Several vanadium alloys have also been studied, and it has been verified that the addition of Ni significantly reduces the solubility, while the addition of Ti increases the solubility. In particular, the ternary alloy $V_{85}Ni_{10}Ti_5$ showed a very high permeability (9.3×10^{-8} mol m^{-1} s^{-1} $Pa^{-0.5}$ at 400°C). Finally, the equiatomic Ti–Ni alloys have also exhibited interesting permeability values.

Although the results of the research on these new materials for hydrogen separation membranes are very promising, no important applications are reported at the present time. The aspects of stability and reliability still have to be addressed. Examples of applications are reported for the refractory metals which exhibit a good resistance to aggressive environments such as hydroiodic acid. A study has also demonstrated that Nb or Ta dense membranes could be applied for hydrogen recovery from the decomposition of HI in the iodine–sulphur process for thermochemical hydrogen production.

Unlike the new metal alloys used as alternatives to Pd, PGMs are extensively applied in several separation processes such as micro- and ultra-filtration, gas separation, demulsification medium and membrane emulsification, gas-liquid contacting and gas dispersion. In particular, the gas dispersion process and membrane emulsification are of interest for the food, pharmaceutical, fine chemicals and cosmetic industries.

4.6 References

1. Lu, G.Q., J.C. Diniz da Costa, M. Duke, S. Giessler, R. Socolow, R.H. Williams, and T. Kreutz, Inorganic membranes for hydrogen production and purification: A critical review and perspective. *Journal of Colloid and Interface Science*, 2007. **314**(2): 589–603.
2. Carrara, A., A. Perdichizzi, and G. Barigozzi, Pd–Ag dense membrane application to improve the energetic efficiency of a hydrogen production industrial plant. *International Journal of Hydrogen Energy*, 2011. **36**(9): 5311–5320.
3. Borgognoni, F., S. Tosti, M. Vadrucci, and A. Santucci, Pure hydrogen production in a Pd–Ag multi-membranes module by methane steam reforming. *International Journal of Hydrogen Energy*, 2011. **36**(13): 7550–7558.

4. Iulianelli, A., T. Longo, and A. Basile, CO-free hydrogen production by steam reforming of acetic acid carried out in a Pd–Ag membrane reactor: The effect of co-current and counter-current mode. *International Journal of Hydrogen Energy*, 2008. **33**(15): 4091–4096.
5. Brunetti, A., G. Barbieri, and E. Drioli, Integrated membrane system for pure hydrogen production: A Pd–Ag membrane reactor and a PEMFC. *Fuel Processing Technology*, 2011. **92**(1): 166–174.
6. Noble, R.D., *Membrane Separations Technology – Principles and Applications*. 1995, Elsevier Science B.V., Amsterdam.
7. Buekenhoudt, A., Stability of Porous Ceramic Membranes, in *Membrane Science and Technology*, E.B. V, Editor 2008, **13**: 1–31
8. Chapter 9 Inorganic membrane reactors—Material and catalysis considerations, in *Membrane Science and Technology*, H.P. Hsieh, Editor. 1996, Elsevier. **3** : 367–410.
9. Massey, B.S., *Mechanics of Fluids*, 6th Edition. 1989: Chapman & Hall, London. ISBN 0412342804.
10. Keizer, K., R.J.R. Uhlhorn, R.J. Vanvuren, and A.J. Burggraaf, Gas separation mechanisms in microporous modified γ-Al_2O_3 membranes. *Journal of Membrane Science*, 1988. **39**(3): 285–300.
11. E.A. Mason, A.P. Malinauskas, *Gas Transport in Porous Media: The Dusty-Gas Model*. 1983, Elsevier, Amsterdam. ISBN 0444421904.
12. Nakamura, K. and K. Matsumoto, Properties of protein adsorption onto pore surface during microfiltration: Effects of solution environment and membrane hydrophobicity. *Journal of Membrane Science*, 2006. **280**(1–2): 363–374.
13. Kuraoka, K., R. Amakawa, K. Matsumoto, and T. Yazawa, Preparation of molecular-sieving glass hollow fiber membranes based on phase separation. *Journal of Membrane Science*, 2000. **175**(2): 215–223.
14. Kukizaki, M. and M. Goto, Demulsification of water-in-oil emulsions by permeation through Shirasu-porous-glass (SPG) membranes. *Journal of Membrane Science*, 2008. **322**(1): 196–203..
15. Nuisin R., S. Omi, and S. Kiatkamjornwong, Synthesis and property behavior of dioctyl phthalate plasticized styrene-acrylate particles by Shirasu porous glass emulsification and subsequent suspension copolymerization. *Journal of Applied Polymer Science*, 2003. **90**: 3037–3050.
16. Kukizaki, M., Porous glass membranes: preparation, characterization and applications, in *Handbook of Membrane Research*, Gorley S.V. (Ed.), 2010, Nova Science Publishers, Hauppauge, New York, 75–217.
17. Kukizaki, M. and M. Goto, Size control of nanobubbles generated from Shirasu-porous-glass (SPG) membranes. *Journal of Membrane Science*, 2006. **281**(1–2): 386–396.
18. Yun, S. and S. Ted Oyama, Correlations in palladium membranes for hydrogen separation: A review. *Journal of Membrane Science*, 2011. **375**(1–2): 28–45.
19. Tosti, S., A. Basile, L. Bettinali, F. Borgognoni, F. Chiaravalloti, and F. Gallucci, Long-term tests of Pd–Ag thin wall permeator tube. *Journal of Membrane Science*, 2006. **284**(1–2): 393–397.
20. Tosti, S., L. Bettinali, and V. Violante, Rolled thin Pd and Pd–Ag membranes for hydrogen separation and production. *International Journal of Hydrogen Energy*, 2000. **25**(4): 319–325.

21. Bhandari, R. and Y.H. Ma, Pd–Ag membrane synthesis: The electroless and electro-plating conditions and their effect on the deposits morphology. *Journal of Membrane Science*, 2009. **334**(1–2): 50–63.
22. Itoh, N., T. Akiha, and T. Sato, Preparation of thin palladium composite membrane tube by a CVD technique and its hydrogen permselectivity. *Catalysis Today*, 2005. **104**(2–4): 231–237.
23. Dolan, M.D., Non-Pd BCC alloy membranes for industrial hydrogen separation. *Journal of Membrane Science*, 2010. **362**(1–2): 12–28.
24. Phair, J.W. and R. Donelson, Developments and design of novel (non-palladium-based) metal membrane for hydrogen separation. *Industrial and Engineering Chemistry Research*, 2006. **45**: 5657–5674.
25. Ockwig, N.W. and T.M. Nenoff, Membranes for hydrogen separation. *Chemical Reviews*, 2007. **107**: 4078–4110.
26. Julbe, A., D. Farrusseng, and C. Guizard, Porous ceramic membranes for catalytic reactors — overview and new ideas. *Journal of Membrane Science*, 2001. **181**(1): 3–20.
27. Tosti, S., Supported and laminated Pd-based metallic membranes. *International Journal of Hydrogen Energy*, 2003. **28**(12): 1445–1454.
28. Uemiya, S., Brief review of steam reforming using a metal membrane reactor. *Topics in Catalysis*, 2004. **29**(1): 79–84.
29. Fan, J., H. Ohya, T. Suga, H. Ohashi, K. Yamashita, S. Tsuchiya, M. Aihara, T. Takeuchi, and Y. Negishi, High flux zirconia composite membrane for hydrogen separation at elevated temperature. *Journal of Membrane Science*, 2000. **170**(1): 113–125.
30. Van Gestel, T., D. Sebold, F. Hauler, W.A. Meulenberg, and H.-P. Buchkremer, Potentialities of microporous membranes for H_2/CO_2 separation in future fossil fuel power plants: Evaluation of SiO2, ZrO2, Y2O3–ZrO2 and TiO2–ZrO2 sol–gel membranes. *Journal of Membrane Science*, 2010. **359**(1–2): 64–79.
31. Gu, Y. and S.T. Oyama, Permeation properties and hydrothermal stability of silica–titania membranes supported on porous alumina substrates. *Journal of Membrane Science*, 2009. **345**(1–2): 267–275.
32. Nakashima T.and Y. Kuroki, Effect of composition and heat treatment on the phase inversion of NaO-B_2O_3-SiO_2-Al_2O_3-CaO glass prepared from volcanic ashes. *Nippon Kagaku Kaishi*, 1981. **8**: 1321.
33. Kukizaki, M. and M. Goto, Preparation and characterization of a new asymmetric type of Shirasu porous glass (SPG) membrane used for membrane emulsification. *Journal of Membrane Science*, 2007. **299**(1–2): 190–199.
34. Hood S.P. and M.E. Nordberg, *Treated borosilicate glass*. US patent 2106744. 1938.
35. Wang, H. and G.R. Gavalas, Mesoporous glass films supported on -Al_2O_3. *Journal of Membrane Science*, 2000. **176**(1): 75–85.
36. Hammel, J.J., Porous inorganic siliceous-containing gas enriching material and process of manufacture and use. US Patent 4853001, cited in Wang H. and G.R. Gavalas, Mesoporous glass films supported on a-Al_2O_3, *Journal of Membrane Science*, 2000. **176**: 75–85.
37. T. Yazawa, H. Tanaka, K. Eguchi, and S. Yokoyama, Novel alkali-resistant porous glass prepared from a mother glass based on the RO-ZrO-B_2O_3-SiO_2 system (R=Mg. Ca, Sr, Ba and Zn). *Journal of Membrane Science*, 1994. **29**: 3433–3440.

38. Dunkerton, S.B., Diffusion bonding – process and applications. *Welding and Metal Fabrication*, 1991. **59**: 132–136.
39. Ryi, S.-K., J.-S. Park, S.-H. Choi, S.-H. Cho, and S.-H. Kim, Fabrication and characterization of metal porous membrane made of Ni powder for hydrogen separation. *Separation and Purification Technology*, 2006. **47**(3): 148–155.
40. Xiong, L., S. Liu, and L. Rong, Fabrication and characterization of Pd/ $Nb_{40}Ti_{31}Ni_{40}$/Pd/porous nickel support composite membrane for hydrogen separation and purification. *International Journal of Hydrogen Energy*, 2010. **35**(4): 1643–1649.
41. Paglieri, S.N., N.K. Pal, M.D. Dolan, S.-M. Kim, W.-M. Chien, J. Lamb, D. Chandra, K.M. Hubbard, and D.P. Moore, Hydrogen permeability, thermal stability and hydrogen embrittlement of Ni–Nb–Zr and Ni–Nb–Ta–Zr amorphous alloy membranes. *Journal of Membrane Science*, 2011. **378**(1–2): 42–50.
42. Dolan, M., N. Dave, L. Morpeth, R. Donelson, D. Liang, M. Kellam, and S. Song, Ni-based amorphous alloy membranes for hydrogen separation at 400°C. *Journal of Membrane Science*, 2009. **326**(2): 549–555.
43. Watanabe, N., H. Yukawa, T. Nambu, Y. Matsumoto, G.X. Zhang, and M. Morinaga, Alloying effects of Ru and W on the resistance to hydrogen embrittlement and hydrogen permeability of niobium. *Journal of Alloys and Compounds*, 2009. **477**(1–2): 851–854.
44. Awakura, Y., T. Nambu, Y. Matsumoto, and H. Yukawa, Hydrogen solubility and permeability of Nb–W–Mo alloy membrane. *Journal of Alloys and Compounds*, 2011. **509**(Supplement 2): S877–S880.
45. Yukawa, H., T. Nambu, and Y. Matsumoto, V–W alloy membranes for hydrogen purification. *Journal of Alloys and Compounds*, 2011. **509**(Supplement 2): S881–S884.
46. Livshits, A., N. Ohyabu, M. Notkin, V. Alimov, H. Suzuki, A. Samartsev, M. Solovyov, I. Grigoriadi, A. Glebovsky, A. Busnyuk, A. Doroshin, and K. Komatsu, Applications of superpermeable membranes in fusion: The flux density problem and experimental progress. *Journal of Nuclear Materials*, 1997. **241–243**: 1203–1209.
47. Alimov, V.N., Y. Hatano, A.O. Busnyuk, D.A. Livshits, M.E. Notkin, and A.I. Livshits, Hydrogen permeation through the Pd–Nb–Pd composite membrane: Surface effects and thermal degradation. *International Journal of Hydrogen Energy*, 2011. **36**(13): 7737–7746.
48. Luo, W., K. Ishikawa, and K. Aoki, High hydrogen permeability in the Nb-rich Nb–Ti–Ni alloy. *Journal of Alloys and Compounds*, 2006. **407**(1–2): 115–117.
49. Dolan, M.D., G. Song, D. Liang, M.E. Kellam, D. Chandra, and J.H. Lamb, Hydrogen transport through $V_{85}Ni_{10}M_5$ alloy membranes. *Journal of Membrane Science*, 2011. **373**(1–2): 14–19.
50. Nishimura, C., M. Komaki, S. Hwang, and M. Amano, V–Ni alloy membranes for hydrogen purification. *Journal of Alloys and Compounds*, 2002. **330–332**: 902–906.
51. Zhang, Y., T. Ozaki, M. Komaki, and C. Nishimura, Hydrogen permeation of Pd–Ag alloy coated V–15Ni composite membrane: effects of overlayer composition. *Journal of Membrane Science*, 2003. **224**(1–2): 81–91.
52. Ozaki, T., Y. Zhang, M. Komaki, and C. Nishimura, Hydrogen permeation characteristics of V–Ni–Al alloys. *International Journal of Hydrogen Energy*, 2003. **28**(11): 1229–1235.

53. Ahmad, A.L. and N.N.N. Mustafa, Sol–gel synthesized of nanocomposite palladium–alumina ceramic membrane for H_2 permeability: Preparation and characterization. *International Journal of Hydrogen Energy*, 2007. **32**(12): 2010–2021.
54. Ahmad, A.L., M.A.T. Jaya, C.J.C. Derek, and M.A. Ahmad, Synthesis and characterization of TiO_2 membrane with palladium impregnation for hydrogen separation. *Journal of Membrane Science*, 2011. **366**(1–2): 166–175.
55. Chai, M., Y. Yamashita, M. Machida, K. Eguchi, and H. Arai, Preparation and characterization of metal-dispersed alumina membranes for selective separation of hydrogen. *Journal of Membrane Science*, 1994. **97**: 199–207.
56. Park, J., T.-W. Hong, and M. Jung, Hydrogen permeation on Al_2O_3-based nickel/cobalt composite membranes. *International Journal of Hydrogen Energy*, 2010. **35**(23): 12976–12980.
57. Ernst, B., S. Haag, and M. Burgard, Permselectivity of a nickel/ceramic composite membrane at elevated temperatures: A new prospect in hydrogen separation? *Journal of Membrane Science*, 2007. **288**(1–2): 208–217.
58. Mazali I.O., A.G. Souza Filho, B.C. Viana, J. Mendes Filho, O.L. Alves, Size-controllable synthesis of nanosized-TiO_2 anatase using porous Vycor glass as template. *Journal of Nanoparticle Research*, 2006. **8**: 141–148.
59. Uemiya, S., W. Kato, A. Uyama, M. Kajiwara, T. Kojima, and E. Kikuchi, Separation of hydrogen from gas mixtures using supported platinum-group metal membranes. *Separation and Purification Technology*, 2001. 22–23: 309–317.
60. Shelekhin, A.B., E.J. Grosgogeat, and S.-T. Hwang, Gas separation properties of a new polymer/inorganic composite membrane. *Journal of Membrane Science*, 1992. **66**(2–3): 129–141.
61. Markovi ,A., D. Stoltenberg, D. Enke, E.U. Schlünder, and A. Seidel-Morgenstern, Gas permeation through porous glass membranes: Part I. Mesoporous glasses—Effect of pore diameter and surface properties. *Journal of Membrane Science*, 2009. **336**(1–2): 17–31.
62. Markovi ,A., D. Stoltenberg, D. Enke, E.U. Schlünder, and A. Seidel-Morgenstern, Gas permeation through porous glass membranes: Part II: Transition regime between Knudsen and configurational diffusion. *Journal of Membrane Science*, 2009. **336**(1–2): 32–41.
63. Uemiya, S., N. Sato, H. Ando, T. Matsuda, and E. Kikuchi, Steam reforming of methane in a hydrogen-permeable membrane reactor. *Applied Catalysis*, 1990. **67**(1): 223–230.
64. Song, J.-Y. and S.-T. Hwang, Formaldehyde production from methanol using a porous Vycor glass membrane reactor. *Journal of Membrane Science*, 1991. **57**(1): 95–113.
65. Trianto, A. and T. Kokugan, Dehydrogenation of pure cyclohexane in the membrane reactor and prediction of conversion by pseudo equilibrium model. *Journal of Chemical Engineering of Japan*, 1998. **31**(4): 596–603.
66. Trianto A. and T. Kokugan, Effects of catalyst amount, membrane tube diameter and permeation rate on the performance of porous membrane reactors. *Journal of Chemical Engineering of Japan*, 2001. **34**(11): 1332–1340.
67. Kumar S., M. Agrawal, S. Kumar, and S. Jilani, The production of syngas by dry reforming in membrane reactor using alumina-supported Rh catalyst: A simulation study. *International Journal of Chemical Reactor Engineering*, 2008. **6**: A109.

68. Tereschenko, G.F., M.M. Ermilova, V.P. Mordovin, N.V. Orekhova, V.M. Gryaznov, A. Iulianelli, F. Gallucci, and A. Basile, New Ti–Ni dense membranes with low palladium content. *International Journal of Hydrogen Energy*, 2007. **32**(16): 4016–4022.
69. Tosti, S., R. Borelli, F. Borgognoni, P. Favuzza, C. Rizzello, and P. Tarquini, Study of a dense metal membrane reactor for hydrogen separation from hydroiodic acid decomposition. *International Journal of Hydrogen Energy*, 2008. **33**(19): 5106–5114.
70. Nomura M., S. Kasahara, and S.-I. Nakao, Silica membrane reactor for the thermochemical iodine–sulphur process to produce hydrogen. *Industrial and Engineering Chemistry Research*, 2004. **43**: 5874–5879.
71. Grena, R., M. Celino, and P. Tarquini, DFT study of interstitial hydrogen in tantalum lattice. *International Journal of Hydrogen Energy*, 2011. **36**(21): 13858–13865.

4.7 Appendix: nomenclature

4.7.1 Notation

$\alpha_{A/B}$	selectivity factor of two components A and B
D	diffusion coefficient
D_0	pre-exponential factor of diffusivity
E_D	apparent activation energy of diffusivity
E_S	apparent activation energy of solubility
J	hydrogen flux
L	thickness
p_h	hydrogen partial pressure on the high pressure side
p_l	hydrogen partial pressure on the low pressure side
P	permeability
S	solubility coefficient
S_0	pre-exponential factor of solubility
x_A and x_B	fractions of the components A and B in the feed side
y_A and y_B	fractions of components A and B in the permeate side

4.7.2 Abbreviations

CVD	chemical vapour deposition
DFT	density functional theory
MR	membrane reactor
PGM	porous glass membrane
PVD	physical vapour deposition
SPG	Shirasu porous glass
TR	traditional reactor

5 Nanocomposite membranes for membrane reactors

A. GUGLIUZZA, ITM-CNR, Italy

DOI: 10.1533/9780857097330.1.218

Abstract: This chapter discusses nanocomposite membranes in relation to their use in membrane reactors. The concept of hybrid nanocomposites is explained, along with the advantages of using them as a challenging alternative to the organic and inorganic systems that are most frequently used. The chapter further analyses fabrication techniques and characterization methods, based on structure–property relationships. Finally, several important applications of nanocomposite membranes are discussed, in the fields of catalysis, biocatalysis and energy.

Key words: nanocomposite, nanoparticles, fillers, polymers, carbon nanotubes, zeolites, membrane reactors, hybrid structures, mixed matrices, gas separation, catalysis, biocatalysis, fuel cells, energy.

5.1 Introduction

Extracellular skeletal biocomposites are a natural example of high-performance nanocomposites, in which minerals and fibrous biopolymers are interlaced in functional mixed matrices (Samir *et al.*, 2005). The perfect integration of (bio)organic and inorganic components results in outstanding properties if the structures and functions are optimized at multiple length scales. Depending on the level of interplay between the components, a high degree of sophistication and miniaturization can be expected.

Hybrid systems represent a real challenge for many scientists, including membranologists looking to overcome recurrent shortcomings that affect molecular separation through membranes. One of the most critical issues is the loss of selectivity that occurs with the increase in permeability and vice versa.

There is a great demand for highly productive and selective processes in many industrial applications, including oil and gas, chemical processing, fuel cells, water purification, bio-separation and clean power generation; it has therefore become necessary to turn to newly developed advanced functional materials, with the aim of establishing valuable alternatives to fully organic and inorganic films. Hybrid nanocomposites have been shown to offer great potential in overcoming the problem of the trade-off curve between selectivity and permeability (Robeson, 1991, 2008).

The nanoscale design of advanced mixed matrices with intricate structures and functions is a challenging but crucial step, especially if we take into account the numerous advantages that these mixed matrices can offer over conventional membranes, listed below:

1. Tenfold improvement in permeability.
2. Preserved or, in some cases, improved selectivity.
3. Greater integrity and absence of defects.
4. Good processability applicable to a broad range of materials.
5. Strengthened mechanics.
6. Adaptability to harsh environments.
7. Ability to scale up (usually difficult in inorganic systems).
8. Ability to scale down, with the major benefit of miniaturization for use in lightweight systems and non-invasive devices.
9. Processability at low pressure.

This chapter discusses the development, breakthrough performance and reliability of nanocomposite membranes, and explores how nanoengineered composition and architecture can influence the separation ability of membrane reactors at the macroscale.

5.2 An overview of fabrication techniques

The high costs associated with fabrication, along with the difficulties involved in both handling and scaling up technologies for the synthesis of defect-free inorganic membranes, constitute major barriers to the processing of this class of materials in many industrial applications. However, it is possible to compensate for another critical deficiency, the low resistance of the majority of polymers to harsh chemical and physical conditions, by synergistically integrating two apparently conflicting classes of materials in a unique hybrid system, as summarized in Table 5.1.

Heterogeneous matrices are usually called 'mixed or hybrid matrices' and are prepared by dispersing guest inorganic fillers in host organic films (Fig. 5.1).

Depending on the final application, various tailoring procedures can be selected to obtain different configurations, ranging from casting techniques for flat membranes to dip-coating for hollow fibre membranes (Fig. 5.2).

In mixed matrices there are a number of crucial factors to be considered, including compatibility, uniformity, adhesion and minimum loading for maximum surface area. Given the strong tendency of nanoinorganic fillers to aggregate, common tailoring procedures, such as melt blending or roller mixing, appear unsuitable. Solution blending (Wara *et al.*, 1995), *in situ* polymerization (Patel *et al.*, 2003) and sol–gel (Gomes *et al.*, 2005) as

Table 5.1 Comparison of properties of inorganic, organic and hybrid membranes

Properties	Inorganic membranes	Organic membranes	Hybrid membranes
Cost	Expensive	Cheap	Moderate
Chemical and thermal stability	High	Moderate	High
Mechanical resistance	Poor	Good	Excellent
Compatibility to solvents	Broad range	Limited	Limited
Swelling	Free	Sensitive	Good resistance
Selectivity	Moderate	Moderate	Exceed Robeson upper boundary
Handling	Brittle	Robust	Robust

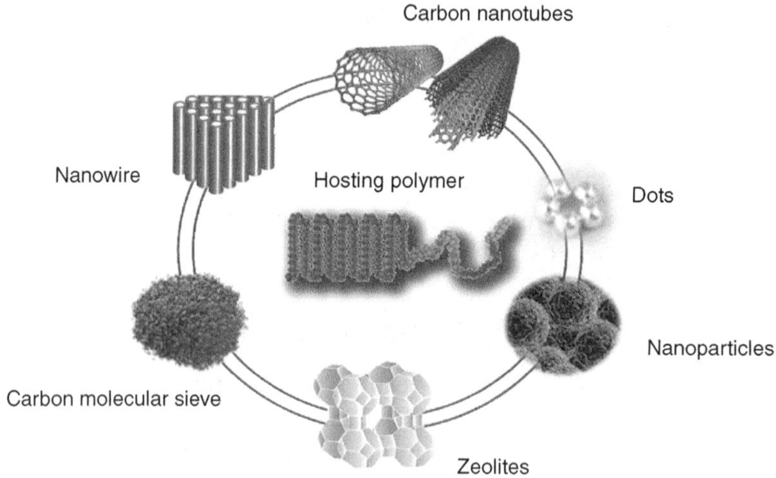

5.1 Classes of inorganic fillers incorporated in polymers.

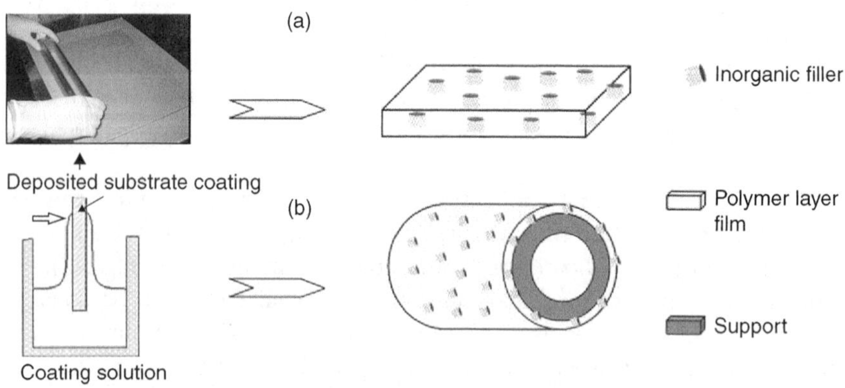

5.2 Representation of nanocomposite membranes: (a) self-standing flat nanocomposite membrane. (b) Supported hollow fibre membrane with mixed coating.

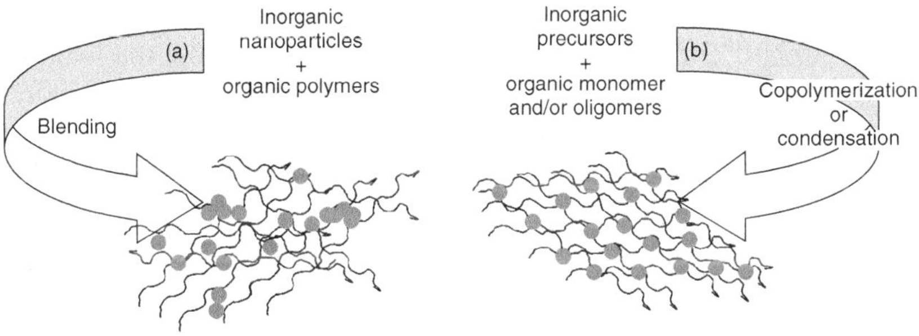

5.3 Nanofillers dispersed in polymer matrices according to different approaches: (a) blending and (b) copolymerization or condensation.

well as directed assembly (Grzelczak *et al.*, 2010) methods are preferable approaches, although the aggregation of the inorganic phase remains a critical issue. This aggregation occurs more frequently when polymer and inorganic particles are blended in the same solution and then cast on a support. Cluster formation and macroscopic phase segregations occur frequently in solution or during solvent evaporation when the materials are lacking in compatibility and miscibility. In this case, the high interconnectivity between polymer chains and the external surface area of the particles depends on the buoyancy of the van der Waals forces and hydrogen bonds established at their interface (Fig. 5.3a).

Functionalized nanoparticles are mixed with organic monomers when *in situ* polymerization techniques are preferred. The monomers are able to form radicals or ions depending on the operating conditions. In this case, phase segregation can be restricted, but not completely prevented. Better interaction of two inorganic and organic systems is instead achieved by mixing oligomers or polymers with inorganic precursors using sol–gel methods. After hydrolysis, the inorganic precursors condense into nanoparticles and are thus uniformly dispersed throughout the polymer matrix (Fig. 5.3b).

In contrast to previous manufacturing approaches, it is proposed that the infusion of volatile precursors of inorganic and metal-oxide nanoparticles into free volume of polymers is a more practical and effective strategy for the dispersal of discrete nanoscale particles. To this end, hybrid films containing Pd nanoparticles have been obtained through infusion; the metal nanoparticles have been shown to have a strong catalytic effect on the gas transport through fluorinated ethylene–propylene copolymer (FEP) (Yu *et al.*, 2004).

Another attractive approach for the fabrication of nanocomposite films is directed assembly, consisting of an intrinsic self-assembly of building blocks aided by templates or external fields or capillary forces. In this case, the self-assembly of nanoparticles can be dragged by straightforward intermolecular interactions established between capping agents and a predefined template. These forces can be controlled by adjusting the temperature, pH, solvent polarity and so on. The template can also be a block copolymer serving as a scaffold for the arrangement of nanoparticles. The spatial distribution of the polymer sites with affinity towards specific building blocks directs the periodicity of the assembly in hierarchical structures.

In combination with thermodynamics, the use of external fields, including electrical, magnetic and flow fields, or capillary forces based on liquid–liquid interfaces, can also drive the self-assembly of nanoparticles and colloidals in nanocomposites (Gast and Zukoski, 1989; Promislow and Gast, 1996; Hermanson *et al.*, 2001; Dijkstra, 2005; Hynninen *et al.*, 2005; Boker *et al.* 2007).

Engineered nanocomposite membranes can also be realized through molecular-level blending processes, covering the organization and composition of various materials at different length scales. In this case, multilayered films with uniformly stratified embedded nanoparticles can be tailored with desirable properties (Kotov, 2003).

With regard to the uniform introduction of nanoparticles into the polymer, mechanical dispersion has often proved to be inadequate. On the other hand, the organic functionalization of nanoparticles by capping agents, ligands and stabilizers seems to be a viable method of enhancing the spatial distribution and inducing ordered nanoaggregations by controlling enthalpic and entropic interactions. As a result, chemical composition and tight placement of the components are achieved at the nanoscale level and the desired functions are mastered on the macroscale (Patel *et al.*, 2004; Balazs *et al.*, 2006; Shevchenko *et al.*, 2006; Skirtach *et al.*, 2007).

Another important issue to be considered is the size of the nanoparticles. The available surface area has been shown to be 2 to 4 orders of magnitude greater than that estimated for more conventional micron-scale particles, resulting in a better dispersion (Wu *et al.*, 2002).

In summary, directed functionalization, combined with nanometer size, allows the construction of complex architectures with the desired predictable structure-property relationships, including reinforced mechanical resistance (Wu *et al.*, 2005), enhanced surface properties (Bae *et al.*, 2006), superior catalytic activity (Sidorov *et al.*, 2001) and improved transport (Merkel *et al.*, 2003).

5.3 Examples of organic/inorganic nanocomposite membranes

Since any inorganic filler can have intrinsic porosity and the ability to discriminate between differently sized and functionalized molecules, membranes that are more selective and more resistant can be achieved by using a combination of rigid adsorptive inorganic phase and processable organic phase (Pal, 2008; Widjojo *et al.*, 2008). MFI type zeolites (ZSM-5 and Silicalite-1), metals and metal-oxide have often been considered for use as the dispersed selective phase; furthermore, zirconium dioxide (ZrO_2) has been added to polysulfone (PSF) solution, yielding membranes with optimized permeability as the content of ZrO_2 was increased (Genne *et al.*, 1996). Polyimides have been mixed with porous layered aluminophosphate (AlPO) in the presence of a long-chain surfactant in order to make the inorganic material compatible with the polymer, and to induce the swelling necessary to allow the polymer to penetrate into the interlayer spacing of AlPO (Jeong *et al.*, 2004). Enhanced selectivity of CO_2 over CH_4 (α = 18–40) has been achieved; this has been attributed to the fact that molecular sieve mechanisms favour smaller molecules. Titanium dioxide (TiO_2) has been dispersed in methacrylic acid monomers, which have been polymerized under microwave radiation (Liu, 1997), while polyacrylonitrile (PAN) containing hydrolysate of tetraethoxysilane (TEOS) has been formed in membranes for gas separation (Iwata *et al.,* 2003). Nafion 117 and silicon alkoxide precursors have been mixed to prepare nanocomposite membranes using the *in situ* sol–gel synthesis technique (Ladewig *et al.*, 2007). In this case, it was shown that polymer free volume changed as a function of the materials incorporated by using Positron Annihilation Lifetime Spectroscopy; an increased selectivity for transport over protons was estimated for mixed matrices with respect to pure Nafion 117. A new class of proton-conducting hybrid membranes has been synthesized by incorporating silica and phosphosilicates in Nafion using the sol–gel technique (Mistry *et al.*, 2008). Specifically, it was shown that the inorganic phase was directed to a specific phase of the membrane, favouring the formation of Si–O–Si and Si–O–P bridges within the membrane structure, thereby increasing the water retention properties up to 200°C. This was higher than can be achieved with unmodified Nafion.

Carbon molecular sieve (CMS) membranes are another promising type of molecular sieve films, a potential alternative to the use of zeolites for various separation processes. They are prepared through the carbonization of polymer precursors at high temperature and in an inert atmosphere, leading to superior selectivity, taken to be the result of the finely tuned pore size and distribution variation (Rao *et al.*, 2008; Ismail *et al.,* 2009). CMS membranes offer numerous advantages over zeolites, because they display a higher affinity for polymers, resulting in improved adhesion and a larger contact area (Vu *et al.*, 2003a, 2003b). The use of carbon nanotubes (CNTs) to replace CMS or other inorganic molecular sieve fillers is a particularly promising approach. The need to match the pore size to the diameter of the target molecule with small dispersion suggests that CNTs would be

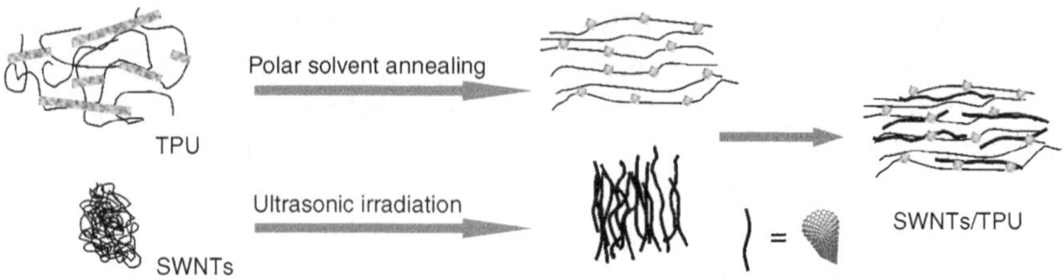

5.4 Alignment of CNTs by using chemical and physical methods (Chen, 2005). TPU, thermoplastic polyurethane; SWNTs, single-walled nanotubes.

extremely useful for tuning porosity in membranes at the nanoscale level (Ago *et al.*, 2000; Kanzow *et al.*, 2001; Belin and Epron, 2005; Ismail *et al.*, 2008). CNTs are regarded as one of the most versatile fillers because of their superior geometrical perfection. They exhibit a number of unique properties, including extraordinary mechanical and thermal resistance, excellent electronic properties, high aspect ratio and surface area, and frictionless surfaces. Their use in nanocomposite membranes is expected to increase the number of adsorption sites for small molecules, such as interstitial channel sites with high binding energy, and pore sites with a large surface area (Ismail *et al.*, 2009). The major challenge is to deliver gases, liquids and ions inside the CNTs in order to improve the discriminating capacity of the membrane based on well-sized channels and/or the ability of the penetrant to interact with the walls of the CNT (Nednoor *et al.*, 2005; Corry, 2008; Guo *et al.*, 2010). The controlling mechanism for the transport of molecules through confined systems is clearly expected to be different from that occurring in the bulk phase (Liu *et al.*, 2005). In this respect, the incorporation and alignment of CNTs by chemical and physical means are two relevant aspects of research into the optimization of trans-membrane flux and selectivity (Cervini *et al.*, 2008). Chen and Tao (2005) have demonstrated how the dimension compatibility of CNTs and segment polymer chains allows the assembly and orientation of CNTs through solvent annealing and ultrasonic irradiation (Fig. 5.4).

Hinds *et al.* (2004) used a spin-coating approach to fill the inter-tube gap with polystyrene (PS), yielding free-standing membranes with a thickness of 5–10 μm and leaving CNT alignment intact from top to bottom of the polymer film.

Oxidation and plasma etching are two processes that commonly occur during purification (Huang and Dai, 2002; Mi *et al.*, 2007), and the creation of open end termini in the structure has been shown to be an important factor in enhancing the reactivity of the nanotube tips.

Within this framework, the integration of experimental and theoretical studies is generally recommended for the construction of predictive models and the design of new materials at multiple length scales. The macroscopic properties of these materials can be programmed and determined by means of a very fine control of the chemistry and tight placement of each single component in predetermined nano- and micro-spaces (Kim *et al.,* 2007; Gethard *et al.*, 2011).

5.4 Structure-property relationships in nanostructured composite membranes

The incorporation of nanofillers into polymer matrices inevitably leads to significant changes in the performance of the native polymers. Entropic and enthalpic interactions can determine the spatial distribution, yielding engineered multicomposite materials with desirable macroscopic functions. However, the absence of guidelines and the small number of databases provide limited information about structure-property relationships in nanocomposites. Further study is therefore required to correlate structural changes with various macroscopic features, including mechanical, viscoelasticity, crystallinity, surface and transport properties (Ajayan *et al.,* 2003). It would be particularly valuable to study examples that show the effect of the interactions between polymer and nanofillers on the performance of the final nanocomposite.

5.4.1 Morphology effects on mechanical properties

Any variation in nanoparticle distribution throughout the matrices can induce significant morphological changes, which are also reflected in a modification of the properties of the native materials. Studies have been carried out examining the motion of nanoparticles during membrane formation (Nunes *et al.*, 1999; Lu *et al.*, 2006; Xiao *et al.*, 2006). The study regarding the distribution of zeolites to outer selective layers of Matrimid 5218 hollow fibres is particularly useful in this regard (Jiang *et al.*, 2005). It is proposed that the parabolic shape of the dope's axial velocity profile determines the migration of the zeolite, with controlling factors suggested to be (a) shear within the spinneret; (b) die swell when exiting from the annulus spinneret; and (c) elongation drawing in the air-gap region.

In every case, the migration or aggregation of nanofillers into polymer matrices, along with the mutual polymer/filler interactions, can induce significant structural changes that are reflected in the intrinsic features of the materials, including mechanical resistance.

Jordan *et al.* (2005) provided a detailed discussion of the effects of the particle size on strain of crystalline, semicrystalline and amorphous polymers. A reduction of the maximum strain is usually estimated for the first two

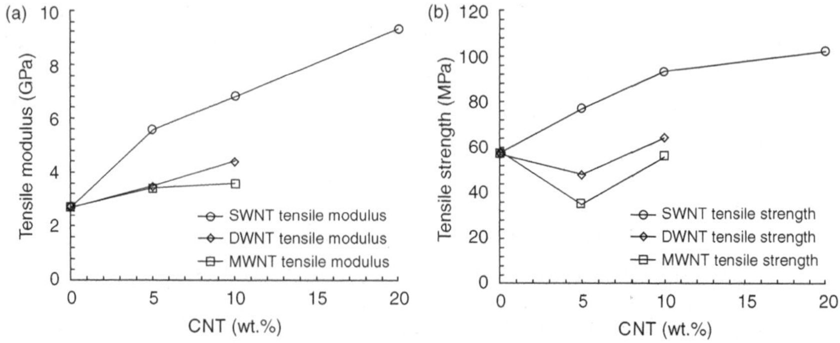

5.5 Effects of SWNT, dual-walled nanotube (DWNT) and MWNT loading on mechanical resistance of polyacrilonitrile (PAN) membranes: comparison of tensile modulus (a) and tensile strength (b).

classes of polymers when nanoparticles are added to the matrix, whereas an increase in strain-to-failure is observed for amorphous polymers as the particle size is reduced. Generally, the Young's modulus is observed to be dependent on the volume fraction of inclusion, even if an increase in this specific viscoelastic property is usually observed with decreasing particle size due to cooperative intermolecular interactions between polymer and filler (Vollenberg and Heikens, 1989; Priya and Jog, 2002; Sheng *et al.,* 2004).

Guo *et al.* (2010) carried out a study into the effects of various classes of CNTs when incorporated at different loading into polyacrylonitrile (PAN), demonstrating that the tensile modulus and tensile strength of PAN/CNT composites improved with rising nanofiller content (Fig. 5.5).

It was also demonstrated that single-walled nanotube (SWNT), in contrast to other types of nanofiller, exhibited variable interactions with PAN depending on metallic impurity, surface area and diameter.

However, the crystallinity of the native polymer is not usually subject to variation in nanocomposite films. For example, it has been shown that the addition of $CaCO_3$ or SiO_2 into polypropylene (PP), or of nano-sized montmorillonite clays into polyamide PA6-matrices or TiO_2 into poly(ethylene oxide-*b*-amide-6) (PEBAX), does not affect the content of crystallinity of the polymer (Zoppi *et al.*, 2000; Rong *et al.*, 2001; Varlot *et al.*, 2002; Sheng *et al.*, 2004).

5.4.2 Effects on surface properties

Song *et al.* (2005) found that inter-domain distance and domain size variation in organoclays incorporated in polyurethane (PU) cause a significant increase in tensile strength (120%) and elongation at break (100%), but also lead to an important reduction in the interfacial surface free energy of the films. Systems based on α,ω-dimethacrylate PEO oligomer (BEMA 1400) combined with methacryloyl-oxypropyltrimethoxysilane (MEMO) and tetraethoxysilane (TEOS) have been used as a coating for PET and were shown to change the surface wettability, because of a selective segregation of MEMO (Si/C) on the surface. In this case, silica nanophases also caused a change in the oxygen permeability due to the length of the path of the molecules through coated PET films (Malucelli *et al.*, 2006). Aulin *et al.* (2009) demonstrated that the build up of hierarchical structures on the surface of silicon after coating with 1H,1H,2H,2H-perfluorooctyltrichlorosilane (PFOTS) induces superoleophobic surface properties. Real time contact angle measurements confirmed the quick spreading of various oils on the traditional cellular planar surface, whereas a remarkable increase in wetting resistance was observed for all surfaces functionalized by PFOTS. A change in the contact angle from 0 to 140° has been estimated for nanocomposite surfaces. This provides an indication of the effect of the chemical environment on the surface free energy and related polar and non-polar components at the interface (Fig. 5.6).

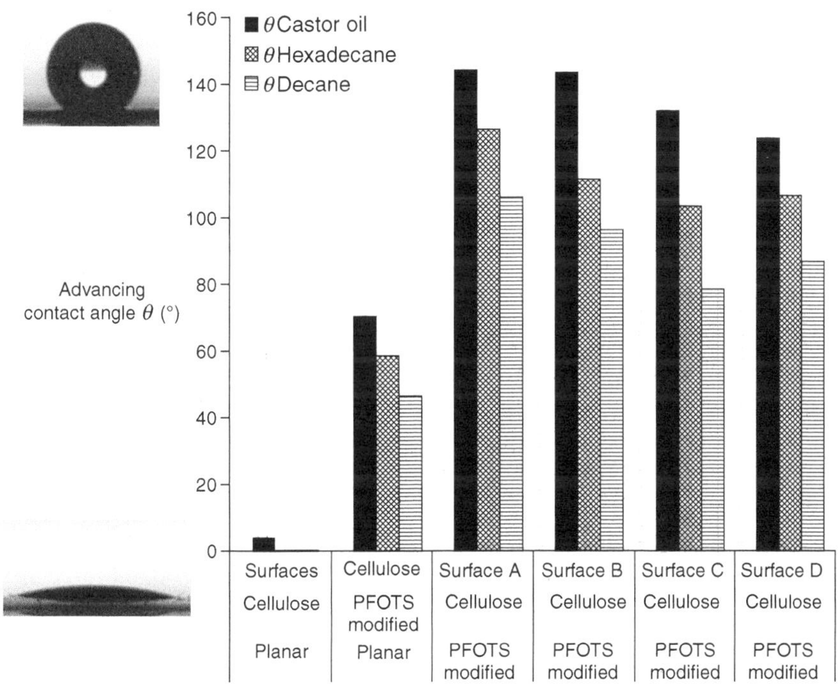

5.6 Advancing contact angles for non-polar liquids on PFOTS-coated and non-coated structured and planar cellulose nanocrystal surfaces.

Hybrid nanostructures have also been shown to control fouling and polarization phenomena during separation processes. Fouling-resistant nanocomposite membranes have been prepared through the electrostatic self-assembly of TiO_2 nanoparticles on an anionic sulfonated polyethersulfone (SPES) surface, for use in membrane bioreactors (Bae *et al.*, 2006; Yang *et al.*, 2007). The fouling was mitigated by the presence of TiO_2 nanoparticles, which limited the hydrophobic interactions between the membrane surface and organic foulants. Similarly, Al_2O_3 particles have been added into polyvinylidene fluoride (PVDF) mixtures, yielding membranes with a more hydrophilic surface and improved antifouling performance (Lu *et al.*, 2005).

5.4.3 Effects on transport

Another significant area of concern is related to the influence of nanofillers on the transport through nanocomposites. The loading of inorganic nanoparticles and the compatibility between inorganic and organic phases play a decisive role in the redistribution of free volume and the modulation of the affinity to permeating molecules.

Poly(vinyl alcohol) (PVA) substrates coated with a non-porous hydrophilic polymer/multiwalled nanotube (MWNT) nanocomposite layer have been put forward for potential use in oil/water emulsion separation (Wang *et al.*, 2005). Specifically, it has been proposed that CNTs could improve the flux rate for the formation of hydrophilic nanochannels for water transfer through nanocomposite membranes. The immobilization of CNTs in hydrophobic membrane pores was shown to have a favourable effect on water–membrane interactions and to promote higher vapour permeability, since liquid was prevented from entering the membrane pores (Gethard *et al.*, 2011). The literature also mentions improvements in water flux and selectivity when polymeric membranes are loaded with hydrophilic nanofillers, including silica (Bottino *et al.*, 2001), ZrO_2 (Bottino *et al.*, 2002) and TiO_2 (Yang *et al.*, 2006).

Xu *et al.* (2009) compared the performance of some nanocomposite membranes containing silica nanoparticles in gas separation processes (Fig. 5.7).

Interesting results have been achieved by blending CO_2-affinity polymers with inorganic compounds. Silica dispersed in PEG and PEO membranes was shown to improve CO_2 permeability, with CO_2/N_2 ratios of 9.1 and 9.03 respectively (Xu *et al.*, 2009), whereas an improvement in CO_2-affinity was achieved using PEBAX/ceramic blends and PEBAX/organic fillers (Gugliuzza and Drioli, 2005), as the solubility process is determined by favourable intermolecular interactions. Improved performances have also been achieved for the separation of H_2 from CO_2 when polysulfone/zeolite 3A mixed matrix membranes (MMM) were prepared by using the amino-propyl-trimethoxysilane agent to improve the adhesion between the organic and inorganic phases; an H_2/CO_2 selectivity of 72 and H_2 permeability of 7.1 Barrer were obtained

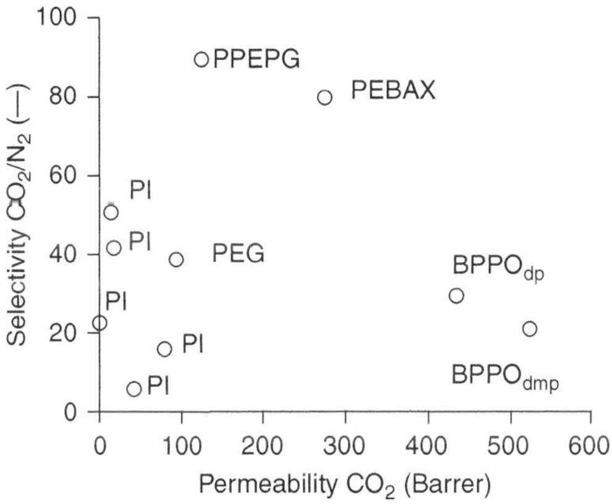

5.7 CO_2 permeability versus CO_2/N_2 selectivity estimated for various composite membranes embedding silica nanoparticles. (Data adapted from Xu, 2009.) Permeability CO_2 1 Barrer = 10^{-10}(cm^3 (STP)·cm)/ cm^2·s·cmHg); PI = polyimide; PEG = polyethylene glycol; PEBAX = poly (ether-b-amide); PPEPG = polyphenylene ether polyglycol; $BPPO_{dp}$ = brominated poly(2,6-diphenyl-1, 4-phenylene oxide); $BPPO_{dmp}$ = brominated poly(2,6-dimethyl-1,4-phenylene oxide)

(Guiver *et al.*, 2002). An increase in H_2/CO_2 separation factor from 1.53 to 3.57 has been observed for membranes containing 40% zeolite (Khan *et al.*, 2010, 2011). Enhanced performance has been achieved for nanocomposites based on Multiwalled Carbon Nanotubes (MWCNTs) in poly(bisphenol A-co-4-nitrophthalicanhydride-co-1,3-phenylene diamine) (PBNPI). Reduced *d*-spacing compared to the pure polymer has yielded H_2 permeability of 14 Barrer, H_2/CH_4 selectivity of 8, and H_2/CO_2 selectivity of 6 at 15% of CNT (Weng *et al.*, 2009).

Nanocomposite membranes based on poly(4-methyl-2-pentyne) (PMP) and nano-sized fumed silica fillers have been put forward as a possible alternative to the less chemically resistant PTMSP for the removal of C_2^+ hydrocarbons (He *et al.*, 2002). Specifically, an *n*-butane/methane separation factor of 26 and *n*-butane permeability of 19 000 Barrer were estimated when 30% hydrophilic silica was added to PMP, resulting in a simultaneous increase in *n*-butane/methane selectivity and *n*-butane permeability, in contrast to the more conventional trade-off relationships between selectivity and permeability in polymers. The addition of fumed silica in PMP enables this polymer to challenge PTMSP in terms of both efficiency and productivity, as it disrupts the molecular packing by fillers, which in turn causes a redistribution of the free volume (Merkel *et al.*, 2002).

It has been observed that nanoparticles smaller than 50 nm cause a significant increase in permeability and, as a result, the highest polymer/particle interfacial area, leading to amplified *d*-spacing. On the other hand, a restriction of the diffusional pathway was observed when cloisite Na^+ nanoclays were added to poly(ethylene-*co*-vinyl acetate) (EVA) for the separation of chloroform from acetone via pervaporation (Anilkumar *et al.*, 2008). In this case, an ultra-thin dispersion of exfoliated silicates promotes strong interactions between the polymer chains and the layered nanoclays, resulting in a tenfold higher chloroform/acetone selectivity: $\alpha = 3.8$ for unfilled membranes and $\alpha = 36$ for membranes containing 3% nanoclays. The reduced permeability is a direct consequence of an increase in tortuosity of the path. In this case, the membrane breaks the chloroform/acetone azeotrope, overcoming the limits of the evaporative selectivity.

5.5 Major application of hybrid nanocomposites in membrane reactors

The term 'membrane reactors' is used to refer to devices in which two distinct events – a chemical reaction and a separation process – take place in a single stage. When biological events control the conversion and separation process, the system is called a 'membrane bioreactor'.

Nanocomposite membranes have practical applications in many fields of membrane reactor technology, including chemicals (Ozdemir *et al.*, 2006), food and environmental safety (Lewis *et al.*, 2011), biomedical (Hule and Pochan, 2007) and renewable power (Lakshminarayana and Nogami, 2009). Here, we illustrate a few recent applications of nanocomposite membranes in the areas of chemistry, biotechnology and energy.

5.5.1 Hybrid membranes in catalytic processes

In catalysis, the immobilization and heterogenization of catalytic nano-compounds on a solid support facilitates the access of reactants to the catalytic sites, prevents aggregation and allows recovery and reuse, as well as prolonging the life of the catalyst. Depending on the specific type of process, the reaction can occur in gas or in liquid phase. Table 5.2 summarizes the various interesting types of applications that use hybrid catalytic nanocomposite membranes for conversion and separation of chemical compounds in a single stage (Ozdemir *et al.*, 2006; Biswas *et al.,* 2011; Domènech *et al.,* 2011).

5.5.2 Nanocomposite bioreactors

Recently, the combination of the oxidative activity of nanoparticles with the catalytic activity of one enzyme in stacked membrane reactors has been successfully proposed for use in water purification (Lewis *et al.*, 2011).

Table 5.2 Examples of hybrid membranes operated in catalytic processes

Membrane type	Reaction type	Nanocomposite membrane
Catalytic reaction in gaseous phase		
Catalytic	CO_2 adsorption and conversion into carbonates	CuO, MgO and SiO_2/PEG
Catalytic	Hydrogenation of cyclopentadiene	PVP–Pd/CA, PVP–Pd/PAN, EC–Pd/CA, AR–Pd/CA
Catalytic	Hydrogenation of butadiene in 1-butene	Mono- and bimetallic polymeric fibres
Catalytic	Hydrogenation of ethylene and propylene	Pd/polyMTD
Catalytic	Hydrogenation of methylene–cyclohexane	$PdPVDF_{20}PVP_{10}$, $PdPVDF_{20}$
Catalytic	MTBE decomposition	$H_3PW_{12}O_{40}$/PPO/Al_2O_3
Catalytic reaction in liquid phase		
Catalytic	Hydration of α-pinene	PVA–HPMo
Catalytic	Epoxidation of cyclohexene and styrene	Cis$[Mn(bpy)_2]^{2+}$–NaY–PDMS
Catalytic	Cyclohexane oxidation	Fe(TPP)Cl–PDMS, Fe(TDCPP) Cl–PDMS, Fe(PCl_8)Cl–PDMS
Catalytic	Benzyl alcohol to benzaldehyde oxidation	(Py, Pb_2Ru_2O6O0)–Nafion 417
Catalytic	Alcohols photooxidation	$W_{10}O_{32}$4_PDMS/PVDF
Catalytic	Reduction of *p*-nitrophenol by $NaBH_4$ to *p*-aminophenol	Pd_0/sulfonated polyethersulfone with Cardo group membranes
Catalytic reaction in pervaporation		
Catalytic	N-butyl alcohol–acetic acid esterification	PVA/PAA–$Zr(SO_4)$2$4H_2O$
Catalytic	Hydrogenation of 4-chlorophenol	PEBA–Pd
Catalytic	Epoxidation of propene to propene oxide	Mn(TPP)Cl/non-polar pervaporation membranes

PVP Poly N-vinyl-2-pyrrrolidone; CA Cellulose Acetate; AR melamine-formaldehyde resin; MTD Methyltetracyclododecene; EC ethyl cellulose; TPP Tetraphenylporphyrin; TDCPP 5,10,15,20-tetrakis(2,6-dichlorophenyl)porphyinato.

Pore-functionalized synthetic membranes have been tailored to bring about toxic organic degradation and detoxification from water without the addition of expensive or harmful chemicals. The proposed system consists of two nanocomposite stacked membranes: one embedding enzyme (glucose oxidase, GOx) for the catalytic production of hydrogen peroxide (H_2O_2) from glucose and oxygen; and the other including iron ions or ferrihydrite/iron oxide nanoparticles for the decomposition of H_2O_2 to form powerful free radical oxidants. The model reaction is the degradation of a chlorinated organic contaminant (Fig. 5.8).

The membrane pores are coated with a pH-responsive PAA network by means of sequential assembly, and bring immobilized reactants into closer

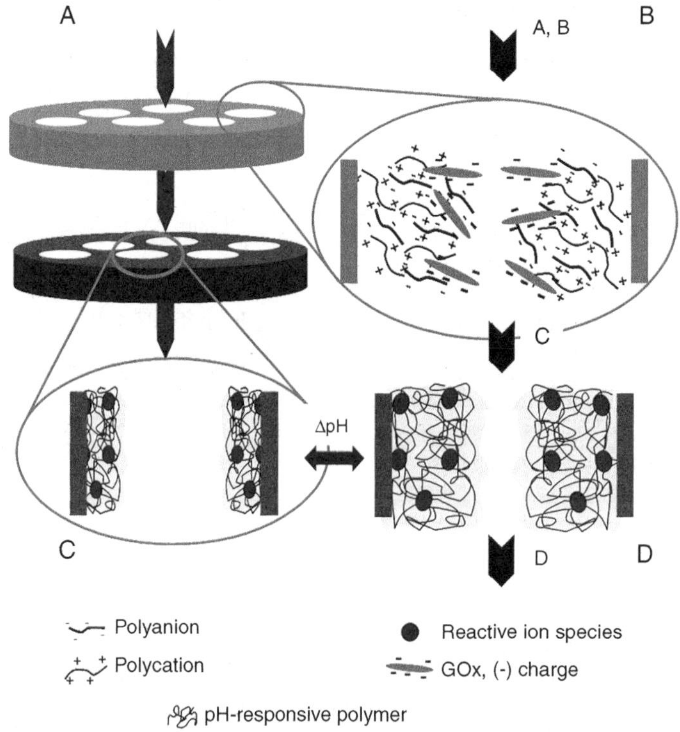

5.8 Biocatalytic nanocomposite membrane system adapted from Lewis *et al.* (2011). (A) Stacked functional membranes operated via convective flow. (B) Pore of the top membrane with immobilized enzyme. (C,D) Pore of the bottom membrane with immobilized iron species in collapsed state and swelled state via pH control.

proximity to the feed solution. The polymeric layers in the membrane pores permit controlled opening and closing of the gates through pH-driven collapse of the network. H_2O_2 is produced in the top membrane and then transported to the bottom membrane, where it reacts with the bound iron species and forms the free radicals necessary for contaminant degradation.

Magnetic nanoparticles of Fe_3O_4 along with alkaline phosphatase (ALP) have also been incorporated into a sol–gel/chitosan biosensor membrane (Loh *et al.*, 2008). The result was a two-fold increase in the sensitivity of the biosensor to the herbicide 2,4-dichlorophenoxyacetic acid (2,4-D) with a recovery of 95–100%. The role played by nanoparticles in enhancing the productivity of the bioreactor and significantly reducing the tendency towards enzyme deactivation/pore-blocking has been discussed by Hicke *et al.* (2006). Epoxy-reactive nanoparticles with a diameter of 200–230 nm have been covalently immobilized on the pore walls of track-etched membranes

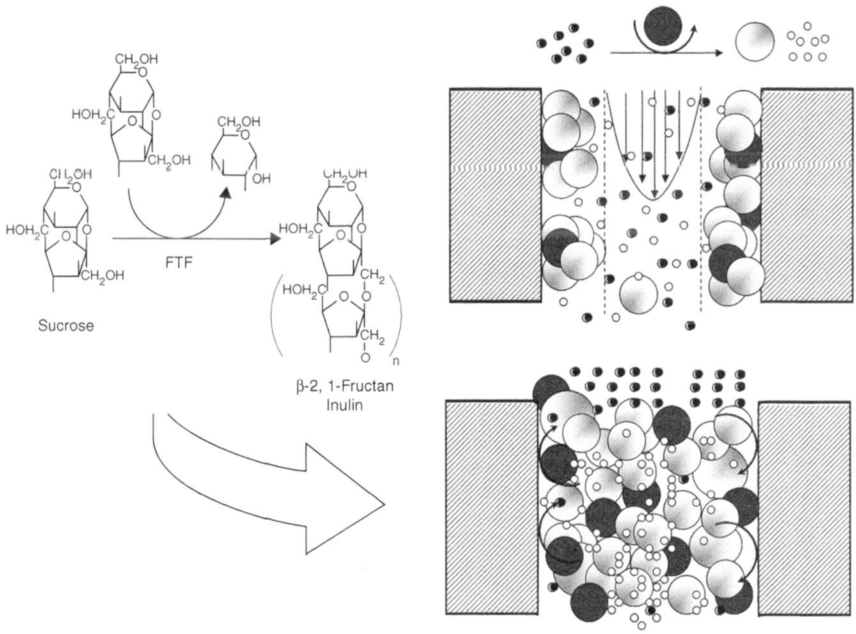

5.9 Nanocomposite bioreactors for the synthesis of poly-(2-1)-fructan (inulin) from sucrose by inulinsucrase (fructosyltransferase, FTF) in the presence of epoxy-reactive nanoparticles. (Scheme adapted from Hicke *et al.*, 2006.)

in order to synthesize high-molecular weight inulin from sucrose using fructosyltransferase (FTF) (Fig. 5.9).

The increased operation time of enzyme–membrane reactors (EMR) and the formation of larger amount of inulin have both been attributed to the ability of nanoparticles to work as turbulence promoters and supports, which supply an increased surface area for covalent enzyme immobilization.

5.5.3 Nanocomposite membranes for power production

Nanocomposite membranes with embedded inorganic nanofillers have been used to construct new functional materials for energy conversion and energy storage. There have been a number of recent advances in organic–inorganic nanocomposite membranes, particularly in the design of structures at a molecular scale for fuel cells, batteries and electrolysers. These electrochemical devices rely on ion-containing polymer electrolytes that facilitate ion transport to and from the electrodes (Fig. 5.10).

In recent years, particular attention has been devoted to the development of nanostructured hybrid membranes with superior ion transport properties; these allow the water to be managed in operating conditions with low

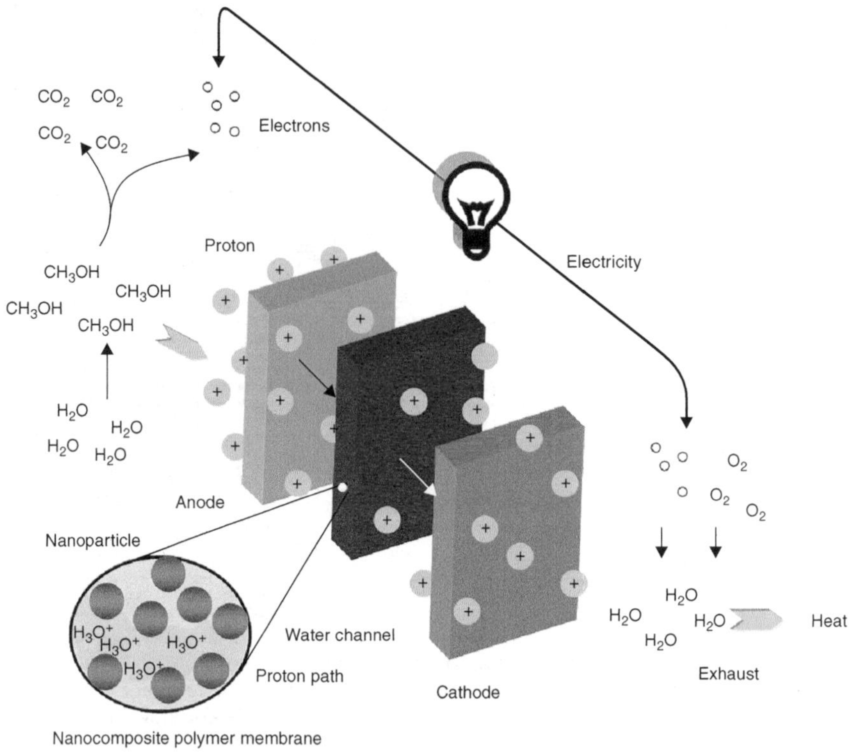

5.10 Scheme of electrolyte membranes

humidity, thereby reducing methanol crossover and improving conductivity and thermal and mechanical stability (Mangiagli *et al.*, 2009; Xu *et al.*, 2009).

As an example, CNTs have been dispersed in sulfonated poly[bis(benzi midazobenzisoquinolinones)] (SPBIB) membranes, yielding higher proton conductivity (102×10^{-3} Scm^{-1} for CNT/SPBIBI(0.5) against 90×10^{-3} S cm^{-1} estimated for Nafion117, measured in water at 20°C.). This result was attributed to the ordered and smaller cluster-like ion domains of the composite layers. Morphological changes were caused by the uniform dispersion of CNTs through pi–pi stacking (π–π) between the pyridinone ring and the side walls of the MWCNTs (Li *et al.*, 2009). The compatibility between CNTs and Nafion was also optimized by grafting poly(oxyalkylene) onto the walls of the CNTs via a covalent amide linkage, forming amine-functionalized tube walls via a sol–gel process in order to establish favourable interactions between the amino moieties of CNTs and -SO_3H groups of Nafion (Chen *et al.*, 2008). Enhanced water retention up to 200°C was achieved for Nafion by incorporating silica and phosphosilicate nanoparticles into the polymer via a sol–gel process (Mistry *et al.*, 2008). An increase in thermal stability due to the interactions between the inorganic component and the polymer membrane was associated with an improvement in the ability of the inorganic filler to bind water molecules, preventing its loss at high temperature. Similarly, the incorporation of inorganic nanoparticles with an affinity for water, such as silica, has allowed water management for different nanocomposite polymers to be improved (Smith and Zharov, 2009). Organic–inorganic nanocomposite zwitterionic polymer electrolyte membranes (PEMs) have been developed for direct methanol fuel cell applications (Tripathi and Shahi, 2009). Low methanol diffusion (3.95×10^{-7} cm s^{-1}) and reasonable proton conductivity (4.85×10^{-2} S cm^{-1}) have been achieved for ZI-70, a zwitterionomer membrane with 70 wt% of PVA of 3-[[3-(triethoxysilyl)propyl]amino]propane-1-sulfonic acid in the membrane matrix.

5.6 Conclusions and future trends

Nanocomposite membranes represent an emerging class of materials in the area of nanotechnology. Hybrid systems can offer enhanced performance compared to more conventional systems, as the features of lightweight polymers are combined with low technology nanofillers. Nanocomposites with associated selective properties, such as separation, adsorption, sensing and catalysis, are expected to create new market opportunities in the fields of pharmaceuticals, water and wastewater, food, medical and healthcare and energy.

Depending on the final intended use of the membrane, the design of nanocomposites is tailored to meet certain criteria, such as enhanced mechanical and chemical resistance, improved selectivity and permeability, biocompatibility and biodegradability. However, the integration of different components into nanocomposites can only meet specific needs through a detailed understanding of the structure-property relations involved. Today, the real challenge is to identify the types of thermodynamic and molecular interactions and other factors that affect the direct ordered assembly of nanofillers into polymer matrices, ensuring efficient scale-up. This involves maintaining close control over the placement and functionalization of each single component forming multifunctional devices. Although various synthetic pathways have been proposed over the years, many unexplored mechanisms are yet to be investigated, and a number of questions about the fabrication, characterization and use of nanocomposites remain unresolved. A multidisciplinary effort is required, along with the development of models describing the formation path of nanostructured hybrid membranes, in order to achieve a more complete picture of the nature and structure of new materials for use in the construction of breakthrough membranes with desirable properties.

5.7 References

Ago, H, Komatsu, T, Ohshima, S, Kuriki, Y and Yumura, Y (2000), 'Dispersion of metal nanoparticles for aligned multiwall carbon nanotube arrays', *Appl Phys Lett* **77**, 79–81.

Ajayan, PM, Schadler, LS and Braun, PV (2003), *Nanocomposite Science and Technology*, Ajayan and Schadle Eds, Wiley VCH, Weinheim, Germany.

Anilkumar, S, Kumaran, MG and Thomas S (2008), 'Characterization of EVA/Clay nanocomposite membranes and its pervaporation performance', *J Phys Chem B*, **112**, 4009–4015.

Aulin, C, Yun, SH, Wågberg, L and Lindstro, T (2009), 'Design of Highly Oleophobic Cellulose Surfaces from Structured Silicon Templates', *ACS Appl Mater Interfaces*, **11**, 2443–2452

Bae, TH, Kim, IC and Tak, TM (2006), 'Preparation and characterization of fouling-resistant TiO_2 self-assembled nanocomposite membranes', *J Membr Sci*, **275**, 1–5.

Balazs, CA, Emrick, T and Russell, TP (2006), 'Nanoparticle Polymer Composites: Where Two Small Worlds Meet', *Science*, **314**(5802), 1107–1110.

Belin, T and Epron, F (2005), 'Characterization methods of carbon nanotubes: A review', *Mater Sci Eng, B*, **119**, 105–118.

Biswas, A, Tokoly, T, Wang, T, Ramidi, P, Ghosh, A, Dervishi, E, Watanabe, F, Biris, AS, Bayer, LS and Norton, MG (2011), 'Design and synthesis of sprayable nanocomposite coatings for carbon capture and direct conversion into environmentally safe stable carbonates', *Chem Phys Lett*, **508**, 276–280.

Boker, A, He, J, Emrick, T and Russell, TP (2007), 'Self-assembly of nanoparticles at interfaces', *Soft Matter*, **3**, 1231–1248.

Bottino, A, Capannelli and G, Asti, VD (2001), 'Preparation and properties of novel organic-inorganic porous membranes', *Sep Purif Technol*, **22–23**, 269–275.

Bottino, A, Capannelli, G and Comite, A (2002), 'Preparation and characterization of novel porous PVDF-ZrO2 composite membranes', *Desalination*, **146**, 35–40.

Cervini, R, Simon, GP, Ginic-Markovic, M, Matisons, JG, Huynh, C and Hawkins, S (2008), 'Aligned silane-treated MWCNTS/ liquid crystal polymer film', *Nanotechnology*, **19**(17) 175602-1–175602-10.

Chen, W and Tao X (2005), 'Self-organizing alignment of carbon nanotubes in thermoplastic polyurethane', *Macromol Rapid Commun*, **26**, 1763–1767

Chen, WF, Wu, JS and Kuo, PL (2008), 'Poly(oxyalkylene)diamine-functionalized carbon nanotube/perfluorosulfonated polymer composites: synthesis, water state, and conductivity', *Chem Mater*, **20**, 5756–5767.

Corry, B (2008), 'Designing carbon nanotube membranes for efficient water desalination', *J Phys Chem B*, **112**(5), 1427–1434.

Dijkstra, M (2005), 'Phase diagram of dipolar hard and soft spheres: manipulation of colloidal crystal structures by an external field', *Phys Rev Lett*, **94**, 138303–138306.

Domènech, B, Muñoz, M, Muraviev, DN and Macanás, J (2011), 'Polymer-stabilized palladium nanoparticles for catalytic membranes: ad hoc polymer fabrication Nanoscale', *Res Lett*, **6**(406), 1–5.

Gast, AP and Zukoski, CF (1989), 'Electrorheological Fluids as Colloidal Suspensions', *Adv Coll Inter Sci*, **30**, 153–202.

Genne, I, Kuypers, S and Leysen, R (1996), 'Effect of the addition of ZrO2 to polysulfone based UF membranes,' *J Membr Sci*, **113**, 343–350.

Gethard, K, Sae-Khow, O and Mitra, S (2011), 'Water Desalination Using Carbon-Nanotube-Enhanced Membrane Distillation', *ACS App Mater Inter*, **3**, 110–114.

Gomes, D, Nunes, SP and Peinemann, KV (2005), 'Membranes for gas separation based on poly(1-trimetylsilyl-1-propyne)-silica nanocomposites', *J Membr Sci*, **246**, 13–25.

Grzelczak, M, Vermant, J, Furst, EM and Liz-Marza´n LM (2010), 'Directed self-assembly of nanoparticles', *ACSNANO*, **4**(7), 3591–3605.

Gugliuzza, A and Drioli, E (2005), 'Evaluation of the CO_2 permeation through functional assembled monolayers: relationships between structure and transport', *Polymer*, **46**(23), 9994–10003.

Guiver, MD, Thi, NL and Robertson, GP (2002), US Patent 20,020,062,737.

Guo, H, Minus, ML, Jagannathan, S and Kumar, S (2010), 'Polyacrylonitrile/carbon nanotube composite films', *ACS Appl Inter Mater*, **2**(5), 1331–1342.

He, Z, Pinnau, I and Morisato, A (2002), 'Nanostructured poly(4-methyl-2-pentyne)/silica hybrid membranes for gas separation', *Desalination*, **146**, 11–15.

Hermanson, KD, Lumsdon, SO, Williams, JP, Kaler, EW and Velev, OD (2001), 'Dielectrophoretic assembly of electrically functional microwires from nanoparticle suspensions', *Science*, **294**, 1082–1086.

Hicke, HG, Becker, M, Paulke BR and Ulbricht, M (2006), 'Covalently coupled nanoparticles in capillary pores as enzyme carrier and as turbulence promoter to facilitate enzymatic polymerizations in flow-through enzyme–membrane reactors', *J Membr Sci*, **282**, 413–422.

Hinds, BJ, Chopra, N, Rantell, T, Andrews, R, Gavalas, V and Bachas, L.G (2004), 'Aligned multiwalled carbon nanotube membranes', *Science*, **303**, 62–65.

Huang, S and Dai, L (2002), 'Plasma etching for purification and controlled opening of aligned carbon nanotubes', *J Phys Chem B*, **106**, 3542–3545.

Hule, RA and Pochan, DJ (2007), 'Polymer nanocomposites for biomedical applications', *MRS Bull*, **32**, 354–358.

Ismail, AF, Goh, PS, Tee, JC, Sanip, SM and Aziz, M (2008), 'Review of purification techniques for carbon nanotubes', *NANO: Brief Reports and Reviews, Nano*, **3**(3), 127–143.

Ismail, AF, Goh, PS, Sanip, SM and Aziz, M (2009), 'Transport and separation properties of carbon nanotube-mixed matrix membrane', *Sep Purif Technol*, **70**, 12–26.

Iwata, M, Adachi, T, Tomidokoro, M, Ohta, M and Kobayashi, T (2003), 'Hybrid sol-gel membranes of polyacrylonitrile–tetraethoxysilane composites for gas permselectivity', *J Appl Polym Sci*, **88**, 1752–1759.

Jeong, HK, Krych, W, Ramanan, H, Nair, S, Marand, E and Tsapatsis, M (2004), 'Fabrication of polymer/selective-flake nanocomposite membranes and their use in gas separation', *Chem Mater*, **16**, 3838–3845.

Jiang, LY, Chung, TS, Cao, C, Huang, Z and Kulprathipanja, S (2005), 'Fundamental understanding of nano-sized zeolite distribution in the formation of the mixed matrix single- and dual-layer asymmetric hollow fiber membranes', *J Membr Sci*, **252**, 89–100.

Jordan, J, Jacob, KI, Tannenbaum, R, Sharaf, MA and Jasiuk, I (2005), 'Experimental trends in polymer nanocomposites – a review', *Mat Sci Eng A*, **393**, 1–11.

Kanzow, H, Lenski, C and Ding, A (2001), 'Single-wall carbon nanotube diameter distributions calculated from experimental parameters', *Phys Rev B*, **63**, 125402–125407.

Khan, AL, Li, X and Vankelecom, IFJ (2011), 'SPEEK/Matrimid blend membranes for CO_2 separation', *J Membr Sci*, **380**, 55–62.

Khan, AL, Cano-Odena, A, Gutiérrez, Minguillón, C and Vankelecom, IFJ (2010), 'Hydrogen separation and purification using polysulfone acrylate-zeolite mixed matrix membranes,' *J Membr Sci*, **350**, 340–346.

Kim, S, Chen, L, Johnson, JK and Maranda, E (2007), 'Polysulfone and functionalized carbon nanotube mixed matrix membranes for gas separation: Theory and experiment', *J Membrane Sci*, **294**, 147–158.

Kotov, NA, (2003), 'Layer-by-layer of nanoparticles and nanocolloids: intermolecular interactions, structure and materials perspectives' in: *Multilayer thin films*, Decher, G, Schlenoff, JB (Eds), Wiley-VCH Verlag GmbH & Co. KGaA, Weinheim, Germany

Ladewig, BP, Knott, RB, Hill, AJ James, Riches, D, White, JW, Martin, DJ, Diniz da Costa, JC and Lu, JQ (2007), 'Physical and electrochemical characterization of nanocomposite membranes of nafion and functionalized silicon oxide', *Chem Mater*, **19**, 2372–2381.

Lakshminarayana, G and Nogami, M (2009), 'Synthesis and characterization of proton conducting inorganic-organic hybrid nanocomposite films from mixed phosphotungstic acid/phosphomolybdicacid/tetramethoxysilane/3glycidoxypropyltrimethoxysilane/phosphoric acid for H_2/O_2 fuel cells', *J Renewable Sustainable Energy*, **1**, 063106, 1–19.

Lewis, RS, Datta, S, Gui, M, Coker, EL, Huggins, FE, Daunertc, S, Bachas, L and Bhattacharyya, D (2011), 'Reactive nanostructured membranes for water purification', *PNAS*, **108**(21), 8577–8582.

Li, N, Zhang, F, Wang, J, Li, S and Zhang, S (2009), 'Dispersions of carbon nanotubes in sulfonated poly[bis(benzimidazobenzisoquinolinones)] and their proton-conducting composite membranes', *Polymer*, **50**, 3600–3608

Liu, H (1997), 'Synthesis of TiO_2 nanopowder enwrapped by organic membrane with microwave induced plasma method', *Huaxue Tongbao*, **10**, 44–46

Liu, Y, Wang, Q, Zhang, L and Wu, T (2005), 'Dynamics and density profile of water in nanotubes as one-dimensional fluid', *Langmuir*, **21**, 12025–12030.

Loh, KS, Lee, YH, Musa, A, Salmah, AA and Zamri, I (2008), 'Use of Fe_3O_4 nanoparticles for enhancement of biosensor response to the herbicide 2,4-dichlorophenoxyacetic acid', *Sensors*, **8**, 5775–5791.

Lu, Y, Yu, SL and Chai, BX (2005), 'Preparation of poly(vinylidene fluoride) (PVDF) ultrafiltration membrane modified by nano-sized alumina (Al_2O_3) and its antifouling research', *Polymer*, **46**, 7701–7706.

Lu, Y, Yu, SL, Chai, BX and Shun, XD (2006), 'Effect of nano-sized Al_2O_3-particle addition on PVDF ultrafiltration membrane performance', *J Membrane Sci*, **276**, 162–167.

Malucelli, G, Priola, A, Amerio, E, Pollicino, A, Di Pasquale, G, Pizzi, D, de Angelis, MG and Doghieri, F (2006), 'Surface and barrier properties of hybrid nanocomposites containing silica and PEO segments', *J Appl Polym Sci*, **103**, 4107–4115.

Mangiagli, PM, Ewing, CS, Xu, K, Wang, Q, Hickner, MA (2009), 'Dynamic water uptake of flexible ion-containing polymer networks', *Fuel Cells*, **9**(4), 432–438.

Merkel, TC, Freeman, BD, Spontak, RJ, He, Z, Pinnau, I, Meakin, P and Hill AJ (2002), 'Ultrapermeable, reverse-selective nanocomposite membranes', *Science*, **296**, 519–521.

Merkel, TC, Freeman, BD, Spontak, RJ, He, Z and Pinnau, I (2003), 'Sorption, transport, and structural evidence for enhanced free volume in poly(4-methyl-2-pentyne)/fumed silica nanocomposite membranes', *Chem Mater*, **15**, 109–123.

Mi, W, Lin, YS and Li, Y (2007), 'Vertically aligned carbon nanotube membranes on macroporous alumina supports', *J Membr Sci*, **304**(1–2), 1–7.

Mistry, MK, Choudhury, NR, Dutta, NK, Knott, R, Shi, Z, and Holdcroft, S (2008), 'Novel organic-inorganic hybrids with increased water retention for elevated temperature proton exchange membrane application', *Chem Mater*, **20**, 6857–6870.

Nednoor, P, Chopra, N, Gavalas, V, Bachas, LG and Hinds, BJ (2005), 'Reversible biochemical switching of ionic transport through aligned carbon nanotube membranes', *Chem Mater*, **17**(14), 3595–3599

Nunes, SP, Peinenmann, KV, Ohlorogge, K, Alpers, A, Keller, M and Pires, ATN (1999), 'Membranes of poly(ether imide) and nanodispersed silica', *J Membr Sci*, **157**, 219–226.

Ozdemir, SS, Buonomenna, MG and Drioli, E (2006), 'Catalytic polymeric membranes: Preparation and application', *Appl Catal A: Gen*, **307**, 167–183.

Pal, R (2008), 'Permeation models for mixed matrix membranes', *J Coll Interf Sci*, **317**, 191–198.

Patel, NP, Miller, AC and Spontak, RJ (2003), 'Highly CO_2-permeable and selective polymer nanocomposite membranes', *Adv Mater*, **15**, 729–733

Patel, NP, Zielinski, JM, Samseth, J and Spontak RJ (2004), 'Effects of pressure and nanoparticle functionality on CO_2-selective nanocomposites derived from cross-linked poly(ethylene glycol)', *Macromol Chem Phys*, **205**, 2409–2419.

Priya, L and Jog, JP (2002), 'Poly(vinylidene fluoride)/clay nanocomposites prepared by melt intercalation: crystallization and dynamic mechanical behavior studies', *J Polym Sci B: Polym Phys*, **40**, 1682–1689.

Promislow, JHE and Gast, AP (1996), 'Magneto rheological fluid structure in a pulsed magnetic field', *Langmuir*, **12**, 4095–4102

Rao, PS, Wey, MY, Tseng, HH, Kumar, IA and Weng, TH (2008), 'A comparison of carbon nanotube molecular sieve membrane with polymer carbon molecular sieve membranes for the gas separation application', *Microporous Mesoporous Mater*, **113**, 499–510

Robeson, LM (1991), 'Correlation of separation factor versus permeability for polymeric membranes', *J Membr Sci*, **61**, 165–185.

Robeson, LM (2008), 'The upper bound revisited', *J Membr Sci*, **320**, 390–400.

Rong, MZ, Zhang, MQ, Zheng, YX, Zeng, HM, Walter, R and Friedrich, K (2001), 'Structure-property relationships of irradiation grafted nanoinorganic particle filled polypropylene composites', *Polymer*, **42**, 167–183.

Samir, MASA, Alloin, F and Dufresne, A (2005), 'Review of recent research into cellulosic whiskers, their properties and their application in nanocomposite field', *Biomacromolecules*, **6**, 612–626

Sheng, N, Boyce, MC, Parks, DM, Rutledge, GC, Abes, JI and Cohen, RE (2004), 'Multiscale micromechanical modeling of polymer/clay nanocomposites and the effective clay particle', *Polymer*, **45**, 487–506.

Sheng, N, Boyce, MC, Parks, DM, Rutledge, GC, Abes, JI and Cohen, RE (2004), 'Multiscale micromechanical modeling of polymer/clay nanocomposites and the effective clay particle', *Polymer*, **45**, 487–506.

Shevchenko, EV, Talapin, Dv, Kotov, NS, O'Brien, S and Murray, BC (2006), 'Structural diversity in binary nanoparticle superlattices', *Nature*, **439**(5), 55–59.

Sidorov, SN, Volkov, IV, Davankov, VA, Tsyurupa, MP, Valetsky, PM, Bronstein, LM., Karlinsey, R, Zwanziger, JW, Matveeva, VG, Sulman, EM, Lakina, NV, Wilder EA and Spontak, RJ (2001), 'Platinum-containing hyper-cross-linked polystyrene as a modifier-free selective catalyst for l-sorbose oxidation', *JACS*, **123**, 10502–10510.

Skirtach, AG, De´jugnat, C, Braun, D, Susha, AS, Rogach, AL and Sukhorukovr, GB (2007), 'Nanoparticles distribution control by polymers: aggregates versus non-aggregates', *J Phys Chem C*, **111**(2), 555–564.

Smith, JJ and Zharov, I (2009), 'Preparation and proton conductivity of sulfonated polymer-modified sintered and self-assembled silica colloidal crystals', *Chem Mater*, **21**, 2013–2019.

Song, M, Xia, HS, Yao, KJ and Hourston, DJ (2005), 'A study on phase morphology and surface properties of polyurethane/organoclay nanocomposite', *Eur Polym J*, **41**, 259–266

Tripathi, BP and Shahi, VK (2009), '3-[[3-(Triethoxysilyl)propyl]amino] propane-1-sulfonic acid-poly(vinyl alcohol) cross-linked zwitterionic polymer electrolyte membranes for direct methanol fuel cell applications', *ACS Inter Mat*, **1**(5), 1002–1012.

Varlot, KM, Reynaud, E, Vigier, G and Varlet J (2002), 'Mechanical properties of clay-reinforced polyamide', *J Polym Sci B: Polym Phys*, **40**, 272–283.

Vollenberg, PHT and Heikens, D (1989), 'Particle size dependence of the Young's modulus of filled polymers: 1. preliminary experiments', *Polymer*, **30**, 1656–1662.

Vu, DQ, Koros, WJ and Miller, SJ (2003a), 'Mixed matrix membranes using carbon molecular sieves. II. Modeling permeation behavior', *J Membr Sci*, **211**(2), 335–348.

Vu, DQ, Koros, WJ and Miller, SJ (2003b), 'Effect of condensable impurity in CO_2/CH_4 gas feeds on performance of mixed matrix membranes using carbon molecular sieves', *J Membrane Sci*, **221**(1–2), 223–239.

Wang, X, Chen, X, Yoon, K, Fang, D, Hsiao, BS and Chu, B (2005), 'High flux filtration medium based on nanofibrous substrate with hydrophilic nanocomposite coating', *Environ Sci Technol*, **39**, 7684–7691

Wara, NM, Francis, LF and Velamakanni, BV (1995), 'Addition of alumina to cellulose acetate membranes', *J Membr Sci*, **104**, 43–49

Weng, TH, Tseng, HH and Weya, MY (2009), 'Preparation and characterization of multi-walled carbon nanotube/PBNPI nanocomposite membrane for H_2/CH_4 separation', *Int J Hydrogen Energ*, **34**, 8707–8715.

Widjojo, N, Chung, TS and Kulprathipanja, S (2008), 'The fabrication of hollow fiber membranes with double-layer mixed-matrix materials for gas separation', *J Membrane Sci*, **325**, 326–335.

Wu, CL, Zhang, MQ, Rong, MZ and Friedrich, K (2002), 'Tensile performance improvement of low nanoparticles filled-polypropylene composites', *Compos Sci Technol*, **62**, 1327–1340.

Wu, CL, Zhang, MQ, Rong, MZ and Friedrich, K (2005), 'Silica nanoparticles filled polypropylene: effects of particle surface treatment, matrix ductility and particle species on mechanical performance of the composites', *Compos Sci Technol*, **65**(3–4), 635–645.

Xiao, YC, Wang, KY, Chung, TS and Tan J (2006), 'Evolution of nano-particle distribution during the fabrication of mixed matrix TiO_2-polyimide hollow fiber membranes', *Chem Eng Sci*, **61**, 6228–6233.

Xu, K, Chanthad, C, Gadinski, MR, Hickner, MA and Wang, Q (2009), 'Acid-functionalized polysilsesquioxane-nafion composite membranes with high proton conductivity and enhanced selectivity', *ACS Inter Mat*, **1**(11), 2573–2579.

Xu, L, Yu, L and Han, LF (2009), 'Polymer-nanoinorganic particles composite membranes: a brief overview', *Front Chem Eng China*, **3**(3), 318–329.

Yang, YN, Wang, P and Zheng, QZ (2006), 'Preparation and properties of polysulfone/TiO_2 composite ultrafiltration membranes', *J Polym Sci Part B: Polym Phys*, **44**, 879–887.

Yang, YN, Zhang, HX, Wang, P, Zheng, QZ and Li J (2007), 'The influence of nano-sized TiO_2 fillers on the morphologies and properties of PSF UF membrane', *J Membr Sci*, **288**, 231–238.

Yu, J, Liu, RYF, Poon, B, Nazarenko, S, Koloski, T, Vargo, T, Hiltner, A, and Baer, E (2004), 'Polymers with palladium nanoparticles as active membrane materials', *J Appl Pol Sci*, **92**, 749–756.

Zoppi, RA, das Neves, S and Nunes, SP (2000), 'Hybrid films of poly(ethylene oxide-b-amide-6) containing sol-gel silicon or titanium oxide as inorganic fillers: effect of morphology and mechanical properties on gas permeability', *Polymer*, **41**(14), 5461–5470.

5.8 Appendix: nomenclature

5.8.1 Notation

α	separation factor
π	ligand

5.8.2 Abbreviations

ALP	alkaline phosphatase
AlPO	aluminophosphate
Barrer	gas permeability coefficient expressed as 10^{-10} (cm^3 (STP) cm)/(cm^2 s cmHg)
BEMA	α,ω-dimethacrylate PEO oligomer
$BPPO_{dp}$	brominated poly(2,6-diphenyl-1,4-phenylene oxide)
$BPPO_{dmp}$	brominated poly(2,6-dimethyl-1,4-phenylene oxide)
CMS	carbon molecular sieve
CNT	carbon nanotube
2,4-D	2,4-dichlorophenoxyacetic acid

DWNT	dual-walled nanotube
EMR	enzyme–membrane reactor
EVA	poly(ethylene-*co*-vinyl acetate)
FTF	fructosyltransferase
GOx	glucose oxidase
MEMO	methacryloyl-oxypropyltrimethoxysilane
MFI	ZSM-5 and silicalite-1
MMM	mixed matrix membranes
MWNT	multiwalled nanotube
PA	polyamide
PAA	polyacrylic acid
PAN	polyacrylonitrile
PBNPI	poly(bisphenol A-*co*-4-nitrophthalicanhydride-*co*-1,3-phenylene diamine)
PDMS	polydimethylsiloxane
PEBAX	poly(ethylene oxide-β-amide)
PEG	polyethylene glycol
PEM	polymer electrolyte membrane
PEO	polyethyleneoxide
PFOTS	1H,1H,2H,2H-perfluorooctyltrichlorosilane
PI	polyimide
PMP	poly(4-methyl-2-pentyne)
PP	polypropylene
PPEPG	polyphenylene ether polyglycol
PS	polystyrene
PSF	polysulfone
PTMSP	poly [1-(trimethyl-silyl) propine]
PU	polyurethane
PVA	poly(vinyl alcohol)
PVDF	polyvinylidene fluoride
SPBIB	sulfonated poly[bis(benzimidazobenzisoquinolinones)]
SWNT	single-walled nanotube
TEOS	tetraethoxysilane
TPU	thermoplastic polyurethane

Part II

Zeolite, ceramic and carbon membranes and catalysts for membrane reactors

6
Zeolite membrane reactors

C. ALGIERI, ITM-CNR, Italy and A. COMITE and
G. CAPANNELLI, University of Genoa, Italy

DOI: 10.1533/9780857097330.2.245

Abstract: Zeolite membranes, with their well-defined pore size, and high thermal and chemical stability, can be used as membrane reactors. However, the development of zeolite membrane reactors requires the preparation of defect-free membranes. Different methods for their preparation have been developed, including *in situ* and secondary growth methods. The secondary growth method has several advantages over the *in situ* method, such as easier operation, higher controllability in crystal orientation, a thicker microstructure and a higher level of reproducibility.

Despite enormous efforts made to improve the quality of these membranes, their industrial application on a large scale is confined to T and NaA zeolites to use in pervaporation and vapour permeation processes. In addition, commercial zeolitic membranes cannot yet be employed in gas separation because of the presence of inter-crystalline defects. This problem, along with other disadvantages such as reproducibility problems in the preparation step, the cost of the membrane modules and their sealing at high temperatures, has meant that there are relatively few papers on the use of zeolite membrane reactors in processes of industrial interest. Therefore, further improvements are necessary to facilitate the introduction of zeolite membrane reactors into industry.

Key words: zeolite, zeolite membranes, zeolite membrane reactors.

6.1 Introduction

Zeolite membranes have attracted a lot of interest for their uniform pore size at molecular scale, which allows the separation of liquid and gaseous mixtures in a continuous way. Because of their thermal and chemical stability, they can also be used in processes at high temperatures and in the presence of organic solvents where polymeric membranes fail. In addition, zeolite materials exhibit intrinsic catalytic properties which clearly suggests the use of zeolite membranes as catalytic membrane reactors (CMRs). In the last two decades, enormous progress on zeolite membrane synthesis has been made, but only 20 structures are used for membrane preparation even if 170 zeolitic structures are available today (Baerlocher *et al.*, 2007). The high cost and poor reproducibility in the synthesis step hinder the application of the zeolite membranes at industrial level (Caro *et al.*, 2005; Mcleary *et al.*, 2006). Until now, only NaA and T-type zeolite membranes have been commercialized and used in alcohol dehydration and solvent dewatering (Kondo and Kita, 2010). Their success in these applications has encouraged a few companies to commercialize other types of zeolite membrane. However, they are still not used for gas separation, since they show very low separation factors due to the presence of inter-crystalline defects (Noack *et al.*, 2001).

The concept of a zeolite membrane reactor is based on the use of zeolite as a membrane implemented in a reactor in order to carry out chemical reactions. Based on this definition, a wide variety of reactor arrangements and configurations can be envisaged. For example:

(a) the availability of different zeolite structures and types,
(b) the positioning of the zeolite membrane in the reactor,
(c) the reactant flow configuration with respect to the zeolite layer.

Another classification of zeolite membrane reactors is based on the aim of the application, in the same way as that generally adopted for inorganic membrane reactors:

(a) Conversion enhancement by:
  (i) product removal
  (ii) removal of catalyst poisons of inhibitor products.
(b) Selectivity enhancement by:
  (i) reactant distribution
  (ii) control of the residence time
  (iii) control of the reactant traffic.

In this chapter the application of zeolite membranes as membrane reactors will be illustrated. A short overview on the methods to synthesize membranes will be given. Basic concepts on mass transport through zeolite membranes will be also shown. Finally, the application of the zeolite membranes for use as reactors will be discussed in more detail.

6.1.1 Synthesis of zeolite membranes

Zeolite membranes are usually synthesized on porous supports of alumina or stainless steel, since a self-standing zeolite layer is very fragile. The most common methods used for the synthesis of zeolite membranes are:

(a) direct *in situ* crystallization
(b) vapour-phase transport
(c) secondary growth.

In the *in situ* method the support is put in contact with a synthesis solution or a gel under hydrothermal conditions. At the same time, under appropriate conditions, zeolite nuclei are formed on the support and they will grow by forming a zeolite layer. Zeolite crystals heterogeneously nucleated on the support surface are also present. This technique does not give the possibility of controlling the nucleation of the crystals which is necessary to form a very uniform and compact zeolite layer (Li *et al.*, 2003).

In the vapour-phase transport method (Matsukata *et al.*, 1994), the precursor is deposited on the support as a gel layer which is converted to zeolite under the action of the heated (water and template) vapour. This method is convenient because it minimizes the use of the structure-directing agent, which can be quite expensive. The disadvantages of this method are the necessity of depositing a very thick gel layer on the support in order to achieve the zeolite film formation and the presence of cracks in the zeolite layer (Tsay and Chiang, 2000).

The secondary growth (Lovallo and Tsapatsis, 1996) method has two steps: seeding and growth. During the first step, zeolite seeds are deposited on the surface of the support before the hydrothermal treatment. The hydrothermal synthesis favours the formation of a dense zeolite layer for the regrowth of the seeds. This method of decoupling the nucleation from the crystal growth makes it possible to optimize the conditions of each state independently. In fact, this method allows for better control of the membrane structure and a higher reproducibility than the *in situ* method. Seeding is a crucial step because it influences the membrane quality. Currently, different seeding procedures are used to deposit the zeolite nuclei. The most frequently used are dip coating (Lovallo and Tsapatsis, 1996; Bernal *et al.*, 2001) and rubbing (Hasegawa *et al.*, 2002). Other seeding techniques are spin coating (Mintova and Bein, 2001) and the use of a cationic polymer to favour the adhesion of the zeolite seeds on the support (Hedlund *et al.*, 1999). However, these procedures present some limitations. For example, the rubbing technique is more appropriate for preparing membranes having the selective layer on the external surface of the tubular supports, the dip coating needs to be repeated more times to create an almost uniform layer and the spin coating permits the preparation of the zeolite layer on planar supports. The seeding procedures that involve filtration are more controllable and are also suitable for sowing the inner surface of tubular supports (Takata *et al.*, 2002; Huang *et al.*, 2004; Pera-Titus *et al.*, 2008). There are two filtration modes: dead-end and cross-flow. The second allows a more uniform and compact zeolite layer because the zeolite slurry is pumped tangentially along the surface of the support. During the filtration, the positioning of tubular support in a horizontal position and without rotation causes an inhomogeneous coverage of the support. To overcome this disadvantage, a new seeding procedure for tubular membranes has been designed, which combines the support tilting and rotation with the cross-flow filtration of the zeolite water suspension through the support (Algieri *et al.*, 2009).

Recently, microwave and continuous-flow synthesis have been proposed for the preparation of zeolite membranes. The microwave route offers some advantages over the conventional hydrothermal synthesis, including a shorter synthesis time and the possibility of synthesizing very small zeolite particles with a very narrow particle size distribution (Xu *et al.*, 2004; Huang and Yang, 2007). For film produced via microwave heating, synthesis can proceed in a manner similar to *in situ* crystallization. Alternatively, the substrate can be dip-coated in the zeolite precursor gel prior to crystallization. Microwave synthesis leads to different membrane characteristics from those synthesized by conventional heating. For example, membranes prepared by microwave heating have a higher permeance (Xu *et al.*, 2000; Cheng *et al.*, 2002).

In continuous-flow synthesis the reactants are continuously supplied on the lumen side of the supports (Pera-Titus *et al.*, 2008). A continuous process is desirable because it would: (1) be energy efficient by eliminating the high energy consumption required for the repeated heat-up and cool-down in batch crystallizers, (2) require smaller equipment and lower capital costs, and (3) produce a more uniform product because of the controlled operating conditions. However, working at low flow rates, mass and heat transfer rates are also low and non-uniform zeolitic layers are synthesized (Ju *et al.*, 2006).

6.1.2 Zeolite membrane characterization

The morphological quality of zeolite membranes can be evaluated using different techniques. The thickness and morphology of the zeolite films are evaluated using scanning electron microscopy (SEM). In particular, a cross-sectional view shows the thickness of the zeolite layer on the support and the top view shows the size and also the shape of the crystals. Figure 6.1 shows the top view of a supported mordenite framework inverted (MFI) zeolite membrane.

Energy dispersive X-ray (EDX) is used to determine the bulk Si/Al ratio and elemental compositions of zeolite membranes. Figure 6.2 shows the SEM cross-section with the elemental composition obtained by an EDX probe.

X-ray analysis is used to verify the topology of the crystals grown on the porous support. Besides, the intensity of the peaks gives information about the orientation of the crystals.

Fluorescence confocal optical microscopy is a powerful tool for the non-destructive analysis of zeolite membranes. The grain boundary network of the zeolite layer can be observed along the thickness of the membranes and so the defects can be clearly visualized (Bonilla *et al.*, 2001).

Nitrogen adsorption experiments are typically used to determine the micropore volume and porosity of the zeolite powders. However, this

6.1 Top view of an MFI supported zeolite membrane prior to calcination. (Reprinted from *Microporous and Mesoporous Materials* 47, Algieri C, Golemme G, Kallus S, Ramsay J D F, 'Preparation of thin supported membranes by *in situ* nucleation and secondary growth', 127–134, Copyright (2001) with permission of Elsevier.)

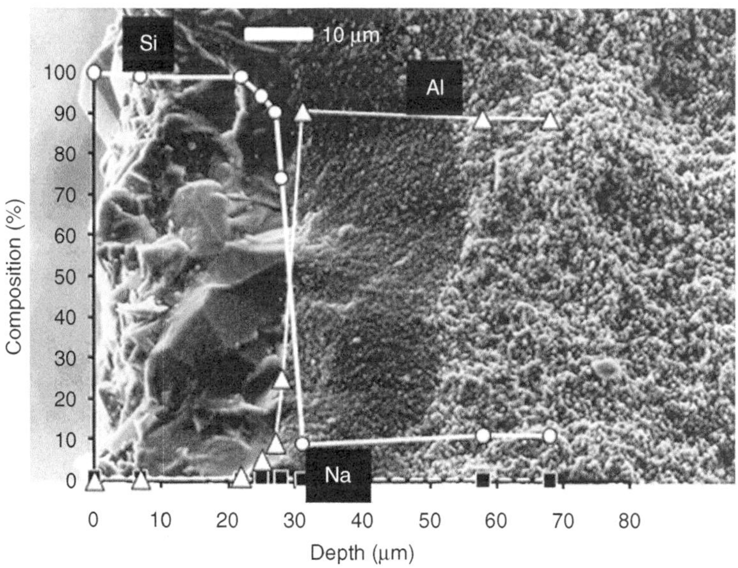

6.2 SEM cross-section of a ZSM-5 membrane obtained at the University of Genoa. The zeolite crystalline layer is about 30 µm thick. The penetration and anchoring of the zeolite layer in the tortuous pore structure of the support is indicated by the presence of Si.

method is difficult to use for supported zeolite membranes, because the supports generally do not fit inside the sample tubes of commercial equipment. An alternative method for the determination of porosity in thin films is the porosimetry, which allows to analyse the contribution of micropores and defects to the overall flux through the membrane (Hedlund *et al.*, 2002). In this method, He permeance through the membrane is measured while the activity of a condensable hydrocarbon is increased stepwise from 0 to 1. The hydrocarbons normally employed are *n*-hexane and *p*-xylene.

6.2 Separation using zeolite membranes

Mass transport through the zeolite layer is the result of five steps (den Exter *et al.*, 1996; Burggraaf, 1999):

1. Adsorption of the species on the external surface of the membrane
2. Mass transport from the external surface into the zeolite pore
3. Intra-crystalline zeolite diffusion
4. Mass transport out of the zeolite pores to the external surface
5. Desorption from the external surface to the bulk.

Steps 1 and 5 depend on the following factors: permeation conditions (temperature and partial pressures), the nature of the chemical species and the type of crystalline material. Steps 1 and 5 are, generally, assumed to be fast processes. Steps 2, 3 and 4 are usually activated processes (Barrer, 1990). Intra-crystalline zeolite diffusion is described as configurational diffusion.

The different situations that can be encountered analysing the permeation of molecules through a porous membrane are reported below. When the pore diameter of a porous solid is in the macropore range, collisions between the molecules will occur much more frequently than collisions with the wall. In this case molecular diffusion is the dominant mechanism. As the size of the pores decreases (mesoporous solid), the number of collisions with the wall increases and can become more frequent and important than the molecule–molecule collisions. At this point, Knudsen diffusion takes over. When the pore diameter becomes comparable to the size of the molecules (microporous solid), the molecules continuously collide with the walls. When this happens, diffusion behaves as an activated process and the term 'configurational regime' is used to describe it.

Intra-crystalline permeation through a zeolite membrane can be described using different approaches (Krishna, 2006). For example, in the Fickian approach, the concentration gradient is the driving force through the zeolite membrane; whereas in the Maxwell–Stefan (MS) approach the gradient of the thermodynamic potential is the driving force. The MS approach allows an approximation of the transport flux for multicomponent mixtures by using information from single components.

Considering the permeation of a single component through an ideal zeolite membrane in a wide temperature range, following the Fickian approach, it can be assumed that the total flux N is a combination of the 'surface flux', N^s, (which takes place at low–medium temperatures) and the 'activated gaseous flux', N^g (which can occur at high temperatures) (Xiao and Wei, 1992; den Exter *et al.*, 1996; Burggraaf, 1999).

$$N = N^s + N^g \quad [6.1]$$

Surface flux can be described using the following equation:

$$N^S = -D^S \frac{\mathrm{d}c}{\mathrm{d}z} \quad [6.2]$$

where D^s is the Fick diffusivity for the surface flux, which can be expressed as:

$$D^S = D_o\Gamma \quad [6.3]$$

where D_o is the 'corrected' or 'intrinsic diffusivity' and Γ is the 'thermodynamic correction factor'

$$\Gamma = \frac{\mathrm{d}\ln p_i}{\mathrm{d}\ln c_i} \quad [6.4]$$

Temperature dependence of the transport diffusivity increases with the temperature. An Arrhenius-type dependence on the temperature is often assumed.

$$D_o = D_\infty e^{-E_{\mathrm{Diff}}/RT} \quad [6.5]$$

In any case, the overall dependence on the temperature will be affected by the adsorption, depending on the pressure and temperature conditions.

The thermodynamic correction-factor value depends on the concentration of the adsorbed phase (loading), and it can be quite large. In turn, the loading depends on the type of adsorption that takes place in the zeolite micropores.

The Langmuir isotherm for a single adsorbed component is represented by the following equation:

$$\vartheta = \frac{q}{q_{\text{sat}}} = \frac{bp}{1+bp} \qquad [6.6]$$

Then

$$\Gamma = \frac{\mathrm{d}\ln p}{\mathrm{d}\ln q} = \frac{1}{1-\vartheta} \qquad [6.7]$$

and the surface flux is more explicitly described by:

$$N^s = -\frac{D_o}{1-\vartheta}\rho_{\text{app}}q_{\text{sat}}\frac{\mathrm{d}\vartheta}{\mathrm{d}z} \qquad [6.8]$$

At high temperatures, the adsorption phenomena can become negligible and molecules can be considered as being in a quasi-gaseous state even within the constrained environment of the zeolite framework. This state is referred to as 'gas translational diffusion', 'activated gaseous diffusion' or 'activated-Knudsen diffusion'.

When activated gaseous diffusion occurs, the flux is described by the next equation:

$$N^g = -\frac{D^g}{RT}\frac{\mathrm{d}p}{\mathrm{d}z} \qquad [6.9]$$

The driving force for permeation is the pressure gradient $\mathrm{d}p/\mathrm{d}z$.

The diffusion coefficient will depend on gas molecular velocity:

$$D^g = \mathrm{d}_p u_m e^{-E_e/RT} \qquad [6.10]$$

For ideal gases, molecular velocity can be calculated through the kinetic theory

$$u_m = \sqrt{\frac{8RT}{\pi M}} \qquad [6.11]$$

From these equations it can be deduced that single gas transport through a zeolite membrane will be strongly affected by the nature of the adsorptive-interaction between the permeating molecule and the zeolite. The behaviour of the permeating flux should increase with the temperature, but a maximum can be observed depending on which diffusion mechanism is dominant. In the presence of defects in the membrane layer, Knudsen diffusion and viscous mass transport can greatly contribute to the overall flux and will strongly influence its expected temperature dependence.

Although a simple approach has been used, the complexity of predicting the mass transport and separation through real zeolite membranes, where inter-crystalline defects also need to be taken into account, is very evident. High-selectivity separations can be achieved by using nearly perfect zeolite membranes. In order to be suitable for an industrial-scale application, beside the high permselectivity, zeolite membranes should have a high permeation flux, which is accomplished with a very low membrane thickness. Unfortunately, by decreasing the membrane thickness the negative influence of inter-crystalline defects on permselectivity can be very detrimental, if proper synthesis procedures are not employed. The thickness of a zeolite layer depends on the synthesis route and conditions and on the number of depositions. For example, a ZSM-5 membrane obtained by direct *in situ* crystallization with a two-step deposition showed a thickness between 30–40 μm. Anyway, at laboratory level, zeolite membranes with a thickness in the order of few microns can be obtained with a sufficient quality (White *et al.*, 2010).

Efforts are ongoing to find a way to avoid, reduce or eliminate the presence of inter-crystalline defects, which, aside from poor synthesis reproducibility, are the main obstacle to the industrial application of zeolite membranes for gas separation.

Nevertheless, if mixtures of gas and vapour of higher molecular mass species, or liquid mixtures of two species with different volatility and surface tension, are considered, the separation factors and permeation fluxes can be very interesting, but such separations cannot be predicted from the pure gas permeance. Silicalite membranes are hydrophobic and preferentially adsorb organic molecules that are small enough to enter the pore openings. Therefore, they can be used to separate hydrocarbon mixtures with relatively high separation factors. The selectivity for *n*-heptane isooctane has a maximum of 138 at 373 K for the ternary mixture of isooctane, *n*-heptane and *n*-hexane (Funke *et al.*, 1996).

The description of the separation of multicomponent mixtures requires a more complex approach, for example by using Maxwell–Stefan methodology. However, the real membrane often assumes a more complex structure, in which, beside the microporous zeolite layer, the mesoporosity of the intra-crystalline-defects and of the underlying support can play an important role, especially when the capillary condensation phenomenon can occur, as in the case of the permeation of vapour. Kondo and Kita (Kondo and Kita, 2010) attempted an interpretation of the dehydration process by including narrow non-zeolitic pores into the support. The water molecules in the feed selectively adsorbed in zeolite pores are then transported to the non-zeolitic pore, where they are released in the permeate side of the membrane.

From a practical point of view, solvent dehydration is the only separation process based on the use of a zeolite membrane that has found industrial application. Mitsui Engineering and Shipbuilding Co. (Japan) introduced its zeolite membrane pervaporation process to the market in 2000. Hermsdorfer Institut für Technische Keramik e.V. with Inocermic GmbH (Germany) (now merged into Fraunhofer-Institut für Keramische Technologien und Systeme IKTS) have developed a NaA zeolite membrane for the vapour permeation and dewatering of ethanol (from 20 kg/(m^2h) of permeate flux at 135°C down to a water concentration of about 0.2%). Hitachi Zosen Corporation (Hitz) has developed a dehydration system, using NaA membrane elements, which was adopted for the first commercial bio-ethanol plant with the capacity of 50 000 L/d in Japan, 2008. In 2008, Sato and co-workers (Sato *et al.*, 2008) achieved the synthesis of industrial-scale Na-Y zeolite membranes. Anyway, due to the high cost of the zeolitic membranes, their application in dehydration processes is limited to applications with very low water contents.

6.3 Zeolite membrane reactors

In the case of equilibrium-limited reactions, the removal of products increases the conversion. In the field of zeolite membrane reactors this concept has been used for hydrogen removal in dehydrogenation reactions. Zeolite membranes having MFI topology were almost always used for the conversion of alkanes to olefins. Dehydrogenation of iso-butane has been investigated in a membrane reactor combining a PtIn/zeolite catalyst and a supported MFI membrane with tubular configuration (Ciavarella *et al.*, 2001). The results put in evidence as the isobutene yield was four times higher than that obtained using a conventional reactor. In another paper the performance of two different membranes (MFI zeolite and Pd membranes) in an extractor-type CMR using a trimetallic Pt-In-Ge catalyst were compared (van Dyk *et al.*, 2003). Both CMRs gave better results than those obtained using conventional reactors. The two CMRs presented different separation properties but very similar yields; this result was attributed to the lack of efficiency of the catalyst in the two membrane reactors. Caro and coworkers (Illigen *et al.*, 2001), studying also the dehydrogenation of iso-butane, showed that the H_2/iso-butane mixture separation factor was near 1 at room temperature and increased to 70 at 500°C. These results can be explained by the fact that, at low temperature, permeation is controlled by the adsorption and the permeate is enriched in butane. Increasing the temperature, diffusion becomes the dominant mechanism because the butane is less adsorbed. Moreover, for the different experimental conditions considered, the membrane reactor showed higher iso-butane conversion with respect to the conversion obtained using a conventional reactor, as shown in Fig. 6.3. The iso-butane

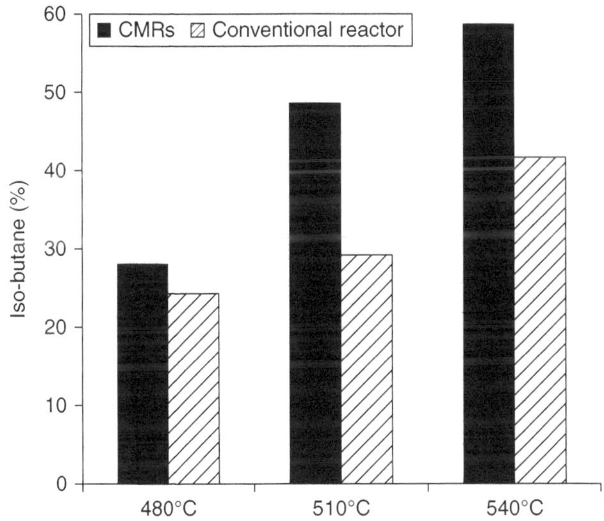

6.3 Iso-butane conversion with and without H_2 removal. Weight-hourly-space-velocity (WHSV) = 0.5 h^{-1}. (Elaborated from Illigen *et al.*, 2001.)

selectivity was found to be less than 96%, while with the packed-bed reactor (PBR) it is under 90%.

Styrene is one of the most important monomers used for the production of synthetic rubbers and plastics. About 90% of this olefin is produced by the catalytic dehydrogenation of ethylbenzene. Different membrane-types have been investigated for this reaction. A good result was obtained using a palladium membrane reactor because the hydrogen selectively permeates through a palladium membrane by the solution–diffusion mechanism. However, these membranes are very expensive and undergo degradation at high temperatures. Recently, a silicalite (MFI topology) membrane was used for carrying out this reaction (Kong *et al.*, 2007). The conversion with the CMR was 7% higher than with the fixed bed reactor – above 600°C without a decrease of the styrene selectivity. As the temperature increased from 580°C to 650°C, the conversion increased. The explanation of this result is the same as that given by Caro and coworkers (Illigen *et al.*, 2001). It was also found that the conversion decreased, increasing the space velocity due to the low residence time in both membrane and fixed bed reactor.

A zeolite membrane having MFI topology was also used to test the reaction of methanol to olefins (Masuda *et al.*, 2003). The authors produced olefins with a high selectivity (80–90%) and high conversion (60–98%) by adjusting the diffusion and chemical reaction rates of the molecular species in the zeolite layer of the membrane.

The *para*-xylene recovery is a very important step in a large petrochemical plant. Xylene has three isomers: *para*-xylene (PX), *meta*-xylene (MX) and *ortho*-xylene (OX), used as solvents or intermediates of different derivatives (Lai *et al.*, 2003). However, the most important isomer is PX because it is the feed for the production of terephthalic acid which in turn is used for the synthesis of polyester resin and fibres (Xomeritakis *et al.*, 2001). Extensive studies of xylene separation using zeolite membranes, particularly MFI topology, have been carried out by different research groups in the past few years. For example, Dittmeyer and coworkers (Haag *et al.*, 2006) studied xylene isomerization and PX separation by using an H-ZSM-5 membrane with a flat configuration. The experimental results indicated that the zeolite membrane improved the catalytic result in comparison to the conventional PBR. In fact, the conversions of MX and OX with the CMR at 400°C were higher at 15% and 4.4%, respectively. This increase is due to the removal of PX from the membrane, which is possible because the MFI zeolite has a pore size closer to the kinetic diameter of the PX (~0.58 nm) than the other two isomers. The membrane, nevertheless, showed a poor separation performance, mainly due to the diffusion of different isomers through the defects which have a negative effect on the conversion. This result provided further evidence of the necessity of preparing defect-free zeolite membranes before they can be applied as CMRs. Recent studies on the isomerization of MX using a MFI membrane loaded with Pt-HZSM-5 catalyst (Daramola *et al.*, 2010) demonstrated the possibility of obtaining ultra-pure PX in the permeate side and with PX selectivity close to 100%. These results showed that it is possible to reduce the operational costs through a reduction in energy consumption during the ultra-pure PX production. It is, however, necessary to prepare defect-free membranes with appreciable PX flux in order to make this technology attractive and competitive with existing technologies.

MFI zeolite membranes which have a partial modification of the zeolite channels are able to achieve a high selectivity and permeance during a hydrogen separation from water gas shift reaction at a high temperature (Tang *et al.*, 2010). The zeolite membrane was modified by the *in situ* catalytic cracking of methyl diethoxysilane (Gu *et al.*, 2008). The modified zeolite membrane exhibited a H_2/CO_2 permselectivity of 68.3 with a hydrogen permeance of 2.94×10^{-7} mol m^{-2} s^{-1} Pa^{-1}. The modified membrane also exhibited a high stability in the temperature range 400–550°C. Furthermore, at 550°C, the membrane reactor achieved a CO conversion of 81.7 %. This is higher than that obtained using a PBR.

The Fischer–Tropsch synthesis (FTS) allows the synthesis of liquid hydrocarbons from various feedstocks, such as coal and natural gas, and has gained increasing interest in recent years. The removal of water, a reaction product, are threefold: to reduce the deactivation of the catalysts; to increase the reactor productivity; and to enhance the conversion of CO_2 to long-chain hydrocarbons by displacing the equilibrium of the water gas shift reaction on the reaction rate (Rohde *et al.*, 2005). Different hydrophilic zeolite membranes were used for the selective removal of water from mixtures of H_2 and CO. For example, ZSM-5 and mordenite membranes were employed for the water removal under conditions typical of the FTS (Espinoza *et al.*, 2000). Mordenite membranes exhibited very high H_2O fluxes and good permselectivities. For the same topic, Lynde Type A (LTA) (NaA) zeolite membrane was used to study the permeation of single components of water, vapour, CO, H_2, CH_4 and their binary mixtures (Zhu *et al.*, 2005). The permeance of water vapour in the binary mixture is almost similar to the value found as single species. However, the permeance of the gas components decreased when the water was present. These results can be explained by the fact that the adsorbed water molecules in the membrane block the other gas species. Increasing the temperature, the amount of water adsorbed in the membrane slightly decreases and the selectivity for water in the binary mixture decreases. Kapteijn and coworkers (Rohde *et al.*, 2008) illustrated the potential of the FT membrane-reactor concept by mathematical study. In particular, two cases were studied using the dimensionless model of the membrane reactor with Co- and Fe-based catalysts. This study demonstrated that the removal of water is very important for the protection of Co-based catalysts and for conversion enhancement in Fe-based catalysts. In the same work, a hydroxyl sodalite membrane on an alumina support was prepared by direct hydrothermal synthesis and exhibited an ideal H_2O/H_2 selectivity of above 10^6. The membrane also exhibited good thermal and mechanical stability making it a promising candidate for FT synthesis.

The zeolite membranes can also act as a distributor in order to control the addition of reactants to the catalyst and to cut down the side reactions. The use of membrane reactors seems to be very relevant for carrying out oxidative dehydrogenation of alkanes to control the oxygen feeding in order to limit the highly exothermic total combustion (Pantazidis *et al.*, 1995). For example, V-MFI membranes were used for the oxidative dehydrogenation of propane (Julbe *et al.*, 2000). These membranes are active for the partial oxidation of propane at 550°C. However, the oxygen distributor configuration did not improve the reactor performance compared with the flow-through that can be attributed to back-mixing effects. In these cases the membrane was catalytically inert and coupled with a conventional fixed bed of catalysts, which were positioned on one side of the membrane. Another possible application is to use the membrane as an active contactor which is catalytically active but not necessarily permselective (Julbe *et al.*, 2001). In particular, a catalytic zeolite membrane, with catalytically active particles dispersed into a very thin zeolite layer, ensures a good contact between the reactants and the active site of the catalyst. This reduces the by-pass problems present in the PBR and reduces the pressure drop (Bernardo *et al.*, 2008). The same authors (Bernardo *et al.*, 2008) also studied carbon monoxide selective oxidation (Selox) from hydrogen-rich gas streams by using

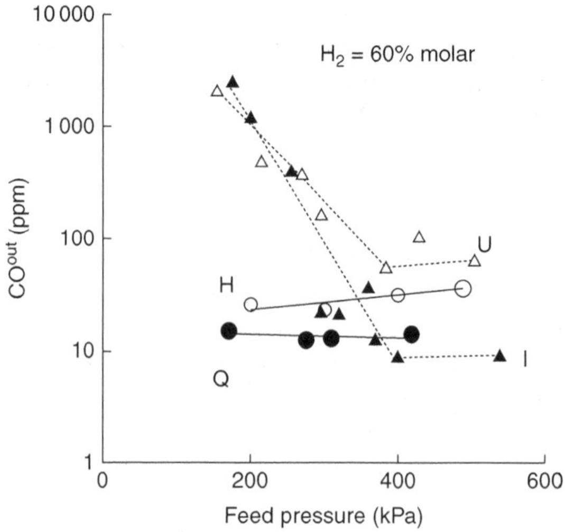

6.4 CO concentration measured at the MR exit versus feed pressure for different Pt/Na-Y membranes. (Reprinted from *Separation and Purification Technology* 62, Bernardo P, Algieri C, Barbieri G, Drioli E, 'Hydrogen purification from carbon monoxide by means of selective oxidation using zeolite catalytic membranes', 629–635, Copyright (2008) with permission of Elsevier.)

catalytic zeolite membranes (Pt/Na-Y). The catalytic membranes reduced an amount of CO from 10 000 ppm down to 10–50 ppm, depending on the operating conditions. The results in terms of CO conversion using Pt loaded zeolite membranes and characterized by different permeation properties are shown in Fig. 6.4. The pressure effect is negligible for the less permeable membranes (H and Q), while it is positive for more permeable membranes (I and U). Membrane Q showed the best permeation and Selox performance, reducing the CO amount down to 10 ppm at a low reaction pressure. The effect of the pressure on CO conversion can be explained by considering different possible paths: into the zeolite channels and through the defects. CO permeating into the zeolite channels has a very high probability of interacting with the catalyst and, thus, to react because of the very small pore size. In this case, at any pressure (in the range investigated) the interaction between reactants and catalytic sites is the same. The reactant molecules through defects have a large volume, owing to the pore size of the support, and the interaction with the catalyst particle depends on the pressure. The experimental results indicate that the membranes H and Q are almost defect-free, while the membranes I and U present defects into the zeolite layer. These results confirm the good potentiality of catalytic zeolite membranes for a deep purification of H_2-rich streams, and with the possibility to use the hydrogen for fuel cell applications.

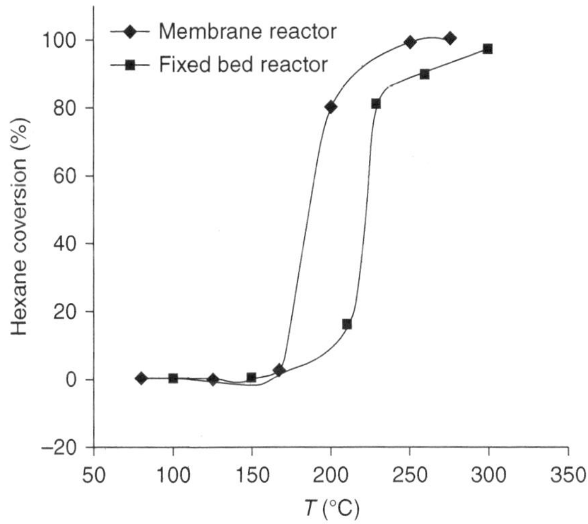

6.5 Comparison of the performance of membrane and fixed bed reactor configurations. Space velocity 300 h^{-1}. (Elaborated from Aguado *et al.*, 2005.)

Santamaria and coworkers (Aguado *et al.*, 2005) prepared Pt/ZSM-5 membrane reactors for the combustion of *n*-hexane present at a low concentration in the air. Experimental results showed that *n*-hexane combustion was achieved at 210°C. A comparison of the conversion of the hexane obtained using the membrane reactor and the fixed bed reactor evidenced the better performance of the membrane reactor with a light-off temperature lower (about 70°C) than that obtained in the fixed bed reactor as illustrated in Fig. 6.5. Another example of the successful application of a Pt/ZSM-5 membrane in the catalytic combustion of volatile organic compounds (VOCs) (Bottino *et al.*, 2001) showed that a high removal efficiency of toluene can be obtained and that the zeolite membrane with a relatively thick layer (40 μm) is clearly affected by mass transfer limitations.

Bottino and coworkers studied propane oxide hydrogenation by membrane reactors by using a V/ZSM5 membrane (Bottino *et al.*, 2002). The yields in oxidative dehydrogenation of propane with this system were low, due to the intrinsic catalytic activity negatively affecting the selectivity of the zeolite towards the intermediate products. The reported data showed that a thick and dense zeolite layer can perform reactant segregation very well and acts as an oxygen distributor. Moreover, the introduction of the two reactants (propane and oxygen) from the two opposite sides of the membrane (tube and shell) leads to a local profile of the hydrocarbon/oxygen ratio in the catalytic layer, which is unfavourable to the over-oxidation of propylene.

Considering the application of a zeolite membrane as a reactor for high temperature processes, therefore, the following aspects should be taken into account:

- Zeolite membranes, due to their microporous nature and hence the low diffusivity coefficients, could work under mass transfer limited regime due to the high thickness of the zeolite layer. When the membrane acts as a reactant distributor the mass transfer should help to obtain the desired reactant profile in the reactor. In a catalytic zeolite membrane reactor, when the intrinsic reaction rate is higher than the diffusion rate, the mass transfer becomes a limiting factor for the conversion.
- Zeolites may show an intrinsic catalytic activity. This on the one hand suggests their application to reactions which are catalysed by zeolites. On the other hand, when the zeolite layer or the membrane reactor hosts a different catalyst, the possible negative effect of the zeolite on the overall selectivity or conversion needs to be evaluated in order to reduce it or to select the most proper zeolite.
- In a catalytic zeolite membrane, especially when the zeolite layer is sufficiently thick, the catalyst distribution can be an important factor in determining the reactor performance due to presence of mass transfer limitations which affect the reactant profile.

6.4 Modeling of zeolite membrane reactors

Many studies have been devoted to the modeling of zeolite membranes in different operations, such as in the permeation of gases and vapours, the separation of liquid mixtures, and in catalytic reactions in combination with separation in membrane reactors. In all these situations the development of a model for permeation behaviour through the zeolite layer is fundamental to its implementation in the design equations. The description of the permeation of gas and vapours through a zeolite membrane carried out using the adsorption–diffusion model described previously (where the permeating component adsorbs into the zeolite micropore and then, under the driving force of the chemical potential, diffuses through the microporosity of the zeolite layer) is considered too simplistic to take account for the coupling effects of multicomponent transport. Therefore, a more general and rigorous multicomponent transport model based on the MS approach is necessary (Salem *et al.*, 2006). Moreover, proper models need to be used for the description of adsorption equilibria of components in the zeolite framework. Various models have been successfully applied to the description of the adsorption equilibrium in zeolite membranes for single component, binary mixture and multicomponents:

- Langmuir isotherm
- dual-site Langmuir (DSL) isotherm
- extended or multicomponent Langmuir isotherm
- adsorbed solution theories (the ideal adsorbed solution theory (IAST), and the real adsorbed solution theory (RAST)).

The behaviour of a single component or of a binary mixture is usually well described by the single and dual site Langmuir isotherms. The Langmuir approach can also be extended to multicomponent mixtures by assuming that the saturation loadings of the mixture component are the same. Configurational-bias Monte Carlo simulations have shown that, at high pressures, the molecules with a high saturation loading will fill up the zeolite pores excluding other molecules even if they are strongly adsorbed (Vlugt *et al.*, 1999). The application of the multicomponent Langmuir isotherm is limited in cases where the mixture components have the same saturation loading. The prediction of the behaviour of multicomponent mixtures is, therefore, often carried out using adsorbed solution theories. The adsorbed solution theories are not limited by the restriction of using a single component isotherm for all the species, meaning that the best isotherm for each component in the mixture can be selected on the basis of experimental data (Calleja *et al.*, 1994; Krishna *et al.*, 1995, 1999, 2002; Krishna and Paschek, 2000a, 2000b). RAST differs from IAST in that the former introduces the activity coefficients in the chemical potential expression to take into account deviations from the ideal condition. Other authors have implemented the MS expressions considering the pure and multicomponent Nitta, Langmuir–Freundlich and Toth isotherms (Lito *et al.*, 2011). The Onsager approach to describe the permeation through the zeolite layers uses irreversible thermodynamics. The molar fluxes are linear functions of the gradient of chemical potential. The proportionality coefficients, defined as Onsager coefficients, can be determined from Kinetic Monte Carlo and Molecular Dynamics simulations. In particular, the modeling of permeation in the presence of heat effects, as for example in pervaporation, has been investigated by using non-equilibrium thermodynamics, since the conventional MS models do not take the coupled local heat effect into account. The theory of non-equilibrium thermodynamics (NET) is based on the assumption of local equilibrium and linear flux-force relations. Its application in a pervaporation process through a zeolite membrane in which liquid–vapour interfaces are present has been demonstrated by Kuhn *et al.* (2009). NET aspects applied to the coupled transport of both heat and mass in a zeolite membrane are discussed in more detail in Chapter 18.

In a zeolite membrane reactor the dual functions of reaction and separation are coupled in a single unit. Zeolite membrane reactors can be classified into two main groups:

- The zeolite membrane plays the role of separator without being involved in the chemical reaction, which takes places on a suitable catalyst close to the membrane (e.g., fixed or fluidized bed).
- The zeolite contains the catalyst or it is intrinsically catalytic.

In the former case, modeling of the membrane reactor is carried out by inserting the permeation model of the zeolite membrane in the design equations of the reactor in order to analyse the addition or extraction of reactants or products through the zeolite layer, respectively. An example is given by Daramola and coworkers where a simple Fickian flux approach has been chosen to investigate an extractor-type zeolite membrane reactor (Daramola *et al.*, 2011). In other cases, typical experimental values of the permeance of single components are introduced in the design equation of a membrane reactor. An example is the study developed by Jeong and coworkers (Jeong *et al.*, 2004) on the catalytic dehydrogenation of cyclohexane in a Faujasite (FAU) type zeolite membrane reactor that worked in isothermal operation and with a plug flow pattern. A similar approach was used by Kumar and coworkers investigating a zeolite membrane reactor for cyclohexane dehydrogenation (Kumar *et al.*, 2009). Fong *et al.* (2008) discussed the use of zeolite membrane and membrane reactors for the production of PX and provided an example of a model development chart which considers the need also to introduce into the equations parameters such as sorption and diffusion coefficients and molecular sieving. During the development of a catalytic zeolite membrane reactor, adsorption, diffusion and reaction should be simultaneously taken into account. An interesting example of the evaluation of the Thiele modulus using IAST compared to the multicomponent Langmuir adsorption approach was reported by Baur and Krishna (2005).

The modeling of zeolite membrane for permeation of single component and mixtures is already in a well-defined stage of development, giving an interesting contribution to the understanding of the involved phenomena and it needs to be better consolidated. On the other hand, the modeling of zeolite membrane reactors is still at an early stage since the permeation is assumed with too simple or empirical equations. Therefore, an effort is necessary in order to introduce in the design equations the models developed for the zeolitic membranes.

6.5 Scale-up and scale-down of zeolite membranes

The scale-up of the zeolite membranes is currently hindered by their cost influenced for at least 70% of the support price (Baker, 2002). The cost of a zeolite membrane module is approximately US \$3000 per m^2 (Caro *et al.*, 2000). The application of a zeolite membrane at industrial level would be economically feasible by reducing the membrane module cost by a factor 10 (Meindersma and De Haan, 2002). This reduction of the cost depends on the use of cheaper supports, such as the mullite (Algieri *et al.*, 2009), and of different configurations, such as hollow fibres (Richter *et al.*, 2003) which are cheaper per unit area than tubular supports. The reproducibility of the synthesis in zeolite membranes is another problem. Even if improved results could be obtained using CMRs, another big problem for their practical application is the lack of a module that will work in extreme temperature conditions. Full-ceramic modules would solve the issue of high temperature sealing (Caro *et al.*, 2005), in this case it would be necessary to seed and grow the zeolite layer after the completion of the housing. The scale-down of the zeolitic membranes seems to be a way to reduce the difficulties mentioned before. In fact, it would give the possibility of preparing defect-free membranes. The combination of membrane reactor and process miniaturization concepts makes it possible to provide a new method of carrying out the chemical reactions using a more efficient, cleaner and safer approach (Caro and Noack, 2008). For example, Yeung and coworkers (Lai *et al.*, 2003) carried out a Knoevenagel condensation of benzaldehyde and ethyl cyanoacetate to produce ethyl 2-cyano-3-3-phenylacrylate, an intermediate stage for fine chemical preparation. Cs\NaX crystals were selected as the catalyst and ZSM-5 as the membrane, which was prepared using the secondary growth method. The membrane was used to remove the water, a by-product of the reaction. The performance of four different reactors (PBR, packed-bed membrane reactor (PBMR), multi-channel reactor and multi-channel membrane microreactor) were studied. The highest yield (30%) obtained with the PBR was lower than that obtained with the microreactor (60%), and the latter presented a lower amount of the catalyst (20%). The poor performance of the PBR is due to the large external mass transfer resistance presented in the catalytic bed and absent with the microreactor. Recently, Santamaria and co-workers (Sebastiàn *et al.*, 2008) prepared various zeolite layers having different topologies (mordenite, Na-Y, ETS10) on the walls of microreactors by using the secondary growth method. The synthesis procedure gave highly reproducible zeolite films well adhered to the support.

The use of membrane microreactors offers many advantages, but at the same time, it is necessary to reduce their cost, to change the stainless steel used to manufacture the microreactors with more economic materials such as silicon.

Zeolites are also considered promising candidates for the production of gas sensors, due to their adsorption capacity, high surface area and porosity, presence of mobile ions and catalytic activity (Sahner *et al.*, 2008). A zeolite with FAU topology having large pores (super-cage of about 13 Å) showed the capacity for encapsulating large dye molecules and complexes. For example, this zeolite loaded with 2-(hydroxymethyl)anthraquinone showed good performance as an oxygen sensor (McGilvray *et al.*, 2006). Zeolite-coated interdigital capacitors revealed impressive hydrophilic properties and good potential as humidity sensors because they were able to detect a water vapour concentration under 0.5 ppm (Urbiztondo *et al.*, 2011).

6.6 Conclusion and future trends

In this chapter it has been demonstrated that zeolite membrane reactors are interesting candidates for carrying out different processes where the zeolite phase has the dual function of separator and catalyst. For this reason, there has, in the last ten years, been increasing research activity to prepare zeolite membranes with high flux and good separation factors. However, industrial application on large scale seems to be far off. Indeed, such application has so far been confined to T and NaA-type membranes used in pervaporation and vapour permeation processes. Their commercialization is hindered by the reproducibility problems in the preparation step, their long-term stability and the necessity to increase the surface area to volume ratio using capillary or multi-channel supports. Besides, when high temperature processes are considered, improvement in the sealing of the membrane modules is also necessary.

Researchers have made enormous efforts and significant progress in understanding the synthesis procedures and the formation mechanism of the zeolite layer in order to improve the quality of the membranes. However, further improvements based on producing reproducible defect-free zeolite membranes and to reduce their manufacturing cost need to be carried out to facilitate their introduction in the industry as membrane reactors.

6.7 References

Aguado S, Coronas J and Santamaria J (2005), 'Use of zeolite membrane reactors for the combustion of VOCs present in air at low concentrations', *Chem Eng Res Des*, **83**(A3), 295–301.

Algieri C, Bernardo P, Barbieri G and Drioli E (2009), 'A novel seeding procedure for preparing tubular NaY zeolite membranes', *Micropor Mesopor Mat*, **119**, 129–136.

Baker R W (2002), 'Future directions of membrane gas separation technology', *Ind Eng Chem Res*, **41**, 1393–1411.

Baerlocher C, McCusker L B and Olson D H (2007), *Atlas of zeolite framework types*, Elsevier, Amsterdam.

Barrer R M (1990), 'Porous crystal membranes', *J Chem Soc Faraday Trans*, **86**, 1123–1130.

Baur R and Krishna R, 2005, 'The effectiveness factor for zeolite catalysed reactions', *Catal Today*, **105**, 173–179.
Bernal M P, Xomeritakis G and Tsapatsis M (2001), 'Tubular MFI zeolite membranes made by secondary (seeded) growth', *Catal Today*, **67**, 101–107.
Baur R and Krishna R (2005), 'The effectiveness factor for zeolite catalysed reactions', *Catal Today*, **105**, 173–179
Bernardo P, Algieri C, Barbieri G and Drioli E (2008), 'Hydrogen purification from Carbon monoxide by means of Selective oxidation using zeolite catalytic membranes', *Sep Purif Technol*, **62**, 629–635.
Bonilla G, Tsapatsis M, Vlachos D G and Xomeritakis G (2001), 'Fluorescence confocal optical microscopy imaging of the grain boundary structure of zeolite MFI membranes made by secondary (seeded) growth', *J. Membrane Sci*, **182**, 103–109.
Bottino A, Capannelli G, Comite A and Novelli N (2001), 'VOC abatement: performance of different ceramic membranes in a catalytic membrane reactor', *La Chim Ind -Milan*, **3**, 200–206.
Bottino A, Capannelli G and Comite A (2002), 'Catalytic membrane reactors for the oxidehydrogenation of propane: experimental and modelling study', *J Membrane Sci*, **197**, 75–88.
Burggraaf A J (1999), 'Single gas permeation of thin zeolite (MFI) membranes: theory and analysis of experimental observations', *J Membrane Sci*, **155**, 45–65.
Calleja G, Jimenez A, Pau J, Dominguez L and Perez P (1994), 'Multicomponent adsorption equilibrium of ethylene, propane, propylene and CO_2 on 13X zeolite', *Gas Sep Purif*, **8**, 247–256.
Caro J, Noack M, Kolsch P and Schafer R (2000), 'Zeolite membranes-state of their development and perspective', *Micropor Mesopor Mat*, **38**, 3–24.
Caro J, Noack M and Kolsch P (2005), 'Zeolite membranes: from laboratory scale to technical applications', *Adsorption*, **11**, 215–227.
Caro J and Noack M (2008), 'Zeolite membranes-recent development and progress', *Micropor Mesopor Mat*, **115**, 215–233.
Cheng Z L, Chao Z S and Wan H L (2002), 'Research of A type zeolite as well as zeolite membrane by microwave heating', *Chin J Inorg Chem*, **18**, 528–532.
Ciavarella P, Casanave D, Moueddeb H, Miachon S, Fiaty K and Dalmon J A (2001), 'Isobutane dehydrogenation in a membrane reactor I influence of the operating conditions on the performance', *Catal Today*, **67**, 177–184.
Daramola M O, Burger A J, Giroir-Fendler A, Miachon S and Lorenzen L (2010), 'Exctractor-type catalytic membrane reactor with nanocomposite MFI-alumina tube as separation unit: prospect for ultrapure para-xylene production from m-xylene isomerisation over Pt-HZSM-5 catalyst', *Appl Catal A-Gen*, **386**, 109–115.
Daramola M O, Burger A J and Giroir-Fendler A (2011), 'Modelling and sensitivity analysis of a nanocomposite MFI-alumina based extractor-type zeolite catalytic membrane reactor for m-Xylene isomerization over Pt-HZSM-5 catalyst', *Chem Eng J*, **171**, 618–627.
den Exter M J, Jansen J C, van de Graaf J M, Kapteijn F, Moulijn J A and van Bekkum H (1996), 'Zeolite-based membranes preparation, performance and prospects', in H. Chon, S.I. Woo, S.-E. Park (eds.), *Recent advances and new horizons in zeo-*

lite science and technology, Studies in surface science and catalysis, Amsterdam, Elsevier, 413–454.

Espinoza R L, du Toit E, Santamaria J, Menèndez M, Coronas J and Irusta S (2000), 'Use of membranes in Fischer-Trops reactors', *Catal Today*, **106**, 143–148.

Fong Y, Abdullah A Z, Ahmad A L and Bhatia S (2008), 'Development of functionalized zeolite membrane and its potential role as reactor combined separator for para-xylene production from xylene isomers', *Chem Eng J*, **139**, 172–193

Funke H H, Argo A M, Baertsch C D, Falconer J L and Noble R D (1996), 'Separation of close-boiling hydrocarbons with silicalite zeolite', *J Chem Soc Faraday Trans*, **92**, 2499–2502.

Gu X, Tang Z and Dong J (2008), 'On-stream modification of MFI zeolite membranes for enhancing hydrogen separation at high temperature', *Micropor Mesopor Mat*, **111**, 441–448.

Haag S, Hanebuth M, Mabande G T P, Avhale A, Schiwieger W and Dittmeyer R (2006), 'On the use of a catalytic H-ZSM-5 membrane for xylene isomerisation', *Micropor Mesopor Mater*, **96**, 168–176.

Hasegawa Y, Sotowa K-I, Kusakabe K and Moroka S (2002), 'The influence of feed composition on CO oxidation using zeolite membranes loaded with metal catalysts', *Micropor Mesopor Mat*, **53**, 37–43.

Hedlund J, Noack M, Kolsch P, Creaser D, Caro J and Sterte J (1999), 'ZSM-5 membranes synthesized without organic templates using a seeding technique', *J Membrane Sci*, **159**, 263–273.

Hedlund J, Sterte J, Anthonis M, Bons A J, Carstensen B, Corcoran N, Cox D, Deckman H, De Gijnst W, de Moor P P, Lai F, Mchenry J, Mortier W, Reinoso J and Peters J (2002), 'High flux MFI membranes', *Micropor Mesopor Mat*, **52**, 179–189.

Huang A, Lin Y S and Yang W (2004), 'Synthesis and properties of A-type zeolite membranes by secondary growth method with vacuum seeding', *J Membrane Sci*, **245**, 41–51.

Huang A and Yang W (2007), 'Hydrothermal synthesis of NaA zeolite membrane together with microwave heating and conventional heating', *Mater Lett*, **61**, 5129–5132.

Illigen U, Schäfer R, Noack M, Kolsh P, Kühnle A and Caro J (2001), 'Membrane supported catalytic dehydrogenation of iso-butane using an MFI zeolite membrane reactor', *Catal Commun*, **2**, 339–345.

Jeong B-H, Sotowa K-I and Kusakabe K (2004), 'Modeling of an FAU-type zeolite membrane reactor for the catalytic dehydrogenation of cyclohexane', *Chem Eng J*, **103**, 69–75

Ju J, Zeng C, Zhang L and Xu N (2006), 'Continuous synthesis of zeolite NaA in a microchannel reactor', *Chem Eng J*, **116**, 115–121.

Julbe A, Farrusseng D, Jalibert J C, Moriodatos C and Guizard C (2000), 'Characteristics and performance in the oxidative dehydrogenation of propane of MFI and V-MFI zeolite membranes', *Catal Today*, **56**, 199–209.

Julbe A, Farrusseng D and Guizaed C (2001), 'Porous ceramic membranes for catalytic reactors-overview and new ideas', *J Membrane Sci*, **181**, 3–20.

Kondo M and Kita H (2010), 'Permeation mechanism through zeolite NaA and T-type membranes for practical dehydration of organic solvents', *J Membrane Sci*, **361**, 223–231.

Kong C, Lu J, Yang J and Wang J (2007), 'Catalytic dehydrogenation of ethylbenzene to styrene in a zeolite silicalite-1 membrane reactor', *J Membrane Sci*, **306**, 29–35.

Krishna R and Van Den Broeke L J P (1995), 'The Maxwell-Stefan description of mass transport across zeolite membranes', *Chem Eng J*, **57**,155–162

Krishna R, Vlugt T J H and Smit B (1999), 'Influence of isotherm inflection on diffusion in silicalite', *Chem Eng Sci*, **54**, 1751–1757.

Krishna R and Paschek D (2000a), 'Separation of hydrocarbon mixtures using zeolite membranes: a modelling approach combining molecular simulations with the Maxwell–Stefan theory', *Sep Purif Technol*, **21**, 111–136.

Krishna R and Paschek D (2000b), 'Permeation of Hexane Isomers across ZSM-5 Zeolite Membranes', *Ind Eng Chem Res*, **39**, 2618–2622.

Krishna R, Calero S and Smit B (2002), 'Investigation of entropy effects during sorption of mixtures of alkanes in MFI zeolite', *Chem Eng J*, **88**, 81–94.

Kriskna R (2006), 'The Maxwell–Stefan formulation of diffusion in zeolite', in Conner W.C. and Fraissard J (Eds), *Fluid transport in nanoporous materials*, Netherland, Springer, 9–39.

Kumar S, Gaba T and Kumar S (2009), 'Simulation of Catalytic Dehydrogenation of Cyclohexane in Zeolite Membrane Reactor', *Int J Chem React Eng*, **7**, A13.

Kusakabe K, Kuroda T and Morooka S (1998), 'Separation of carbon dioxide from nitrogen using ion-exchanged faujasite-type zeolite membranes formed on porous support tubes', *J Membrane Sci*, **148**, 13–23.

Kuhn J, Stemmer R, Kapteijn F, Kjelstrup S and Gross J (2009), 'A non-equilibrium thermodynamics approach to model mass and heat transport for water pervaporation through a zeolite membrane', *J Membrane Sci*, **330**, 388–398.

Lai Z, Bonilla G, Diaz I. Nery J G, Sujaoti K, Kokkoli E, Terasaki O, Thompson R W, Tsapatsis M and Vlachos D G (2003), 'Microstructural Optimization of a Zeolite Membrane for Organic Vapour Separation', *Science*, **300**, 456–460.

Lai S M, Ng C P, Martin-Aranda R and Yeung K L (2003), 'Knoevenagel condensation reaction in zeolite membrane microreactor', *Micropor Mesopor Mat*, **66**, 239–252.

Li G, Kikuchi E and Matsukata M (2003), 'The control of phase and orientation in zeolite membranes by the secondary growth method', *Micropor Mesopor Mat*, **62**, 211–220.

Lito P F, Santiago A S, Cardoso S P, Figueiredo B R and Silva C M (2011), 'New expressions for single and binary permeation through zeolite membranes for different isotherm models', *J Membrane Sci*, **367**, 21–32.

Lovallo M C and Tsapatsis M (1996), 'Preferentially Oriented Submicron Silicalite Membranes', *AICHE J*, **42**, 3020–3029.

Masuda T, Asanuma T, Shouji M, Mukai S R, Kawase M and Hashimoto K (2003), 'Methanol to olefins using ZSM-5 zeolite catalyst membrane reactor', *Chem Eng Sci*, **58**, 649–656.

Matsukata M, Nishiyama N and Ueyama K (1994), 'Zeolitic membrane synthesized on a porous alumina support', *Chem Soc Chem Commun*, **3**, 339–340.

McGilvray K, Chretien M, Lukeman M and Scaiano J (2006), 'Simple and smart oxygen sensor based on the intrazeolite reactions of a substituted anthraquinone', *Chem Commun*, **42**, 4401–4403.

McLeary E, Jansen J C and Kapteijn F (2006), 'Zeolite based films, membranes and membrane reactors: Progress and prospects', *Micropor Mesopor Mat*, **90**, 198–220.

Meindersma G W and de Haan A B (2002), 'Economical feasibility of zeolite membranes for industrial scale separation of aromatic hydrocarbons', *Desalination*, **149**, 29–34.

Mintova S and Bein T (2001), 'Microporous films prepared by spin-coating stable colloidal suspensions of zeolites', *Adv Mater*, **13**, 1880–1883.

Noack M, Kolsch P, Schafer R, Toussaint P and Caro J (2001), 'Molekularsieb-Membranen für industrielle Anwendungen – Probleme, Fortschritte, Lösungen', *Chem Ing Tech*, **73**, 958–967.

Pantazidis A, Dalmon J A and Mirodtos C (1995), 'Oxidative dehydrogenation of propane on catalytic membrane reactors', *Catal Today*, **25**, 403–408.

Pera-Titus M, Llorens J, Cunill F, Mallada R and Santamarıa J (2005), 'Preparation of zeolite NaA membranes on the inner side of tubular supports by means of a controlled seeding technique', *Catal Today*, **104**, 281–287.

Pera-Titus M, Bausach M, Llorens J and Cunill F (2008), 'Preparation of inner-side tubular zeolite NaA membranes in a continuous flow system', *Sep Purif Technol*, **59**, 141–150.

Richter H, Voight I, Fischer G and Puhlfürß P (2003), 'Preparation of zeolite membranes on the inner surface of ceramic tubes and capillaries', *Sep Purif Technol*, **32**, 133–138.

Rohde M P, Unruh D and Schaub G (2005), 'Membrane application in Fischer–Tropsch synthesis reactors–Overview of concepts', *Catal Today*, **106**, 143–148.

Rohde M P, Schaub G, Khajavi S, Jansen J C and Kapteijn F (2008), 'Fischer–Tropsch synthesis with in situ H_2O removal – Directions of membrane development', *Micropo Mesopor Mater*, **115**, 123–136.

Sahner K, Hagen G, Schonauer D, Reiß S and Moos R (2008), 'Zeolites-versatile materials for gas sensors', *Solid State Ionics*, **179**, 2416–2423.

Salem A, Ghoreyshi A A and Jahanshahi M (2006), 'A multicomponent transport model for dehydration of organic vapors by zeolite membranes', *Desalination*, **193**, 35–42.

Sato K, Sugimoto K and Nakane T (2008), 'Synthesis of industrial scale NaY zeolite membranes and ethanol permeating performance in pervaporation and vapor permeation up to 130 °C and 570 kPa', *J Membrane Sci*, **310**, 161–173.

Sebastiàn V, de la Iglesia O, Casado l, Kolb G, Hessel V and Santamaria J (2008), 'Preparation of zeolite films as catalytic coatings on microreactor channels', *Micropor Mesopor Mater*, **115**, 147–155.

Takata Y, Tsuru T, Yoshioka T and Asaeda M (2002), 'Gas permeation properties of MFI zeolite membranes prepared by the secondary growth of colloidal silicalite and application to the methylation of toluene', *Micropor Mesopor Mater*, **54**, 257–268.

Tang Z, Kim S J, Reddy G K, Dong J H and Smirniotis P (2010), 'Modified zeolite membrane reactor for high temperature water gas shift reaction', *J Membrane Sci*, **354**, 114–122.

Tsay C S and Chiang S T (2000), 'Supported zeolite membrane by vapour-phase regrowth toluene', *AICHE J*, **46**, 616–625.

Urbiztondo M, Pellejero I, Rodriguez A, Pina M P and Santamaria J (2011), 'Zeolite-coated interdigital capacitors for humidity sensing', *Sensor Actuat B: Chem*, **157**, 450–459.

Van Dyk L, Miachon S, Lorenzen L, Torres M, Fiaty K, Dalmon J A (2003), 'Comparison of microporous MFI and dense Pd membrane performances in an extractor-type CMR', *Catal Today*, **82**, 167–177.

Vlugt T J H, Krishna R and Smit B (1999), 'Simulations of adsorption isotherms for linear and branched alkanes and their mixtures in silicalite', *J Phys Chem B*, **103**, 1102–1118.

White J C, Dutta P K, Shqau K and Verweij H (2010), 'Synthesis of ultrathin zeolite Y membranes and their application for separation of carbon dioxide and nitrogen gases', *Langmuir*, **26**, 10287–10293.

Xiao J and Wei J (1992), 'Diffusion mechanism of hydrocarbons in zeolites – I. theory', *Chem Eng Sci*, **47**, 1123–1141.

Xomeritakis G, Lai Z and Tsapatsis M (2001), 'Separation of xylene isomer vapors with oriented MFI membranes made by seeded growth', *Ind Eng Chem Res*, **40**, 544–552.

Xu X, Bao Y, Song C, Yang W, Liu J and Lin L (2004), 'Microwave-assisted hydrothermal synthesis of hydroxy-sodalite zeolite membrane', *Micropor Mesopor Mater*, **75**, 173–181.

Xu X C, Yang W S, Liu J and Lin L W (2000), 'Synthesis of a high –performance NaA zeolite membrane by microwave heating', *Adv Mater*, **12**, 195–198.

Zhu W, Gora L, van den Berg A W C, Kapteijn F, Jansen J C and Moulijn J A (2005), 'Water vapour separation from permanent gases by a zeolite-4A membrane', *J Membrane Sci*, **235**, 57–66.

6.8 Appendix: nomenclature

6.8.1 Notation

b Langmuir adsorption parameter (Pa^{-1})
c concentration (mol m^{-3})
d_p pore diameter (m)
D^g activated gas translational diffusivity ($m^2\ s^{-1}$)
D^S surface diffusivity ($m^2\ s^{-1}$)
D_o corrected diffusivity ($m^2\ s^{-1}$)
D_∞ Arrhenius-type pre-exponential factor for the surface diffusivity ($m^2\ s^{-1}$)
E_a^g activation energy for activated gas translational diffusion (J mol^{-1})
E_{diff} activation energy for surface diffusion (J mol^{-1})
M molecular mass (mol kg^{-1})
N molar flux in a stationary coordinate frame of reference (mol $s^{-1}\ m^{-2}$)
N^g activated gaseous flux (mol $s^{-1}\ m^{-2}$)
N^s surface flux (mol $s^{-1}\ m^{-2}$)
P, p pressure (Pa)

q	adsorbed molar loading (mol kg^{-1})
q_{sat}	monolayer molar loading (mol kg^{-1})
R	universal gas constant = 8.3145 (J mol^{-1} K^{-1})
T	temperature (K)
u_m	average velocity (m s^{-1})
z	axial coordinate (m)
ϑ	fractional coverage (–)
ρ_{app}	apparent density (kg m^{-3})
Γ	thermodynamic correction factor (–)
i,j	subscript, component in mixture

7
Dense ceramic membranes for membrane reactors

X. TAN, Tianjin Polytechnic University, China and
K. LI, Imperial College London, UK

DOI: 10.1533/9780857097330.2.271

Abstract: Dense ceramic membrane reactors are made from composite oxides, usually having perovskite or fluorite structure with appreciable mixed ionic (oxygen ion and/or proton) and electronic conductivity. They combine the oxygen or hydrogen separation process with the catalytic reactions into a single step at elevated temperatures (>700°C), leading to significantly improved yields, simplified production processes and reduced capital costs. This chapter mainly describes the principles of various types of dense ceramic membrane reactors, and the fabrication of the membranes and membrane reactors.

Key words: membrane reactors, mixed ionic-electronic conductors (MIECs), perovskite, oxygen permeable membrane, proton conducting membrane.

7.1 Introduction

A membrane reactor is a device that combines the membrane separation process with chemical reactions in a single step. Due to the integration of reaction and separation, the chemical process becomes much simpler and the operational costs can thereby be reduced drastically. As illustrated in Fig. 7.1, the membrane reactor is capable of promoting a reaction process by: (1) selectively removing one of the products through the membrane from the reaction zone, making the equilibrium reaction shift to the product side; (2) creating a well-defined reaction interface (or region) between two reactant streams; and (3) supplying a particular reactant to the reaction zone, thus establishing an optimum concentration profile in the reactor (Sirkar *et al.*, 1999). As a result, the yield can be increased (even surpassing the thermodynamic limitations for equilibrium reactions) and/or the selectivity can be improved by suppressing other undesired side-reactions or the secondary reaction of products. In addition, membrane reactors are energy efficient and relatively safe in operation, and may avoid formation of hot spots as encountered in conventional co-feed reactors.

Dense ceramic membranes are made from composite oxides usually having perovskite or fluorite crystalline structure (Bouwmeester, 2003; Liu

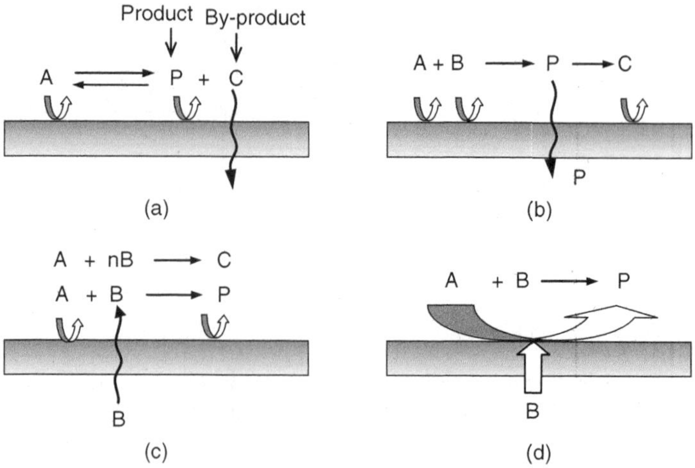

7.1 Principle of membrane reactor to promote reactions: (a) selective permeation of by-product of an equilibrium-limited reaction; (b) selective permeation of an intermediate product; (c) dosing a reactant through the membrane; and (d) supplying a well-defined reaction interface. In (a) and (b) the membrane functions as a product extractor, and as a reactant distributor in (c). (In the figure, A and B stand for two different reactants, C for a by-product, P for the desired product and nB for n mole of the reactant B.)

et al., 2006). A large number of oxygen vacancies are presented in the membrane, mostly generated by doping strategy, leading to ionic conductivity. Depending on the mediator of the electronic flux, the ceramic membranes may be divided into three different classes (Hendriksen *et al.*, 2000):

- A single material membrane based on a material exhibiting both ionic and electronic conductivity (MIEC membrane).
- A dual-phase composite consisting of percolating phases of an ion conductor and an electron conductor (dual-phase membrane).
- A pure ionic electrolyte conductor, with suitable electrodes connected to an external circuit for the electronic current (electrolyte membrane).

Under an electrochemical potential gradient, oxygen or hydrogen can be permeated through the dense ceramic membrane in a dissociated or ionized form rather than through conventional molecular diffusion, and thus an extremely high permselectivity (up to 100%) can be achieved. For this unique property, as well as the high thermal resistance, dense ceramic membranes have been extensively applied in a variety of oxidation and dehydrogenation reactions where the membrane functions as either a product extractor or a reactant distributor (Liu *et al.*, 2001).

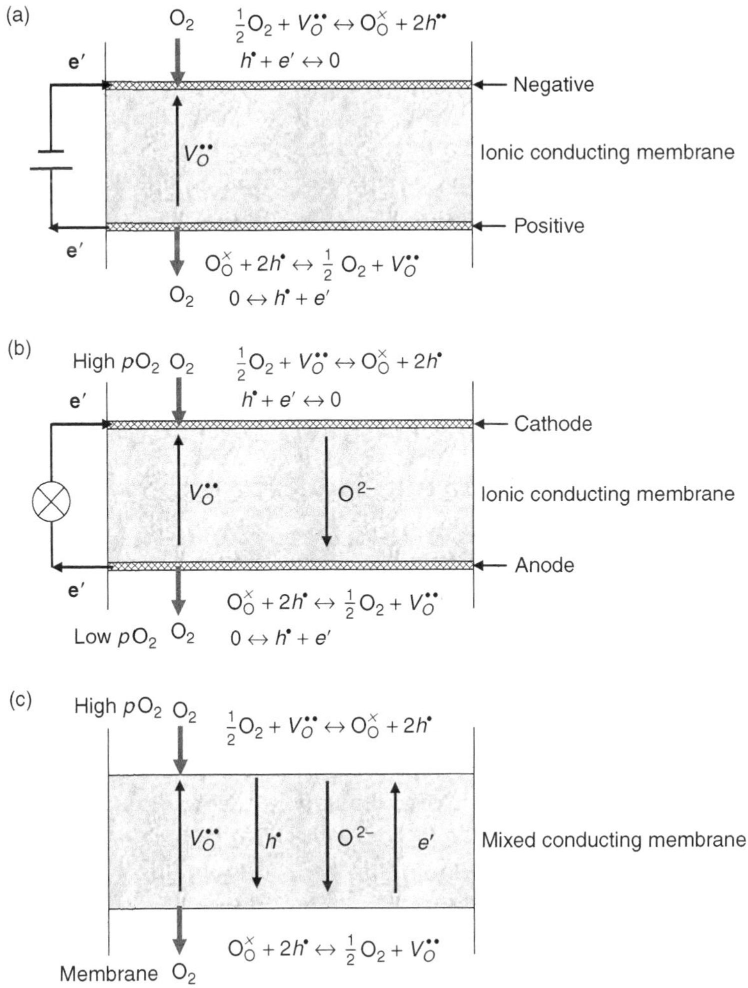

7.2 Oxygen permeation modes in the dense ceramic membranes: (a) electrochemical oxygen pump; (b) oxygen permeation together with production of electricity; and (c) oxygen permeation in mixed conducting membrane.

The electrochemical potential gradient for oxygen/hydrogen permeation can be generated by an oxygen/hydrogen partial pressure gradient or by using an external power source across the membrane. Figure 7.2 describes schematically the oxygen permeation modes associated with different membranes. For the electrolyte membrane, an external circuit is required for electron transport with two electrodes attached to the membrane surfaces to distribute/collect electrons. When an external power source is applied, the oxygen can be electrochemically pumped from the negative side to the positive side of the membrane regardless of the oxygen partial pressures on each side, as shown in Fig. 7.2a. When an oxygen partial pressure gradient is present across the membrane, the oxygen can also be permeated through the electrolyte membrane by the aid of the external circuit to transport electrons, as shown in Fig. 7.2b. As a result, electrical powder can be co-generated along with the oxygen permeation. For the MIEC and dual-phase membranes, the oxygen can be permeated through the membrane under the oxygen partial pressure gradient at high temperatures without the need of electrodes or an external circuit. This makes the oxygen permeation through the membrane much simplified and the operational cost can accordingly be reduced. The dense ceramic membrane reactor is usually referred to as without an external power source. Oxygen/hydrogen permeation takes place under an oxygen/hydrogen partial pressure gradient across the membrane.

7.2 Principles of dense ceramic membrane reactors

7.2.1 Oxygen-separation membrane

Dense ceramic membranes are made from composite oxides in which there are a large number of oxygen vacancies in the crystalline lattice. These oxygen vacancies allow the oxygen ions with enough energy to move from one site to another, leading to an appreciable oxygen ionic conductivity. On the other hand, in the presence of oxygen gas, the oxygen vacancies on the membrane surface can be filled by the oxygen atoms accompanied by consumption of electrons (n-type conductor) or formation of electron holes (p-type conductor) according to the following reactions,

$$\frac{1}{2}\mathrm{O}_2 + V_{\mathrm{O}}^{\bullet\bullet} + 2e' \xleftrightarrow{k_f / k_r} \mathrm{O}_{\mathrm{O}}^{x} \quad (\text{for } n\text{-type conductor}) \qquad [7.1a]$$

or

$$\frac{1}{2}\mathrm{O}_2 + V_{\mathrm{O}}^{\bullet\bullet} \xleftrightarrow{k_f / k_r} \mathrm{O}_{\mathrm{O}}^{x} + 2h^{\bullet} \quad (\text{for } p\text{-type conductor}) \qquad [7.1b]$$

where the charged defects are defined using the Kröger–Vink notation, that is, O_{O}^{x} stands for lattice oxygen, $V_{\mathrm{O}}^{\bullet\bullet}$ for oxygen vacancy and $h^{\bullet}$ for electron hole. k_f and k_r are the forward and the reverse reaction rate constants for the surface exchange reaction, respectively. It should be noted that the above surface exchange reaction may involve many sub-steps, such as oxygen adsorption, dissociation, recombination and charge transfer, but all of them can be combined into a single surface exchange reaction.

The equilibrium between the electrons and holes is given by,

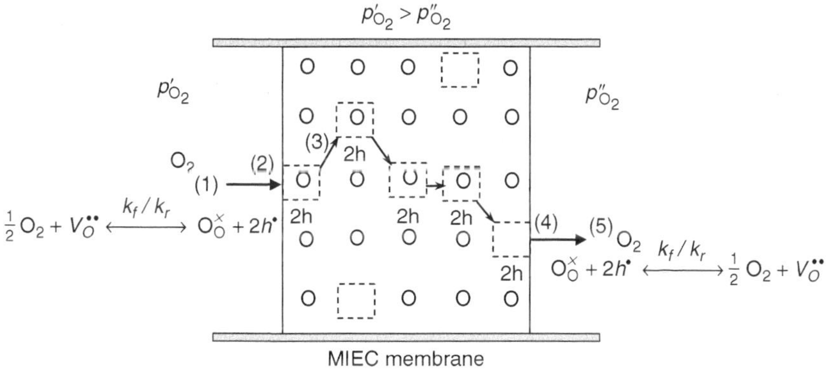

7.3 Oxygen permeation through a mixed ionic-electronic conducting membrane: dashed boxes represent oxygen vacancy.

$$0 \longleftrightarrow e' + h^{\bullet} \qquad [7.2]$$

In the meantime, the ceramic membrane may also demonstrate high electronic conductivity due to the capability of the transition metals existing in different valence states in the structure (MIEC membrane). When the MIEC membrane is located under an oxygen partial pressure gradient at elevated temperature, the oxygen may permeate from the high oxygen partial pressure side to the low oxygen partial pressure side, yielding a net oxygen flux. Figure 7.3 demonstrates schematically the oxygen permeation process through a MIEC membrane, including the following steps in series: (1) oxygen molecular diffusion from the gas stream to the membrane surface (high pressure side); (2) reaction between molecular oxygen and oxygen vacancy on the membrane surface (high pressure side); (3) bulk diffusion of oxygen vacancy across the membrane; (4) reaction between lattice oxygen and electron hole on the membrane surface (low pressure side); and (5) mass transfer of oxygen from the membrane surface to the gas stream (low pressure side). Generally, the gaseous transfer resistances of steps (1) and (5) can be negligible, and the oxygen permeation is controlled by diffusion of oxygen ions in the membrane and/or surface oxygen exchange kinetics on either or both sides of the membrane. In the oxygen permeable MIEC membrane, the common charge defects are oxygen vacancies and electron holes, and the oxygen permeation flux can be given by (Xu and Thomson, 1999; Tan *et al.*, 2005b):

$$J_{O_2} = \frac{1}{2\delta}\int_{(I)}^{(II)} \frac{(C_h + 4C_V)D_V D_h}{C_h D_h + 4C_V D_V} \times \mathrm{d}C_V \qquad [7.3]$$

where δ is the membrane thickness, C_i and D_i are the concentration and the diffusion coefficient of defect i. The subscripts h and V represent hole and oxygen vacancy, respectively. The oxygen permeability of the MIEC membrane is determined by the inherently conducting properties of the membrane material and is inversely proportional to the membrane thickness. For a high oxygen flux, the membrane has to possess high and equivalent oxide ionic and electronic conductivities.

When the flux is governed by bulk oxygen diffusion, it is generally described by Wagner's equation (Chen *et al.*, 1997),

$$J_{O_2} = -\frac{RT}{16F^2\delta}\int_{(I)}^{(II)} t_e \sigma_{\text{ion}} \,\text{d}\ln p_{O_2} \qquad [7.4]$$

where σ_{ion} is the ionic conductivity, t_e is the electronic transference number, R is the ideal gas constant, T is the temperature and F is the Faraday constant.

Mostly, the MIEC membranes have a perovskite structure with the general composition of $A_{1-x}A'_xB_{1-y}B'_yO_{3-\delta}$ ($0 \leq x \leq 1$; $0 \leq y \leq 1$), where the A-site ion is a lanthanide, A = La (Pr, Nd, Sm, Gd) or an alkaline earth, A′ = Ca, Sr, Ba, and the B-site is occupied by one or two different transition metals, B = Cr, Mn, Fe, Co and B′ = Co, Ni, Cu. Such perovskite oxides have demonstrated high structural and phase stability and excellent mixed ionic–electronic conduction properties, which stems from the fact that they may exhibit high degrees of oxygen deficiency and high mobility of the oxide ion vacancies (Hendriksen *et al.*, 2000). The high electronic conductivity is due to the capability of the transition metals to exist in different valence states in the structure. In most cases, the electronic conductivity in the perovskite membranes overwhelms the oxygen ionic conductivity and the oxygen flux can be derived in terms of the oxygen partial pressures as (Xu and Thomson, 1999; Tan *et al.*, 2005b):

$$J_{O_2} = \frac{k_r\left[\left(p'_{O_2}\right)^{0.5} - \left(p''_{O_2}\right)^{0.5}\right]}{\left(p''_{O_2}\right)^{0.5} + \frac{2\delta k_f}{D_V}\cdot\left(p'_{O_2}p''_{O_2}\right)^{0.5} + \left(p'_{O_2}\right)^{0.5}} \text{(for disc membrane)} \qquad [7.5a]$$

or

$$J_{O_2} = \frac{k_r\left[\left(p'_{O_2}\right)^{0.5} - \left(p''_{O_2}\right)^{0.5}\right]}{\frac{R_m}{R_o}\cdot\left(p''_{O_2}\right)^{0.5} + \frac{2k_f\left(R_o - R_{\text{in}}\right)}{D_V}\cdot\left(p'_{O_2}p''_{O_2}\right)^{0.5} + \frac{R_m}{R_{\text{in}}}\cdot\left(p'_{O_2}\right)^{0.5}} \qquad [7.5b]$$

(for tubular membrane)

where p'_{O_2} and p''_{O_2} are the oxygen partial pressures on the outer and the inner surfaces of the tubular membrane, respectively; R_m the algorithm radius, $R_m = (R_o - R_{in}) / \ln(R_o / R_{in})$, in which R_o and R_{in} are respectively the outer and the inner radius of the fibre (cm); D_V is the diffusion coefficient of oxygen vacancy.

Apart from the perovskite oxides, some non-perovskite-type oxides also exhibit mixed oxygen ionic and electronic conducting properties and are used as the oxygen-separation membrane, which has been discussed in detail elsewhere (Liu *et al.*, 2006).

7.2.2 Hydrogen separation membrane

For composite oxides with a general formula of $AB_{1-x}M_xO_{3-\delta}$, where A is Ca, Sr or Ba; B is Ce or Zr; M is taken from the group consisting of Tm, Nd, Ga, Y, Yb, In and Tb, protons can be formed in the water vapour or hydrogen-containing environment at high temperature according to the following reactions:

$$H_2O + V_O^{\bullet\bullet} \longleftrightarrow O_O^{\times} + 2H^{\bullet} \quad [7.6]$$

$$\frac{1}{2}H_2 + h^{\bullet} \longleftrightarrow H^{\bullet} \quad [7.7a]$$

or

$$\frac{1}{2}H_2 \longleftrightarrow H^{\bullet} + e' \quad [7.7b]$$

where $H^{\bullet}$ stands for proton. The protons are also considered to bond to oxygen ions, forming substitutional hydroxyl, $OH_O^{\bullet}$ namely

$$H_2O + V_O^{\bullet\bullet} + O_O^{x} \longleftrightarrow 2OH_O^{\bullet} \quad [7.8a]$$

$$H_2 + 2O_O^{x} + 2h^{\bullet} \longleftrightarrow 2OH_O^{\bullet} \quad [7.8b]$$

Figure 7.4 shows the formation of charged defects in the proton conducting membranes exemplified by $SrCe_{0.95}Yb_{0.05}O_{3-\delta}$. As can be seen, four types of charged defect, including oxygen vacancy, proton, electron and hole, may coexist in the membrane at high temperature. Two mechanisms are

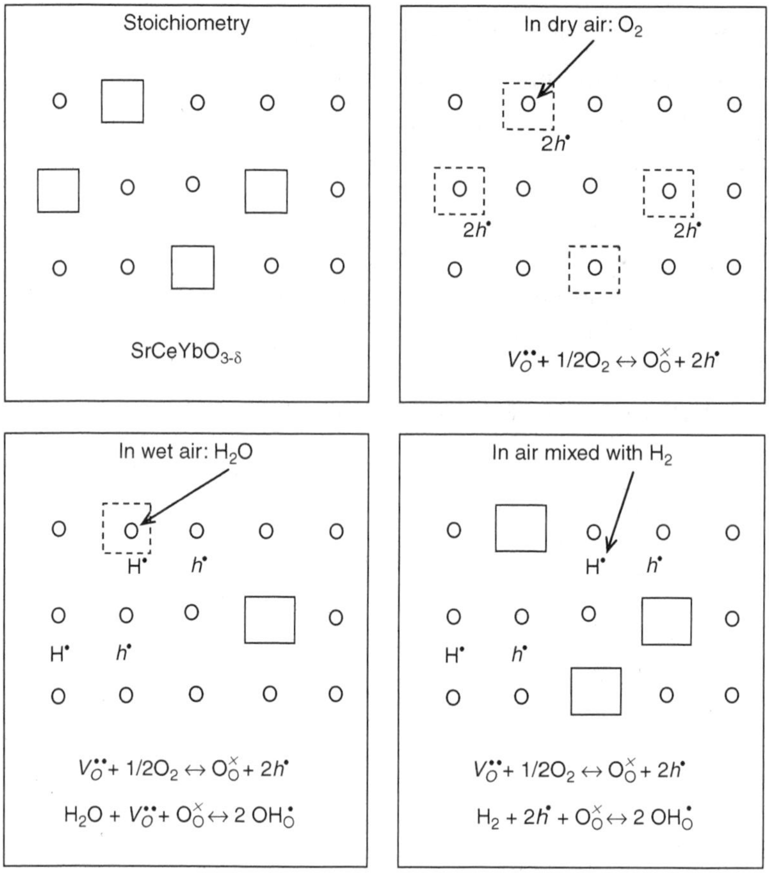

7.4 Formation of the charged defects in the proton conducting ceramic membranes.

usually applied to describe the proton transport in the proton conducting membranes: the free proton transport mechanism and the vehicle proton transport mechanism. In the free proton transport, protons jump between stationary oxygen ions. Each jump is followed by a rotation around the oxygen ion to get in a position for the next jump. The jump is normally considered the rate-limiting step and the rotation is easier. In vehicle mechanism, the proton moves as a passenger on a larger species such as oxygen ions. The vehicle mechanism can only account for dominant long-range transport processes in oxides when proceeding on interstitial lattice sites. Therefore, the hydroxyl ion transport cannot provide a dominant conductivity but accomplishes minority oxygen transport in a proton conductor. The mobility of proton in the oxide is limited by the proton transfer between fixed oxygen sites, but is facilitated by thermal fluctuation of the oxygen ion separation.

Therefore, the proton transport has close activation energies to (and generally somewhat lower than) those for oxygen vacancy mobility.

The most well-known proton ceramic conductors with appreciable protonic conductivity are the doped $SrCeO_3$, $SrZrO_3$, $CaZrO_3$ and $BaCeO_3$ perovskite-type oxides (Iwahara, 1996; Iwahara *et al.*, 2004). Their protonic conductivities in hydrogen atmosphere are in the order of 10^{-3}–10^{-2} S cm^{-1} at 600–1000°C. Among these oxides, the $BaCeO_3$ based perovskites exhibit much higher protonic conductivity than the others, but oxygen ions contribute to the conduction as the temperature is increased. Although the conductivity of $SrCeO_3$ based oxides is lower, the transport number of protons is higher than that of $BaCeO_3$ based ones. The zirconate based oxides such as $SrZrO_3$ or $CaZrO_3$ show good chemical stability and mechanical strength and they are more stable against carbon dioxide gas, which reacts with cerate materials below 800°C, but their conductivity is lower than the cerates.

As for the oxygen-separation membrane, hydrogen permeation through the proton conducting membrane can be achieved by applying a hydrogen partial pressure gradient or by using an external power source across the membrane. For the mixed conducing membrane, it is also required to possess high and equivalent protonic and electronic conductivity so as to achieve high hydrogen flux. But it is usually not true, especially for the proton conducting membranes. Therefore, dual-phase membranes are often used to achieve appreciable hydrogen permeation fluxes (Siriwardane *et al.*, 2000).

7.2.3 Oxygen permeable membrane reactor

The oxygen permeable membrane reactor combines the oxygen-separation with chemical reactions in a single unit. It is usually used for a variety of oxidation reactions. The oxidation reaction takes place on one side of the membrane while the oxygen as a reactant is introduced through the membrane from the other side into the reaction zone. Figure 7.5 illustrates the principle of oxygen permeable membrane reactors exemplified by the partial oxidation of methane to syngas. Methane and air are fed to separate chambers of the membrane reactor. On the reaction side, the methane is oxidized by surface O^{2-} and the surface oxygen is depleted; bulk O^{2-} diffuses from the oxygen-rich side to fill in the oxygen vacancies. On the oxygen-rich side, gaseous O_2 is first reduced to O^{2-}, which diffuses towards the reaction side. The driving force is the oxygen partial pressure gradient across the membrane. In this case, the dense ceramic membrane introduces oxygen, not in the gaseous state but in atomic form, to the reaction chamber. This type of operation allows complete control over the contact mode of the reactants with each other, and with the catalytically active surface. On the other hand, if the membrane surface is not sufficiently active to catalyse the oxidation

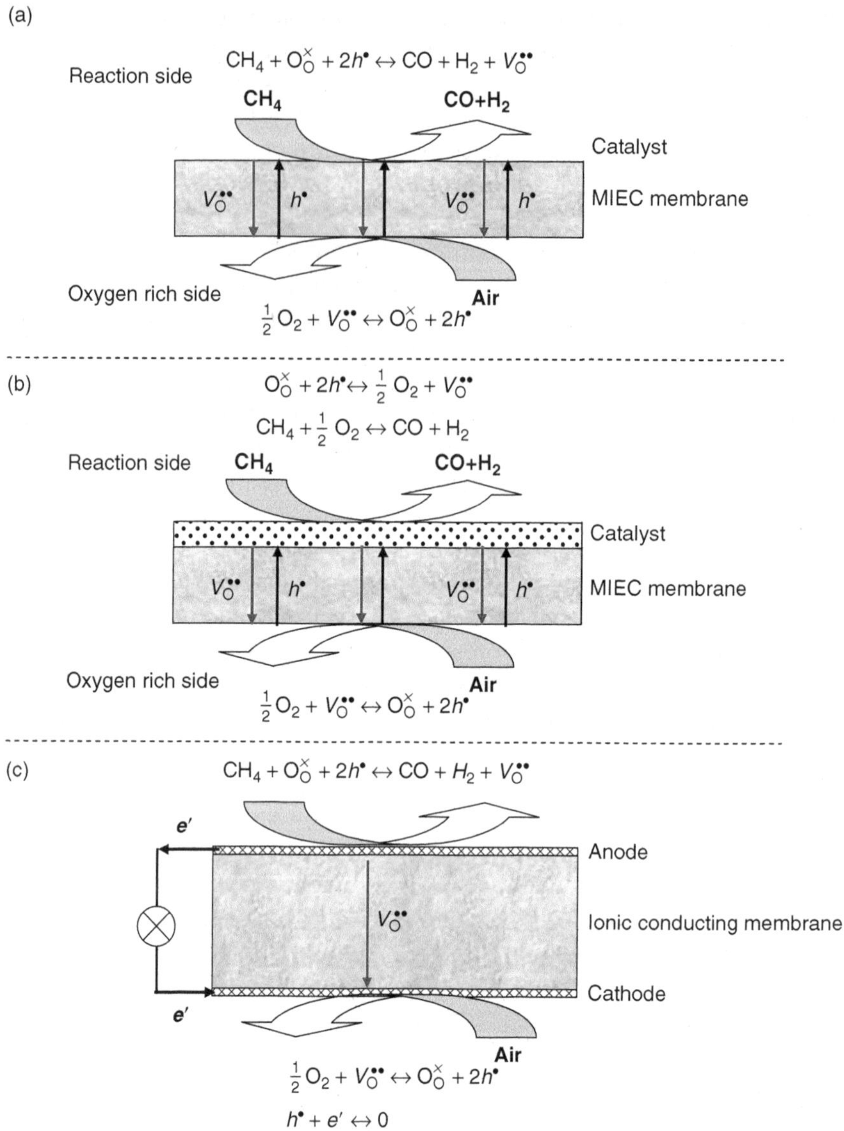

7.5 Principles of the oxygen permeable membrane reactor for partial oxidation of methane to syngas: (a) catalytic membrane reactor; (b) packed membrane reactor; and (c) SOFC-type membrane reactor.

reaction, the oxygen can also be permeated through the membrane in the form of oxygen ions, and forms again into gaseous O_2 on the reaction side. Therefore, catalyst is required to consume the permeated oxygen, and the reactor may be thus called a packed membrane reactor (Fig. 7.5b).

For the pure oxygen ion conducting membrane, the electrons released from the chemical reactions have to be transported to the oxygen-rich side via an external circuit so as to precede the reaction and the oxygen permeation, as shown in Fig. 7.5c. In this case, electrical power is co-generated along with the production of valuable chemicals. Therefore, the membrane reactor operating in this mode is also called a solid oxide fuel cell type membrane reactor (SOFC-MR).

As can be seen, air instead of pure oxygen can be used directly as the oxidant in the dense ceramic membrane reactors because of the 100% oxygen permselectivity of the membrane. This may lead to a dramatic decrease in the operation cost of the membrane reactor. Furthermore, the oxygen can be introduced into the reaction side in a controllable mode. Thus, the excessive temperature excursion can be avoided, making operation of the reactor more safe.

7.2.4 Hydrogen permeable membrane reactor

The hydrogen permeable membrane reactor combines hydrogen separation with chemical reactions in a single unit. It is usually used for a variety of dehydrogenation reactions. Figure 7.6a illustrates the principle of mixed proton-hole/electron conducting membrane reactor, exemplified by the dehydrogenation coupling of methane. The methane and oxygen are introduced into each side of the membrane, respectively. On the reaction side, methane is adsorbed and transformed to methyl radical on the surface of the membrane with the form of protons; coupling occurs between two methyl radicals in the gas phase. The protons permeate through the membrane at high temperature towards the oxygen-side surface and react with the adsorbed oxygen into water molecules with the formation of electron holes. In this case, the membrane acts as an internal circuit for electron conduction and the oxygen does not take part in the methane-coupling reaction. If there is no oxygen present on the cathode side of the membrane, the hydrogen produced by the reaction also can be permeated through the membrane by using a sweep gas. But very few cases have been studied, because the hydrogen permeability in the proton conducting membranes is too small.

When the membrane is a pure proton conductor, the electrons released from the chemical reactions have to be transported to the oxygen-rich side via an external circuit so that the reaction can proceed, as shown in Fig. 7.6b. Therefore, electrical power can be co-generated along with the production of C_2 compounds, and the membrane reactor becomes a SOFC-type one. Since the catalytic activity of the membrane surface is poor, a catalyst layer is often required in the proton conducting membrane reactors.

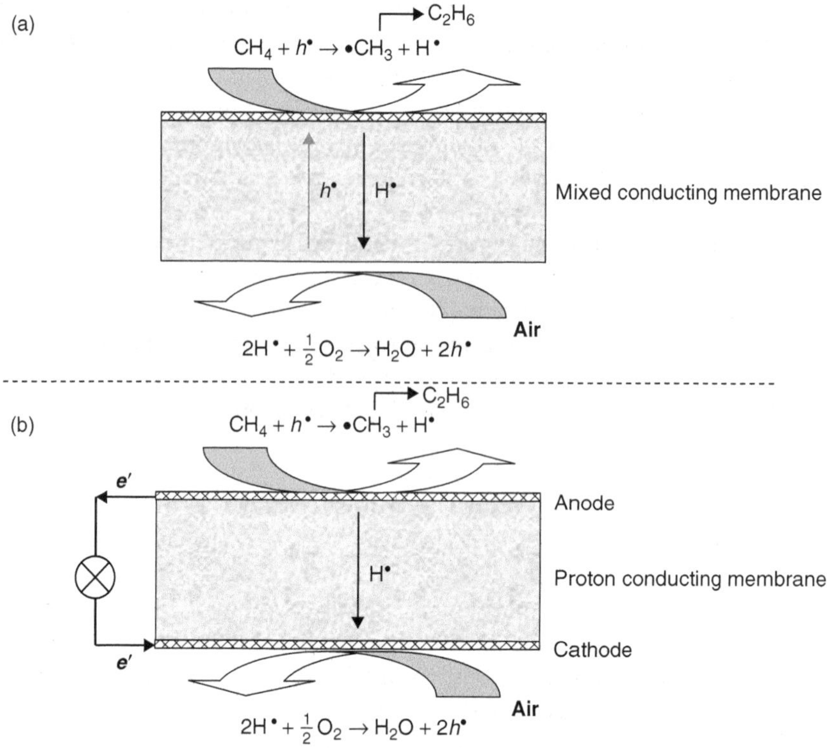

7.6 Principles of hydrogen permeable membrane reactor for methane coupling: (a) catalytic membrane reactor; and (b) SOFC-type membrane reactor.

7.3 Membrane preparation and catalyst incorporation

7.3.1 Preparation of dense ceramic membranes

Dense ceramic membranes can be fabricated in three configurations: disc/flat-sheet membrane, tubular membrane and hollow fibre membrane. Their preparation processes are described respectively in details as follows.

Disc/flat-sheet membrane

Disc/flat-sheet-shaped membranes are mostly applied in dense ceramic membrane reactors due to the ease of the fabrication process: the ceramic material powder is pressed into discs in a stainless steel mould under an isostatic or hydraulic pressure, followed by sintering at a high temperature. Such disc-shaped membranes usually have a thickness of about 1 mm so as to provide enough mechanical strength. The disc membrane provides a very limited effective area (~a few cm^2) for separation and reactions and low permeation fluxes due to its large thickness.

In order to obtain high permeation fluxes, the membrane thickness has to be reduced. Flat-sheet thin-film membranes can be fabricated by tape casting and lamination on porous supports and subsequent co-sintering (Juste *et al.*, 2010; Kaiser *et al.*, 2011). A shear-thinning slurry is prepared from attrition-milled ceramic powder by adding a binder (e.g. methyl methacrylate), a plasticizer (e.g. dibutyl phthalate) and a dispersing agent (e.g. CP 213) in an azeotropic solvent of butanone-2 and ethanol. A cohesive and flexible green tape with a controlled thickness is obtained using a doctor blade tape casting apparatus. The green tape is then applied on the porous tape casted support by lamination (application of pressure and heat between two rollers). Before sintering a very slow debindering cycle is applied to avoid damage to the laminated structure. The laminated structure is co-sintered in air at an elevated temperature into dense membranes.

Tubular membrane

Tubular membranes are usually prepared by a plastic extrusion method (Lu *et al.*, 2000). To prepare for extrusion, the calcined ceramic powder is mixed with several additives to make a slip with enough plasticity to be easily formed into tubes, while retaining satisfactory strength in the green state. The additives include a solvent (butanol or xylene), a dispersant, a binder and a plasticizer. After the slip is prepared, some of the solvent is allowed to evaporate to yield a plastic mass that is forced through a die at a high pressure to produce hollow tubes. The dimensions of the extruded tube can be controlled by the orifice diameters of the die. This is heated to a temperature range of 150–400°C to facilitate the removal of the gaseous species formed during decomposition of the organic additives. After that the tube is sintered at a high temperature in stagnant air for a given time to obtain the membrane tubes.

Tubular membranes with one dead-end can be prepared by cold isostatic pressing (CIP) with green machining method (Akin and Lin, 2002). The ceramic powder is first pressed into cylindrical rods. A carbide bit is used to drill the rods into dead-end geometry. The dead-end tubes are then sintered and annealed at elevated temperatures into dense membranes.

Hollow fibre membrane

Hollow fibre membranes have a micro-tubular configuration with outer diameter of less than 2 mm. The above extrusion-sintering method can also be used to fabricate hollow fibre membranes. But the method mostly used to fabricate dense ceramic hollow fibre membranes is the phase

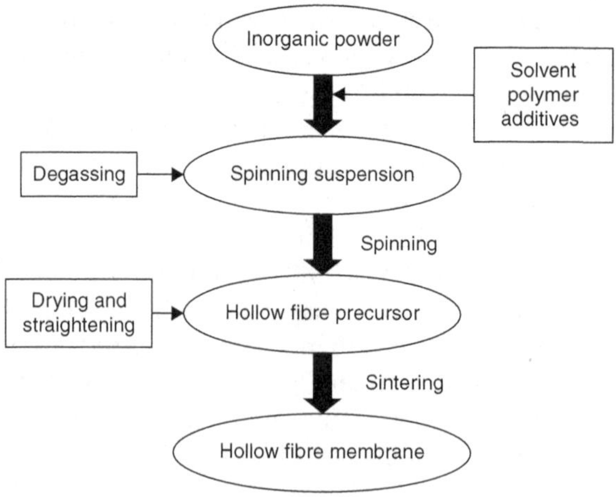

7.7 Phase inversion-sintering process to fabricate ceramic hollow fibre membranes.

inversion-sintering technique (Tan *et al.*, 2005a; Li, 2010). It is a three-step process including: (1) preparation of a spinning suspension; (2) spinning of hollow fibre precursors; and (3) high-temperature sintering, as illustrated in Fig. 7.7. The spinning suspension is composed of a polymer binder, ceramic powders, solvent and additives. The solvent(s) must dissolve the polymer binder and the additives and must show a high exchange rate with non-solvent (coagulants). All the components except inorganic powders are used to pre-design the membrane morphology (cross-sectional structure) and to facilitate the spinning and phase inversion. The suspension is pressurized through a tube-in-orifice spinneret (Figure 7.8a) into a coagulating bath to form hollow fibre precursors. Due to the phase inversion occurring in the nascent membranes, an asymmetric structure consisting of porous support and relatively dense layers can be formed depending on the suspension composition and the spinning conditions (Fig. 7.8b). Sintering of the hollow fibre precursors is to remove the organic components and to bond the ceramic particles into fibre products. The morphology of the hollow fibres can be generally retained but the microstructure may be changed during the sintering process. Since the porous and the intermediate structures are formed in a single step, the preparation is much simplified and the membrane cost can be accordingly decreased.

The properties of the ceramic membranes are strongly dependent on their morphology and microstructure (i.e., the size and shape of grains, the porosity, the pore size, and the pore-size distribution, etc.) of the hollow fibres. Theoretically, the hollow fibre morphology can be tailored as expected for different applications by modulation of the suspension composition and the

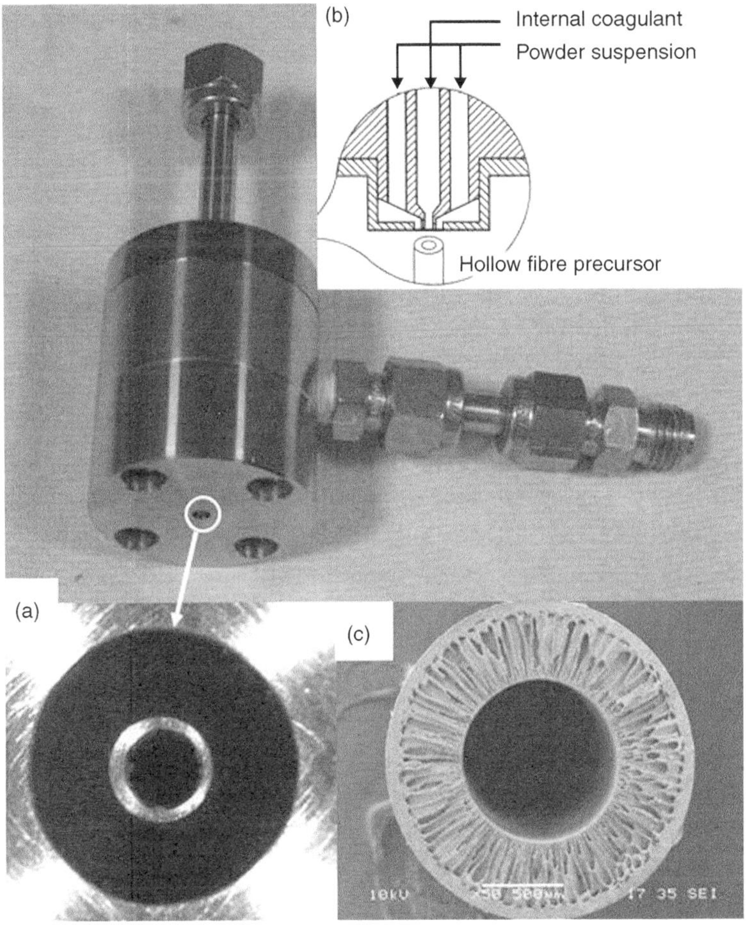

7.8 Structure of the spinneret (a and b) and the hollow fibre membranes.

spinning parameters (Kingsbury and Li, 2009; Tan *et al.*, 2011). However, it is still a great challenge to precisely design and control the macro- and micro-structure of the ceramic hollow fibre membranes, because too many factors, such as the particle size and its distribution, the shape and the surface property of ceramic powders, the composition and viscosity of the spinning suspension, the spinning conditions (spinning rate, air gap, internal coagulant, etc.) and the sintering parameters (sintering temperature, dwelling time, heating rate), can solely or jointly affect the formation of membrane structures. Therefore, a well-designed phase inversion/sintering process coupled with an optimal spinning suspension is the key to obtaining asymmetric hollow fibres with desired permeation characteristics as well as excellent mechanical strength.

7.3.2 Incorporation of a catalyst in membrane reactors

Inherently catalytic properties of the membrane

As described above, dense ceramic membranes are made of composite oxides with a large number of oxygen vacancies in the crystalline lattice. Such materials are inherently catalytic to the oxidation and dehydrogenation reactions. Therefore, dense ceramic membrane may serve as both catalyst and separator, and catalyst is not required in the membrane reactor. As shown in Fig. 7.5a, the lattice oxygen directly takes part in the chemical reactions. Since the chemical reactions take place on the membrane surface, it is required to have a very porous membrane surface so as to contain a sufficient quantity of active sites. This can be achieved in the membrane preparation process, or by coating a porous membrane material after the preparation. The main potential problems for this are that the membrane may not have sufficient catalytic activity, and the catalytic selectivity cannot be modulated with respect to the considered reactions.

Catalyst packed on or next to the membrane

In order to improve the catalytic activity and selectivity of the membrane reactor, a catalyst has to be used. A simple way is to place the catalyst pellets on/next to the membrane, as shown in Fig. 7.9a. The membrane mainly functions as either a product extractor or a reactant distributor, although it also plays some role in the reaction. The reaction selectivity is mainly determined by the catalyst. This incorporation mode is most popular in practical use and is easily operated. Since the catalyst is physically separated from the membrane, only the separation function of the membrane needs to be controlled. The high selectivity of the dense ceramic membranes leads to highly attractive results (pure H_2 extraction in dehydrogenation reactions and direct use of air in partial oxidation reaction). But the permeability of the membrane has to be improved as high as possible.

Catalyst incorporated with the membrane

A more efficient way for incorporation of catalyst with the membrane is to coat the catalyst on the membrane surface. The catalyst layer is generally porous and integrated with the membrane into a single body, as shown in Fig. 7.9b. The catalyst coating layer can be formed by a screen painting method. A catalyst serigraphy ink is first prepared from attrition-milled powders dispersed in an organic medium with the addition of rice starch as pore forming agent. A thin layer of the ink is screen printed onto one side surface of the dense membrane and dried at room temperature. After heat treatment to remove the organics, a porous catalyst layer with the porosity

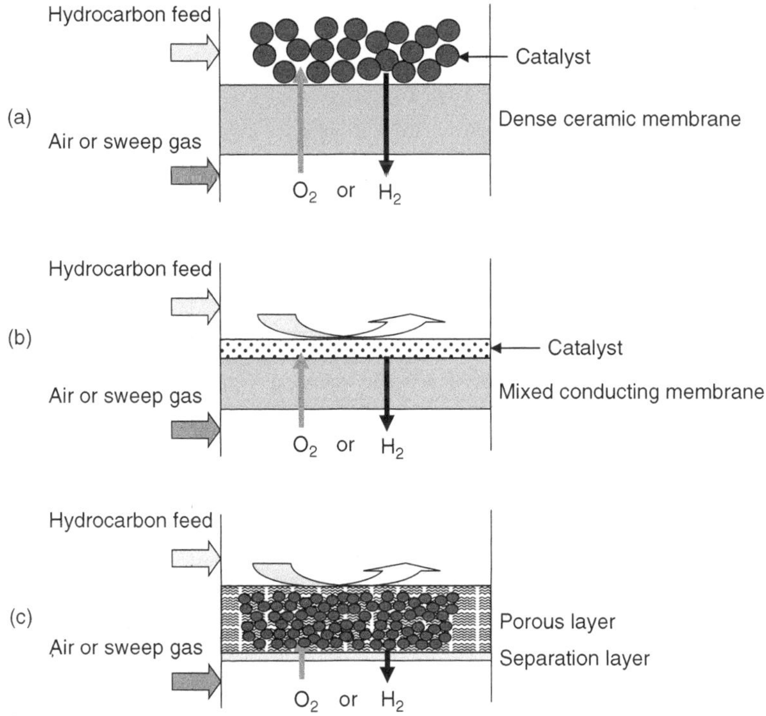

7.9 Membrane/catalyst combination modes: (a) catalyst separated by the membrane; (b) catalyst coated on the membrane; and (c) catalyst deposited in porous layer.

depending on the concentration of pore former agent will be formed on the membrane surface (Juste *et al.*, 2010).

The catalyst layer can also be formed together with the membrane in a single step (Wu *et al.*, 2010). The catalyst and membrane powders are made into the spinning suspension with different compositions. The two suspensions are co-extruded through a triple orifice spinneret into a coagulant bath for solvent exchange to form hollow fibre precursors. A high-temperature co-sintering is performed to obtain the catalyst/membrane dual-layer hollow fibres. Figure 7.10 shows the morphology of the LSM–YSZ/NiO–YSZ dual-layer hollow fibres where the LSM–YSZ serves as the oxygen-separation membrane and the NiO–YSZ layer as the catalyst for methane conversion. This method requires that the catalyst and the membrane layers should have matching sintering behaviours to avoid crack and to achieve great adhesion between the two layers.

In Fig. 7.9c, the catalyst is dispersed in the porous layer of the membrane to form a membrane catalyst. One of the reactants or the products traverses the membrane into or out of the reaction zone. In this case, the

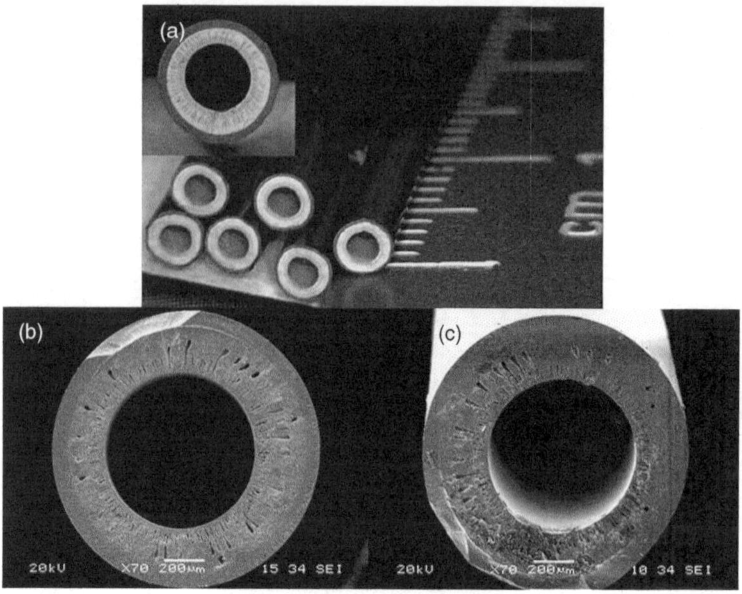

7.10 Morphology of the catalyst/membrane double-layer hollow fibres: (a) hollow fibre precursors; (b) LSM–YSZ/NiO–YSZ hollow fibre; and (c) reduced LSM–YSZ/Ni–YSZ fibre.

access of reactants to the catalyst is improved and the catalyst efficiency can be increased greatly. It is estimated that a membrane catalyst could be ten times more active than in the form of pellets provided that the membrane thickness and porous texture, as well as the quantity and location of the catalyst in the membrane, are adapted to the kinetics of the reaction (Dittmeyer *et al.* 2001). For catalytic membranes, the membrane composition, activity and porous texture have to be optimized for each reaction and kept stable upon use. In addition, if the catalyst is reactive with the separation layer of the membrane, direct contact between them should be avoided.

7.4 Fabrication of membrane reactors

7.4.1 Disc/flat-sheet membrane reactor

Disc/flat-sheet membrane reactors are mostly applied in research work because they can be fabricated easily in laboratory with a small amount of membrane material. Figure 7.11 shows the structure of the disc/flat-sheet membrane reactor. The membrane disc is mounted between two vertical ceramic or quartz tubes, and then placed in a bigger quartz tube. Pyrex or gold gaskets are used to obtain an effective seal between the disk and the walls of the inner tubing at high temperatures by placing the assembly in

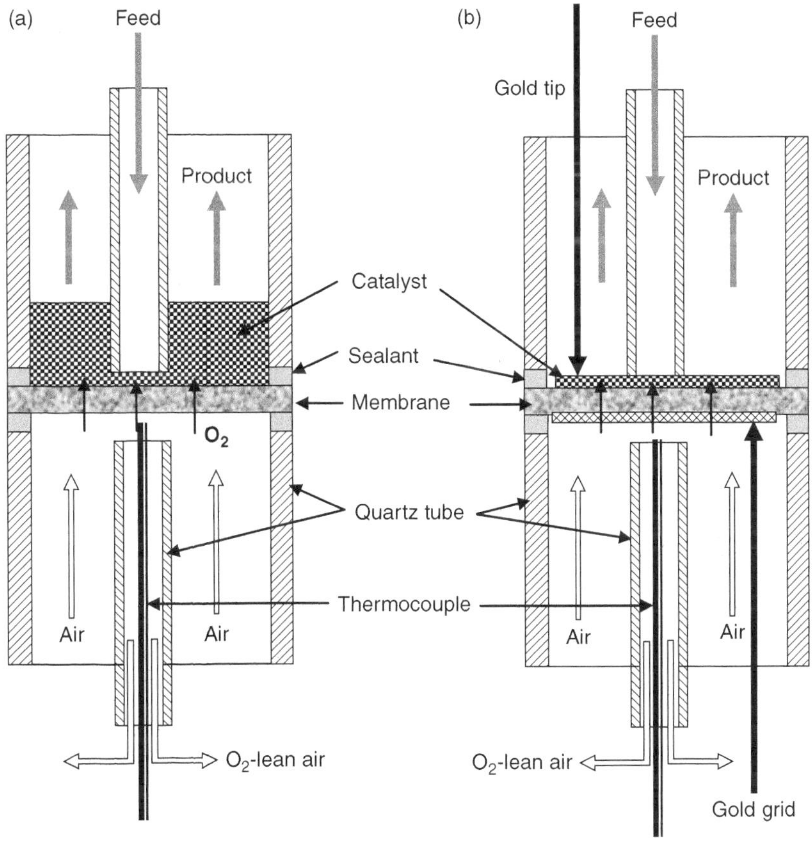

7.11 Structure of the disc/flat-sheet membrane reactor: (a) catalyst packed on membrane; and (b) catalyst coated on membrane with electrodes and external circuit.

compression with the use of spring clamps. The catalyst is usually packed on the membrane or coated on the permeate side surface.

Sometimes, an external circuit is established by fixing a gold lead on each side of the membrane with gold or silver paint, which are fired at an elevated temperature. The setup of Fig. 7.11b can also be used to measure the polarization of the membrane surface and to evaluate the activity of oxygen at the surface of the membrane and its evolution during catalytic reactions (Löfberg *et al.*, 2004).

7.4.2 Tubular membrane reactor

In order to increase the membrane areas for separation and reaction, tubular membrane reactors can be applied. Figure 7.12(a) shows the structure of

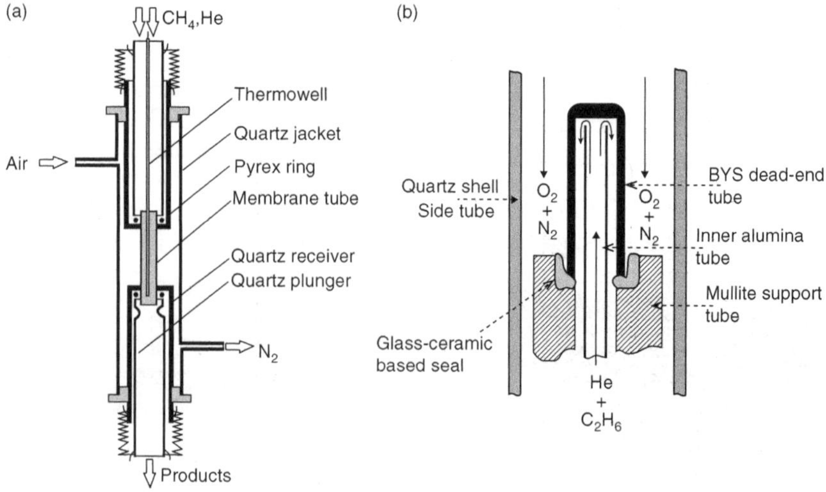

7.12 Schematic diagram showing (a) the tubular membrane reactors and (b) the dead-end tubular membrane reactor for the partial oxidation of ethane.

a typical tubular membrane reactor (Lu *et al.*, 2000). The membrane tube is connected to two quartz tubes with Pyrex rings as the sealant, and then pressed tight by the quartz plungers with springs. The seal is achieved by heating the reactor slowly to about 1000°C in a stagnant atmosphere so that the Pyrex rings become soft. The group is then placed in quartz tube serving as the shell side the membrane shell-and-tube reactor. A quartz thermowell is positioned along the axis of the membrane tube and allows a sliding thermocouple to profile the temperature along the reactor. The catalyst can be packed in the tube side or on the shell side of the membrane reactor.

The use of a dead-end ceramic membrane tube can make the assembly of the membrane reactor much simpler. As shown in Fig. 7.12(b), a dead-end ceramic membrane tube is sealed onto the mullite tube in which a smaller dense alumina tube is coaxially placed. The whole system is sited in a quartz tube, which serves as the shell side in the reactor operation. The short step in the mullite tube serves as a holder for both the seal material and the membrane. For the $Bi_{1.5}Y_{0.3}Sm_{0.2}O_3$ (BYS) membrane mounted to the mullite tube, the sealant consisting of 45 wt% BYS, 25 wt% $SrCe_{0.95}Tb_{0.05}O_3$, 10 wt% Pyrex glass, 10 wt% $NaAlO_2$ and 10 wt% B_2O_3 can be used (Akin and Lin, 2002).

7.4.3 Hollow fibre membrane reactor

In general, the hollow fibre membrane reactor can be assembled into reactors following the same procedures as those of tubular membranes. The small diameter of the hollow fibres allows a high membrane packing density in the

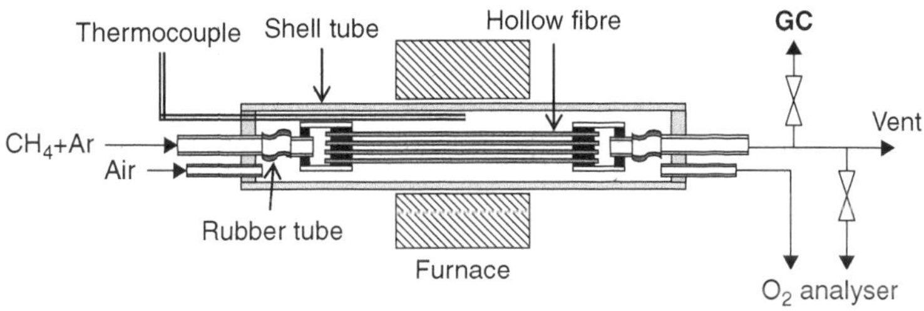

7.13 Configuration of the dense ceramic hollow fibre membrane reactor. In the figure GC refers to gas chromatography.

membrane reactor. Figure 7.13 schematically illustrates the configuration of the hollow fibre membrane reactor (Tan *et al.*, 2007; 2008). A quartz shell is used to house the hollow fibres, two pairs of gas inlet/outlet fittings and alumina thermocouple sleeve assembly. These are housed in MACOR end caps that are fitted closely to the inner wall of the shell tubing. (MACOR is the trademark name of the products made by Marco Products Company, San Fernando, California.) The hollow fibres are placed in the module using a pair of MACOR tubes, each tube closed at one end with holes drilled through to accommodate the ends of the hollow fibres. The other end of each MACOR tube is completely open. The open ends of these tubes are attached to the lumen-side gas inlet/outlet housed in the end caps by a flexible silicone tubing to offset the thermal expansion of the hollow fibres. The sealing is achieved by using a high-temperature water-based glass–ceramic sealant ('Light Grade' Fortafix, UK). A K-type thermocouple which could be moved along the length of the module is inserted in an alumina sleeve, the tip of which is positioned close to the centre of the hollow fibres, allowing the temperature profile to be recorded during operation. A custom made furnace with a short heat zone can be used to heat the reactor so that the sealing points are kept off from the high-temperature zone. The catalyst, if used, is usually packed in the fibre lumen for the convenient operation.

7.5 Conclusion and future trends

Dense ceramic membranes allow oxygen or hydrogen permeation in a dissociated or ionized form other than the conventional molecular diffusion, and thus exhibit extremely high selectivity (up to 100%). They can be incorporated into membrane reactors for a variety of oxidation and dehydrogenation reactions where the membrane functions as either a product extractor or a reactant distributor. Three configurations, that is, disc/flat-sheet, tubular and hollow fibre membranes have been applied in membrane reactors. They exhibit respective advantages/disadvantages in terms of the ease and cost of fabrication, the effective membrane area/volume ratio, and mechanical robustness, and can be used for different purposes. The catalysts can be incorporated with the membrane reactor by packing on/next to the membrane, by coating on the membrane surface, or by dispersing in the porous structure of the composite membrane. The fabrication of intact dense membrane with high performance and a membrane reactor in large scale is the main challenge for commercializing dense ceramic membrane reactor technology. For this purpose, future work has to be focused on both the material and engineering solutions including:

- Developing novel membrane materials with high thermal and chemical stability especially in CO_2- and water vapour containing atmosphere.
- Developing new fabrication technologies to improve membrane permeation rate (with higher permeability and higher membrane area/volume ratio) and mechanical strength, and also to reduce membrane costs.
- Deepening theoretical and experimental investigation into the design and control of membrane permeability to match the catalytic activity of catalysts in membrane reactors.
- Developing novel high-temperature sealants and sealing techniques for membrane reactor fabrication.

7.6 Acknowledgements

The authors gratefully acknowledge the research funding provided by the National Natural Science Foundation of China (NNSFC, No. 20976098, No. 21176187) and the Royal Academy of Engineering Research Exchanges with China and India Scheme.

7.7 References

Akin F T and Lin Y S (2002), 'Selective oxidation of ethane to ethylene in a dense tubular membrane reactor', *J Membr Sci*, **209**, 457–467.

Bouwmeester H J M (2003), 'Dense ceramic membranes for methane conversion', *Catal Today*, **82**, 141–150.

Chen C H, Bouwmeester H J M, van Doorn R H E, Kruidhof H and Burggraaf A J (1997), 'Oxygen permeation of $La_{0.3}Sr_{0.7}CoO_{3-\delta}$', *Solid State Ionics*, **98**, 7–13.

Dittmeyer R, Hollein V and Daub K (2001), 'Membrane reactors for hydrogenation and dehydrogenation processes based on supported palladium', *J Mol Catal A: Chem*, **173**, 135–184.

Hendriksen P V, Larsen P H, Mogensen M, Poulsen F W and Wiik K (2000), 'Prospects and problems of dense oxygen permeable membranes', *Catal Today*, **56**, 283–295.

Iwahara H (1996), 'Proton conducting ceramics and their applications', *Solid State Ionics*, **86–88**, 9–15.

Iwahara H, Asakura Y, Katahira K and Tanaka M (2004), 'Prospect of hydrogen technology using proton-conducting ceramics', *Solid State Ionics*, **168**, 299–310.

Juste E, Julian A, Geffroy P-M, Vivet A, Coudert V, Richet N, Pirovano C, Chartier T and Del Gallo P (2010), 'Influence of microstructure and architecture on oxygen permeation of $La_{(1-x)}SrxFe_{(1-y)}(Ga, Ni)_YO_{3-\delta}$ perovskite catalytic membrane reactor', *J Eur Ceramic Soc*, **30**, 1409–1417.

Kaiser A, Foghmoes S, Chatzichristodoulou C, Søgaard M, Glasscock J A, Frandsen H L and Hendriksen P V (2011), 'Evaluation of thin film ceria membranes for syngas membrane reactors—Preparation, characterization and testing', *J Membr Sci*, **378**, 51–56.

Kingsbury B F K and Li K (2009), 'A morphological study of ceramic hollow fibre membranes', *J Membr Sci*, **328**, 134–140.

Li K (2010), 'Ceramic hollow fiber membranes and their applications', in Enrico Drioli and Lidietta Giorno, *Comprehensive Membrane Science and Engineering*, Oxford: Academic Press, v **1**: 253–273.

Liu S, Tan X, Li K and Hughes R (2001), 'Methane coupling using catalytic membrane reactors', *Catal Rev*, **43**, 147–198.

Liu Y, Tan X and Li K (2006), 'Mixed conducting ceramics for catalytic membrane processing', *Catal Rev*, **48**, 145–198.

Löfberg A, Boujmiai S, Capoen E, Steil M C, Pirovano C, Vannier R N, Mairesse G and Bordes-Richard E (2004), 'Oxygen permeation versus catalytic properties of bismuth-based oxide ion conductors used for propene oxidation in a catalytic dense membrane reactor', *Catal Today*, **91–92**, 79–83.

Lu Y, Dixon A G, Moser W R, Ma Y H and Balachandran U (2000), 'Oxygen-permeable dense membrane reactor for the oxidative coupling of methane', *J Membr Sci*, **170**, 27–34.

Sirkar K K, Shanbhag P V and Kovvali A S (1999), 'Membrane in a reactor: A functional perspective', *Ind Eng Chem Res*, **38**, 3715–3737.

Siriwardane R V, Poston Jr. J A, Fisher E P, Lee T H, Dorris S E and Balachandran U (2000), 'Characterization of ceramic hydrogen separation membranes with varying nickel concentrations', *Appl Surf Sci*, **167**(1–2), 34–50.

Tan X, Liu Y and Li K (2005a), 'Preparation of $La_{0.6}Sr_{0.4}Co_{0.2}Fe_{0.8}O_{3-\alpha}$ hollow fiber membranes for oxygen production by a phase-inversion/sintering technique', *Ind Eng Chem Res*, **44**, 61–66.

Tan X, Liu Y and Li K (2005b), 'Mixed conducting ceramic hollow fiber membranes for air separation', *AIChE J*, **51**, 1991–2000.

Tan X, Pang Z, Gu Z and Liu S (2007), 'Catalytic perovskite hollow fiber membrane reactors for methane oxidative coupling', *J Membr Sci*, **302**, 109–114.

Tan X, Li K, Thursfield A and Metcalfe I S (2008), 'Oxyfuel combustion using a catalytic ceramic membrane reactor', *Catal Today*, **131**, 292–304.

Tan X, Liu N, Meng B and Liu S (2011), 'Morphology control of perovskite hollow fibre membranes for oxygen separation using different bore fluids', *J Membr Sci*, **378**, 308–318.

Wu Z, Wang B and Li K (2010), 'A novel dual-layer ceramic hollow fibre membrane reactor for methane conversion', *J Membr Sci*, **352**, 63–70.

Xu S J and Thomson W J (1999), 'Oxygen permeation rates through ion-conducting perovskite membranes', *Chem Eng Sci*, **54**, 3839–3850.

7.8 Appendices

7.8.1 Appendix 1: oxygen permeation flux through the MIEC membrane

The transfer flux of charged species in electrolyte membranes can be described by the Nernst–Planck equation:

$$J_i = -\frac{\sigma_i}{z_i^2 F^2}\nabla\mu_i + C_i\upsilon \qquad [A1]$$

where $\sigma_i, \mu_i, z_i, C_i$ are the conductivity, electrochemical potential, charge number and concentration of species i, respectively; υ is the local velocity of inert marker; F is the Faraday constant. The electrochemical potential for each charged species consists of a chemical potential or an activity term and a local electrostatic potential term, ϕ.

$$\mu_i = \mu_i^0 + RT\ln a_i + z_i F\phi \qquad [A2]$$

where μ_i^0 a_i, ϕ, R and T are the standard chemical potential, activity and the Galvanic (internal) potential, the gas constant and temperature, respectively. For the ideal state, the activity of defect can be replaced by its concentration (activity coefficient is unit).

The conductivity of defect can be correlated to its concentration and diffusivity, which is a measure of the random motion of the species i in the lattice, by the Nernst–Einstein equation:

$$\sigma_i = \frac{z_i^2 F^2}{RT} C_i D_i \qquad [A3]$$

where D_i is the diffusion coefficient of charged species i.

The relationship between the flux of mobile ions and the current density through the membrane can be expressed as:

$$I = \mathrm{F}\sum_i z_i J_i \qquad [A4]$$

For the MIEC membranes without external circuit, the overall charge balance is applied or $\sum z_i J_i = 0$ and the local velocity of inert marker is negligible, $\upsilon = 0$. Accordingly, the transport flux of charged defects in the MIEC membrane at steady state can be derived (one-dimensional model) from Equations [A1] to [A3] as:

$$J_i = -D_i C_i \left[\frac{1-t_i}{C_i} \cdot \frac{\mathrm{d}C_i}{\mathrm{d}x} - \sum_{j \neq i} \frac{z_i}{z_j} \cdot \frac{t_j}{C_j} \cdot \frac{\mathrm{d}C_j}{\mathrm{d}x} \right] \quad \text{[A5]}$$

where t_j is the transport number of defect j and t_i is the transport number of defect i:

$$t_i = \frac{\sigma_i}{\sum \sigma_j} = \frac{z_i^2 D_i C_i}{\sum_j z_j^2 D_j C_j} \quad \text{[A6]}$$

In oxygen permeable MIEC membranes, oxygen vacancy and electron hole are the primary mobile charge carriers, and the oxygen vacancy flux may be derived from Equation [A5] as:

$$J_V = -\frac{(C_h + 4C_V) D_V D_h}{C_h D_h + 4C_V D_V} \cdot \frac{\mathrm{d}C_V}{\mathrm{d}x} \quad \text{[A7]}$$

where the subscripts h and V represent hole and oxygen vacancy, respectively.

Based on the stoichiometric relation between oxygen and vacancy, the oxygen bulk diffusion flux in the membrane can be given by:

$$J_{O_2} = -\frac{1}{2} J_V = \frac{1}{2\delta} \int_{(I)}^{(II)} \frac{(C_h + 4C_V) D_V D_h}{C_h D_h + 4C_V D_V} \cdot \mathrm{d}C_V \quad \text{[A8]}$$

It can be seen that oxygen permeation flux through the MIEC membranes is determined by the defect concentrations on the membrane surfaces and is inversely proportional to the membrane thickness.

For most MIEC perovskite membranes, electronic conductivity usually overwhelms ionic conductivity, that is, $C_h D_h >> C_V D_V$ and $C_h >> C_V$. Therefore, Equation [A8] may reduce to:

$$J_{O_2} = \frac{D_V \left(C_V'' - C_V' \right)}{2\delta} \quad \text{[A9]}$$

On the other hand, the surface exchange reaction rates (Equations [7.1a] and [7.1b]) integrated with all the sub-steps on the O_2-rich and the O_2-lean side may be written, respectively, as:

$$J_{O_2} = k_f \left(p_{O_2}' \right)^{0.5} C_V' - k_r \quad \text{[A10a]}$$

$$J_{O_2} = k_r - k_f \left(p''_{O_2}\right)^{0.5} C''_V \quad \text{[A10b]}$$

where k_f and k_r are, respectively, the forward and reverse reaction rate constants for the surface reactions; p'_{O_2} and p''_{O_2} are the partial pressures of oxygen on the O_2-rich and O_2-lean side, respectively. It is noted that Equation [A10] is based on the fact that the electron holes are essentially constant on membrane surfaces due to the overwhelming electronic conductivity in perovskites, and thus the exchange reaction rates may be pseudo zero-order with respect to electron hole concentration at steady state under isothermal condition.

Combination of Equations [A9] and [A10] gives the overall oxygen permeation flux through perovskite membranes in terms of the partial pressures and the membrane thickness as:

$$J_{O_2} = \frac{k_r\left[\left(p'_{O_2}\right)^{0.5} - \left(p''_{O_2}\right)^{0.5}\right]}{\left(p''_{O_2}\right)^{0.5} + \frac{2\delta k_f}{D_V}\cdot\left(p'_{O_2}p''_{O_2}\right)^{0.5} + \left(p'_{O_2}\right)^{0.5}} \quad \text{[A11]}$$

7.8.2 Appendix 2: notation

C_i	concentration of charged defect i (mol/m^3)
D_i	diffusion coefficient of charged defect i (m^2/s)
F	Faraday constant
k_r	the reverse reaction rate constant for oxygen surface exchange reaction (mol/(cm^2 s))
k_f	the forward reaction rate constant for oxygen surface exchange reaction (m/(Pa$^{0.5}$ s))
J_{O_2}	oxygen permeation flux (mol/(m^2 s))
p'_{O_2}, p''_{O_2}	oxygen partial pressure in the upstream and downstream side (Pa)
R	ideal gas constant (8.314 J/(mol.K))
R_m	algorithmic radius of fibre, $R_m = (R_o - R_{in})/\ln(R_o/R_{in})$
R_o, R_{in}	outer and inner radius of the tubular membrane (m)
T	temperature (K)
t_i	transport number of defect i t_j transport number of defect j
z_i	charge number of defect i
δ	membrane thickness (m)
μ_i	electrochemical potential of defect i

μ_i^0 standard chemical potential
σ_i conductivity of defect i
ϕ local electrostatic potential or Galvanic (internal) potential
υ local velocity of inert marker

Subscripts

e electron
h electron hole
V oxygen vacancy

8
Porous ceramic membranes for membrane reactors

S. SMART, The University of Queensland, Australia, S. LIU, Curtin University, Australia, J. M. SERRA, Universidad Politécnica de Valencia, Spain, J. C. DINIZ DA COSTA, The University of Queensland, Australia and A. IULIANELLI and A. BASILE, ITM-CNR, Italy

DOI: 10.1533/9780857097330.2.298

Abstract: This chapter discusses the research and development of porous ceramic membranes and their application as membrane reactors (MRs) for both gas and liquid phase reaction and separation. The most commonly used preparation techniques for the synthesis of porous ceramic membranes are introduced first followed by a discussion of the various techniques used to characterise the membrane microstructure, pore network, permeation and separation behaviour. To further understand the structure-property relationships involved, an overview of the relevant gas transport mechanisms is presented here. Studies involving porous ceramic MRs are then reviewed. Of importance here is that while the general mesoporous nature of these membranes does not allow excellent separation, they are still more than capable of enhancing reaction conversion and selectivity by acting as either a product separator or reactant distributor. The chapter closes by presenting the future research directions and considerations of porous ceramic MRs.

Key words: mesoporous ceramic membranes, alumina, titania, zirconia, membrane reactors for dehydrogenation reactions.

8.1 Introduction

A membrane is a selective barrier that allows passage to certain species, while rejecting or hindering the passage of other species. They were first described in the scientific literature in 1748 by Jean-Antonie Nollet (Nollet, 1748) and were used by the research community to develop physical and chemical theories such as osmosis and the kinetic theory of gases. However, despite the concept of membrane separation, originally proposed by Graham in 1866, for separating a solution into its components (Graham, 1866) these early examples were unreliable, exhibited poor selectivity and were ultimately too expensive to be applied to industrial or commercial applications until the 1970s where polymeric membranes found applications in both water desalination and natural gas processing (Laverty and O'Hair,

Table 8.1 Pore size classification

Type	Pore diameter (nm)	Transport mechanism	Applications field
Microporous	<2	Micropore diffusion	Gas separation
Mesoporous	2–50	Knudsen diffusion	Gas separation, ultrafiltration, nanofiltration
Macroporous	>50	Sieving	Ultrafiltration, nanofiltration

Source: Rouquerol *et al.* (1994).

1990). By contrast porous ceramic membranes had found application since the 1960s in the field of large-scale gas diffusion processes for uranium isotope separation. It was only in the 1980s that porous ceramic membranes found other non-nuclear industrial applications, mainly oriented towards microfiltration and ultrafiltration water treatment processes.

A porous ceramic membrane is, for the purposes of this chapter, defined as an inorganic, non-metallic porous solid formed into the desired membrane geometry, fired at high temperatures and subsequently cooled. The pore size of the membrane is classified as per the International Union of Pure and Applied Chemistry (IUPAC) standards as outlined in Table 8.1.

Ceramic membranes may consist of a single morphology with a narrow pore distribution, but are more commonly composed of several layers of different ceramic materials forming an asymmetric composite membrane. Generally, they possess a macroporous support, one or two mesoporous intermediate layers and a microporous top layer. In particular, the macroporous layer provides the mechanical support, whereas the middle layer bridges the pore size differences between the support layer and the top layer, where the actual separation takes place. The most commonly used materials for ceramic membranes are Al_2O_3, TiO_2, ZrO_2 and SiO_2. The majority of commercial ceramic membranes are available in plate, monolith or tubular geometries. Conventionally, they are assembled as a plate and frame module using sheet membranes or as a tubular module by means of either a porous monolith with multiple channels or as closely packed membrane tubes. Several research groups have constructed ceramic membranes in a hollow fibre form, useful for overcoming the physical brittleness often associated with ceramic materials but still retaining a high surface area to volume ratio.

There has been considerable growth in the research and development of porous ceramic membranes in the past two decades as demonstrated in Fig. 8.1 where the total number of papers focused on ceramic membranes as compared with the total number of scientific papers for inorganic membranes as a whole. Figure 8.2 further illustrates the percentage distribution

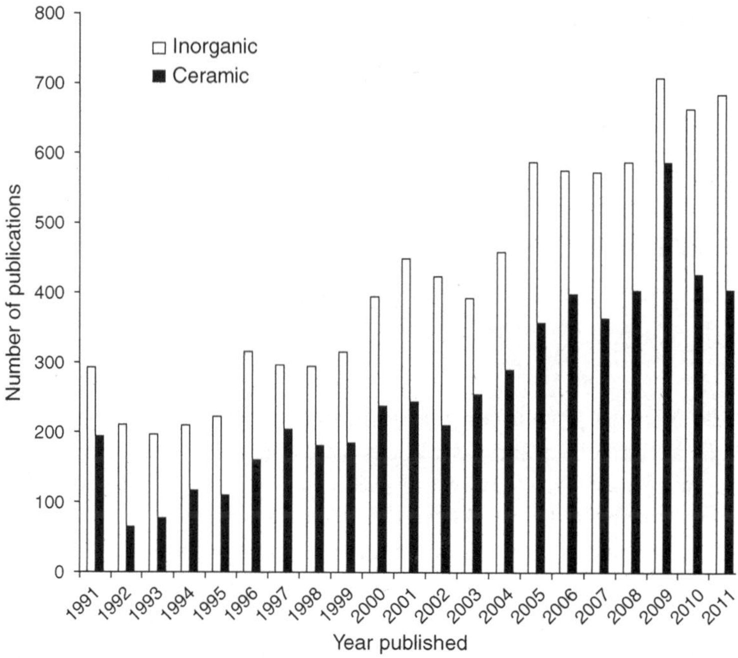

8.1 Number of published papers per year for the last two decades (Elsevier Scopus database).

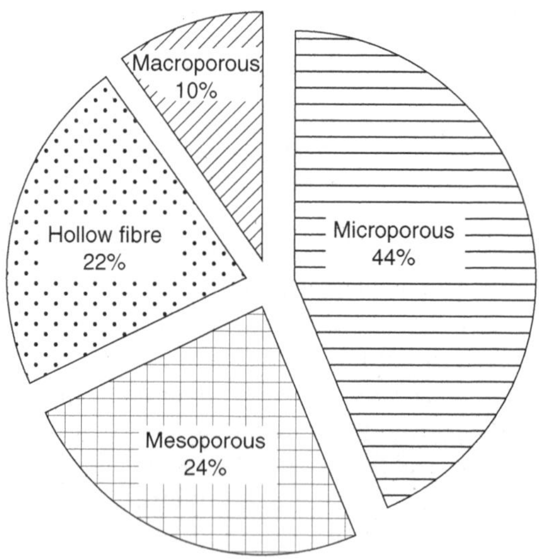

8.2 Percentage distribution of the studies carried out on porous ceramic membranes.

of the studies carried out on ceramic membranes, based on their different geometries. The majority of these studies have focused on using porous ceramic membranes for separation applications with only one-third utilising them in a MR configuration. The incorporation of a membrane into a reactor can serve several functions, the most common of which is to separate a product from the reaction mixture to enhance conversion and reduce downstream processing requirements. The scope of this chapter is to examine the preparation and characterisation techniques for porous ceramic membranes and to highlight their potential when applied to the field of MR technology.

8.2 Preparation of porous ceramic membranes

There are a wide variety of techniques used for the preparation of porous ceramic membranes; however, they all share the following common steps:

- formation of ceramic particles or membrane precursors;
- packing of the particles or coating of the precursor into a particular geometric shape such as flat sheet, monolith or tube;
- consolidation of the membrane by sintering or calcination at high temperature.

Conventionally, the production of symmetric macroporous ceramic filter elements, which also double as support structures for asymmetric composite membranes, is achieved through physical, shaping techniques which are designed to shape a suspension or slurry of ceramic particles into a solid membrane 'green' (unsintered) body (Fig. 8.3). The production of the final porous

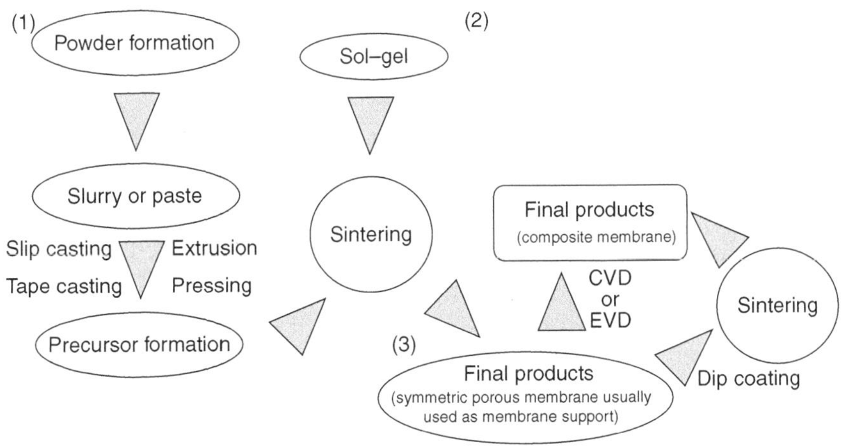

8.3 Preparation scheme of a porous ceramic membrane using conventional techniques. CVD, chemical vapour deposition; EVD, electrochemical vapour deposition.

ceramic filter element or membrane support is achieved by firing or sintering the 'green' body at high temperatures. The final pore size, porosity and therefore performance obtained is primarily a function of both the original ceramic particle size and the sintering conditions. By contrast, the 'green' body shaping process primarily controls the final geometry of the membrane but not its final microstructure. Asymmetric composite membranes (multilayer membranes) can be further produced on a macroporous support through a coating or deposition step followed by further firing. In this case, the formation of the coating solution, coating technique and firing steps all influence the pore structure of the final membrane. Each of these techniques, included sintering, will be briefly reviewed in the following subsections of this chapter.

8.2.1 Casting

Casting is a commonly used technique for the preparation of ceramic membranes, whereby wet slurry of ceramic particles is formed into the desired shaped as the solvent is removed either by capillary suction (slip casting) or evaporation (tape casting). Tape casting is a process used for producing large quantities of flat sheet ceramic membranes. The process, shown schematically in Fig. 8.4 involves a casting knife, a reservoir for the slurry of ceramic powder and a drying zone. Briefly, ceramic slurry is placed in the reservoir chamber which has a small 'gap' controlled by the casting knife or 'doctor blade'. Underneath this gap is a polymer tape, and movement of either the tape (most common) or the knife/reservoir set-up results in the formation of a slurry layer with the thickness of the 'doctor blade gap'. The slurry and tape then pass through an oven or drying zone where the slurry solvent is evaporated forming a solid ceramic film on the polymer backing. This can then be wound onto a spool for storage before sintering.

The thickness of porous ceramic membranes prepared in this manner is typically in the range of a few millimetres (Larbot, 1996), although thinner membranes have been realised (Sahibzada *et al.*, 2000). Tape casting is a high volume production technique for flat sheet membranes and is employed on both a research and commercial scale.

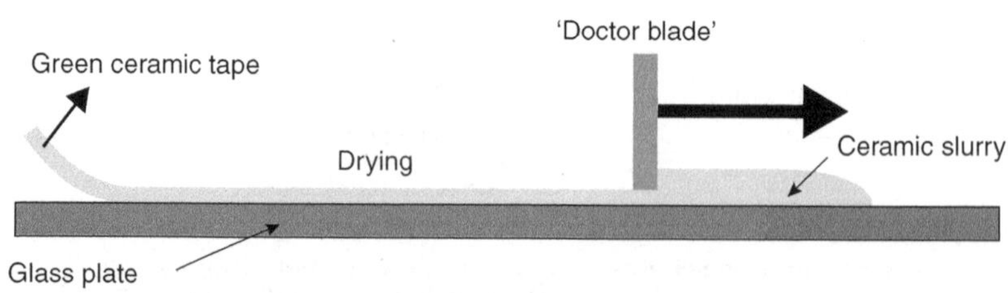

8.4 Tape casting technique.

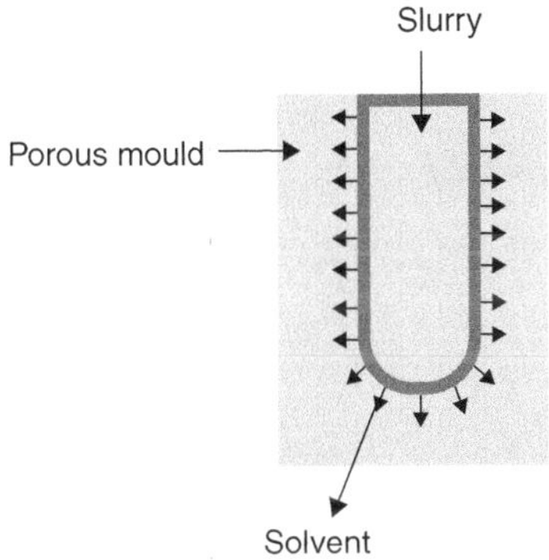

8.5 Slip casting method for ceramic membrane preparation.

Slip casting involves pouring the ceramic slurry into a porous mould, of the desired shape, which imparts a capillary suction force to extract the solvent from the suspension (Fig. 8.5). It is crucial that the capillary-induced solvent removal takes place quickly to avoid penetration of the particles into the pores of the mould. In this way the particles form a layer on the mould surface to form the 'green' membrane. The use of a mould allows both simple and complex shapes to be produced, conferring a significant advantage over tape casting; however, the rate of production is limited by the number of available moulds, which ensures the process is only commercially effective for small to medium production volumes. The parameters controlling the 'green' body production process are primarily the particle size, shape, loading and slurry viscosity (Leenaars and Burggraaf, 1985c, Uhlhorn *et al.*, 1989).

8.2.2 Extrusion

Extrusion is an important and productive technique used extensively for the preparation of porous ceramic tubes. The extrusion process bares significant similarities to both tape casting and fibre spinning in that a slurry or paste of ceramic particles is forced through a die to form the final 'green' membrane. However, extrusion relies on plastic deformation of the slurry and evaporation of any remaining solvent to keep the membrane in its desired final shape (Fig. 8.6), whereas fibre spinning relies on a coagulation bath to remove the solvent and fix the membrane structure. The other crucial difference lies in the membrane 'green' body microstructure, with extruded membranes displaying a homogeneous cross-sectional microstructure, in contrast to spun membranes where the homogeneity of the cross-sectional microstructure is governed by the interactions between the solvent, binder and coagulation bath (Isobe *et al.*, 2006) as discussed further in Section 8.2.3.

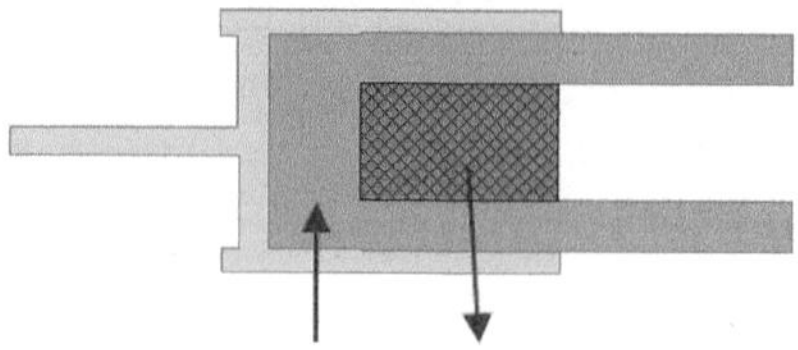

8.6 Extrusion technique.

8.2.3 Hollow fibre ceramic membranes

Polymeric hollow fibre membranes are currently in use for a number of process applications, including filtration, desalination, gas separation and in MRs. Ceramic membranes have been developed for similar applications, especially where high temperature operation or harsh feeds containing high concentrated alkaline, acid or organic solvents exclude the use of existing polymeric membranes. However, currently produced ceramic membranes are usually in the form of flat discs, finite sized tubes with diameters of at least several millimetres, or multi-channel monoliths and consequently have low surface area/volume ratios ranged from 30 to 250 m^2 m^{-3} (Hsieh, 1989, 1991; Saracco *et al.*, 1994). These low area/volume ratios compare unfavourably with polymeric hollow fibre modules where area/volume ratios of several thousands are obtainable. This limits the application of current inorganic monolithic, tubular and disc membranes. This limitation is most evident in catalytic MRs, where it is desirable to maximise the area of the membrane module to increase the permeation rate to remove the product species from the reaction zone (Armor, 1998). Traditional inorganic hollow fibres have been made by the extrusion of the melting inorganic materials such as glass or steel, which requires very expensive high temperature extrusion equipment (Kuraoka *et al.*, 2002). Recently, the well-known phase-inversion method, commonly employed for spinning polymeric hollow fibre membranes, has been successfully modified to prepare the ceramic hollow fibres (Liu, 2007; Liu and Gavalas, 2004; Liu *et al.*, 2001, 2003, 2006; Tan *et al.*, 2001). Because of the phase-inversion characteristics, the prepared inorganic hollow fibres possess an asymmetric structure, which provides a better permeability for a given thickness. Thus, they can be used directly not only in many separation processes, but also to be served as a porous support for composite membrane formation (Liu *et al.*, 2005). More importantly, this synthesis skill is very versatile and can be expanded directly to prepare any kinds of inorganic or ceramic membrane materials as long as these inorganic materials are not water-soluble and the particle size ranges from 100 microns to 50 nanometres to make a suitable spinning dope. The physical properties, particularly the pore size, porosity and mechanical strength of the resultant ceramic hollow fibre membranes, are jointly affected by many factors such as the individual material, particle size or size distribution, the composition of the spinning dope, the spinning and the sintering conditions. The three main steps in this synthesis process are briefly illustrated below.

Preparation of a spinning suspension

A typical polymeric solution, where the weight ratio of the polymer (typically polyethersulfone (PESf)) and the organic solvent (1-methyl-2-pyrrolidinone (NMP)) is 1: 4, forms the basis for the ceramic hollow fibre spinning solution. This composition is traditionally used to prepare polymeric hollow fibre membranes; however, due to the good binding capability of PESf, it is now also used as the organic binder to prepare ceramic hollow fibres. Other sulphur-containing polymers such as polysulfone (PSF) and sulphur-free polymers such as polyetherimide (PEI), can also be used for this purpose. The difference between the usage of binders that contain sulphur and those that do not is that some inorganic materials can be contaminated by the presence of sulphur during the subsequent sintering process (Leo *et al.*, 2011). After the polymer solution is formed, the ceramic powder is then added, followed by extensive mixing for 3 to 10 h to ensure that all the ceramic powder is uniformly dispersed in the polymer solution. The required energy of mixing is dependent on the potential of the particles to aggregate (Smart *et al.*, 2010b). The weight ratio of ceramic particles to polymer can be controlled from 3 to 12 depending on the particle size and the requirement of the final resultant membranes. The typical viscosity of the spinning dope measured at room temperature by Physica UDS-200 rheometer at shear rates of 3 rpm is around 55 Pa s^{-1}. When the viscosity is too low, up to 2% (by weight) of polyvinylpyrrolidone (PVP) can be introduced into the spinning solution to modulate its viscosity. Finally, the spinning mixture or dope is degassed at room temperature for a few hours by applying vacuum to remove the air bubbles generated during mixing. The existence of these gas bubbles will result in defective hollow fibres.

Spinning of the ceramic hollow fibre precursor

The degassed spinning dope containing the dispersed ceramic powders is transferred to a stainless steel reservoir and pressurised to 20–50 psig using nitrogen or air. The spinning apparatus is schematically shown in Fig. 8.7. The details of the spinning apparatus and the procedures on hollow fibre spinning have been described elsewhere (Deshmukh and Li, 1998). Briefly, a tube-in-orifice spinneret (the inset photograph in Fig. 8.7) with orifice diameter/inner diameter of the tube of approximately 2.0/0.72 mm is used to obtain 'green' hollow fibres. The air-gap can be kept at 2–5 cm during the

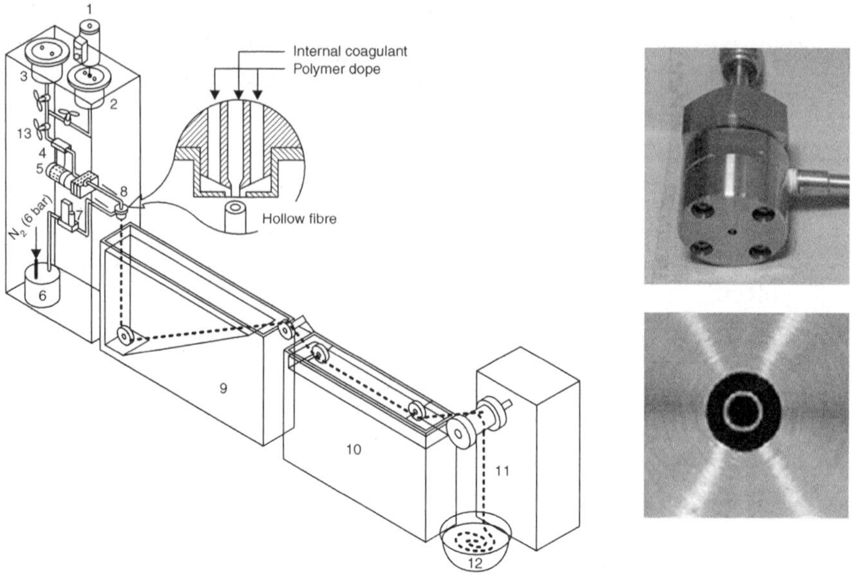

8.7 Ceramic hollow fibre membrane precursor spinning apparatus (inset: photograph of a spinneret). (1) Mechanical stirrer; (2) mixing tank; (3) solution tank; (4) filter; (5) gear pump; (6) internal coagulant; (7) mass flow controller; (8) spinnerette; (9) coagulation bath; (10) washing bath; (11) motor guide; (12) storage tank; (13) valve. Reproduced from Li (2010), with permission from Elsevier.

spinning process, although this is highly dependent on the desired microstructure of the hollow fibres (Boom *et al.*, 1994). The nascent 'green' hollow fibre is passed through a coagulation bath to immediately solidify the hollow fibre shape, although to complete the full phase-inversion process the 'green' fibres are soaked inside the bath for at least 24 h during which time the solvent stops diffusing from the polymer solution phase to the water phase. The 'green' hollow fibre is then dried at room temperature to remove any water prior to sintering. It should be noted that many parameters of the spinning process like dope composition, the bore liquid composition, the air-gap distance and the hollow fibre extruding speed will have significant influence on the physical properties of the resultant hollow fibre membranes (Tan *et al.*, 2011).

Hollow fibre sintering

The sintering process of the 'green' hollow fibres is similar to the sintering processes utilised in the manufacture of porous ceramic membranes via other techniques. The major concern during sintering of the hollow fibres is the removal of the binder and subsequent sintering without collapsing the hollow fibre shape. Traditionally, this is accomplished via heating in a programmable furnace at lower temperatures ranged from 400 to 800°C for 4 h to remove the organic polymer binder and then sintering at higher temperatures for about 10 h to allow the fusion and bonding to occur to make the hollow fibre robust. One empirical rule is that the sintering temperature should be chosen at around three-fourth of the material's melting point. Otherwise, the ceramic particles would not sufficiently fuse or bond during sintering and the mechanical strength of the resultant hollow fibre is very low. In the case of Al_2O_3, the melting point is 2054°C and thus it suggests that preparation of the Al_2O_3 hollow fibre membranes with sufficient mechanical strength are possible at the sintering temperature of 1540°C. Sintering of the hollow fibre at higher temperatures that is, 1600°C shows, of course, an increase in mechanical strength; however, membrane porosity and gas permeability are decreased considerably. Therefore, there is a trade-off between the mechanical strength, membrane porosity and gas permeability (Liu *et al.*, 2003).

8.2.4 Sintering

In the production of porous ceramic membrane, sintering is used to eliminate any remaining binder or solvent and most importantly to interconnect the particles contained in the green body in order to create a solid body with the desired characteristics. Ultimately, the driving force for the sintering process is a reduction in the total interfacial energy of the particles of which the surface area associated with the pores is easily the largest contributor. Porosity is maintained therefore through careful control of the sintering parameters and crucially by halting the sintering process before complete densification is achieved. Sintering can be achieved through two main mechanisms: solid-state sintering, where the green body is densified wholly in a solid state at the sintering temperature and liquid-state sintering, where a liquid phase is present in the green body during sintering (Kang, 2005). A stylised schematic of the sintering process is displayed in Fig. 8.8, although the production of a porous ceramic membrane would necessitate halting the sintering process during one of the intermediate steps.

In general, the densification rate during sintering increases with decreasing particle size and increasing sintering temperature and time, as shown schematically in Fig. 8.9. Since sintering is a thermally activated process, the variables affected by the change in temperature (viscosity, diffusivity, etc.) are expressed as an exponential function of the temperature, but the exact relation is different for each of the different sintering mechanisms. An increase in sintering temperature leads into an enhancement of the densification rate relative to particle growth rate. To obtain porous materials the sintering process commonly takes place at a lower temperature or

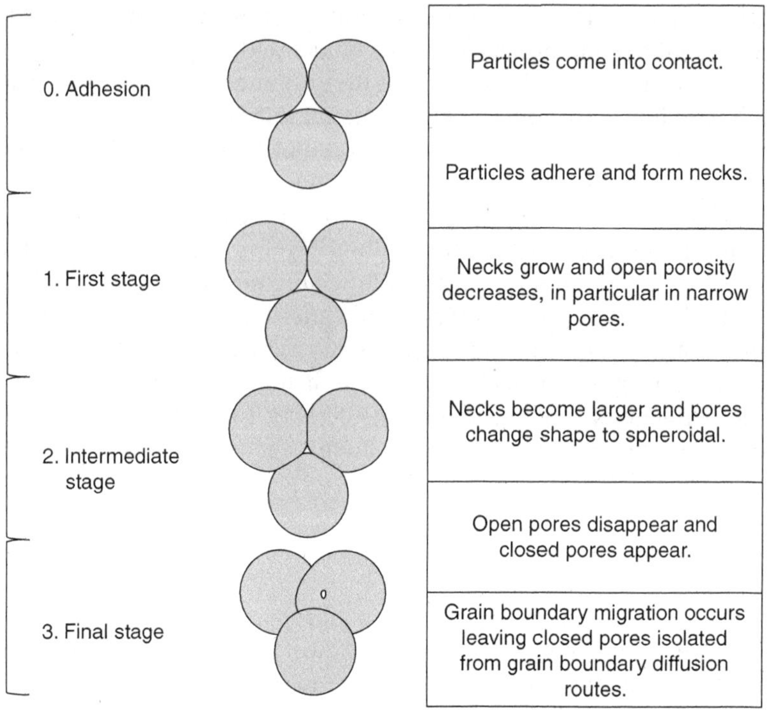

8.8 Scheme of densification stages in the sintering process.

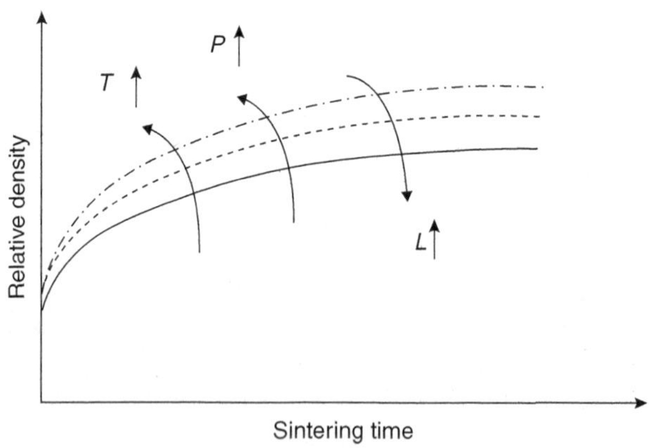

8.9 Effect of sintering parameters on densification. *T*: temperature; *P*: pressure; *L*: particle size.

Table 8.2 Influence of powder properties on porous materials

Properties of powder	Properties of porous materials
Particle shape	Pore shape, porosity, pore size distribution, tortuosity, surface area
Particle size	Pore size, mechanical strength, surface area
Particle size distribution	Pore size, pore size distribution, porosity
Porosity	Porosity, pore size distribution, surface area
Agglomeration	Pore size, pore size distribution, porosity, pore shape

Source: Ishizaki *et al.* (1998).

for a shorter time than those for dense materials. These conditions allow particle bonding without significant densification; this is because at lower temperatures surface diffusion is more dominant than volume diffusion. As temperatures increase, grain boundary diffusion predominates and volume diffusion becomes the main route for mass transport, which suggests that sintering at low temperatures leads to produce porous materials with well-developed necks and open porosity, with the caveat that it takes longer periods of time to sinter owing to the low diffusivity at low temperatures (Falamaki *et al.*, 2004, Levänen and Mäntylä, 2002). The sintering atmosphere also has a significant effect on the final sintered body, and atmospheric or higher oxygen partial pressures are typically employed for sintering porous ceramics.

While the sintering parameters have a significant effect on the final membrane pore size distribution and overall porosity, the initial properties of the ceramic particles also strongly influence the properties of the final membrane. In general, the pore size of porous solids is determined mainly by the particle size and the pore shape is governed by the shape of the starting powder (Ishizaki *et al.*, 1998) although additional relationships are summarised in Table 8.2.

The effect of particle size is such that when powders with similar shapes, but different sizes, are sintered at the same conditions using the same sintering mechanism, the required time to obtain the same degree of densification of the largest particle is directly proportional to a coefficient between the radii of both particles (Herring, 1950). It is noteworthy that the law proposed by Herring is generally not satisfied in sintering of powders because the mechanism of grain growth is different from the mechanism of densification. Nevertheless, this scale law can be used as a good approximation and clearly demonstrates the effect of particle size on the sintering process. It is evident that the particle shape is a very important parameter to consider, given that it affects the largest number of properties in the resulting sintered porous body. However, it is much more difficult to control particle size and is often neglected.

8.2.5 Sol–gel coating

Sol–gel chemistry is a well-studied technique in ceramic engineering that was first applied to membranes by Leenaars for developing ceramic ultrafiltration membranes (Leenaars and Burggraaf, 1985a, Leenaars and Burggraaf, 1984b; Leenaars *et al.*, 1984). It is currently considered one of the most important techniques for the production of meso- and microporous membranes from a variety of ceramic precursors (Anderson *et al.*, 1988; Brinker and Scherer, 1990; Brinker *et al.*, 1985, 1988; Gieselmann *et al.*, 1988; Klein and Gallagher, 1988; Larbot *et al.*, 1989; Moosemiller *et al.*, 1989). Ultimately, the sol–gel process is controlled by a series of simultaneous hydrolysis and condensation reactions involving a metal salt or metal–organic precursor, water and the corresponding alcohol that ultimately form an interconnected network or gel. As shown in Fig. 8.10, controlling the rate and selectivity of the hydrolysis and condensation reactions leads to two main routes through which the sol–gel membrane can be fabricated, each with its own respective pore size:

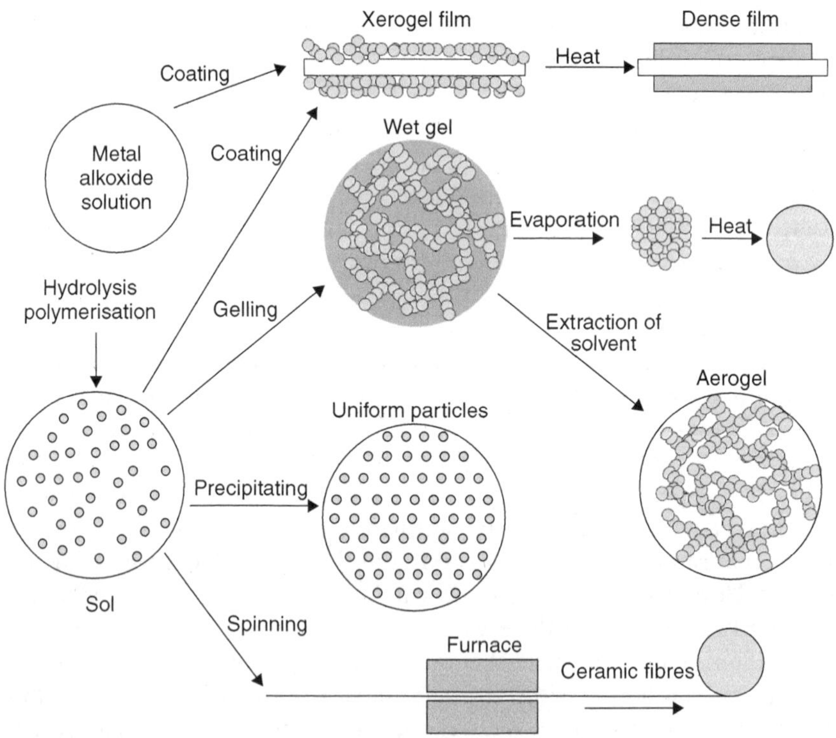

8.10 Sol–gel technologies and their products.

1. The colloidal way, in which the sol consists of discrete, dense nanoparticles, that is, a colloid of oxide particles. The sol is then coated onto a membrane support where the nanoparticles polymerise to form a colloidal gel. The pore sizes are dictated by the inter-particle gaps and are therefore most commonly mesoporous (Brinker and Scherer, 1990).
2. The polymeric way, in which the sol consists of long chain, networked, metal-organic polymers. The sol is then coated onto a membrane support where it undergoes limited interconnecting to form a polymeric gel. The pore sizes are dictated by the spaces in between the polymer chains and are therefore often microporous in nature (Brinker and Scherer, 1990).

The sol–gel reactions and therefore the ultimate pore size of the membrane can be finely controlled by adjusting the synthesis conditions of the hydrolysis and condensation reactions. To be effective in the fields of gas separation and reaction, porous ceramic membranes must have pores sizes on a molecular scale (i.e., less than 1 nm) to enable separation based on kinetic diameters of the gaseous species. Therefore the polymeric sol–gel route is the most widely applied for the synthesis of the final membrane layer. However, in order to form a suitable membrane, free from pin-holes and defects, the final membrane layer must be coated on an ultra-smooth interlayer with pore sizes on the same order of magnitude. The colloidal sol–gel route is ideal for creating interlayers with the desired structure and colloidal alumina, zirconia and titania sols are utilised extensively for this purpose (Verweij, 2003). Ceramic membranes manufactured in this manner have proven highly effective at gas separation (Smart *et al.*, 2010a, 2011) and microporous silica membranes have been investigated in a variety of MR designs (Battersby *et al.*, 2009; Galuszka *et al.*, 2011; Oklany *et al.*, 1998; Tsuru *et al.*, 2004, 2008).

Dip coating (as shown schematically in Fig. 8.11) is the most common coating technique utilised with sol–gel chemistry. Dip-coating parameters including the viscosity of the sol, dipping speed, immersion and drying time affect both the membrane thickness and its pore size distribution (Brinker and Scherer, 1990). The drying process is critical in sol–gel chemistry governing both the pore size and densification through both reaction chemistry and capillary forces (Brinker *et al.*, 1985). In dip coating the drying process takes place immediately following the withdrawal of the membrane from the bulk solution. As the liquid meniscus recedes, the sol gels on the membrane surface and excess liquid is simultaneously removed via evaporation. The thickness of the layer is governed by the withdrawal speed and the viscosity of the sol with a faster withdrawal speed and thicker viscosity resulting in thicker layers. The gel is then fired or calcined to form the final porous ceramic layer; however, if the layer is too thick, thermal stresses will

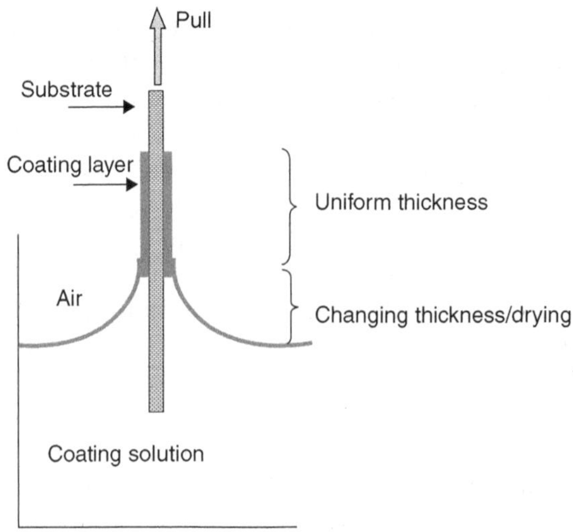

8.11 Dip-coating technique.

cause cracking during calcination. Therefore, in order to create a defect-free membrane, the process should be performed in a clean-room environment (de Vos and Verweij, 1998a, 1998b) and the dip-coating process should be repeated a number of times to create the desired membrane structure (de Lange *et al.*, 1995a; Kusakabe *et al.*, 2003).

8.2.6 Chemical vapour deposition

Chemical vapour deposition (CVD) is a technique that allows the coating of the membrane support by depositing a porous ceramic layer by means of chemical reactions in a gaseous medium, surrounding the membrane support at an elevated temperature. The use of gaseous phase reactants means that different reactions to typical sol–gel techniques can be utilised giving tighter control over the final nanostructure and pore size distribution. CVD membranes typically have a very controlled and narrow pore size distribution and often exhibit higher selectivities than comparable sol–gel processes. As such, they are best suited to gas separation applications and are only employed for the production of the final membrane layer. Figure 8.12 shows a simplified CVD system set-up, including an apparatus for metering a mixture of reactive and carrier gases, a heated reaction chamber and a system for the treatment and disposal of exhaust gases (Li *et al.*, 2002; Xomeritakis and Lin, 1996). The gas mixture (typically consisting of hydrogen, nitrogen or argon and reactive gases) is fed into the reaction chamber and heated to the desired temperature. The pore size and structure of the final membrane are controlled by the choice of reactive precursor gases as

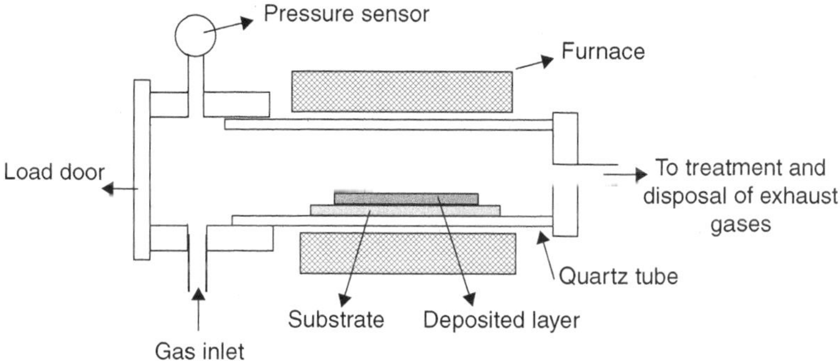

8.12 CVD technique.

well as the reaction conditions. The thickness of the membrane is primarily determined by the deposition time.

Despite the tighter pore size control and often superior membrane separation, CVD techniques are costly and difficult to scale to high production volumes. They are currently commercially employed only in high-value end products such as the semi-conductor industry. By comparison sol–gel techniques, while not as precise, utilise less costly precursors and are more readily scaled up to high production volumes.

8.3 Characterisation of ceramic membranes

The performance of porous ceramic membranes is typically expressed by the permeate flux (throughput) and the selectivity (separation ability), which in turn are governed by the pore size distribution, porosity and intrinsic membrane surface properties, and as such there are a variety of direct and indirect characterisation techniques used to evaluate the potential of a membrane and predict its performance. In this way new membranes can be efficiently and effectively screened and later optimised without the need for lengthy permeation experiments, reducing overall development time. Similarly, these techniques can be used to understand and/or verify a new membrane's transport and separation mechanisms.

8.3.1 Characterisation of membrane morphology and microstructure

The separation characteristics and performances of the composite ceramic membranes are ultimately determined by the morphology and underlying microstructure of the membrane, whether it be a symmetric macroporous filter element or an asymmetric microporous composite membrane for gas

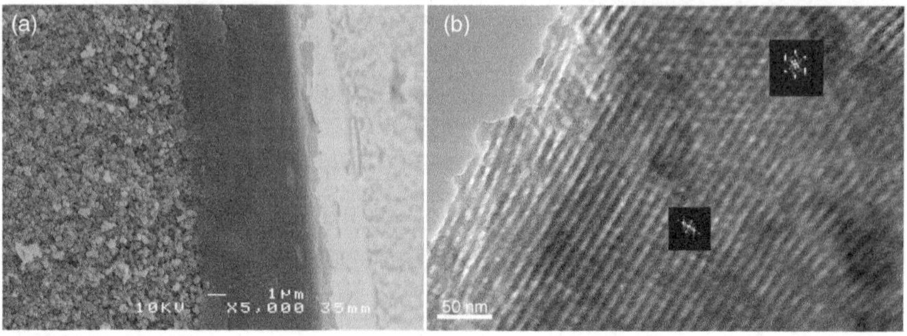

8.13 Electron microscopy images of porous ceramic membranes taken with SEM (a) and TEM (b).

separation. The morphology of membrane surfaces and their cross sections is routinely analysed via techniques such as optical or electron microscopy. Optical microscopy is conventionally used to observe large defects on the membrane surface in a non-destructive manner. Otherwise, for high magnification of the membrane separation layer or its surface texture, electron microscopy may be employed. Therefore, some of the most common techniques are scanning electron microscopy (SEM) or field emission scanning electron microscopy (FESEM) (Flegler *et al.*, 1995), transmission electron microscopy (TEM) (Flegler *et al.*, 1995) and atomic force microscopy (AFM) (Quate, 1994). These techniques allow information about surface roughness and grain size (or shape) of ceramics to be provided at the membrane surface, as well as layer continuity and thickness of the membrane cross section (Fig. 8.13). It is also possible to estimate pore size and shape using electron microscopy, especially for mesoporous and macroporous membrane structures; however, the limited field of vision, lengthy preparation and analysis times and heterogeneity of most porous ceramics, limit its quantitative application to ordered mesoporous structures as shown in Fig. 8.13b.

8.3.2 Characterisation of pore structure and network

The characterisation of porous membranes is, for the most part, no different from the characterisation of porous solids, although in the case of membranes, the nature of the percolative network of pores is crucial for understanding transport of the desired species through the membrane and rejection of undesired species. The most common and accurate method for determining pore size and structure is by employing probe molecules to adsorb the pore walls and/or condense within the pores themselves.

The technique of gas adsorption/desorption is perhaps the most widely used for characterisation of porous materials for the determination of pore size, specific surface area and pore size distribution. The adsorption/

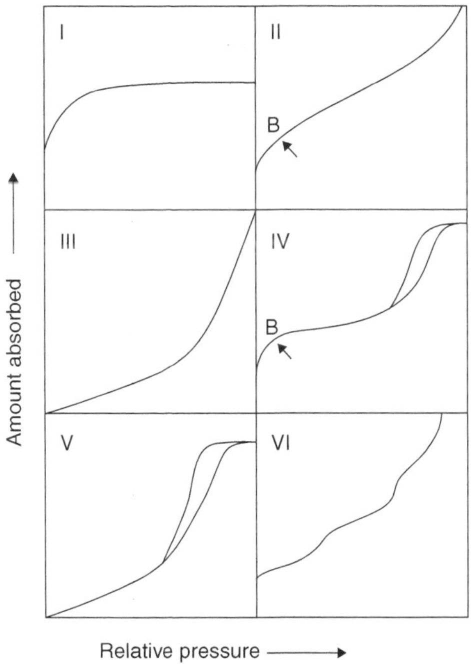

8.14 Types of physisorption isotherms. The point B refers to monolayer adsorption capacity from which the surface area of the sample can be determined. (Reprinted from Sing *et al.*, 1985.)

desorption isotherm of a non-reactive gas is presented as the quantity adsorbed as a function of the relative pressure, typically the ratio of the applied pressure and the saturation pressure of the gas. The shape of the isotherm (as shown in Fig. 8.14) is used in conjunction with the pore geometry (typically cylindrical, slit or spherical), surface energy and a variety of mathematical models to determine the specific surface area, average pore size and pore size distribution of the porous material (Do, 1998).

In practice, the majority of the physisorption isotherms can be divided into six groups (Sing *et al.*, 1985) as shown in Fig. 8.14, each of which has characteristic pore size regimes and pore surface energies. There is a plethora of literature reviews, book chapters and entire books (Do, 1998) dedicated to the understanding of adsorption analysis and as such a full discussion is outside the scope of this chapter. However, the most important isotherms for porous ceramic membrane materials are type I, which corresponds to microporous solids, and types IV and V, which are characteristic of mesoporous solids (especially ceramics) undergoing capillary condensation and hysteresis during desorption.

Despite its widespread usage and relatively well-understood analysis techniques, the use of adsorption/desorption isotherms to characterise porous ceramic membrane materials is not without its drawbacks. Firstly, the technique treats all pores the same, whether they be part of a percolative network of interconnecting pathways or dead-ends, which may lead to misleading and inaccurate results and ultimately flawed transport mechanisms. Secondly, the technique does not deal effectively with multiple pore sizes across several orders of magnitude, as found in asymmetric composite ceramic membranes. It is therefore impractical and often impossible to obtain useful results from performing the analysis on a piece of the composite membrane. Instead research groups routinely make the membrane materials in bulk and analyse the pore structure of each layer separately. As the drying conditions (which influence pore structure) are invariably different between bulk materials and thin films, the analysis can only serve as a qualitative guide (de Vos and Verweij, 1998b), although studies of silica materials have shown the results to be reasonably analogous (Brinker *et al.*, 1985).

Thirdly, the final pore size distribution is critically dependent on the mathematical and physical models used in the analysis of the isotherms. Incorrectly applied assumptions of pore geometry and surface energy can significantly alter the results. Most authors resolutely report results using the Brunauer–Emmett–Teller (BET) model, due to its simplicity and general applicability; however, with the advent of more powerful computers the application of the fundamentally more accurate Density Functional Theory is becoming more common. In summary, gas adsorption/desorption is a simple method for obtaining pore size and pore size distribution of membrane materials, in particular when the isotherm model used is sufficiently correspondent to the pore geometry of the membranes.

Permporometry is a characterisation technique for determining active pores of ceramic membranes that makes use of the capillary condensation phenomena. The method was first developed by (Eyraud *et al.*, 1984) as gas–liquid permporometry and has subsequently been improved to the gas–vapour permporometry method (Cao *et al.*, 1993). Permporometry is based on controlled blocking of pores by capillary condensation with a condensable vapour combined with a simultaneous measurement of counter-diffusion of non-condensable gases through the pores that are not blocked by the condensed vapour. The relationship between the real pore diameter, d_p and the Kelvin radius r_K is given by:

$$d_p = 2(r_K + t) \qquad [8.1]$$

where t is the thickness of the layer formed on the inner surface of the pores, usually 0.3–0.5 nm. To determine the number of pores that are open at a relative vapour pressure, the counter-diffusion of, for example, nitrogen and oxygen is applied. In the absence of an overall mechanical pressure difference across the membrane, diffusion via the Knudsen mechanism can be adopted to describe flux across the membrane as given by Equation [8.6] in Section 8.5.2. Based on the Kelvin equation, its relationship between the Kelvin radius and real pore size and the Knudsen diffusion equation, the pore size and pore size distribution of a membrane can be obtained with a stepwise reduction (or increase) of relative pressure.

Permporometry has size limitations relating to the fact that micropores do not experience capillary condensation and cannot therefore be studied using this technique. However, the choice of condensable vapour influences the Kelvin radius and thus the size ranges of pores that can be analysed. Cao and co-workers used the permporometry technique to study the pore size distribution of γ-alumina membranes with a pore radius ranging from about 2 to 10 nm (Cao *et al.*, 1993). Their results indicated that the permporometry technique can effectively measure the active pores which show a sharp pore size distribution with an average Kelvin radius, r_K of 9 nm. More recently other groups have utilised water vapour to push the limits of permporometry and investigate microporous ceramic membranes with pore size diameters between 0.5–30 nm (Tsuru *et al.*, 2001).

Thermoporometry is a technique that allows the pore structures of materials in the liquid state to be studied to determine the pore size distribution in porous media by means of the Gibbs–Thomson equation (Brun *et al.*, 1977). The principle of the method is based on the decrease of the triple point temperature of a liquid filling a porous material. Importantly phase transitions, such as crystallization or melting, for a liquid confined within a pore can shift to lower temperatures which can be correlated to pore size. This difference in transition temperature between confined and bulk solvent can be analysed by differential thermal analysis or differential scanning calorimetry techniques.

In contrast to porometry techniques which rely on the liquid wetting or vapour condensing within the pores, porosimetry is a fast and popular technique based on the non-wetting properties of mercury. Briefly, mercury is forced into a dry membrane with the volume of mercury recorded at each applied pressure. The largest pores fill with mercury at a certain minimum pressure and, as the pressure is increased, smaller and smaller pore sizes are filled with mercury until a maximum intrusion value is reached, where it is assumed that all pores are filled. Thus, the pore size distribution of the membrane can be determined because every applied pressure is related to one specific pore size. The disadvantage is that small pore sizes require high pressures, which may damage the membrane pore structure. In addition, the technique is considered destructive, as once the pores are filled with mercury it is very difficult to remove and the contamination is more or less permanent. Furthermore, like adsorption/desorption studies, this method determines all the pores presented in the membrane including those non-percolating, dead end pores, which results in an overestimation of the membrane permeation characteristics.

8.3.3 Permeation measurements

The most accurate way of determining membrane performance is to test the membrane in a permeation set-up, as this mimics the real-life operation of the membrane. Permeation measurements can be made with either a liquid or gas and can consist of a single component or a mixture. Single component tests are most common in evaluating gas separation membranes as they avoid competitive interactions between the mixture components. The throughput or production rate is referred to as the permeability of a membrane or the permeability coefficient, which is the flux normalised against the cross membrane driving force (pressure difference) and the membrane thickness ($mol \cdot m \cdot m^{-2} \cdot s^{-1} \cdot Pa^{-1}$). However, the membrane thickness is not readily available for meso- and microporous ceramic membranes, and so permeance, P ($mol \cdot m^{-2} \cdot s^{-1} \cdot Pa^{-1}$), is normally employed instead. It is possible to back calculate an average pore size from the permeation measurements, although the mathematical relationships used are dependent on the transport mechanism of the membrane. For example, analysis of macroporous membranes is conventionally performed by using either the Hagen–Poiseuille equation for capillary pore or the Carman–Kozeny equation for pores formed between packed spheres. Microporous membranes typically obey an activated transport or molecular sieving mechanism while mesoporous membranes exist within a transition region between viscous flow and Knudsen diffusion, and the dominant mechanism is highly dependent on pore size and the permeating molecule. These transport mechanisms are further explored in Section 8.4.

The separation performance of a membrane is typically reported as its selectivity. For single component tests, permselectivity, or ideal selectivity $\left(S_{ij}^*\right)$, is given as the ratio of the permeance of two different gas species as given by:

$$S_{ij}^* = \frac{P_i}{P_j} \qquad [8.2]$$

where P_i and P_j are the single gas permeability of component i and j, respectively. By testing with multiple components with a variety of kinetic diameters it is possible to establish an approximate average pore size of the porous ceramic membrane. Briefly, the permeance of each component is plotted against the relevant kinetic diameter (for gases) or the molecular weight (for liquids). A sudden decrease in permeance (often several orders of magnitude) indicates that the average pore size of the membrane is in this range. However, the cut-off point is ambiguous as molecules, especially larger ones, can orient themselves in a variety of configurations and 'squeeze' through pores that are smaller than the relevant molecular dimension. Furthermore, the technique is strongly influenced by the nature of the interactions between the probe molecule and the pore walls.

For mixed components both the interactions between the gas molecules and membrane as well as interactions between the various components will affect the transport. Thus, the true separation ability of membrane materials for a mixture will deviate from the permselectivity (typically less) and is commonly described by the separation coefficient (S_{ij}) as given by (Koros *et al.*, 1996):

$$S_{ij} = (x_i/x_j)/(y_i/y_j) \tag{8.3}$$

where x_i and y_i are the composition of component i in the permeate and feed streams, respectively.

8.4 Transport and separation of gases in ceramic membranes

The diffusing flux through different membranes can be adequately described by Fick's law (Equation [8.4]), indicating that gas transport through porous membranes is driven by a cross membrane pressure gradient. Based on the differences in partial pressures, gas diffusivities, molecular sizes and shapes, gases can be separated when they flow through a membrane.

$$J = -\tilde{D}\frac{\mathrm{d}c}{\mathrm{d}z} \tag{8.4}$$

where, J is the membrane flux, $\tilde{D}$ is the chemical diffusion coefficient, c is the local concentration and z relates to the direction of permeate flux.

In principle, the separation properties of a multilayer porous ceramic membrane, such as permselectivity, should be dependent only on the pore size distribution of the top separation layer. However, they can be compromised if resistances in the intermediate layers and the macroporous support become significant. For transport through macro- and meso-pores, molecular diffusion, Knudsen diffusion and viscous flow all contribute to the total transport, while the activated surface flow of the adsorbed phase will affect microporous transport. Therefore, any theoretical models used in analysing the transport data of gases through a porous ceramic membrane with a distributed pore size must take the following contributions into consideration: (1) viscous flow, (2) Knudsen flow, (3) surface flow and (4) molecular sieving effects. As the majority of porous ceramic MR work has focused on separation of gaseous components, only gas transport models will be considered in this section.

The ratio of average pore radius of the membrane to the mean free path of the gas molecules is a useful parameter for determining which transport mechanism is dominant under the specified conditions. For example, when the mean free path is much smaller than the pore size, that is, $(r_p/\lambda) > 3$ (Kong and Li, 2001), gas transport through the membrane occurs via molecule–molecule collisions and viscous flow. Therefore, the molar flux of gases can be described by the Hagen–Poiseuille law as given by (de Lange *et al.*, 1995b):

$$F_{P,0} = \frac{\varepsilon_p \bar{r}^2}{8\tau\eta RT\delta} p_m \qquad [8.5]$$

where $F_{P,0}$ is the Poiseuille permeation (mol·s^{-1}·m^{-2}·Pa^{-1}), ε_p the porosity (–), τ the tortuosity (–), η the gas viscosity (N·s·m^{-2}), δ the thickness (m) of the porous layer, and R the gas constant (J·mol^{-1}·K^{-1}), T the absolute temperature (K), $\bar{r}$ the model pore radius (m) and p_m is the mean pressure (Pa).

Knudsen diffusion is relevant when the mean free path of the gas molecules is greater than the pore size of the membrane, that is, $(r_p/\lambda) < 0.05$ (Liepmann, 1961). In this case, the permeating molecule is more likely to collide with the pore wall than with another molecule inside the pore, and the Knudsen diffusivity D_k can be described by:

$$D_k = \frac{d_p}{3}\sqrt{\frac{8RT}{\pi M}} \qquad [8.6]$$

where M is the molecular weight of the diffusing molecule (Bird *et al.*, 2002). This mechanism is often predominant in mesoporous and macroporous membranes (Fain, 1994; Kärger and Ruthven, 1992; Koros and Fleming, 1993; Sotirchos and Burganos, 1999). There is some controversy as to whether the effect of the Knudsen model neglecting van der Waals forces is significant and whether adsorption effects need to be accounted for in small pore size, low temperature situations. The overall effect of using the unmodified Knudsen model appears to be an overestimation of the diffusion coefficient at moderate temperatures (Bhatia and Nicholson, 2011; Bonilla and Bhatia, 2011), although experimental work from other research groups does not appear to corroborate these simulation results (Higgins *et al.*, 2009; Ruthven *et al.*, 2009).

Surface flow occurs when the permeating species exhibit a strong affinity for the membrane surface and adsorb along the pore walls at sufficiently low temperature and/or high pressure. This flow regime is generally only significant for strongly adsorbed gas molecules on very small pores, such as below 2 nm. In such cases the surface diffusivity D_s should be considered as given by:

$$D_s = D_{s,0}\Gamma \frac{\mathrm{d}C_{\mathrm{ads}}}{\mathrm{d}p} \qquad [8.7]$$

where $D_{s,0}$ is the intrinsic diffusivity of the adsorbed phase, C_{ads} the adsorbed concentration and p the pressure. Γ is a function that takes into account the thermodynamic effect determined by the adsorption isotherm. Furthermore, surface flow often occurs in parallel with other transport mechanisms, such as Knudsen or viscous flow (Kärger and Ruthven, 1992), depending on the pore characteristics of the membranes. Therefore, the total diffusivity in the porous structure can be obtained by combining the gas-phase diffusion and surface diffusion. However, the surface transport models are distinguished in the specialised literature in the following three categories:

- The hydrodynamic model: developed with the consideration that the adsorbed gas is in a form of liquid film (condensed gas) (Gilliland *et al.*, 1958).
- The random walk mode: developed based on the two-dimensional form of Fick's law (Ash *et al.*, 1963).
- The hopping model: developed with the assumption that molecules can hop over the surface (Okazaki *et al.*, 1981).

Taken to its logical conclusion, capillary condensation is therefore a form of surface flow and is routinely observed for certain zeolite membranes where highly adsorbing hydrocarbons can condense inside the pores, blocking the flow of smaller molecules and overriding the initial size based sieving. If carefully controlled this phenomena can also be utilised to reach very high selectivities, as the formation of the liquid-like dense layer of the condensable gas blocks and prevents the flow of the non-condensable gas (Sotirchos and Burganos, 1999, Uhlhorn *et al.*, 1992, Uhlmann *et al.*, 2010).

Flux through microporous membranes incorporates both adsorption and diffusion characteristics and as such the equations developed are modified based on the membrane material and pore structure. For example, the following expression (Equation [8.8]) for permeating flux through microporous silica membranes is accepted as an appropriate description of molecular sieving or activated transport (de Lange *et al.*, 1995c):

$$J = \frac{\mu}{\delta} D_0 K_0 q_s \exp\left[\frac{E_d - Q_a}{RT}\right] \Delta p \qquad [8.8]$$

where μ is a correction factor related to the porous structure of the silica membrane, δ is the membrane thickness, D_0 and K_0 the pre-exponential factor for the diffusivity and adsorption constant, respectively, while E_d and Q_a the activation energy for diffusion and isosteric heat of adsorption, respectively, q_s the adsorbed amount at saturation and Δp is the pressure gradient. Gas adsorption is assumed to take place according to Henry's law.

For the transport of gas mixtures, the generalised Maxwell–Stefan equation (Krishna and Wesselingh, 1997) has been widely adopted to describe multi-component diffusion. Although quantitative descriptions of gas diffusion in various microporous or mesoporous ceramic membranes based on statistical mechanics theory (Oyama *et al.*, 2004) or molecular dynamic simulation (Krishna, 2009) have been reported, the prediction of mixed gas permeation in porous ceramic membranes remains a challenging task, due to the difficulty in generating an accurate description of the porous network of the membrane.

8.5 Ceramic membrane reactors

The use of porous ceramic membranes in MR configurations has received considerable attention in the last few decades. This is due in part to the renewed interested in MRs as a process engineering concept and also because many of the reactions being explored occur at high temperatures where polymeric membranes are not suitable (Hsieh, 1989; Lu *et al.*, 2007; Shu *et al.*, 1991). While there are a myriad of important functions that can be achieved through the combination of membrane separation with reactor technology (Sirkar *et al.*, 1999); the majority of studies investigating the use of porous ceramic membranes in MRs have typically utilised their ability to remove a product and enhance a thermodynamically limited equilibrium (Mohan and Govind, 1988) or to control the introduction of a reactant into the reaction zone (Kölsch *et al.*, 2002).

There have also been several studies into the use of porous ceramic membranes in a membrane bioreactor (MBR) set-up (Cicek *et al.*, 1999; Gander *et al.*, 2000; Grzeoekowiak-Przywecka and Slomi ska, 2005; Takahiro, 1996; Xu *et al.*, 2002; Xu *et al.*, 2003; Zhang *et al.*, 2009). The basic principle here is to use a conventional suspended solids bioreactor to process municipal and industrial waste waters in combination with a porous membrane of the correct pore size range to filter out the suspended solids and other microorganisms, producing clean effluent. This market is dominated by polymeric MBRs, due to their lower production costs and overall improved commercial viability; however, in certain niche applications especially those with aggressive feed waters or cleaning regimes, porous ceramic MBRs may find applications. In the last decades ceramic MBRs have found commercial application in treating cosmetic processing effluents at the Lancome plant in France (Cicek *et al.*, 1999), reclamation of water from pulp and paper industries in Japan (Kimura, 1991), as well as several smaller municipal waste water treatment facilities around the world.

8.5.1 Porous ceramic membranes as product separators

MRs where the membrane acts as a product separator represent a process coupling membrane separation with a chemical reaction step in one unit. Owing to the integration of reaction and separation, the number of process vessels is reduced along with the overall plant footprint, lowering capital costs. Furthermore, an MR in this set-up may also promote the reaction process by selectively removing at least one of the products from the reaction zone, shifting equilibrium limited reactions further to the product side and increasing conversion (Fogler, 2006).

Mesoporous ceramic membranes have been applied in this way to a variety of dehydrogenation reactions owing to their selectivity of hydrogen, and a representative summary of these studies can be found in Table 8.3. Typically, these MRs are operated in a packed bed membrane reactor (PBMR) configuration with the catalyst packed inside the feed stream. In laboratory scale experiments this takes the form of a catalyst-in-tube configuration which allows for easy assembly; however, it is not suitable for industrial applications as it would require complete shut-down and disassembly of the MR module to replace either the catalyst or a broken membrane. The majority of reported cases clearly show better yields and higher conversions can be achieved in PBMRs compared to conventional packed bed reactors under the same operating conditions. In addition, continuous removal of hydrogen by the membrane shows additional benefits in slowing down undesirable side reactions, resulting in an increase in the reaction selectivity.

The performance of porous ceramic MRs functioning as product separators is controlled by both the membrane and the catalyst and can be predicted by examining the product of the *Damkohler* and *Peclet* numbers (Battersby *et al.*, 2006). For example, in a silica MR for C_3H_8 dehydrogenation, the increase in propene yield is only significant for a relatively small value of propane feed loading, because at high propane feed loading, the hydrogen formed cannot be removed fast enough through the membrane and conversion is again limited by the thermodynamic equilibrium (Weyten *et al.*, 2000). Therefore, in order to achieve an enhancement of net performance in MRs, the hydrogen production and the hydrogen extraction rate must be equivalent, that is the *Damkohler-Peclet* number must be 1 (Battersby *et al.*, 2006). This can be realised by adjusting parameters such as the reactant feed flow rate and the membrane area (Bernstein *et al.*, 1996).

Table 8.3 Dehydrogenation reactions in porous ceramic membrane reactors

Reaction	Membrane	Catalyst	*T* (°C)	Conversion (χ)	Yield (Y) (%)	Reference
Propane to propene	Silica/Al_2O_3	Cr_2O_3/Al_2O_3	535	–	Y_{MR} (propene) = 28% Y_{FBR} (propene) = 24%	Schäfer *et al.*, 2001
Propane to propene	Silica/ Al_2O_3 Pd–Ag/ Al_2O_3	Cr_2O_3/Al_2O_3	500	–	Y_{MR} (propene) = 33% Y_{MR} (propene) = 60% Y_{FBR} (propene) = 20%	Weyten *et al.*, 2000
Cyclohexane to benzene	Silica/Al_2O_3	Pt/Al_2O_3	250	χ_{MR} = 88% χ_{EQ} = 45%		Akamatsu *et al.*, 2008
Methylcyclohexane to toluene	Silica/Al_2O_3	Pt/Al_2O_3	240	χ_{MR} = 90% χ_{EQ} = 60%		Oda *et al.*, 2010

In order to increase the permeation through the membrane laboratory, studies routinely increase the partial pressure difference of hydrogen across the membrane by sweeping the permeate side with another gas. Ideally the sweep gas should not permeate and be non-reactive like nitrogen, but steam is also commonly trialled, although this can also prematurely degrade some ceramic membranes with poor hydrothermal stability (Yu *et al.*, 2005) 2008). With regards to dehydrogenation reactions specifically, the dehydrogenation of isobutene was studied theoretically in a packed bed zeolite MR with a Pt–In catalyst (Casanave *et al.*, 1999). In this case, the dehydrogenation reaction was limited by the transport properties of the membrane in co-current mode. However, when the sweep was run in a counter-current mode, kinetics limitations predominated. This led to the interesting finding that, despite the separation factor being higher for the counter-current sweep flow, the reaction yield was constant regardless of sweep direction. In some instances if the membrane exhibits low hydrogen selectivity, the equilibrium can still be enhanced by introducing a sweep gas on the reaction or feed side. This results in the dilution of the feed but has been shown to also lead to an increase in conversion. However, one of the main problems with microporous membranes is represented by the high sensitivity to steam and coke in the presence of light alkanes and olefins. On the contrary, one of the advantages of microporous membranes is the incorporation of catalyst, making them a true catalytic membrane.

8.5.2 Porous ceramic membranes as reactant distributors

Porous ceramic membranes have also found application in MRs as reactant distributors, particularly for catalytic oxidative dehydrogenation reactions where the intermediate or final products are more reactive towards oxygen than the original hydrocarbons, which limits reaction selectivity. As a result, total rather than partial oxidation of the original hydrocarbon is often achieved. Systems which can control the oxygen concentration along the reactor length, such as membranes by means of an MR, can increase the reaction selectivity towards the intermediate or final products (Coronas *et al.*, 1995).

The most preferable configuration is the catalytic membrane reactor (CMR) where the membrane also acts as a catalyst providing a reactive interface for the reaction to take place. The recent surge in research related to dense, ionic oxygen transport membranes has fuelled a significant body of research into partial oxidation MR designs using these materials (Caro *et al.*, 2007); however, PBMRs employing porous ceramic membranes are still being investigated due to their higher mechanical, thermal and chemical stability. Even without the molecular sieving effects present in silica-based membranes, where dosing pure oxygen from an air feed would be possible with the correct pore size tuning, researchers have observed enhanced reaction selectivity, enhanced conversion and decreased total oxidation using mesoporous γ-alumina membranes as summarised in Table 8.4. For CMRs, the structural characteristics of the membrane and, in particular, the relative laminar and Knudsen contributions to permeation, have a strong influence on the reactor performance (Alfonso *et al.*, 2002).

There have also been several theoretical studies comparing oxidative dehydrogenation MRs with traditional packed bed reactors (Rodríguez *et al.*, 2010). These studies show that, as well as reaction selectivity and conversion enhancement, the use of PBMRs can prevent oxygen accumulation, reduce the occurrence of 'hot spots' within the reaction bed (Assabumrungrat *et al.*, 2002) and improve the inherent production rates and safety of the process vessel.

8.6 Conclusions and future trends

MR research for industrial application is driven by the desire to reduce both capital and operational expenditure, by combining reaction and separation processes into a single unit. The physical coupling of reaction and separation processes reduces the plant footprint and reduces the need for costly upstream and downstream separation unit operations. The MR offers considerable advantages during operation, including the ability to react and separate at the same temperature and pressure achieving significant energy savings. Conventional separation technologies often operate at reduced temperatures and pressures which are often in direct contrast to conventional reactors where temperature and pressure are used to enhance production. However, these benefits come at the cost of increased process complexity and, in some cases, reduced operational control and so balancing the MR operation is a major requirement. This involves optimising the performances of both membrane and reactor that can achieve maximum feedstock conversion and yield of desired products while minimizing costs through reduced reactor volume and use of cheaper construction materials.

Porous ceramic membranes are of particular interest in this respect as they exhibit high mechanical, chemical and thermal robustness and are therefore excellent candidates for high temperature and pressure MR applications. Furthermore, the wide variety of cheap, commercially available porous ceramic membrane geometries and large-scale production techniques will ensure that porous ceramic MRs will remain under the research spotlight despite several better performing, but otherwise expensive and temperamental materials that have come to light in the last decade. Indeed, the major drawback with porous ceramic membranes is the low selectivity offered by the, primarily mesoporous, materials for gas-phase separations. This can limit

Table 8.4 MRs for oxidative dehydrogenation reactions

Reaction	Membrane	Catalyst	MR	*T* (°C)	Conversion (χ)	Yield (Y)	Selectivity (S)	Reference
Butane to butadiene and butene	Porous γ-Al_2O_3	V/MgO	CMR	550	–	–	$S_{(MR)\ (olefin)} = 53\%$ $S_{(FBR)\ (olefin)} = 46\%$	Alfonso *et al.*, 2002
Butane to butadiene and butene	Porous γ-Al_2O_3	V/MgO	PBMR	600	$\chi_{MR} = 69\%$ $\chi_{FBR} = 62\%$	$Y_{MR\ (C_4)} = 28\%$ $Y_{FBR\ (C_4)} = 24\%$	$S_{MR\ (C4)} = 61\%$ $S_{FBR\ (C4)} = 51\%$	Ge *et al.*, 2001
Ethane to ethylene	Porous γ-Al_2O_3	VO_x/γ-Al_2O_3	PBMR	630	$\chi_{MR} = 58\%$ $\chi_{FBR} = 47\%$	$Y_{MR\ (ethyl)} = 33\%$ $Y_{FBR\ (ethyl)} = 22\%$	–	Klose *et al.*, 2004
Ethane to ethylene	Porous stainless steel	V_2O_5/Al_2O_3	FLBMR	675	$\chi_{MR} = 69\%$ $\chi_{FBR} = 71\%$	$Y_{MR\ (ethyl)} = 37\%$ $Y_{FLBR\ (ethyl)} = 23\%$	$Y_{FLBMR\ (ethyl)} = 55\%$ $Y_{FLBR\ (ethyl)} = 32\%$	Ahchieva *et al.*, 2005
Propane to propylene	Porous γ-Al_2O_3	V_2O_5/Al_2O_3	PBMR	600	$\chi = 67\%$	–	$S_{(propylene)} = 11\%$	Ramos *et al.*, 2000
Methanol to hydrogen	Porous γ-Al_2O_3	Pt	CMR	275	$\chi = 84\%$	–	$S_{(hydrogen)} = 55\%$	Brinkmann *et al.*, 2001
Methane to C_2	Porous γ-Al_2O_3	Mn–W–Na/ SiO_2	PBMR	805	$\chi_{MR} = 46\%$ $\chi_{FBR} = 46\%$	$Y_{MR\ (C_2)} = 27\%$ $Y_{FBR\ (C_2)} = 14\%$	$S_{MR\ (C2)} = 60\%$ $S_{FBR\ (C2)} = 29\%$	Lu *et al.*, 2000

FBR, fixed bed reactor; FLBMR, fluidized bed membrane reactor; FLBR, fluidized bed reactor.

the opportunities offered by the MR set-up in that the processing plants will still need to employ downstream product separation technologies in order to meet product purity requirements. However, the enhancement of conversion and reaction selectivity has been clearly demonstrated both in the laboratory and with modelling simulations. From an industrial point of view the high production rates (i.e., permeation) offered by ceramic membranes may significantly outweigh the lower selectivity. Finally, there is significant scope to utilise the pores with these ceramic membranes as micro or indeed even nano-channel reactors by incorporating catalysts inside the pores of the membrane itself. This forms a catalytic membrane which can be used to reduce the reactor volume (or increase the membrane surface area) by effectively removing conventional catalyst packing volumes. Thus porous ceramic membranes show significant potential for utilisation in MR configurations.

8.7 Acknowledgements

The authors would like to acknowledge Mr Diego Schmeda for his work in producing several of the figures used in the chapter and for discussions relating to sintering behaviour.

8.8 References

Ahchieva, D., Peglow, M., Heinrich, S., Mörl, L., Wolff, T. and Klose, F. (2005) Oxidative dehydrogenation of ethane in a fluidized bed membrane reactor. *Applied Catalysis A: General*, **296**(2), 176–185.

Akamatsu, K., Ohta, Y., Sugawara, T., Hattori, T. and Nakao, S. (2008) Production of hydrogen by dehydrogenation of cyclohexane in high-pressure (1–8 atm) membrane reactors using amorphous silica membranes with controlled pore sizes. *Industrial & Engineering Chemistry Research*, **47**(24), 9842–9847.

Alfonso, M. J., Menéndez, M. and Santamaría, J. (2002) Oxidative dehydrogenation of butane on V/MgO catalytic membranes. *Chemical Engineering Journal*, **90**, 131–138.

Anderson, M. A., Gieselmann, M. J. and Xu, Q. (1988) Titania and alumina ceramic membranes. *Journal of Membrane Science*, **39**, 243–258.

Armor, J. N. (1998) Applications of catalytic inorganic membrane reactors to refinery products. *Journal of Membrane Science*, **147**, 217–233.

Ash, R., Barrer, R. M. and Pope, C. G. (1963) Flow of Adsorbable Gases and Vapours in a Microporous Medium. I. Single Sorbates. *Proceedings of the Royal Society of London. Series A. Mathematical and Physical Sciences*, **271**, 1–18.

Assabumrungrat, S., Rienchalanusarn, T., Praserthdam, P. and Goto, S. (2002) Theoretical study of the application of porous membrane reactor to oxidative dehydrogenation of n-butane. *Chemical Engineering Journal*, **85**, 69–79.

Battersby, S., Smart, S., Ladewig, B., Liu, S., Duke, M.C., Rudolph, V. and Diniz da Costa, J.C. (2009) Hydrothermal stability of cobalt silica membranes in a water gas shift membrane reactor. *Separation and Purification Technology*, **66**, 299–305.

Battersby, S., Teixeira, P. W., Beltramini, J., Duke, M.C., Rudolph, V. and Diniz da Costa, J.C. (2006) An analysis of the Peclet and Damkohler numbers for dehydrogenation reactions using molecular sieve silica (MSS) membrane reactors. *Catalysis Today*, **116**, 12–17.

Bernstein, L. A., Reo, C. M. and Lund, C. R. F. (1996) A batch membrane reactor for laboratory studies. *Journal of Membrane Science*, **118**, 93–100.

Bhatia, S. K. and Nicholson, D. (2011) Some pitfalls in the use of the Knudsen equation in modelling diffusion in nanoporous materials. *Chemical Engineering Science*, **66**, 284–293.

Bird, R. B., Stewart, W. E. and Lightfoot, E. N. (2002) *Transport phenomena*, New York, Wiley.

Bonilla, M. R. and Bhatia, S. K. (2011) The low-density diffusion coefficient of soft-sphere fluids in nanopores: Accurate correlations from exact theory and criteria for applicability of the Knudsen model. *Journal of Membrane Science*, **382**, 339–349.

Boom, R. M., van den Boomgaard, T. and Smolders, C. A. (1994) Mass transfer and thermodynamics during immersion precipitation for a two-polymer system: Evaluation with the system PES–PVP–NMP–water. *Journal of Membrane Science*, **90**, 231–249.

Brinker, C. J., Hurd, A. J. and Ward, K. J. (1988) Fundamentals of sol–gel thin-film formation. In *Ultrastructure Processing of Advanced Ceramics*. Mackenzie, J. D. and Ulrich, D. R. (Eds.). New York, Wiley, 223–253.

Brinker, C. J. and Scherer, G. W. (1990) *Sol-Gel Science: The Physics and Chemistry of Sol-Gel Processing*, Boston, Academic Press.

Brinker, C. J., Scherer, G. W. and Roth, E. P. (1985) Sol-Gel-Glass: II. Physical and structural evolution during constant heating rate experiments. *Journal of Non-Crystalline Solids*, **72**, 345–368.

Brinkmann, T., Perera, S. P. and Thomas, W. J. (2001) An experimental and theoretical investigation of a catalytic membrane reactor for the oxidative dehydrogenation of methanol. *Chemical Engineering Science*, **56**(6), 2047–2061.

Brun, M., Lallemand, A., Quinson, J.-F. and Eyraud, C. (1977) A new method for the simultaneous determination of the size and shape of pores: the thermoporometry. *Thermochimica Acta*, **21**, 59–88.

Cao, G. Z., Meijernik, J., Brinkman, H. W. and Burggraaf, A. J. (1993) Permporometry study on the size distribution of active pores in porous ceramic membranes. *Journal of Membrane Science*, **83**, 221–235.

Caro, J., Caspary, K. J., Hamel, C., Hoting, B., Kolsch, P., Langanke, B., Nassauer, K., Schiestel, T., Schmidt, A., Schomacker, R., Seidel-Morgenstern, A., Tsotsas, E., Voigt, I., Wang, H. H., Warsitz, R., Werth, S. and Wolf, A. (2007) Catalytic membrane reactors for partial oxidation using perovskite hollow fiber membranes and for partial hydrogenation using a catalytic membrane contactor. *Industrial and Engineering Chemistry Research*, **46**, 2286–2294.

Casanave, D., Ciavarella, P., Fiaty, K. and Dalmon, J. A. (1999) Zeolite membrane reactor for isobutane dehydrogenation: Experimental results and theoretical modelling. *Chemical Engineering Science*, **54**, 2807–2815.

Cicek, N., Dionysiou, D., Suidan, M.T., Ginestet, P. and Audic, J. M. (1999) Performance deterioration and structural changes of a ceramic membrane bioreactor due to inorganic abrasion. *Journal of Membrane Science*, **163**, 19–28.

Coronas, J., Menéndez, M. and Santamaría, J. (1995) The porous-wall ceramic membrane reactor: an inherently safer contacting device for gas-phase oxidation of hydrocarbons. *Journal of Loss Prevention in the Process Industries*, **8**, 97–101.

De Lange, R. S. A., Hekkink, J. H. A., Keizer, K. and Burggraaf, A. J. (1995a) Formation and characterization of supported microporous ceramic membranes prepared by sol-gel modification techniques. *Journal of Membrane Science*, **99**, 57–75.

De Lange, R. S. A., Hekkink, J. H. A., Keizer, K. and Burggraaf, A. J. (1995b) Permeation and separation studies on microporous sol-gel modified ceramic membranes. *Microporous Materials*, **4**, 169–186.

De Lange, R. S. A., Keizer, K. and Burggraaf, A. J. (1995c) Analysis and theory of gas transport in microporous sol-gel derived ceramic membranes. *Journal of Membrane Science*, **104**, 81–100.

De Vos, R. M. and Verweij, H. (1998a) High-selectivity, high-flux silica membranes for gas separation. *Science*, **279**, 1710–1711.

De Vos, R. M. and Verweij, H. (1998b) Improved performance of silica membranes for gas separation. *Journal of Membrane Science*, **143**, 37–51.

Deshmukh, S. P. and Li, K. (1998) Effect of ethanol composition in water coagulation bath on morphology of PVDF hollow fibre membranes. *Journal of Membrane Science*, **150**, 75–85.

Do, D. D. (1998) *Adsorption Analysis: Equilibria and Kinetics*, London, Imperial College Press.

Eyraud, C., Betemps, M., Quinson, J. F., Chatelut, F., Brun, M. and Rasneur, B. (1984) Determination of the pore-size distribution of an ultrafilter by gas-liquid permporometry measurement comparison between flow porometry and condensate equilibrium porometry. *Europe-Japan Congress on Membranes and Membrane Processes*. Stresa, Italy.

Fain, D. E. (1994) Membrane gas separation principles. *MRS Bulletin*, **19**, 40–43.

Falamaki, C., Afarani, M. S. and Aghaie, A. (2004) Initial sintering stage pore growth mechanism applied to the manufacture of ceramic membrane supports. *Journal of the European Ceramic Society*, **24**, 2285–2292.

Flegler, S. L., Heckman, J. W. and Klomparens, K. L. (1995) *Scanning and Transmission Electron Microscopy: An Introduction*, New York, Oxford University Press.

Fogler, H. S. (2006) *Elements of Chemical Reaction Engineering*, Upper Saddle River, NJ, Prentice Hall PTR.

Galuszka, J., Giddings, T. and Iaquaniello, G. (2011) Membrane assisted WGSR – Experimental study and reactor modeling. *Chemical Engineering Journal,* **213**, 363–370.

Gander, M., Jefferson, B. and Judd, S. (2000) Aerobic MBRs for domestic wastewater treatment: a review with cost considerations. *Separation and Purification Technology*, **18**, 119–130.

Ge, S. H., Liu, C. H. and Wang, L. J. (2001) Oxidative dehydrogenation of butane using inert membrane reactor with a non-uniform permeation pattern. *Chemical Engineering Journal*, **84**(3), 497–502.

Gieselmann, M. J., Anderson, M. A., Moosemiller, M. D. and Hill, C. G. (1988) Physicochemical properties of supported and unsupported Gamma-Al_2O_3 and TIO_2 ceramic membranes. *Separation Science and Technology*, **23**, 1695–1714.

Gilliland, E. R., Baddour, R. F. and Russell, J. L. (1958) Rates of flow through microporous solids. *AIChE Journal*, **4**, 90–96.

Graham, T. (1866) On the absorption and dialytic separation of gases by colloid septa. *Philosophical Magazine Series 4*, **4**, 401–420.

Grzeœkowiak-Przywecka, A. and Slomi ska, L. (2005) Continuous potato starch hydrolysis process in a membrane reactor with tubular and hollow-fiber membranes. *Desalination*, **184**, 105–112.

Herring, C. (1950) Effect of Change of Scale on Sintering Phenomena. *Journal of Applied Physics*, **21**, 301.

Higgins, S., Desisto, W. and Ruthven, D. (2009) Diffusive transport through mesoporous silica membranes. *Microporous and Mesoporous Materials*, **117**, 268–277.

Hsieh, H. P. (1989) Inorganic Membrane Reactors – a Review. *Membrane Reactor Technology*. Washington, DC, USA, AIChE.

Hsieh, H. P. (1991) General characteristics of inorganic membranes. In *Book of Inorganic Membranes Synthesis, Characteristics and Applications*. Bhave, R. R. (Ed.). New York, Springer, 65–93.

Ishizaki, K., Komarneni, S. and Nanko, M. (1998) *Porous Materials: Process Technology and Applications*, Dordrecht, Netherlands, Kluwer Academic.

Isobe, T., Tomita, T., Kameshima, Y., Nakajima, A. and Okada, K. (2006) Preparation and properties of porous alumina ceramics with oriented cylindrical pores produced by an extrusion method. *Journal of the European Ceramic Society*, **26**, 957–960.

Kang, S.-J. L. (2005) *Sintering: Densification, Grain Growth, and Microstructure*, Amsterdam, Elsevier Butterworth-Heinemann.

Kärger, J. and Ruthven, D. M. (1992) *Diffusion in Zeolites and Other Microporous Solids*, New York, Wiley.

Kimura, S. (1991) Japan's aqua renaissance '90 project. *Water Science and Technology*, **23**, 1573–1582.

Klein, L.C. and Gallagher, D. (1988) Pore structures of sol-gel silica membranes. *Journal of Membrane Science*, **39**, 213–220.

Klose, F., Wolff, T., Thomas, S. and Seidel-Morgenstern, A. (2004) Operation modes of packed-bed membrane reactors in the catalytic oxidation of hydrocarbons. *Applied Catalysis A: General*, **257**(2), 193–199.

Kölsch, P., Noack, M., Schäfer, R., Georgi, G., Omorjan, R. and Caro, J. (2002) Development of a membrane reactor for the partial oxidation of hydrocarbons: direct oxidation of propane to acrolein. *Journal of Membrane Science*, **198**, 119–128.

Kong, J. and Li, K. (2001) An improved gas permeation method for characterising and predicting the performance of microporous asymmetric hollow fibre membranes used in gas absorption. *Journal of Membrane Science*, **182**, 271–281.

Koros, W. J. and Fleming, G. K. (1993) Membrane-based gas separation. *Journal of Membrane Science*, **83**, 1–80.

Krishna, R. (2009) Describing the diffusion of guest molecules inside porous structures. *Journal of Physical Chemistry C*, **113**, 19756–19781.

Krishna, R. and Wesselingh, J. A. (1997) The Maxwell-Stefan approach to mass transfer. *Chemical Engineering Science*, **52**, 861–911.

Kuraoka, K., Hirano, T. and Yazawa, T. (2002) High-selectivity, high-flexibility glass hollow-fiber membrane for gas separation. *Chemical Communications*, **38**(6), 664–665.

Kusakabe, K., Shibao, F., Zhao, G., Sotowa, K. I., Watanabe, K. and Saito, T. (2003) Surface modification of silica membranes in a tubular-type module. *Journal of Membrane Science*, **215**, 321–326.

Larbot, A. (1996) Ceramic processing techniques of support systems for membranes synthesis. In *Fundamentals of Inorganic Membrane Science and Technology*. Burggraaf, A. J. and Cot, L. (Eds.). Amsterdam, Elsevier.

Larbot, A., Julbe, A., Guizard, C. and Cot, L. (1989) Silica membranes by the sol-gel process. *Journal of Membrane Science*, **44**, 289–303.

Laverty, B. and O'hair, G. (1990) The applications of membrane technology in the natural gas industry. In *The Membrane Alternative: Energy Implications for Industry*. Howell, J. A. (Ed.). London, Elsevier.

Leenaars, A. F. M. and Burggraaf, A. J. (1985a) The preparation and characterization of alumina membranes with ultra-fine pores: Part 3. The permeability for pure liquids. *Journal of Membrane Science*, **24**, 245–260.

Leenaars, A. F. M. and Burggraaf, A. J. (1985b) The preparation and characterization of alumina membranes with ultra-fine pores: Part 4. Ultrafiltration and hyperfiltration experiments. *Journal of Membrane Science*, **24**, 261–270.

Leenaars, A. F. M. and Burggraaf, A. J. (1985c) The preparation and characterization of alumina membranes with ultrafine pores. 2. The formation of supported membranes. *Journal of Colloid and Interface Science*, **105**, 27–40.

Leenaars, A. F. M., Keizer, K. and Burggraaf, A. J. (1984) The preparation and characterization of alumina membranes with ultra-fine pores. *Journal of Materials Science*, **19**, 1077–1088.

Leo, A., Smart, S., Liu, S. and Diniz Da Costa, J.C. (2011) High performance perovskite hollow fibres for oxygen separation. *Journal of Membrane Science*, **368**, 64–68.

Levänen, E. and Mäntylä, T. (2002) Effect of sintering temperature on functional properties of alumina membranes. *Journal of the European Ceramic Society*, **22**, 613–623.

Li, K. (2010) Ceramic hollow fiber membranes and their applications. In *Comprehensive Membrane Science and Engineering*. Drioli, E. and Lidietta, G. (Eds). Oxford, Elsevier, pp. 253–273.

Li, Y. -Y., Nomura, T., Sakoda, A. and Suzuki, M. (2002) Fabrication of carbon coated ceramic membranes by pyrolysis of methane using a modified chemical vapor deposition apparatus. *Journal of Membrane Science*, **197**, 23–35.

Liepmann, H. W. (1961) Gas kinetics and gas dynamics of orifice flow. *Journal of Fluid Mechanics*, **10**, 65–79.

Liu, L. H., Tan, X. Y. and Liu, S. M. (2006) Yttria stabilized zirconia hollow fiber membranes. *Journal of the American Ceramic Society*, **89**, 1156–1159.

Liu, S. (2007) Thin porous glass tubes. *Materials Chemistry and Physics*, **103**, 147–152.

Liu, S. and Gavalas, G. R. (2004) Glass–carbon composite hollow fibers. *Industrial and Engineering Chemistry Research*, **43**, 3137–3140.

Liu, S., Li, K. and Hughes, R. (2003) Preparation of porous aluminium oxide (Al_2O_3) hollow fibre membranes by a combined phase-inversion and sintering method. *Ceramics International*, **29**, 875–881.

Liu, S., Tan, X., Li, K. and Hughes, R. (2001) Preparation and characterization of $SrCe_{0.95}Yb_{0.05}O_{2.975}$ hollow fibre membranes. *Journal of Membrane Science*, **193**, 249–260.

Liu, S., Teo, W. K., Tan, X. and Li, K. (2005) Preparation of PDMSvi–Al_2O_3 composite hollow fibre membranes for VOC recovery from waste gas streams. *Separation and Purification Technology*, **46**, 110–117.

Lu, G. Q., Diniz Da Costa, J.C., Duke, M., Giessler, S., Socolow, R., Williams, R. H. and Kreutz, T. (2007) Inorganic membranes for hydrogen production and purification: A critical review and perspective. *Journal of Colloid and Interface Science*, **314**, 589–603.

Lu, Y., Dixon, A. G., Moser, W. R. and Hua Ma, Y. (2000) Oxidative coupling of methane in a modified γ-alumina membrane reactor. *Chemical Engineering Science*, **55**(21), 4901–4912.

Mohan, K. and Govind, R. (1988) Analysis of equilibrium shift in isothermal reactors with a permselective wall. *AIChE Journal*, **34**, 1493–1503.

Moosemiller, M. D., Hill, C. G. and Anderson, M. A. (1989) Physicochemical properties of supported Gamma-Al_2O_3 And TIO_2 ceramic membranes. *Separation Science and Technology*, **24**, 641–657.

Nollet, J. A., Geurin, H.-L. and Delatour, L.-F. (1748) *Lecons de physique experimentale*. London, printed for S. Austen.

Oda, K., Akamatsu, K., Sugawara, T., Kikuchi, R., Segawa, A. and Nakao, S. (2010) Dehydrogenation of methylcyclohexane to produce high-purity hydrogen using membrane reactors with amorphous silica membranes. *Industrial & Engineering Chemistry Research*, **49**(22), 11287–11293.

Okazaki, M., Tamon, H. and Toei, R. (1981) Interpretation of surface flow phenomenon of adsorbed gases by hopping model. *AIChE Journal*, **27**, 262–270.

Oklany, J. S., Hou, K. and Hughes, R. (1998) A simulative comparison of dense and microporous membrane reactors for the steam reforming of methane. *Applied Catalysis A: General*, **170**, 13–22.

Oyama, S. T., Lee, D., Hacarlioglu, P. and Saraf, R. F. (2004) Theory of hydrogen permeability in nonporous silica membranes. *Journal of Membrane Science*, **244**, 45–53.

Quate, C. F. (1994) The AFM as a tool for surface imaging. *Surface Science*, **299–300**, 980–995.

Ramos, R., Menéndez, M. and Santamarìa, J. (2000) Oxidative dehydrogenation of propane in an inert membrane reactor. *Catalysis Today*, **56**(1–3), 239–245.

Rodríguez, M. A. L., Ardissone, D. E., López, E., Pedernera, M. N. and Borio, D. O. (2010) Reactor designs for ethylene production via ethane oxidative dehydrogenation: comparison of performance. *Industrial and Engineering Chemistry Research*, **50**, 2690–2697.

Rouquerol, J., Avnir, D., Fairbridge, C.W., Everett, D. H., Haynes, J. H., Pernicone, N., Ramsay, J. D. F., Sing, K. S. W. and Unger, K. K. (1994) Recommendations for the characterization of porous solids. *Pure and Applied Chemistry*, **66**(8), 1739–1758.

Ruthven, D. M., Desisto, W. J. and Higgins, S. (2009) Diffusion in a mesoporous silica membrane: validity of the Knudsen diffusion model. *Chemical Engineering Science*, **64**, 3201–3203.

Sahibzada, M., Steele, B.C. H., Hellgardt, K., Barth, D., Effendi, A., Mantzavinos, D. and Metcalfe, I. S. (2000) Intermediate temperature solid oxide fuel cells operated with methanol fuels. *Chemical Engineering Science*, **55**, 3077–3083.

Saracco, G., Vesteeg, G. F. and Van Swaaij, W. P. M. (1994) Current hurdles to the success of high-temperature membrane reactors. *Journal of Membrane Science*, **95**, 105–123.

Schäfer, R., Noack, M., Kölsch, P., Thomas, S., Seidel-Morgenstern, A., and Caro, J. (2001) Development of a H_2-selective SiO_2-membrane for the catalytic dehydrogenation of propane. *Separation and Purification Technology*, **25**(1–3), 3–9.

Shu, J., Grandjean, B. P. A., Van Neste, A. and Kaliaguine, S. (1991) Catalytic palladium-based membrane reactors. a review. *Canadian Journal of Chemical Engineering*, **69**, 1036–1060.

Sing, K. S. W., Everett, D. H., Haul, R. A. W., Moscou, L., Pierotti, R. A., Rouquerol, J. and Siemieniewska, T. (1985) Reporting physisorption data for gas solid systems with special reference to the determination of surface-area and porosity (Recommendations 1984). *Pure and Applied Chemistry*, **57**, 603–619.

Sirkar, K. K., Shanbhag, P. V. and Kovvali, A. S. (1999) Membrane in a reactor: a functional perspective. *Industrial and Engineering Chemistry Research*, **38**, 3715–3737.

Smart, S., Ding, L. P. and Diniz Da Costa, J.C. (2011) Inorganic membranes for synthesis gas processing. In *Advanced membrane science and technology for sustainable energy and environmental applications*. Basile, A. and Nunes, S. P. (Eds.). Cambridge, Woodhead Publishing Ltd, 215–254.

Smart, S., Lin, C. X.C., Ding, L., Thambimuthu, K. and Diniz Da Costa, J.C. (2010a) Ceramic membranes for gas processing in coal gasification. *Energy and Environmental Science*, **3**, 268–278.

Smart, S. K., Edwards, G. A. and Martin, D. J. (2010b) Thermoplastic polyurethane nanocomposites. In *Rubber Nanocomposites*. John Wiley and Sons, Ltd, 239–254.

Sotirchos, S. V. and Burganos, V. N. (1999) Transport of gases in porous membranes. *MRS Bulletin*, **24**, 41–45.

Takahiro, S. (1996) A dense cell culture system for microorganisms using a stirred ceramic membrane reactor incorporating asymmetric porous ceramic filters. *Journal of Fermentation and Bioengineering*, **82**, 264–271.

Tan, X., Liu, N., Meng, B. and Liu, S. (2011) Morphology control of the perovskite hollow fibre membranes for oxygen separation using different bore fluids. *Journal of Membrane Science*, **378**, 308–318.

Tan, X., Liu, S. and Li, K. (2001) Preparation and characterization of inorganic hollow fiber membranes. *Journal of Membrane Science*, **188**, 87–95.

Tsuru, T., Hino, T., Yoshioka, T. and Asaeda, M. (2001) Permporometry characterization of microporous ceramic membranes. *Journal of Membrane Science*, **186**, 257–265.

Tsuru, T., Morita, T., Shintani, H., Yoshioka, T. and Asaeda, M. (2008) Membrane reactor performance of steam reforming of methane using hydrogen-permselective catalytic SiO^2 membranes. *Journal of Membrane Science*, **316**, 53–62.

Tsuru, T., Yamaguchi, K., Yoshioka, T. and Asaeda, M. (2004) Methane steam reforming by microporous catalytic membrane reactors. *AIChE Journal*, **50**, 2794–2805.

Uhlhorn, R. J. R., Keizer, K. and Burggraaf, A. J. (1989) Gas and surface diffusion in modified γ-alumina systems. *Journal of Membrane Science*, **46**, 225–241.

Uhlhorn, R. J. R., Keizer, K. and Burggraaf, A. J. (1992) Gas transport and separation with ceramic membranes. Part I. Multilayer diffusion and capillary condensation. *Journal of Membrane Science*, **66**, 259–269.

Uhlmann, D., Smart, S. and Diniz Da Costa, J.C. (2010) High temperature steam investigation of cobalt oxide silica membranes for gas separation. *Separation and Purification Technology*, **76**, 171–178.

Verweij, H. (2003) Ceramic membranes: morphology and transport. *Journal of Materials Science*, **38**, 4677–4695.

Weyten, H., Luyten, J., Keizer, K., Willems, L. and Leysen, R. (2000) Membrane performance: the key issues for dehydrogenation reactions in a catalytic membrane reactor. *Catalysis Today*, **56**, 3–11.

Xomeritakis, G. and Lin, Y. S. (1996) Fabrication of a thin palladium membrane supported in a porous ceramic substrate by chemical vapor deposition. *Journal of Membrane Science*, **120**, 261–272.

Xu, N., Xing, W., Xu, N. and Shi, J. (2002) Application of turbulence promoters in ceramic membrane bioreactor used for municipal wastewater reclamation. *Journal of Membrane Science*, **210**, 307–313.

Xu, N., Xing, W., Xu, N. and Shi, J. (2003) Study on ceramic membrane bioreactor with turbulence promoter. *Separation and Purification Technology*, **32**, 403–410.

Yu, W., Ohmori, T., Kataoka, S., Yamamoto, T., Endo, A., Nakaiwa, M. and Itoh, N. (2008) A comparative simulation study of methane steam reforming in a porous ceramic membrane reactor using nitrogen and steam as sweep gases. *International Journal of Hydrogen Energy*, **33**, 685–692.

Yu, W., Ohmori, T., Yamamoto, T., Endo, A., Nakaiwa, M., Hayakawa, T. and Itoh, N. (2005) Simulation of a porous ceramic membrane reactor for hydrogen production. *International Journal of Hydrogen Energy*, **30**, 1071–1079.

Zhang, F., Jing, W., Xing, W. and Xu, N. (2009) Experiment and calculation of filtration processes in an external-loop airlift ceramic membrane bioreactor. *Chemical Engineering Science*, **64**, 2859–2865.

8.9 Appendix: nomenclature

8.9.1 Notation

c local concentration ($mol \cdot m^{-3}$)
C_{ads} adsorbed concentration ($mol \cdot m^{-2}$)
$\tilde{D}$ chemical diffusion coefficient ($m^2 \cdot s^{-1}$)
D_0 mean intrinsic diffusion coefficient for micropore diffusion ($m^2 \cdot s^{-1}$)
d_p real pore diameter (m)
D_s surface diffusivity ($m^2 \cdot s^{-1}$)
$D_{s,0}$ intrinsic diffusivity of the adsorbed phase ($m^2 \cdot s^{-1}$)
$F_{P,0}$ Poiseuille permeation ($mol \cdot s^{-1} \cdot m^{-2} \cdot Pa^{-1}$)
J_i gas flux through the membrane ($mol \cdot m^{-2} \cdot s^{-1}$)
K_0 intrinsic Henry constant ($mol \cdot m^{-3} \cdot Pa^{-1}$)
M gas molecular weight ($g \cdot mol^{-1}$)
p pressure (Pa)
P_i single gas permeability of component i ($mol \cdot s^{-1} \cdot m^{-2} \cdot Pa^{-1}$)
p_m mean pressure (Pa)
Δp_i gas partial pressure difference across the membrane (Pa)
Q_a isosteric heat of adsorption ($J \cdot mol^{-1}$)

R	gas constant ($J \cdot mol^{-1} \cdot K^{-1}$)
$\bar{r}$	model pore radius (m)
r_K	Kelvin radius (m)
r_p	real pore radius (m)
S^*_{ij}	permselectivity or ideal selectivity of component *i* in comparison to component *j*
S_{ij}	separation coefficient of component *i* from component *j* in a gas mixture
T	absolute temperature (K)
T	thickness of the *t* layer formed on the inner surface of the pores during adsorption (m)
x_i	composition of component *i* in the permeate
y_i	composition of component *i* in the feed stream
z	position within the membrane in the direction of permeate flux (m)
Γ	function for surface diffusivity that takes into account thermodynamic effects as determined by adsorption isotherm
δ	membrane thickness (m)
ε_p	porosity
η	viscosity ($N \cdot s \cdot m^{-2}$)
λ	mean free path of the gas molecules
μ	correction factor for porous structure of microporous membrane
τ	tortuosity

9

Microporous silica membranes: fundamentals and applications in membrane reactors for hydrogen separation

S. SMART, J. BELTRAMINI, J. C. DINIZ DA COSTA, The University of Queensland, Australia and A. HARALE, S. P. KATIKANENI and T. PHAM, Saudi Aramco, Saudi Arabia

DOI: 10.1533/9780857097330.2.337

Abstract: This chapter discusses the research and development of membrane reactors, incorporating microporous silica-based membranes, specifically for hydrogen production. Microporous silica membranes are first introduced alongside a discussion of relevant gas transport mechanisms, membrane performance parameters, membrane reactor designs and membrane reactor performance metrics. This is followed by an in-depth analysis of the various research investigations where silica membrane reactors have been used to produce hydrogen and/or syngas from hydrocarbon reforming reactions. Of particular importance here is the hydrothermal instability of silica-based membranes at the required operating temperatures and so the chapter closes by presenting the future research trends and industrial design challenges and considerations of silica-based membrane reactors.

Key words: microporous silica, membrane reactors, water–gas shift reaction, autothermal reforming, steam reforming.

9.1 Introduction

The incorporation of a membrane into a reactor can serve several functions, the most common of which is to separate a product from the reaction mixture to enhance conversion and reduce downstream processing requirements. However, the concept of a membrane reactor was not investigated until the latter half of the twentieth century. Only in recent decades has there been an intense worldwide effort on membrane reactors and membrane catalysis to bring this technology to a demonstrated commercial application. Membranes can be classified as organic, inorganic or hybrid (i.e., organic/inorganic composites) (Lu *et al.*, 2007b). Organic membranes can be further classified as polymeric and biological, while inorganic membranes can be further divided into metallic (dense phase) and ceramic (both porous and non-porous) membranes. Membranes made of dense materials include palladium membranes, which are permeable only to hydrogen, and solid oxide electrolyte dense membranes such as modified zirconia and perovskites, which permeate oxygen and hydrogen at high temperatures. Porous inorganic membranes can be divided into macroporous (pore diameter, $d_p > 50$ nm), mesoporous ($50 > d_p > 2$ nm) and microporous ($d_p < 2$ nm). Macroporous materials, such as α-alumina membranes, provide no separation function, but may be used to support layers of smaller pore size to form composite membranes, or in applications where a well-controlled reactive interface is required (Dixon, 2003). Generally, mesoporous materials for membranes have pore sizes in the range 4–5 nm, thus permeation and selectivity are governed by Knudsen or bulk diffusion. Microporous membranes such as carbon molecular sieves, porous silica and zeolites serve as molecular sieves, separating molecules based on kinetic diameters with very high separation factors.

Membranes have been used commercially for many years for the separation of gases, but their application to catalysis is driven by a number of features. The main idea behind incorporating a membrane into a reaction process is to separate a particular reactant or product from a stream and thus drive a reaction further, providing an enhancement to process efficiencies. Membranes also offer a barrier to prevent two incompatible reactants, such as H_2 and O_2 from being on the same side of the reactor. In addition, the concept of producing a purified product provides the added feature of enhancing the value of the product as well as reducing the size and capital cost of the downstream purification-unit operations. Since separation and purification are key steps in the production of chemicals, there is keen interest in incorporating a membrane into a reactor in order to reduce costs and the number of operating units in a production plant. To achieve this goal, there are critical operational features of a membrane that need to be analysed. These can be summarized as:

- The selectivity or separation ability of the membrane. In most cases this must exceed 10 to provide any benefit.
- The production rate or flux of the membrane. This must be sufficient to either add or remove desired components at a rate compatible with the reaction.
- The membrane material must be stable over, ideally, several years of industrial operation, and the manufacturing process must consistently produce high-quality and defect-free membranes.

9.2 Microporous silica membranes

Molecular sieving silica (MSS) membranes are amorphous, microporous ceramic membranes that separate gas species based on their adsorption

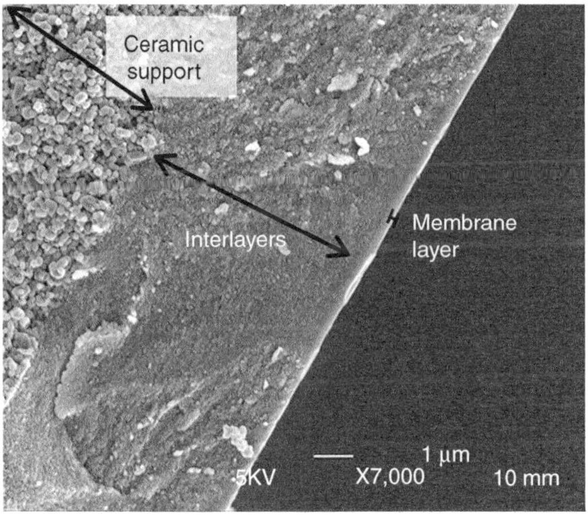

9.1 Asymmetric membrane structure showing relative thicknesses of ceramic support, interlayers and top membrane layer.

behaviour and kinetic diameters (de Lange *et al.*, 1995b). The pore size can be tailored depending on the silica synthesis, which in combination with their excellent thermal, mechanical and chemical stability has seen MSS materials attract significant research interest as inorganic membranes, particularly for the production and separation of H_2. The performance of MSS membranes is dependent on the synthesis conditions and overall membrane morphology. MSS membranes are commonly derived from sol–gel synthesis techniques (Cao *et al.*, 1996; de Vos and Verweij, 1998b, 1998a) or chemical vapour deposition (CVD) (Alsyouri *et al.*, 2003; Kanezashi *et al.*, 2008). The sol–gel route is the most studied, as it is the easiest and most economical method to produce high-quality membranes (Smart *et al.*, 2010). Conventionally, silica membranes for gas separation are manufactured with an asymmetric structure of reducing pore size as shown in Fig. 9.1. Here the thick, macroporous substrate provides the mechanical strength required for the membrane to operate at high pressure.

In order to generate a thin, uniform and defect-free MSS membrane for gas separation on top of the substrate, interlayers are used for transition from the large pores and rough surface. Sol–gel silica materials for MSS membranes are typically synthesized by hydrolysing monomeric tetrafunctional alkoxide precursors in the presence of a mineral acid (Brinker and Scherer, 1990). The manufacture of the silica membrane is done by dip coating, in a clean environment, a membrane support in the silica sol–gel and calcining the dried film to fix the silica structure for use as a membrane. This synthesis process, which is often repeated several times to repair defects (de Lange *et al.*, 1995a; Kusakabe *et al.*, 2003; Gu and Oyama, 2007),

Table 9.1 Silica-based membranes for H_2/CO_2 separation

Membrane	H_2 permeance (mol m^{-2} s^{-1} Pa^{-1})	Permselectivity	Testing conditions
Silica (de Vos and Verweij, 1998a)	2×10^{-6}	64 (H_2/N_2)	200°C, ΔP = 50–300 kPa
Template silica (Pakizeh *et al.*, 2007)	4.2×10^{-6}	35 (H_2/CO_2)	400°C, ΔP = 20–100 kPa
Metal silica (Igi *et al.*, 2008)	$2–4 \times 10^{-6}$	250–730 (H_2/N_2)	500°C, P_{H_2O} = 300 kPa

produces silica structures with pores that can be tuned as finely as 3–5 Å. Examples of the performance of several silica-derived membranes are listed in Table 9.1.

The major failing of pure MSS membranes is their chemical and structural instability when exposed to steam, which typically manifests as a dramatic reduction in surface area and corresponding increase in average pore diameter (Leboda and Mendyk, 1991; Fotou *et al.*, 1995; Leboda *et al.*, 1995). The degradation process occurs through the sorption of water onto the pore surface via silanol groups (Si–OH), followed by the breaking of available siloxane bonds (Si–O–Si) to form more silanol groups (Michalske and Freiman, 1982; Brinker and Scherer, 1990). This hydrolysis results in a 'mobile' phase of silica oligomers inside the pore, which migrate to the most thermodynamically favourable sites within the pore before undergoing further condensation to form a dense silica structure (Duke *et al.*, 2006). This degradation leads to the densification of small pores and the widening of large ones, ultimately destroying the finely controlled pore sizes of the silica matrix and the ability of membrane to effectively separate gases. This hydrothermal instability ultimately makes pure MSS membranes unsuitable for use in any membrane reactor that processes steam or humid gas streams.

The majority of the research work into microporous silica membranes over recent decades has been focussed on improving their hydrothermal stability. The initial approach was to produce water repelling or hydrophobic membranes, either by using organic–inorganic precursors to functionalize the silica network with alkyl side chains (Raman and Brinker, 1995; Cao *et al.*, 1996; Raman *et al.*, 1996; de Vos *et al.*, 1999), or by incorporating short chain surfactants into the silica structure (Tsai *et al.*, 2000; Giessler *et al.*, 2001; Duke *et al.*, 2004c, 2006), which when calcined in a non-oxidizing atmosphere preserve the template, carbonizing it within the silica. This hybrid organic–inorganic approach has demonstrated improved hydrothermal stability of the silica network by either decreasing the hydrophilicity of the silica surface or inhibiting the movement of the 'mobile' silica phase once formed. However, the hybrid organic–inorganic structures often possess a larger pore size and a correspondingly lower selectivity towards H_2 (de Vos *et al.*, 1999; Duke *et al.*, 2004b).

Alternative efforts at improving silica membrane performance have focussed on modifying the silica structure through the addition of metals and metal oxides, either through sol–gel (Kanezashi *et al.*, 2005; Kanezashi and Asaeda, 2005, 2006; Ikuhara *et al.*, 2007; Battersby *et al.*, 2009a, 2009b; Uhlmann *et al.*, 2009, 2010, 2011) or CVD techniques (Gu *et al.*, 2008; Gu and Oyama, 2009). Various metal oxides such as TiO_2, ZrO_2, Fe_2O_3, Al_2O_3, NiO and Co_3O_4 have been studied in an attempt to produce membranes with a higher flux, lower cost and enhanced stability (Yoshida *et al.*, 2001; Asaeda *et al.*, 2003; Kanezashi and Asaeda, 2005; Kanezashi *et al.*, 2005; Uhlmann *et al.*, 2010). Several studies have shown that membranes doped with cobalt (Igi *et al.*, 2008; Uhlmann *et al.*, 2009) or nickel (Kanezashi and Asaeda, 2006) and their oxides (Uhlmann *et al.*, 2010) show very high selectivity and flux, with enhanced hydrothermal stability up to temperatures of 200°C. The enhanced flux and selectivity of these membranes was ascribed to the affinity of the dopants to H_2 (Ikuhara *et al.*, 2007). The stability afforded by the incorporation of metals and metal oxides is thought to be the result of reduced mobility of the 'mobile' silica phase, in a manner similar to the organic–inorganic structures (Igi *et al.*, 2008). Furthermore, the surface chemistry of metal and metal oxide doped silica membranes may minimize access to silanol groups, preventing adsorption and eventual hydrolytic cleavage of the silica matrix.

9.2.1 Membrane transport and performance metrics

The transport and separation of gases through a silica membrane is controlled by the overarching porous structure of the silica network (de Lange *et al.*, 1995b). As a typical silica membrane consists of a mechanical support and interlayers, as well as the topmost separation layer, there are a number of different transport mechanisms that should be considered when analysing the total flow through the membrane, including molecular diffusion, Knudsen diffusion and viscous flow. However, only the topmost silica separation layer will be considered here as it contributes the greatest resistance to permeate flow and actively enables separation of various components. Microporous silica membranes exhibit a molecular sieving or activated diffusion transport mechanism wherein the flux through the membrane can be expressed by (de Lange *et al.*, 1995b):

$$J = \frac{\mu}{\delta} D_0^* K_0 q_s \exp\left[\frac{E_d - Q_a}{RT}\right] \Delta P \qquad [9.1]$$

where μ is a correction factor related to the porous structure of the silica membrane and δ is the membrane thickness. D_0 and K_0 are pre-exponential factors for the diffusivity and adsorption constants, while E_d and Q_a are the activation energies for diffusion and isosteric heat of adsorption, respectively; finally, q_s is the adsorbed amount at saturation and ΔP is the pressure gradient across the membrane. Here gas adsorption is assumed to take place according to Henry's law and both the diffusivity and the Henry's law constants are expressed by the Arrhenius form. The situation for gas mixtures is far more complicated, due to the inherent difficulties in accurately describing the nature of porous silica matrix. The generalized Maxwell–Stefan equation is widely used, although statistical mechanics and molecular simulations have also been reported (Krishna and Wesselingh, 1997; Oyama *et al.*, 2004; Krishna, 2009).

The performance of all membranes is typically described by the permeability and permselectivity parameters. Permeability, or the permeability coefficient, is the flux normalized against the cross-membrane driving force (pressure difference) and the membrane thickness ($mol \cdot m \cdot m^{-2} \cdot s^{-1} \cdot Pa^{-1}$). However, the membrane thickness is not readily available for silica membranes, and so permeance, Q ($mol \cdot m^{-2} \cdot s^{-1} \cdot Pa^{-1}$), is normally employed instead. Permselectivity, or ideal selectivity S^{ij}, is the ratio of the permeability of two different gas species as given by:

$$S^{ij} = \frac{\mathrm{Perm}_i}{\mathrm{Perm}_j} \qquad [9.2]$$

where Perm_i and Perm_j are the single gas permeabilities of components i and j, respectively. For mixed components, both the interactions, between the gas molecules and membrane as well as interactions between the various components, will affect the transport. Thus, the true separation ability of membrane materials for mixed components will deviate from the permselectivity and is typically described by the separation coefficient, α^{ij}, as given by (Koros *et al.*, 1996):

$$\alpha^{ij} = \frac{x_i / x_j}{y_i / y_j} \qquad [9.3]$$

where x_i and y_i are the compositions of component i in the permeate and feed streams, respectively. Finally, the gas flow rate through silica membranes can be calculated by:

$$F = QA\left(P_f - P_p\right) \quad [9.4]$$

where A is the membrane area. P_f and P_p are the pressures at the feed and permeate sides, respectively. This suggests that the only operational parameter that can affect the flux through a membrane module is the pressure gradient across the membrane. However, the flux at the design stage of a membrane or membrane reactor module can also be altered by the intrinsic properties of the membrane material chosen, permeance, and the total membrane area.

9.3 Membrane reactor function and arrangement

The incorporation of a membrane into a reactor unit operation can serve a myriad of important functions including (Sirkar *et al.*, 1999):

1. Separation of products from the reaction mixture.
2. Separation of a reactant from a mixed stream for introduction into the reactor.
3. Controlled addition of one reactant or two reactants.
4. Non-dispersive phase-contacting.
5. Segregation of a catalyst in a reactor.
6. Immobilization of a catalyst in a membrane.
7. Membrane acting as the catalyst.
8. Membrane acting as the reactor.
9. Transfer of heat.
10. Immobilizing the liquid reaction medium.

Silica membranes are clearly not capable of performing all functions; however, under appropriate operating conditions, they may serve more than one. Ultimately, however most of these functions can easily be categorized into the three main concepts: extraction, distribution and contact. The most common type of membrane reactor works according to the extractor principle. In this case the function of the membrane is to continuously and selectively remove a product from the reaction environment. This is the most common way that silica-based membranes have been incorporated into membrane reactors in order to remove H_2 from the reaction products. Therefore a reversible reaction, thermodynamically limited to equilibrium conversion under a given set of operating conditions, can be enhanced by the removal of H_2. Additional advantages to the equilibrium shift are a reduction in undesired side or sequential reactions. If the reaction rate of the undesired secondary reaction is higher than that of the primary reaction, the reaction selectivity can be significantly increased by removing the desired intermediate species. A further advantage of selectively removing the product lies in avoiding or reducing the downstream separation unit operation size and complexity through *in situ* separation.

The second main function of the membrane in a membrane reactor system is as a distributor for the reactant feed. In this case a reactant is specifically added to the reaction mixture across a membrane to assist with even distribution of a limiting reagent, in order to prevent hot spots and side reactions from developing. In addition, the membrane can be used as an upstream separation unit, selectively dosing the desired component from a mixture for use in a separate reaction. This functionality is of prime importance for processes with competing or cascading reactions. This can reduce both the upstream and downstream processing requirements by reducing the need for additional separation units before the reactor and by controlling possible side reactions and thus unwanted products in the reactor exit stream.

Finally, the two-sided geometry of membranes allows alternative means of bringing reactants into contact as compared to a conventional packed bed reactor. In a conventional reactor with a catalyst bed, the reactants must contact each other as they diffuse into the catalyst pellet. The products must then diffuse out of the catalyst pellet and may contact new reactants entering, prompting both diffusional limitations and possible unwanted side reactions. A membrane provides an alternate option for the reactants to contact the active catalyst sites in that two reactants can be fed from opposite sides of the membrane coming into contact either at the edge of the membrane or in the case of porous membranes, mixtures of reactants can be forced through pores coated with catalyst, like a nano-fluidic channel. In this case the pore can be tailored to optimize the contact time and permeation regime in a superior way to conventional fixed bed reactors.

The main building block of a membrane reactor system is the module, which as a unit operation bears more resemblance to a membrane unit than to a traditional reactor. For commercial applications, it is desirable to fabricate membranes into a modular form that maximizes both productivity and selectivity. The most common commercial membrane geometries are flat sheet and tubular, and currently there are five module types namely, plate-and-frame and spiral-wound modules, based on flat membranes, and tubular, capillary and hollow-fibre modules, based on tubular membrane geometries. The module types of most relevance to silica membrane modules and membrane reactors are the plate-and-frame and tubular types. The inflexibility of silica membranes prevents them from being used in a spiral-wound set-up and the technology to produce capillary and hollow-fibre membrane geometries has not been adequately explored for silica membranes. Plate-and-frame modules are easily constructed by

Table 9.2 Common acronyms used in the literature

Acronym	Explanation
CMR	Catalytic membrane reactor
PBMR	Packed bed membrane reactor
PBCMR	Packed bed catalytic membrane reactor
FBMR	Fluidized bed membrane reactor
FBCMR	Fluidized bed catalytic membrane reactor
CNMR	Catalytic non-permselective membrane reactor
SLPCMRS	Supported liquid-phase catalytic membrane reactor-separator
PFR (PBR, FBR)	Plug flow reactor (packed bed reactor, fixed bed reactor)

placing flat membranes parallel to each other and are commonly employed in laboratory studies. A spacer plate is used to separate the feed flows running alongside different membranes in the module. For the plate-and-frame type modules the membrane surface per module volume, also known as the packing density, is roughly around 100–400 m^2/m^3. Tubular membranes consist of a thin selective membrane layer deposited on a tubular support coated either on the inner or outer surface of the tube. The number of tubes incorporated into the module varies anywhere from a single tube as used in most laboratory studies to several hundred depending on the industrial surface area requirements. The maximum packing density for tubular membranes is approximately 300 m^2/m^3.

The selection of a module shape depends on a number of factors, including cost, heat management, manufacturability, maintainability, operability, efficiency and membrane replacement. Membrane modules and thus membrane reactors can be combined into number of stages. The options for membrane system layout are virtually endless, as stages can be combined in various ways incorporating both compressors and recycle streams. Furthermore, membrane reactors inevitably contain catalysts and there are several ways in which the catalyst can be incorporated (Tsotsis *et al.*, 1993). The classification of membrane reactors that incorporate catalysts is mainly based on the location of the catalyst with respect to the membrane as shown in Table 9.2.

Firstly, the catalyst may be deposited within the membrane structure; this includes the case where the membrane itself may be intrinsically catalytically active, as in the CMR. Secondly, the catalyst particles may be packed adjacent to the membrane, as in the PBMR, either on a single side as for flat sheet geometries or, for tubular geometries, the catalyst may be placed inside or outside the membrane tube. Thirdly, combining these two options yields catalyst both in the membrane pores and packed as particles in or around the membrane, as in the PBCMR. Similar configurations are available for the cases where the membranes are used with fluidized beds, such as the FBMR and the FBCMR. Researchers employing silica membranes have most commonly chosen a PBMR as the most practical configuration, as it allows them to employ commercial catalysts in non-specialized reaction rigs (Battersby *et al.*, 2009a). However, novel CMR configurations have also been investigated (Tsuru *et al.*, 2004, 2006, 2008).

The PBMR functions to change the composition in contact with the catalyst particles, either by intermediate or product removal, or by a gradual supply of reactant species. However, the CMR also provides an alternative way to contact the reactants and catalyst. Microporous membranes such as silica can be especially highly effective, due to their high surface area per unit mass if a catalyst can be well dispersed within their pores (Dixon *et al.*, 2003). The microporous membrane geometry allows products to leave the catalytic site without having to diffuse against reactants, and possibly react further. The membrane also provides effective contact, since all molecules can be forced to pass through the membrane unlike the less-contacting packed bed. However, the disadvantages include operating in a diffusion-controlled regime, and difficulties in obtaining sufficient catalyst loading in the membrane (Tsuru *et al.*, 2006).

9.4 Membrane reactor performance metrics and design parameters

Membrane reactors are complex unit operations and in most respects more than simply the sum of their parts. As such, the means of designing membrane reactor systems and then analysing their performance requires the use of additional concepts and parameters. In modelling simulations and other mathematical treatments of silica-based membrane reactors, the reactor section is commonly considered as a packed bed reactor, operating under isothermal, plug flow conditions. Under these conditions there are two important parameters that govern reactor performance, namely the reaction rate and the space velocity (Dixon *et al.*, 2003). The Damköhler number (Da) is a dimensionless ratio of these two parameters and provides insight as to whether the residence time of the feed in the reactor is sufficient to reach equilibrium. Membrane performance is commonly established by comparing the relative size of convective transport of the feed *past* the membrane against the permeation of the desired species *through* the membrane. The *Peclet* number (Pe) is a dimensionless ratio of these two parameters and evaluates the membrane yield or the effectiveness of the membrane to remove all the desired product from the reaction zone.

The performance of a membrane reactor is then a combination of the reactor performance and the membrane performance. The product of Da and Pe numbers, the $DaPe$ number, gives the ratio of the maximum reaction rate per reactor volume over the maximum permeation per reactor volume. The $DaPe$ number is defined as:

$$DaPe = \frac{r_0 W_{\text{cat}}}{Q_{\text{H}} A \Delta P} \qquad [9.5]$$

where W_{cat} is the catalyst mass, Q_H is hydrogen permeance, A the membrane area, ΔP is the cross-membrane pressure difference and r_0 is the reaction rate constant. If the membrane permeance is low relative to the reaction rate then the *DaPe* number is greater than 1, implying the amount of the product removed is small in comparison to the total amount of product being generated by the reaction, or the reactor is oversized, and the membrane reactor operates essentially as a packed bed reactor. Conversely, if the membrane permeance is high relative to the reaction rate then the *DaPe* number is less than 1, implying the space velocity is too low and that the membrane reactor is not taking full advantage of the capabilities of the membrane, or the reactor is undersized.

A silica membrane reactor will be efficient and effective therefore, when the membrane, working at full capacity, is able to process all the H_2 produced by the reaction. This situation arises when the $DaPe = 1$ and simulations for silica membrane reactors for the water gas shift (WGS) reaction have indeed demonstrated that maximum CO conversion was achieved at *DaPe* close to 1 (Battersby *et al.*, 2006; Ikuhara *et al.*, 2007). Thus the *DaPe* number is a valuable metric to evaluate the potential performance of a membrane reactor and a valuable, yet simple, design tool to ensure that both the reactor and membrane components work together for maximum efficacy. However, the *DaPe* number does not take into account the selectivity of the membrane which obviously does affect the membrane reactor performance. Both experimental and simulation studies have shown that higher permeation results in higher conversion and product yield enhancements (Battersby *et al.*, 2006; Boutikos and Nikolakis, 2010; Lim *et al.*, 2010). That is not to say that a membrane with a low selectivity cannot be successfully utilized in a membrane reactor set-up. Provided the membrane has nominal selectivity for the desired products over reactants, the conversion of equilibrium-limited reactions will be enhanced in a membrane reactor system. However, the product purity will remain dilute and thus additional operational and capital expenditure will be required for further downstream processing. If the membrane is unable to separate gases then the system behaves as a packed bed reactor.

Additional metrics for assessment of the performance of a catalytic membrane reactor, specifically to generate H_2 from liquid hydrocarbon fuels, are provided in Equations [9.6][9.8]. Similar equations can be applied to processes other than H_2 generation as well. The liquid hydrocarbon feed conversion is given by:

$$X = 1 - \frac{f_{p,H_2}}{f_{H_2}^{O}} \qquad [9.6]$$

where $f_{H_2}^{O}$ is the equivalent feed of molecular hydrogen in the inlet stream to the reactor and f_{p,H_2} is the molar flow rate of H_2 in the permeate. The H_2 utilization or yield is the percentage of the H_2 feed that is removed in the permeate stream as given by:

$$Y_{H_2} = 100\frac{f_{p,H_2}}{f_{H_2}^{O}} \qquad [9.7]$$

Hydrogen productivity, P_{H_2}, is defined as the yield of H_2 in the permeate stream per reactor volume per time as given by Equation [9.8], assuming a tubular reactor:

$$P_{H_2} = \frac{f_{p,H_2}}{\pi L D^2/4} \qquad [9.8]$$

where d is the reactor diameter and L is the reactor length.

9.5 Catalytic reactions in a membrane reactor configuration

Microporous silica membranes operate on a molecular sieving principle and are usually tailored to separate H_2 from other, larger, gases. Hence their incorporation into membrane reactor technology is most suited to those reactions that either consume or produce H_2 and especially, given their thermal stability, for those reactions involving H_2 that occur at temperatures up to 800°C. This section explores the membrane reactor performance of several of the catalytic reactions where silica-based membranes have been effectively demonstrated at the research level.

9.5.1 H_2 production from direct reforming of hydrocarbons

There are three main industrial processes for the production of H_2 namely: steam reforming (SR), partial oxidation (POX) and autothermal reforming (ATR). All three methods transform a hydrocarbon source into synthesis gas (syngas), a mixture of H_2, CO and CO_2, before further reaction and purification to produce a pure H_2 product. Natural gas or methane is currently the most utilized hydrocarbon source for H_2 production; however, with the push for cleaner fuels and a decarbonized economy, other hydrocarbon sources such as liquid petroleum fuels (gasoline, diesel and heavy oils), biomass and coal are all being explored (Navarro *et al.*, 2007).

Steam reforming

SR converts hydrocarbons into syngas through the addition of steam and heat (Kolb, 2008). The main reactions that take place are the strongly endothermic reforming Reaction [9.9] and [9.10] and the moderately exothermic WGS Reaction [9.11]:

$$C_nH_m + nH_2O \rightarrow nCO + (n + 1/2m)H_2 \qquad \Delta H > 0 \qquad [9.9]$$

$$CH_4 + H_2O \leftrightarrow CO + 3H_2 \qquad \Delta H = 206 \text{ kJ/mol} \qquad [9.10]$$

$$CO + H_2O \leftrightarrow CO_2 + H_2 \qquad \Delta H = -41.2 \text{ kJ/mol} \qquad [9.11]$$

A multitude of side reactions are possible, depending on the hydrocarbon source, with coke formation being the most common and least desirable. Total hydrocarbon conversion is typically achieved in industrial practice, so the reaction product composition is determined by the thermodynamic equilibrium of the reactions which are in turn affected by the operating conditions at which the process takes place. Of the three industrial reforming processes, SR produces the highest hydrogen concentration in the product stream (Kolb, 2008). Nevertheless, the endothermic nature of this reaction makes the process energy intensive and so thermal integration within the hydrocarbon reforming process is essential (Elnashaie *et al.*, 1988). It should be emphasized that CO_2 is not only produced via the shift reaction, but also directly via the SR reaction. Due to its endothermic nature, SR is favoured at high temperature and because reforming is accompanied by a volume expansion, it is also favoured by low pressure. In contrast, the exothermic shift reaction is favoured at low temperatures but is unaffected by changes in pressure. Increasing the amount of steam will enhance the CH_4 conversion, but this constitutes an additional energy input into the process. In practice steam to carbon ratios (S/C) of around 3 are applied to balance conversion, energy requirements and to suppress coke formation during the reaction (Rostrup-Nielsen *et al.*, 2002).

The SR process is divided into two sections: a high-temperature and pressure section (typically 800–1000°C and 30–40 bar) in which the reforming and WGS reactions occur, followed by an additional (two-step) shift section at lower temperatures (typically 200–400°C) in order to maximize the CO conversion to H_2 through Reaction [9.11]. In principle, silica membranes can be incorporated into the first stage of the SR process where, by the application of Le Chatelier's principle, conversion of the hydrocarbons can be enhanced through use of a membrane reactor. In addition to the advantage of producing separate H_2 and CO_2 streams, the removal of H_2 from the reaction zone shifts the equilibrium of the reforming reactions to the product side. As a consequence, high conversions can be reached at relatively low temperatures, perhaps even without the need of a separate WGS reactor, although this does preclude the use of highly active, low-temperature SR catalysts. Furthermore, the SR reaction under standard conditions is accompanied by a volume expansion of the reaction mixture, which can negatively affect the equilibrium hydrocarbon conversion and decrease H_2 production. A membrane reactor arrangement, however, will facilitate the removal of H_2 from the reaction mixture, decreasing the overall reaction volume and enhancing hydrocarbon conversion.

The majority of the published work on SR membrane reactors has utilized dense palladium (Pd) or Pd-alloy membranes (Lu *et al.*, 2007a); however, Pd-based membrane reactors have been reviewed in detail elsewhere (Uemiya, 2004) and will be covered in this handbook in other chapters. Indeed, there are only a handful of published studies that have utilized silica-based membranes for an SR membrane reactor. Sogge and Strøm made the first economic assessment on catalytic membrane systems based on mesoporous silica membranes, ultimately finding that the lack of selectivity meant the system was not cost effective (Sogge and Strøm, 1997). However, the advent of highly selective microporous silica-based membranes demonstrated the potential of the technology (Oklany *et al.*, 1998; Lee *et al.*, 2004; Tsuru *et al.*, 2004, 2006, 2008; Yu *et al.*, 2005, 2008; Hacarlioglu *et al.*, 2006; Lim *et al.*, 2010). Figure 9.2 shows a catalytic membrane reactor consisting of a microporous silica membrane and a catalyst-impregnated membrane support (Tsuru *et al.*, 2004). The authors of this study conducted the methane SR

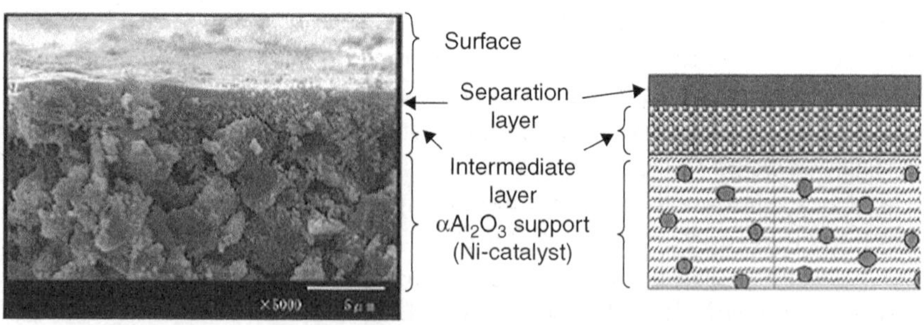

9.2 Cross section of a catalytic membrane (reprinted from Tsuru *et al.*, 2004).

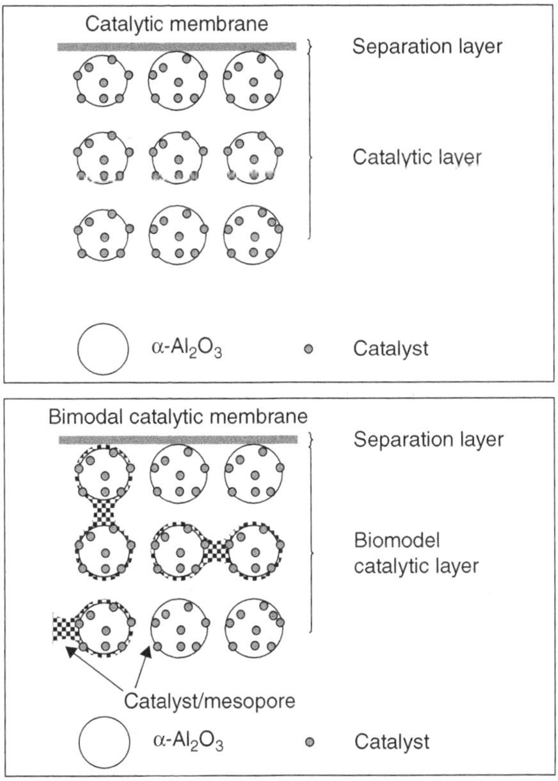

9.3 Modified catalytic membrane support (reprinted from Tsuru *et al.*, 2006).

reaction at 500°C, and the feed and permeate pressure were maintained at 100 and 20 kPa, respectively. It was found that methane conversion increased up to approximately 0.8, beyond the equilibrium conversion of 0.44 at the prevailing temperature and pressure by extracting H_2 in the permeate stream. Modelling simulations performed by the authors, which were in good agreement with the experimental results, also showed that methane conversion increased with increasing H_2 permeability and H_2 selectivity, reaching a maximum when the flow rate of H_2 through the membrane was 10–30 times greater than the inlet flow rate of methane (Tsuru *et al.*, 2004).

In later studies the authors modified the catalytic membrane support (Fig. 9.3) to enhance catalyst dispersion, noting that higher methane conversions were obtained at faster space velocities and lower catalyst loadings than for the original catalyst-impregnated supports (Tsuru *et al.*, 2006). Interestingly, when the silica membrane layer was applied and the system operated as a catalytic membrane reactor the highest methane conversion reached was only 0.7, as compared to the previously obtained 0.8 at 500°C, despite utilizing a membrane with similar H_2 permeance. This still, however, represented an enhancement above the equilibrium value of 0.44. Results from the work of Tsuru *et al.* highlight the complexity of membrane reactor systems as, intuitively, a more active catalyst should result in a higher conversion. However, this intuitive analysis neglects the valuable insight offered by the dimensionless Damköhler and Peclet numbers as discussed previously. Indeed, as the catalytic activity of the support was increased for the same volume of catalyst, the Damköhler number for the second study would have been larger in comparison to the original system. Yet the Peclet number remained unchanged, as the permeation properties of the membrane were comparable. Thus the product of the Damköhler and Peclet numbers, *DaPe,* was increased in the second study, bringing the membrane reactor system closer to the performance of a conventional reactor and thus a lower conversion was to be expected.

Further simulations and studies from the Hiroshima University group showed that methane conversions of 0.8 and even higher are possible with membranes that exhibit H_2/N_2 selectivities higher than 100 and H_2/H_2O selectivities higher than 15, provided the *DaPe* is appropriately controlled (Tsuru *et al.*, 2008). Tsuru *et al.* further showed experimentally that increasing the reaction pressure resulted in a higher H_2 yield and more significantly an increased methane conversion for a given feed flow rate. This is in direct contrast to a conventional reactor wherein the equilibrium conversion is lowered as the pressure increases due to the volume expansion of the reaction mixture under standard conditions. The authors clearly demonstrated that an increased reaction pressure is advantageous for the membrane reactor performance for two reasons. Firstly, the shift in equilibrium is greater at higher pressures which, in this case, were a shift of methane conversion from 0.22 to 0.8 at a feed pressure of 400 kPa as compared to a shift of 0.44 to 0.6 at a feed pressure of 100 kPa (Tsuru *et al.*, 2008). Secondly, as the permeate flux is directly influenced by the partial pressure difference across the membrane, an increase in reaction pressure, coupled with an increase in methane conversion, results in a higher H_2 production rate and higher H_2 yield.

This finding mirrored the earlier work of Hacarlioglu *et al.* who also showed that an increase in reaction pressure resulted in an increase in H_2 yield from the SR of methane using a silica membrane reactor (Hacarlioglu *et al.*, 2006). The authors also found that the membrane reactor set-up gave higher methane conversions as the pressure increased in comparison to a conventional packed bed reactor, and that the conversion enhancement also increased with pressure. However, they did not observe that methane conversion increased with increasing reaction pressure as per the work of Tsuru *et al.* (Tsuru *et al.*, 2008). Rather, the methane conversion decreased with increasing reaction pressure in the same manner as per equilibrium and the packed bed reactor, but the conversion value was higher and the decrease was at a lower rate (Hacarlioglu *et al.*, 2006). For example at 923 K and 100 kPa the membrane reactor conversion was > 0.975. This is an improvement of 10% over the equilibrium-limited packed bed reactor at 0.9. At a higher pressure of 2000 kPa, the membrane reactor conversion was 0.55 representing an improvement of 80% over the packed bed reactor at 0.3. Increasing reaction temperature was also observed to increase methane conversion in the membrane reactor set-up; however, the enhancement offered by the membrane reactor decreased as reaction temperature increased (Hacarlioglu *et al.*, 2006).

The effect of membrane selectivity was further explored by Lim *et al.* who compared membrane reactors with similar H_2 permeance but with different H_2 selectivity in the SR of an ethanol feedstock (Lim *et al.*, 2010). The authors observed that the membrane with the highest H_2/CH_4 selectivity gave the largest enhancement in H_2 yield over a packed bed reactor (23% as compared to 8%). However, they postulated that the effect would have been more significant if not for the modest H_2 permeance of the membranes which was lower than the H_2 production rate from the reforming reaction (i.e., $DaPe > 1$).

Finally, simulation studies using the properties of silica membranes in a SR membrane reactor arrangement have shown that the use of a sweep gas on the permeate side can enhance the hydrocarbon conversion, as a result of the increased driving force across the membrane and the greater subsequent H_2 removal from the reaction zone (Yu *et al.*, 2005). However, in practical terms, the use of a sweep gas serves to dilute the H_2 product purity, increasing downstream processing requirements and negating several of the attractive aspects of the membrane reactor concept. Steam is commonly employed as an appropriate sweep gas due to the relative ease of removal from the H_2 product stream (Oklany *et al.*, 1998; Yu *et al.*, 2008). There are, however, several problems with the use of a steam as a sweep gas. Firstly, the back-permeation of steam from the permeate side to the reaction zone is possible, depending on the operating conditions, in particular the partial pressure difference of steam across the membrane. This phenomenon is not commonly considered in simulations (Yu *et al.*, 2008) and yet the impact may be significant. In some instances the back-permeation of steam may serve to increase the S/C ratio of the reaction mixture and actually increases the reaction conversion (Oklany *et al.*, 1998). However, it may also serve to reduce H_2 permeation through the membrane by competing for access to the silica pore network (Uhlmann *et al.*, 2010). Secondly, the direction of the sweep gas flow may have significant impact on both the reaction conversion and product yield. Counter-current flow appears to provide a more uniform driving force from the reactor inlet to the reactor outlet resulting in higher reaction conversions and H_2 yields (Yu *et al.*, 2008). Thirdly, the separation processes required for removing the water from the H_2 product stream will be highly dependent on the final application of the H_2. Complete dehydration of the H_2 product for compression or liquefaction from storage and transport will require significant capital investment, although this is somewhat tempered by the fact that all silica-based membrane reactor units will produce at best a saturated H_2 stream and the dehydration step will be required in any case. Finally, while steam is commonly available in most industrial processes, using a steam sweep gas does constitute an additional energy load for the process.

Autothermal reforming

Oxidative SR or ATR is the production of syngas via the combination of SR and complete oxidation of the hydrocarbon feedstock in a single reaction chamber. Here, the O_2 feed is balanced so that the energy produced via the complete oxidation of the hydrocarbon is equivalent to the energy required in the SR reaction, making the process self-sustaining (Kolb, 2008). The complete oxidation of the hydrocarbon can be accomplished using a burner or a catalyst. The flameless combustion process over a suitable catalyst (typically containing Pt) is generally preferred in order to keep the reaction temperature at a reasonable level. If methane is used as a feedstock then Equation [9.12] applies.

$$4CH_4 + O_2 + 2H_2O \rightarrow 4CO + 10H_2 \qquad [9.12]$$

The catalytic ATR process operates at approximately 850–900°C and can be adjusted to give a H_2/CO ratio that ranges from 1 to 2.5, supplying less H_2 per unit feed than the SR process. As a result, the ATR is not generally favoured for hydrogen production. However, modern large scale methanol or gas-to-liquid production facilities are shifting to the ATR configuration as it can produce a H_2/CO ratio close to 2 without CO_2 recycling, tolerates a low S/C and is compact (Smart *et al.*, 2011). The potential benefits from employing a membrane reactor for the ATR process are greater when fast dynamic response is a critical requirement. Typical application of the dynamically responsive ATR is for hydrogen production on board a vehicle for use in a hydrogen fuel cell power train.

There have been very few experimental studies examining membrane reactors for ATR, and all of them have dealt with Pd-based membranes. Recent collaborations, between the Films and Inorganic Membrane Laboratory at The University of Queensland in Australia and Saudi Aramco's Research and Development Center, on the use of silica-based membrane reactors for the ATR of liquid fuels has yielded some promising experimental results. Cobalt-doped silica membranes (Uhlmann *et al.*, 2009) were incorporated with a commercial catalyst in an ATR catalytic membrane reactor to process a gasoline feed. In comparison to the fixed bed reactor (employing the same

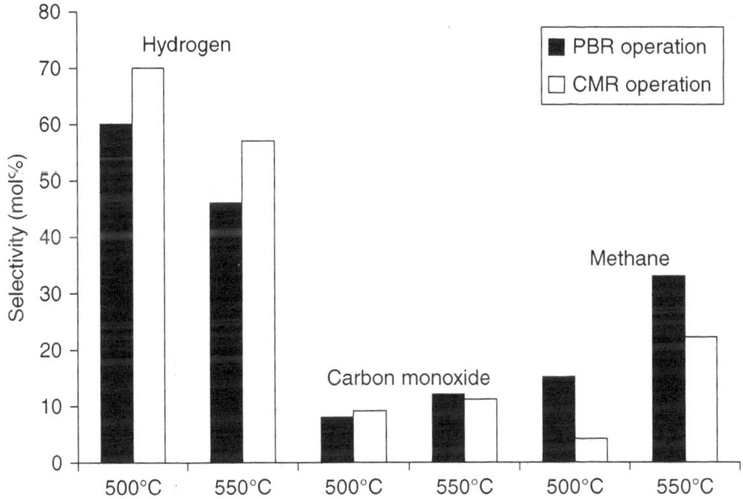

9.4 Product selectivity comparison for CMR and PBR at 500°C and 550°C.

catalyst), the ATR CMR increased total gasoline conversion to over 89% and enhanced the H_2 product yield from 62% to 74%. Of particular interest, the ATR reaction in this work was carried out at much lower temperatures, between 500 and 550°C. The catalytic membrane reactor consistently performed better than the fixed bed reactor, as shown in Fig. 9.4, delivering higher hydrogen and lower methane selectivities.

9.5.2 Hydrogen production from syngas (water–gas shift (WGS) reaction)

The WGS reaction (Equation [9.11]) is a well-known exothermic, equilibrium-limited reaction used in the production of hydrogen and ammonia. It is conventionally conducted in a two-stage operation with excess steam where each stage is operated at different temperatures to overcome the equilibrium limitations. In the first stage, often referred to as the high-temperature WGS (HT-WGS) reactor, the shift reaction takes place at temperatures from 300–450°C using iron-based catalysts to take advantage of fast reaction kinetics. The second stage, or low-temperature WGS (LT-WGS) reaction, occurs at temperatures from 200–250°C, using copper-based catalysts to maximize conversion. The combination of the two-stage reaction process with the additional separation requirements for H_2 production means that the WGS reaction is an ideal candidate for the application of membrane reactor technology. Not surprisingly, there has been significant research into WGS membrane reactors, with a specific focus on employing H_2 separation membranes. Most of the published literature has focussed on palladium-based membranes, due to their ability to produce pure H_2; however, the high cost of Pd and chemical instability of Pd-based alloys to impurities in the feed stream have seen silica-based membranes receive more attention in the past decade.

The first published scientific study to successfully use silica-based membranes in a membrane reactor configuration for the LT-WGS reaction was conducted by Giessler *et al.* (Giessler *et al.*, 2003). Previous attempts to use mesoporous ceramic membranes failed as the lack of H_2 selectivity provided by the membrane coupled with poor productivity prevented the membrane reactor overcoming equilibrium limitations regardless of reaction conditions (Criscuoli *et al.*, 2000). Instead Giessler *et al.* incorporated novel carbon templated microporous silica membranes with varying degrees of hydrophilicity and structural stabilization (Giessler *et al.*, 2001) and critically demonstrated that molecular sieving membranes could enhance CO conversion above equilibrium at temperatures from 220–350°C in a membrane reactor configuration. In addition, the authors demonstrated that the highest conversions were achieved with the least hydrophilic (and thus most hydrostable) membrane, with permeate sweep gas flow rates greater than or equal to half the feed flow rate and with H_2O to CO feed molar ratios of <1.

More recent studies have investigated the use of metal silica membranes in the LT-WGS reaction (Battersby *et al.*, 2008, 2009a, 2010). These membranes have demonstrated improved hydrothermal stability, especially below 250°C when exposed to wet gas streams (Battersby *et al.*, 2009b; Uhlmann *et al.*, 2009). These membranes gave high CO conversion of approximately 95% at 250°C with high H_2 selectivity for over 200 hours of operation. These studies expanded the understanding of the effects of the operational parameters on the membrane reactor performance. While the highest conversions can be obtained with low space velocities, the reactor production is low and so the economic advantages offered by the membrane reactor are outweighed by the added process complexity. The benefits of the membrane reactor were therefore only realized with higher space velocities, which resulted in a greater shift away from the equilibrium limitations and higher rates of H_2 production and separation.

The largest knowledge gap in the use of silica membrane reactor for the WGS is that all but a limited number of published studies have focussed their investigations on the LT-WGS reaction. However, the LT-WGS reaction is already more thermodynamically favoured to high CO conversions and so the benefits of employing a membrane reactor are limited. The HT-WGS reaction on the other hand has a lower CO conversion and so there is a greater potential enhancement if a silica membrane reactor is employed. In the only study published so far, Galuszka *et al.* employed a

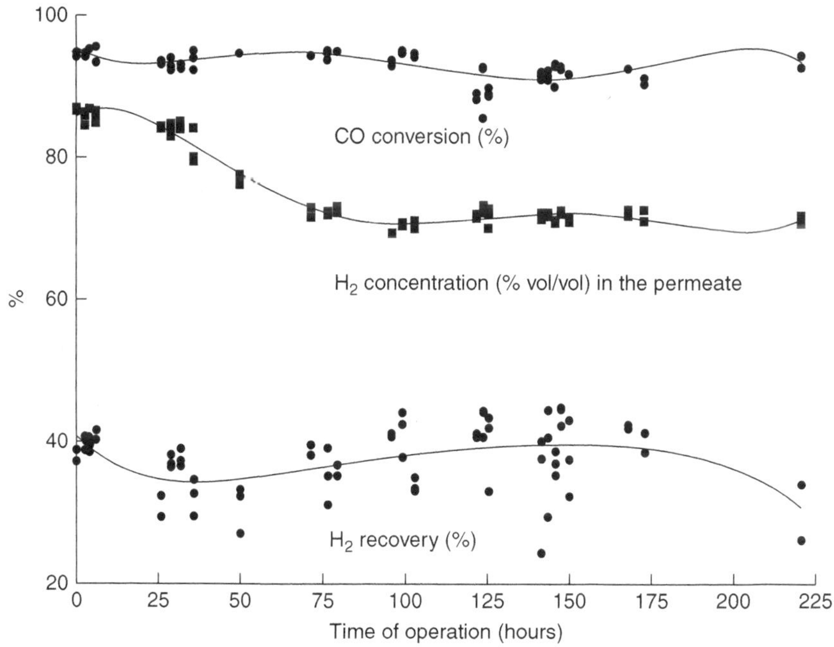

9.5 Combined results of the HT-WGS membrane reactor operating at 400°C.

silica membrane reactor for the WGS reaction using a simulated gasification mixture at 450–500°C for approximately 40 h (Galuszka *et al.*, 2011). The CO conversion was approximately 7% higher than the thermodynamic equilibrium, although modelling simulations conducted by the same authors showed that small changes in the reactor parameters could bring about conversion enhancements of up to 15%. Indeed, Diniz da Costa and co-workers have demonstrated that this is the case for operating a HT-WGS membrane reactor for 225 h, as CO conversions were maintained 10–15% above the thermodynamic equilibrium (Diniz da Costa *et al.*, 2009).

Figure 9.5 shows the combined results for CO conversion, H_2 purity in the permeate stream and H_2 recovery for a HT-WGS membrane reactor. These results stemmed from an attempt to scale up a single-tube membrane reactor to a four-tubes-in-series membrane module. Sud Chemie HT-WGS catalysts were assembled in the inner shell of the membranes. In order to optimize the membrane reactor process operation, the ratio of membrane area per catalyst weight was kept at 25 $cm^2\ g^{-1}$, 2 g of catalyst for a 50 cm^2 of membrane area. For the first 100 h of operation, CO conversion varied from 92.5–95%. For the remainder of operating time, the CO conversion reduced to 90–93%. Hence, CO conversion improved from 10–15% as compared to the thermodynamic equilibrium at testing conditions of 80% (Diniz da Costa *et al.*, 2009).

Large variations in H_2 recovery are observed in Fig. 9.5. These variations were associated with blockages in the flow, which also affected the reaction and partial pressure of H_2 along the membrane tube. However, the recovery was ~40%, which implies that 60% of H_2 went to the retentate stream and required downstream separation, suggesting that the membrane system was undersized. Finally, the concentration of H_2 in the permeate stream decreased from a high initial value of 84–87% down to 69–72% after 100 h operation (Fig. 9.5). The reduction was attributed to gas leaking through the seals between the membrane and module interface (Diniz da Costa *et al.*, 2009). For the remainder of the operation, the membrane reactor delivered constant H_2 purity varying between 70–73%.

In summary, there is significant potential for the utilization of silica membranes in catalytic membrane reactors. To maximize the competitive advantage of catalytic membrane reactors over traditional packed bed reactors, the silica membranes must ideally possess a high H_2 permeance and high H_2 selectivity. Furthermore, the production rate of H_2 from the reaction must be balanced against the permeation performance of the membrane to gain the optimal conversion and H_2 yield enhancement. For hydrocarbon feedstocks, additional concerns, such as the formation of coke on both the catalyst and the membrane itself, must be carefully considered. Coke formation on the membrane may serve to hinder H_2 flow across the membrane, further reducing H_2 permeation and the effectiveness of the membrane reactor system.

9.6 Industrial considerations

The industrial design of a membrane reactor using silica-derived membranes requires consideration of several distinct constraints. Firstly, the membranes themselves must be suitable for processing industrial gas streams. The major performance criteria for commercial applications of membranes are high flux rates, high selectivity for the desired component (H_2), long-term stability at industrial operating temperatures, pressures and in the presence of other gases and contaminants (e.g., CO_2, CO, CH_4, H_2O, H_2S), ease of fabrication and low production costs. Table 9.3 summarizes the stringent 2015 targets set by the US Department of Energy (DOE) for hydrogen separation membranes (US DOE, Hydrogen Production Interim Update, 2011). The majority of these targets are achievable within the time frame given the growth of research in this area, although the durability/stability of the membranes remains the most serious concern.

In addition to the membrane performance, the optimal membrane reactor design will have to address the practical aspects of the chosen industrial applications. One important consideration will be the impact of thermal

Table 9.3 US DOE technical targets for microporous membranes for hydrogen separation and purification

Performance criterion	Units	2006 Status	2015 Target
Flux rate	$scfh/ft^2$	150	300
Membrane material and all module costs	$\$/ft^2$ of membrane	200	<100
Durability	Hours	1100	>43 800
Operating capability	psi	500	400–1000
Hydrogen recovery	%	80	>90
Hydrogen quality	% of total (dry) gas	>95	99.99

stresses on the structural integrity. Firstly, metal pressure vessels are necessary in order to cope with high temperatures and pressures, but they must also accommodate ceramic membranes. Metals and ceramics have different thermal expansion coefficients (TEC) and variations in the operating temperature will cause several parts of the system to experience different degrees of expansion. So matching the TECs of all the components of the membrane reactor becomes a very important mechanical design feature. Indeed, heat management during normal operation, as well as start-up and shut-down procedures, is a critical area of design consideration for membrane reactors. In the case of exothermic reactions, hot spots during normal operation may induce heat stresses either in the membrane film and substrate, or at the membrane/seal interface. Novel designs, such as in Fig. 9.6, allow for superior heat management in the membrane reactor module. For instance, the metallic coils allow for gas heat exchange and act as an expander to match the coefficient of expansion between the pressure vessel shell and the ceramic tubes in the inner shell. In addition, catalysts can be placed inside the metallic coils, which function as pre-reaction chambers. The ceramic membranes are the solid black cylinders within the module and are located between the Swagelok™ connectors. Feed streams enter from the left and retentate stream for each set of tubes exit to the right. The permeate stream is via the shell side of the module.

The geometry of the membrane reactor and the relative locations and flow directions of the feed, permeate and retentate streams all play important roles in the reactor performance as discussed in Section 9.3. The disk-shaped membrane is the most common membrane configuration and has been extensively used to study the fundamental properties in laboratory settings. The disadvantages of the disk-shaped membrane configuration include challenges associated with sealing at high temperatures, small membrane area, and scale up for commercial sized applications. For industrial set-ups, tubular configurations offer easier sealing options, as high-temperature sealing issues can be addressed by using long membranes

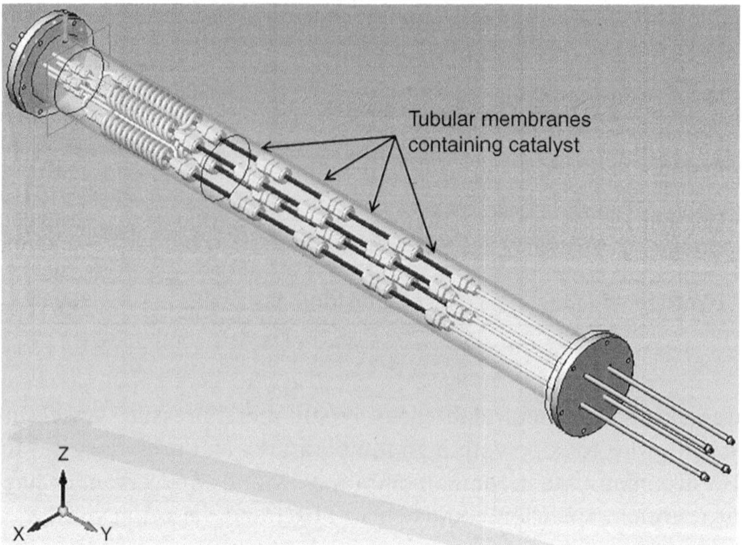

9.6 Membrane reactor design with integrated heat management systems and optimal feed flow/catalyst contact with the catalyst placed inside the membranes (Diniz da Costa *et al.*, 2009).

and keeping the two sealing ends away from the high-temperature zone. However, the membrane reactor length will dictate the pressure drop across the unit. Shorter modules will reduce the pressure drop, but will also lead to more seals per module. Furthermore, it is difficult to operate a membrane reactor constructed from a long tube because thermal variations along the reactor length create mechanical instability in the membrane.

The catalyst placement in the reactor will also play an important role in the optimal membrane reactor design. The simplest catalyst arrangement for researchers to design and operate in small bench top scale experiments is the placement of the catalyst inside the feed stream. This corresponds with either the catalyst being placed inside the tubular membrane supports, as in Fig. 9.6, or being placed inside the shell of the module, which then becomes the feed/retentate side. The catalyst-in-tube arrangement allows for easy assembly; however, it is not suitable for industrial applications as it would require complete shut-down and disassembly of the membrane reactor module to replace either the catalyst or a broken membrane. The catalyst-in-shell arrangement, by contrast, requires significantly more catalyst to fill the shell and is thus more susceptible to oversizing; however, it does offer the ability to replace the catalyst without impacting the membranes. More complex catalyst placements include coating the catalyst on the membrane top layer or inside the membrane itself; however, in such cases replacing the catalyst usually means replacing the complete membrane.

Another complex problem facing membrane reactor technology employing ceramic membranes or ceramic substrates is sealing membranes at high pressures and temperatures as reported elsewhere (Armor, 1995; Saracco *et al.*, 1999; Duke *et al.*, 2004a; Battersby *et al.*, 2009a; Smart *et al.*, 2010). Silica-derived membranes are generally coated on ceramic supports. High-temperature, heat induced stress, results in membrane tubes cracking at the sealing interface of the membrane tube and the metal module (Duke *et al.*, 2004a). Hence, ceramic tubes are particularly susceptible to mechanical failure and novel, robust designs are required. One example is silica membranes coated on porous stainless steel, which departs from the primary weakness associated with ceramic tubes. Another approach involves the design of special graphite ferrules integrated into tube connectors (Yacou *et al.*, 2012). Under these conditions it was shown that graphite sealed ceramic tubes operated extremely well up to 500°C for 2000 h under several temperature cycles.

Finally, there are potential limitations related to materials. Pure silica membranes tend to densify at operating temperatures in excess of 500–600°C. At these temperatures, the pores in the silica matrix close, resulting in loss of H_2 flux through the membrane. Hence, SR and ATR membrane reactors operating in excess of these temperatures are beyond the limit of pure silica membranes. In addition, steam has well-known detrimental effects on silica. However, metal oxide silica membranes have been proven to cope with steam (Kanezashi and Asaeda, 2006), attributed to the ability of the metal oxide to reduce the thermally induced molecular motion of silica during hydrothermal degradation (Igi *et al.*, 2008). Cobalt oxide as a dopant reduces water adsorption, thus decreasing the hydrophilicity of silica and increasing hydrothermal resistance (Battersby *et al.*, 2009a). Metal oxide dopants have been shown to oppose densification, possibly allowing metal oxide silica membranes to operate at temperatures well beyond 500–600°C. Finally, impurities in industrial gas streams are also a concern, though cobalt oxide silica membranes have also proven stable for gases containing H_2S (Uhlmann *et al.*, 2011).

9.7 Future trends and conclusions

Membrane reactor research for industrial application is driven by the desire to reduce production costs. The key competitive advantage here is the ability to couple reaction and separation in a single unit, a membrane reactor. This avoids the construction of two unit operations, a reactor followed by downstream separation unit. In terms of process operation, a membrane reactor offers considerable advantages. First, membrane reactors can realize significant energy savings by reacting and separating desired products at the same temperature. For instance, conventional separation processes (e.g., absorption or adsorption) require low temperature (<50°C) for optimal operation. As reaction process are generally carried out at high temperatures (>200°C), cooling down a gas stream to meet the low temperature requirements of the separation process will attract energy penalties and reduce efficiencies. Second, high yields and conversions can be realized in the case of equilibrium-limited reactions as per the SR, ATR and WGS reactions where more H_2 is produced by the reaction if H_2 is concomitantly withdrawn from the reaction chamber by a membrane.

Nevertheless all these advantages must be balanced against industrial realities. Currently silica-based membranes are not hydrostable enough to warrant industrial deployment and, as a result, the majority of research into silica membrane reactors has focussed on improving the hydrothermal stability of the membranes themselves. This is most prevalent in the WGS membrane reactor field, where both carbon templated and metal oxide doped membranes have been trialled. However, despite the success of carbon templated silica at improving the performance and moderate hydrothermal stability of silica membranes for H_2 separation from wet gas streams (Duke *et al.*, 2006), they have not been utilized in other WGS membrane reactor studies. There are a number of factors behind the lack of technology uptake, even at the research level. Firstly, there is an inherent problem in using a stabilizer for the silica matrix that can itself be consumed by the reaction components. Carbon, whether it is covalently bonded to the silica network or simply present as amorphous carbon inside the micropores, can react with either H_2O to produce CO and H_2, or with CO_2 to produce CO. Thus, the initial protection provided by the carbon templating will diminish over time, relative to the temperature of membrane reactor operation. Secondly, the fine pore size control necessary to achieve high H_2 selectivity, and thus exceptional membrane reactor performance, is more difficult to achieve with carbon templated silica as the addition of organic ligands or surfactant templates is well correlated with an increased pore size (van Bommel *et al.*, 1991; Li *et al.*, 2011).

Studies have certainly shown that carbon templated silica membranes are stable enough to be used under LT-WGS conditions (Giessler *et al.*, 2003); however, the industrial practicality of such a set-up is questionable (Smart *et al.*, 2010). In this case, CO conversion is generally high, in excess of 90%, using a conventional packed bed reactor (PBR). The use of membrane reactors for the LT-WGS have shown CO conversion improvement of up to 7% as compared to PBR for a feed H_2O/CO molar ratio of 1/1 (Battersby *et al.*, 2009a). If the production rates are very high, then 7% improvement can be viewed as profitable. However, by changing the H_2O/CO to 2/1 or higher, the CO conversion will be as high in a PBR as in a membrane reactor (Battersby *et al.*, 2009a). This means that adding a cheap reactant (e.g., water) to the reaction in excess of the required stoichiometry, high CO conversions can be attained at a much lower capital and operation cost. This case is however, different for the HT-WGS which has fast kinetics but lower CO conversions. CO conversions have been increased up to 7% (Galuszka *et al.*, 2011) and 15% (Diniz da Costa *et al.*, 2009) at H_2O:CO ratios >1, hence the HT-WGS may be more attractive for membrane reactor industrial operation.

The embedding of metal oxides in silica membrane has opened a window of opportunities to confer hydrothermal, thermal and chemical stability otherwise not available in pure silica membranes. This is an area of future research which is likely to take metal oxide silica membranes to operate at temperatures beyond the WGS reaction and to cope with SR and ATR operating conditions. Novel mechanical designs of membrane reactors are currently imperative to enable the next step of shifting membrane reactors from small laboratory scale to large pilot trials. Sealing still remains a major weakness, while advanced, mechanically robust, metal substrates may provide an elegant solution to this problem. Finally, balancing the membrane reactor operation is a major requirement to optimize the performances of both membrane and reactor that can achieve maximum feedstock conversion and yield of desired products while minimizing costs through reduced reactor volume and use of cheaper construction materials.

9.8 Acknowledgements

The authors acknowledge financial support from the Centre for Low Emission Technology, the Queensland Government via the NIRAP funding scheme, the Australian Research Council (DP110101185), and Saudi Aramco under the International Research Contract 6600023043.

9.9 References

Alsyouri, H. M., Langheinrich, C., Lin, Y. S., Ye, Z. B. and Zhu, S. P. (2003) Cyclic CVD modification of straight pore alumina membranes. *Langmuir*, **19**, 7307–7314.

Armor, J. N. (1995) Membrane catalysis: Where is it now, what needs to be done? *Catalysis Today*, **25**, 199–207.

Asaeda, M., Kanezashi, M., Yoshioka, T. and Tsuru, T. (2003) Gas permeation characteristics and stability of composite silica-metal oxide membranes. In V. N. Burganos, R. D. Noble, M. Asaeda, A. Ayral and Leroux, J. D. (Eds) *MRS Fall Meeting, Symposium AA – Membranes – Preparation, Properties and Applications*. Boston, Massachusetts.

Battersby, S., Duke, M.C., Liu, S., Rudolph, V. and Dinizda Costa, J.C. (2008) Metal doped silica membrane reactor: Operational effects of reaction and permeation for the water gas shift reaction. *Journal of Membrane Science*, **316**, 46–52.

Battersby, S., Ladewig, B. P., Duke, M., Rudolph, V. and da Costa, J.C. D. (2010) Membrane reactor modelling, validation and simulation for the WGS reaction using metal doped silica membranes. *Asia-Pacific Journal of Chemical Engineering*, **5**, 83–92.

Battersby, S., Smart, S., Ladewig, B., Liu, S., Duke, M.C., Rudolph, V. and Dinizda Costa, J.C. (2009a) Hydrothermal stability of cobalt silica membranes in a water gas shift membrane reactor. *Separation and Purification Technology*, **66**, 299–305.

Battersby, S., Tasaki, T., Smart, S., Ladewig, B., Liu, S., Duke, M.C., Rudolph, V. and Dinizda Costa, J.C. (2009b) Performance of cobalt silica membranes in gas mixture separation. *Journal of Membrane Science*, **329**, 91–98.

Battersby, S., Teixeira, P. W., Beltramini, J., Duke, M.C., Rudolph, V. and Dinizda Costa, J.C. (2006) An analysis of the Peclet and Damkohler numbers for dehydrogenation reactions using molecular sieve silica (MSS) membrane reactors. *Catalysis Today*, **116**, 12–17.

Boutikos, P. and Nikolakis, V. (2010) A simulation study of the effect of operating and design parameters on the performance of a water gas shift membrane reactor. *Journal of Membrane Science*, **350**, 378–386.

Brinker, C. J. and Scherer, G. W. (1990) *Sol-Gel Science: the Physics and Chemistry of Sol-Gel Processing,* Boston, Academic Press.

Cao, G. Z., Lu, Y. F., Delattre, L., Brinker, C. J. and Lopez, G. P. (1996) Amorphous silica molecular sieving membranes by sol-gel processing. *Advanced Materials*, **8**, 588–591.

Criscuoli, A., Basile, A. and Drioli, E. (2000) An analysis of the performance of membrane reactors for the water-gas shift reaction using gas feed mixtures. *Catalysis Today*, **56**, 53–64.

de Lange, R. S. A., Hekkink, J. H. A., Keizer, K. and Burggraaf, A. J. (1995a) Formation and characterization of supported microporous ceramic membranes prepared by sol-gel modification techniques. *Journal of Membrane Science*, **99**, 57–75.

de Lange, R. S. A., Keizer, K. and Burggraaf, A. J. (1995b) Analysis and theory of gas transport in microporous sol-gel derived ceramic membranes. *Journal of Membrane Science*, **104**, 81–100.

de Vos, R. M., Maier, W. F. and Verweij, H. (1999) Hydrophobic silica membranes for gas separation. *Journal of Membrane Science*, **158**, 277–288.

de Vos, R. M. and Verweij, H. (1998a) High-selectivity, high-flux silica membranes for gas separation. *Science*, **279**, 1710–1711.

de Vos, R. M. and Verweij, H. (1998b) Improved performance of silica membranes for gas separation. *Journal of Membrane Science*, **143**, 37–51.

Diniz da Costa, J. C., Lu, G. Q., Rudolph, V., Smart, S., Gopalakrishnan, S., Ladewig, B. P. and Tasaki, T. (2009) *Proof of Concept Engineering Membranes and Catalytic Membrane Reactors*. Centre for Low Emission Technology, Brisbane.

Dixon, A. G. (2003) Recent research in catalytic inorganic membrane reactors. *International Journal of Chemical Reactor Engineering*, **1**, 1–35.

Dixon, A. G., Nijemeisland, M. and Stitt, E. H. (2003) CFD simulation of reaction and heat transfer near the wall of a fixed bed. *International Journal of Chemical Reactor Engineering*, **1**, A22.

Duke, M., Rudolph, V., Lu, G. Q. and Dinizda Costa, J.C. (2004a) Scale-up of molecular sieve silica membranes for reformate purification. *AIChE Journal*, **50**, 2630–2634.

Duke, M.C., da Costa, J.C. D., Lu, G. Q., Petch, M. and Gray, P. (2004b) Carbonised template molecular sieve silica membranes in fuel processing systems: permeation, hydrostability and regeneration. *Journal of Membrane Science*, **241**, 325–333.

Duke, M.C., Dinizda Costa, J.C., Do, D. D., Gray, P. G. and Lu, G. Q. (2006) Hydrothermally robust molecular sieve silica for wet gas separation. *Advanced Functional Materials*, **16**, 1215–1220.

Duke, M.C., Dinizda Costa, J.C., Lu, G. Q., Petch, M. and Gray, P. (2004c) Carbonised template molecular sieve silica membranes in fuel processing systems: permeation, hydrostability and regeneration. *Journal of Membrane Science*, **241**, 325–333.

Elnashaie, S. S. E. H., Al-Ubaid, A. S. and Soliman, M. A. (1988) Steam reforming: kinetics, catalyst deactivation and reactor design. In D.M. Bibby, C. D. C. R. F. H. and Yur chak, S. (Eds.) *Studies in Surface Science and Catalysis*. Elsevier, Amsterdam.

Fotou, G. P., Lin, Y. S. and Pratsinis, S. E. (1995) Hydrothermal stability of pure and modified microporous silica membranes. *Journal of Materials Science*, **30**, 2803–2808.

US Department of Energy (2011) Fuel Cell Technologies Program Multi-Year Research, Development and Demonstration Plan – 3.1 – Hydrogen Production, 2011 Interim Update.

Galuszka, J., Giddings, T. and Iaquaniello, G. (2011) Membrane assisted WGSR – Experimental study and reactor modeling. *Chemical Engineering Journal,* **213**, 363–370.

Giessler, S., Dinizda Costa, J.C. and Lu, G. Q. (2001) Hydrophobicity of templated silica xerogels for molecular sieving applications. *Journal of Nanoscience and Nanotechnology*, **1**, 331–336.

Giessler, S., Jordan, L., Dinizda Costa, J.C. and Lu, G. Q. (2003) Performance of hydrophobic and hydrophilic silica membrane reactors for the water gas shift reaction. *Separation and Purification Technology*, **32**, 255–264.

Gu, Y., Hacarlioglu, P. and Oyama, S. T. (2008) Hydrothermally stable silica-alumina composite membranes for hydrogen separation. *Journal of Membrane Science*, **310**, 28–37.

Gu, Y. and Oyama, S. T. (2007) High molecular permeance in a poreless ceramic membrane. *Advanced Materials*, **19**, 1636–1640.

Gu, Y. and Oyama, S. T. (2009) Permeation properties and hydrothermal stability of silica-titania membranes supported on porous alumina substrates. *Journal of Membrane Science*, **345**, 267–275.

Hacarlioglu, P., Gu, Y. and Oyama, S. T. (2006) Studies of the methane steam reforming reaction at high pressure in a ceramic membrane reactor. *Journal of Natural Gas Chemistry*, **15**, 73–81.

Igi, R., Yoshioka, T., Ikuhara, Y. H., Iwamoto, Y. and Tsuru, T. (2008) Characterization of co-doped silica for improved hydrothermal stability and application to hydrogen separation membranes at high temperatures. *Journal of the American Ceramic Society*, **91**, 2975–2981.

Ikuhara, Y. H., Mori, H., Saito, T. and Iwamoto, Y. (2007) High-temperature hydrogen adsorption properties of precursor-derived nickel nanoparticle-dispersed amorphous silica. *Journal of the American Ceramic Society*, **90**, 546–552.

Kanezashi, M. and Asaeda, M. (2005) Stability of H_2-permselective Ni-doped silica membranes in steam at high temperature. *Journal of Chemical Engineering of Japan*, **38**, 908–912.

Kanezashi, M. and Asaeda, M. (2006) Hydrogen permeation characteristics and stability of Ni-doped silica membranes in steam at high temperature. *Journal of Membrane Science*, **271**, 86–93.

Kanezashi, M., Fujita, T. and Asaeda, M. (2005) Nickel-doped silica membranes for separation of helium from organic gas mixtures. *Separation Science and Technology*, **40**, 225–238.

Kanezashi, M., O'Brien-Abraham, J., Lin, Y. S. and Suzuki, K. (2008) Gas permeation through DDR-type zeolite membranes at high temperatures. *AIChE Journal*, **54**, 1478–1486.

Kolb, G. (2008) *Fuel processing for fuel cells*, Weinheim, Germany, Wiley-VCH.

Koros, W. J., Ma, Y. H. and Shimidzu, T. (1996) Terminology for membranes and membrane processes. *Pure and Applied Chemistry*, **68**, 1479–1489.

Krishna, R. (2009) Describing the diffusion of guest molecules inside porous structures. *Journal of Physical Chemistry C*, **113**, 19756–19781.

Krishna, R. and Wesselingh, J. A. (1997) The Maxwell-Stefan approach to mass tranfer. *Chemical Engineering Science*, **52**, 861–911.

Kusakabe, K., Shibao, F., Zhao, G., Sotowa, K. I., Watanabe, K. and Saito, T. (2003) Surface modification of silica membranes in a tubular-type module. *Journal of Membrane Science*, **215**, 321–326.

Leboda, R. and Mendyk, E. (1991) Hydrothermal modification of porous structure of silica adsorbents. *Materials Chemistry and Physics*, **27**, 189–212.

Leboda, R., Mendyk, E., Gierak, A. and Tertykh, V. A. (1995) Hydrothermal modification of silica gels (xerogels).2. Effect of the duration of treatment on their porous structure. *Colloids and Surfaces a-Physicochemical and Engineering Aspects*, **105**, 191–197.

Lee, D., Hacarlioglu, P. and Oyama, S. T. (2004) The effect of pressure in membrane reactors: Trade-off in permeability and equilibrium conversion in the catalytic reforming of CH_4 with CO_2 at high pressure. *Topics in Catalysis*, **29**, 45–57.

Li, G., Kanezashi, M. and Tsuru, T. (2011) Preparation of organic–inorganic hybrid silica membranes using organoalkoxysilanes: the effect of pendant groups. *Journal of Membrane Science*, **379**, 287–295.

Lim, H., Gu, Y. and Oyama, S. T. (2010) Reaction of primary and secondary products in a membrane reactor: Studies of ethanol steam reforming with a silica–alumina composite membrane. *Journal of Membrane Science*, **351**, 149–159.

Lu, G. Q., Dinizda Costa, J.C., Duke, M., Giessler, S., Socolow, R., Williams, R. H. and Kreutz, T. (2007a) Inorganic membranes for hydrogen production and purification: A critical review and perspective. *Journal of Colloid and Interface Science*, **314**, 589–603.

Lu, G. Q., Dinizda Costa, J.C., Duke, M., Giessler, S., Socolow, R., Williams, R. H. and Kreutz, T. (2007b) Inorganic membranes for hydrogen production and purification: A critical review and perspective. *Journal of Colloid and Interface Science*, **314**, 589–603.

Michalske, T. A. and Freiman, S. W. (1982) A Molecular interpretation of stress-corrosion in silica. *Nature*, **295**, 511–512.

Navarro, R. M., Peña, M. A. and Fierro, J. L. G. (2007) Hydrogen production reactions from carbon feedstocks: Fossil fuels and biomass. *Chemical Reviews*, **107**, 3952–3991.

Oklany, J. S., Hou, K. and Hughes, R. (1998) A simulative comparison of dense and microporous membrane reactors for the steam reforming of methane. *Applied Catalysis A: General*, **170**, 13–22.

Oyama, S. T., Lee, D., Hacarlioglu, P. and Saraf, R. F. (2004) Theory of hydrogen permeability in nonporous silica membranes. *Journal of Membrane Science*, **244**, 45–53.

Pakizeh, M., Omidkhah, M. R. and Zarringhalam, A. (2007) Synthesis and characterization of new silica membranes using template-sol-gel technology. *International Journal of Hydrogen Energy*, **32**, 1825–1836.

Raman, N. K., Anderson, M. T. and Brinker, C. J. (1996) Template-based approaches to the preparation of amorphous, nanoporous silicas. *Chemistry of Materials*, **8**, 1682–1701.

Raman, N. K. and Brinker, C. J. (1995) Organic "template" approach to molecular sieving silica membranes. *Journal of Membrane Science*, **105**, 273–279.

Rostrup-Nielsen, J. R., Sehested, J. and Nørskov, J. K. (2002) Hydrogen and synthesis gas by steam- and C02 reforming. *Advances in Catalysis*. Academic Press.

Saracco, G., Neomagus, H. W. J. P., Versteeg, G. F. and Swaaij, W. P. M. V. (1999) High-temperature membrane reactors: potential and problems. *Chemical Engineering Science*, **54**, 1997–2017.

Sirkar, K. K., Shanbhag, P. V. and Kovvali, A. S. (1999) Membrane in a reactor: a functional perspective. *Industrial and Engineering Chemistry Research*, **38**, 3715–3737.

Smart, S., Ding, L. P. and Dinizda Costa, J.C. (2011) Inorganic membranes for synthesis gas processing. In Basile, A. and Nunes, S. P. (Eds) *Advanced membrane science and technology for sustainable energy and environmental applications*. Cambridge, Woodhead Publishing Ltd.

Smart, S., Lin, C. X.C., Ding, L., Thambimuthu, K. and Diniz da Costa, J.C. (2010) Ceramic membranes for gas processing in coal gasification. *Energy and Environmental Science*, **3**, 268–278.

Sogge, J. and Strøm, T. (1997) Membrane reactors—a new technology for production of synthesis gas by steam reforming. In De Pontes, M., R. L. E. C. P. N. J. H. S. and Scurrell, M. S. (Eds) *Studies in Surface Science and Catalysis*. Elsevier, Amsterdam.

Tsai, C.-Y., Tam, S.-Y., Lu, Y. and Brinker, C. J. (2000) Dual-layer asymmetric microporous silica membranes. *Journal of Membrane Science*, **169**, 255–268.

Tsotsis, T. T., Minet, R. G., Champagnie, A. M. and Liu, P. K. T. (1993) Catalytic membrane reactors. In Becker, E. R. and Pereira, C. J. (Eds) *Computer-Aided Design of Catalysts*. New York, Marcel Dekker.

Tsuru, T., Morita, T., Shintani, H., Yoshioka, T. and Asaeda, M. (2008) Membrane reactor performance of steam reforming of methane using hydrogen-permselective catalytic SiO2 membranes. *Journal of Membrane Science*, **316**, 53–62.

Tsuru, T., Shintani, H., Yoshioka, T. and Asaeda, M. (2006) A bimodal catalytic membrane having a hydrogen-permselective silica layer on a bimodal catalytic support: Preparation and application to the steam reforming of methane. *Applied Catalysis A: General*, **302**, 78–85.

Tsuru, T., Yamaguchi, K., Yoshioka, T. and Asaeda, M. (2004) Methane steam reforming by microporous catalytic membrane reactors. *AIChE Journal*, **50**, 2794–2805.

Uemiya, S. (2004) Brief review of steam reforming using a metal membrane reactor. *Topics in Catalysis*, **29**, 79–84.

Uhlmann, D., Liu, S., Ladewig, B. P. and Dinizda Costa, J.C. (2009) Cobalt-doped silica membranes for gas separation. *Journal of Membrane Science*, **326**, 316–321.

Uhlmann, D., Smart, S. and Dinizda Costa, J.C. (2010) High temperature steam investigation of cobalt oxide silica membranes for gas separation. *Separation and Purification Technology*, **76**, 171–178.

Uhlmann, D., Smart, S. and Dinizda Costa, J.C. (2011) H_2S stability and separation performance of cobalt oxide silica membranes. *Journal of Membrane Science*, **380**, 48–54.

van Bommel, M. J., Bernards, T. N. M. and Boonstra, A. H. (1991) The influence of the addition of alkyl-substituted ethoxysilane on the hydrolysis—condensation process of TEOS. *Journal of Non-Crystalline Solids*, **128**, 231–242.

Yacou, C., Smart, S. and Dinizda Costa, J.C. (2012) Long term performance cobalt oxide silica membrane module for high temperature H_2 separation. *Energy and Environmental Science*, **5**, 5820–5832.

Yoshida, K., Hirano, Y., Fujii, H., Tsuru, T. and Asaeda, M. (2001) Hydrothermal stability and performance of silica-zirconia membranes for hydrogen separation in hydrothermal conditions. *Journal of Chemical Engineering of Japan*, **34**, 523–530.

Yu, W., Ohmori, T., Kataoka, S., Yamamoto, T., Endo, A., Nakaiwa, M. and Itoh, N. (2008) A comparative simulation study of methane steam reforming in a porous ceramic membrane reactor using nitrogen and steam as sweep gases. *International Journal of Hydrogen Energy*, **33**, 685–692.

Yu, W., Ohmori, T., Yamamoto, T., Endo, A., Nakaiwa, M., Hayakawa, T. and Itoh, N. (2005) Simulation of a porous ceramic membrane reactor for hydrogen production. *International Journal of Hydrogen Energy*, **30**, 1071–1079.

9.10 Appendix: nomenclature

9.10.1 Notation

A	membrane area
Da	Damköhler number
$DaPe$	Damköhler Peclet number
D_0	pre-exponential factor for diffusivity
d	reactor diameter
E	activation energy
E_d	activation energy of diffusion
F	flow rate
$f_{H_2}^{O}$	the equivalent feed of molecular hydrogen in the inlet stream to the reactor
f_{p,H_2}	the molar flow rate of H_2 in the permeate
J_i	flux of component i
K_0	pre-exponential factor for adsorption
L	reactor length
P_i	partial pressure of component i
P_f	pressure of feed stream
P_p	pressure of permeate stream
P_{H_2}	yield of H_2 in the permeate stream per reactor volume per time
$Perm_i$	permeability of component i

Pe	Peclet number
Q_a	isosteric heat of adsorption
Q_i	permeance of component i
q_s	adsorbed amount at saturation
R	gas constant
S^{ij}	permselectivity or ideal selectivity of component i from component j
W_{cat}	mass of catalyst
X	liquid hydrocarbon feed conversion
x_i	composition of component i in the permeate
Y_{H_2}	H_2 utilization
y_i	composition of component i in the feed
α^{ij}	separation coefficient of component i from component j
δ	membrane thickness
μ	correction factor related to the porous structure of the silica membrane

10
Carbon-based membranes for membrane reactors

K. BRICEÑO, Universitat Rovira i Virgili, Spain,
A. BASILE, ITM-CNR, Italy,
J. TONG, Colorado School of Mines, USA and
K. HARAYA, National Institute of Advanced Industrial Science and Technology (AIST), Japan

DOI: 10.1533/9780857097330.2.370

Abstract: Membrane reactor research has been focused on new membrane materials to be integrated in a compact configuration. Carbon membranes have scarcely been explored in the past because of mechanical drawbacks. For this reason, it is recommended that carbon membranes are supported. However, this can cause the formation of defects which are disadvantageous in membrane reactor (MR) applications. This chapter explores the main variables to be considered in the development of carbon membranes, mainly focusing on when the carbon material has to be supported. Some applications are revised for macro and micro reactors.

Key words: membrane reactors, microfabrication, carbon membranes, gas separation, microreactors, carbon molecular sieve membranes (CMSM).

10.1 Introduction

Membrane reaction processes are systems where separation and reaction are carried out simultaneously, and the continuous extraction of one of the products can shift the equilibrium, enhancing yield and selectivity as compared with a traditional system.[1] The development of membrane reactors has gone hand-in-hand with innovations in membrane materials and catalysts. Specifically, in the case of membranes, the same type of materials used to obtain them can also be adapted to support different catalysts. In terms of the separation and catalysis functions, porous membranes with permeance superior to dense membranes are the preferred candidates for use in membrane reactors; these include porous oxide, zeolite, glass, metal, and, more recently, carbon membranes. Although carbon membranes are still in their infancy and have some serious challenges, such as weak mechanical strength as unsupported membranes and bad controllability and reproducibility of manufacture as supported membranes, they are believed to be promising candidates for porous membrane based membrane reactors because of their ease of manufacture, low costs of both manufacture and raw materials, molecular-sieve separation effect and high permeance.[2–7]

Carbon membranes can be manufactured through the pyrolysis of polymer precursors. Their permeance and separation factors are better than corresponding polymeric membranes from the same polymer precursors. The investigation of carbon membranes have mostly focused on the sheet, capillary, hollow fiber and composite configurations. However, for large-scale applications the capillary or hollow fiber membranes are more important, due to their high packing density. For this reason, the first attempts to test carbon membrane reactors have been made using hollow fiber membranes for dehydrogenation reactions.[8] In fact, the integration of carbon membranes in membrane reactors developed out of the same need to find alternatives to silica and zeolite membranes. However, very few attempts have been reported in the literature, due to the early development state of carbon membranes. There are still some problems to be overcome, especially when carbon membranes are supported on a second substrate with different composition and structure in order to improve their mechanical strength. The crack, hydrothermal and mechanical resistance inevitably result. Moreover, the properties measured for the carbon membranes cannot be simply transferred to the carbon-based membrane reactors like their competitive counterparts. Therefore, the fabrication of high-quality carbon membrane with controllable and reproducible processes, and the exploration of the appropriate chemical reactions for carbon-based membrane reactors are the two key tasks for the development of the promising carbon-based membrane reactors.[5]

In this chapter, we first give an overview of carbon membrane materials (Section 10.2) and the classification of carbon membranes (Section 10.3). Then, unsupported carbon membranes, based on planar membranes and asymmetric hollow fiber membranes are discussed (Section 10.4). In Section 10.5, the supported CMSMs are reviewed in detail in terms of precursors, supports, fabrications and problems. In Section 10.6, carbon-based membrane reactors are discussed in detail, based on the topics of dehydrogenation reactions, hydration reactions, hydrogen production reactions, H_2O_2 synthesis, bio-diesel synthesis, and new carbon membranes for carbon membrane reactors (CMRs). In the end, the new concept of using carbon membranes in microscale devices (microcarbon-based membrane reactor) is outlined (Section 10.7).

10.1.1 Carbon membrane materials

Carbon is normally stabilized in various multi-atomic structures with different molecular configurations called allotropes. The three best known allotropes of carbon are amorphous carbon, graphite and diamond. Once considered exotic, fullerenes are nowadays commonly synthesized and

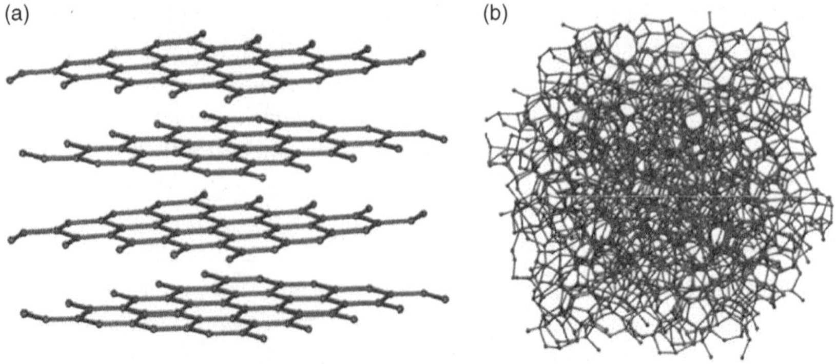

10.1 Molecular configurations of (a) graphite structure and (b) amorphous carbon structure.

used, including buckyballs, carbon nanotubes, carbon nanobuds and nanofibers. Several other exotic allotropes have also been discovered, such as lonsdaleite, glassy carbon, carbon nanofoam and linear acetylenic carbon. Although most of the carbon allotropes can be used as materials for the preparation of carbon membranes, the amorphous carbon and graphite related molecular configurations (as shown in Fig. 10.1) have been most extensively investigated as carbon membrane materials.

Graphite has a layered, planar structure (Fig. 10.1a). In each layer, the carbon atoms are arranged in a hexagonal lattice with separation of 0.142 nm, and the distance between planes is 0.335 nm. The two known forms of graphite, α (hexagonal) and β (rhombohedral), have very similar physical properties (except that the graphene layers stack slightly differently). The hexagonal graphite may be either flat or buckled. The α form can be converted to the β form through mechanical treatment, and the β form reverts to the α form when it is heated above 1300°C. At normal pressures carbon takes the form of graphite, in which each atom is bonded trigonally to three others in a plane composed of fused hexagonal rings, just like those in aromatic hydrocarbons. The resulting network is two-dimensional, and the resulting flat sheets are stacked and loosely bonded through weak van der Waals forces. After carbonization, an organic precursor can be converted into a material with higher carbon content. Below 1500 K there is a parallel alignment of molecules even though the structure of each hexagonal layer is not regular, which is favorable for the formation of holes.

Amorphous carbon (Fig. 10.1b) is an assortment of carbon atoms in a non-crystalline, irregular and glassy state, which is essentially graphite but not held in a crystalline macrostructure. As with all glassy materials, some short-range order can be observed. It is the main constituent of substances such as charcoal, lampblack (soot) and activated carbon. In a crystallographic sense, however, these materials are not truly amorphous, but are polycrystalline or nanocrystalline materials of graphite or diamond within an amorphous carbon matrix.

For the purposes of this chapter we are focusing carbon membrane materials with a turbostratic structure. This structure was considered in the past as amorphous carbon, which is not completely true. Nowadays, it is considered that there are fundamental building blocks of so-called amorphous carbon and microcrystalline domains where hexagonal planes are not aligned but displaced from each other or overlapped, giving a highly disordered character. In fact, there is a folding of hexagonal sheets which brings about holes of size smaller than 2 nm.[9] Carbon molecular sieves (CMSs) are a kind of turbostratic carbon membrane material with nanoporous structure that allows separation of molecules based on adsorption rate differences given by its size and shape.[3] Similarly, some obvious distinctions exist between CMSs and zeolite materials. For example, CMSs are mainly amorphous materials, but zeolites are crystallized materials. The structure of zeolites contains water of hydration that limits their application at high temperatures due to possible structure collapse. On the contrary, CMSs do not have this characteristic, which makes it possible to apply CMSs at high temperatures.[3] Furthermore, in comparison with zeolites, CMSs have other advantages such as good shape selectivity to planar molecules, high hydrophobicity, electrical neutrality and synthesis feasibility.[5]

CMSs can be obtained from controlled pyrolysis of synthetic and natural precursors. In the case of synthetic precursors, a large variety of polymers has been employed in the past. Although high-quality CMSs can be easily obtained from synthetic polymer, the cost of such synthetic polymers is very high compared with natural polymers, which makes commercialization more intricate and expensive.[3] For this reason, in the case of air recovery through pressure swing adsorption, the adsorbents of CMSs are obtained from cost-effective natural polymers instead of synthetic polymers. However, the advantages originating from pyrolysis of high-quality synthetic polymer, as for example, polyimides, drive the use of these polymers over supports. Lower quantities of polymer can be used when the polymer is supported than in the case of self-supported membranes. The fabrication process of self-supported membranes, as in the case of hollow fibers, must imply the use of a large quantity of polymer for dope formulation, which must be extruded for producing the asymmetric membrane (composed by the membrane itself but also by the self-supporting part). From this point of view, the application of CMSs to carbon membrane separation and (macro- and micro-) CMR holds more promise.

10.1.2 Carbon membrane classifications

Carbon membranes can be obtained after pyrolysis of corresponding polymer precursor membranes. The carbon membrane prepared by Koresh and

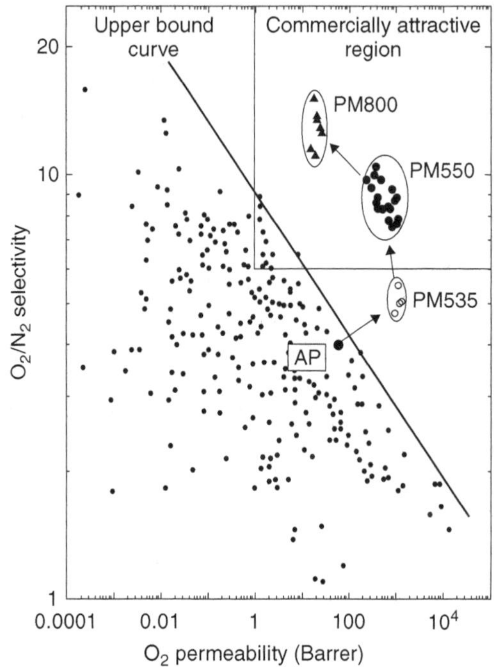

10.2 Robeson's plot: literature data for O_2/N_2 selectivity versus O_2 permeability related to polymeric and carbon membranes. (AP represents polymer precursors of PM carbon membranes. PM stands for polymeric membrane. The number that follows indicates the temperature (in °C) reached at the end of the heating).[11]

Sofer *et al.*[10] is one of the earliest works in this topic. These membranes are important due to the improved trade-off upper limit between permeability and selectivity compared with their polymer precursor membranes. Singh-Ghosal and Koros[11] reported the Robeson's plot (O_2/N_2 selectivity versus O_2 permeability) for some carbon membranes and corresponding polymer membranes (Fig. 10.2). It is obvious that the performance of carbon membranes is much better than that of corresponding polymer membranes. Moreover, the permeance of the carbon membranes depends on the surface characteristics and the interactions between pores and gas molecules rather than on the bulk properties as for the polymer membranes. When carbon membranes separate molecules based on the molecular-sieving mechanism, the molecules have to overcome an energetic barrier created by the differences between pore dimension and gas molecules.

Classification by transport mechanism

Carbon membranes have been extensively reported for gas separations,[5] which can be classified on the basis of the main transport mechanisms. In the case of gas transport depending on the effective sizes of the gas molecules rather than on the adsorption effect, we talk about molecular sieves carbon membranes (MSCMs) with pore sizes in the range of 3–5 Å. When gas permeation is governed by adsorption effects rather than the sizes of the gas molecules, we talk about adsorption selective carbon membranes (ASCMs) with pore sizes in the range of 5–7 Å.[12] On the other hand, depending on both the pore size and the pore population, it is possible to find alternative transport mechanisms such as Knudsen diffusion and adsorption-surface diffusion, etc. For Knudsen diffusion, the mean free path of gas molecules is larger than the pore sizes. Gas molecules diffuse through the pores in agreement with the concentration gradient while colliding with the pore walls. The diffusion flux is inversely dependent on the square root of molecular weight. This Knudsen diffusion mechanism occurs more often when pore sizes are in mesoporous range (2–50 nm) or the pore population is between meso- and macro-porosity. As for the adsorption-surface diffusion mechanism, the transport behavior of gas molecules is mainly dominated by the properties of the pore surface. During transport, gas molecules can be adsorbed on the pore walls and diffuse on the surface, which depends on how strong one or more components are adsorbed on the surface. The adsorbed molecules can plug the pores avoiding the passage of their own molecules but allowing the passage of other species.

Classification by configuration

An alternative classification of carbon membranes is based on their configuration, which is described in Fig. 10.3. It is clear that carbon membranes can

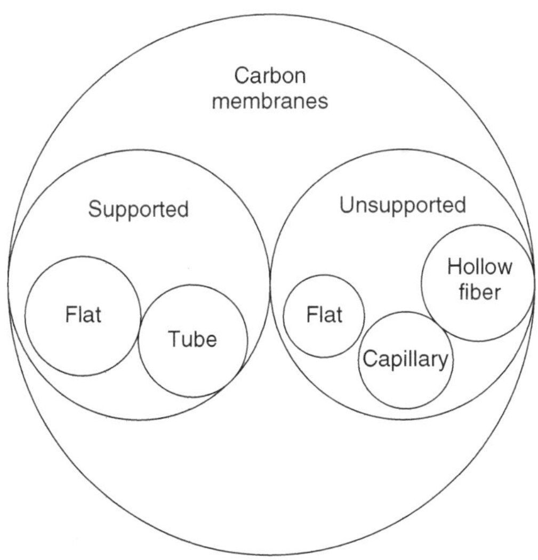

10.3 Classification of carbon membranes.

be divided into supported and unsupported carbon membranes, like most of the other membranes. The unsupported carbon membranes include the configurations of flat, capillary and hollow fiber, etc. The carbon membranes with these three configurations, especially the hollow fiber, can sometimes be fabricated with asymmetric structure with a thin separation layer (small pore size) and thick mechanical support layer (large pore size), which can also be called self-supported carbon membranes. In this chapter, we describe this asymmetric structure as supported because at least two types of materials are involved. The unsupported carbon membranes are easy to fabricate and the membrane quality is easy to control and reproduce. However, the intrinsic weak mechanical strength of these kind of membranes greatly limits their practical applications. On the other hand, for the supported carbon membranes, we mainly focus on the configurations with thin top carbon separation layer supported on a porous substrate with high mechanical strength and non-carbon compositions such as ceramic and glass substrates. The supported carbon membranes can be further divided into macro-flat, micro-flat, macro-tube and micro-tube, etc. It is easy to see that this kind of configuration can greatly improve the mechanical strength and increase the membrane packing volume density. However, the fabrication of these kind of supported membranes still has a lot of challenges. It is difficult to obtain well-controlled and reproducible high-quality carbon membranes on foreign porous substrates.

10.2 Unsupported carbon membranes

10.2.1 Symmetric flat carbon membranes

In the case of unsupported flat carbon membranes, Suda and Haraya[13] reported that the pyrolysis of Kapton polyimide at 1273 K can successfully result in high-quality carbon membranes. Kapton is one of the most studied materials for the fabrication of carbon membranes. The flat carbon membranes, obtained by pyrolyzing Kapton at the temperatures range of 1073–1273 K, showed molecular-sieving behavior. They also observed the global amorphous nature of their carbon membrane and the increasing pyrolysis temperature increased crystalline domains, which was related to the decrease of permeability and the increase of permselectivity. These changes were coupled with the decrease of interplanar spacing (d_{002}). They concluded that the contribution of the decreased interplanar spacing, which decreased amorphous zone, is the main reason for the increased microporosity in their carbon membranes. Suda and Haraya[14] explained the influences of pyrolysis conditions on tailoring the microstructure of their carbon membranes. The experimental parameters of pyrolysis temperature, pyrolysis atmosphere and heating rate were confirmed to be the main variables for controlling the formation of pore structures. In addition, they also studied the effects of the three experimental variables on the pore population of the carbon membranes. This is another important property to affect the gas separation performance for carbon membranes. In fact, it is possible that a pore population obtained at specific pyrolysis condition can bring benefit to the separation for a couple of gases but not for some other gases with similar effective sizes.

10.2.2 Asymmetric carbon membranes

Symmetric flat carbon membranes are good prototypes for studying the relationship between experimental conditions and pore formation behavior and property in the carbon membranes. However, they are limited for industrial applications because of the low permeance and volume density, which can be improved by an alternative asymmetric geometry such as asymmetric capillary and hollow fiber membranes. Haraya *et al.*[2] fabricated asymmetric capillary CMS. They obtained asymmetric structure using the phase inversion process through controlled coating polyamic acid (PA) membrane on a polytetrafluoroethylene (PTFE) micro-tube. After polymer gelification in several liquid baths, the PTFE micro-tube was removed and the PA capillary membrane was left, which was further imidized to the Kapton membrane and pyrolyzed at 1223 K to form the final capillary carbon membranes. The structures of the resultant membranes were highly dependent on coagulation baths. Therefore, the thickness of dense layer and the pore sizes of membranes can be controlled using different baths. For example, the membranes prepared in a methanol bath showed a H_2/N_2 selectivity up to 1080, while the lower H_2/N_2 selectivity of 95.9 was obtained for the membranes prepared in a water bath. However, the phase inversion technique for fabricating capillary carbon membranes showed a relatively low reproducibility. One of the reasons is that it is difficult to control the rapid formation of the dense layer, which is responsible for the separation performance. Moreover, this method allows the formation of large defects, which can decrease the separation performance of the asymmetric membranes. It should also be mentioned that this asymmetric configuration of capillary carbon membranes requires an important consideration of the technical machinery.

10.3 Supported carbon membranes

As an alternative to unsupported carbon membranes, the carbon separation layer can be supported on flat or tubular substrates to fabricate supported carbon membranes. In the case of supported membrane configuration, we have to consider the influence of the same experimental variables reported in symmetric unsupported flat carbon membranes (pyrolysis temperature,

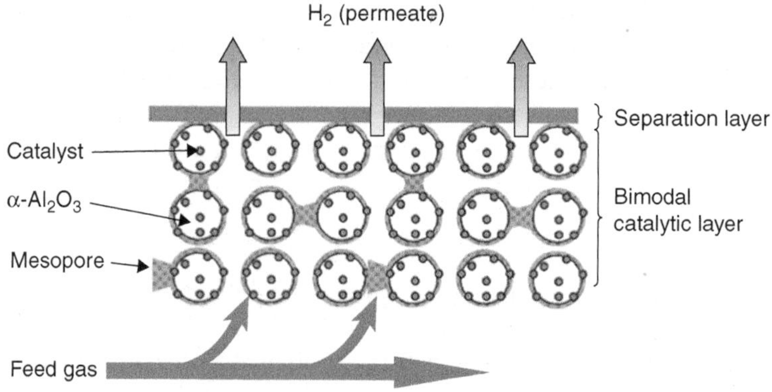

10.4 Schematic concept of a bimodal catalytic membrane (microporous top layer coated on a bimodal catalytic support).[16]

pyrolysis atmosphere and heating rate). In addition, another experimental parameter of fabrication method has to be considered carefully, because it can greatly influence the membrane nanostructure.[15] It is obvious that the supported carbon membranes have very good mechanical stability because porous substrates of ceramic or metal with high mechanical strength are employed. The special structure of the substrate makes membranes suitable for all combinations of catalysts and carbon separation layers, which can bring great benefit to the application of membrane reactors. This concept is schematically described in Fig. 10.4, which includes a microporous top layer and a bimodal catalytic support.[16] Therefore, both gas separation and catalytic reaction functions are combined to form membrane reactors. In the following parts we will introduce the most popular supported carbon membranes of CMS membrane on tubular substrates based on polymer precursors.

10.3.1 Polymer precursors

As mentioned before, the carbon membranes can be fabricated through pyrolyzing the corresponding polymer precursor at a high temperature in a controlled atmosphere. In selecting polymer precursors for the fabrication of carbon membranes, the preference must be for those polymers that show graphitization behavior. For example, commercial polyimides are chosen as polymer precursors for carbon membrane because they can change into highly crystallized graphite films after carbonization. Concerning this idea, Shiflett and Foley[17] reported that differences in polymer structure can affect the degree of cross-linking in the char. At higher temperatures, the highly cross-linked polymer will not be suitable for graphitizing. This implies that some amorphous carbon can be available in the graphite domains in the

10.5 Schematic of folded graphite-like layers.[3]

turbostratic carbon structure as represented in Fig. 10.5.[3] The control of graphite domains is very important in tailoring the pore size distribution of carbon membrane. Moreover, in addition to the polymer structure itself, the organization of the polymer during casting can also affect the final structure of carbon membrane. In this sense, Shao *et al.*[18] found that the polyimide films coated from polyimide solution with N,N-dimethylformamide as solvent have a crystallized structure. On the other hand, those polyimide films coated from polyimide solution with 1-methyl-2-pyrrolidone and dichloromethane as solvents have an amorphous structure. For carbon membranes obtained from amorphous polymer films after pyrolyzing at 823°K, the polymer structure determines the permeance of the resultant carbon membrane. However, at higher pyrolysis temperatures, the permeance of the resulted carbon membrane will be more influenced by pyrolysis conditions.

The most commonly used polymer precursors for carbon membranes have been reported to be polyimides, polyfurfuryl alcohol, phenol formaldehyde resins and cellulose. Their common characteristic is that they do not melt during pyrolysis at high temperature, which keeps their original shape and structure during the thermal heating and decomposition process. In this sense, the commercially available Matrimid and Kapton are the fully imidized polyimides with high T_g values. They do not abruptly change their structure during pyrolysis. This important characteristic of these two polyimides has been extensively appreciated by some investigators in the fabrication of CMSM.[2,18–20]

As is the case with unsupported carbon membranes, pyrolysis temperature also influences the carbon structure of supported carbon membranes. Centeno *et al.*[21] studied the effect of pyrolysis temperature on the permeance of phenolic resins supported on a ceramic tube. They showed how permeance decreases with an increase of pyrolysis temperature after

973 K. Between 973°K and 1023°K, the membranes were highly effective in the separation of absorbable and non-absorbable species which were considered as ASCM instead of carbon molecular-sieving membranes. Over 1073 K the carbon structure becomes more ordered, which implies decrease of pore size causing molecular-sieving behavior.

Kusakabe *et al.*[22] coated a PA membrane on the outer surface of a porous alumina tube, which was later pyrolyzed at 873–1173 K and post-oxidized at 673–773 K in series. They showed that combining the post-oxidation treatment with the normal thermal pyrolysis step increased the gas permeance of composite membranes.

Instead of the PA precursors, Hayashi *et al.*[23] deposited a polyimide film on the outer surface of a porous alumina tube by dip-coating three times. After imidization and pyrolyzation at 973–1073 K, the carbon membranes were fabricated on porous alumina tube. The enhancement of the volume of micropores accessible to smaller molecules has been observed. Hayashi *et al.* obtained an optimal pyrolysis temperature of 973 K and maximum permeance was achieved. In order to improve selectivity, a carbon layer was further deposited on the resultant supported carbon membrane by chemical vapor deposition (CVD) of propylene at 923 K. The CVD process favors the deposition of carbon in micropores, which explains the increase of the selectivity of CO_2/N_2 from 47 to 73.

However, polyimide precursors are very expensive. In order to explore alternative, cost-effective polymer precursors, Fuertes and Centeno[24] spin-coated a polyetherimide film on a porous carbon disk. They used isopropanol as a coagulation bath for the gelification of polyetherimide and obtained a defect-free membrane. After further plasticization by thermal treatment, the polymeric membrane minimally penetrated the porous substrate. The supported carbon membrane obtained through this process showed molecular-sieving behavior. In a successive work, the same investigators[25] started to use PA in the preparation of coated tubes. In that work, they reported the addition of an intermediate layer in the macroporous substrate in order to improve the substrate surface towards defect-free carbon membranes. The introduction of this intermediate layer allowed decreasing the pore size in the substrate. After 1–3 coats, all the resultant membranes showed ideal selectivity factors against Knudsen theoretical values. Moreover, permeance also increased with increasing operational temperature, which indicated all the membranes were showing molecular-sieving behavior.

10.3.2 Porous substrates

In the design of the carbon-supported membrane, the porous substrates have to be considered too. The most popular commercial porous substrates are porous stainless steel (SS)[26] and porous alumina.[27] The selection of substrate depends on the type of tight, or connection of the membrane into the module. However, in both cases the macroporous structure can result in large defects on the supported carbon membrane. For this reason, some strategies have been considered to reduce this effect. Therefore, Liu *et al.*[27] modified the surface morphology of an alumina macroporous substrate by dip-coating in boehmite sol several times. Merrit *et al.*[28] incorporated nanofillers into the pores of a porous SS substrate to reduce pore size, and much thinner carbon membrane was obtained on the modified substrate. The oxygen permeance increased to some degree without sacrificing selectivity.

10.3.3 Fabrication techniques

Once the polymer precursor and the porous substrate have been decided, we can go to the most important step in the fabrication of CMSM on the substrate. The fabrication technique must guaranty reproducibility and defect minimization. The separation carbon layer has to be placed over a defect-free substrate, otherwise all substrate defects can be copied to the coated carbon membrane, which in most cases will decrease the selective properties of the composite membrane.

In spite of the different strategies that can be considered to minimize defects on substrates, fabrication techniques can also reduce or minimize this problem. Table 10.1 summarizes the most important methods reported in fabrication of CMSM supported as flat or tubes.

10.3.4 Technical challenges

A common problem in the fabrication of supported carbon membranes is related to cracks formation and the minimization of defects arising from pyrolysis. Figure 10.6 shows a supported carbon membrane obtained from a commercial polyimide Matrimid coated on a porous substrate after pyrolyzing at 973 K with a ramp rate of 2.5 K/min in N_2 atmosphere. The differences of thermal expansion coefficients between substrate and coated polymer film during pyrolysis created a lot of cracks in the resultant carbon membrane. Therefore, the selection of polymer precursor, the optimization of pyrolysis temperature and the deposition parameters of the polymer layer are believed to be very important for avoiding such cracks.

10.4 Carbon membrane reactors (CMRs)

Due to their molecular-sieving effect, carbon membranes show a high H_2 selectivity, and for this aspect they are considered a good candidate for

Table 10.1 Methods for the preparation of supported CMSM

Method	Support	Coating time (s)	Thickness (μm)	Max. pyrolysis temperature (K)	Membrane position	Reference
Ultrasonic nozzle	SS tube	3	15 ± 2	873	Outside	17
Dip-coating	Carbon support	1	35	1073	Inside	20
Dip-coating + spinning	Ceramic	–	2	973–1273	Inside	21
Dip-coating + polypropylene pyrolysis	α-alumina	3	–	973	Outside	23
Spinning coating	Macroporous carbon disc	1 or 3	2	1073	Outside	24
Ultrasonic spray system + rotation	Sintered SS	1–3	–	723	–	26
Sponge soaked + spinning coating	α-Al_2O_3 modified by boehmite sol	1–3	4.7–13.9	873–973	Outside	27
Vapor deposition polymerization	γ-Al_2O_3/α-Al_2O_3	1–2	6	873	Outside	29

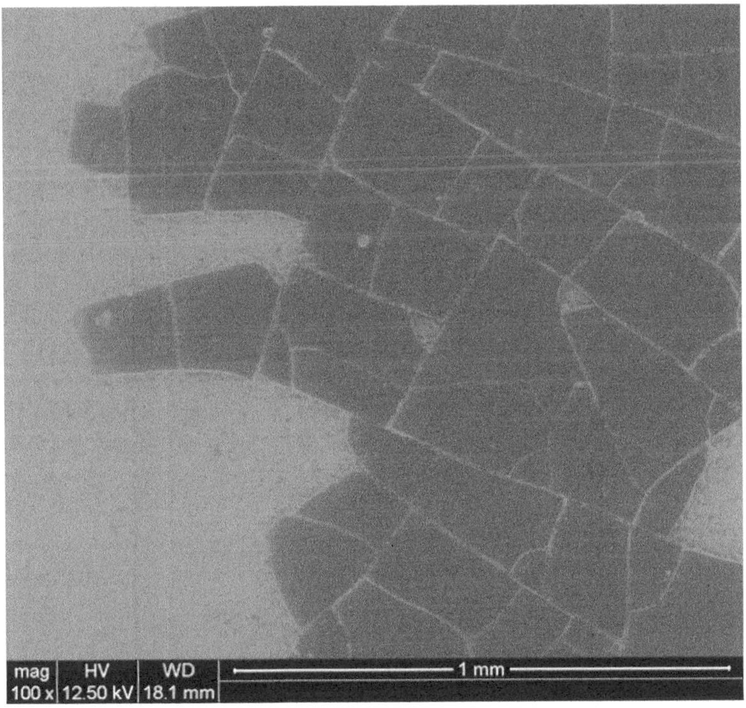

10.6 Crack formation on carbon-supported membrane.

hydrogen production via dehydrogenation reactions in CMRs. In fact, CMRs are one of the most promising applications of carbon membranes. Nevertheless, despite the very high potential of CMRs, to the best of our knowledge only a few works have been published in this specialized and very interesting field.

10.4.1 Dehydrogenation reactions

In 2000, Itoh and Haraya[8] constructed the first CMR and experimentally examined the performance of a dehydrogenation reaction. Asymmetric polyimide hollow fibers were pyrolyzed in a vacuum oven at 1023 K in order to obtain hollow fiber carbon membranes. Their CMR consisted of SS in which 20 carbonized hollow fibers (0.295 mm diameter and 128 mm long) and catalyst pellets (0.5 wt% Pt/Al_2O_3) were allocated. The reactor, used for cyclohexane dehydrogenation to benzene at 468 K, showed a fair improvement over equilibrium conversions. In detail, the temperature dependency of the permeation rates showed that the carbon membrane had micropores with an average diameter close to those of the gas molecules and therefore the permeation process was molecular-sieving controlled. The ideal H_2/Ar

separation factor (i.e., the ratio of permeation rates) showed a value of 53 at 448 K and decreased with increasing temperature.

Compared with the traditional reactor, obviously higher cyclohexane to benzene conversion has been observed in the CMR. The conversion of 25–37% at 468 K was obtained while keeping the pressure in the permeated side (H_2 side) relatively high. On the other hand, while keeping the pressure in the permeated side relatively low, conversion as high as 30–70% was obtained, which was ascribed to the larger amount of H_2 permeated through the carbon membrane. The experimental results also matched a simulated mathematical model, derived on the assumption of ideal flow. This assumption effectively explains the experimental data within a limited range of the reaction condition. Beyond this range, however, the simulated predictions deviated significantly from the experimental results, which probably can be ascribed to the radial concentration polarization of H_2 that reduced the conversion and a possible decrease of the catalyst activity at a lower H_2 atmosphere. This preliminary simulated work evidenced that not only a more precise model, but also additional studies on the integration of carbon membrane and catalyst, were needed.

The dehydrogenation reaction has also been attempted by Sznejer and Sheintuch in 2004.[30] They constructed a CMR using a molecular-sieve carbon membrane and carried out dehydrogenation reactions. Their CMR was composed of 100 fibers with diameter of 100 μm and thickness of 10 μm, which had a total membrane area of 150 cm^2. The catalyst pellets were loaded between the tube and the membrane module. In particular, the isobutane dehydrogenation on a chromia-alumina catalyst was chosen as a model reaction. Following the simulated results summarized in Table 10.2, the conversion of isobutane in the plug flow reactor (PFR) mode and all the three CMR modes increased gradually with the increasing temperature. At a certain temperature, the operation in the counter-current mode showed higher isobutane conversion than the operation in the co-current mode. The conversion values in the vacuum mode are quite close to those in the counter-current mode. On the other hand, the experimental results for PFR mode and CMR counter-current and vacuum modes are studied at different operational temperatures. The temperature really showed a great effect on the isobutane conversion, in that almost no reaction occurred at relatively low temperatures. Although the higher temperature showed obvious enhancement of isobutane conversion, the reaction in CMR operation modes was not stable. Furthermore, the effect of feed flux on the isobutane conversion was experimentally studied at 773 K, and the results are shown in Fig. 10.7. For comparison, the simulated results for the CMR counter-current operation mode are also plotted in the same figure. It is evident that all the experimental and simulated conversions decrease gradually with increasing feed flow, which can be ascribed to the decreasing contact

Table 10.2 Simulated conversion results at various temperatures and operation modes

Temperature (K)	PFR (%)	CMR (%)		
		Co-current	Counter-current	Vacuum
723	18.6	48.7	52.1	53.9
773	32.4	74.1	79.9	80.8
623	49.6	89.7	95.1	95.1

Source: Adapted from Sznejer and Sheintuch (2004).

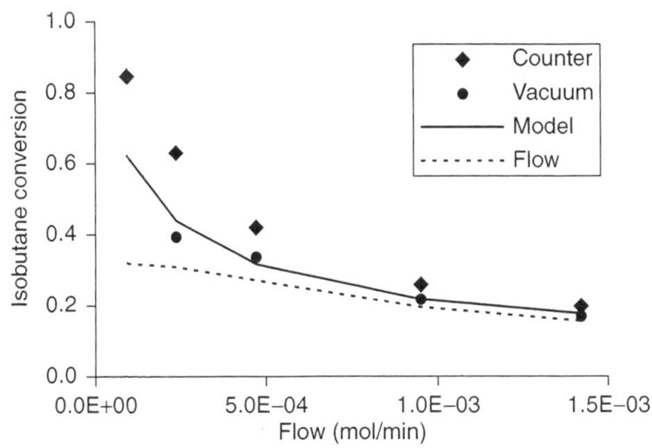

10.7 Results from an experimental study of the effect of feed flux on the isobutane conversion at 773 K.[30]

time between the catalyst and the membrane because of the increasing feed flow (space velocity). As expected, the conversion of the PFR (i.e., no separation) shows the lowest values for each feed flow studied. The highest isobutane conversion was achieved in the sweep gas counter-current flow mode with a maximum of 85% at 773 K. In the vacuum mode (no nitrogen transport and dilution effect), the isobutane conversions are lower than those in the sweeping gas mode (but still higher than that of 30% in the PFR) with a maximum of 40% at 773 K. Nevertheless, in this case, no dilution occurs, so that all the improvement is due only to the H_2 separation. Moreover, the discrepancy between the experimental results and the simulated predictions may result from the experimental factors or the assumed constant transport selectivities in the simulation. A much deeper investigation of these aspects is required because, as stressed by the authors, the most important improvement should be in enhancing and understanding transport selectivity. Following the experimental results, the highly selective and relatively inexpensive molecular-sieve carbon membrane used by the authors can be considered a good candidate to be used in a MR at relatively high temperatures.

Simultaneously, based on the same experimental studies, Sheintuch *et al.* studied the dehydrogenation of isobutane in CMRs from a modeling point of view.[31] In particular, they considered the separation in a membrane module either by maintaining the shell-side under lower pressure (or under vacuum) or by sweeping it with an inert gas. The main purpose of their work was to derive multi-component transport expressions from thermodynamics of molecular adsorption into, and diffusion within, the pores of the carbon membranes. Simultaneously, by comparing the simulation results with the experimental ones, the probability analysis of both transport multi-component single-file transport and single-component single-file transport in co- and counter-diffusion modes was applied in the design of CMRs. Furthermore, in order to explain the dramatic change in permeabilities measured in counter-diffusion mode and those measured as single-component, the measured pore size distribution factor was also included in their simulation work. The analysis was repeated for a family of parallel pores.

Simulated predictions and experimental results showed good consistency for the CMR performance obtained for a reaction coupled with separation by sweeping the hydrogen with nitrogen, but large discrepancy for a reaction coupled with vacuum-driven separation. CMR performance in the former mode is better due to excellent transport selectivity, which was attributed to mutual blocking of counter-diffusion by nitrogen and hydrocarbons.

10.4.2 Hydration reactions

Later, the CMRs were also used in an attempt to carry out homogeneous catalytic reactions for example, hydration of propene. Lapkin *et al.*[32] prepared a carbon membrane from a macroporous phenolic resin and constructed a CMR for the hydration reaction. In this gas phase continuous catalytic membrane reactor, the flat carbon membrane was used as a contactor for carrying out reactions at high temperature and pressure. In particular, the hydration of propene, catalyzed by an aqueous solution of phosphoric acid, was selected as a suitable model reaction. Olefin and water were fed separately in order to have the additional benefit of an increased alcohol concentration in the product stream because of the absence of steam in the propene feed.

The membrane contactor type of reactor, fabricated from a porous carbon artifact, was used for solving the difficulties associated with the supported liquid phase (SLP) hydration catalyst, without introducing deleterious mass-transfer resistance. In fact, the porous contactor is able to provide complete retention of the liquid catalyst, whereas the membrane reactor configuration serves to enable the retention of catalyst. As is well known, thermodynamic equilibrium between gas and liquid at the interface is influenced by chemical reaction, diffusion of products from the reaction zone, and convective flow on either side of the membrane. In such a contactor-reactor with a porous membrane, an increase in conversion is expected. In particular, the system was operated in a mass-transfer limited regime. Moreover, periodic oscillation of transmembrane pressure reduced mass-transfer resistance and improved the overall reactor performance. Another important aspect is related to the possibility of achieving a stable reactor operation at high operating pressure by tailoring the porous structure of the carbon membrane and coupling the reactor with an on-line feedback pressure controller.

The flat carbon membranes (diameter 31 mm; thickness 2–4 mm), used for carrying out experiments at 403 K and 2 MPa, were allocated in an SS reactor. The experimental results of both CMR and a conventional SLP reactor were compared using the apparent rate of production (*Rp* in $kg_{alcohol}/m^3_{membrane\ pore\ volume}\cdot h$) of propan-2-ol. *Rp* depends on: the rate of reaction in the liquid phase; the rate of gas–liquid mass transfer; the mass transfer within the membrane; and the convective transport in the gas space above the membrane. Therefore, *Rp* does not directly correspond to the intrinsic reaction rate. The best experimental measure of '*Rp*' was referred to equilibrium. In particular, *Rp* was estimated as the percentage of alcohol concentration in the vapor phase divided by the concentration of alcohol at equilibrium.

In the absence of any mass-transfer resistance induced by the porous membrane, the continuous sweep of products in the gaseous phase effected a shift in thermodynamic equilibrium towards formation of alcohol. This was also demonstrated by the authors by introducing a convective mass transfer to the batch reaction model. In fact, a dynamic model of the reactor was also developed, and the results of simulations compared favorably with experiment and the performance of a commercially operated conventional reactor.

Both experimental and simulation results showed that oscillation of transmembrane pressure significantly reduced the liquid-phase mass-transfer resistance. This aspect is considered to be very important for slow reactions, for which the volume of liquid catalyst is a more important parameter than the gas–liquid interfacial area, as indeed is the case for the hydration of propene.

Feeding reagents separately to the contactor (propene fed as gas, and water fed as liquid, on opposite sides of membrane) results in a lower concentration of water in the vapor products, which decreases the cost of downstream product separation.

Among the various advantage of this hybrid configuration (e.g., the rates of alcohol production are satisfactory; the retention of catalyst and separation of product from the reaction mixture are important for industrial-scale operation; the carbon membranes used for the contactor are robust and are unaffected by the presence of a strong acid, such as phosphoric acid catalyst), also must be added that the membrane preparation and fabrication techniques can be optimized for reducing the mass-transfer resistance in the liquid-filled pores and consequently increase the yield of desired product.

For commercial applications, the development of membranes in the form of monoliths or hollow fiber modules with considerably thinner walls than that of flat disk membranes is considered necessary, especially for a better shift in thermodynamic equilibrium.

10.4.3 Hydrogen production reactions

Zhang *et al.*[33] carried out the methanol steam reforming reaction, in both an MR and a conventional reactor for hydrogen production. The fixed bead reactor (FBR) and CMR consisted of an SS tube and a tubular carbon membrane, respectively, with the same inner diameter (i.d.) of 6 mm. The carbon membrane tube, a pinhole-free carbon composite membrane layer (thickness 20–30 μm), was sealed inside an SS tube (length 30 cm, i.d. 2 cm). In both reactors, 1.05 g of $Cu/ZnO/Al_2O_3$ catalyst pellets was packed in the reaction zone. Experimental parameters such as temperature, flow rate of carrier gas, and feed ratio were investigated to better understand the separation effect of hydrogen in the CMR on the methanol conversion and the product selectivity. The experimental results confirmed that both the methanol conversion and the H_2 selectivity in the CMR are much better than those in FBR under all the studied experimental conditions. On the other hand, if the overall yields of hydrogen were kept same in both the reactors, higher hydrogen purity can be obtained using CMR. The main results are summarized in the following:

1. In both the CMR and FBR, methanol conversion increased gradually with the increase in either the operational temperature or the feed ratio of H_2O/CH_3OH. Methanol conversion in the CMR was always higher than that in the FBR throughout the studied temperature range, and the methanol conversion at 523 K in the CMR was as high as 99.9%.
2. An increase in the carrier gas flow rate (only in CMR) increased the methanol conversion up to a plateau.
3. With regard to permeation selectivity in the CMR, hydrogen selectivity was 97%.
4. The CO yield in the CMR was always lower than that in the FBR throughout the studied temperature range.

Recently, the research group including Professor Adelio Mendes (Portugal)[6,7] studied two types of membranes, a CMS and a Pd one, for carrying out methanol steam reforming in a MR. The potential advantages of CMS membranes over palladium membranes to conduct the methanol steam reforming are related to very high permeabilities and hydrogen recovery, whereas Pd membranes are more expensive but exhibit much higher selectivity towards hydrogen. The authors' main objective was to analyze in which conditions each of the membranes performs better than the other, and also how both membranes can be integrated simultaneously in the same reactor in order to get a synergy. For these purposes, a 1-D comprchensive mathematical model of a packed-bed membrane reactor was developed for determining the effect of various parameters, such as the *Da* number, the reaction temperature, and so on, on methanol conversion, carbon monoxide concentration in the hydrogen rich stream, and hydrogen recovery. The goal was to maximize methanol conversion and hydrogen recovery, while keeping the CO concentration at the permeate side below 10 ppm. The authors emphasized that carbon membrane allows the permeation of water depending on pressures and residence time, whereas Pd membrane allows this under any conditions. Depending on hydrogen pressure in retentate and membrane permeance, water depletion can affect the final methanol conversion. In addition, they noticed that different properties of the carbon membranes can affect hydrogen purity based on the different reaction rates. For example, if the reaction rates of the reactants in the retentate side are slow, the consumption of reactants can be affected by increasing water partial pressure. *Da* number analysis affirmed that it is possible to have an excess of steam and shift the steam reforming and water gas shift reactions towards the production of hydrogen and consumption of carbon monoxide for high hydrogen selectivity. Moreover, it was also established that, depending on *Da* numbers, the reaction rates could be too high to change the gas mixture composition at the reaction side due to the increase of methanol conversion. In the particular case of the reverse water gas shift reaction, they also mentioned that carbon monoxide production and hydrogen selectivity reduction are both determined by the contact time and the characteristics of the carbon membranes. The H_2/CO reaction selectivity increases for the Pd-MR, whereas for the CMS-MR the opposite effect is observed. Hydrogen recovery increases with *Da* numbers for both MR, although the effect is almost unnoticeable for the Pd-MR and for low *Da* numbers. The CMS-MR presents higher hydrogen recovery than does the Pd-MR at high hydrogen concentrations, and the Pd-MR proves to be more advantageous for lower hydrogen production rates. Finally, the Pd-MR performance is enhanced by high retentate pressures and low permeate pressures, while the CMS-MR performance is enhanced for intermediate values. Concerning the combination of 'CMS + Pd membrane reactor', it shows some advantages for the CMS-MR; specifically, higher hydrogen recovery is achieved, keeping the CO concentration at the permeate side below 10 ppm. In comparison to the Pd-MR, this membrane combination allows the use of smaller membranes and higher feed flow rates, without prejudice to membrane reactor performance.

The reaction of methanol steam reforming was studied in a CMR over a commercial $CuO/ZnO/Al_2O_3$ catalyst by Sá *et al.*[7] A commercial CMSM was allocated in a CMR, which was operated at atmospheric pressure and with vacuum (15 mbar) at the permeate side, at 473 K. High methanol conversion and hydrogen recovery were obtained with low carbon monoxide permeate concentrations. A sweep gas configuration was simulated with a 1-D model. The experimental mixed-gas permeance values at 473 K were used in a mathematical model, which showed good agreement with the experimental data. The advantages of using water as sweep gas for increasing both methanol conversion and hydrogen recovery were investigated.. The concentration of carbon monoxide at the permeate side was under 20 ppm in all simulation runs. These results indicate that the permeate stream can be used to feed a polymer electrolyte membrane fuel cell (PEMFC). Good agreement between the mathematical model and the experimental data was observed. It was found that methanol conversion, hydrogen recovery and hydrogen yield are all enhanced by lower feed flow rates, due to higher residence times, with the drawback of higher carbon monoxide production. The simulation study showed that using water as sweep gas brings several advantages. In addition to an increase in both methanol conversion and hydrogen recovery, the production of carbon monoxide decreases drastically. The results presented in this study confirm the potential of using methanol steam reforming in a CMR to produce humidified hydrogen directly usable for PEMFC applications.

Briceño *et al.*[34] prepared a carbon membrane starting from a polyimide material coated and pyrolyzed under N_2 atmosphere on TiO_2–ZrO_2 macroporous tubes. The supported carbon membrane was used both to determine its permeation for low molecular weight gases such as H_2, CH_4, CO, N_2 and CO_2, and also for carrying out the methanol steam reforming in MR. As for the steam reforming of methanol, both the H_2 yields and the methanol conversions at different operation temperatures were compared respectively for both TR and CMR reactors (Table 10.3). The experimental data obtained in this preliminary study on the methanol steam reforming demonstrates the possible use of this type of carbon membrane in MR applications: both the yield and the methanol conversion of CMR are above those of the TR system.

10.4.4 H_2O_2 synthesis

The research group coordinated by Professor Gabriele Centi prepared a series of tubular catalytic membranes (TCM) and tested in the direct synthesis of H_2O_2.[35–37] Such TCMs are asymmetric α-alumina mesoporous

Table 10.3 H_2 yield and methanol conversion versus temperature in both TR and CMR for steam reforming of methanol

Temperature (K)	Reactor	H_2 yield (%)	Methanol conversion (%)
473	TR	64	44
	CMR	68	48
503	TR	70	45
	CMR	71	52
523	TR	72	49
	CMR	75	55

Source: Adapted from Briceño *et al.* (2012).

membranes supported on macroporous α-alumina, either with or without a subsequent carbon coating, which were respectively named as carbon coated alumina membranes (CAMs) and α-alumina asymmetric membranes (AAMs). For AAM, a dense Pd film was deposited on such supports by electroless plating deposition. After further thermal treatment in inert atmosphere at 773 K, a surface with well-developed large and ordered crystallites was obtained. For CAM, Pd catalyst was introduced by the deposition–precipitation technique. The carbon membrane layer of CAM possesses a high surface area and micro- to meso-pores. Catalytic tests were carried out in a semi-batch re-circulating mode under very mild conditions. Concentrations as high as 250–300 ppm H_2O_2 were commonly achieved with both carbon membrane deposition and without carbon membrane deposition after 6–7 h on stream. However, the H_2O_2 decomposition rate was particularly high in the presence of H_2. CAM's catalytic activity depended strongly on the Pd particle size, whereas well-developed crystallites were necessary to improve catalytic activity of AAM's. These features seem to indicate that a smooth metallic surface was necessary to improve catalytic activity.

The preparation of a novel catalytic membrane system to be used in multiphase H_2O_2 production has also been discussed in detail by Tennison *et al.* in 2007.[38] In their review, it was shown that it is possible to produce a membrane system that is potentially suitable for use in both multiphase and gas phase membrane reactor systems based on a 2-layer ceramic substrate. Moreover, the performance is sensitive to the degree of perfection of the support. The carbon membrane deposited within the nanoporous layer of the substrate has the structure and surface area to enable high dispersions of catalyst metals to be achieved when oxidized in carbon dioxide that have shown good performance in the direct synthesis of H_2O_2. When prepared under nitrogen, despite the simple production route, the carbon membrane shows excellent gas separation characteristics.

10.4.5 Bio-diesel synthesis

Dubé *et al.*[39] studied the production of bio-diesel in a semi-batch two-phase CMR using an acid- and a base-catalyzed transesterification of canola oil. Tubular carbon membranes (pore size 0.05 μm, i.d. 6 mm, O.D. 8 mm and length 1200 mm) were used in the reactor for their stability at high temperatures and resistance to chemical attack. The methanol and sulfuric acid catalyst, after pre-mixing, were charged into the membrane module. The authors demonstrated that CMR is able to alleviate many of the difficulties related to the transesterification of triglycerides to FAME (fatty acid methyl esters). The tests were carried out in the CMR in semi-batch mode at 333, 338 and 343 K and at different catalyst concentrations and feed flow rates. Increases in temperature, catalyst concentration and feedstock (methanol/oil) flow rate significantly increased the conversion of oil to bio-diesel. The novel reactor enabled the separation of reaction products (FAME/glycerol in methanol) from the original canola oil feed. The two-phase membrane reactor was particularly useful in removing unreacted canola oil from the FAME product yielding high purity bio-diesel and shifting the reaction equilibrium to the product side. In this work it was found that maintaining a 2-phase (emulsified) system in the MR inhibits the transfer of triglyceride and non-reacting lipids to the product stream. The production of a triglyceride-free FAME leads to the production of high-quality FAME. For this reason, the MR allowing a phase barrier which limits the presence of triglyceride and non-reacting lipids in the product is highly desirable in maintaining quality assurance in the production of bio-diesel.

10.4.6 Development of new carbon membranes for CMRs

More recently, Coutinho *et al.*[4] prepared carbon membranes by pyrolysis of polyetherimide hollow fibers to be used as a catalyst or catalyst support as well as separation medium, raising reaction yield and selectivity, and reducing the need for further separation steps. The authors investigated the pyrolysis process parameters as well as stabilization parameters to find an optimum condition for preparing polyetherimide-based carbon membranes. The experimental results permitted the production of thermostable carbon hollow fibers and selection of best treatment conditions. In particular, at the stabilization temperature >773 K, an intensive degradation of the fiber was observed. An initial exposure to an oxidizing atmosphere seems to be crucial in controlling the final membrane properties. In this atmosphere, heating rates as low as 1 $K \cdot min^{-1}$ during stabilization reduce cracks in the surface of final membranes.

An excellent review concerning the fabrication of carbon membranes for gas separation as well as for membrane reactors was published by Saufi and Ismail.[5] Among the various aspects, some important key-points are summarized in the following:

1. The production of carbon membranes currently involves a very high cost (between one and three orders of magnitude greater than that of a typical polymeric membrane). Therefore, carbon membranes must achieve a superior performance in order to compensate for their higher cost.
2. Optimization of fabrication parameters during the pyrolysis process is arguably the best way to achieve this goal and because of the large number of parameters involved, computer simulation will help in optimizing the pyrolysis process.
3. Once a high performance carbon membrane is produced, it is important to determine the effect of its exposure to water vapor, which becomes important when the carbon membrane is commercialized.
4. Moreover, the ambient humidity found in the atmosphere can also have an adverse effect on carbon membrane performance. Therefore, the study of storage conditions for carbon membranes are also an important aspect to be considered.

10.5 Micro carbon-based membrane reactors

As mentioned in the previous sections, porous carbon membrane materials can be supported on different substrates to form a supported carbon membrane, which can be used for gas separation or the construction of membrane reactors for the improvement of reaction performance. The pyrolysis of the polymer precursor membrane with specific characteristics of graphitization and tunable pore size distributions for the fabrication of the carbon membranes can also be used in other important applications such as carbon-based microporous devices. De Jong *et al.*[40] posed the need to find alternative materials for the fabrication of porous microfluidic devices, which should provide the rapid screening of toxic or dangerous reactions in microscale. Moreover, the micro-devices can economize on high-value reactants because of the high compactness of the device. Therefore, in the following parts of this section, we are going to give a simple overview of carbon-based micro membrane reactors, a kind of micro device.

Micro-devices with microchannels consisting of ceramic, polymer and silicon, etc. have been extensively fabricated to provide spaces (the walls of microchannels) for the contact of the reactant and product. However, for further enhancing the reactor, the integration of the separation function through the channel walls is believed to be very attractive. Therefore, we

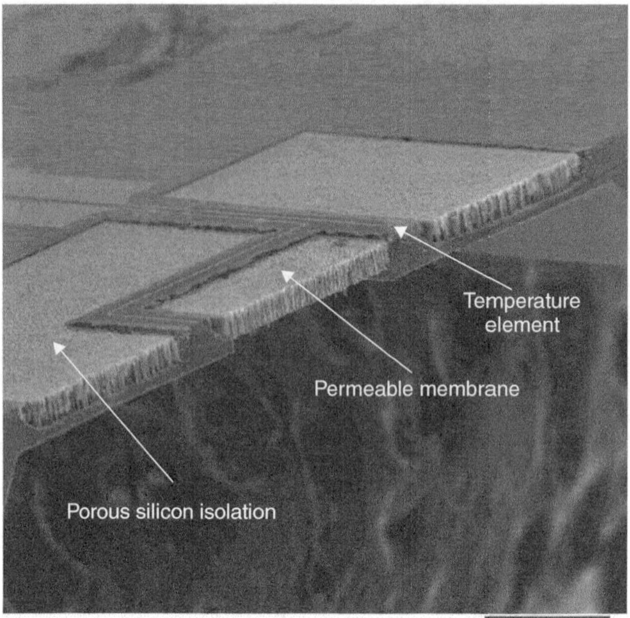

10.8 SEM cross-section of porous membrane reactor.[41]

think that, by controlling the same experimental variables as for the fabrication of the supported carbon membranes, it is also possible to design microporous layers as the walls of the microchannels in micro-devices for specific applications. De Jong *et al.* also pointed to the application of porous materials in membrane-based-gas-liquid contacting and porous channel emulsification for production of mono-dispersed droplets. In fact, Splinter *et al.*[41] have explored the fabrication of a porous silicon membrane doped with palladium for conversion of CO to CO_2 at 140°C as component in gas sensor devices as shown in Fig. 10.8, which is a very good example of the integration of porous membrane into the microreactors.

In the case of carbon-based micro membrane reactors, the fabrication of the glassy carbon films has been reported by Schueller *et al.*[42] through molding a furfuryl alcohol modified resin. Polymer precursor of polydimethylsiloxane (PDMS) is put in contact with a substrate with defined cavities or microchannels. Two techniques of MIMIC (micro molding in capillaries) and μTM (micro transfer molding) have been used for this fabrication. After a drop of polymer was added to one of these structures, the fluid filled the channels. In both techniques, the polymer was cured at temperature and PDMS peeled off from the substrates leaving the tridimensional feature. After further treatments, the resin was converted to glassy carbon while keeping its shape. Moreover, these authors also reported that the use of the

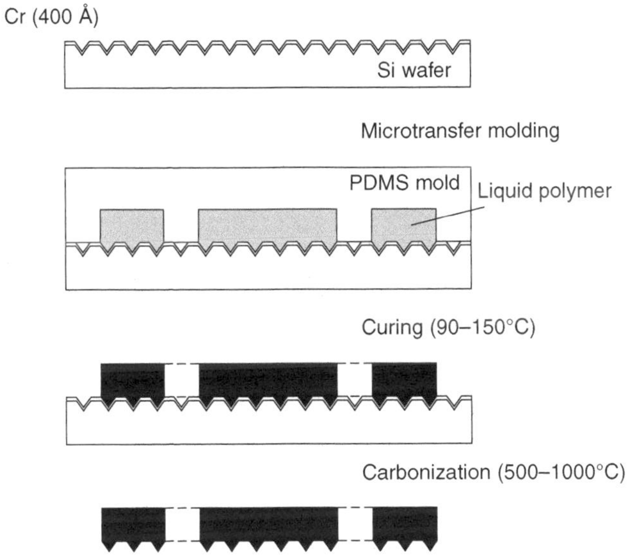

10.9 Schematic describing the preparation of a glassy carbon microstructure using a PDMS mold.[42]

reactive ion etching technique to generate patterns on a glassy carbon surface. Figures 10.9 and 10.10 show a schematic view of the process and some details of the microstructure. These methods allow the fabrication of microstructures of large area and high aspect ratio in short periods of time.

Fabrication of carbonaceous microstructures has been reported also by Malladi *et al.*[43] by spin-coating an SU-8 photoresistant material on a Si wafer. The photoresist material is coated on a Si wafer under a UV mask that transfers the pattern, which undergoes pyrolysis at 900°C in N_2 atmosphere. This process allows the fabrication of suspended carbon microelectromechanical systems (C-MEMS), which are highly appreciated on sensor applications.

To the best knowledge of these authors, there is no micro device integrating carbon glassy carbon membranes. The nearest in membrane fabrication is the fabrication of a composite material proposed in the past by Carretero *et al.*[44] They fabricated a microporous glassy carbon membrane infiltrated inside a mesoporous Al_2O_3 support. After dip-coating a phenolic resin on the inner face of Al_2O_3 tubes, the composite membrane was cured in air at 170°C and later at 700°C under N_2 atmosphere to obtain the microporous structure. The X-ray diffraction analysis (XRD) pattern evidenced formation of graphite and graphite layers stacked in a disorder structure in unsupported samples of the pyrolyzed resin. The obtained membrane showed a decrease of permeability with molecular size of gases tested (H_2, CO_2, O_2, N_2) with values of 5×10^{-7}–10×10^{-7} mol/m^2·s·Pa.

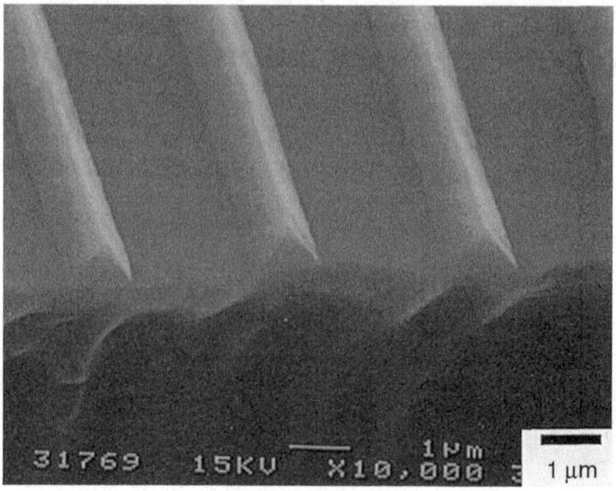

10.10 Close up of glassy carbon surface.[42]

10.6 Conclusions and future trends

Although carbon membranes are still in their infant stage, and have some serious challenges such as weak mechanical strength for unsupported membranes and bad controllability and reproducibility of the fabrication for supported membranes, they are believed to be promising candidates for porous membrane based membrane reactors because of their ease of fabrication, low cost of both fabrication and raw materials, molecular-sieve separation effect and high permeance.

A turbostratic carbonaceous structure allows obtaining carbon membranes suitable for gas separation and to be also used in membrane reactors. The pore size distribution and porosity of these kind of carbon membrane materials can be easily tailored by controlling the fabrication variables. There is an extended work developed on non-supported carbon membrane, such as planar and asymmetric hollow fiber configurations. However, in order to consider the practical industrial applications they have to solve the mechanical stability problem. In this sense, the fabrication of carbon thin films can be developed over substrates that bring them higher mechanical stability. However, the goal to achieve supported carbon membranes has then to face the problem of crack formation. During pyrolysis, the resultant carbon membrane from its polymer precursor has to achieve the turbostratic structure without cracks or defects. For example, for membrane fabrication the selection of the carbon membrane material has to consider graphitization behavior. Pyrolysis conditions such as temperature and heating rate will determine the pore structure. It is also very important to consider the type of substrate, because defects from the substrate's surface can be transmitted to the supported polymer and later carbon membrane. On minimization of defects, the modification of the support and several coating steps are strategies to follow.

The most important application for carbon membrane is to construct a carbon-based membrane reactor to carry out some special chemical reactions. Dehydrogenation reactions, hydration reactions and hydrogen production reactions were commonly carried out in CMRs. The concept of separation of hydrogen, one of the products, in the dehydrogenation and hydrogen production reactions, really shifted the reaction equilibrium to the product side by increasing the conversions. However, hydrogen selectivity and membrane stability certainly need further improvement to meet the reactions' demand. Moreover, the simulated work was also tried. One of the conclusions was that a new and more rigorous mathematical model needs to be developed for better describing these membrane reactions process. The flat carbon membrane was also used as a contactor for carrying out hydration of propene. The carbon membrane contactor has the following advantages. The rates of alcohol production are satisfactory; the retention of catalyst and separation of product from the reaction mixture are important for industrial-scale operation; the carbon membranes used for the contactor are robust and are unaffected by the presence of a strong acid, such as phosphoric acid catalyst. For commercial applications, the development of carbon membranes in the form of monoliths or hollow fiber modules, with considerably thinner walls than those of flat disk membranes, is considered necessary, especially for a better shift in thermodynamic equilibrium. Furthermore, the reactions of H_2O_2 and bio-diesel synthesis were also tried in the CMRs and exciting preliminary results have been obtained. Certainly more work needs to be done in this research area. Simultaneously, the research showing how to improve the fabrication of carbon membranes and make them suitable for application in CMRs has also been tried and summarized.

Finally, the flexibility that supported carbon membranes can bring to membrane reactors at a high scale can be transferred to the micro scale. For this reason, similarly to silica materials, it is reported possible to achieve carbonaceous microstructures applying traditional microfabrication techniques using other materials. It is possible to achieve integration of carbonaceous microporous structures in small chips for micro reactions. Using PDMS molding allows casting polymeric materials towards defined structures using a mask that allow features with dimensions over 20 μm.

10.7 Acknowledgements

The authors wish to thank Prof. Naotsugu Itoh (Utsunomiya University, Japan) for his help in reviewing the draft of the chapter.

10.8 References

1. J. Zaman, A. Chakma, Inorganic membrane reactor, *J. Membrane Sci.*, 1994, **92**, 1–28.
2. K. Haraya, H. Suda, H. Yanagishita, S. Matsuda, Asymmetric capillary membrane of a carbon molecular sieve, *J. Chem. Soc., Chem. Commun.*, 1995, **17**, 1781–1782.
3. H.C. Foley, Carbogenic molecular sieves: synthesis, properties and applications, *Microporous Mater.*, 1995, **4**, 407–433.
4. M.B. Coutinho, V.M.M. Salim, C.P. Borgees, Preparation of carbon hollow fiber membranes by pyrolysis of polyetherimide, *Carbon*, 2003, **41**, 1707–1714.
5. S.M. Saufi, A.F. Ismail, Fabrication of carbon membranes for gas separation—a review, *Carbon*, 2004, **42**, 241–259.
6. S. Sá, H. Silva, J. M. Sousa, A. Mendes, Hydrogen production by methanol steam reforming in a membrane reactor: palladium vs carbon molecular sieve membranes, *J. Membrane Sci.*, 2009, **339**, 160–170.
7. S. Sá, J. M. Sousa, A. Mendes, Steam reforming of methanol over a CuO/ZnO/Al_2O_3 catalyst part II: a carbon membrane reactor, *Chem. Eng. Sci.*, 2011, **66**, 5523–5530.
8. N. Itoh, K. Haraya, A carbon membrane reactor, *Catal. Today*, 2000, **56**, 103–111.
9. José Miguel Martín Martínez, Adsorción física de gases y vapores por carbones, Universidad de Alicante, ISBN: 84–86809–33–9.
10. J.E. Koresh, A. Sofer, *Separ. Sci. and Technol.*, 1983, **18**, 723–734.
11. A. Singh-Ghosal, W.J. Koros, Air separation properties of flat sheet homogeneous pyrolytic carbon membranes, *J. Membrane Sci.*, 2000, **174**, 177–188.
12. A.B. Fuertes, Adsorption-selective carbon membrane for gas separation, *J. Membrane Sci.*, 2000, **177**, 9–16.
13. H. Suda, K. Haraya, Molecular sieving effect of carbonized Kapton polyimide membrane, *J. Chem. Soc. Chem. Commun.*, 1995, **11**, 1179–1180
14. H. Suda, K. Haraya, Gas permeation through micropores of carbon molecular sieve membranes derived from Kapton polyimide, *J. Phys. Chem. B*, 1997, **101**, 3988–3994.
15. M.S. Strano, A.L. Zydney, H. Barth, G. Wooler, H. Agarwal, H.C. Foley, Ultrafiltration membrane synthesis by nanoscale templating of porous carbon, *J. Membrane Sci.*, 2002, **198**, 173–186.
16. T. Tsuru, T. Morita, H. Shintani, T. Yoshioka, M. Asaeda, Membrane reactor performance of steam reforming of methane using hydrogen-permselective catalytic SiO2 membranes, *J. Membrane Sci.*, 2008, **316**, 53–62.
17. M.B. Shiflett, H.C. Foley, Ultrasonic deposition of high-selectivity nanoporous carbon membranes, *Science*, 1999, **285**, 1902–1905.
18. L. Shao, T.S. Chung, G. Wensley, S.H. Goh, K.P. Pramoda, Casting solvent effects on morphologies, gas transport properties of a novel 6FDA/PMDA–TMMDA copolyimide membrane and its derived carbon membranes, *J. Membrane Sci.*, 2004, **244**, 77–87.
19. K.M. Steel, W.J. Koros, Investigation of porosity of carbon materials and related effects on gas separation properties, *Carbon*, 2003, **41**, 253–266.
20. W. Wei, G. Qin, H. Hu, L. You, G. Chen, Preparation of supported carbon molecular sieve membrane from novolac phenol–formaldehyde resin, *J. Membrane Sci.*, 2007, **303**, 80–85.

21. T.A. Centeno, J.L. Vilas, A.B. Fuertes, Effects of phenolic resin pyrolysis conditions on carbon membrane performance for gas separation, *J. Membrane Sci.*, 2004, **228**, 45–54.
22. K. Kusakabe, M. Yamamoto, S. Morooka, Gas permeation and micropore structure of carbon molecular sieving membranes modified by oxidation, *J. Membrane Sci.*, 1998, **149**, 59–67.
23. J. Hayashı, H. Mizuta, M. Yamamoto, K. Kusakabe, Sh.Morooka, Pore size control of carbonized BPDA-pp ODA polyimide membrane by chemical vapor deposition of carbon, *J. Membrane Sci.*, 1997, **124**, 243–251.
24. A.B. Fuertes, T.A. Centeno, Carbon molecular sieve membranes from polyetherimide, *Micropor. Mesopor Mat.*, 1998, **26**, 23–26.
25. A.B. Fuertes, T.A. Centeno, Preparation of supported carbon molecular sieves membranes, *Carbon*, 1999, **37**, 679–684.
26. M.B. Shiflett, H.C. Foley, Reproducible production of nanoporous carbon membranes, *Carbon*, 2001, **39**, 1421–1446.
27. B. S. Liu, N. Wang, F. He, J. X. Chu, Separation performance of nanoporous carbon membranes fabricated by catalytic decomposition of CH_4 using Ni/polyamideimide templates, *Ind. Eng. Chem. Res.*, 2008, **47**, 1896–1902.
28. A. Merritt, R. Rajagopalan, H.C. Foley, High performance nanoporous carbon membranes for air separation, *Carbon*, 2007, **45**, 1267–1278.
29. H. Wang, L. Zhang, G. R. Gavalas, Preparation of supported carbon membranes from furfuryl alcohol by vapor deposition polymerization, *J.. Membrane Sci.*, 2000, **177**, 25–31
30. G. Sznejer, M. Sheintuch, Application of a carbon membrane reactor for dehydrogenation reactions, *Chem. Eng. Sci.*, 2004, **59**, 2013–2021.
31. M. Sheintuch, I. Efremenko, Analysis of a carbon membrane reactor: from atomistic simulations of single-file diffusion to reactor design, *Chem. Eng. Sci.*, 2004, **59**, 4739–4746.
32. A. A. Lapkin, S. R. Tennison, W. J. Thomas, A porous carbon membrane reactor for the homogeneous catalytic hydration of propene, *Chem. Eng. Sci.*, 2002, **57**, 2357–2369.
33. X. Zhang, H. Hu, Y. Zhu, S. Zhu, Methanol steam reforming to hydrogen in a carbon membrane reactor system, *Ind. Eng. Chem. Res.*, 2006, **45**, 7997–8001.
34. K. Briceño, A. Iulianelli, D. Montané, R. Garcia-Valls, A. Basile, Carbon molecular sieve membranes supported on non-modified ceramic tubes for hydrogen separation in membrane reactors, *Int. J. Hydogen Eneg.*, 2012, **37**(18), 13536–13544.
35. S. Abate, G. Centi, S. Melada, S. Perathoner, F. Pinna, G. Strukul, Preparation, performances and reaction mechanism for the synthesis of H_2O_2 from H_2 and O_2 based on palladium membranes, *Catal. Today*, 2005, **104**(2–4), 323–328.
36. S. Melada, F. Pinna, G. Strukul, S. Perathoner, G. Centi, Direct synthesis of H_2O_2 on monometallic and bimetallic catalytic membranes using methanol as reaction medium, *J. Catal.*, 2006, **237**, 213–219.
37. S. Melata, F. Pinna, G. Strukul, S. Perathoner, G. Centi, Palladium-modified catalytic membranes for the direct synthesis of H_2O_2: preparation and performance in aqueous solution, *J. Catal.*, 2005, **235**, 241–248.
38. S. R. Tennison, K. Arnott, H. Richter, Carbon ceramic composite membranes for catalytic membrane reactor applications, *Kinet. Catal.*, 2007, **48**(6) 864–876.
39. M.A. Dubé, A.Y. Tremblay, J. Liu, Biodiesel production using a membrane reactor, *Bioresource Technol.*, 2007, **98**, 639–647.

40. J. de Jong, B. Ankone, R. G. H. Lammertink, M. Wessling, New replication technique for the fabrication of thin polymeric microfluidic devices with tunable porosity, *Lab. Chip*, 2005, **5**, 1240–1247.
41. A. Splinter, J. Stürman, O. Bartels, W. Benecke, Micro membrane reactor: a flow a through-membrane for gas pre-combustion, *Sensor Actuat. B-Chem.*, 2002, **83**, 169–174
42. O.J.A. Schueller, S.T. Brittain, G.M. Whitesides, Fabrication of glassy carbon microstructures by soft lithography, *Sensor Actuat. B-Phys.*, 1999, **72**, 125–139.
43. K. Malladi, C. Wang, M. Madou, Fabrication of suspended carbon microstructures by e-beam writer and pyrolysis, *Carbon*, 2006, **44**, 2602–2607.
44. J. Carretero, J.M. Benito, A. Guerrero-Ruiz, I. Rodriguez-Ramos, M.A. Rodriguez, Infiltrated glassy carbon membranes in γ-Al_2O_3 supports, *J. Membrane Sci.*, 2006, **281**, 500–507.

10.9 Appendix: nomenclature

10.9.1 Abbreviations

AAM	α-alumina asymmetric membrane
ASCM	adsorption selective carbon membranes
CAM	coated alumina membrane
C-MEMS	carbon microelectromechanical systems
CMR	carbon membrane reactor
CMS	carbon molecular sieves
CVD	chemical vapor deposition
FAME	fatty acid methyl esters
FBR	fixed bead reactor
HAMR	hybrid adsorbent-membrane reactor
MIMIC	micro molding in capillaries
MR	membrane reactor
CMSM	carbon molecular sieves membranes
LDH	layered double hydroxides
PA	polyamic acid
PEMFC	proton exchange membrane fuel cells
PSA	pressure swing adsorption
Pd	palladium
PDMS	polydimethylsiloxane
PFR	plug flow reactor
PTFE	polytetrafluoroethylene
SRM	steam reforming of methanol
XRD	X-ray diffraction
μTM	micro transfer molding

11

Advances in catalysts for membrane reactors

M. HUUHTANEN, P. K. SEELAM, T. KOLLI, E. TURPEINEN and R. L. KEISKI, University of Oulu, Finland

DOI: 10.1533/9780857097330.2.401

Abstract: Membrane reactors with a catalyst bed are designed to be used in various reactions, such as hydrogenation, dehydrogenation, oxidation and reforming reactions. The catalyst can be introduced into the reactor as a bed in several ways in the form, for example, of pellets, extrudates or tablets; or it can be incorporated in the reactor as a catalytic membrane wall. However, in many cases, the studies concentrate on the membrane itself, the development of catalysts is ignored, and commercial catalysts are used in the experiments. Most of the catalysts tested are aluminium oxide (alumina, Al_2O_3) based, as alumina is a mature support and already well proven in convectional reactors. However, some new catalyst materials such as carbon nanotubes (CNTs), carbon black, gels and anodic aluminium oxide (AAO) are developed as innovative catalyst supports and catalysts, since there is also a need for new catalysts for membrane reactors.

Key words: catalytic membrane, packed-bed membrane reactor, catalyst distribution, deactivation, hydrogen spill-over, sustainability.

11.1 Introduction

Membranes are used nowadays typically for the purification and separation of gases in several industrial processes. The role of catalysts in the development of membrane reactors (MRs) has in many cases been ignored in some levels, as the research has been concentrated on the membrane and development of new membranes. There is, however, a need for a catalyst in the MR to perform the desired reaction. The MRs containing a catalyst have been classified in three categories, namely extractor, distributor and contactor reactors (Miachon and Dalmon, 2004; van Dyk *et al.*, 2003). These reactor types are presented in Fig. 11.1.

In the extractor-type reactor one of the reaction products is removed through the membrane from the reactor. The distributor type is used for reactant delivery along the catalyst bed. In the contactor type the flow either goes through the membrane and catalyst bed, or is introduced from both sides of the membrane to the catalyst bed. It is also worth noting that the amount of catalyst needs to be sufficient for the reactor type, and that both the membrane and the catalyst have an influence on the efficiency of

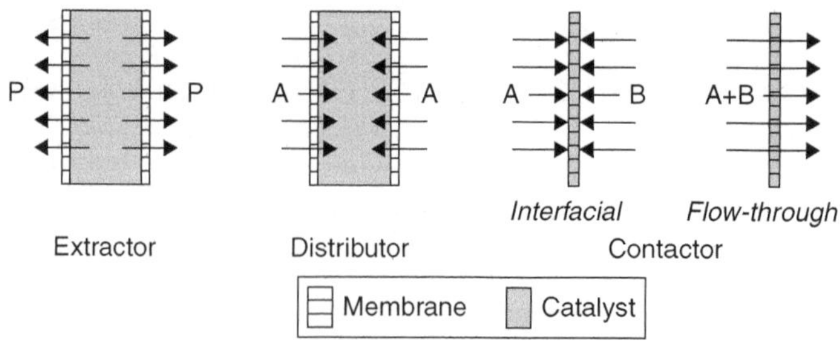

11.1 Catalyst placement in extractor, distributor and two contactor-type membrane reactors (A, B: reactants; P: product) (Miachon and Dalmon, 2004).

the system. Especially in the packed-bed membrane reactor (PBMR) the membrane usually acts only as a separator unit, and the catalyst particles as a bed are placed inside or in the annulus of the MR (Miachon and Dalmon, 2004).

Figure 11.1 illustrates how the catalyst and the membrane unit can be combined in various ways. According to Miachon and Dalmon (2004), general overall analysis is almost impossible for classifying the combination of a catalyst and a membrane as there are so many various possibilities. The MR with a catalyst can be either a (catalytic) packed-bed membrane reactor ((C)PBMR), a (catalytic) fluidised-bed membrane reactor ((C)FBMR) (where the catalyst is part of the set-up inside the free space reactor), or a catalytic membrane reactor (CMR) (where the catalytic material is part of the membrane-based wall) (Dudukovic, 1999; Gallucci *et al.*, 2011). This chapter mainly concentrates on the catalytic materials used for hydrogen production from hydrocarbons and alcohols in MRs.

11.1.1 Reaction and permeation kinetics – *Da* and *Pe* description

The ratio of reaction and permeation rates is critical in designing an MR. Dimensionless numbers are important in parametric analysis of engineering problems. They allow comparison of two systems that are vastly different by combining the parameters of interest. Dimensionless numbers are used to simplify the meaning of the information in scaling-up the reactor for real flow conditions and to determine the relative significance of the terms in the equations. The Damköhler number (*Da*) is the ratio of characteristic fluid motion or residence time to the reaction time, and the Peclet number (*Pe*) defines the ratio of transport rate by convection to diffusion or dispersion (Basile *et al.*, 2008a; Battersby *et al.*, 2006; Moon and Park, 2000; Tosti *et al.*, 2009). In the case of an MR, *Da* and *Pe* are defined in Equations [11.1] and [11.2].

$$Da = \frac{(\text{reactor volume}) \times (\text{reaction rate}) \times (\text{pressure})}{\text{inlet flow rate}} \qquad [11.1]$$

$$Pe = \frac{\text{inlet flow rate}}{(\text{permeability}) \times (\text{membrane area}) \times (\text{pressure})} \qquad [11.2]$$

The combination of the dimensionless numbers *Da* and *Pe* (*DaPe*) gives information about the reaction-permeation kinetic analysis in MRs:

$$DaPe = \frac{(\text{reaction rate}) \times (\text{reactor volume})}{(\text{permeability}) \times (\text{membrane area})} \qquad [11.3]$$

If the desired product formed during the reaction is totally permeated through the membrane then $DaPe = 1$. This implies a well-designed and efficient MR. If the permeation is too low it results in a very high *DaPe* and the membrane behaves as a plug flow reactor. A high permeation rate means a low *DaPe* and thus the tube and shell sides will achieve the equilibrium state very fast. The optimal value of *DaPe* should be in the range of $0.1 < DaPe < 10$. The catalyst activity that is, reaction kinetics should also be near the optimal value in order to achieve high MR performance (Basile *et al.*, 2008a; Tosti *et al.*, 2009). So, the catalyst must be active enough to carry out the reaction when the reaction equilibrium is shifted to the product side.

11.1.2 Hydrogen spill-over effects

By definition, the reverse hydrogen spill-over is the migration of nascent hydrogen atoms, which have formed during the reaction, that is, $H_2 \geq [H] + [H]$, from weakly chemisorbed sites (e.g., on the catalyst support) to the strong metal sites (e.g., Pd) and finally desorbed as a hydrogen molecule (Fig. 11.2). The hydrogen spill-over effects are widely studied in heterogeneous catalysis and their effects in metal catalysed reactions such as hydrogenation, hydrocracking, dehydrogenation, aromatisation, water-gas shift (WGS), and steam reforming (SR) reactions. In catalysis, the spill-over of hydrogen behaviour follows the dissociation–adsorption–diffusion–desorption route over the active metal sites (Cheng *et al.*, 2008; Rei *et al.*, 2011).

The hydrogen spill-over effect is also an important phenomenon in hydrogen separating MRs. Rei *et al.* (2011) have investigated the spill-over effects of hydrogen in SR of methanol and *n*-hexane. A set of different positions that is, distance between the catalyst bed and the Pd-membrane

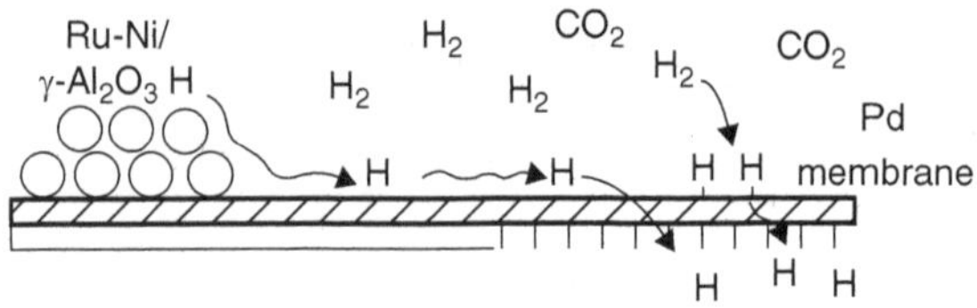

11.2 Hydrogen spill-over during steam reforming reaction in catalytic PBMR (Rei *et al.*, 2011).

configurations were tested in SR reactions. The contact of the membrane and the catalyst bed should be close enough in order to allow the nascent hydrogen atoms from the catalyst surface site to migrate to the surface Pd site and then finally permeate through the membrane. In a conventional fixed or packed-bed reactor (PBR) the hydrogen spill-over takes place also on the catalyst. Nascent hydrogen atoms may both recombine, forming molecular hydrogen and then desorbing into the gas phase, or they may inhibit or slow down the desired product formation. In the case of MRs, the backward reaction is reduced due to a nascent H atom permeation through the membrane. Moreover, MR shifts the reaction towards the product side that is, to the forward direction. Due to rapid removal of nascent hydrogen atoms by spill-over and surface diffusion of atoms, the reaction rate is enhanced and higher conversions are achieved. Thus the spill-over effect is a more important phenomenon to exhibit higher membrane performances than the reduction in the hydrogen partial pressure by membrane permeation.

11.2 Requirements of catalysts for membrane reactors

In CMRs, the catalyst and the membrane have two distinct functions. Here the discussion is concentrated mainly on the catalytic function that is, enhancing the reaction rate. Information about the membrane in terms of separation function is provided in the other chapters of this Handbook. The main requirements of catalytic materials in MRs are stability (that is, coke and poison resistance), mechanical and thermal stability (no sintering), attrition resistance, high activity per unit volume in the reactor, high selectivity, long operation life, reproducibility and economy (preparation and materials costs) (Sanchez Marcano and Tsotsis, 2002). Regeneration and activation processes for the catalysts are easier in the PBMR than in the CMR. In the inorganic MR in gas phase, pressure is the driving force, thus the catalyst needs to withstand this pressure in the case of a CMR. The shape and size of catalyst particles are also critical when applying pressure in membrane assisted catalytic processes. An important consideration is the complete utilisation of the active catalyst and the membrane material during the reaction and separation processes.

There have not been many studies done which deal with catalysis phenomena in PBMRs and CMRs. In the case of a PBMR or a MR, the membrane itself is not acting as a catalyst. Generally, after a catalytic reaction the membrane is involved in shifting the thermodynamic limitations. There are several physico-chemical properties required to prepare the most active and selective catalysts for certain reactions. The catalyst and membrane should function hand-in-hand. The catalyst should exhibit the desired activity of forcing the reaction towards the equilibrium (Miachon and Dolmon 2004). If any one of the functions deteriorates, the overall efficiency of the process will be reduced. However, many studies on heterogeneous and homogenous catalysis are devoted to developing new catalysts for various chemical reactions performed in a conventional fixed-bed reactor (FBR) and a PBR. The catalyst behaviour is slightly different in the MRs compared with the conventional reactor systems. The stability of catalyst materials is more critical in CMRs than in the FBR or the PBR. Furthermore, catalyst behaviour has a strong effect on the membrane function. An important aspect is the incorporation of the active catalyst phase into a CMR. The most common type discussed in the literature is the packed catalyst bed in a tubular MR, that is, the PBMR module in shell and/or lumen side. Sometimes the membrane is a catalyst itself and can then act as both a catalyst and a separating layer, henceforth termed the CMR. The three different types of catalyst placements in a MR are presented in Fig. 11.3 (Miachon and Delmon, 2004; Sanchez Marcano and Tsotsis, 2002).

11.2.1 Optimal catalyst bed pattern and catalyst distribution inside packed-bed membrane reactor (PBMR)

The catalyst distribution inside the PBMR is the critical parameter in enhancing MR performance. The utilisation of catalyst's active phase during the reaction in accordance with the membrane separation of desired components will affect positively. For example, to have well-mixed fluid phases no interphase mass transfer or diffusional resistance during the reaction and separation should occur (Yeung *et al.*, 1994). In the PBMR the active catalyst pellets are distributed inside the membrane module on the feed side as in the conventional FBR. In the CMR the active catalyst particles and sites are located within the membrane. Morbidelli *et al.* (2001) have reported a non-uniform catalyst distribution and its optimal bed location in the PBMR or an inert membrane reactor with a catalyst on the feed side (IMRCF) and in the CMR. In the case of a non-uniform catalyst distribution, that is,

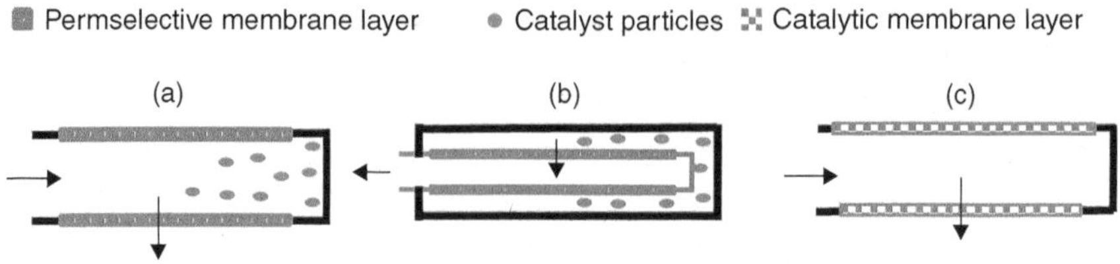

11.3 Common methods of introducing catalyst materials into the MR. Catalyst is packed inside the tubular membrane module (a) in the shell side or (b) in the tube (annulus) side, or (c) catalyst is dispersed along the membrane during *in-situ* preparation.

changing the active sites of pellets from the centre to the surface, the conversion is found to be higher than for the uniform distribution. This may be due to the reduction in mass transfer resistance as the non-uniform distribution of active sites is located closer to the pellet surface. In terms of the reactor performance, the non-uniform catalyst distribution is more positive in the case of the CMR, whereas minimal effects in the case of the PBMR can be detected. In the case of a uniform distribution, the backward reaction competes with the selective removal of the product in the CMR. Thus, non-uniform catalyst distribution maximises the reactant concentration at the active site of a catalyst and thus no diffusional limitations through the pellet or the membrane are present (Yeung *et al.*, 1994). Moreover, the radial distribution and the concentration of the catalyst will influence catalyst activity and selectivity (Grünewald *et al.*, 2004). In the FBR, longer residence times are required to obtain higher conversion per pass, whereas with shorter contact times between the catalyst and the reactant the same conversion can be obtained, as in the case of the FBR.

Szegner *et al.* (1997) have studied the effect of a non-uniform catalyst distribution in a PBMR for dehydrogenation of ethane using Pt–Sn supported on a γ-alumina composite membrane. In this study, a non-uniform Pt distribution within the catalyst pellets was prepared by impregnation, and later Sn was introduced uniformly into the pellets. Over catalyst with narrow surface step distribution, the conversion was high, that is, 80% beyond the equilibrium.

Caravella *et al.* (2008) have proposed a numerical model for the reactive–permeative stage of the MR, to investigate the importance of the catalyst and the membrane area axial distribution in methane steam reforming (MeSR). By optimising the number of reactive–permeative stage lengths one can achieve better performance than by the conventional MR, by maximising the hydrogen recovery factor or conversion. In another study, the decomposition of ammonia from coal gasification was studied in an integrated CMR (Abashar, 2002). In this study, the effect of different catalyst bed patterns was evaluated and the optimal condition to achieve better reactor performance was reported in order. Two different types of catalyst layers were proposed, a single layer (single catalyst layer and two catalysts-well-mixed pattern) and double layers (spatially layered bed pattern). It was concluded that the spatially layered catalyst bed pattern performed better and achieved higher ammonia conversion at a lower temperature and with a reduced reactor volume (i.e., reduced mass of the catalyst). In the PBMR, reactor performance is also based on the catalyst size, shape, structure, pellet distribution and layer configuration. A well-mixed catalyst bed pattern can be made by combining two catalysts physically mixed in a spatial layer (Abashar, 2002). Similar studies on the effect of bed pattern have been done, for example, by Basile *et al.* (2008a). Two catalysts, Ni and Ru-based catalyst beds packed on the lumen side of the Pd–Ag MR, performed better than a single Ni-based catalyst in acetic acid SR.

11.3 Catalyst design, preparation and formulation

11.3.1 General aspects in catalyst design and selection

Catalyst design is recognised as a part of the integral reactor design at the reactor level. There are several characteristics which the catalyst used must fulfil, regardless of the reaction or reactor in question. Some of these characteristics are listed, for example, by Liu (2007) and Richardson (1989):

- hydrothermal stability,
- mechanical strength,
- resistance against poisoning (especially coking),
- activity,
- selectivity,
- low pressure drop,
- high catalyst loading per reactor volume, and
- high internal and external mass transfer rates.

In addition, some general principles can be used in the selection of catalysts (Richardson 1989):

1. Before selection of the catalyst, the operating conditions in the reactor and reactor system must be carefully considered.
2. Thermodynamic consideration such as appropriate temperature and pressure range, as well the compositions for the reactants in the inlet and products in the outlet are defined.
3. Competing side reactions and product degradation reactions are identified.

4. Practical considerations, such as the risk of explosion or corrosion problems, should be determined.
5. Economic factors such as optimum energy utilisation, costs of feed material and the reactor itself, separation phenomena and possible by-products and their value should be evaluated.

However, it should be noticed that the preparation and formulation methods presented in the literature are in many cases only general overviews, as very often the details of the catalyst design and the formulation's manufacturing properties are confidential industrial information.

Catalytic materials for MRs have some particular requirements compared to a conventional tube flow reactor. The catalytic material should be in a form that can be inserted easily into the membrane reactor, and the catalyst should not have any mechanical failure or properties which are not suitable for a MR. Very fine powder form catalysts cannot be used, as the small particulates may block the pores of the membrane; however, small particulates (>0.2 mm) have been considered (e.g., Li *et al.*, 2010). Thus, in many cases, the catalysts generally used in MRs are pellets, extrudates or tablets. In addition to these forms, novel fibre type or foam catalysts have been studied as support materials for active metals. Li *et al.*, (2010) have presented in their study one kind of a method of encapsulating the catalyst particles (diameter 0.2–1.7 mm) which combines a catalyst particulate with a membrane layer. This has been reported to increase the selectivity of the reaction, and thus the separation process is much easier.

11.3.2 Combining catalysts with membranes

As discussed earlier, the concept of coupling a catalyst with a membrane into one layer/module represents the CMR. There are not many studies with this kind of a catalytic membrane layer, where reaction and separation occur simultaneously on the same material. Unlike in the PBMR, catalyst material is dispersed on the membrane surface (supported or self-supported membranes) (Specchia *et al.*, 2006), or the membrane itself is catalytically active such as a Ni membrane in the CMR (Ryi *et al.*, 2009). Moreover, the membrane itself can exhibit some catalytic properties as in the case of Pd/Al_2O_3, Pd–Ag alloy, etc. This could be more advantageous in reducing reactor volume and costs. However, in most of cases, metal membranes can be utilised to remove selectively the desired components such as H_2 using a Pd-based membrane, but they are not active enough or do not participate in chemical reactions due to low activity. In the CMR, the catalyst should be more robust than the catalyst used in the PBMR. It is considered more efficient to design a catalytic membrane layer which is also effective in the separation having suitable permeation properties for the desired components. There

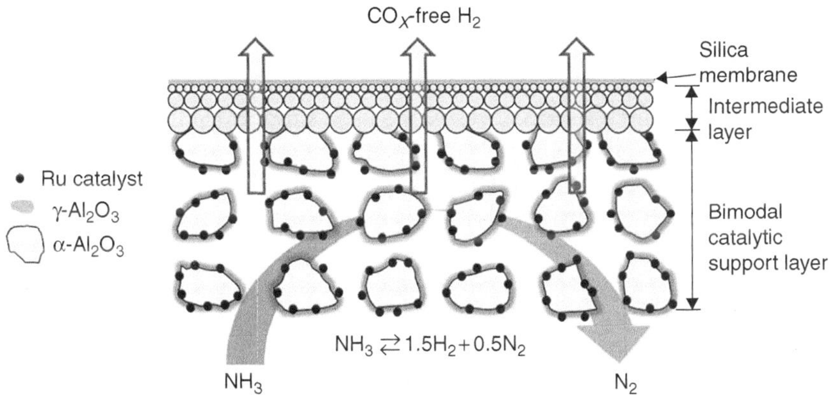

11.4 A BCM reactor for hydrogen production (Li *et al.*, 2011).

are many approaches to preparing catalytic membrane materials which can be self-supported or supported once. Ryi *et al.* (2009) have studied MeSR using a novel Ni catalytic membrane for hydrogen production. A porous Ni catalytic membrane was shown to perform reaction and hydrogen permeation which was 2.5 times that of CO and CO_2 at high space velocities. The deposited catalyst layer should not have detrimental effects on the permeation properties of the membrane. One of this type of a membrane is the bimodal catalytic membrane (BCM), illustrated in Fig. 11.4, designed and tested in ammonia decomposition for CO_x free hydrogen production (Li *et al.*, 2011). In the BCM there is a silica membrane layer for hydrogen permeation, a SiO_2–ZrO_2 intermediate layer, and catalytic bimodal supports γ-Al_2O_3/α-Al_2O_3, where Ru is deposited using an impregnation method. A comparison was made between bimodal (Ru/γ-Al_2O_3/α-Al_2O_3) and monomodal (Ru/α-Al_2O_3) catalytic supports in ammonia decomposition. For the BCM, stability and activity are much higher than for the monomodal catalytic support. The main reason for better performance of the BCM is that, after the γ-Al_2O_3 incorporation, the Ru particles are less than 40 nm in size, compared with 100 nm on monomodal as the support material (i.e., mesoporous γ-Al_2O_3). In the case of the BCM, the specific surface area is significantly increased, and thus Ru dispersion is improved.

The length scale plays a vital role in heterogeneous catalysis to control the structure, active phase and overall framework of the catalyst system. In the CMR, catalysis by nano-scale materials will improve the physico-chemical properties of the catalytic membrane. A uniform nanostructured catalytic membrane was prepared using AAO and atomic layer deposition techniques and tested in oxidative dehydrogenation of cyclohexane. Nanostructured CMs have excellent properties, such as uniform pores, contact time control, uniform diffusion paths and active site isolation (no sintering), compared to conventional scaffold based membranes. A vanadium pentoxide (V_2O_5) coated membrane on the AAO support, that is, V_2O_5/Al_2O_3/AAO, was tested in oxidative dehydrogenation reaction and it exhibited better performance with controlled and uniform pores of the used support (Stair *et al.*, 2006).

In a review (*Anonymous*, 1993) the platinum group metals (PGM) were reported to be used in a CMR. The total yield, selectivity and conversion per pass can be improved by coupling the catalyst with a permselective membrane layer as in the case of a porous Ni membrane, Ru on Al_2O_3, etc. The PGMs are widely studied in catalysis and applied in many industries due to their robustness. In membrane separations for example, Pd metal is used for hydrogen separation and purification. The cost of PGMs is, however, the major factor that would make them unfeasible in some commercial applications, in that sense; non-noble metals which are cheap, such as Ni, Co, in the SR reactions, Cu, Fe in WGS reactions, are the most commonly used active materials. Most of the studies are dedicated to developing non-noble metal catalysts due to their lower cost. A membrane can act as a support material for the catalyst. In addition, it is important to control the catalytic physico-chemical properties of the active metal for example, dispersion, morphology, on the membrane materials.

Deposition of Rh particles from the $Rh(acac)_2$ solution on the microporous silica separation layer using a wet impregnation method results in a layer of 4 nm Rh nanoparticles on the composite membrane structure. Depositing the Rh particles in the micropores of the silica reduced the permeability (i.e., He/N_2) compared with the bare silica membrane. Thus, depositing metal particles on the membrane can be crucial, without affecting the membrane properties. New methods and approaches can be developed in order to produce catalytically active membrane materials for the reaction and separation processes. Sometimes, during the reduction and regeneration steps, microcracks or defects can be formed in the silica layer. An optimal metal loading with controlled metal particles should be deposited, otherwise it will affect the permeability of the membrane layer. It is also observed that the activity of the Pt/Al_2O_3 catalytic membrane in the dehydrogenation of iso-butane was reduced due to coke formation inside the reactor. Adding Sn to the Pt on alumina catalyst reduced coke formation and increased the total yield of iso-butene (*Anonymous*, 1993).

11.3.3 Novel catalysts for membrane reactors

New catalytic materials and supports for catalysts have been studied to improve catalyst's activity and selectivity, and also to bring new knowledge on how new catalysts will act in and influence the membrane structure. Development work has been undertaken with AAO, nanostructured carbons as carbon black and CNT based supports decorated with, for example,

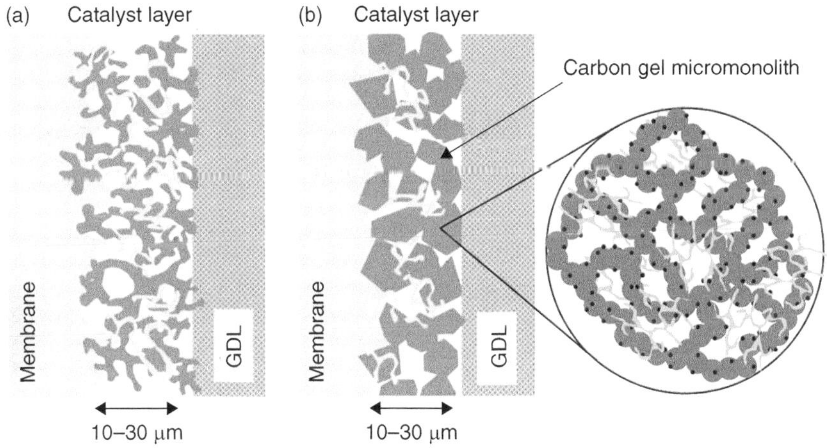

11.5 PEM fuel cell with membrane, catalyst layer and gas diffusion layer (GDL): (a) carbon black and (b) carbon gel supports with Pt particles (black spots) (Job *et al.*, 2009).

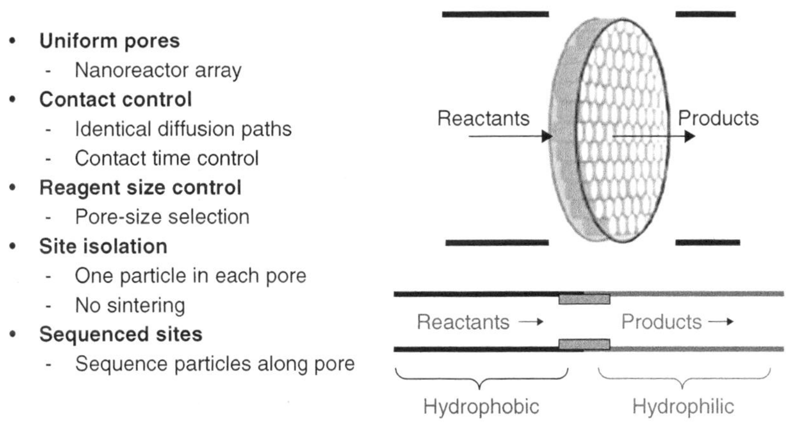

11.6 Nanostructured membrane framework (Stair *et al.*, 2006).

PGMs such as Pt and Pd (Figs 11.5 and 11.6). The membranes developed have been tested with various reactions, such as in proton exchange membrane (PEM) fuel cells and in hydrocarbon hydrogenation (e.g., Halonen *et al.*, 2010; Job *et al.*, 2009; Stair *et al.*, 2006). The nanostructured membrane AAO framework studied by Stair *et al.* (2006) is presented in Fig. 11.6. A similar type of a framework has been used also with the CNT support framework (Kordás *et al.*, 2006). This kind of a structure provides more uniform contact time and controlled reagent flow, as well as decreased sintering phenomena (Stair *et al.*, 2006). By designing new catalyst systems by

Table 11.1 Experimentally reported studies for membrane assisted SR, partial oxidation and dry reforming using gaseous feedstock

Process	Catalyst	Catalytic properties	Reaction conditions	Membrane reactor type	References
SR of methane	Commercial Ni/Al_2O_3	Ni 47.2 wt%	$T = 500–550°C$, GHSV = 5600 h^{-1}	Pd	Tong and Matsumura, 2006
	Ru/alumina	Ru 2 wt%, density of the catalyst 2.3 g cm^{-3}	$T = 500°C$, SV = 2000–3000 h^{-1},	Pd/PSS	Matsumura and Tong, 2008
	Ni:Mg:Al:La at 0.43:0.30:1.06:0.015 ratio	$S_{BET} = 103.4\ m^2g^{-1}$, pore vol. 0.44 $cm^3\ g^{-1}$, pore size distribution = 5.0–20.0 nm	$T = 450–550°C$, $p = 300–900$ kPa, GHSV = 4000–8000 $mLg^{-1}{}_{cat}h^{-1}$	Composite Pd	Chen *et al.*, 2008
	Commercial Ni/Al_2O_3	Ni 9.9 and 47.2 wt%	$T = 527°C$, S/C = 3, GHSV = 1120 h^{-1}	Pd	Tong and Matsumura, 2005
	Ni-based catalyst, La, B and Rh promoters	Ni 16–29 wt% promoter 0–9.65 wt%, dispersion = 9.8–78%, Ni particle size = 1.4–10.3 nm	$T = 600°C$, $p = 1.2$ bar		Ligthart *et al.*, 2011
	Commercial noble metal POx catalyst	$S_{BET} = 155\ m^2g^{-1}$	$T = 550–650°C$, $p = 2–4$ bar, S/C = 4	Pd	Patil *et al.*, 2007
	Commercial catalyst	Mean particle diameter = 85–90 μm	$T = 500–600°C$, $p = 1500–2600$ kPa S/C = 2–3.5	Fluidised bed Pd–Ag	Chen *et al.*, 2007b
	Commercial Ni/Al_2O_3		$T = 400–500°C$, $p = 1–3$ bar, S/C = 2, GHSV = 3710 h^{-1}	Pd–Ag dense*	Iulianelli *et al.*, 2010

	Commercial Ni-based catalyst		S/C = 3, T = 530–590°C, p = 25–42 bar	Pd/Al_2O_3	Pieterse *et al.*, 2010
SR of higher hydrocarbons	Ni/La_2O_3-Al_2O_3	S_{BET} = 174.0 $m^2\ g^{-1}$, pore volume = 0.40 $cm^3\ g^{-1}$, pore size = 5.6 nm, pore size distribution = 3.0–9.2 nm	T = 550°C, S/C = 2.7	Composite Pd	Chen *et al.*, 2006
SR of propane	Commercial RK-212 catalyst	Mean particle diameter = 179 μm	T = 475–550°C	Fluidised-bed $Pd_{77}Ag_{23}$	Rakib *et al.*, 2010
SR of *n*-hexane	$NiRuCe/Al_2O_3$	Ni 15, Ru 1, Ce 1 wt%, S_{BET} = 140 $m^2\ g^{-1}$, particle size = 8–16 nm	T = 500°C, S/C = 2.5, GHSV= 1000 h^{-1}.	Pd	Rei *et al.*, 2011
Partial oxidation of methane	Rh/Al_2O_3	Catalyst dispersion 47%, S_{BET} = 12.7 $m^2\ g^{-1}$	CH_4/O_2 = 1, T = 500°C	Packed-bed Pd–Ag	Cheng *et al.*, 2009b
	Ni/α-Al_2O_3 catalysts	Ni 5% and 17%, S_{BET} = 1 $m^2\ g^{-1}$	T = 700°C and 750°C CH_4/O_2 = 2	Fluidised-bed silicalite and Pd membrane	Ostrowski *et al.*, 1998
	Pd/γ-Al_2O_3	Pd 5.0 wt%	T = 350–550°C, CH_4/O_2 = 3	Fixed-bed double tubular Pd	Galuszka *et al.*, 1998
	Commercial NiO/Al_2O_3; Ni/Al_2O_3	17.5% NiO/γ-Al_2O_3, S_{BET} = 97.38 and 97.62 $m^2\ g^{-1}$.	T = 825–885°C	Perovskite-type $La_{0.6}Sr_{0.4}C_{0.2}Fe_{0.8}O_{3-\gamma}$	Jin *et al.*, 2000a and 2000b
Dry reforming of methane	Rh/La_2O_3; Rh/La_2O_3-SiO_2	S_{BET} = 180 $m^2\ g^{-1}$, Rh = 0.2 and 0.6 wt%, dispersion = 0.08–0.79%	T = 550°C W/F = 1.05×10^{-5} gh/ml	Pd–Ag	Irusta *et al.*, 2005
	Ni/Al_2O_3; $Ni/Ce_{0.5}Zr_{0.5}O_2$; Ni/ZrO_2; Ni/ZrO_2–CeO_2	Ni = 5 wt%	T = 510°C, p = 1 atm, CO_2/CH_4 = 1–3.8	Pd	Ferreira-Aparicio *et al.*, 2005

(*Continued*)

Table 11.1 Continued

Process	Catalyst	Catalytic properties	Reaction conditions	Membrane reactor type	References
	$Pd/\gamma\text{-}Al_2O_3$	Pd 5.0 wt%	T= 550–675°C, CH_4/CO_2 = 1.2	Fixed-bed double tubular Pd	Galuszka *et al.*, 1998
	Rh/La_2O_3; $Rh/La_2O_3\text{-}SiO_2$	Rh 0.6 wt%	T = 550°C, CH_4/CO_2 = 1	Pd–Ag	Coronel *et al.*, 2011
	$Ru/La_2O_3\text{–}SiO_2$	Ru 0.6 wt%, Ru dispersion 24, 29, 36%, 15, 27, 40 and 50 wt% of La_2O_3, particle size = 2.5, 3.11, 3.75 nm	T = 550°C, CO_2/CH_4 = 1–1.9, W/F = 4.5×10^{-6} g h/ml	Pd–Ag dense* double tubular	Faroldi *et al.*, 2011
	Pt/La_2O_3, Rh/La_2O_3	Rh 0.2 and 0.6 wt%; Pt = 0.41, 0.79, 0.93 and 2.35 wt%	T = 550°C	Pd/Ag	Munera *et al.*, 2003
	$Ni_{0.03}Mg_{0.97}O$, Ni/La_2O_3	Ni 5 and 2 wt%, S_{BET} = 25 and 9 m^2/g^{-1}	T = 625–750°C, CH_4/CO_2 = 1	Commercial Vycor glass	Prabhu *et al.*, 1999
Partial oxidation of simulated hot coke oven gas	Ru–Ni/Mg(Al)O	Particle size = 20–40 μm	T = 875°C	$BaCo_{0.7}Fe_{0.2}Nb_{0.1}O_3\text{–}\delta$ (BCFNO)	Cheng *et al.*, 2009a
Partial oxidation of methane in coke oven gas	La_2O_3 or CeO_2 modified $LiNi/\gamma\text{-}Al_2O_3$ catalysts	Ni 9 wt%, Li 0.5 wt%, La and Ce 1–15 wt%	T = 875°C	BCFNO	Zhibin Yang *et al.* 2009

*SSP = self-supported palladium membrane; S_{BET} = BET specific surface area; W_{cat} = catalyst weight; T = reaction temperature; p = reaction pressure; SV = space velocity; WHSV = weight-hourly-space-velocity; GHSV = gas-hourly-space-velocity.

nanotechnology and other novel approaches, one can reduce the mass of the catalyst or metal loading (g/m^2). Thus, the overall cost of the process can be reduced and the material and energy efficiency be improved.

11.4 Case studies in membrane reactors

The role of hydrogen as an energy carrier is gaining significant worldwide interest due to its clean, renewable, and non-polluting nature. SR, partial oxidation (POx), and CO_2 reforming have been widely investigated in hydrogen production from hydrocarbons and alcohols. In the following chapters the membranes in reforming of hydrocarbons and alcohols are presented.

11.4.1 Reforming of gaseous hydrocarbons

Traditionally, reforming has been performed using a tubular FBR technology. Catalysts belonging to the group VIII in the Periodic Table such as nickel have been found to exhibit promising catalytic performance for the reforming of hydrocarbons.

A membrane reactor configuration is attractive because reaction and separation steps can be combined in one unit. Moreover, the continuous withdrawal of hydrogen from the system shifts the equilibrium of the reforming reaction to the product side and enhances yields. This makes it possible to operate reforming at moderate temperatures. In Table 11.1 studies related to membrane assisted steam and dry reforming, and POx of gaseous feedstocks such as methane and hydrocarbons for hydrogen production are listed. In addition, a few POx studies with coke oven gas have been done (e.g., Cheng *et al.*, 2009a; Yang *et al.*, 2009). As shown in Table 11.1 the typical membranes reported in the literature are dense palladium membranes or silver–palladium membranes, and the catalysts used are mainly PGMs and base metals on aluminium and transition metal oxides. The catalysts studied, especially in the case of MeSR, are commercial ones almost invariably. This indicates clearly that the focus of studies has been in testing of membranes instead of catalyst development.

11.4.2 Reforming of alcohols

SR of alcohols has been widely studied in conventional reactor systems for hydrogen production. Bio-derived alcohols can be utilised as the raw materials for hydrogen production. Alcohols have high H/C ratio and hydrogen can be produced at low operating temperatures compared to methane reforming. In scientific literature most studies deal with catalyst development and catalyst performance in SR reactions. Moreover, SR is an equilibrium limited reaction. In MRs, conversion values beyond the equilibrium conversion can be achieved through the 'shift effect'. Most of the studies related

Table 11.2 Catalyst used in methanol, ethanol and other SR reactions performed in PBMR

Process	Catalyst	Catalytic properties	Reaction conditions	Membrane reactor type	References
MSR	Commercial $Cu/ZnO/Al_2O_3$ MDC-3, G66B	$CuO = 31–38\%$, $ZnO = 41–60\%$, $Al_2O_3 = 9–21\%$	$T = 350°C$, $p = 6–15$ atm, $H_2O/CH_3OH = 1.2$, WHSV = 1 h^{-1}	Double-jacketed supported Pd	Lin and Rei, 2001
	Commercial $Cu/ZnO/Al_2O_3$ modified ZrO_2	$W_{cat} = 4.5–25.5$ g, 0.5 mm particle diameter	$T = 250–300°C$, $p = 1–5$ bar, $H_2O/CH_3OH = 1:1$	Single-fibre packed-bed Pd–Ag/Al_2O_3	Israni and Harold, 2011
	Commercial Cu/Zn/Mg	Cu-based	$T = 300°C$, $p = 1.5–3.5$ bar, $H_2O/CH_3OH = 3/1$ molar ratio, WHSV = 0.36–1.82 h^{-1}	Dense Pd–Ag*	Iulianelli *et al.*, 2008a
	$CuOAl_2O_3ZnOMgO$ precipitation method	CuO 64%, Al_2O_3 10%, ZnO 24%, MgO 2%	$T = 350°C$, $p = 1.3$ bar, $H_2O/CH_3OH = 6:1$, $W_{cat} = 3$ g	Dense Pd–Ag*	Basile *et al.*, 2008b
	$LaNi_{0.95}Co_{0.05}O_3/Al_2O_3$ co-precipitation method	Alumina as binder	$T = 400°C$, $p = 0.05$ bar	Pd–Ag CMR	Wilhite *et al.*, 2006
	Commercial $CuO/ZnO/Al_2O_3$		$T = 310°C$	Pd	Fu *et al.*, 2007
	Commercial Ru–Al_2O_3	$W_{cat} = 3$ g 5 wt% Ru	$T = 400°C$, $p = 1.3$ bar	Pd–Ag supported and dense	Gallucci *et al.*, 2007
	Commercial $CuO/ZnO/Al_2O_3$ (G66MR)	Particle size 250–355 μm	$T = 200°C$, $p = 1$ bar	Carbon molecular sieve	Sá *et al.*, 2011
	$ZnO/CuO/Al_2O_3$ MDK-20		$T = 200–260°C$, $p = 1.2$ bar, $H_2O/CH_3OH = 1.3$	Dense $Pd–Ag_{20\%}$*	Basile *et al.*, 2006

ESR	Commercial Ni/ZrO_2 and Co/Al_2O_3	Ni 25 wt%, Co 15 wt%	$T = 400°C$, $p = 8–12$ bar, GHSV = 800–3200 h^{-1}	Pd/PSS composite	Seelam *et al.*, 2012
	Commercial Ru/Al_2O_3	$W_{cat} = 4–7$ g Ru 5 wt%, size 1–2 mm	$T = 400–450°C$, $p = 1.5–2.0$ bar $H_2O:C_2H_5OH$ = 8.4–13.0	Dense Pd–Ag*	Tosti *et al.*, 2008a
	Commercial Pt and Ni-based catalyst	$W_{cat} = 6.3$g, Pt/Al_2O_3, Pt 0.5%, spheres 2 mm size, NiO/SiO_2, $W_{cat} = 6$ g, Ni 25%, cylinders, 2 mm size	$T = 400–450°C$, $p = 1.5–2.0$ bar, $H_2O:C_2H_5OH$ = 8.4–13.0	Dense Pd–Ag*	Tosti *et al.*, 2008b
	Rh/SiO_2 and Pt/TiO_2 Degussa P25 for WGS	Prepared using synthetic method	$T = 300–600°C$	Composite Pt/PSS	Yu *et al.*, 2009
	Commercial MDC-3 $Zn–Cu/Al_2O_3$		$T = 320–450°C$, $p = 3–10$ atm	Pd–Ag/PSS composite	Lin *et al.*, 2008
	Commercial Ru/Al_2O_3	$W_{cat} = 3$ g, Ru 0.5 wt%	$T = 400°C$, $H_2O:C_2H_5OH$ =11:1, GHSV = 2000 h^{-1}	Dense Pd–Ag*	Iulianelli *et al.*, 2009
	Pt/Al_2O_3	$W_{cat} = 7.2$ g, spheres of diameter 2–3 mm	$T = 400$ and $450°C$, $p = 150–200$ kPa, $H_2O:C_2H_5OH$ = 4, 10 and 13	Dense Pd–Ag*	Santucci *et al.*, 2011
	Commercial Ru/Al_2O_3	$W_{cat} = 3$ g, Ru 5 wt%	$T = 400°C$, $p = 1.3$ bar	Dense Pd–Ag*	Gallucci *et al.*, 2007
	Commercial Co/Al_2O_3	$W_{cat} = 3$ g, Co 15 wt%	$T = 400°C$, $p = 3–8$ bar	Pd/PSS composite	Basile *et al.*, 2011 and Iulianelli *et al.*, 2010

(*Continued*)

Table 11.2 Continued

Process	Catalyst	Catalytic properties	Reaction conditions	Membrane reactor type	References
	Commercial Ni and Rh/La–Al_2O_3	W_{cat} = 3 g, Co 15 wt% La 3.1 wt%, Rh 4 wt%, S_{BET} = 143m^2 g^{-1}, particle size 150–250 mm	T = 400°C, p = 3–8 bar	Dense Pd–Ag	Papadias *et al.*, 2010
	Na–Co/ZnO	W_{cat} = 0.45 g, Na 0.2–2 wt%, Co 12.5 wt%, particle size: 0.6–0.85 mm	T = 600–750°C, p = 7–70 atm, S/C = 3–12, GHSV = 8500–83 000 h^{-1}	Silica–alumina composite MR	Lim *et al.*, 2010
Acetic acid SR	Commercial Ni and Ru	W_{cat} = 4 g of Ni W_{cat} = 2 g of 5% Ru	T = 400–450°C, p = 1.5–2.5 bar	Dense Pd–Ag*	Basile *et al.*, 2008c
	Commercial Ni		T = 400°C, p = 1.5–4.0 bar H_2O/CH_3COOH = 10:1	Pd–Ag dense*	Iulianelli *et al.*, 2008b
Glycerol SR	Commercial Ru/Al_2O_3	W_{cat} = 3 g, Ru 0.5 wt%	T = 400°C, p = 1.5–4.0 bar, H_2O/$C_3H_8O_3$ = 6:1 WHSV = 0.1–1 h^{-1}	Pd–Ag dense*	Iulianelli *et al.*, 2011
Naphtha reforming	Pt and Re	Pt 0.3 wt% Re 0.3 wt%	T = 505°C, p = 34–37 bar	Fluidised bed Pd–$Ag_{23\%}$	Rahimpour, 2009

*SSP = self-supported palladium membrane; S_{BET} = BET specific surface area; W_{cat} = catalyst weight; T = reaction temperature; p = reaction pressure; WHSV = weight-hourly-space-velocity; GHSV = gas-hourly-space-velocity.

to MRs emphasise the membrane and MR development. In Table 11.2 the catalysts used for SR reactions in a MR are presented. Most of the catalysts studied are commercial and not much information can be found about them. In methanol steam reforming (MSR), Cu and Pd are found to be the most active and selective elements. Cu and Pd-based catalysts are widely studied and the catalysts based on these compounds are commercially available. For ethanol steam reforming (ESR) reaction, Ni and Co based catalysts are found to be the most active and selective. In addition to that, noble metals, such as Rh and Pt-based catalysts, are recognised also to be active and selective, and are widely studied in ESR. Due to their cost, noble metals or PGMs are less preferred for commercial use, thus non-noble metals (i.e., base metals) are the preferred ones. Only a few articles are found where the new catalysts are designed and tested in the MR. It is important to design a novel and robust catalyst in accordance with membrane characteristics. For example, CNT based catalysts were studied in ESR (Seelam *et al.*, 2010) with metal and metal oxide nanoparticles decorated on CNTs. These kind of scaffold supports can be used in the preparation of novel catalytic membrane materials as well, instead of conventional membranes.

11.5 Deactivation of catalysts

In industrial processes, catalyst deactivation is a serious problem. The loss of catalyst activity and/or selectivity cost time and money depending on the deactivation type (chemical, mechanical or thermal) and whether the deactivation phenomena are reversible. To avoid damage such as poisoning by impurities from feedstock, fouling, thermal degradation, chemical degradation, and mechanical failure, it is vital to know the mechanism of deactivation (Bartholomew, 2003). Catalyst deactivation is dependent on reaction, concentrations of reactants, the ratio of inlet reactants, catalyst material, reactor (e.g., membrane), reaction conditions (temperatures, pressures) and so on. Hydrocarbon reforming technology still has disadvantages to overcome. The processes require a high operation temperature, above ca. 700°C, due to the equilibrium of the endothermic reactions. In addition, hydrocarbon decomposition over a catalyst is well known to be associated with carbon deposition (coking), which may lead to serious deactivation of the catalyst. The benefit of using the MR, compared with the conventional commercial reactor, is the possibility of using lower temperatures with higher conversions and yields. In addition, unwanted side reactions as well as coking of the catalyst are suppressed (Deshmukh *et al.*, 2007). Figure 11.7 presents an example of sintering, poisoning (sulphur) and coke formation on the Ni/RhPd/Al_2O_3 catalyst during SR of jet fuel. Nickel is known to deactivate the heavy hydrocarbon reactions in the presence of sulphur. Adding a noble metal on the Ni-based catalyst inhibits the poisoning by sulphur as well as

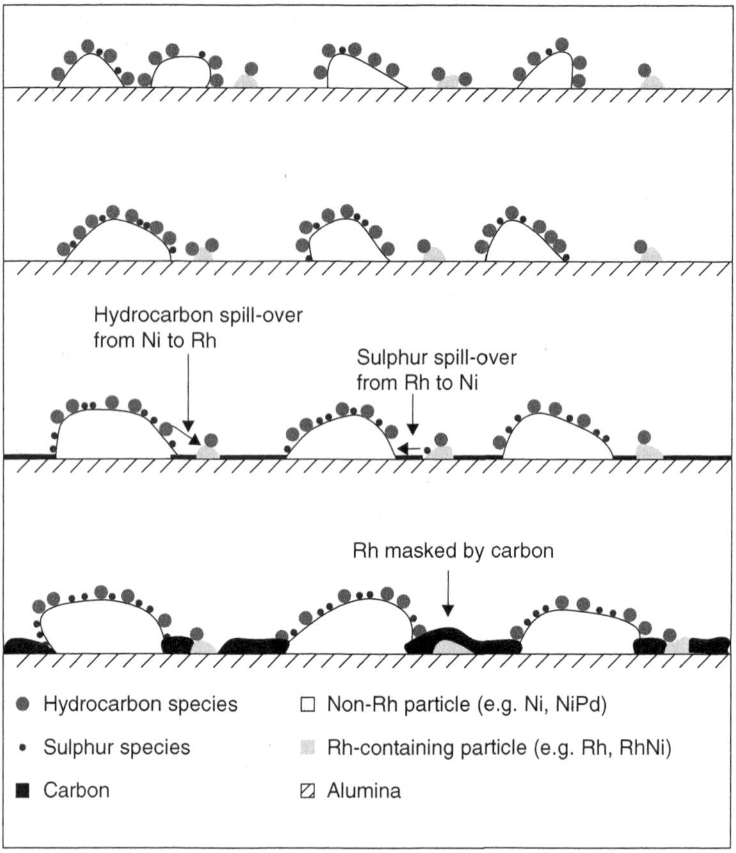

11.7 Mechanism of catalyst deactivation during SR of sulphur containing liquid fuel (Lakhapatri and Abraham, 2011).

adsorbed hydrocarbon migration (carbon deposition) from the sulphur poisoned Ni on the support material (Lakhapatri and Abraham, 2011).

11.5.1 Coke formation and regeneration

Coke formation is the most well-known deactivation mechanism in the membrane catalysed reactions (Reitz *et al.*, 2000). The three reversible reactions that contribute to the coke formation and form the basis for removal of carbon in the regeneration step by gasification (Dömök, 2007; Trimm, 1997, 1999) are as follows:

$$2CO \leftrightarrow C + CO_2 \quad [11.4]$$

(Boudouard reaction, at low temperature)

$$CO + H_2 \leftrightarrow H_2O + C$$
$$\text{(reverse of carbon gasification, at low temperature)} \quad [11.5]$$

$$CH_4 \leftrightarrow 2H_2 + C$$
$$\text{(methane decomposition, at high temperature)} \quad [11.6]$$

$$C_nH_{m+2} \rightarrow nC + (n+1)H_2 \quad [11.7]$$

Carbon can be deposited on the catalyst surface with three different structures: *whisker-like carbon* or *filamentous carbon, encapsulating carbon* and *pyrolytic carbon* (Chiappetta *et al.*, 2010; Christensen *et al.*, 2006). Filament growth causes no immediate deactivation, but may result in a mechanical breakdown of the catalyst pellets (Trimm, 1997). It has been found that there are many ways to minimise or limit the formation of coke: (1) the ratio of carbon-to-stream in the feedstock, (2) the use of a different catalyst or continuous regeneration of the catalyst, and finally (3) control of surface reactions by sulphur or metals or support (Deshmukh *et al.*, 2007; Trimm, 1997). Since coke formation reactions [11.4][11.6] are reversible, it is possible to calculate the best possible carbon/stream ratio in the process and thus avoid coke formation. Using modelling gives useful information about the reaction conditions. However, high steam/carbon ratio means higher production costs. The best solution is to add an additional component to the reactant mixture to avoid the coke formation, for example, the oxygen addition in the case of MeSR or the dehydrogenation of ethane, propane and butane. The problem is that an additional compound can promote the formation of undesired chemical species (Deshmukh *et al.*, 2007).

It has been shown that a small trace of sulphur on the inlet stream can reduce coke formation on the nickel catalyst. Deactivated sulphur–nickel sites inhibit the active carbon ($C\alpha$) polymerisation/isomerisation to less active carbon ($C\beta$) due to the lack of free sites. In other words, SR needs three to four nickel sites while carbon formation requires six to seven nickel sites. Thus, the SR catalyst loses some of its activity, but coke formation is minimal (Trimm, 1997, 1999).

Support materials or promoters such as ceria (CeO_2) and zirconia (ZrO_2) have an effect on coke formation during hydrogen production from methane in a membrane reformer. Cerium-zirconium oxides ($Ce_xZr_{x-1}O_2$) are known to behave as an oxygen storage compound (OSC). This means that $Ce_xZr_{x-1}O_2$ compounds have a high mobility of lattice oxygen through Ce^{3+}/Ce^{4+} redox properties and thus continuously release oxygen from the lattice to oxidise carbon species. Due to this effect, carbon deposition is lower on the Ni/$Ce_xZr_{x-1}O_2$ catalyst than on the Ni/Al_2O_3 and Ni/ZrO_2 catalysts. However, if the carbon deposition is high, but no deactivation is found, the reason for this phenomenon is that the active metal sites remain free for reactant activation. This kind of an effect was found for the Ni/ZrO_2–CeO_2 catalyst (Ferreira-Aparicio *et al.*, 2005). Seelam *et al.* (2012) have also reported that the ZrO_2 supported catalysts were more prone to coke formation, which leads to difficulties in catalyst regeneration. However, it has been found that regeneration over Al_2O_3 is much more effective than over ZrO_2.

11.5.2 Poisoning

When fuel contains heavier hydrocarbons than methane, or it is biofuel, or contains alcohols, the feedstock often contains additional compounds such as sulphur and phosphorus, that is, fertiliser impurities. In the petrochemical industry, gas-borne reactive species (i.e., sulphur, arsenic, chlorine, mercury, zinc, etc.) or unsaturated hydrocarbons (i.e., acetylene, ethylene, propylene and butylene) may act as contaminating agents (Deshmukh *et al.*, 2007). These impurities cause catalyst deactivation by poisoning. The effect of a poison on an active surface is seen as site blockage or atomic surface structure transformation (Babita *et al.*, 2011). Therefore, it is important to choose poisoning-resistant catalyst materials. For example, nickel is not the most effective MSR catalyst although it is widely used in industry due to its low market price compared to ruthenium and rhodium. Both Ru and Rh are more effective in MSR and less carbon is formed in these systems, than in the case of Ni. However, due to the cost and availability of precious metals, these are not widely used in industrial applications.

11.5.3 Sintering

In MSR, the catalysts used are based on Cu, Pd, Ru, Ni, Zn or on a combination of these compounds (Basile *et al.*, 2008b). The choice of the active metal in the process is dependent on the reaction temperature. The Cu-based catalyst can be a good solution at low temperatures $T < 300°C$ for MSR membrane reactors. However, Cu catalysts have been found to suffer thermal deactivation above the 300–350°C range, mainly due to sintering of Cu particles. (Twigg and Spencer, 2001).

For the WGS reaction, Fe, Cu, Co, Au and Pt-based catalysts have been studied and some of them have been commercially tested. In WGS reaction, the deactivation of these catalyst materials have been studied following the loss in surface area, formation of carbonates/formates, chemical transformation of support materials, chemical poisoning, blocking the surface redoxability, CO_2 inhibition, sintering and finally carbon deposition. (Babita *et al.*, 2011) Besides high temperatures (HTs), the increased H_2/HC molar ratio has also been found to deactivate the Pd–Ag membrane by sintering in catalytic naphtha reforming. Hydrogen may induce catalyst surface reconstruction, including sintering, particle migration, growth or coalescence and particle shape changes with various supported and unsupported metals. Due to these effects of hydrogen, the molar ratio of H_2/HC should be held at a reasonable value (Rahimpour, 2009).

11.6 The role of catalysts in supporting sustainability

The importance of catalysis in the support of sustainability is enormous (Centi and Perathoner, 2008; Keiski *et al.*, 2010; Sheldon, 2008). New heterogeneous catalysts for improved reaction selectivity, and catalytic activity and stability, are presently designed, that is, new catalytic materials, CNTs, ligands for selective and efficient catalysis and catalysts (Pan and Bao, 2008). This also concerns catalysts used in CMRs (Halonen *et al.*, 2010; Kordás *et al.*, 2006; Seelam *et al.*, 2010, 2012). The use of alternative reaction media and alternative activation of catalytic reactions also offers new ideas to enhance catalytic reactions, for example, the use of supercritical fluids, ionic liquids (IL) and microwaves as reaction media and activation agents for example, preventing carbon formation during reactions, for targeted energy transfer and in synthesis of new materials, fuels and chemicals (Ballivet-Tkatchenko *et al.*, 2003; Reichardt and Welton, 2011; Welton, 2008). Catalysis has already for a long time fulfilled the principles of Green chemistry and Green engineering in industrial applications (Anastas and Zimmerman, 2003). Combining chemical and biocatalysis to produce hydrogen from renewable feedstocks is an innovative and timely target in catalysis research (Sheldon, 2008). In process engineering, microstructured catalytic reactors, including CMRs with phenomena integration are merging into the market and thus need new and innovative catalytic materials to enhance sustainability in the chemical industry. New functionalised materials and catalysts as well as biomass-based innovations are the key driving forces today and for the future.

Catalysis is forecast to have a great impact on chemicals manufacturing and process intensification in the near future via new reaction routes (chemo-enzymatic) and via designing catalysts combining the best features of biocatalysis, homogeneous and heterogeneous catalysis (Sheldon, 2008; Somorjai, 2004). To develop clean and sustainable chemical processes, the transfer of the catalytic efficiency shown by enzymes in nature to chemical processes is a timely challenge (Lozano *et al.*, 2011). Enzymes provide the most important routes for Green organic synthesis. One innovative approach in this area comes from the combination of biochemical catalysis (e.g., enzymes and microbes) with the use of neoteric solvents, such as supercritical carbon dioxide ($scCO_2$) and ILs. Neoteric solvents help stabilisation and easy recycling of biocatalysts, and the use of continuous flow conditions

(Lozano *et al.*, 2011). These improvements apply not only to biocatalysis but also chemical catalysis. The most innovative idea is to mimic the metabolic pathways found in the nature. Thus, continuous flow multi-enzymatic and multi-chemo-enzymatic processes in multiphase systems for synthesising new products are underway.

The design of catalysts based on nanomaterials offers new tools for chemical process integration at nano-scale (Somorjai and Park, 2007a, 2007b; Somorjai and Rioux, 2005). With nanomaterials the design of industrial processes can be done with a better integration of units. This approach also enhances sustainability via allowing more efficient use of noble metals and rare earth elements used as active and support materials in many catalysis fields. This all guarantees sustainable use of raw materials and energy, since catalysis by nanomaterials offers selective and optimised chemical conversion of raw materials into desired products with minimal use of expensive materials. It also opens ideas in seeking substituting materials for the expensive and critical metals and elements used as active components in heterogeneous catalysis. This results in following the key principles of Green chemistry and engineering approaches (Anastas and Warner, 1998; Anastas and Williamson, 1998). These approaches will change the way that chemicals and other products are manufactured, distributed and used in our society in the future and will enhance sustainability.

Combining the best features of homogeneity, heterogeneity and biocatalysis into one reactor is considered to have a key role in tomorrow's society (Somorjai, 2004). In a nanometre scale, better control of activity, selectivity and deactivation of catalysts is attained (Somorjai and Park, 2007a, 2007b). The design, synthesis, characterisation and manipulation of nano-scale catalysts have already been a keen thematic in research. Catalysis nanoscience takes advantage of for example, self-assembly of catalytic sites in predetermined two- and three-dimensional configurations. Research activities and innovations are needed and underway in this area of research.

Thus, it is evident that catalysis has a key role in sustainable development (Keiski *et al.*, 2010) and that a new era in catalysis has been opened. Catalysts have an enormous impact on industry and everyday life in making processes more efficient, increasing the operating profit and making processes environmentally friendly, economic and safe.

11.7 Conclusions and future trends

Membrane reactors can be used in many industrial reactions such as hydrogenation, oxidation and reforming. Nowadays, mainly commercial catalysts are introduced in MRs to improve the desired reactions such as hydrogen production from hydrocarbons and alcohols. For example, a typical reactor for the reforming of hydrocarbons is a tubular FBR. A promising catalytic material for reforming of hydrocarbons has been found in group VIII in the Periodic Table, especially nickel (Ni). In SRE, besides nickel, cobalt has been found to be active and selective. Noble metals have also been studied in an SRE reactor, but due to high market prices, these metals are not widely used in industry.

Catalysts in the MR can be placed in three different ways. The classifications of these types of membranes are extractor, distributor and contactor. The last one can be either interfacial or flow-through. Both the catalyst and membrane have an influence on the reaction. Depending on the MR, the catalyst is dispersed on the membrane, the membrane can be catalytically active, or the membrane exhibits some catalytic properties. In addition, how the catalyst pellets, extrudates, tables, fibres or foams are distributed inside the MR is a critical parameter.

The demands for the catalyst inside the MR are high activity, stability and selectivity. The deactivation of the catalyst due to coke formation during hydrocarbon decomposition has been found to have an effect on the catalyst performance. Other mechanisms for catalyst deactivation in the case of membranes are poisoning and sintering. Poisoning can be avoided by using impurity-free fuels. The chemical processes at HTs may cause sintering, and therefore the design of catalysts for lower temperatures is important.

Studies to find new catalysts for MRs are relatively few. Recently, researchers have been concentrating on work with CNTs, nanostructured carbons such as carbon black, and AAO as supports decorated with PGMs, such as platinum (Pt) or palladium (Pd). The aim of new catalyst materials is to reduce process costs and to improve material and energy efficiency. A successful approach in chemical manufacturing seems to be the intensification of processes via integration of several catalytic steps and downstream processes together, for example, CMRs. It can well be concluded that catalysis plays a key role as a driver for sustainability though improving the quality of life and protecting human health and the environment. This approach will also enhance the use of CMRs and catalytic membranes in chemical and related industries. To find new sustainable catalytic processes for industrial or academic purposes in the future is very important.

11.8 References

Abashar M E E (2002) Integrated catalytic membrane reactors for decomposition of ammonia. *Chem Eng Proc*, **41**, 403–412.

Anastas P T, Williamson T C (Eds) (1998) *Green Chemistry; Frontiers in Benign Chemical Syntheses and Processes*. Oxford University Press, Inc, New York, USA, 364 p.

Anastas P T, Warner J C (1998) *Green Chemistry, Theory and Practice*. Oxford University Press, Inc., New York, USA, 135 p.

Anastas P T, Zimmerman J B (2003) Design through the 12 principles green engineering. *Env Sci Techn*, **37**, 94A–101A.

Anonymous (1993) Membrane catalysis and the J Platinum Group Metals, a review of two recent publications, *Platinum Metals Rev*, **37**(3), 153–158.

Babita K, Sridhar S, Raghavan K V (2011) Review: membrane reactors for fuel cell quality hydrogen though WGSR – review of their status, challenges and opportunities. *Int J Hydrogen Energ*, **36**, 6671–6688.

Ballivet-Tkatchenko D, Picquet M, Solinas M, Franciò G, Wasserscheid P, Leitner W (2003) Acrylate dimerisation under ionic liquid–supercritical carbon dioxide conditions. *Green Chem*, **5**, 232–235.

Basile A, Tereschenko G F, Orekhova N V, Ermilova M M, Gallucci F, Iulianelli A (2006) An experimental investigation on methanol steam reforming with oxygen addition in a flat Pd–Ag membrane reactor, *Int J Hydrogen Energ*, **31**, 1615–1622.

Basile A, Gallucci F, Tosti S (2008a) Synthesis, characterization, and applications of palladium membranes, In: Mallada R and Menéndez M (Eds), *Membrane Science and Technology*, Vol. 13, Elsevier, Amsterdam, pp. 255–323

Basile A, Parmaliana A, Tosti S, Iulianelli A, Gallucci F, Espro C, Spooren J, (2008b) Hydrogen production by methanol steam reforming carried out in membrane reactor on Cu/Zn/Mg-based catalyst. *Catal Today*, **137**, 17–22.

Basile A, Gallucci F, Iulianelli A, Borgognoni F, Tosti S (2008c) Acetic acid steam reforming in a Pd–Ag membrane reactor: The effect of the catalytic bed pattern, *J Membrane Sci*, **311**, 46–52.

Basile A, Pinacci P, Iulianelli A, Broglia M, Drago F, Liguori S, Longo T, Calabrò V (2011) Ethanol steam reforming reaction in a porous stainless steel supported palladium membrane reactor. *Int. J. Hydrogen Energ*, **36**, 2029–2037.

Bartholomew C (2003) *Catalyst Deactivation and Regeneration. Kirk-Othmer Encyclopedia of Chemical Technology*. John Wiley & Sons, Inc. New York.

Battersby S, Teixeira P W, Beltramini J, Duke M C, Rudolph V, Diniz da Costa J C (2006) An analysis of the Peclet and Damköhler numbers for dehydrogenation reactions using molecular sieve silica (MSS) membrane reactors. *Catal Today*, **116**, 12–17.

Caravella A, Maio F P D, Renzo A D (2008) Optimization of membrane area and catalyst distribution in a permeative-stage membrane reactor for methane steam reforming. *J Membrane Sci*, **321**, 209–221

Centi G, Perathoner S (2008) Catalysis, a driver for sustainability and societal challenges. *Catal. Today*, **138**, 69–76.

Chen Y, Wang Y, Xu H, Xiong G (2008) Efficient production of hydrogen from natural gas steam reforming in palladium membrane reactor. *Appl Catal B: Env*, **80**, 283–294.

Chen Y, Xu H, Wang Y, Xiong G (2006) Application of Coprecipitated Nickel Catalyst to Steam Reforming of Higher Hydrocarbons in Membrane Reactor. *Chinese J Catal*, **27**(9), 772–777.

Chen Z, Grace J R, Lim C J, Li A (2007b) Experimental studies of pure hydrogen production in a commercialized fluidized-bed membrane reactor with SMR and ATR catalysts. *Int J Hydrogen Energ*, **32**, 2359–2366.

Cheng H, Chen L, Cooper A C, Sha X, Pez G P (2008), Hydrogen spillover in the context of hydrogen storage using solid-state materials. *Energy Environ Sci*, **1**, 338–354

Cheng H, Lu X, Liu X, Zhang Y, Ding W (2009a) Partial oxidation of simulated hot coke oven gas to syngas over Ru-Ni/Mg(Al)O catalyst in a ceramic membrane reactor. *J Natural Gas Chem*, **18**, 467–473.

Cheng Y S, Peña M A, Yeung K L (2009b) Hydrogen production from partial oxidation of methane in a membrane reactor. *J Taiwan Inst Chem Eng*, **40**, 281–288.

Chiappetta G, Barbieri G, Drioli E (2010) Pd/Ag-based membrane reactors on small scale: Assessment of the feed pressure and design parameters effect on the performance. *Chem Eng Proc: Proc Intensific*, **49**, 722–731.

Christensen K O, Chen D, Lødeng R, Holmen A (2006) Effect of supports and Ni crystal size on carbon formation and sintering during steam methane reforming. *Appl Catals A: Gen* **314**, 9–22.

Coronel L, Múnera J F, Lombardo E A, Cornaglia L M (2011) Pd based membrane reactor for ultra pure hydrogen production through the dry reforming of methane. Experimental and modeling studies. *App Catal A: Gen*, **400**, 185–194.

Deshmukh S A R K, Heinrich S, Mörl L, van Sint Annaland M, Kuipers J A M (2007) Membrane assisted fluidized bed reactors: potential and hurdles. *Chem Eng Sci*, **62**, 416–436.

Dömök M, Tóth M, Raskó J, Erdohelyi A (2007) Adsorption and reactions of ethanol and ethanol–water mixture on alumina-supported Pt catalysts. *Appl Catal B: Env*, **69**, 262–272.

Dudukovic M P (1999) Trends in catalytic reaction engineering. *Catal Today*, **48**, 5–15.

van Dyk L, Miachon S, Lorenzen S, Torres M, Fiaty K, Dalmon J-A (2003) Comparison of microporous MFI and dense Pd membrane performance in an extractor-type CMR. *Catal Today*, **82**, 167–177.

Faroldi B M, Lombardo E A, Cornaglia L M (2011) Ru/La_2O_3–SiO_2 catalysts for hydrogen production in membrane reactors. *Catal Today*, **172**, 209–217.

Ferreira-Aparicio P, Benito M, Kouachi K, Menad S (2005) Catalysis in membrane reformers: a high-performance catalytic system for hydrogen production from methane. *J Catal*, **231**, 331–343.

Fu C-H, Wu J C S (2007) Mathematical simulation of hydrogen production via methanol steam reforming using double-jacketed membrane reactor. *Int J Hydrogen Energ*, **32**, 4830–4839.

Gallucci F, van Sint Annaland M and Kuipers J A M (2011) Chapter 10, Modeling of membrane reactors for hydrogen production and purification. In: Drioli E and Barbieri G (Eds), *Membrane Engineering for theTtreatment of Gases, Vol. 2: Gas-Separation Problems Combined with Membrane Reactors*. Royal Society of Chemistry, Cambridge, UK, pp. 1–39.

Gallucci F, Basile A, Tosti S, Iulianelli A, Drioli E (2007) Methanol and ethanol steam reforming in membrane reactors: An experimental study. *Int J Hydrogen Energ*, **32**, 1201–1210.

Galuszka J, Pandey R N, Ahmed S (1998) Methane conversion to syngas in a palladium membrane reactor. *Catal Today*, **46**, 83–89.

Grünewald M, Agar D W (2004) Enhanced catalyst performance using integrated structured functionalities. *Chem Eng Sci*, **59**, 5519–5526.

Halonen N, Rautio A, Leino A-R, Kyllönen T, Tóth G, Lappalainen J, Kordás K, Huuhtanen M, Keiski R L, Sápi A, Szabo M, Kukovecz Á, Kónya Z, Kiricsi I, Ajayan P M, Vajtai, R (2010) Three-dimensional carbon nanotube scaffolds as particulate filters and catalyst support membranes. *ACS Nano*, **4**, 2003–2008

Irusta S, Munera J, Carrara C, Lombardo E A, Cornaglia L M (2005) A stable, novel catalyst improves hydrogen production in a membrane reactor. *Appl Catal A: Gen*, **287**, 147–158.

Israni S H, Harold M P (2011) Methanol steam reforming in single-fiber packed bed Pd-Ag membrane reactor: Experiments and modeling. *J Membrane Sci*, **369**, 375–387.

Iulianelli A, Longo T, Basile A (2008a) Methanol steam reforming reaction in a Pd–Ag membrane reactor for CO-free hydrogen production. *Int J Hydrogen Energ*, **33**, 20, 5583–5588.

Iulianelli A, Longo T, Basile A (2008b) CO-free hydrogen production by steam reforming of acetic acid carried out in a Pd–Ag membrane reactor: The effect of co-current and counter-current mode. *Int J Hydrogen Energ*, **33**, 4091–4096.

Iulianelli A, Longo T, Liguori S, Seelam P K, Keiski R L, Basile A (2009) Oxidative steam reforming of ethanol over Ru–Al_2O_3 catalyst in a dense Pd–Ag membrane reactor to produce hydrogen for PEM fuel cells. *Int J Hydrogen Energ*, **34**(20), 8558–8565.

Iulianelli A, Manzolini G, De Falco M, Campanari S, Longo T, Liguori S, Basile A (2010) H_2 production by low pressure methane steam reforming in a Pd/Ag membrane reactor over a Ni-based catalyst: Experimental and modelling. *Int J Hydrogen Energ*, **35**, 11514–11524.

Iulianelli A, Liguori S, Longo T, Tosti S, Pinacci P, Basile A (2010) An experimental study on bio-ethanol steam reforming in a catalytic membrane reactor. Part II: Reaction pressure, sweep factor and WHSV effects. *Int J Hydrogen Energ*, **35**, 3159–3164.

Iulianelli A, Seelam P K, Liguori S, Longo T, Keiski R, Calabrò V, Basile A (2011) Hydrogen production for PEM fuel cell by gas phase reforming of glycerol as byproduct of bio-diesel. The use of a Pd–Ag membrane reactor at middle reaction temperature. *Int J Hydrogen Energ*, **36**, 3827–3834.

Jin W, Gu X, Li S, Huang P, Xu N, Shi J (2000a) Experimental and simulation study on a catalyst packed tubular dense membrane reactor for partial oxidation of methane to syngas. *Chem Eng Sci*, **55**, 2617–2625.

Jin W, Li S, Huang P, Xu N, Shi J, Lin Y S (2000b) Tubular lanthanum cobaltite perovskite-type membrane reactors for partial oxidation of methane to syngas. *J Membrane Sci*, **166**, 13–22.

Job N, Berthon-Fabry S, Chatenet M, Marie J, Brigaudet M, Pirard J-P (2009) Nanostructured carbons as platinum catalyst support for proton exchange membrane fuel cell electrode, *Topics Catal*, **52**, 2117–2122.

Keiski R L, Kolli T, Huuhtanen M, Ojala S, Pongrácz E. (2010) Chapter 7. Challenges in catalysis for sustainability. In: Cybulski, Moulijn and Stankiewi cz (Eds) *Novel Concepts in Catalysis and Chemical Reactors, Improving the Efficiency for the Future*. Wiley-VCH Verlag GmbH & Co. KGaAP, Weinheim. pp. 143–162.

Kordás K, Tóth G, Levoska J, Huuhtanen M, Keiski R, Härkönen M, George T G, Vähäkangas J (2006) Room temperature chemical deposition of palladium nanoparticles in anodic aluminium oxide templates. *Nanotechnology*, **17**, 1459–1463.

Lakhapatri S L, Abraham M A (2011) Analysis of catalyst deactivation during steam reforming of jet fuel on Ni-(PdRh)/γ-Al_2O_3 catalyst. *Appl Catal A Gen*, **405**, 149–159.

Li G, Kanezashi M, Tsuru T (2011) Highly enhanced ammonia decomposition in a bimodal catalytic membrane reactor for COx-free hydrogen production, *Catal Comm*, **15**, 60–63.

Li X, Zhang Y, Meng M, Yang G, San X, Takahashi M, Tsubaki N (2010) Silicalite-1 membrane encapsulated Rh/activated-carbon catalyst for hydroformulation of 1-hexene with high selectivity to normal aldehyde. *J Membrane Sci*, **347**, 220–227.

Ligthart D A J M, Pieterse J A Z, Hensen E J M (2011) The role of promoters for Ni catalysts in low temperature (membrane) steam methane reforming. *Appl Catal A Gen*, **405**, 108–119.

Lim H, Gu Y, Oyama S T (2010) Reaction of primary and secondary products in a membrane reactor: studies of ethanol steam reforming with a silica–alumina composite membrane. *J Membrane Sci*, **351**, 149–159.

Lin Y-M, Rei M-H (2001) Study on the hydrogen production from methanol steam reforming in supported palladium membrane reactor. *Catal Today*, **67**, 77–84.

Lin W-H, Liu Y-C, Chang H-F (2008) Hydrogen production from oxidative steam reforming of ethanol in a palladium–silver alloy composite membrane reactor. *J Chinese Inst Chem Eng*, **39**(5), 435–440.

Liu W (2007) Multi-scale catalyst design. *Chem Eng Sci*, **62**, 3502–3512.

Lozano P, Garcia-Verdugo E, Luis S V, Pucheault M, Vaultier M (2011) (Bio)Catalytic continuous flow processes in $scCO_2$ and/or ILs: towards sustainable (Bio)catalytic synthetic platforms. *Curr Organic Synthesis*, **14**, 810–823.

Matsumura Y, Tong J (2008) Methane steam reforming in hydrogen-permeable membrane reactor for pure hydrogen production, *Topics Catal*, **51**(1–4), 123–132.

Miachon S, Dalmon J-A (2004) Catalysis in membrane reactors: what about the catalyst? *Topics Catal*, **29**(1–2), 59–65.

Moon W S, Park S B (2000) Design guide of a membrane for a membrane reactor in terms of permeability and selectivity. *J Membrane Sci*, **170**, 43–51.

Morbidelli M, Gavrilidis A, Varma A (Eds) (2001) *Catalyst Design: Optimal Distribution of Catalyst in Pellets, Reactors, and Membranes*. Cambridge University Press, Cambridge, UK, Chapter 5, pp. 95–98.

Múnera J, Irusta S, Cornaglia L, Lombardo E (2003) CO_2 reforming of methane as a source of hydrogen using a membrane reactor. *Appl Catal A: Gen*, **245**, 383–395.

Ostrowski T, Giroir-Fendler A, Mirodatos C, Mleczko L (1998) Comparative study of the partial oxidation of methane to synthesis gas in fixed-bed and fluidized-bed membrane reactors. Part II: Development of membranes and catalytic measurements. *Catal Today*, **40**, 191–200.

Pan X, Bao X (2008) Reactions over catalysts confined in carbon nanotubes. *Chem Comm*, 6271–6281.

Papadias D D, Lee S H D, Ferrandon M, Ahmed S (2010) An analytical and experimental investigation of high-pressure catalytic steam reforming of ethanol in a hydrogen selective membrane reactor. *Int J Hydrogen Energ*, **35**, 2004–2017.

Patil C S, van Sint Annaland M, Kuipers J A M (2007) Fluidised bed membrane reactor for ultrapure hydrogen production via methane steam reforming: experimental demonstration and model validation. *Chem Eng Sci*, **62**, 2989–3007.

Pieterse J A Z, Boon J, van Delft Y C, Dijkstra J W, van den Brink R W (2010) On the potential of nickel catalysts for steam reforming in membrane reactors. *Catal Today*, **156**, 153–164.

Prabhu A K, Radhakrishnan R, Oyama S T (1999) Supported nickel catalysts for carbon dioxide reforming of methane in plug flow and membrane reactors. *Appl Catal A: Gen*, **183**, 241–252.

Rahimpour M R (2009) Enhancement of hydrogen production in a novel fluidized-bed membrane reactor for naphtha reforming. *Int J Hydrogen Energ*, **34**, 2235–2251.

Rakib M A, Grace J R, Lim C J, Elnashaie S S E H B, Ghiasi B (2010) Steam reforming of propane in a fluidized bed membrane reactor for hydrogen production. *Int J Hydrogen Energ*, **35**, 6276–6290.

Rei M-H, Yeh G-T, Tsai Y-H, Kao Y-L, Shiau L-D (2011) Catalysis-spillover-membrane. III: The effect of hydrogen spillover on the palladium membrane reactor in the steam reforming reactions. *J Membrane Sci*, **369**, 299–307.

Reichardt C, Welton T (Ed.) (2011) *Solvents and Solvent Effects in Organic Chemistry*. Wiley-VCH Verlag & Co, KGaA, Weinheim, 718 pp.

Reitz T L, Ahmed S, Krumpelt M, Kumar R, Kung H H (2000) Characterization of CuO/ZnO under oxidizing conditions for the oxidative methanol reforming reaction. *J Mol Catal A: Chem*, **162**, 275–285.

Richardson J T (1989) *Principles of Catalyst Development*. New York, Plenum Press, 288 pp.

Ryi S-K, Park J-S, Kim D-K, Kim T-H, Kim S-H (2009) Methane steam reforming with a novel catalytic nickel membrane for effective hydrogen production. *J Membrane Sci*, **339**(1–2), 189–194.

Sá S, Sousa J M, Mendes A (2011) Steam reforming of methanol over a CuO/ZnO/Al_2O_3 catalyst part II: A carbon membrane reactor. *Chem Eng Sci*, **66**(22), 5523–5530.

Sanchez Marcano J G, Tsotsis Th T (2002) *Catalytic Membranes and Membrane Reactors*. Wiley-VCH Verlag GmbH, Weinheim, 251 p.

Santucci A, Annesini M C, Borgognoni F, Marrelli L, Rega M, Tosti S (2011) Oxidative steam reforming of ethanol over a Pt/Al_2O_3 catalyst in a Pd-based membrane reactor. *Int J Hydrogen Energ*, **36**(2), 1503–1511.

Seelam P K, Huuhtanen M, Sápi A, Szabó M, Kordás K, Turpeinen E, Keiski R L (2010), CNT-based catalysts for H_2 production by ethanol reforming. *Int J Hydrogen Energ*, **35**, 12588–12595.

Seelam P K, Liguori S, Iulianelli A, Pinacci P, Calabrò V, Huuhtanen M, Keiski R, Piemonte V, Tosti S, De Falco M, Basile A (2012), Hydrogen production from bio-ethanol steam reforming reaction in a Pd/PSS membrane reactor, *Catal Today,* **193**(1), 42–48. doi:0.1016/j.cattod.2012.01.008 (in press).

Sheldon R A (2008) E factors, green chemistry and catalysis: an odyssey. *Chem Commun*, 3352–3365.

Somorjai G A (2004), On the move. *Nature*, **430**, 730.

Somorjai G A, Rioux R M (2005) High technology catalysts towards 100% selectivity: Fabrication, characterization and reaction studies. *Catal Today*, **100**, 201–215.

Somorjai G A, Park J Y (2007a) Frontiers of surface science. *Physics Today*, **60**(10), 48–53.

Somorjai G A, Park Y A (2007b) The impact of surface science on the commercialization of chemical processes. *Catal Lett*, **115**(3–4), 87–98.

Specchia S, Fino D, Saracco G, Specchia V (2006) Inorganic membrane reactors. In: Cybulski A, Moulijn J A (Eds) *Structured Catalysts and Reactors*, 2nd ed. CRC Press, Boca Raton, FL, USA, Chapter 18, pp. 621–622.

Stair PC, Marshall C, Xiong G, Feng H, Pellin M J, Elam J W, Curtiss L, Iton L, Kung H, Kung M, Wang H-H (2006) Novel, uniform nanostructured catalytic membranes. *Topics Catal*, **39**, 181–186.

Szegner J, Yeung K L, Varma A (1997) Effect of catalyst distribution in a membrane reactor: Experiments and model. *AIChE J*, **43**, 2059–2072.

Tong J, Matsumura Y (2005) Effect of catalytic activity on methane steam reforming in hydrogen-permeable membrane reactor *Appl Catal A: Gen* **286**, 226–231.

Tong J, Matsumura Y (2006) Pure hydrogen production by methane steam reforming with hydrogen-permeable membrane reactor. *Catal Today*, **111**, 147–152.

Tosti S, Basile A, Borgognoni F, Capaldo V, Cordiner S, Di Cave S, Gallucci F, Rizzello C, Santucci A, Traversa E (2008a) Low temperature ethanol steam reforming in a Pd-Ag membrane reactor: Part 1: Ru-based catalyst. *J Membrane Sci*, **308**(1–2), 250–257.

Tosti S, Basile A, Borgognoni F, Capaldo V, Cordiner S, Di Cave S, Gallucci F, Rizzello C, Santucci A, Traversa E (2008b) Low temperature ethanol steam reforming in a Pd-Ag membrane reactor: Part 2. Pt-based and Ni-based catalysts and general comparison. *J Membrane Sci*, **308**, 258–263.

Tosti S, Basile A, Borelli R, Borgognoni F, Castelli S, Fabbricino M, Gallucci F, Licusati C (2009), Ethanol steam reforming kinetics of a Pd–Ag membrane reactor. *Int J Hydrogen Energ*, **34**, 4747–4754.

Trimm D L (1997) Coke formation and minimization during steam reforming. *Catal Today*, **37**, 233–238.

Trimm D L (1999) Catalysts for the control of coking during steam reforming. *Catal Today*, **49**, 3–10.

Twigg M V, Spencer M S (2001) Deactivation of supported copper metal catalysts for hydrogenation reactions. *Appl Catal A: Gen*, **212**, 161–174.

Welton T (2008) Is catalysis in ionic liquids a potentially green technology? *Green Chem*, **10**, 483.

Wilhite B A, Weiss S E, Ying J Y, Schmidt M A, Jensen K F (2006), High-purity hydrogen generation in a microfabricated 23 wt% Ag–Pd membrane device integrated with 8:1 $LaNi_{0.95}Co_{0.05}O_3/Al_2O_3$ catalyst. *Adv Mater*, **18**, 1701–1704.

Yang Z, Zhang Y, Ding W, Zhang Y, Shen P, Zhou Y, Liu Y, Huang S, Lu X (2009) Hydrogen production from coke oven gas over LiNi/γ-Al_2O_3 catalyst modified by rare earth metal oxide in a membrane reactor. *J Natural Gas Chem*, **18**, 407–414.

Yeung K L, Aravind, R, Zawada R J X, Szegner J, Cao G, Varma A (1994), Nonuniform catalyst distribution for inorganic membrane reactors: theoretical considerations and preparation techniques. *Chem Eng Sci*, **49**, 4823–4838.

Yu C-Y, Lee D-W, Park S-J, Lee K-Y, Lee K-H (2009) Study on a catalytic membrane reactor for hydrogen production from ethanol steam reforming. *Int J Hydrogen Energ*, **34**, 2947–2954.

11.9 Appendix: nomenclature

11.9.1 Abbreviations

AAO	anodic aluminium oxide
BCM	bimodal catalytic membrane
BET	Brunauer–Emmett–Teller
CMR	catalytic membrane reactor
CNT	carbon nanotube
ESR	ethanol steam reforming
FBR	fixed-bed reactor
HC	hydrocarbon
GDL	gas diffusion layer
GHSV	gas-hourly-space-velocity
HT	high temperature
IL	ionic liquid
IMRCF	inert membrane reactor with a catalyst on the feed side
MeSR	methane steam reforming
MSR	methanol steam reforming
MR	membrane reactor
OSC	oxygen storage compound
PBR	packed-bed reactor
PBMR	packed-bed membrane reactor
PEM	proton exchange membrane
PGM	precious group metal
POx	partial oxidation
PSS	porous stainless steel
PVD	physical vapour deposition method
$scCO_2$	supercritical carbon dioxide
SS	stainless steel
SSP	self-supported Pd-based membranes
SR	steam reforming
SV	space velocity
WGS	water-gas shift
WHSV	weight-hourly-space-velocity

Part III

Membrane reactor modelling, simulation and optimisation

12

Mathematical modelling of membrane reactors: overview of strategies and applications for the modelling of a hydrogen-selective membrane reactor

M. DE FALCO, University of Rome 'Campus Bio-Medico', Italy and A. BASILE, ITM-CNR, Italy

DOI: 10.1533/9780857097330.3.435

Abstract: The present chapter deals with the main issues for membrane reactor modelling. Reactor modelling strategies, together with a classification of the main mathematical model categories, are presented and discussed. Then the membrane reactor concept is illustrated and, specifically, dense hydrogen-selective membranes are presented. As an example, a natural gas steam reforming Pd-based membrane reactor is completely modelled. Model equations, derived from mass, energy and momentum balances, are reported and explained together with boundary conditions.

Key words: membrane reactors, reactor modelling, selective membrane, natural gas steam reforming, hydrogen production, Pd-based membrane.

12.1 Introduction

Modelling of reactors is a crucial topic in process design. Throughout the ages, researchers have understood the importance of modelling and have tackled the challenge of developing more and more accurate algorithms. The development of software and hardware has increased the computational capacity of mathematical modelling and nowadays the behaviour of reactors and all chemical and physical phenomena occurring inside the reaction environment can be deeply analysed.

Practically, reactor modelling can be used to:

1. Improve understanding of both the process and the reaction. Computer simulation allows useful information on dynamic and steady-state reactor and process behaviour to be obtained even before the reactor is fabricated.
2. Optimize reactor operating conditions. By developing a reliable mathematical model, designers have the availability of a tool that can simulate the behaviour of the reactor, change operating conditions without the need of carrying out a long and expensive experimental phase.

Here, two model classifications are presented. Firstly, models can be classified based on how they are obtained:

- Theoretical models are developed from chemistry and physics principles.
- Empirical models are obtained by fitting experimental data.
- Semi-empirical models are a combination of theoretical and empirical models. They are theoretical models in which one or more empirical parameters are calculated from experimental data.

Theoretical models have two main benefits: they are applicable within wide ranges of operating conditions, and they offer a physical insight into process behaviour. On the other hand, they are complex to develop, time-consuming and require suitable hardware.

Empirical models are much simpler but they have a serious drawback, that is, usually they are valid only within the range in which experimental data are obtained. In fact, the lack of physical and chemical bases disallows extrapolating data and predicting reactor behaviour if operating conditions change.

Certainly, most reactor models are semi-empirical. Starting from strong physical and chemical bases, model equations are obtained and, typically, an ordinary differential equations (ODE) or partial differential equations (PDE) set has to be solved. In an equation set, parameters usually need to be fitted (for example reaction-kinetics rate coefficients, catalyst adsorption coefficients, heat-transfer coefficients, etc.), and they are calculated using experimental data. Semi-empirical models have three important advantages (Seborg *et al.*, 1989):

1. they incorporate theoretical knowledge;
2. they can be extrapolated over a wider range of operating conditions than empirical models;
3. they require less effort than theoretical models.

In the next section, the equation structure is described for theoretical and semi-empirical of membrane reactors.

Another mathematical model classification is proposed by Froment and Bischoff (1990) and is based on the level of complexity, and consequently of accuracy.

A heterogeneous catalytic reactor, composed, for example, of a solid catalyst and a fluid feedstock, has been described by (De Falco *et al.*, 2011a):

- a heterogeneous model. Fluid and solid phases are modelled separately, imposing balance equations for each phase. Mass and heat fluxes between the solid and fluid phases are expressed in terms of particle-to-fluid mass and heat transport coefficients.

Table 12.1 Pseudo-homogeneous and heterogeneous reactor model classification

Pseudo-homogeneous models	One-dimensional	Ideal
		+ axial mixing
	Two-dimensional	+ radial mixing
Heterogeneous models	One-dimensional	Interfacial gradients
		+ intra-particle gradients
	Two-dimensional	+ radial mixing

- a pseudo-homogeneous model. Fluid and solid phases are considered as a single pseudo-phase and the balances are imposed for only one phase. Heat and mass transport coefficients inside the bed are calculated by expressions which account for the simultaneous presence of two phases.

For each category, models can be classified in order of their growing complexity, as reported in Table 12.1 (Froment and Bischoff, 1990). For a presentation of each model type, refer to De Falco *et al.* (2010a, 2011a).

The simplest model is pseudo-homogeneous, one-dimensional and ideal, where solid and gas phases are analysed as a one single phase and no axial and radial mixing is taken into account. Obviously, this type of model is very easily implemented but, on the other hand, it can only give a first approximation assessment.

The most complex model is heterogeneous, with both axial and radial mixing, intra-particle and interfacial gradients. Such models give extremely accurate results, but they are much more complex to formulate and implement and, usually, such a high accuracy grade is not required for industrial process assessments.

It is clear that a rigorous formulation of all phenomena occurring inside a reaction environment leads to a level of complexity that is difficult to manage. The skill of reactor designers is mainly in introducing proper assumptions, reducing formulation complexity, but at the same time not making unacceptable approximations.

12.2 Membrane reactor concept and modelling

A membrane reactor (MR) is a system combining reaction and separation of one or more products, with the separation operation performed by a selective membrane.

The MR concept is based on the removal of a reaction product in order to avoid the reaction equilibrium conditions to be achieved and promoting reaction kinetics. Most of industrial chemical reactions are reversible

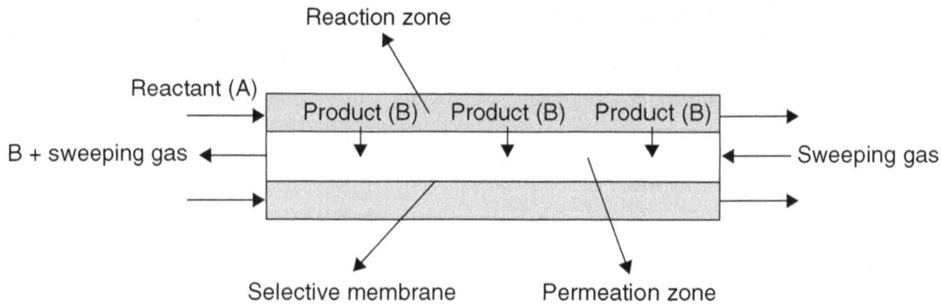

12.1 IMR layout.

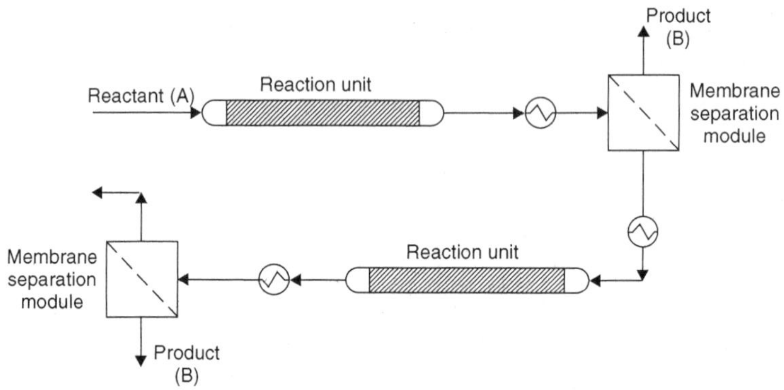

12.2 SMR (two stages) layout.

reactions, thermodynamically limited since equilibrium conditions cannot be overcome in the reacting mixture. From a thermodynamic point of view, equilibrium is represented by a constraint, that is, the equilibrium constant, on mole concentrations, temperature and pressure, derived from the Second Law of Thermodynamics. When equilibrium conditions are achieved, no net change in state variables is observed. From a kinetics point of view, it means that at equilibrium the reaction rate of the direct reaction is equal to the reaction rate of the inverse reaction.

An MR can be designed in one of the following two configurations:

- Integrated membrane reactor (IMR), where the selective membrane is integrated directly in the reaction environment and the reaction product is removed as it is produced. A sweeping gas is fed counter-currently (as shown in Fig. 12.1) or co-currently to carry out the product permeated through the membrane.
- Staged membrane reactor (SMR) or reactor and membrane module (RMM), where the selective membrane is placed outside the reactor, in a proper unit located downstream (Fig. 12.2).

The following two figures show both the configurations (De Falco *et al.*, 2011b), imposing an elementary reaction:

$$A \Leftrightarrow B$$

In the present chapter, only the IMR shown in Fig. 12.1 is modelled. The SMR modelling is simply composed of the series of a packed bed reactor model (equations described in Section 12.2.2) followed by a membrane separation unit, where only the hydrogen stream permeated through the selective membrane (Equation [12.8]) has to be evaluated.

12.2.1 MR modelling strategies

An IMR is divided into two zones:

- the reaction zone, where a catalyst to promote the reactions is packed;
- the permeation zone, where the product permeated through the membrane is collected and carried out, by a sweeping gas or by reducing the pressure under atmospheric.

The catalyst can be packed in the annular zone, as shown in Fig. 12.1, or in the inner tube, depending on the type of reaction and on the process requirements.

Modelling an IMR involves integrating a catalytic reactor-zone model and a permeation-zone model. The two models are then linked by boundary conditions.

With respect to reaction-zone models, all types listed in Table 12.1 can be implemented. In the following, for sake of simplicity, only pseudo-homogeneous models are presented. IMR models can be one- or two-dimensional. Usually, axial mixing phenomena are neglected and therefore two model types can be implemented to describe IMR behaviour:

1. ideal model (1D);
2. ideal + radial mixing model (2D).

Catalytic bed reactors are usually in turbulent conditions, making the 1D model absolutely reliable for a sufficiently deep assessment of reactor performance. On the other hand, the 2D models add information on radial temperature and concentration profiles and, in some cases, are necessary for reactor design. Particularly in membrane reactors, there is the problem of membrane overheating, for example the hydrogen-selective Pd-based dense membrane, which can be integrated in many hydrogen production processes (Dittmeyer *et al.*, 2001; Howard *et al.*, 2004; Peachey, *et al.* 1996;

Tosti *et al.*, 2002). Due to stability-related problems, it generally has to operate at a temperature below 500°C (Bredesen, 2008). Considering that in a catalytic reactor the temperature profiles are usually steep (De Falco *et al.*, 2008), if a 2D model is implemented, the reactor zone temperature is calculated in every point inside the reactor, and both axial and radial profiles are available, giving the reactor designer a much more reliable assessment of the membrane temperature profile.

The permeation zone model consists of a simple fluid-dynamics tube model, and a 1D model can be used. The energy and mass boundary conditions enable connection between the two models, as described in Section 12.2.4.

12.2.2 Reaction-zone models

As reported above, the reaction-zone model can be two-fold: an ideal model or a radial mixing model.

The 1D ideal model is a plug-flow model, which assumes that concentrations, temperature and pressure vary only in the axial direction. Moreover, no axial mixing is considered; therefore the only transport mechanism is the axial convective flux.

Mass, energy and momentum balances for an elementary reaction $A \Leftrightarrow B$ are:

$$\frac{d\tilde{C}_A}{d\tilde{z}} = -\frac{\rho_B \times L}{u_s \times C_{A,\text{in}}} r_A \qquad [12.1]$$

$$\frac{d\tilde{T}}{d\tilde{z}} = -\frac{L}{u_s \times \rho_g \times c_p \times T_{\text{in}}}\left[(-\Delta H)\rho_B r_A - 4\frac{U}{d_t}(T - T_r)\right] \qquad [12.2]$$

$$\frac{d\tilde{P}}{d\tilde{z}} = \frac{f \times G \times \mu_g \times L}{\rho_g d_p^{\,2} \times P_{\text{in}}} \times \frac{(1-\varepsilon)^2}{\varepsilon^3} \qquad [12.3]$$

where $\tilde{z}$, $\tilde{C}_A$, $\tilde{T}$ and $\tilde{P}$ are the dimensionless axial coordinate, A-component concentration, reactor temperature and pressure, respectively; $C_{A,\text{in}}$, T_{in} and P_{in} are the inlet concentration, temperature and pressure, respectively; u_s is the gas superficial velocity; ρ_B and ρ_g are the catalytic bed and gas density, respectively; r_A is the A-component reaction rate; c_p is the gas mixture specific heat; $(-\Delta H)$ is the heat of reaction; d_t and d_p are the internal tubular reactor diameter and the equivalent particle diameter, respectively; L is the reactor length; U is the overall heat transport coefficient between the external energy source and the reaction bed; T_r is the temperature of energy source; f is the friction factor; is superficial mass flow velocity; μ_g is the gas mixture viscosity and ε is the void fraction of the packed bed.

The overall heat transport coefficient U is calculated as the sum of different heat resistances in a series. For a packed bed reactor, the heat transport resistances are:

- the external heat source;
- the reactor tube wall;
- the first layer of gas mixture, in which the heat transport only occurs by molecular conduction (Tsotsas and Schlünder, 1990);
- the pseudo-homogeneous phase (gas + solid phases).

The expression for the calculation of U is:

$$U = \left(\frac{1}{h_{\text{ex}}} \cdot \frac{A_i}{A_o} + \frac{t}{k_{\text{met}}} \cdot \frac{A_i}{A_m} + \frac{1}{h_W} + \frac{\text{d}_t}{8\lambda_{\text{er}}} \right)^{-1} \qquad [12.4]$$

where h_{ex}, k_{met}, h_W and λ_{er} are the heat transport coefficients in the external energy source side, the tube wall conductivity, the heat transport coefficient in the first layer near the tube wall (Dixon and Cresswell, 1979; Li and Finlayson, 1977) and effective radial thermal conductivity of the pseudo-homogeneous phase (De Wasch and Froment, 1972; Elnashaie and Elshishini, 1993), respectively; while A_i, A_o and A_m are the bed size heat-exchanging surface, the heating fluid media heat-exchanging surface and the log mean of them, respectively.

If a 2D mode is considered, the radial mixing phenomenon has to be added.

Heat and mass diffusive fluxes in radial direction are superimposed on the convective axial transport, leading to the following equations:

$$\frac{4 \times \varepsilon \times D_{\text{er}}}{\text{d}_t^2} \left(\frac{\partial^2 \tilde{C}_A}{\partial \tilde{r}^2} + \frac{1}{\tilde{r}} \frac{\partial \tilde{C}_A}{\partial \tilde{r}} \right) - \frac{u_s}{L} \frac{\text{d}\tilde{C}_A}{\text{d}\tilde{z}} = \frac{\rho_B r_A}{C_{A,\text{in}}} \qquad [12.5]$$

$$-\frac{4 \cdot \lambda_{\text{er}}}{\text{d}_t^2} \left(\frac{\partial^2 \tilde{T}}{\partial \tilde{r}^2} + \frac{1}{\tilde{r}} \frac{\partial \tilde{T}}{\partial \tilde{r}} \right) - \frac{u_s \rho_g c_p}{L} \frac{\text{d}\tilde{T}}{\text{d}\tilde{z}} = \frac{(-\Delta \text{H}) \rho_B r_A}{T_{\text{in}}} \qquad [12.6]$$

where the effective radial mass diffusivity D_{er} should be calculated by evaluating experimentally the radial mass Peclet number: $Pe_{\text{mr}} = (u_s d_p / D_{\text{er}})$.

But, assuming that the Reynolds number $\left(\mathrm{Re}_p = G\mathrm{d}_p / \mu_g\right)$ is greater than 1000, which is probable in industrial conditions, a simplification could be made since Pe_{mr} reaches a constant value depending only on geometric features (Kulkarni and Doraiswamy, 1980):

$$Pe_{\mathrm{mr}} = 8.8 \times \left(2 - \left(1 - \frac{2 \times d_p}{d_t}\right)^2\right) \quad [12.7]$$

12.2.3 Permeation zone model

The permeation zone model consists of a simple fluid-dynamics tube model and a 1D model can be always implemented since radial profile is usually of no interest in reactor design. The mass balance is expressed as follows:

$$\frac{\mathrm{d}Y_B}{\mathrm{d}\tilde{z}} = \pm \frac{J_B 2 \cdot r_{o,i} \cdot L}{u_s c_{A,\mathrm{in}} \cdot \left(r_{i,\mathrm{o}}^2 - r_{\mathrm{o},i}^2\right)} \quad [12.8]$$

where $r_{\mathrm{o},i}$ is the external radius of the inner tube, $r_{\mathrm{i},o}$ the internal radius of outer tube, J_B is B flux through the membrane, calculated depending on the type of membrane and permeation mechanism and

$$Y_{\mathrm{H}_2} = \frac{F_{B,\mathrm{perm}}}{F_{A,\mathrm{in}}} \quad [12.9]$$

with $F_{B,\mathrm{perm}}$ and $F_{A,\mathrm{in}}$ equal to the permeated B flow and inlet A flow-rate, respectively. The sign + or − depends on co-current (+) or counter-current (−) configurations.

The energy balance is:

$$\frac{\mathrm{d}\tilde{T}_P}{\mathrm{d}\tilde{z}} = \frac{L}{F_{\mathrm{TOT},p} \times c_{p,p} \times T_{\mathrm{in}}} \times \left[U_1 \times 2\pi \times r_{i,i} \times T_{\mathrm{in}} \times \left(\tilde{T} - \tilde{T}_P\right) + J_B 2\pi \times r_{o,i} \times (h_{B,r} - h_{B,p})\right] \quad [12.10]$$

where $F_{\mathrm{TOT},p}$ is the total permeation zone flow-rate; $c_{p,p}$ is the specific heat of permeation zone fluid mixture; U_1 is the overall heat-transfer coefficient between reaction and permeation zone; $r_{i,i}$ is the internal radius of the inner tube; and $h_{B,r}$ and $h_{B,p}$ the B component specific enthalpy in the reaction zone and permeation zone, respectively.

The term U_1 is calculated by:

$$U_1 = \left[\frac{1}{h_W} + \frac{\delta}{k_{\text{mem}}} + \frac{r_{o,i}}{r_{i,i}} \cdot \frac{1}{h_{\text{perm}}} \right]^{-1} \quad [12.11]$$

where h_W is the heat-transfer coefficient corresponding to the membrane wall; δ and k_{mem} are the thickness and the thermal conductivity (respectively) of the membrane with its support; and h_{perm} is the forced convection heat transport coefficient in the permeation zone given by (turbulence conditions):

$$h_{w,p} = \frac{\lambda_{g,\text{perm}}}{r_{i,i}} \times 0.023 \times \text{Re}_{\text{perm}}^{0.8} \times \text{Pr}_{\text{perm}}^{1/3} \quad [12.12]$$

where Pr_{perm} is the Prandtl number $\left(= c_p \mu_g / k_g\right)$; $r_{i,i}$ is the internal radius of the inner tube; and $\lambda_{g,\text{perm}}$ is the conductivity of gas stream in the permeation zone.

12.2.4 Boundary conditions

If a 1D reaction-zone model is implemented, the boundary conditions have to be imposed only in the inlet section. Boundary conditions for a 1D IMR model are defined as follows:

$$\tilde{z} = 0:$$

$$\tilde{C}_A = 1$$

$$\tilde{T} = 1$$

$$\tilde{P} = 1$$

$$Y_{H_2} = 0 \text{ Co-current configuration}$$

$$\tilde{T}_P = \frac{T_{P,\text{in}}}{T_{\text{in}}} \quad [12.13]$$

The equations set [12.1–12.3, 12.8, 12.10] is an ODEs set, to be solved by a numerical procedure.

If a sweeping gas counter-current configuration is imposed, the boundary conditions on mass and energy in permeation zone have to be imposed in the outlet section:

$$\tilde{z} = 1:$$

$$Y_{H_2} = 0 \text{ Counter-current configuration}$$

$$\tilde{T}_P = \frac{T_{P,\text{in}}}{T_{\text{in}}} \quad [12.14]$$

Consequently, the equation set is a boundary value problem (BVP), which can be solved by a shooting method.

If a 2D reaction-zone model is imposed, two boundary conditions on radius ($\tilde{r} = 0$ and $\tilde{r} = 1$) have to be defined as well:

$$\tilde{z} = 0, \forall \tilde{r}:$$

$$\tilde{C}_A = 1$$

$$\tilde{T} = 1$$

$$\tilde{P} = 1$$

$$Y_{H_2} = 0 \text{ Co-current configuration}$$

$$\tilde{T}_P = \frac{T_{P,\text{in}}}{T_{\text{in}}} \quad [12.15]$$

$$\tilde{z} = 1:$$

$$Y_{H_2} = 0 \text{ Counter-current configuration}$$

$$\tilde{T}_P = \frac{T_{P,\text{in}}}{T_{\text{in}}} \quad [12.16]$$

$\tilde{r} = 1, \forall \tilde{z}$:

$$\frac{\partial(\tilde{C}_A)}{\partial \tilde{r}} = 0$$

$$\lambda_{\text{er}} \frac{\partial \tilde{T}}{\partial \tilde{r}} = \frac{q_r \times r_{i,\text{o}}}{T_{\text{in}}} = \frac{U \times r_{i,\text{o}}}{T_{\text{in}}} \left(T_{w,0} - T_{|r_{i,\text{o}}} \right) \quad [12.17]$$

where q_r is the heat flux from the external source to the reactor packed bed and λ_{er} is the effective radial conductivity of the catalytic bed and the gas stream system.

$\tilde{r} = \frac{r_{\text{o},i}}{r_{i,\text{o}}}, \forall \tilde{z}$ (boundary conditions on selective membrane):

$$\frac{\partial(\tilde{C}_A)}{\partial \tilde{r}} = 0$$

$$\frac{\text{d}_p}{Pe_{\text{mr}}} \times \frac{\partial(\tilde{C}_B)}{\partial \tilde{r}} = J_B \times \frac{r_{i,\text{o}}}{u_s c_{A,\text{in}}} \text{ with } \tilde{C}_A + \tilde{C}_B = 1$$

$$\lambda_{\text{er}} \frac{\partial \tilde{T}}{\partial \tilde{r}} = q_m \times \frac{r_{i,\text{o}}}{T_{\text{in}}} = U_1 \times \frac{r_{i,\text{o}}}{T_{\text{in}}} \left(T_{|r_{\text{o},i}} - T \right) \quad [12.18]$$

where q_m is the heat flux from the reactor packed bed to the permeation zone.

12.3 A hydrogen-selective membrane reactor application: natural gas steam reforming

Among chemical processes, hydrogen production processes are the most interesting for membrane reactor application for two main reasons:

1. Hydrogen production reactions are typically endothermic and limited by thermodynamic equilibrium conditions. The integration of a selective membrane would allow the thermodynamics thresholds to be overcome, since equilibrium conditions would never be achieved.
2. Hydrogen-selective membranes are well-known and many research groups and industries have tested them for many years (Basile *et al.*, 2005; Chen *et al.*, 2010; De Falco *et al.*, 2011c; Kikuchi, 1995; Mendes *et al.*, 2010). Among hydrogen-selective membranes, the dense membranes, particularly Pd-based ones, are the most characterized and tested, as a wide scientific literature attests.

The process assessed in this section, in order to illustrate to the readers an IMR modelling case study, is the natural gas steam reforming process, being the most applied process in producing hydrogen industrially.

12.3.1 Methane steam reforming (MSR) process

Among many methods, MSR is the main process used for producing a mixture of hydrogen and carbon monoxide (synthesis gas) and, currently, it contributes about 50% of the world's hydrogen production (Armor, 1999; Scholz, 1993).

MSR is a catalytic process involving a reaction between methane and steam, and it is characterized by (1) a multi-step of steam reforming, water–gas shift and H_2 purification, and (2) severe reaction conditions. A schematic flow sheet for a conventional MSR process is shown in Fig. 12.3 (Barelli *et al.*, 2008).

Methane reformation is the first step, in which methane and steam react at 800–1000°C, 14–20 bar over Ni-based catalyst in endothermic reactions in order to produce syngas, a mixture mostly made up of H_2 and CO:

$$\text{MSR: } CH_4 + H_2O \leftrightarrow CO + 3H_2 \quad \Delta H^0_{298} = 206 \text{ kJ/mol} \qquad [12.19]$$

$$\text{Water–gas shift: } CO + H_2O \leftrightarrow CO_2 + H_2 \quad \Delta H^0_{298} = -41 \text{ kJ/mol} \qquad [12.20]$$

which, taken together, yield:

$$\text{Global reaction: } CH_4 + 2H_2O \leftrightarrow CO_2 + 4H_2 \quad \Delta H^0_{298} = 165 \text{ kJ/mol} \qquad [12.21]$$

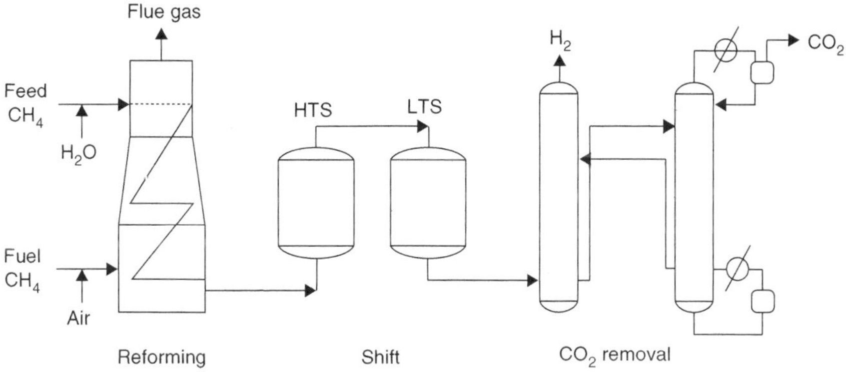

12.3 Flowsheet for a conventional MSR process.

The global reaction is strongly endothermic and very fast over an Ni-based catalyst, so that equilibrium conditions are quickly reached and a significant hydrogen yield is achieved only at high temperature (850–950°C).

In the second step, H_2 and CO_2 are produced by water–gas shift (WGS) reaction (12.20) and this process occurs in two stages consisting of high temperature shift (HTS) and low pressure shift (LTS) reactors. The HTS is loaded with high temperature catalyst, generally chromium-promoted iron oxide which operates at 350–400°C, whereas the LTS is loaded with low-temperature catalyst of copper-promoted zinc oxide, which operates at 200°C (Ledjeff-Hey *et al.*, 2000).

In order to obtain high-purity H_2 flow, the stream coming out from the shift reactors, containing about 86% H_2, 12% CO_2, 0.4% CO and 1.6% CH_4 on a dry basis (Kirk and Othmer, 1999), requires further purification steps.

Several techniques can be used for H_2 separation from mixtures with other gases, as shown in Table 12.2. Nevertheless, pressure swing adsorption (PSA), cryogenic distillation and membrane separation techniques are the main hydrogen separation processes (Adhikari and Fernando, 2006).

The cryogenic distillation process is usually carried out at very low temperature. It uses the difference in the boiling temperatures of the feed components to effect the separation and this process consumes a considerable amount of energy (Hinchliffe and Porter, 2000).

Very high hydrogen purity is not practicable with a cryogenic system. As a consequence, the PSA process is the most commonly selected hydrogen purification technology.

Table 12.2 Techniques used for H_2 purification

Technique	Principle	H_2 recovery (%)	H_2 purity (%)
Cryogenic separation	Partial condensation of gas mixtures at low temperatures	Up to 98	90–98
Polymer membrane	Differential rate of diffusion of gases through a permeable membrane	>85	92–98
Metal hybrid separation	Reversible reaction of hydrogen with metals to form hydrides	75–95	99
Solid polymer electrolyte cell	Electrolytic passage of hydrogen ions across a solid polymer membrane	95	99.8
PSA	Selective adsorption of impurities from gas stream	70–85	99.99
Palladium membrane	Selective diffusion of hydrogen through a palladium alloy membrane	Up to 99	>99.999%

The PSA process is based on the principle that adsorbents are capable of adsorbing more impurities at a higher gas-phase partial pressure than at a lower partial pressure (Adhikari and Fernando, 2006). This process consists of a series of beds filled with molecular sieves or active carbon, in which all components of the gaseous stream, except H_2, are preferentially adsorbed. The high-purity H_2 production, up to 99.99%, and the CO and CO_2 reduction at low ppm, are the main benefits of this process. On the other hand, the hydrogen recovery range of this process is between 70% and 80%, depending on the desired H_2 purity, and therefore at least 20% of the hydrogen is lost with impurities (Dalle Nogare *et al.*, 2007).

Hydrogen can also be separated from the gas mixture with the use of membranes. For instance, the selective surface flow (SSF) membrane, developed by Air Products and Chemicals Inc., can be integrated into the PSA process to increase the overall H_2-recovery. Indeed, it has been demonstrated that the integrated process can increase the net hydrogen recovery to 84–5% from 77–8% (Sircar *et al.*, 1999). This membrane type can separate gas mixtures by the selective adsorption–surface diffusion–desorption mechanism. In particular, when the gas coming out of the PSA unit passes through the high pressure side of the SSF membrane, the CO, CO_2 and CH_4 molecules are selectively adsorbed on the membrane, and then diffuse toward the low pressure side of the membrane, where they are desorbed into the permeate stream. As a consequence, a H_2-rich stream is produced from the high pressure side of the membrane.

Three research lines are ongoing in order to improve the MSR process. The first approach is to improve catalyst performance in terms of activity, mechanical strength, better resistance to carbon formation and sulphur poisoning, and better catalyst efficiency by improving the pellet configuration (Adris *et al.*, 1996).

The second approach is to enhance the properties of the reactor tube material to be able to withstand higher stresses at high temperature and thermal flux.

The third research line deals with reforming the reactor configuration. In particular, three major areas are being developed: (1) transferring from a fixed-bed reactor to a fluidized-bed reactor; (2) use of membrane technology; and (3) changing from external firing to internal heat supply.

Numerous attempts have been made to achieve radical improvements in the reforming process performance through configuration changes. In particular, the most significant attempt has been given to employing hydrogen permselective membranes during the MSR reaction.

MSR conversion is limited by the thermodynamic equilibrium, and this is one of the most serious constraints related to the MSR reaction. Indeed, in order to achieve complete conversion of methane in conventional fixed-bed reformers, the reaction temperature has to be in the range of 800–900°C. At this elevated temperature the catalyst undergoes deactivation due to carbon formation; it also results in blockage of reformer tubes and increased pressure drops (Trimm, 1997).

In order to avoid the problems associated with catalyst fouling, high process energy requirements and poor energy integration, the MSR reaction can be carried out in an MR. In particular, the use of hydrogen permselective MRs allows the MSR reaction in milder operative conditions, conversion enhancement of equilibrium limited reactions, combining the chemical reaction and hydrogen separation in a single system and, as a consequence, improving hydrogen yield and selectivity.

In other words, the integration of a hydrogen-selective membrane could allow the removal of a product of the reactions (H_2), and consequently equilibrium conditions would never be achieved and, in comparing an MRS process using hydrogen-selective membranes with a conventional one:

- greater methane conversion would be achieved at the same operating temperature, or
- the same methane conversion would be achieved at lower operating temperature.

In fact, many works report that, if a Pd-based membrane is integrated in the reaction environment, the methane conversion (>90%) achieved in a traditional reactor at 850–1000°C could be obtained at a much lower temperature (about 500°C) (De Falco *et al.*, 2011a).

Research in SMR reaction modelling carried out on conventional reactors has been focused on the development of reaction kinetic mechanisms and evaluation of kinetic parameters. Indeed, in the past, much attention has been placed on the preparation of catalysts and the evaluation of the MSR processes and equipment, with little work being done on the kinetics and mechanisms of the reaction. As a result, kinetic data were lacking and contradictory mechanisms had been proposed up until 1970. After that, some groups, such as Temkin (1979) and Xu and Froment (1989a), worked on MSR and investigated the kinetics, mainly with Ni-based catalysts.

In particular, Temkin carried out the MSR on nickel foil at atmospheric pressure, studying reforming kinetics. These experiments were conducted in the temperature range of 470–900°C. At 900°C the rate of the reforming reaction is a first-order equation. Later, Xu and Froment improved the modelling of the reforming rate. These researchers focused on a detailed reaction kinetic mechanism to model the intrinsic kinetics of SMR reactions on $Ni/MgAl_2O_4$-spinel catalyst. Their proposed reaction scheme consisted of 21 sets of rate equations. The number of possible reaction mechanisms was reduced, taking into account a thermodynamic analysis and using the Langmuir equilibrium relation.

These works identified the main three reactions [12.19], [12.20] and [12.21] that occur during SMR reaction and derived their rate expressions [12.22], [12.23] and [12.24]. These rate expressions identify the rate-limiting step in the absence of mass-transfer limitations:

$$r_1 = k_1\left(\frac{p_{CH_4} \times p_{H_2O}}{p_{H_2}^{2.5}} - \frac{p_{H_2}^{0.5} \times p_{CO}}{K_1}\right) / \mathrm{DEN}^2 \qquad [12.22]$$

$$r_2 = k_2\left(\frac{p_{CO} \times p_{H_2O}}{p_{H_2}} - \frac{p_{CO_2}}{K_2}\right) / \mathrm{DEN}^2 \qquad [12.23]$$

$$r_3 = k_3\left(\frac{p_{CH_4} \times p_{H_2O}^2}{p_{H_2}^{3.5}} - \frac{p_{H_2}^{0.5} \times p_{CO_2}}{K_3}\right) / \mathrm{DEN}^2 \qquad [12.24]$$

where r_1, r_2 and r_3 are the reaction rates of the steam reforming reaction, the WGS and the global reaction, respectively, and DEN is calculated as:

$$DEN = 1 + K_{CH_4} \times p_{CH_4} + K_{H_2O} \times \frac{p_{H_2O}}{p_{H_2}} + K_{H_2} \times p_{H_2} + K_{CO} \times p_{CO}$$

where K_i (i = CH_4, H_2O, H_2, CO) are adsorption equilibrium constants (reported in Section 12.3.3) and p_i are the partial pressures of components i.

Xu and Froment (1989b) developed the most widely accepted kinetic model, deriving the intrinsic parameters and incorporating diffusional limitations through the evaluation of the tortuosity factor, effective diffusivities, and the effectiveness factor. These parameters were used in the simulation of commercial reactors and industrial steam reformers with satisfactory results.

12.3.2 Modelling analysis of the MSR reaction

Industrial SMR is a mature technology. As a result, there are numerous mathematical models in the academic and commercial literature that simulate steam methane reformers. These models differ in their simplifying assumptions and in the reactor configuration type, whether fixed-bed or fluidized-bed.

Fixed-bed reactors models can be classified by their dimensionality (one-dimensional or two-dimensional) and by their complexity (pseudo-homogeneous or heterogeneous). In a one-dimensional model gradients are considered only in the axial direction, whereas in a two-dimensional model gradients are assumed in both the axial and radial directions.

In pseudo-homogeneous models, the process gas and catalyst are assumed to be at the same temperature and to be almost in contact. The pseudo-homogeneous assumption simplifies mass-transfer modelling, since external and internal diffusion are not considered explicitly. An effectiveness factor is applied to reaction rates to model the lower concentration of reactants at the catalyst sites. Since the process gas and catalyst are assumed to be at the same temperature, an overall heat-transfer coefficient can be used to describe heat transfer from the inner tube wall to the catalyst and process gas.

In heterogeneous models, separate material (and energy) balances are performed on the bulk-process gas and on the process gas diffusing through the catalyst particle. Unlike pseudo-homogeneous models, the material balance on the bulk-process gas does not contain a reaction-rate expression.

In Table 12.3 some fixed-bed reactor models for the MSR reaction are reported. As shown, several one-dimensional fixed-bed reactor models make pseudo-homogeneous assumptions (Murty and Murthy, 1988; Singh and Saraf, 1979; Yu *et al.*, 2006) and other heterogeneous assumptions (Plehiers

Table 12.3 Several SMR reaction models

Authors	Reactor model	Reaction kinetic
Singh and Saraf (1979)	-1D -pseudo-homogeneous -no effectiveness factors used -assumed diffusion limitations accounted for in kinetics -plug-flow	Used first-order kinetic rate expressions developed by Haldor Topsoe (shown in Singh and Saraf, 1979)
Murty *et al.* (1988)	-1D -pseudo-homogeneous -diffusion limitations accounted for in kinetics -plug-flow	Used first-order kinetic rate expressions developed by Haldor Topsoe (shown in Singh and Saraf, 1979)
Plehiers and Froment (1989)	-1D -heterogeneous -plug-flow	Xu and Froment (1989a) diffusion limitations
Alhabdan *et al.* (1992)	-1D heterogeneous -plug-flow (not stated) - derived a material balance on a catalyst pellet using characteristic length	Xu and Froment (1989a)
Pedernera *et al.* (2003)	-2D heterogeneous -PDEs from momentum balances	Xu and Froment (1989a)
Yu *et al.* (2006)	-1D -pseudo-homogeneous	Yu *et al.* 2006 -reaction kinetics derived from stoichiometric equations -1D pseudo-homogeneous
Wesenberg and Svendsen (2007)	-2D heterogeneous	Xu and Froment (1989a)
Ebrahimi *et al.* (2008)	-1D -pseudo-homogeneous -plug-flow	Xu and Froment (1989a) diffusion limitations

and Froment, 1989; Soliman *et al.*, 1988). Many two-dimensional heterogeneous models exist in the literature but only a few (Pedernera *et al.*, 2003; Wesenberg and Svendnsen, 2007) are shown in Table 12.3. This table also shows that the Xu and Froment kinetic model is widely used.

However, as reported above, conventional fixed-bed steam methane reformers have some disadvantages, such as low heat-transfer rates, diffusional resistance and lack of uniformity of temperature within the reactor. In order to avoid these drawbacks, a fluidized-bed reactor can be used for carrying out the MSR reaction. Historically, two classes of models have been proposed to describe the performance of fluidized-bed reactors; one is based on a pseudo-homogeneous approach and the other on a two-phase approach (Mahecha-Botero *et al.*, 2009). Recently, Ho (2003) classified reactor models as pseudo-homogeneous, two phased, and with multiple regions, giving also a general overview of each model. The author highlighted the main aspect of multiphase modelling, indicating that no single model is likely to be applicable in all cases.

12.3.3 Mathematical model of the Pd-based dense membrane MSR reaction

To properly describe the reaction-zone model, a reaction-kinetics scheme is needed. There are many kinetics expressions reported in the literature, to which the Xu-Froment Equations [12.22], [12.23] and [12.24] are applied (Xu and Froment, 1989).

All the constants appearing in the reaction rates equations are shown in the following:

- Rate coefficients:

$$k_1 = 9.49 \times 10^{16} \exp\left(\frac{-28879}{T}\right) \frac{\text{kmol} \cdot \text{kPa}^{0.5}}{\text{kg} \cdot \text{h}}$$

$$k_2 = 4.39 \times 10^{4} \exp\left(\frac{-8074.3}{T}\right) \frac{\text{kmol} \cdot \text{kPa}^{-1}}{\text{kg} \cdot \text{h}}$$

$$k_3 = 2.29 \times 10^{16} \exp\left(\frac{-29336}{T}\right) \frac{\text{kmol} \cdot \text{kPa}^{0.5}}{\text{kg} \cdot \text{h}}$$

- Adsorption equilibrium constants:

$$K_{\text{CH}_4} = 6.65 \times 10^{-6} \exp\left(\frac{4604.28}{T}\right) \text{kPa}^{-1}$$

$$K_{\text{H}_2\text{O}} = 1.77 \times 10^{5} \exp\left(\frac{-10666.35}{T}\right)$$

$$K_{\text{H}_2} = 6.12 \times 10^{-11} \exp\left(\frac{9971.13}{T}\right) \text{kPa}^{-1}$$

$$K_{CO} = 8.23 \times 10^{-7} \exp\left(\frac{8497.71}{T}\right) \quad \text{kPa}^{-1}$$

- Equilibrium constants:

$$K_1 = 10266 \cdot 76 \exp\left(\frac{-26830}{T} + 30.11\right) \quad \text{kPa}^2$$

$$K_2 = \exp\left(\frac{4400}{T} - 4.063\right)$$

$$K_3 = K_1 \times K_2 \quad \text{kPa}^2$$

Among the hydrogen-selective membranes that can be assembled in a steam reforming membrane reactor, the Pd-based dense membranes are surely the most interesting.

Progress in the field of palladium-based membrane has been driven by their capacity to produce a pure hydrogen stream, owing to infinite hydrogen perm-selectivity with respect to all other gases. On the other hand, some drawbacks limit their commercialization, mainly:

- palladium-based membranes suffer from poisoning from various compounds;
- the cost of palladium is high and the lack of a real industrialization of manufacturing processes means the Pd-based membrane cost is still uncompetitive.

For a better understanding of Pd-based membrane technology, refer to (De Falco *et al.*, 2011a).

Hydrogen permeation flux is described by the following expression

$$J_{H_2} = \frac{P_H}{\delta} \times \left(p^n_{H_2,r} - p^n_{H_{2,p}}\right) \qquad [12.25]$$

where δ is the membrane thickness, $p_{H_2,r}$ and $p_{H_2,p}$ are the hydrogen partial pressures in the reaction and permeation zone, and P_H is the membrane permeability, derived by an Arrhenius-type law:

$$P_H = P_0 \times \exp\left(-\frac{E_a}{R \cdot T}\right) \qquad [12.26]$$

where the pre-exponential factor P_0 and the activation energy E_a depend on the type of membrane and can be obtained for a specific membrane if experimental data are available. The value of the n exponent in Equation [12.25] depends on which is the leading phenomenon in the permeation mechanism. If the membrane is very thin, the dissociative adsorption of the molecular hydrogen on the membrane surface dominates the permeation mechanism and n is equal to 1; if the membrane is thick (some tens of micron), the diffusion of atomic hydrogen is the leading phenomenon and n is equal to 0.5. There are, as well, mid-span situations.

In the following, an integrated steam reforming Pd-based membrane reactor is modelled by a 2D approach. All equations and boundary conditions are defined and explained.

12.3.4 Mathematical model implementation

In the present section, only the 2D mathematical model is reported. For the 1D model formulation, please refer to De Falco *et al.* (2007a).

The model is composed of mass balances (one for each of the five reaction components), energy balance and momentum balance. The momentum balance is usually formulated ignoring radial mixing.

Mass balances

Reaction zone:

$$\frac{\partial\left(\widetilde{u_s}\widetilde{c_i}\right)}{\partial\tilde{z}}=\frac{\mathrm{d}_p\times L}{Pe_{\mathrm{mr}}\times r_{i,\mathrm{o}}^{2}}\times\left(\frac{\partial^2\left(\widetilde{u_s}\widetilde{c_i}\right)}{\partial\tilde{r}^2}+\frac{1}{\tilde{r}}\times\frac{\partial\left(\widetilde{u_s}\widetilde{c_i}\right)}{\partial\tilde{r}}\right)-\frac{\eta\times\rho_b\times L}{u_{s,0}c_{\mathrm{CH_4,in}}}\times r_i \qquad [12.27]$$

where $r_{i,o}$ is the internal radius of outer tube; $\tilde{u}_s$ and $u_{s,0}$ are the dimensionless and the inlet gas velocity, respectively; $\tilde{c}_i$ and $c_{\mathrm{CH_4,in}}$ are the dimensionless i-component concentration and the inlet methane concentration, respectively; and r_i is the reaction rate of component i.

It should be noted that the gas velocity changes along the reaction, since the total mole number is not constant, and consequently the $\tilde{u}_s$ term has to be inside the differential.

With respect to the mass effective radial Peclet (Pe_{mr}) number, if the Reynolds number is assumed to be greater than 1000, Pe_{mr} reaches the constant value depending only on geometric features calculated by Equation [12.7].

Otherwise, if Re < 1000, the Pe_{mr} has to be evaluated, for example using the Wilke-Hougen method (Wilke and Hougen, 1945).

Permeation zone:

$$\frac{dY_{H_2}}{d\tilde{z}} = \pm \frac{J_{H_2} 2 \times r_{o,i} \times L}{u_s c_{CH_4,in} \times \left(r_{i,o}^2 - r_{o,i}^2 \right)} \quad [12.28]$$

where J_{H_2} is H_2 flux through the Pd-based membrane, calculated by Equation [12.25] and $Y_{H_2} = F_{H_2,perm} / F_{CH_4,in}$, with $F_{H_2,perm}$ and $F_{CH_4,in}$ equal to the permeated hydrogen flow-rate and inlet methane flow-rate.

Energy balances

Reaction zone:

$$\frac{\partial \tilde{T}}{\partial \tilde{z}} = \frac{\lambda_{er} \times L}{(u_s c_{tot}) \times c_{p,mix} \times r_{i,o}{}^2} \times \left(\frac{\partial^2 \tilde{T}}{\partial \tilde{r}^2} + \frac{1}{\tilde{r}} \times \frac{\partial \tilde{T}}{\partial \tilde{r}} \right) + \frac{\eta \times \rho_b \times L \times \sum_{i=1}^{3} (-\Delta H_i) \times r_i}{(u_s c_{tot}) \times c_{p,mix} \times T_{in}} \quad [12.29]$$

where $c_{p,mix}$ is the gas mixture specific heat, η is the catalyst effectiveness factor and c_{tot} is the total mixture concentration. The reactions rates are calculated according to Equations [12.21][12.24]

Permeation zone:

$$\frac{d\tilde{T}_P}{d\tilde{z}} = \frac{L}{(u_{s,perm} \times c_{tot,perm}) \times c_{p,mix,perm} \times T} \times \left[U_1 \times 2\pi r_{i,i} \times T_{in} \times \left(\tilde{T} - \tilde{T}_P \right) + J_{H_2} 2\pi r_{o,i} \times \left(h_{H_2,reac} - h_{H_2,perm} \right) \right] \quad [12.30]$$

where $u_{s,perm}$ and $c_{tot,perm}$ are the gas velocity and the total mixture concentration in the permeation zone, respectively.

Momentum balance

The momentum balance is imposed only in the reaction zone, since it is assumed that the pressure drop is negligible in the permeation zone:

$$\frac{\mathrm{d}\tilde{P}}{\mathrm{d}\tilde{z}} = \frac{f \times G \times \mu_g \times L}{\rho_g \times \mathrm{d}_p{}^2 \times P_{\mathrm{in}}} \times \frac{(1-\varepsilon)^2}{\varepsilon^3} \quad [12.31]$$

Boundary conditions

Assuming a counter-current sweeping gas configuration, the following boundary conditions are imposed:

$$\tilde{z} = 0, \forall \tilde{r}:$$

$$\tilde{u}_s \, \tilde{c}_{\mathrm{CH}_4} = 1$$

$$\tilde{u}_s \, \tilde{c}_i = \frac{u_s c_i}{u_{s,\mathrm{in}} c_{\mathrm{CH}_4,\mathrm{in}}} (i = \mathrm{H_2O},\ \mathrm{H_2}, \mathrm{CO}, \mathrm{CO_2})$$

$$\tilde{T} = 1$$

$$\tilde{P} = 1$$

$$\tilde{T}_P = \frac{T_{P,\mathrm{in}}}{T_{\mathrm{in}}} \quad [12.32]$$

$$\tilde{z} = 1, \forall \tilde{r}:$$

$$Y_{\mathrm{H_2}} = 0 \text{ Counter-current configuration} \quad [12.33]$$

$$\tilde{r} = 1, \forall \tilde{z}:$$

$$\frac{\partial \left(\tilde{u}_s \tilde{c}_i \right)}{\partial \tilde{r}} = 0$$

$$\lambda_{\mathrm{er}} \frac{\partial \tilde{T}}{\partial \tilde{r}} = \frac{q_r \times r_{i,o}}{T_{\mathrm{in}}} = \frac{U \times r_{i,\mathrm{o}}}{T_{\mathrm{in}}} \left(T_{w,\mathrm{o}} - T_{|r_{i,\mathrm{o}}} \right) \quad [12.34]$$

where q_r is the heat flux from the external source to the reactor packed bed and $T_{w,o}$ is the external tube wall temperature.

$$\tilde{r} = \frac{r_{o,i}}{r_{i,o}}, \forall \tilde{z}:$$

$$\frac{\partial\left(\tilde{u}_s \tilde{c}_i\right)}{\partial \tilde{r}} = 0$$

$$\frac{\mathrm{d}_p}{Pe_{\mathrm{mr}}} \times \frac{\partial\left(\tilde{u}_s \tilde{c}_{\mathrm{H}_2}\right)}{\partial \tilde{r}} = J_{\mathrm{H}_2} \times \frac{r_{i,o}}{u_{s,0} c_{\mathrm{H}_2,0}}$$

$$\lambda_{\mathrm{er}} \frac{\partial \tilde{T}}{\partial \tilde{r}} = q_m \times \frac{r_{i,o}}{T_{\mathrm{in}}} = U_1 \times \frac{r_{i,o}}{T_{\mathrm{in}}}\left(T_{|r_{o,i}} - T_P\right) \qquad [12.35]$$

where q_m is the heat flux from the reactor packed bed to the permeation zone.

The 2D steam reforming mathematical model here presented has been widely validated and tested. Refer to De Falco *et al.* (2007b; 2008; 2011a) for model results.

12.4 Conclusions

Modelling a reactor means developing a tool for designers to improve understanding of the process and the reaction and to optimize reactor operating conditions, without the need for long and expensive experimental phases.

Specifically, IMRs can be modelled using various modelling strategies, classified in decreasing order of complexity, from complex 2D heterogeneous models to simple 1D pseudo-homogeneous models. The selection of mathematical model category depends on the quality of information needed by designers to properly dimensioning the reactor.

To develop a membrane reactor design tool, designers need:

1. a reliable reaction-kinetics scheme;
2. expression of membrane permeability and of hydrogen flux through the membrane;

3. expression of all components appearing in the overall heat-transfer coefficients between the external and the reaction zone and between the reaction zone and permeation zone;
4. physical properties of components, such as heat capacity, viscosity, density, etc.When all information is available, the designers can properly implement the model, developing a suitable design tool.

12.5 Acknowledgements

The authors wish to thank both Prof. Josè Sousa and Dr Simona Liguori for their help in improving the manuscript.

12.6 References

Adhikari S. and Fernando S., (2006), 'Hydrogen membrane separation techniques', *Ind. Eng. Chem. Res.*, **45**, 875–881.

Adris A.M., Pruden B.B., Lim C.J. and Grace J.R., (1996), 'On the reported attempts to radically improve the performance of the steam methane reforming reactor', *Canadian J. Chem. Eng.*, **74**, 177–186.

Armor J.N., (1999), 'Review: the multiple roles for catalysis in the production of H2', *Appl. Catal. A: Gen*, **176**, 159–176.

Barelli L., Bidini G., Gallorini F. and Servili S., (2008), 'Hydrogen production through sorption enhanced steam methane reforming and membrane technology: A review', *Energy*, **33**, 554–570.

Basile A., Gallucci F. and Paturzo L., (2005), 'A dense Pd/Ag membrane reactor for methanol steam reforming: experimental study', *Catal. Today*, **104**, 244–250.

Bredesen R., (2008), 'Thin Pd-23w%Ag membranes for hydrogen separation', Lyon, CASTOR/CACHET/DYNAMIS/ENCAP workshop, 22 – 24.

Chen W., Hu X., Wang R. and Huang Y., (2010), 'On the assembling of Pd/ceramic composite membranes for hydrogen separation', *Sep. Pur. Tech.*, **72**, 92–97.

Dalle Nogare D., Baggio P., Thomas C., Mutri L. and Canu P., (2007), 'A thermodynamic analysis of natural gas reforming processes for fuel cell application', *Chem. Eng. Sci.*, **62**, 5418–5424.

De Falco M., Di Paola L. and Marrelli L., (2007a), 'Heat transfer and hydrogen permeability in modelling industrial membrane reactors for methane steam reforming', *Int. J. Hy. En.*, **32**, 2902–2913

De Falco M., Di Paola L., Marrelli L. and Nardella P., (2007b), 'Simulation of large-scale membrane reformers by a two-dimensional model', *Chem. Eng. J.*, **128**, 115–125.

De Falco M., Nardella P., Marrelli L., Di Paola L., Basile A. and Gallucci F., (2008), 'The effect of heat flux profile and of other geometric and operating variables in designing industrial membrane steam reformers', *Chem. Eng. J.*, **138**, 442–451.

De Falco M., Marrelli L. and Basile A., (2010a), 'An industrial application of membrane reactors: modelling of the methane steam reforming reaction', in *Simulation of Membrane Reactors*, Basile A. and Gallucci F.(Eds), Nova Science Publisher, New York.

De Falco M., Marrelli L. and Iaquaniello G., (2011a), *Membrane Reactors for Hydrogen Production Processes*, New York, Springer.
De Falco M., Salladini A., Palo E. and Iaquaniello G., (2011b), 'Reformer and membrane modules (RMM) for methane conversion powered by a nuclear reactor: experimental assessment and perspectives of such novel architecture', in Editor: Pavel Tsvetkov *Nuclear Power*, InTech Open Access Publisher.
De Falco M., Iaquaniello G., and Salladini A., (2011c), 'Experimental tests on steam reforming of natural gas in a reformer and membrane modules (RMM) plant', *J. Membrane Sci.*, **368**, 264–274.
De Wasch A. and Froment G., (1972), 'Heat transfer in packed beds', *Chem. Eng. Sci.*, **27**, 567–576.
Dittmeyer R., Höllein V. and Daub K., (2001), 'Membrane reactors for hydrogenation and dehydrogenation processes based on supported palladium', *J. Mo.l Catal.: A Chem.*, **173**, 135–184.
Dixon A. G. and Cresswell D. L., (1979), 'Theoretical prediction of effective heat transfer parameters in packed beds', *AIChE J.*, **25**(4), 663–675.
Elnashaie S. and Elshishini S., (1993), *Modelling, simulation and optimization of industrial fixed bed catalytic reactors*, New York, Gordon and Breach Science Publisher.
Froment G. F. and Bischoff K. B., (1990), *Chemical Reactor Analysis and Design*, New York, Wiley.
Hinchliffe A.B. and Porter K.E., (2000), 'A comparison of membrane separation and distillation', *Trans. Inst. Chem. Eng.*, **78**, 255–268.
Ho T.C., (2003), 'Modeling', in Handbook of fluidization and fluid-particle system, Wen-Ching Yang (ed.), ch. 9, Routledge ISBN: 9780203912744
Howard B. H., Killmeyer R. P., Rothenberger K. S., Cugioni A. V., Morreale B. D., Enick R. M. and Bustamante F., (2004), 'Hydrogen permeance of palladium-copper alloy membranes over a wide range of temperatures and pressures', *J. Membrane Sci.*, **241**, 207–218.
Kikuchi E., (1995), 'Palladium/ceramic membranes for selective hydrogen permeation and their application to membrane reactor', *Catal. Today*, **25**, 333–337.
Kirk R.E. and Othmer D.F., (1999), *Concise Encyclopedia of Chemical Technology*, 4th Edition, March 1999, Wiley, New York, V12, 950.
Kulkarni B. D. and Doraiswamy L. K. (1980), 'Estimation of effective transport properties in packed bed reactors', *Catal. Rev. Sci. Eng.*, **22**(3), 431–483.
Ledjeff-Hey K., Roes J. and Wolters R., (2000), 'CO_2-scrubbing and methanation as purification system for PEFC', *J. Power Sou.*, **86**, 556–561.
Li C. and Finlayson B. (1977), 'Heat transfer in packed beds – a re-evaluation', *Chem. Eng. Sci.*, **32**, 1055–1066.
Mahecha-Botero A., Chen Z., Grace J.R., Elnashaie S.S.E.H., Lim C.J., Rakib M., Yasuda I. and Shirasaki Y., (2009), 'Comparison of fluidized bed flow regimes for steam methane reforming in membrane reactors: A simulation study', *Chem. Eng. Sci.*, **64**, 3598–3613.
Mendes D., Mendes A., Madeira L. M., Iulianelli A., Sousa J. M. and Basile A., (2010), 'The water-gas shift reaction: from conventional catalytic systems to Pd-based membrane reactors – a review', *Asia-Pacific J. Chem. Eng.*, **5**, 111–137.
Murty C.V.S. and Murthy M.V., (1988), 'Modeling and simulation of a top-fired reformer', *Ind. Eng. Chem. Res.*, **27**, 1832–1840.

Peachey N. M., Snow R. C. and Dye R. C., (1996), 'Composite Pd/Ta metal membranes for hydrogen separation', *J. Membrane Sci.*, **111**, 123–133.
Pedernera M.N., Piña J., Borio D.O. and Bucalá V., (2003), 'Use of a heterogeneous two-dimensional model to improve the primary steam reformer performance', *Chem. Eng. J.*, **94**, 29–40
Plehiers P.M. and Froment G.F., (1989), 'Coupled simulation of heat transfer and reaction in a steam reforming furnace', *Chem. Eng. Technol.*, **12**, 20–26.
Scholz W.H., (1993), 'Processes for industrial production of hydrogen and associated environmental effects', *Gas Sep. Purif.*, **7**, 131–139.
Seborg D. E., Edgar T. F. and Mellichamp D. A., (1989), *Process Dynamics and Control*, New York, Wiley.
Singh C. P. P. and Saraf D. N., (1979) 'Simulation of side fired steam-hydrocarbon reformers', *Ind. Eng. Chem. Proc. Des. Devel.*, **18**, 1–7.
Sircar S., Waldron W.E., Rao M.B. and Anand M., (1999) 'Hydrogen production by hybrid SMR–PSA–SSF membrane system', *Sep. Pur. Techn.*, **17**, 11–20.
Soliman M.A. and El-Nashaie S.S.E.H., (1988), 'Simulation of steam reformers of methane', *Chem. Eng. Sci*, **43**, 1801–1806.
Temkin M.I., (1979), 'Industrial heterogeneous catalytic reactions', *Adv. Catal.*, **28**, 175–292.
Trimm D.L., (1997), 'Coke formation and minimization during steam reforming reactions', *Catal. Today*, **37**, 233–238.
Tosti S., Bettinali L., Castelli S., Sarto F., Scaglione S. and Violante V., (2002), 'Sputtered, electroless and rolled palladium-ceramic membranes', *J. Membrane Sci.*, **196**, 241–249.
Tsotsas E. and Schlünder E., (1990), 'Heat transfer in packed beds with fluid flow: remarks on the meaning and the calculation of a heat transfer coefficient at the wall', *Chem. Eng. Sci.*, **45**, 819–837.
Wesenberg M.H. and Svendnsen H.F., (2007), 'Mass and heat transfer limitations in a heterogeneous model of a gas-heated steam reformer', *Ind. Eng. Chem. Res.*, **46**, 667–676.
Wilke C. R. and Hougen D. A., (1945), 'Mass transfer of gas mixture', *Trans. Am. Ins. Chem. Eng.*, **41**, 445.
Xu J. and Froment G.F., (1989a), 'Methane steam reforming, methanation and water-gas shift: I. Intrinsic kinetics', *AIChE J.*, **35**, 88–96.
Xu J. and Froment G.F., (1989b) 'Methane steam reforming. II. Diffusional limitations and reactor simulation', *AIChE J.*, **35**, 97–103.

12.7 Appendix: nomenclature

12.7.1 Notation

A_i	bed size heat-exchanging surface
A_o	heating fluid media heat-exchanging surface
$\tilde{C}_i$	dimensionless concentration of component *i*
$C_{i,\mathrm{in}}$	inlet concentration of component *i*, temperature and pressure
c_p	gas mixture specific heat
$c_{p,p}$	specific heat of permeation zone fluid mixture

c_{tot}	total mixture concentration
D_{er}	effective radial mass diffusivity
d_p	equivalent particle diameter
d_t	the internal tubular reactor diameter
f	friction factor
$F_{\text{TOT},p}$	total permeation zone flow-rate
G	superficial mass flow velocity
$h_{B,p}$	B component specific enthalpy in permeation zone
$h_{B,r}$	B component specific enthalpy in reaction zone
h_{ex}	the heat transport coefficient in the external energy source side
h_{perm}	forced convection heat transport coefficient in the permeation zone
h_W	heat transport coefficient in the first layer near the tube wall
$(-\Delta H)$	heat of reaction
J_i	i-component flux through the membrane
k_{mem}	thermal conductivity of the membrane
k_{met}	tube wall conductivity
L	reactor length
$\tilde{P}$	dimensionless reactor pressure
P_{H}	membrane permeability
$p_{\text{H}_2,p}$	hydrogen partial pressures in the permeation zone
$p_{\text{H}_2,r}$	hydrogen partial pressures in the reaction zone
P_{in}	inlet pressure
Pe_{mr}	radial mass Peclet number
q_m	heat flux from the reactor packed bed to the permeation zone
q_r	heat flux from the external source to the reactor packed bed
$r_{i,i}$	internal radius of inner tube
$r_{i,o}$	internal radius of outer tube
$r_{o,i}$	the external radius of the inner tube
Re_p	Reynolds number
r_i	i-component reaction rate
$\tilde{T}$	dimensionless reactor temperature
T_{in}	inlet temperature
T_r	temperature of energy source
U	overall heat-transport coefficient between external energy source and reaction bed
U_1	overall heat-transfer coefficient between reaction and permeation zone
u_s	gas superficial velocity

Greek symbols

δ	thickness of the membrane
ε	packed bed void fraction
η	catalyst effectiveness factor
λ_{er}	effective radial thermal conductivity of the pseudo-homogeneous phase
$\lambda_{g,perm}$	conductivity of gas stream in permeation zone.
μ_g	gas mixture viscosity
ρ_B	catalytic bed density
ρ_g	gas density

13
Computational fluid dynamics (CFD) analysis of membrane reactors: simulation of single- and multi-tube palladium membrane reactors for hydrogen recovery from cyclohexane

N. ITOH, Utsunomiya University, Japan and
K. MIMURA, Chiyoda Corporation, Japan

DOI: 10.1533/9780857097330.3.464

Abstract: A palladium membrane reactor has been applied as a promising chemical hydrogen carrier to recover the hydrogen from cyclohexane. However, it has been found that increasing feed rate resulted in an increasing deviation from the ideal analytical model assuming plug-flow and isothermal conditions. This chapter, therefore, presents a computational fluid dynamics (CFD) model development, which takes into account the concentration, temperature and velocity distributions due to mass, heat transfer and flow resistance in the membrane reactor. The model is verified for the dehydrogenation of cyclohexane in a shell-and-tube type palladium membrane reactor, as well as for a multi-tube type.

The CFD model clearly shows that large temperature and concentration distributions are formed both in the radial and axial directions. Simulation results with cyclohexane dehydrogenation were in good agreement with the experimental data for the both membrane reactors. Furthermore, it is demonstrated that the multi-tube model developed is applicable for changing the reactor design, for instance the membrane dimension, the length of catalyst-packed layer, and the operation conditions such as temperature, feed rate and pressure.

Key words: CFD model, membrane reactor, palladium membrane, single tube, multi-tube, dehydrogenation, hydrogen.

13.1 Introduction

Hydrogen is expected to be a clean secondary energy, despite some problems in its transportability and storability, for instance the necessity of high pressure containers for 35–70 MPa of compressed hydrogen gas. There is hydrogen liquefaction alternatively, which is thought to be unrealistic because that would imply an extremely low operating temperature, as low as 20 K. Therefore, more convenient hydrogen carriers are highly desirable. In this context, liquid chemical hydrogen carriers such as cyclohexane are considered suitable from the following viewpoints:

1. A higher hydrogen content (e.g., 7.1 wt%) is very attractive compared with metal hydrides (at most 3 wt%).
2 The dehydrogenated product, benzene, can be reversibly hydrogenated and then reused.
3. Benzene and cyclohexane are liquid at ordinary temperatures, meaning that both are highly suitable for transport and storage.

Dehydrogenation of cyclohexane is an equilibrium reaction, and thus its conversion is limited thermodynamically. It is well known that dehydrogenation can be significantly enhanced by using a composite reactor incorporating a membrane permselective for hydrogen (so-called membrane reactor). Palladium is believed to be one of the best materials for a hydrogen separation membrane because of its infinite permselectivity for hydrogen and comparatively large permeability over a wide range of temperature. To ensure the driving force for hydrogen permeation through the membrane, the reaction-side hydrogen partial pressure must be kept higher than that of the permeation side, which is usually at atmospheric pressure or less (Itoh *et al.*, 2003). Because such a complicated reaction system has many operating parameters, the operation conditions and design of the membrane reactor must be optimized to recover as much hydrogen as possible.

Mathematical modelling is very useful in making a good membrane reactor design. Several membrane reactor models have been proposed in the literatures (Itoh,1987; Barbieri and Di Maio, 1997; Kim *et al.*, 1999; Basile *et al.*, 2001; Assabumrungrat *et al.*, 2002; Nair and Harold, 2006; Smit *et al.*, 2007; De Falco *et al.*, 2007, 2008). Most of the proposed models are one-dimensional or pseudo one-dimensional for simple configurations such as a single membrane tube reactor. The flow inside the reactor is usually assumed to be plug-flow, the species conservation equation is basically in one dimension form, and the energy conservation equation is given in one or two-dimensional form. Two-dimensional models have been proposed by Tiemersma *et al.* (2006) and De Falco *et al.* (2007, 2008). However, their models are described using cylindrical coordinates for simplicity. Therefore, it would not be easy to apply their models to a more complex geometry, such as a multi-tubular membrane reactor or to a full three-dimensional model. If a membrane reactor has simple geometry, a one-dimensional or pseudo one-dimensional assumption can be applied. However, the real membrane reactor geometry, even in a double tube membrane reactor, (see Fig. 13.1) is complicated, and thus the applicability of these simple models is significantly limited.

Although many full three-dimensional analyses using CFD technology have been applied to various engineering disciplines, few attempts have been made in membrane reactor development.

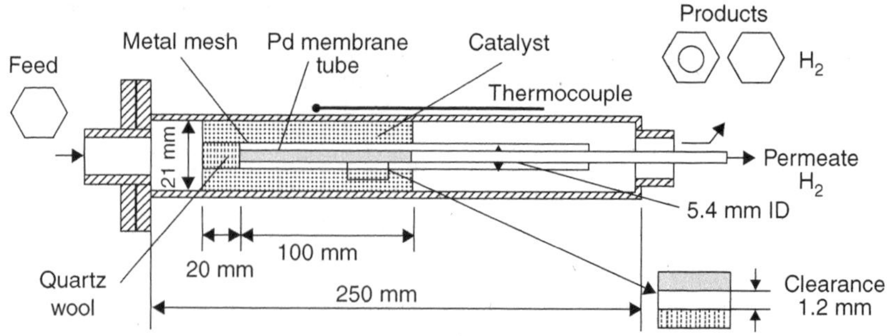

13.1 Cross-sectional view of the palladium membrane reactor (catalyst: 0.5 wt% Pt/Al_2O_3 pellets, membrane: 3 mm OD, 5 μm-thick Pd/Al_2O_3 tube).

In this text, a rigorous modelling method of using CFD technology for a single palladium membrane tube reactor for dehydrogenation of cyclohexane, as well as a multi-tube membrane reactor, is presented. The proposed CFD model will be validated by comparison with experimental data. Also, it will be shown that the model becomes a useful tool in analysing and understanding in detail the phenomena that take place and how they change inside the reactor.

13.2 Single palladium membrane tube reactor

First, the basic CFD model for a single palladium membrane tube reactor with the simplest configuration is developed. The model established, which takes account of a lot of information on the physical properties, the chemical kinetics, the membrane properties and so on, will be proved by comparing with the practical reactor performances.

13.2.1 Experimental data

The single palladium membrane tube reactor used is shown in Fig. 13.1, where the shell tube is stainless-steel with a length of 250 mm and an inside diameter (ID) of 21 mm. A palladium membrane tube was inserted into the centre of the reactor and was covered with a stainless-steel mesh tube to prevent contact with the catalyst pellets, such that the clearance between the membrane tube and stainless-steel mesh was 1.2 mm. The 5 μm-thick palladium membrane supported on a porous α-Al_2O_3 tube was 100 mm long and 3 mm in outer diameter (OD), and showed almost infinite selectivity to hydrogen; that is, the ideal permselectivity for H_2/N_2 observed was more than 10 000 (Itoh *et al.*, 2007). 0.5 wt% Pt/Al_2O_3 cylindrical pellets (3.2 mm diameter and 3.6 mm height) were packed around the membrane tube over 120 mm. The reactor was installed in a constant temperature air bath and the wall temperature was monitored by a thermocouple (TC). Experiments for gas separation using a gaseous mixture (75% H_2 and 25% N_2) and cyclohexane dehydrogenation were carried out at 573 K. Detailed experimental results have been presented elsewhere (Itoh *et al.* 1997, 2003).

13.2.2 Numerical model for the single palladium membrane tube reactor

The numerical model for analysing the membrane reactor was developed using a commercial CFD code, Star-CD v3.2. The modelling procedure is described below (Mimura *et al.*, 2010a).

Governing equations

Governing equations are defined for both the reaction and permeation sides, being represented with suffix 1 and 2 respectively. Steady-state assumption is applied to all the conservation equations. Mass and heat exchange through the membrane is considered. A turbulence model is not used in here because the flow rate is very low in the experimental conditions. The clearance region existing between the membrane tube and the stainless-steel mesh tube is treated as an open area.

Reaction side

(a) Mass conservation equation

$$\nabla \cdot (\rho_1 u_1) = S_{m,1} \qquad [13.1]$$

(b) Momentum conservation equation

- Catalyst bed:

$$\nabla P_1 = -K u_1 \qquad [13.2]$$

$$K = \alpha |u_1| + \beta \qquad [13.3]$$

- Open area:

$$\nabla (\rho_1 u_1 u_1) = \nabla P_1 + \nabla (\mu_1 \nabla u_1) + S_{u,1} \qquad [13.4]$$

(c) Energy conservation equation

- Catalyst bed:

$$\nabla \cdot (\rho_1 u_1 h_1) = \nabla \cdot (\lambda_{\text{eff}} \nabla T_1) + S_{h,1} \quad [13.5]$$

- Open area:

$$\nabla \cdot (\rho_1 u_1 h_1) = \nabla \cdot (\lambda_1 \nabla T_1) + S_{h,1} \quad [13.6]$$

$$h_1 = \sum_i m_{1,i} \left(h_{f_i} + \int_{298}^{T_1} C_{pi} \mathrm{d}T \right) \quad [13.7]$$

(d) Species mass conservation equation

- Catalyst bed:

$$\nabla \cdot (\rho_1 m_{1,i} u_1) = \nabla \cdot (D_{\text{eff},i} \rho_1 \nabla m_{1,i}) + S_{C,1,i} \quad [13.8]$$

- Open area:

$$\nabla \cdot (\rho_1, m_{1,i} u_1) = \nabla \cdot (D_{1,i} \rho_1 \nabla m_{1,i}) + S_{C,1,i} \quad [13.9]$$

Permeation side

(a) Mass conservation equation

$$\nabla \cdot (\rho_2 u_2) = S_{m,2} \quad [13.10]$$

(b) Momentum conservation equation

$$\nabla \cdot (\rho_2 u_2 u_2) = -\nabla P_2 + \nabla \cdot (\mu_2 \nabla u_2) + S_{u,2} \quad [13.11]$$

(c) Energy conservation equation

$$\nabla \cdot (\rho_2 u_2 h_2) = \nabla \cdot (\lambda_2 \nabla T_2) + S_{h,2} \quad [13.12]$$

$$h_2 = \sum_i m_{2,i} \left(h_{f_i} + \int_{298}^{T_2} C_{pi} \mathrm{d}T \right) \quad [13.13]$$

(d) Species mass conservation equation

$$\nabla \cdot \left(\rho_2 m_{2,i} u_2 \right) = \nabla \cdot \left(D_{2,i} \rho_2 \nabla m_{2,i} \right) + S_{C,2,i} \quad [13.14]$$

$S_{u,1}$ and $S_{u,2}$, which represent the momentum exchange across the membrane, are neglected because the momentum exchange across the membrane is comparatively small. $S_{m,1}$ and $S_{m,2}$ show the total mass exchange across the membrane. $S_{c,1,i}$ and $S_{c,2,i}$ represent the species mass exchange across membrane or species mass source/sink due to the reactions occurring in the catalyst bed. In this model, it is assumed that the reaction does not take place on the cells adjacent to the membrane, in order to carry out a stable numerical calculation. Therefore, mass and species mass conservation can be written as $S_{m,1} = -S_{m,2}$ and $S_{c,1,i} = -S_{C,2,i}$ across the membrane. $S_{h,1}$ and $S_{h,2}$ are the exchanged amounts of heat generated due to permeation across the membrane or reaction heat generated in the catalyst bed. With heat conduction across the membrane, the relation, $S_{h,1} = -S_{h,2}$, is assumed to hold.

Reaction, heat and mass transfer in the catalyst bed

The catalyst bed is modelled by porous medium in the CFD code. The following physical models are incorporated utilizing the user subroutines.

Reaction kinetics

The following reaction kinetics model is used for the catalytic reaction of cyclohexane dehydrogenation (Itoh and Wu, 1997).

1. Reaction

$$C_6H_{12} = C_6H_6 + 3H_2 \quad [13.15]$$

2. Reaction rate equation

$$r_v = \frac{k_r \left(K_p \dfrac{P_{C_6H_{12}}}{P_{H_2}^3} - P_{C_6H_6} \right)}{\left(1 + K_D K_P \dfrac{P_{C_6H_{12}}}{P_{H_2}^3} \right)} \left(\text{mol m}^{-3}\ \text{s}^{-1} \right) \quad [13.16]$$

$$k_r = 0.223 \exp\left(-\frac{4270}{T} \right) \left(\text{mol m}^{-3}\ \text{s}^{-1}\ \text{Pa}^{-1} \right) \quad [13.17]$$

$$K_P = 5.06\times 10^{35}\exp\left(-\frac{26490}{T}\right)\left(\mathrm{Pa}^{-3}\right) \quad [13.18]$$

$$K_D = 2.03\times 10^{-10}\exp\left(\frac{6270}{T}\right)\left(\mathrm{Pa}^{-1}\right) \quad [13.19]$$

Effective thermal conductivity

The heat transfer parameters in the catalyst bed are estimated by the following equations (Yagi and Kunii, 1957; Kunii and Smith, 1960).

Effective thermal conductivity λ_{eff} is calculated by the following equations.

$$\frac{\lambda_{\mathrm{eff}}}{\lambda} = \frac{\lambda_{\mathrm{eff}}^0}{\lambda} + (\alpha\beta)\times \mathrm{Pr}\times \mathrm{Re}_p \quad [13.20]$$

$$\mathrm{Re}_p = \frac{\mathrm{d}_p u\rho}{\mu} \quad [13.21]$$

$$\mathrm{Pr} = \frac{Cp\mu}{\lambda} \quad [13.22]$$

$$(\alpha\beta) = -0.298\left(\frac{\mathrm{d}_p}{\mathrm{d}_t}\right) + 0.126 \quad [13.23]$$

λ_{eff}^0 is the effective thermal conductivity without flow and is calculated by the following equation.

$$\frac{\lambda_{\mathrm{eff}}^0}{\lambda} = \varepsilon\left(1 + \frac{\mathrm{d}_p \mathrm{h}_{rv}}{\lambda}\right) + \frac{1-\varepsilon}{\dfrac{1}{(1/\Phi)+(h_{rs}\mathrm{d}_p/\lambda)} + \dfrac{2}{3}\dfrac{\lambda}{\lambda_s}} \quad [13.24]$$

$$h_{rv} = \frac{0.227}{1 + \dfrac{\varepsilon}{2(1-\varepsilon)}\dfrac{1-e}{e}}\left(\frac{T}{100}\right)^3 \left(\mathrm{W\,m^{-2}\,K^{-1}}\right) \quad [13.25]$$

$$h_{rs} = 0.227\frac{e}{2-e}\left(\frac{T}{100}\right)^3 \left(\mathrm{W\,m^{-2}\,K^{-1}}\right) \qquad [13.26]$$

Φ in Equation [13.24] is a function of λ_S / λ, which is given by Kunii and Smith (1960). In this modelling, Φ is approximated by the polynomial function of λ_S / λ.

Effective diffusion coefficient

The effective diffusion coefficient D_{eff} is calculated using the empirical relation postulated for mass transfer in the catalyst bed (Itoh *et al.*, 1994).

$$\frac{1}{\mathrm{Pe}_r} = \frac{0.4}{(\mathrm{Re}_P Sc)^{0.8}} + \frac{0.09}{\left[1+10/(\mathrm{Re}_P\, Sc)\right]} \qquad [13.27]$$

$$Sc = \frac{\mu}{\rho D} \qquad [13.28]$$

$$\mathrm{Re}_P = \frac{\rho u \mathrm{d}_P}{\mu} \qquad [13.29]$$

$$\mathrm{Pe}_r = \frac{u \mathrm{d}_P}{D_{\text{eff}}} \qquad [13.30]$$

Pressure drop

Ergun's equation is used for pressure drop ΔP (Pa m^{-1}) in the catalyst bed (Ergun, 1952).

$$\frac{\Delta P}{\Delta L} = -\frac{150\mu(1-\varepsilon)^2 u}{\varepsilon^3 \mathrm{d}_p^2} - \frac{1.75\rho(1-\varepsilon)u^2}{\varepsilon^3 \mathrm{d}_p} \qquad [13.31]$$

Comparing Equation [13.31] and Equations [13.2] and [13.3] gives the following equations.

$$\alpha = \frac{1.75\rho(1-\varepsilon)}{\varepsilon^3 \mathrm{d}_p} \qquad [13.32]$$

$$\beta = \frac{150\mu(1-\varepsilon)^2}{\varepsilon^2 \mathrm{d}_p^2} \qquad [13.33]$$

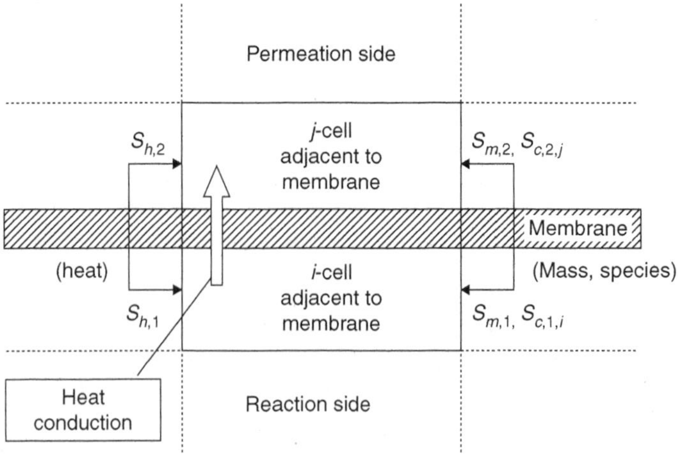

13.2 Model concept for the mass and heat transfer through the membrane.

13.2.3 Mass and heat transfer through the membrane

The membrane is modelled by a heat conducting wall in the CFD code, whereas the mass and heat exchange occurring by permeation through the membrane are represented by the source terms (S_m, S_h) in the basic equations shown above. Figure 13.2 shows the reaction-side cell (*i*-cell) and the permeation-side cell (*j*-cell) to illustrate the model concept. Mass and heat transfer through the membrane are given by mass and heat exchanges between the *i* and *j*-cells. Calculated flux is used for the mass source and sink in the governing equations. These procedures are coded by user subroutines.

Permeation rate is calculated according to the following equation (Itoh and Wu, 1997).

For H_2:

$$J_i = K_{m,i}\left(P_{1,i}^{0.5} - P_{2,i}^{0.5}\right)\left(\text{mol m}^{-2}\text{s}^{-1}\right) \qquad [13.34a]$$

$$K_{m,i} = \frac{Pm_i}{t_m}(\text{mol m}^{-2}\ \text{s}^{-1}\ \text{Pa}^{-0.5}) \qquad [13.35a]$$

For other species:

$$J_i = K_{m,i}\left(P_{1,i} - P_{2,i}\right)\left(\text{mol m}^{-2}\text{s}^{-1}\right) \qquad [13.34b]$$

Table 13.1 Permeation rate parameters

Species	Pm_i in Equations [13.35a] and [13.35b]
H_2	$Pm_H = 2.44 \times 10^{-7} \exp\left(-\frac{1890}{T}\right)$
C_6H_{12}	$Pm_C = 0.0$
C_6H_6	$Pm_D = 0.0$
N_2	$Pm_N = 0.0$

$$K_{m,i} = \frac{Pm_i}{t_m} \ (\text{mol m}^{-2}\text{s}^{-1}\text{Pa}^{-1}) \qquad [13.35b]$$

The permeation rate parameter used in this study is shown in Table 13.1. The composite palladium membrane, which was prepared by the chemical vapour deposition (CVD) technique in this study, was found to have a very large selectivity for hydrogen (ca. 10 000 of ideal selectivity for H_2/N_2 at 300°C), so that the permeabilities of the other components were assumed to be zero (Itoh *et al.*, 2007). Heat transfer through the membrane takes place in two ways: that is, heat conduction, and heat exchange by the permeation. Heat conduction through the membrane is modelled by the CFD code function. Thermal conductivity of 5 W/(m K) is applied to the membrane and support material. Heat exchange through the membrane is calculated as the sum of the permeation flux and enthalpy for each species.

13.2.4 CFD model description of single membrane tube reactor

Figure 13.3 shows the two-dimensional CFD model, which can be reduced to 1/24th of the experimental reactor (Fig. 13.1) because of its cylindrical symmetry. There are 8700 and 1150 meshes for the reaction and permeation sides, respectively,totalling 9850 meshes. Boundary conditions are shown in Fig. 13.4.

13.2.5 Simulation results

Simulation for the gas separation test

Theoretical hydrogen recovery

In order to verify the CFD model, first, simulation was carried out when the membrane reactor was used as a gas separator, in which a mixture

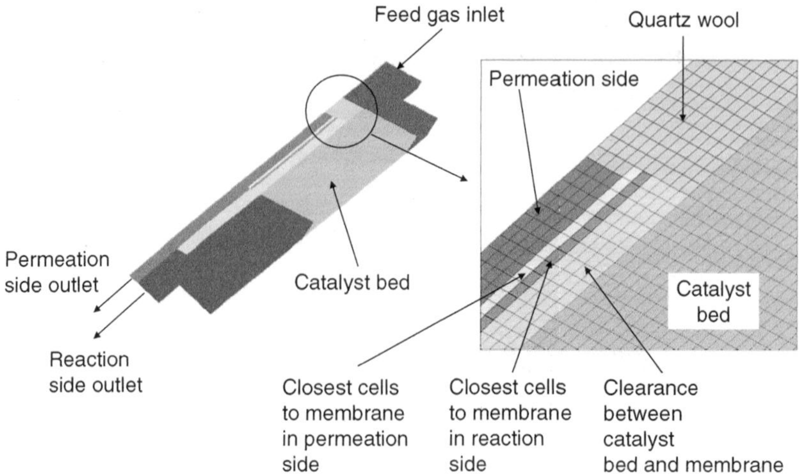

13.3 CFD model.

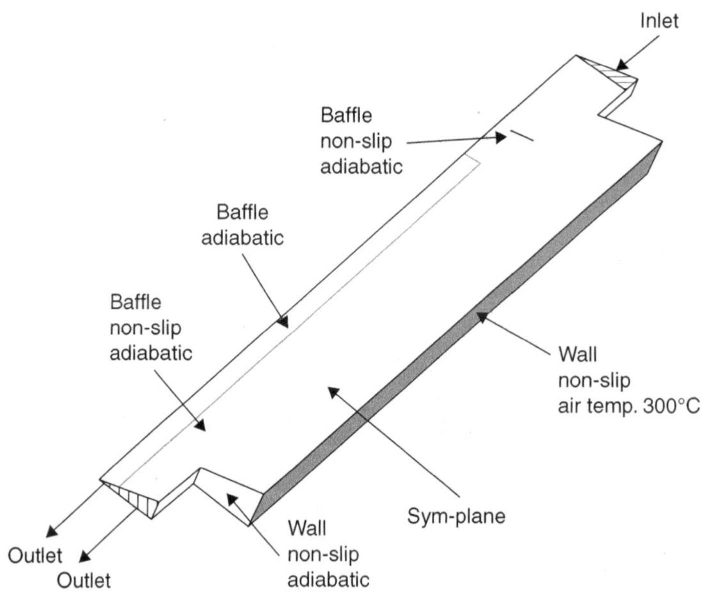

13.4 Boundary condition.

containing hydrogen is fed and the amount of hydrogen permeated through the palladium membrane was measured. Table 13.2 shows the experimental conditions of six cases (Cases 10–15) studied.

'Hydrogen recovery' as used here is defined by the following equation.

Table 13.2 Experimental conditions for the gas separation test at 300°C, where the permeation side is kept 1 bar

Case	Reaction-side pressure, Pr (bar)	Feed rate (cm^3/min) at 25°C, 1 atm		R_{lim} (%)
		H_2	N_2	
10	3	100	33	83.3
11		150	50	
12		200	67	
13	4	100	33	88.8
14		150	50	
15		200	67	

$$\text{Hydrogen recovery} = \frac{R_{\mathrm{H}}}{Q_{\mathrm{H}}^{0}} \times 100 \qquad [13.36]$$

Here, R_{H} (cm^3/min) is hydrogen recovery rate and Q_{H}^{0} (cm^3/min) is hydrogen feed rate.

Hydrogen recovery should reach a limit when the hydrogen partial pressures on the reaction and permeation sides are the same. The theoretical limit, R_{lim} (%), can be given as:

$$R_{\mathrm{lim}} = \frac{1 - \left(P_p / P_r \right)\left(1 / m_{\mathrm{H}}^{0} \right)}{1 - \left(P_p / P_r \right)} \qquad [13.37]$$

Here, m_{H}^{0} is the molar fraction of hydrogen in feed, and P_r and P_p are total pressures on the reaction and permeation sides, respectively. Values for R_{lim} calculated in practice are presented in Table 13.2.

Simulation result

Figure 13.5 shows the experimental and simulation results together with the theoretical hydrogen recovery limit, R_{lim}. All experimental data were found to be close to the CFD simulation results. It is obvious that both decrease as the feed rate increases. This is because the actual hydrogen recovery decreases with decreasing residence time.

This means that the CFD model developed for the membrane reactor in this study is valid in terms of utilization as a gas separator without reaction. It will be tested in the next section whether the model is applicable to the membrane reactor.

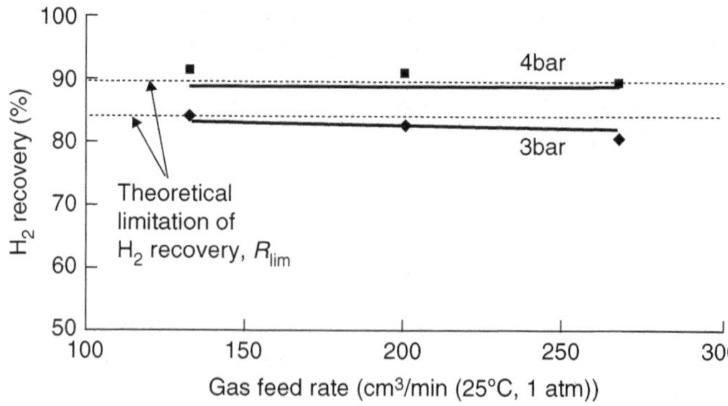

13.5 Experimental and simulation results for the gas separation test using 75%H_2–25%N_2 mixture.

Table 13.3 Experimental conditions for the cyclohexane dehydrogenation

Case	Reaction side				Permeation side
	Inlet temperature (°C)	Inlet pressure (bar)	Inlet flow rate (ml/min)	C_6H_{12} concentration (%)	Pressure (bar)
30	300	3	0.080	100	0.1
31	300	3	0.161	100	0.1
32	300	3	0.241	100	0.1
33	300	3	0.322	100	0.1
34	300	3	0.080	100	1
35	300	3	0.241	100	1
36	300	3	0.322	100	1

Validity of the CFD model for the membrane reactor

Reaction conditions

In order to verify the reactor model, a numerical simulation was carried out for the cyclohexane dehydrogenation experiment. Table 13.3 shows the conditions of the seven cases (Cases 30–36) studied.

Simulation result

1. Changes in temperature and concentration along the reactor

   Since the membrane reactor used in this study was placed in an air oven at 300 °C, the heat-flux from air to wall, q_{wall}, is given by the following equation.

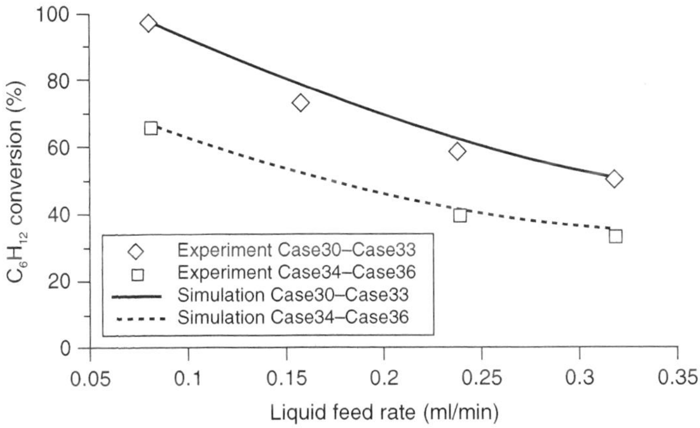

13.6 Experimental and simulation results for the cyclohexane dehydrogenation.

$$q_{\text{wall}} = h_{\text{wall}} \cdot \left(T_{\text{air}} - T_{\text{wall}}\right) \tag{13.38}$$

Here, h_{wall} is the heat transfer coefficient, T_{air} is the air temperature (300°C) and T_{wall} is the wall temperature.

A value of 10 W/(m^2 K) for h_{wall} here, which was determined so as to meet the experimental data, is considered to be reasonable when referring to the empirical data. Figure 13.6 shows the comparison between experimental data and simulation results for cyclohexane conversion.

The CFD simulation results can be recognized to be in good agreement with experimental data. From these results, it can be attested that the developed model had sufficient accuracy to describe the reaction and separation occurring in the membrane reactor. If the reactor has simple geometry, a one-dimensional (1-D) model can be applied to such a simulation. However, the 1-D simulation, assuming a uniform temperature profile and plug-flow in the catalyst bed, is likely to lead to a prediction with large deviation from the experimental result. If the reactor geometry is complicated, such as the multi-tubular membrane reactor, the CFD model will be a more useful and extendable tool.

Figures 13.7–13.10 show the simulation results for temperature and cyclohexane conversion profiles along the catalyst bed, which are averaged in the radial direction at each catalyst bed position. It was found that the temperature suddenly drops by *ca.* 30°C near the inlet of the catalyst bed, due to a large endothermic heat, and then gradually increases along the length of the reactor. However, as the feed supply rate increases, the temperature recovery was found to be later. This is because the heat supply rate from the shell-tube wall is insufficient in the air-circulating oven,

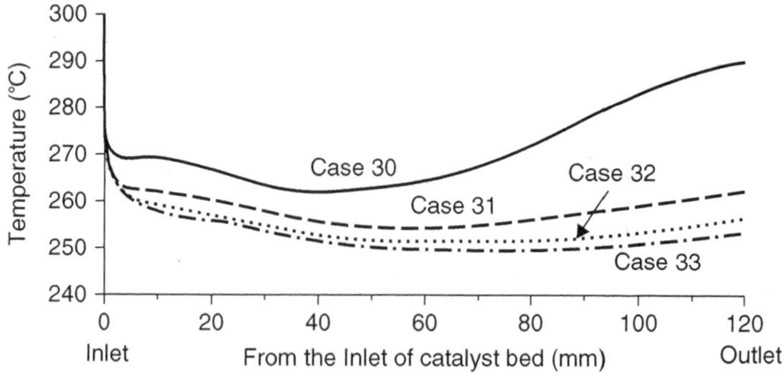

13.7 Temperature profile along the catalyst bed for Case 30 to Case 33 (permeation-side pressure: 0.1 bar).

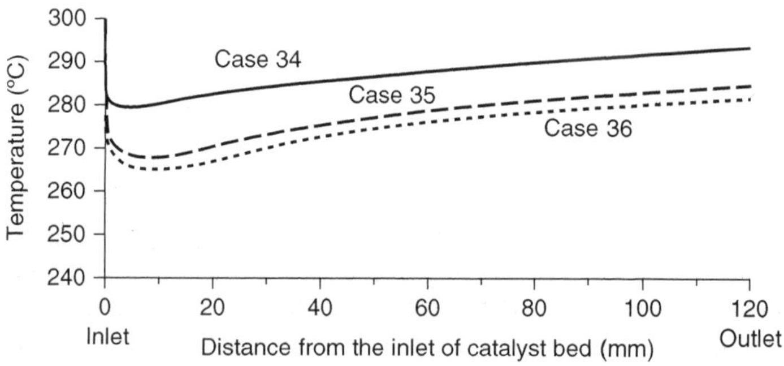

13.8 Temperature profile along the catalyst bed for Case 34 to Case 36 (permeation-side pressure: 1 bar).

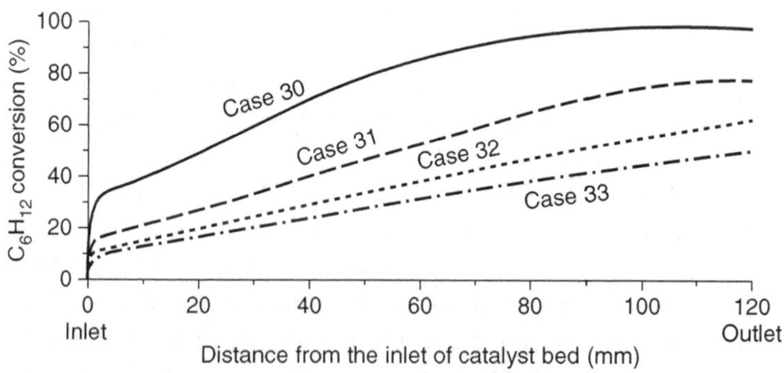

13.9 Cyclohexane conversion profile along the catalyst bed for Case 30 to Case 33 (permeation-side pressure: 0.1 bar).

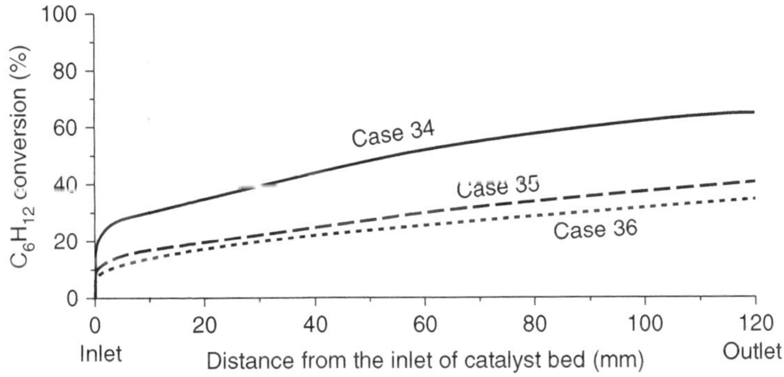

13.10 Cyclohexane conversion profile along the catalyst bed for Case 34 to Case 36 (permeation-side pressure: 1 bar).

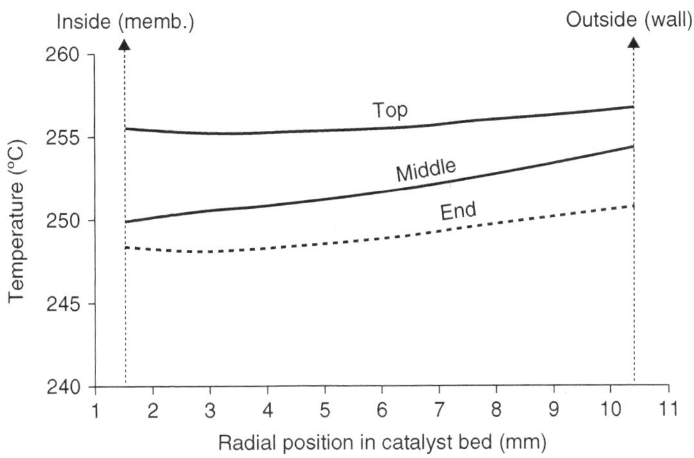

13.11 Typical radial temperature profiles (Case 33).

where the heat transfer coefficient, h_{wall}, is as small as 10 W/(m^2 K). If the heating manner were improved, for instance using a sand bath, the reaction performance would be greatly improved.

In the catalyst-bed temperatures as above, ranging from 250°C to 290°C, the equilibrium cyclohexane conversion at 3 bar was calculated to be 12% to 36%. Therefore, it has been shown that the experimentally obtained conversions can exceed the equilibrium ones in all cases.

2. Radial profile of temperature and concentration

Figures 13.11 and 13.12 show the radial temperature and concentration profile at three positions along the reactor (top middle, and end of the membrane). As can be seen in these results, the catalyst-bed temperature profile has some variations in the radial direction of the catalyst bed,

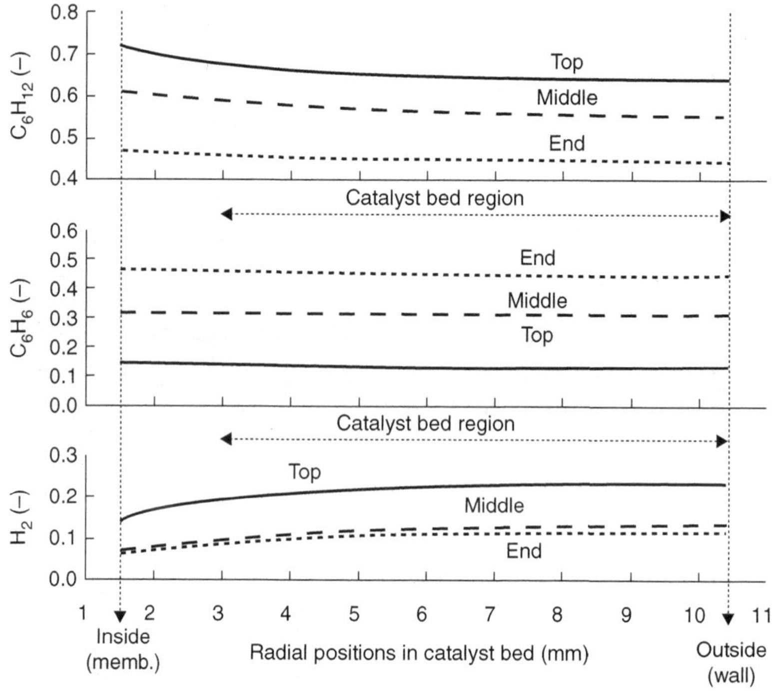

13.12 Typical radial concentration (mol fraction) profile (Case 33).

while there is no significant concentration distribution in the radial direction except for the clearance region.

In the experimental conditions, Re_p in Equation [13.29] is a relatively low value (less than 3.6), so that Pe_r calculated by Equation [13.27] becomes low. Pe_r of hydrogen is lower than other species because its molecular diffusion coefficient is much higher than for other species. This means that hydrogen with a higher effective diffusivity, D_{eff}, can move faster than cyclohexane and benzene in the radial direction. From Fig. 13.12, it can be clearly seen that the radial concentration profile of hydrogen is almost flat in the radial direction, although hydrogen is moving toward the membrane.

The typical temperature and species-concentration contour map obtained by the CFD simulation are shown in Fig. 13.13. This map emphasizes that the heat supply to the catalyst bed is very important for membrane reactor performance.

3. Effect of the clearance on conversion

It has also been found that the clearance region between the membrane and the catalyst bed (1.2 mm) has some effect on the temperature and

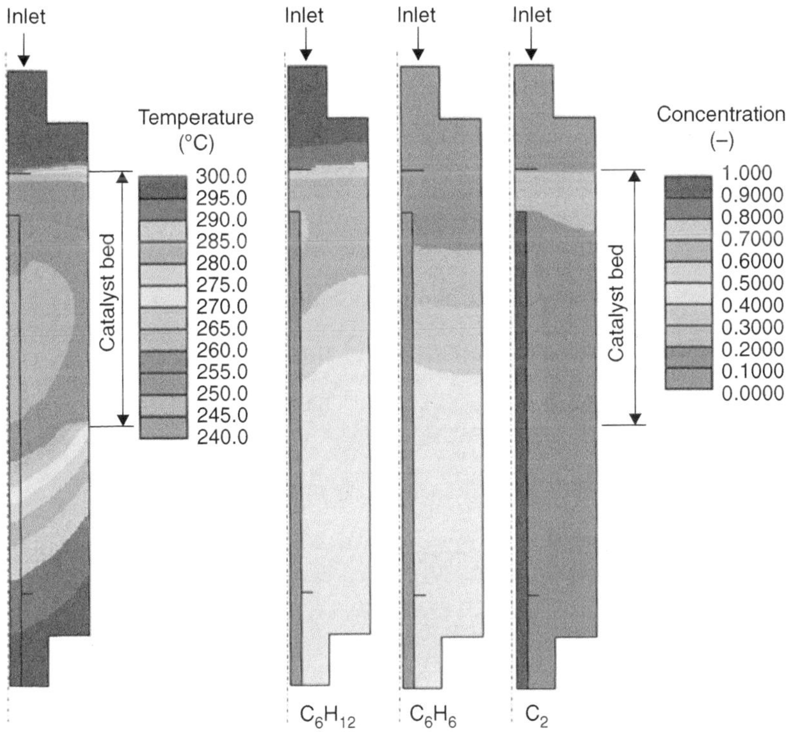

13.13 Typical temperature and concentration (mol. fraction) contour map (Case 33).

species-concentration profile. A part of the reacting gas enters the clearance space near the top of membrane because the pressure drop in the hollow region is smaller than that of the catalyst bed. Such so-called short-circuiting should make the hydrogen concentration near the membrane low, thereby reducing the hydrogen permeation rate.

Additional runs (Cases 40 and 41), and their CFD analyses, were carried out to examine the short-circuiting effect due to the clearance under the conditions shown in Table 13.4. Figure 13.14 shows the contour map of H_2 concentration and gas velocity for comparing the difference between a small clearance with 0.25 mm and a large one with 1.2 mm. The small clearance was obtained by changing from a stainless-steel mesh type protector to a spring type one. According to Fig. 13.14, one can see a higher velocity in the clearance region (near the membrane), that is, a short-circuiting effect. Simultaneously, a lower hydrogen concentration region appears near the membrane and extends in the downstream direction in the large clearance case (Case 40).

Simulation and experimental results for cyclohexane conversion are also shown in Table 13.4, From these results, it was found that they are

Table 13.4 Experimental conditions for examining the effect of clearance between membrane and catalyst bed

Case		Case 40	Case 41
Clearance between membrane and catalyst bed		Large 1.2 mm	Small 0.25 mm
Reaction side	Inlet temperature (°C)	300	
	Inlet pressure (bar)	4	
	Inlet liquid feed rate (ml/min)	0.322	
	Inlet concentration of C_6H_{12} (%)	100	
Permeation side	Pressure (bar)	0.1	
C_6H_{12} conversion (%)	Experiment	59	64
	Simulation	58.2	63.6

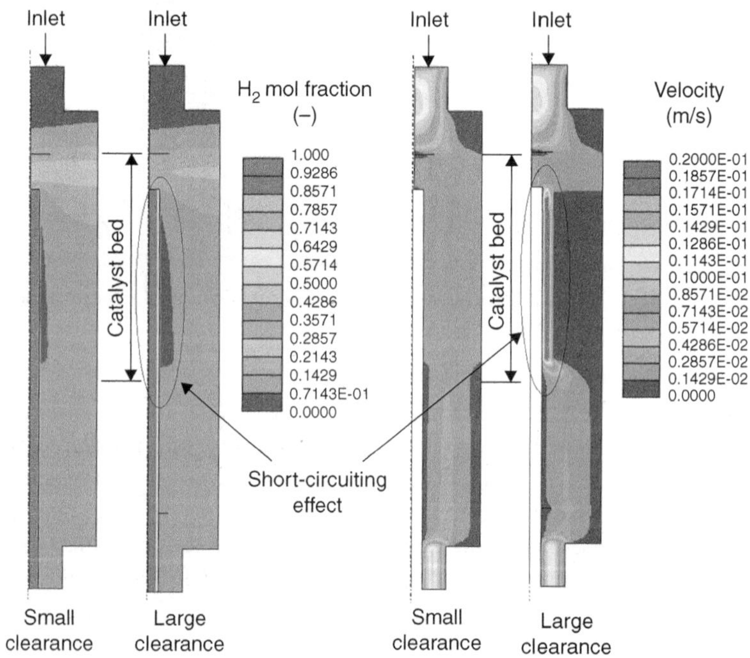

13.14 Effect of the clearance between membrane and catalyst bed on the H_2 mol fraction (left) and gas velocity (right).

in a good agreement and the experimental cyclohexane conversion can be improved from 59 to 64% by reducing the clearance. Hence the CFD model developed can also describe the effect of clearance on the cyclohexane conversion. These results suggest that minimizing the clearance is very important, although the membrane tube itself must be protected from contact with catalyst particles.

Summary for single tube membrane reactor

A palladium membrane reactor has been applied as a promising chemical hydrogen carrier to recover hydrogen from cyclohexane. An elaborate model for the membrane reactor is required to analyse the performance, to optimize the membrane reactor design, and to obtain information for scale-up. In this study, a CFD model using a commercial code (Star-CD v.3.2) was developed by taking into account the mass and heat transfer, the fluid flow, and hydrogen permeation, and its validity was tested for the dehydrogenation of cyclohexane in a single-tube type palladium membrane reactor.

The simulation results were in good agreement with the experimental data. It was found that the heat transfer to the catalyst-packed bed played an important role, especially for endothermic dehydrogenation with a large amount of heat. The clearance between the membrane tube and the protector was found to decrease the conversion of cyclohexane because of the short-circuiting effect, suggesting that one should introduce a different type of protector or a kind of catalyst block to minimize the clearance.

In conclusion, it was proved that the CFD model could be applied to a shell-and-tube membrane reactor. All of the methods developed and the results obtained are expected to be easily extended to the design and analysis of more complicated reactors such as multi-tubular membrane reactors.

13.3 Multi-tube palladium membrane reactor

In order to realize a future hydrogen society, various R&D projects related to hydrogen are ongoing all over the world. As one of national projects in Japan, hydrogen storage and transport systems using chemical hydrides such as cyclohexane and methyl-cyclohexane, which have a comparatively large content of hydrogen exceeding 6 wt%, are being widely investigated. As mentioned above, to recover high purity hydrogen from the chemical carriers, a palladium membrane reactor is expected to be more efficient than the conventional fixed bed type of equilibrium reactor. It is now time to discuss the possibility of scale-up from a single membrane tube to a multi-tube type, in terms of increasing the productivity for practical application.

Similarly to the single-tube type, a CFD model for analysing and designing a multi-tube type of palladium membrane reactor will be developed, where the three-dimensional mass and heat transfer for the dehydrogenation of cyclohexane occurring in the catalyst-packed bed is taken into account (Mimura *et al.*, 2010b).

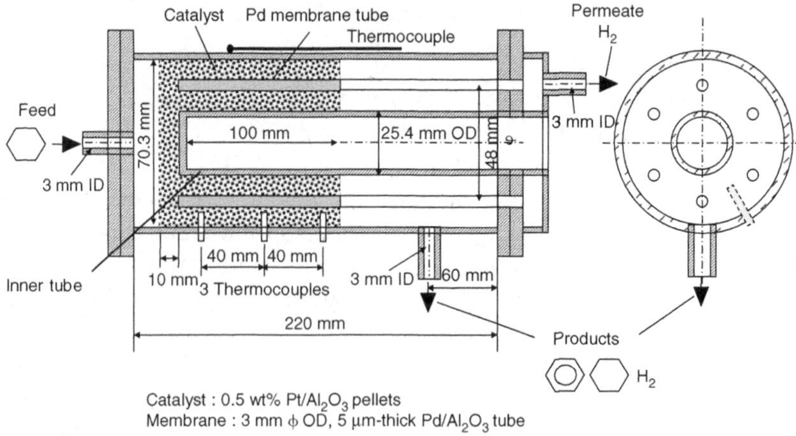

13.15 Cross-sectional view of the multi-tubular palladium membrane reactor.

13.3.1 Experimental data

The multi-tube palladium membrane reactor used in this study is shown in Fig. 13.15. The shell tube is made of stainless-steel with a length of 220 mm and an inside diameter of 70.3 mm, at the centre of which an inner tube (25.4 mm in outer diameter) is inserted to increase the heat transfer area. The 5 μm-thick palladium membrane supported on a porous α-Al_2O_3 tube 100 mm long and 3 mm in outer diameter was prepared by means of CVD, having a high permselectivity of more than 10 000 for H_2/N_2. The six palladium membrane tubes, each of which was covered with a stainless-steel mesh tube to prevent contact with the catalyst pellets, were inserted into the concentric area of the reactor, where 0.5 wt% Pt/Al_2O_3 cylindrical pellets (3.2 mm diameter and 3.6 mm height) were packed uniformly along 110 mm of reactor length. The clearance between the membrane tube and stainless-steel mesh was 1.2 mm. The reactor was heated with a tape heater winded around the shell wall and the wall temperatures along the length were monitored by three TCs.

The experimental set-up is shown in Fig. 13.16. Liquid cyclohexane as feed is vaporized through a pre-heater, then sent to the membrane reactor. The reaction pressure is regulated in the range 2–3 bars with a back-pressure regulator, while the perm-side is evacuated to 0.1 bars with a vacuum pressure controller. The permeation rate of hydrogen is measured with an accumulative flow meter. Experiments for the gas separation using a gaseous mixture (75% H_2 and 25% N_2) as well as cyclohexane dehydrogenation were carried out at 573 K.

Practical conditions and results are summarized in Table 13.5.

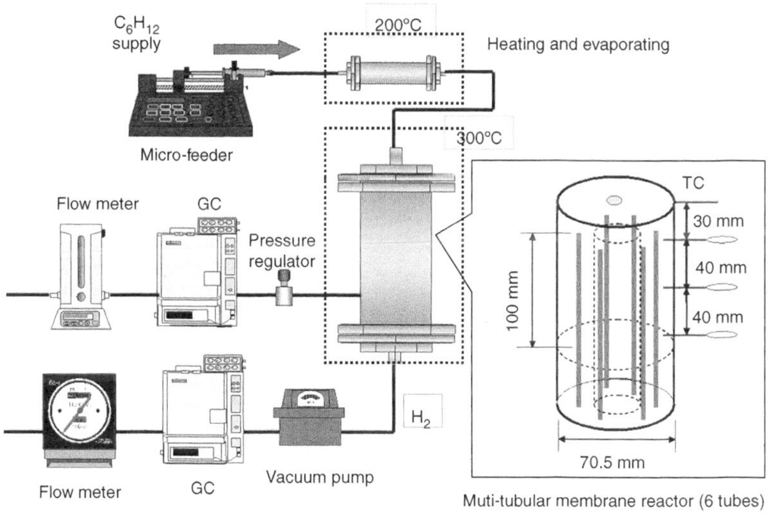

13.16 Experimental apparatus. (GC, gas chromatograph.)

Table 13.5 Experimental conditions and results for the cyclohexane dehydrogenation using the multi-tubular palladium membrane reactor

Case	Units	Case 21	Case 22	Case 23	Case 24	Case 25
Reaction side						
Inlet temperature	°C	300	300	300	300	300
Inlet pressure	Bar (abs.)	2	2	2	2.5	3
Liquid feed rate	ml/min	2.20	1.80	1.55	2.10	2.10
Feed composition						
C_6H_{12}	Vol%	100	100	100	100	100
Permeation side						
Pressure	Bar (abs.)	0.1	0.1	0.1	0.1	0.1
Heating condition						
Shell wall temperature	°C	340	340	340	340	340
Inner pipe wall		300	300	300	300	300
C_6H_{12} conversion						
Experiment	%	80	83	89	76	74
Simulation	%	79.7	85.2	89.9	77.4	75.4
Permeated H_2						
Experiment	Ncc/min	800	733	702	788	782
Simulation	Ncc/min	801	755	718	804	811

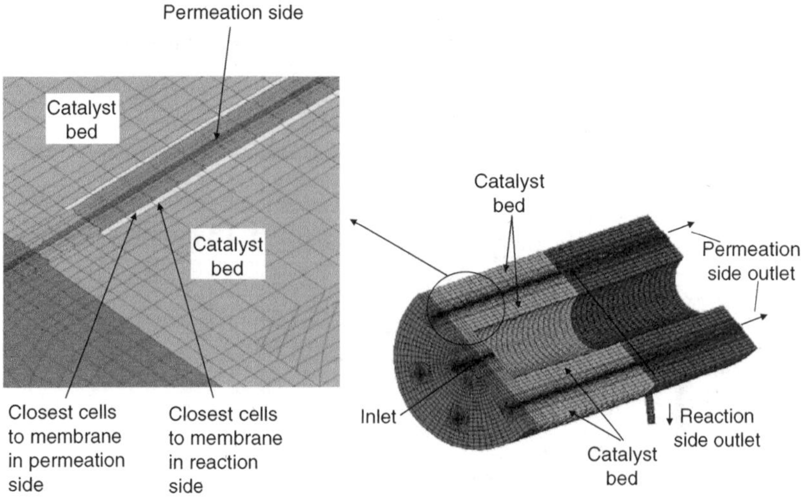

13.17 Model for the mass and heat transfer through the membrane.

13.3.2 Three-dimensional CFD model for the multi-tube palladium membrane reactor

The numerical model for analysing the membrane reactor was developed using a commercial CFD code, Star-CD v3.2. The modelling procedure is almost the same as that for the single tube described above. Equations [13.1–13.35a] were also used for the mass, momentum, species and energy conservation on the reaction and permeation sides, the reaction rate, heat and mass transfer rates, pressure drop and permeability of hydrogen.

Figure 13.17 shows the three-dimensional CFD model constructed, with a total of 140 000 meshes (107 000 for the reaction side and 33 000 for the permeation side). The boundary conditions are shown in Fig. 13.18.

13.3.3 Simulation results

Conversion of cyclohexane

Figure 13.19 shows a comparison between the experimental (plots) and simulated (lines) results for Cases 21–23 in Table 13.5, where both conversion and hydrogen permeation rate can be found to be in good agreement. This indicates that the CFD model developed here is valid for analysing the multi-tubular membrane reactor designed in this text.

Changes of temperature and composition along the reactor length

The contour map of temperature and molar fractions of three components, C_6H_{12}, C_6H_6 and H_2, in the axial cross-section of the membrane reactor

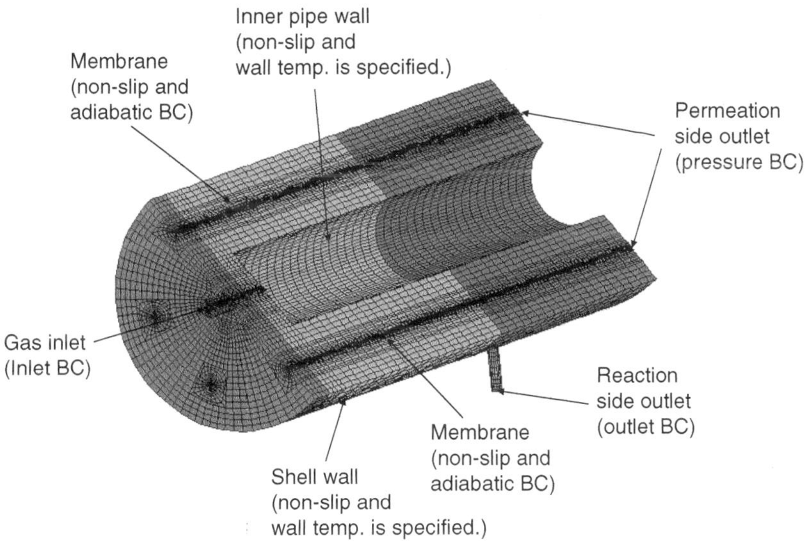

13.18 Boundary conditions (BCs).

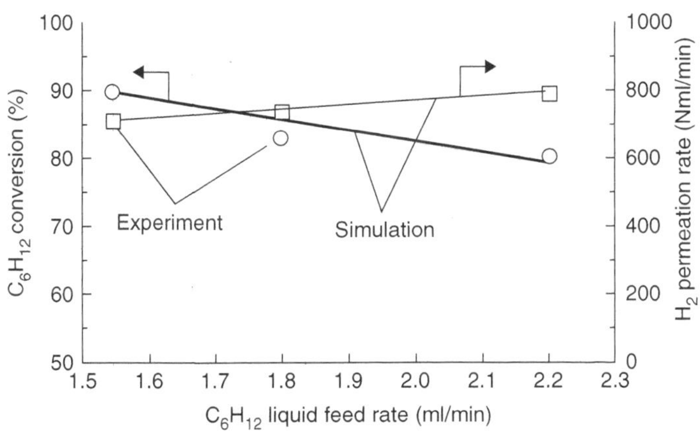

13.19 Simulation and experimental results for the cyclohexane dehydrogenation using the multi-tubular palladium membrane reactor.

(Case 21) is presented in Fig. 13.20. It is found that the temperature drops quickly near the inlet and then gradually rises toward the outlet. This is because the heat transfer rate is not large enough to supply the endothermic reaction heat. With reference to the composition inside the palladium membrane tube, it is clearly seen that the hydrogen mole fraction inside the membrane tube becomes 1 and those for cyclohexane and benzene are zero. Interestingly, one can see that the hydrogen fraction around the membrane

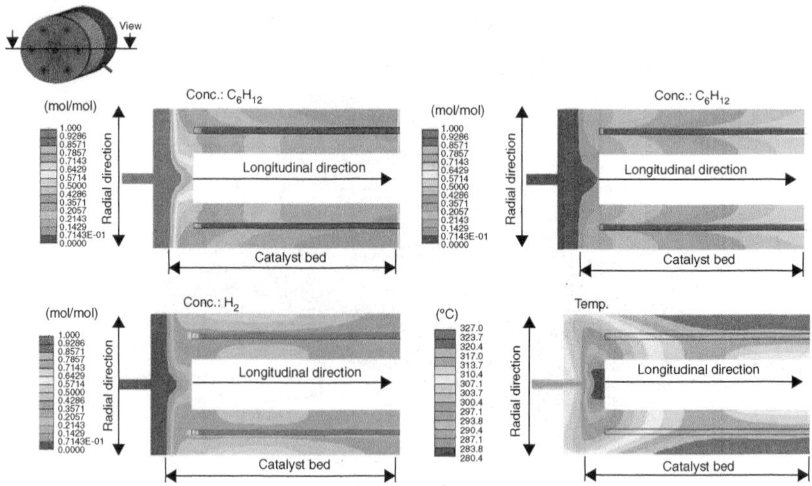

13.20 Temperature and concentration contour map of the longitudinal plane for Case 21.

tube is small, due to its permeation, and instead the fractions of cyclohexane and benzene become relatively high. Therefore, one can expect that the dehydrogenation will be enhanced more near the membrane if the heat transfer method is improved.

A profile of temperature averaged over radial section along the reactor length, which is almost same as that at the position of the TC except near the inlet, is shown by the solid line together with the square plots measured with three TCs in Fig. 13.21. The simulation can almost explain the temperature changes. Next, changes in each composition along the catalyst bed are simulated in Fig. 13.22. From this, it can be seen that initial large changes followed by gradual changes occur. Especially, although hydrogen fraction seems to change only a little from the position 0 mm to 100 mm, the selective separation of hydrogen evolved can enhance dehydrogenation; that is, the conversion at 0 mm of the reactor length is 0.3 and 0.78 at 100 mm. This suggests that much more enhancement of the reaction will be realized if the hydrogen permeation rate is increased for instance, by increasing the membrane area (using thicker membrane tubes).

Radial profile of temperature and composition

Radial contour maps with the catalyst-bed temperature are shown in Fig. 13.23. In all the maps, as the radial position approaches the centre, the temperature drops, because the shell wall is heated but the inner tube wall is not. The temperature near the membrane tube is found to be somewhat lower compared with the same radial position apart from the membrane.

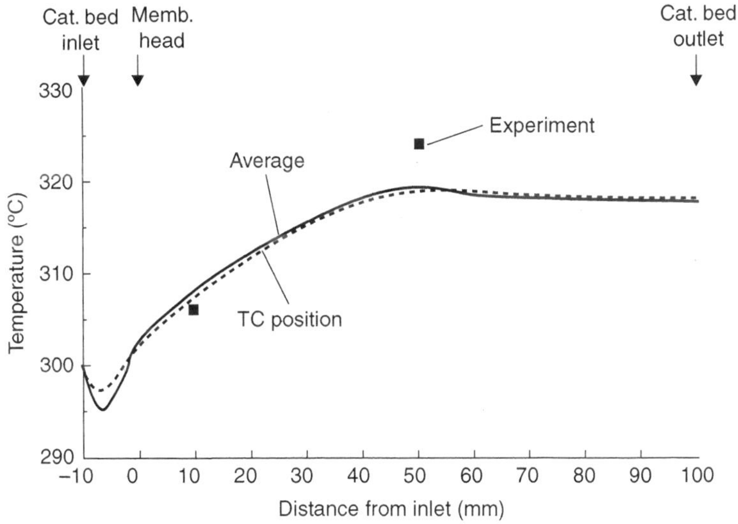

13.21 Temperature profile along the catalyst bed for Case 21. (Thermocouple, TC, is set at 20 mm inside from the shell wall.)

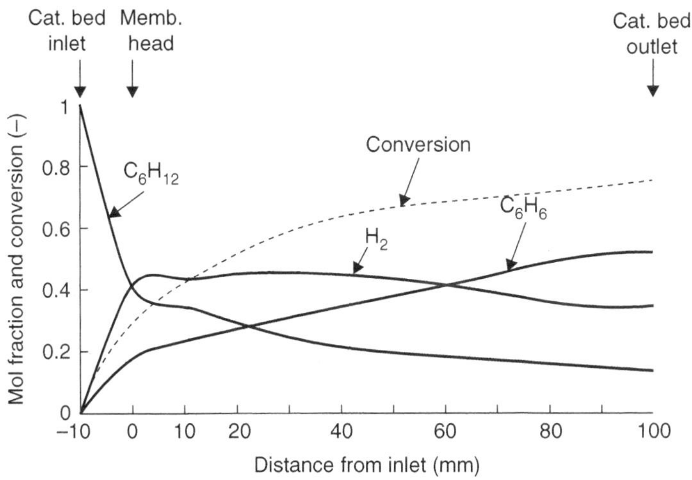

13.22 Concentration profile along the catalyst bed for Case 21.

This is, as mentioned above, because the reaction with an endothermic heat can be accelerated in a lower hydrogen atmosphere around the membrane tube. Actually, it is recognized from Fig. 13.24 that the hydrogen molar fraction around the tube is considerably lower. Further, when one inspects the contour lines of hydrogen around the tube carefully, it becomes clear that they are not formed in concentric circles. Figure 13.25 suggests that such phenomena are attributable to heat transfer in the packed layer.

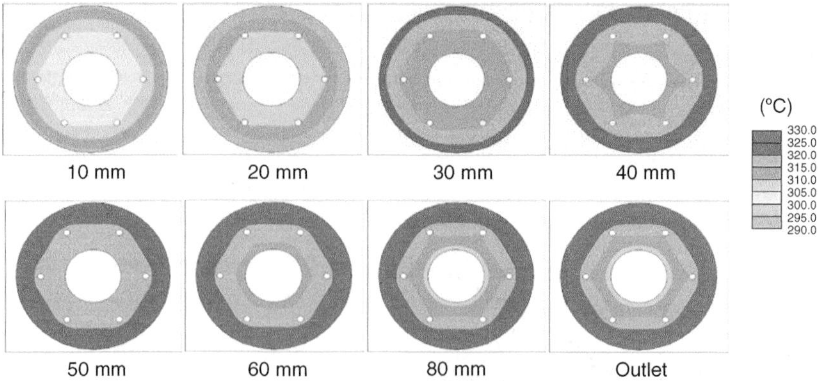

13.23 Temperature contour map of the several transverse planes for Case 21.

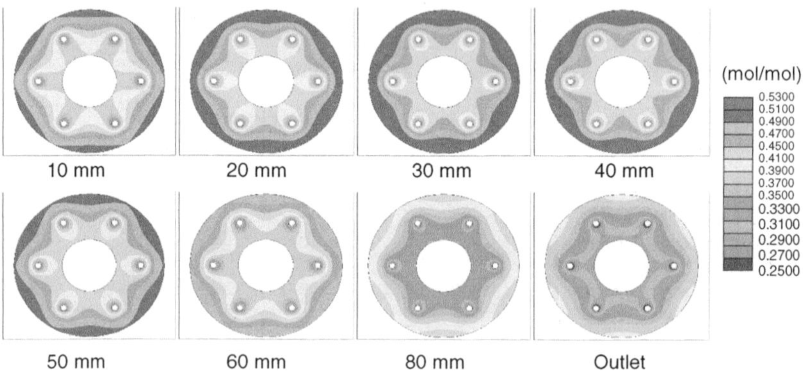

13.24 Hydrogen concentration contour map of the several transverse planes (Case 21).

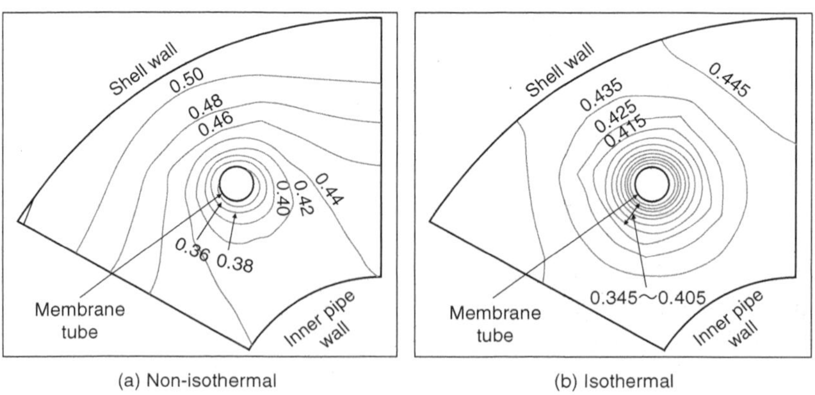

13.25 Hydrogen concentration (mole fraction) contour map for (a) non-isothermal and (b) isothermal conditions at the 20 mm distance from the membrane tube head (Case 21).

13.3.4 Scheme for improving the membrane reactor performance

It has been shown so far that the CFD model describing the multi-tube membrane reactor very well can be constructed successfully. Therefore, in this section, further simulation using the developed CFD model has been conducted to improve the performance of the membrane reactor in terms of heat transfer, amount of catalyst and membrane area.

Heating from the inner tube

The inner tube functioned actually as a heat sink leading heat loss, rather than heat pipe originally designed in the above experiments, because heating was only carried out by the tape heater bound around the reactor shell wall. So, if the heat transfer condition with the inner tube is changed, some improvement in the reactor performance will be expected.

First, let us consider the wall of the inner tube as a heat insulator, which can keep the adiabatic condition without any heat loss. The simulation result is shown as Case 51 in Table 13.6. When comparing with Case 21 (non-adiabatic), the conversion of cyclohexane is found to increase from 79.7% to 84.7%, that is, 5% up.

A second simulation was made on the assumption that the wall temperature of inner tube is the same as that of the shell wall, 340°C (Case 52). The conversion increases 13.7% more than that in Case 21. Also, the hydrogen recovery on the perm-side increases from 800 to 954 ml/min.

Amount of catalyst packed

The catalyst was packed over the 10 mm length, called a pre-reaction zone, between the inlet and the head of membrane tube, as shown in Fig. 13.15. In the pre-reaction zone, the partial pressure of hydrogen is elevated before entering into the membrane installed zone. By doing so, the back permeation of hydrogen from the perm-side to the reaction side will be prevented, which takes place when the hydrogen pressure on the perm-side is higher than that on the reaction side. However, such a pre-reaction zone may be unnecessary if the reaction can proceed fast at a higher temperature. Case 53 shows the results of removing the catalyst in the pre-reaction zone, where the other conditions are the same as those in Case 51. The two are consistent, so that one can say that the removal of the catalyst in the pre-reaction zone is possible.

Diameter of membrane tube

The tubular membrane area is given by $\pi d_m L$, so that it is directly proportional to the diameter of tubular membrane, d_m, and therefore the hydrogen

Table 13.6 Conditions selected for improving the reactor performance by the CFD model developed

	Units	Case 51	Case 52	Case 53	Case 54
Improvement method		No heat loss from inner pipe wall for Case 21	Heating for inner pipe wall for Case 21	Catalyst reduction for Case 21	Membrane tube diameter 6 mm for Case 21
Reaction side					
Feed temperature	°C	300	300	300	300
Feed pressure	bar (abs.)	2	2	2	2
Feed flow rate (liq.)	ml/min	2.20	2.20	2.20	2.20
Feed composition					
C_6H_{12}	vol%	100	100	100	100
Permeation side					
Pressure	bar (abs.)	0.1	0.1	0.1	0.1
Heating condition					
Shell wall temperature	°C	340	340	340	340
Inner pipe wall		Adiabatic	340	Adiabatic	300
C_6H_{12} conversion					
Simulation	%	84.7	92.4	84.7	94.9
Permeated H_2					
Simulation (Case 21:800)	Ncc/min	840	954	840	1154

permeation rate is directly proportional to the membrane area. Accordingly, using thicker tube should lead to higher conversion by separating more hydrogen evolved.

When the present diameter of 3 mm is increased to 6 mm (Case 54), the conversion increases by 15.2% and the hydrogen recovery by 354 ml/min. In this case, although the catalyst volume decreased by 4% by increasing the tube diameter, there was no particular influence on the reaction progress.

Comparing four cases (Cases 51–54), it is made clear that the membrane area is the most effective for improving the performance, as anticipated in the Section 2.3.4.

Table 13.6 shows conditions selected for improving the reactor performance by the CFD model developed.

13.4 Conclusions and future trends

A CFD model for analysing and designing a multi-tube type of palladium membrane reactor was presented, where the three-dimensional mass and heat transfer for the dehydrogenation of cyclohexane occurring in the catalyst-packed bed were taken into account. Simulation results were in good agreement with experimental data. In addition, it was demonstrated that the developed model could be used for further simulation, such as improvement of the reactor performance by an increase in the diameter of membrane tube. The CFD model thus developed will be applicable to understanding the internal phenomena and evaluation of the reactor performance improvement quantitatively, and therefore will be a very useful tool for testing novel ideas, scale-up and optimizing the membrane reactor design.

13.5 References

Assabumrungrat, S., K. Suksomboon, P. Praserthdam, T. Tagawa and S. Goto 'Simulation of a palladium membrane reactor for dehydrogenetion of ethylbenzene', *J. Chem. Eng. Japan*, **35**, 263–273 (2002).

Barbieri, G. and F. P. Di Maio 'Simulation of the methane steam re-forming process in a catalytic Pd-membrane reactor', *Ind. Eng. Chem. Res.*, **36**, 2121–2127 (1997).

Basile, A., L. Paturzo and F. Lagana 'The partial oxidation of methane to syngas in a palladium membrane reactor: simulation and experimental studies', *Catal. Today*, **67**, 65–75 (2001).

De Falco, M., L. Di Paola, L. Marrelli and P. Nardella 'Simulation of large-scale membrane reformers by a two-dimensional model', *Chem. Eng. J.*, **128**, 115–125 (2007).

De Falco, M., P. Nardella, L. Marrelli, L. Di Paola, A. Basile and F. Gallucci 'The effect of heat-flux profile and of other geometric and operating variables in designing industrial membrane methane steam reformers', *Chem. Eng. J.*, **138**, 442–451 (2008).

Ergun, S. 'Fluid flow through packed columns', *Chem. Eng. Prog.*, **48**, 89–94 (1952).

Itoh, N. 'A membrane reactor using palladium', *AIChE J.*, **33**, 1576–1578 (1987).

Itoh, N. and T. Wu 'An adiabatic type of palladium membrane reactor for coupling endothermic and exothermic reactions', *J. Membrane Sci.*, **124**, 213–222 (1997).

Itoh, N., W. C. Xu and K. Haraya 'Radial mixing diffusion of hydrogen in a packed-bed type of palladium membrane reactor', *Ind. Eng. Chem. Res.*, **33**, 197–202 (1994).

Itoh, N., E. Tamura, S. Hara, T. Takahashi, A. Shono, K. Satoh and T. Namba 'Hydrogen recovery from cyclohexane as a chemical hydrogen carrier using a palladium membrane reactor,' *Catal. Today*, **82**, 119–125 (2003).

Itoh, N., T. Akiha and T. Sato 'Formation process of thin palladium membrane on a porous support by MOCVD method', *J. Chem. Eng. Jpn.*, **33**, 211–217 (2007).

Kim, J. H., B. S. Choi and J. Yi 'Modified simulation of methane steam reforming in Pd-membrane/packed-bed type reactor', *J. Chem. Eng. Jpn.*, **32**, 760–769 (1999).

Kunii, D. and J. M. Smith 'Heat transfer characteristics of porous rocks', *AIChE J.*, **6**, 71–78 (1960).

Mimura, K., D. Oka, T. Sato and N. Itoh 'CFD analysis of a single palladium membrane tube reactor for the dehydrogenation of cyclohexane as a chemical hydrogen carrier', *J. Chem. Eng., Jpn.*, **43**, 757–766 (2010a).

Mimura, K., N. Yoshida, T. Sato and N. Itoh 'CFD analysis and design of multi-tubular membrane reactor for dehydrogenation of cyclohexane', *J. Jpn. Petrol. Inst.*, **53**, 283–291 (2010b).

Nair, B. K. R. and M. P. Harold 'Hydrogen generation in a Pd membrane fuel processor: productivity effects during methanol steam reforming', *Chem. Eng. Sci.*, **61**, 6616–6636 (2006).

Smit, J., G. J. Bekink, M. van Sint Annaland and J. A. M. Kuipers 'Experimental demonstration of the reverse flow catalytic membrane reactor concept for energy efficient syngas production. Part 2: model development', *Chem. Eng. Sci.*, **62**, 1251–1262 (2007).

Tiemersma, T. P., C. S. Patil, M. van Sint Annaland and J.A. M. Kuipers 'Modelling of packed bed membrane reactors for autothermal production of ultrapure hydrogen', *Chem. Eng. Sci.*, **61**, 1602–1616 (2006).

Yagi, S. and D. Kunii 'Studies on effective thermal conductivities in packed beds', *AIChE J.*, **3**, 373–381 (1957).

13.6 Appendix: nomenclature

13.6.1 Notation

A_{cell}	area of cell adjacent to membrane (m²/cell)
C_p	specific heat (J/(kg K))
D	molecular diffusion coefficient (m²/s)
D_{eff}	effective diffusion coefficient of catalyst bed (m²/s)
d_m	diameter of membrane tube (m)
d_p	catalyst equivalent diameter (m)
d_t	reactor equivalent diameter (m)
e	emissivity of catalyst bed (—)
h	enthalpy (J/kg)
h_{f}	heat of formation (J/kg)
h_{rs}	radiant heat transfer coefficient between solid surfaces ($\text{W m}^{-2}\ \text{K}^{-1}$)
h_{rv}	radiant heat transfer coefficient between voids ($\text{W m}^{-2}\ \text{K}^{-1}$)
h_{wall}	heat transfer coefficient between air and reactor wall ($\text{W m}^{-2}\ \text{K}^{-1}$)
J	permeation rate (mol/(m²s))
K	constant in Equations [13.2] and [13.3] (Pa sm^{-2})
K_{D}	adsorption constant (Pa^{-1})
K_{P}	equilibrium constant (Pa^{-3})
K_m	gas permeance defined by Equations [13.35a] and [13.35b] ($\text{mol m}^{-2}\ \text{s}^{-1}\ \text{Pa}^{-0.5}$) or ($\text{mol m}^{-2}\ \text{s}^{-1}\ \text{Pa}$)
k_{r}	rate constant ($\text{mol m}^{-3}\ \text{s}^{-1}\ \text{Pa}^{-1}$)
L	length of membrane tube (m)
M_w	molecular weight (kg/kmol)
m	mass fraction of species (wt/wt)

m_H^0	molar fraction of hydrogen in feed (-)
P	pressure (Pa)
Pe	Peclet number (-)
P_m	gas permeability (mol/(m s $Pa^{0.5}$)) or (mol/(m s Pa))
Pr	Prandtl number (-)
$P_{1,i}$	reaction-side partial pressure of component i (Pa)
$P_{2,i}$	permeation-side partial pressure of component i (Pa)
Q_H^0	hydrogen feed rate (cm^3/min)
Re	Reynolds number (-)
R_H	hydrogen recovery rate (cm^3/min)
R_{lim}	theoretical limit (%)
r_v	reaction rate (mol/(m^3-bed s))
Sc	Schmit number (-)
S_c	source term for species mass conservation equation (kg/(m^3 s))
S_h	source term for energy conservation equation (W/m^3)
S_m	source term for mass conservation equation (kg/(m^3 s))
S_u	source term for momentum conservation equation (N/m^3)
T	temperature (K)
t_m	thickness of membrane (m)
u	velocity or specific velocity in the catalyst bed (m/s)
V_{cell}	volume of cell adjacent to membrane (m^3/cell)

Greek symbols

α	coefficient defined by Equations [13.3] and [13.32] (Pa s^2/m^3)
β	coefficient defined by Equations [13.3] and [13.33] (Pa s/m^2)
ε	porosity of catalyst bed (−)
ΔL	length of catalyst bed (m)
ΔP	pressure drop of catalyst bed (Pa)
Φ	effective coefficient with heat transfer (−)
λ	thermal conductivity (W/(m K))
λ_{eff}	effective thermal conductivity of catalyst bed (W/(m K))
λ_{eff}^0	effective thermal conductivity without flow of catalyst bed (W/(m K))
λs	thermal conductivity of catalyst (W/(m K))
μ	viscosity (Pa s)
ρ	density (kg/m^3)

Subscripts and superscripts

1	reaction side
2	permeation side
i	species i

14

Computational fluid dynamics (CFD) analysis of membrane reactors: simulation of a palladium-based membrane reactor in fuel cell micro-cogenerator system

L. ROSES, S. CAMPANARI and
G. MANZOLINI, Politecnico di Milano, Italy

DOI: 10.1533/9780857097330.3.496

Abstract: This chapter presents a bi-dimensional CFD simulation applied to a steam methane reformer coupled with a palladium-based hydrogen-permeable membrane, the so-called 'membrane reformer' (MREF). The interest of this configuration relies on the possibility of implementing this technology within a polymer electrolyte membrane fuel cell (PEMFC)-based micro-cogenerator (also micro-Combined Heat and Power (CHP)) with a net electrical power output in the range of 1–2 kW. A bi-dimensional CFD simulation is important to correctly predict MREF performances due to the coupling of several phenomena in the reactor, such as reaction rate, heat and mass transfer: the reforming reaction heat is supplied by a stream of hot gas coming from the combustion of the unconverted fuel and the unpermeated hydrogen. Through detailed analyses of temperatures, species concentration and reaction rate profiles along the membrane and within the catalyst bed, we study their impact on reactor performance.

Key words: CFD, membrane reformer (MREF), hydrogen production, micro-CHP, palladium membrane.

14.1 Introduction

As discussed in previous chapters, research activity on innovative membrane reactors for hydrogen production has been increasing in the last decade due to energy savings and CO_2 emission reduction targets which have been set in most industrialized countries. One of the most investigated applications of hydrogen energy is distributed small scale CHP. Among different CHP technologies, fuel cell based systems allow the highest net electric efficiency, with significant potential economic benefits for end users. However, the design of a compact and efficient fuel processor is still a relevant issue.

This chapter deals with a bi-dimensional CFD simulation of a MREF, designed for an integrated PEMFC-based micro-CHP system which can be fed with natural gas (NG) or biogas mixtures containing high methane fractions. NG is selected because of its capillary distribution network in the cities of most developed countries, while biogas represents a possible way to exploit renewables in CHP applications. The design net electricity output of the investigated system is in the range of 1–2 kW, which represents the average electric load of a single residential customer in Europe.

The simulations aim to study the feasibility of thermally coupling an MREF with a stream of hot gas resulting from the combustion of the unconverted fuel and the unpermeated hydrogen coming from the fuel processor exhaust streams. The MREF uses a palladium-based membrane, which is permselective to hydrogen, and is designed to operate at a hydrogen recovery factor ranging from 60 to 70%.

Among one-dimensional membrane reactor models available in literature, some assume isothermal conditions (Oklany, 1998; Lin *et al.*, 2003) and others do not (Madia *et al.,* 1999; De Falco *et al.*, 2007; Patel and Sunol, 2007). The main drawback of one-dimensional simulations is their overestimation of hydrogen permeation, because the radial distribution of hydrogen partial pressure is neglected (e.g., the hydrogen permeation across the membrane depends on the difference of hydrogen partial pressures between both sides). Recent studies have outlined the different results obtained by one-dimensional and two-dimensional simulations, emphasizing the importance of correct heat and mass transfer calculation in the radial direction (De Falco *et al.*, 2007; Oyama and Hacarlioglu, 2009; Gallucci *et al.*, 2010).

In this chapter, a bi-dimensional CFD simulation is applied to model a system based on a realistic architecture, analysing the feasibility of thermally integrating an endothermic fixed-bed MREF with a stream of hot gas. The effects of adopting different configurations for the heat supply are initially discussed. Then the links between several operational and technology dependent variables with reactor performance are considered; in particular, feed pressure, steam to carbon ratio, sweep gas flow, catalyst load and membrane permeance are studied. The analysis puts attention on the temperature profiles along the membrane and hydrogen permeation fluxes. Moreover, an analysis is made of the mechanisms and kinetics of carbon deposition over the catalyst, and finally, comments are made regarding the use of appropriate kinetic models for simulation.

14.2 Polymer electrolyte membrane fuel cell (PEMFC) micro-cogenerator systems and MREF

Membrane reactor technology can be applied to hydrogen production due to its capability to deliver high-purity output. An interesting application of this fuel processor is its integration with FC-based power generation systems; in particular, a configuration based on low temperature PEMFC for micro-CHP applications, fuelled with NG and incorporating a membrane reactor for hydrogen production has already been addressed by diverse authors (Salemme *et al.*, 2010; Roses *et al.*, 2011) and is shown in Fig. 14.1. In this layout, a mixture of NG and superheated steam enters the reforming section, where, over a catalyst bed, the NG conversion takes place. The combination of reforming reaction and membrane reactor is usually referred to as an MREF, and this terminology we will adopt. Another input stream shown in the figure is a sweep gas (indicated as 'Sweep' in the figure), whose purpose is explained in Section 14.3.4. For this kind of application, a typical sweep stream is steam. The MREF required in this application is not intended for complete conversion of methane (or complete hydrogen separation), but rather the system is designed so that a fraction of the fuel exits the retentate stream. The amount of unconverted methane and unseparated hydrogen is determined in order to balance the heat necessary to support the reforming reaction. The resulting hydrogen permeated is in the range of 60 to 70% of the maximum that could be produced, and after cooling and separation of condensed water, feeds the anode of the PEMFC for production of electric energy. Thermal energy on the other hand is recovered through heat exchange with the fuel cell and the exhaust gas. Considering that in a typical PEMFC operating temperatures are limited to 75–80°C, heat can be recovered at a maximum temperature of 60–70°C.

The micro-CHP unit based on MREF has several advantages compared to the case of a unit adopting a conventional fuel processor (typically based on a steam reformer, a two stage WGS reactor and preferential oxidation): it has the potential to reach higher net electric efficiency (about 43%), which

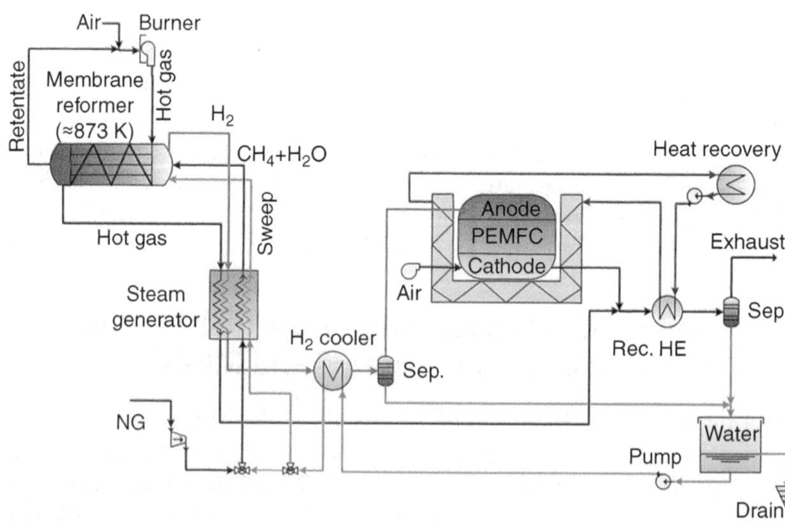

14.1 Layout of PEMFC micro-CHP unit with MREF fuel processor.

is around nine to ten percentage points higher than a conventional system, featuring also less system complexity (Roses *et al.*, 2010a).

14.3 Model description and assumptions

The CFD simulation of the MREF can be run over by a commercial code, where appropriate user-defined routines need to be added in order to model chemical reactions and permeation mechanism. In this case we have adopted ANSYS Fluent® version 6.3.26 (ANSYS Inc., 2010); the setup for the numerical model is 2D axisymmetric calculator, with double precision, pressure-based solver and steady-state condition.

In this section, the mathematical model built behind the simulation is described, together with certain characteristics such as membrane permeance and catalyst load for a configuration set as base case. Further on, results are shown for different simulated scenarios in comparison to the defined base case. In all cases modelled, the fuel fed at the inlet of the system is assumed to be pure methane. The validation of the model, already discussed in Roses *et al.* (2010b), showed good agreement with experimental data and also compared with similar simulation work.

14.3.1 Physical properties and mixture treatment

The thermodynamic and physical properties of the gas mixture reacting in the fuel processor are described in detail because of the strong dependency of reaction kinetics on temperature and composition, hence linked to thermal conductivities and diffusion velocity.

Specific heat capacity of the gas mixture is calculated as mass fraction average of pure species:

$$c_{p,\text{mix}} = \sum_{sp=i} x_i \times c_{p,i} \quad \left(\frac{\text{kJ}}{\text{kg K}}\right) \qquad [14.1]$$

Density, thermal conductivity and dynamic viscosity properties for gas mixtures are calculated with the assumption of ideal gas mixing, expressed as follows:

$$\rho_g = \frac{pM_m}{RT} \quad \left(\frac{\text{kg}}{\text{m}^3}\right) \qquad [14.2]$$

$$\lambda_g = \sum_{sp=i} \frac{y_i \times \lambda_i}{\Sigma_j \left(y_j \times \phi_{ij}\right)} \quad \left(\frac{\text{kW K}}{\text{m}}\right) \text{and } \mu_g = \sum_{sp=i} \frac{y_i \times \mu_i}{\Sigma_j \left(y_j \times \phi_{ij}\right)} \quad \left(\frac{\text{kg}}{\text{m h}}\right) \qquad [14.3]$$

where the term ϕ_{ij} represents the mutual interaction between two species i,j and is defined as:

$$\phi_{ij} = \frac{\left[1+\left(\frac{\mu_i}{\mu_j}\right)^{1/2} \times \left(\frac{M_j}{M_i}\right)^{1/4}\right]^2}{\left[8\times\left(1+\frac{M_i}{M_j}\right)\right]^{1/2}} \qquad [14.4]$$

Specific heat capacities, thermal conductivities and viscosities of single species in equations [14.1–14.4] are given by appropriate fourth degree polynomial expressions:

$$c_{p,i}(T) = \sum_{j=0}^{4} Ac_{p,i,j} \times T^j \quad \left(\frac{\text{kJ}}{\text{kg K}}\right) \qquad [14.5]$$

$$\lambda_i(T) = \sum_{j=0}^{4} A_{\lambda,i,j} \times T^j \quad \left(\frac{\text{kW}}{\text{m K}}\right) \qquad [14.6]$$

$$\mu_i(T) = \sum_{j=0}^{4} A_{\mu,i,j} \times T^j \quad \left(\frac{\text{kg}}{\text{m h}}\right) \qquad [14.7]$$

where default coefficients $Ac_{p,i,j}, A_{\lambda,i,j}$ and $A_{\mu,i,j}$ are assumed as available in the ANSYS Fluent® code.

The reactor shell and the inner pipes are made of austenitic stainless steel with thermal conductivity expressed as follows:

$$\lambda_{\text{steel}}(T) = 38.52 + 5.04\times10^{-2}T \quad \left(\frac{\text{kJ}}{\text{h m K}}\right) \qquad [14.8]$$

For the diffusivity calculation, Maxwell–Stefan equations are used (Merk, 1959), assuming ideal gas behaviour and negligible pressure diffusion; then calculation results can be expressed by the following equations:

$$\vec{J_i} = -\sum_{j=1}^{N-1} \rho_g \times \mathrm{D}_{ij} \times \nabla x_j - \mathbb{D}_{T,i}\frac{\nabla T}{T} \quad \left(\frac{\text{kg}}{\text{h m}^2}\right) \qquad [14.9]$$

Table 14.1 Fick's law diffusion coefficients

D_{ij} (m^2/h)	CH_4	CO	CO_2	O_2	N_2	H_2O
H_2	1.60	1.70	1.40	1.76	1.70	1.55
CH_4	–	0.50	0.40	0.50	0.50	0.43
CO		–	0.36	0.50	0.43	0.43
CO_2			–	0.35	0.36	0.31
O_2				–	0.47	0.43
N_2					–	0.43

Source: Incropera and DeWitt (2007), Taylor and Krishna (1993) and Branan (2002).

$$\mathbb{D}_{ij} = [\mathbb{D}] = [\mathbb{A}]^{-1}[\mathbb{B}] \qquad [14.10]$$

$$\mathbb{A}_{ii} = -\left(\frac{y_i}{D_{iN}}\frac{M_m}{M_N} + \sum_{j \neq i}^{N} \frac{y_j}{D_{ij}}\frac{M_m}{M_i}\right)\left(\frac{\text{h}}{\text{m}^2}\right) \qquad [14.11]$$

$$\mathbb{A}_{ij} = y_i\left(\frac{1}{D_{ij}}\frac{M_m}{M_j} - \sum_{j \neq i}^{N} \frac{1}{D_{iN}}\frac{M_m}{M_N}\right)\left(\frac{\text{h}}{\text{m}^2}\right) \qquad [14.12]$$

$$\mathbb{B}_{ii} = -\left(y_i\frac{M_m}{M_N} + \frac{(1-y_i)M_m}{M_i}\right) \qquad [14.13]$$

$$\mathbb{B}_{ij} = y_i\left(\frac{M_m}{M_j} - \frac{M_m}{M_N}\right) \qquad [14.14]$$

The binary diffusion coefficients D_{ij} related to Fick's law are assumed as shown in Table 14.1; these coefficients correspond to their values at a defined temperature representative of the working range considered in this work (around 873 K).

14.3.2 Reformer section

The reaction taking place in the reformer is developed over a fixed bed made of Ni-based catalyst having density $\rho_s = 1170\ \text{kg/m}^3$; the catalyst particle shape gives a void fraction $\varepsilon = 0.5$ leading to a bed density $\rho_B = 585\ \text{kg/m}^3$. As mentioned above, this model considers the case of pure methane as a fuel. For simplicity, only three fundamental reactions taking place in the reformer are taken into account:

- Reaction 1: $CH_4 + H_2O \Rightarrow CO + 3H_2$ $\quad \Delta H^0_{298K} = 206\,kJ/mol$
- Reaction 2: $CO + H_2O \Rightarrow CO_2 + H_2$ $\quad \Delta H^0_{298K} = -41\,kJ/mol$
- Reaction 3: $CH_4 + 2H_2O \Rightarrow CO_2 + 4H_2$ $\quad \Delta H^0_{298K} = 165\,kJ/mol$

Reaction kinetics are described by the following equations, expressing the respective reaction rates according to the kinetic model of Xu and Froment (1989a).

$$r_1 = \frac{\frac{k_1}{p_{H_2}^{2.5}} \times \left(p_{CH_4} \times p_{H_2O} - \frac{p_{H_2}^3 \times p_{CO}}{K_{eq_1}} \right)}{DEN^2} \quad [14.15]$$

$$r_2 = \frac{\frac{k_2}{p_{H_2}} \times \left(p_{CO} \times p_{H_2O} - \frac{p_{H_2} \times p_{CO_2}}{K_{eq_2}} \right)}{DEN^2} \quad [14.16]$$

$$r_3 = \frac{\frac{k_3}{p_{H_2}^{3.5}} \times \left(p_{CH_4} \times {p_{H_2O}}^2 - \frac{p_{H_2}^4 \times p_{CO_2}}{K_{eq_3}} \right)}{DEN^2} \quad [14.17]$$

$$DEN = 1 + K_{H_2} \times p_{H_2} + K_{CH_4} \times p_{CH_4} + K_{CO} \times p_{CO} + K_{H_2O} \times \frac{p_{H_2O}}{p_{H_2}} \quad [14.18]$$

where:

$$k_1 = 4.225 \times 10^{16} \times e \frac{-2.401 \times 10^5\ \mathbf{kJ/kmol}}{RT} \left(\frac{kmol\,kPa^{0.5}}{kg_{cat}\,h} \right) \quad [14.19]$$

$$k_2 = 1.955 \times 10^4 \times e \frac{-6.713 \times 10^4\ \mathbf{kJ/kmol}}{RT} \left(\frac{kmol}{kPa\,kg_{cat}\,h} \right) \quad [14.20]$$

$$k_3 = 1.020 \times 10^{16} \times e \frac{-2.439 \times 10^5 \text{ kJ / kmol}}{RT} \left(\frac{\text{kmol kPa}^{0.5}}{\text{kg}_{\text{cat}}\text{h}} \right) \quad [14.21]$$

$$K_{H_2} = 6.12 \times 10^{-11} \times e \frac{8.290 \times 10^4 \text{ J/mol}}{RT} \left(\text{kPa}^{-1} \right) \quad [14.22]$$

$$K_{CH_4} = 6.65 \times 10^{-6} \times e \frac{3.828 \times 10^4 \text{ J/mol}}{RT} \left(\text{kPa}^{-1} \right) \quad [14.23]$$

$$K_{CO} = 8.23 \times 10^{-7} \times e \frac{7.065 \times 10^4 \text{ J/mol}}{RT} \left(\text{kPa}^{-1} \right) \quad [14.24]$$

$$K_{H_2O} = 1.77 \times 10^5 \times e \frac{-8.868 \times 10^4 \text{ J / mol}}{RT} \quad [14.25]$$

$$K_{eq_1} = 1.198 \times 10^{17} \times e \frac{-26830K}{T} \left(\text{kPa}^2 \right) \quad [14.26]$$

$$K_{eq_2} = 1.767 \times 10^{-2} \times e \frac{4400K}{T} \quad [14.27]$$

$$K_{eq_3} = K_{eq_1} \times K_{eq_2} \quad [14.28]$$

The pseudo-effectiveness factor, which is sometimes adopted to take into account specific effects related to the bed geometry, was not considered in the kinetics equations presented above, as suggested in Xu and Froment (1989b). This is due to the fact that for an MREF applied to a residential application, compactness is desired; therefore, the catalyst would be packed into small particles ($d_p \leq 0.5$ mm) in order to avoid having diffusion limitations as the rate determining step for the reaction and allowing a more compact reactor.

Pressure loss along the packed bed is calculated adopting Ergun's equation (Ergun, 1952) with a particle size $d_p = 0.5$ mm and uniform void fraction $\varepsilon = 0.5$:

$$\left| \frac{\Delta p}{\Delta z} \right| = \frac{150 \mu_g (1 - \varepsilon)^2}{d_p{}^2 \varepsilon^3} \times u_s + \frac{1.75 \times \rho_g (1 - \varepsilon)}{d_p \varepsilon^3} \times u_s{}^2 \quad [14.29]$$

Heat conduction along the reforming section depends on the catalyst bed, where thermal conductivity of a bi-phase material must be considered. This effect is taken into account by modifying the effective thermal conductivity as a result of the following calculation:

$$\lambda_{\mathrm{er}} = \lambda_{\mathrm{er}}^{0} + 0.111\lambda_g \times \frac{Re_p Pr^{1/3}}{1+46(d_p / d_{\mathrm{eq}})^2}\left(\frac{\mathrm{kJ}}{\mathrm{h\ m\ K}}\right) \qquad [14.30]$$

$$\lambda_{\mathrm{er}}^{0} = \varepsilon \times \left(\lambda_g + 0.95\,\alpha_{\mathrm{ru}}\,d_p\right) + \frac{0.95(1-\varepsilon)}{\dfrac{2}{3\lambda_s} + \dfrac{1}{\left(10\,\lambda_g + \alpha_{\mathrm{rs}}\,d_p\right)}}\left(\frac{\mathrm{kJ}}{\mathrm{h\ m\,K}}\right) \qquad [14.31]$$

$$\alpha_{\mathrm{ru}} = \frac{0.8171\left(\dfrac{T}{100}\right)^3}{1+\left(\dfrac{\varepsilon}{2.(1-\varepsilon)}\left(\dfrac{1-e}{e}\right)\right)}\left(\frac{\mathrm{kJ}}{\mathrm{h\ m^2 K}}\right) \qquad [14.32]$$

$$\alpha_{\mathrm{rs}} = 0.8171\left(\frac{e}{2-e}\right)\left(\frac{T}{100}\right)^3\left(\frac{\mathrm{kJ}}{\mathrm{h\ m^2\ K}}\right) \qquad [14.33]$$

In the above equations, e is the emissivity of the catalyst surface, here assumed equal to 0.8. The term λ_s is the thermal conductivity of the catalyst, which is considered equal to 1.26 kJ m^{-1} h^{-1} K^{-1} (De Wasch and Froment, 1972) and $\lambda_{\mathrm{er}}^{0}$ is the static contribution to the effective conductivity (thermal conductivity of catalytic bed in absence of fluid flow). The same approach was also proposed by other authors (De Wasch and Froment, 1972; Elnashaie and Elshishini, 1993).

14.3.3 Palladium membrane

Palladium-based membranes for hydrogen separation have been deeply investigated in the latest decades. The possibility of separating a pure hydrogen stream makes them significant for applications requiring uncontaminated hydrogen, such as low temperature PEMFC (Roses *et al.*, 2011).

Mass transport across a membrane is called permselective if only some components from a stream mixture permeate through it. Dense palladium-based membranes are permselective to hydrogen transport. Hydrogen permeation across the membrane is based on a solution–diffusion mechanism whose steps could be described as: (i) molecular adsorption on the Pd surface on the feed side of the membrane; (ii) dissociation into atomic hydrogen; (iii) hydrogen atoms entry into the palladium lattice and diffusion across it, while the electrons are interacting with the metal lattice; (iv) on the permeation side of the membrane, hydrogen atoms leaving the lattice and recombining to be desorbed as hydrogen molecules (Berkheimer and Buxbaum, 1985). The overall permeation process across the membrane is modelled using the following equation:

$$J_{H_2} = \frac{P_{H_2}}{\delta} \times \left(p^n_{H_2,\text{feed}} - p^n_{H_2,\text{perm}} \right) \qquad [14.34]$$

The term J_{H_2} is the hydrogen molar flow per membrane surface (mol s^{-1} m^{-2}) and P_{H_2} is the *permeability*, which divided by the membrane thickness δ gives the *permeance;* the term inside brackets represents the *driving force of hydrogen permeation*, where n is the hydrogen partial pressure exponent and depends on the rate-limiting step of the permeation process. When the diffusion of hydrogen atoms through the palladium is rate limiting, which usually happens for membrane thicker than 7–10 μm, the exponent n is equal to 0.5, leading to the so-called Sievert's law, otherwise for thinner membranes (around 3–5 μm), the value of n could be higher, up to 1.0 (Dittmeyer, 2001).

Assuming that n is invariant with the temperature, the permeability can be described by an Arrhenius-type law as follows:

$$P_{H_2} = P^0_{H_2} \times e^{(-E_a/RT)} \qquad [14.35]$$

For this work, the use of a Pd–Ag dense membrane is considered because of its high selectivity (for simplicity, but with good approximation of experimental results, the selectivity is assumed infinite) and for being perceived as the most promising solution for the integration with PEMFC in a medium-term period. Selectivity is critical in low temperature PEMFC applications where the CO concentration in the fuel feeding the cell must be below 10 ppm. The simulation was run focusing on the membrane tested by Gallucci *et al.* (2004) and its characteristics are summarized in Table 14.2.

14.3.4 MREF configuration and base case setup

The configuration of the MREF here discussed is shown in Fig. 14.2. It consists of four coaxial surfaces with a total length of 640 mm: sweep steam

Table 14.2 Membrane characteristics

Variable	Units	Value
δ	m	50×10^{-6}
E_a	kJ kmol^{-1}	14 620
$P^0_{H_2}$	kmol m^{-1} h^{-1} kPa$^{-0.5}$	3.60×10^{-5}
n	–	0.5

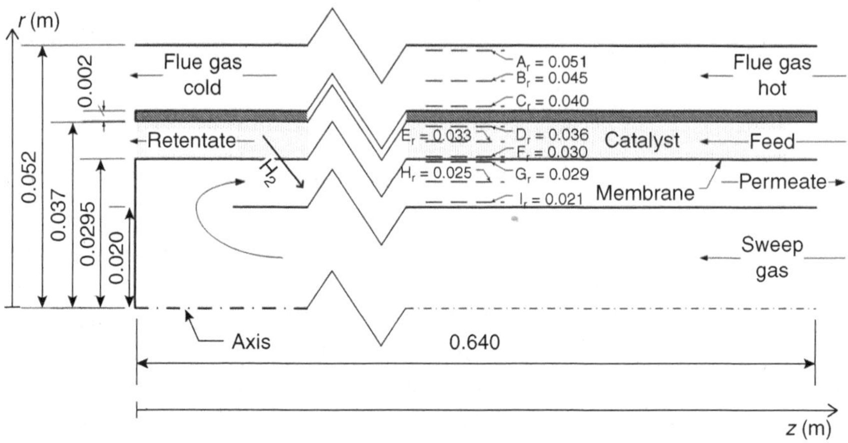

14.2 Layout of MREF configuration (measures and coordinates in m) and definition of sections A–I.

enters from an inner pipe of 20 mm radius to the permeate annular section of 20.0–29.5 mm radius. The feed is a mixture of reactant gas and steam flowing in the annular section limited by 29.5–37.0 mm radius where the reforming reaction takes place over a packed bed of catalyst, running counter-currently to the permeate flow. Finally, in the outer section of 39.0–52.0 mm radius, the external stream provides the heat required by the reforming reaction. The wall separating the heating stream and reformate section is an austenitic stainless steel pipe with 2 mm thickness.

This configuration represents a design of an MREF producing a flow of hydrogen required for feeding a PEMFC with power output in the 1–2 kW range, applicable to the micro-CHP setup described above. Accordingly, initial stream conditions for the simulation as well as thermodynamic conditions of the feed, sweep and hot gases were taken from overall system heat and mass balances. A critical factor which influences the membrane reactor configuration is the capability of limiting the membrane temperature. In palladium-based membranes, depending on the preparation technique,

Table 14.3 Input setup for base case configuration

Parameter	Units	Value
Feed pressure	kPa	800
Feed flow	kmol/h	9.90×10^{-2}
Feed inlet composition	($\%_{molar}$)	0.026 H_2 25.005 CH_4 74.969 H_2O
Feed inlet temperature	K	873
Permeate side pressure	kPa	101
H_2O sweep flow	kmol/h	1.26×10^{-2}
H_2O sweep inlet temperature	K	873
Hot gas flow	kmol/h	1.81×10^{-1}
Hot gas composition	($\%_{molar}$)	13.5 CO_2 4.0 O_2 49.8 N_2 32.7 H_2O
Hot gas inlet temperature	K	1598
Membrane surface	m^2	0.12
Catalyst mass	kg_{cat}	0.55
WHSV	h^{-1}	0.72

thickness and alloys used in the membrane layer, there is a defined maximum temperature above which membrane stability and selectivity performance are prone to severe degradation. This limit is typically placed from 723 to 973 K. For the system here presented we set particular attention where the membrane temperature surpasses 873 K.

In Fig. 14.2, the direction of the hot gas is the same as the retentate, leading to a co-current configuration of the heating stream. In a counter-current configuration, hot gases would move from the left to the right side of the reactor represented in the figure.

To start the simulation, even if the considered case uses a pure steam methane feed, a small amount of hydrogen was added to the inlet stream because of kinetics equation numerical convergence. An inlet hydrogen fraction equal to zero would lead to divergence of kinetics equations for Reactions 1 and 3, as can be seen from Equations [14.15] and [14.17].

Table 14.3 summarizes the input parameters, temperature, pressure and composition of inlet flows. WHSV is the weight hourly space velocity, defined as CH_4 inlet mass flow (kg/h) over catalyst weight (kg_{cat}).

Performance of the membrane reactor will be evaluated in terms of methane conversion $\left(X_{CH_4}\right)$, permeation ratio (PR) and hydrogen recovery factor (HRF), which are defined as follows:

$$X_{CH_4} = \frac{F_{CH_4,r,\text{in}} - F_{CH_4,r,\text{out}}}{F_{CH_4,r,in}} \qquad [14.36]$$

$$\mathrm{PR}_{\mathrm{H}_2} = \frac{F_{\mathrm{H}_2,p,\mathrm{out}}}{F_{\mathrm{H}_2,r,\mathrm{out}} + F_{\mathrm{H}_2,p,\mathrm{out}}} \qquad [14.37]$$

where PR is the ratio between the permeated hydrogen and the produced hydrogen, and

$$\mathrm{HRF} = \frac{F_{\mathrm{H}_2,p,\mathrm{out}}}{4F_{\mathrm{CH}_4,r,in} + F_{\mathrm{H}_2,r,in}} \qquad [14.38]$$

where HRF is the ratio of the permeated hydrogen to the maximum hydrogen flow that can be theoretically produced.

In the MREF, the methane reacts forming CO, CO_2, H_2O and H_2, where the latter can permeate across the membrane and feed the fuel cell. The injection of a sweep gas (in this case steam) decreases the partial pressure of hydrogen on the permeate side of the membrane enhancing the permeation driving force (see Equation [14.34]). The permeation flow is counter-current to the reacting mix in order to have the highest partial pressure difference along the membrane. The assumed inlet velocity profiles are parabolic for sweep gas and hot gas streams, and flat for the feed section.

14.4 Simulation results and discussion of modelling issues

Results of the CFD methodology applied to the membrane reactor integrated in a micro-CHP layout are presented here. First of all, the base case configuration are analysed and discussed. Then, variation of various parameters such as pressure, catalyst loading, membrane permeance, steam to carbon ratio and layout comparison are considered, in order to outline the potential of CFD simulation and the impact of each parameter on membrane reactor performance. The reactor base case configuration has the heating stream in co-current configuration: feed gas and hot gases flow in the same direction (e.g., they both enter from the right side in Fig. 14.2). Temperature and hydrogen molar fraction profiles inside the membrane reactor are presented in Fig. 14.3 and Fig. 14.4, while a reactor overview is summarized in Fig. 14.5. In the same figure, conditions at the inlet and outlet of the membrane reactor are also highlighted.

Starting from Fig. 14.3, simulations show that methane and steam quickly react at the inlet of the reactor producing hydrogen and CO_2. Since the reforming reaction is endothermic, significant heat demand has to be supplied by hot gases. However, hot gases cannot balance the heat of the reaction, leading to a temperature drop inside fixed bed (about 110 K) impacting

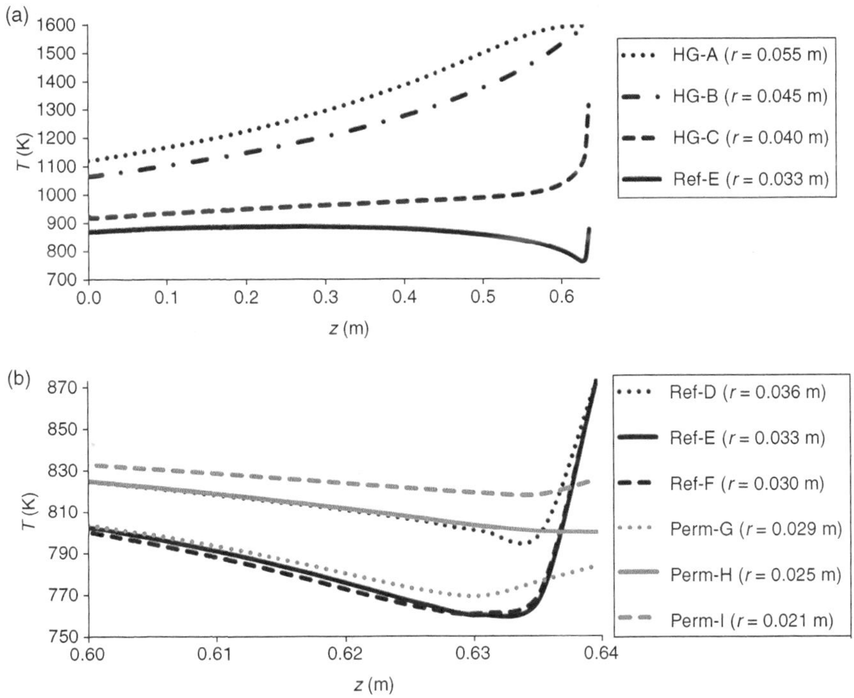

14.3 Temperature profiles along the different sections of the MREF, at various radial positions. (a) Along the total length of the reactor for reforming section and hot gas section at different radial positions; and (b) detailed analysis of the profiles in the reformer inlet region, where the feed inlet is subjected to a steep temperature drop (HG, hot gas; Ref, reformer; Perm, permeate).

on hydrogen conversion and partial pressure. This temperature trend outlines the importance of supplying heat in the first section of the reactor, thus supporting the choice of a co-current configuration. Of course, the adoption of a pre-reforming section or a fluidized bed would have reduced the temperature drop, but increasing the fuel processor complexity and bringing about different additional issues.

Similarly to feed gas trend, hot gases show a marked temperature decrease in the first part of the reactor (stream C). In addition, the temperature profile of hot gases along the radius varied significantly with a positive gradient for higher radius, which suggests predominant conductive heat flux with limited turbulence of hot gases. At the permeate side, temperature in section I is higher than the bed temperature because sweep gas temperature is 873 K at the inlet, which is also higher than fixed-bed temperature. A lower sweep gas temperature could have been assumed with penalties on membrane permeance. After this inlet zone, the temperatures in reformate

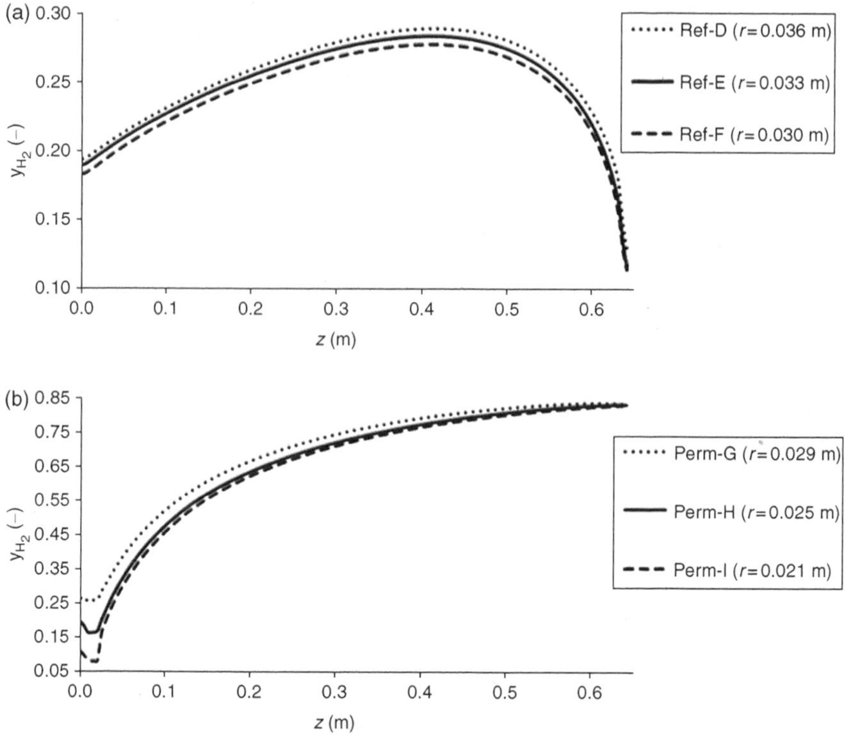

14.4 Hydrogen molar fraction profiles along the MREF at various radial positions. (a) On the reformer section; and (b) on the permeate section. (Ref, reformer; Perm, permeate).

and permeate sides along the radius begin to approach, because of the lower heat consumed by the reactions.

Figure 14.4 outlines the variation of hydrogen fraction along radial and axial directions. For axial variation, the hydrogen concentration in the reforming side has a maximum at about z = 0.4 m: from the inlet to z = 0.4 m, the reaction rate overcomes hydrogen permeation across the membrane. Variation of hydrogen concentration along radius outlines the importance of 2D simulations: average hydrogen partial pressure at the membrane interface, which affects permeation driving force, is about 3% lower than the average in the whole reacting volume. This variation could be due to mass transfer limitations and, in the case of the reforming section, also due to higher reaction rates towards larger radius where the temperature is higher (closer to the hot gas section). Moreover, a 2D simulation allows a better analysis of the temperature profile along the membrane, with critical importance for the safeguard of the membrane, as already addressed.

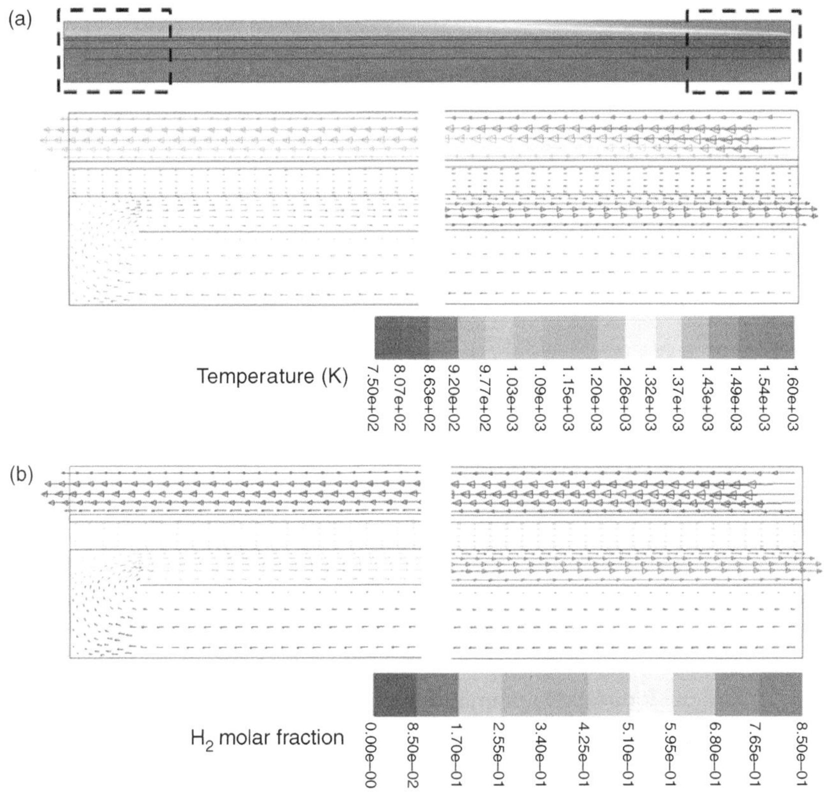

14.5 (a) Temperatures along the MREF: whole semi-axial section of the system and close-up for left and right edges; (b) hydrogen molar fraction for left and right edges of the reactor.

Despite the presence of the packed bed, the calculated pressure loss on the reformer section is very low, accounting for approximately 4.6 kPa. This is due to the small velocities of the reforming stream (i.e., around 0.16 m/s). On permeate and hot gas sides the pressure losses are negligible.

14.4.1 Heating stream: co-current flow versus counter-current flow

This section presents the results calculated for counter-current flow reactor compared with the co-current case. The interest on simulating the case of counter-current flow of the hot gas with respect to the reformate is to assess the most reasonable and feasible configuration in a real unit for (i) obtaining sufficient hydrogen production, (ii) keeping the temperature profiles as flat as possible and (iii) avoiding local overheating and hot spot formation. For this

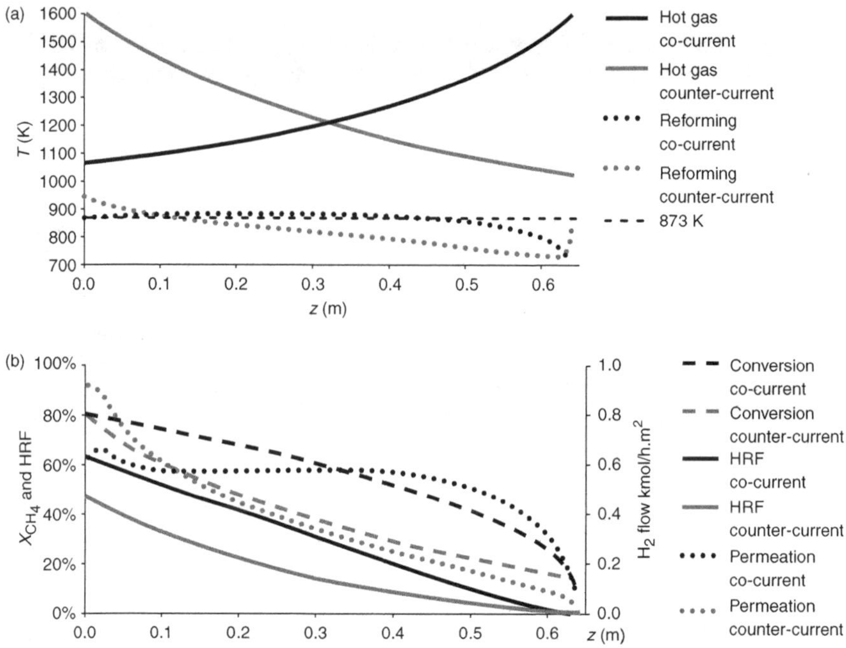

14.6 Comparison between co-current and counter-current reactor configurations. (a) Temperature profiles of hot gas and reforming sections; and (b) methane conversion, HRF and permeation flow.

purpose, permeation results are compared, as well as the membrane temperature profile, with particular attention to the maximum temperature values.

Temperature profiles, methane conversion, HRF and permeated flow are shown in Fig. 14.6. Figure 14.7 shows the membrane temperature profile and the permeation driving force along the reactor. Table 14.4 shows the permeation results and the product outcome, outlining that total hydrogen permeated is lower (26%) for the counter-current configuration. The calculated HRF and methane conversion $\left(X_{CH_4}\right)$ in the counter-current flow are 47% and 81%, compared to 63% and 80% in the co-current case. It can be observed that X_{CH_4} is higher for the counter-current flow case, but the HRF is much smaller. This peculiar behaviour is because in the counter-current flow configuration, hot gases supply heat to the reformate mainly at the end of the reactor, converting methane but with limited impact on permeated hydrogen. As shown in Fig. 14.6, permeated flow in the co-current flow is constant for most of membrane length in the range of 1.2 kmol h^{-1} m^{-2}, while in the counter-current case, it has a constant rise from the beginning and it rises above 1.2 kmol h^{-1} m^{-2} just before the final 0.1 m of the reactor.

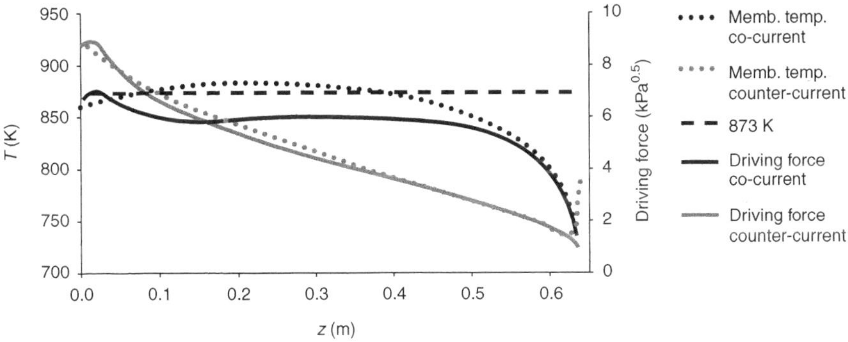

14.7 Comparison of membrane temperature and permeation driving force between co-current and counter-current configurations.

Table 14.4 Permeation and flow results for hot gas flow in co-current and counter-current configuration

Parameter	Units	Co-current value	Counter-current value
Permeated hydrogen flow	kmol/h	6.27×10^{-2}	4.64×10^{-2}
PR_{H_2}	%	81	63
X_{CH_4}	%	80.4	81.3
HRF	%	63.3	46.8
Average driving force	$kPa^{0.5}$	5.6	4.5
Maximum membrane temperature	K	881	920
Average membrane temperature	K	861	814
Average reforming section temperature	K	867	820
H_2/H_2O molar ratio at permeate outlet		5.0	3.7
Retentate outlet composition	$\%_{molar}$	19.0 H_2 6.4 CH_4 3.2 CO 22.9 CO_2 48.5 H_2O	30.6 H_2 5.0 CH_4 6.1 CO 15.6 CO_2 42.7 H_2O
Hot gas outlet temperature	K	1043	989
Retentate outlet temperature	K	867	944
Permeated flow outlet temperature	K	797	761

An important result is related to membrane maximum temperature. As noted above, the membrane stability is compromised if temperature rises much over 873 K. As seen in Fig. 14.7, in the counter-current case, the maximum temperature of 920 K is reached towards the hot gas inlet side, going far over the limits. On the other hand, in the co-current case, the temperature recovers from the steep drop on the reforming inlet section, and then stabilizes within a close ±10 K range around the recommended temperature for most of the reactor length.

To summarize, the adoption of the counter-current configuration leads to an inefficient use of the membrane, featuring also a higher risk of overheating, in comparison with the co-current option.

14.4.2 Feed pressure, steam to carbon ratio, sweep gas and catalyst load

The flexibility given by simulation codes allows easy evaluation of the sensitivity of the reactor performance to changes to specific operational variables. In this section we study the effect of feed pressure, S/C ratio, sweep gas flow and catalyst load on the overall performance variables HRF and X_{CH_4}. Starting from the conditions of the base case configuration, each of the variables was varied in the range reported in Table 14.5.

The change in S/C is performed by modulating the H_2O flow in the feed stream, while keeping the CH_4 flow invariant. With respect to sweep gas flow, this is typically indicated with respect to the fuel feed flow using the sweep gas ratio i_{sw}, defined as the molar ratio between sweep gas flow and methane feed flow:

$$i_{\mathrm{sw}} = \frac{F_{\mathrm{sweep}}}{F_{\mathrm{CH_4}}} \qquad [14.39]$$

With regards to catalyst load, this can be adjusted by changing the dilution of the active catalyst (typically Ni-based) with an appropriate material (typically alumina particles); the higher the catalyst load, the lower the WHSV.

The driving force for hydrogen permeation across the membrane depends on hydrogen partial pressure difference from feed to permeate side (Equation [14.34]). Considering that permeate pressure is set by the fuel cell downstream, only the feed pressure can be varied to see the influence of the driving force on reactor performances. Figure 14.8 shows the positive trend in methane conversion and HRF at higher pressure. Changing the pressure from 600 to 1200 kPa, the HRF increases from 53.9 to 74.1%, with a relative increase of about 40%, while X_{CH_4} shows a slight increase from

Table 14.5 Studied range for sensitivity analysis

	Base case	Range studied
Reacting pressure (kPa)	800	600–1200
S/C	3.0	2.0–3.5
Sweep gas ratio i_{sw}	0.51	0.00–1.02
WHSV (h^{-1})	0.72	0.25–1.50

78.5% to 83.6%. This result differs from the case of a traditional reforming reactor, where methane conversion decreases with pressure because of Le Chatelier's principle. In membrane reactors, there is a dynamic equilibrium consequence of continuous hydrogen subtraction that shifts reaction towards products. It can be summarized that increasing feed pressure has a positive influence on reactor performance, thus reducing membrane surface area. However, a higher pressure has negative consequences on overall system performance (higher compression work, etc.), therefore, the optimization of the micro-CHP system should take into account both thermodynamic and economic aspects, which is however beyond the scope of this chapter.

The increase of steam to carbon ratio leads to higher methane conversion from a thermodynamic point of view and its result is evident in the figure. However, regarding HRF, on the other hand steam dilutes hydrogen reducing its partial pressure. From Fig. 14.8, it can be observed that for the given configuration there is a sort of balance between these two opposite effects, leading to a modest variation of HRF in the range of S/C addressed. It is expected that either extreme condition would be harmful for the system: steam content being too high would result in low hydrogen partial pressure hindering permeation, and a limited amount of steam would favour methane cracking and consequent carbon deposition with possible blockage of the catalyst, as will be discussed further on.

An increase of sweep gas has of course a directly positive effect on hydrogen separation due to the increase in permeation driving force, this effect in some way being analogous to the case discussed above for the increase of reacting pressure. However, taking into consideration the operation of the CHP system as a whole, the criteria for evaluating the benefit or not of modulating the sweep gas flow are much more complex. For example, if in a particular system, the MREF configuration already allows the injection of a sweep gas and there is available thermal energy for production of additional steam, then an increase in the sweep gas flow can be implemented in order to increase separation. However, if there is no way of producing additional steam without increasing the NG consumption, a comprehensive efficiency criterion should be used for selecting the best solution.

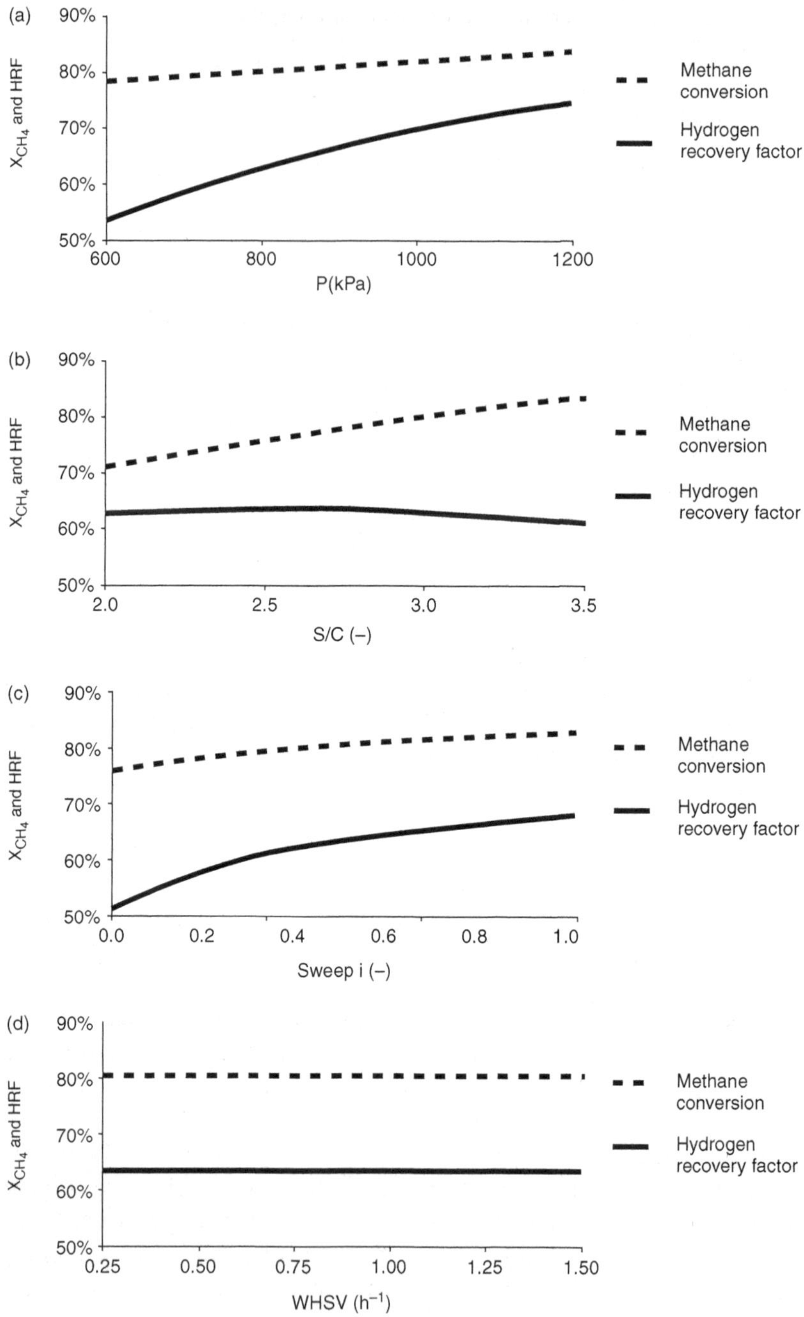

14.8 Effect of (a) pressure, (b) S/C, (c) sweep gas and (d) catalyst load on HRF and X_{CH_4}.

An interesting aspect follows the last figure, dealing with the effect of catalyst load (here shown as WHSV). In the range investigated the variations in HRF and X_{CH_4} are almost negligible. With WHSV changing from 1.50 to 0.25 h^{-1} (with unchanged feed conditions), the HRF increases from 62.9 to 63.7% and X_{CH_4} from 80.0 to 80.8%. These results clearly indicate that the rate-limiting step is much more the H_2 permeation rather than the reaction kinetics, which are influenced by WHSV.

14.4.3 Membrane permeance

The permeance of the membrane becomes an issue when evaluating capital costs for a MREF. Therefore, scientific efforts are strongly focusing on improving permeance without leaving selectivity aside, and for this reason, the impact of this parameter was investigated. We present here the results for the MREF operated under reference case feed conditions, but using hypothetical membranes with permeance two and five times larger than in the reference case. Membranes which have experimentally proven to possess a much higher permeance than that of the technology considered in our base case are reported in literature; however, presently they operate only at lower temperatures: for instance two times higher at temperature below 773 K (Chen *et al.*, 2008; Pizzi *et al.*, 2008; Chang *et al.*, 2010; Catalano *et al.*, 2011) and five times higher at a temperature below 573 K (Chen *et al.*, 2008; Pizzi *et al.*, 2008; Tucho *et al.*, 2009; Catalano *et al.*, 2011).

Figure 14.9 shows the profiles for reforming temperature, methane conversion and HRF. As expected, the higher permeation enhances the methane conversion, but on the other hand it affects the temperature profile due to the higher heat demand of the endothermic reaction. The figure also shows the radial profiles for H_2 molar fraction in the range of radius from 0.020 to 0.037 m (including permeate and reforming sections; see Fig. 14.2) and at defined axial coordinates 'z' of the reactor (corresponding to 10, 50 and 90% of the reactor length). It is possible to see that as permeation becomes larger, the profiles become more pronounced both in absolute and relative terms. The capability given by a 2D simulation for identifying radial profiles of the variables rather than taking it from the bulk reaction allows some limitations of 1D simulation to be overcome. In 1D models, the calculation of the driving force is taken from average values at a certain axial coordinate; neglecting the effect of species diffusion, the partial pressure of hydrogen on the reacting side results are higher, and on the permeate side results are lower with respect to their actual values over the membrane surface. Therefore the driving force would overestimate results using a 1D model. Using the results given by the 2D simulation, it is possible to estimate the average driving force using a cross-sectional averaging method (a 1D proxy) and compare them to the average driving force from 2D modelling.

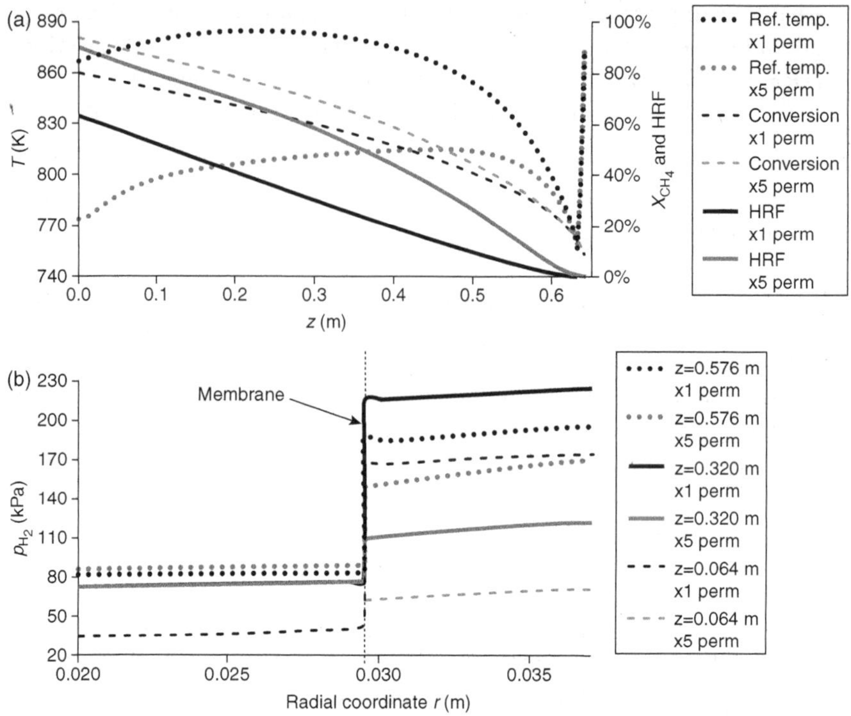

14.9 Comparison of results between base case permeance (x1) and a five times higher value: (a) reforming temperature profiles, methane conversion and HRF along reactor length; (b) radial pressure profiles at different lengths of the reactor.

We can define a cross-sectional averaged H_2 partial pressure:

$$\hat{p}_{H_2(z)} = \frac{\int_{r_{inner}}^{r_{outer}} r p_{H_2(z,r)}\, dr}{\frac{1}{2}(r_{outer}^2 - r_{inner}^2)} \qquad [14.40]$$

then the average driving force in the 1D proxy method results:

$$\int_{z=0}^{z=L} \left(\hat{p}^n_{\mathrm{H_2,feed}(z)} - \hat{p}^n_{\mathrm{H_2,perm}(z)} \right) \mathrm{d}z \frac{1}{L} \qquad [14.41]$$

instead of the calculation used in the 2D method:

$$\int_{z=0}^{z=L} \left(p^n_{\mathrm{H_2,feed}\left(z, r^+_{\mathrm{mem}}\right)} - p^n_{\mathrm{H_2,perm}\left(z, r^-_{\mathrm{mem}}\right)} \right) \mathrm{d}z \frac{1}{L} \qquad [14.42]$$

Table 14.6 Results for enhanced permeance cases

	Permeance base case	Permeance base case *x*2	Permeance base case *x*5
Average reforming temperature (K)	867	840	807
X_{CH_4} (%)	80.0	87.3	93.7
HRF (%)	62.9	78.9	89.6
Average driving force ($kPa^{0.5}$) (area-weighted)	5.6	3.8	1.9
Overestimation error in driving force using 1D calculation (%)	6.3	11.6	28.1

Table 14.6 shows the results for simulations using permeances two and five times that of the base case configuration (Table 14.2) and at WHSV of 1.50 h^{-1}. It is evident that higher membrane permeance (five times) increases HRF from about 63% to 90%, together with methane conversion, which increases from 80% to 94%. On the other hand, the average reforming temperature decreases by 60 K, because the higher permeation flow increases reforming reaction rates and heat demand, which cannot be completely supplied by hot gas. With respect to driving force, it can be seen that for the base case system a 1D model would imply moderate overestimation of the driving force. However, with membrane technologies with higher permeance the error becomes quite significant, up to 28% for the system with five times the base case permeance.

Another interesting effect to discuss is the effect of membrane permeance on reaction rate. In Section 14.4.2 we saw that for a certain variation in catalyst load, the overall results of the reactor were hardly affected, indicating that the rate-limiting step was the permeation of hydrogen. Going back to the kinetic model, if we analyse the reaction rate for methane reforming 'Reaction 1', we see from Equation [14.15] that the reaction rate is faster when the term between brackets increases, which is equivalent to saying that the reaction rate is faster the further the species are from equilibrium. Therefore it is useful to define the equilibrium quotient for reaction rate r_1 as:

$$\frac{K_1}{K_{eq_1}} = \frac{p_{H_2}^3 \times p_{CO}}{p_{CH_4} \times p_{H_2O}} \times \frac{1}{K_{eq_1}} \qquad [14.43]$$

which is further from 1 the further the reaction is from equilibrium. Figure 14.10 shows the plots for the system with base case permeance and also for that five times higher. The results are shown at a radius corresponding to

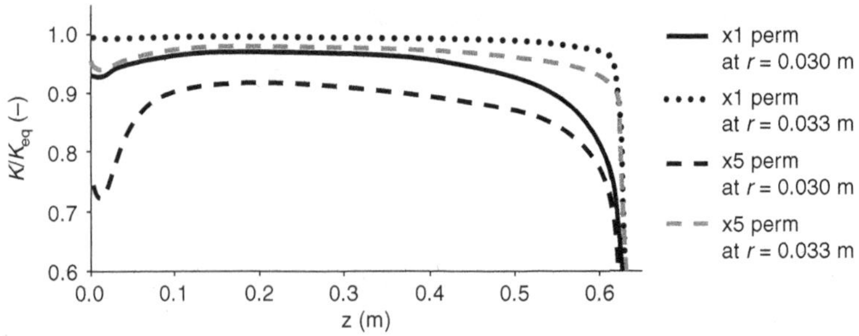

14.10 Equilibrium quotient for reaction r_1 in the reacting section core ($r = 0.033$ m) and in the vicinity of the membrane ($r = 0.030$ m) for permeance equal to the base case and a five times higher value (WHSV 1.50 h^{-1}).

the core of the reacting section ($r = 0.033$ m) and at a radius in the vicinity of the membrane ($r = 0.030$ m). In the base case, the reaction remains very close to equilibrium all along the reactor, and even very close to the membrane, agreeing with what previously was discussed about the rate-limiting step. However, for higher permeance this differs. Therefore, it is fair to say that with the advancement of technology in hydrogen permselective membranes, and its consequent improvement in permeance, an adequate design and selection of the catalyst would no longer be a minor issue in these kinds of application.

14.4.4 Carbon deposition

Carbon deposition deactivates the supported metal catalyst reducing its effectiveness and in extreme cases can even block the reformer (Twigg, 1996). In steam methane reforming, the most likely mechanisms for carbon formation are:

Methane decomposition (or methane cracking):

- Reaction 4: $CH_4 \Rightarrow C + 2H_2 \quad \Delta H^0_{298K} = 75\,kJ/mol$

Boudouard reaction:

- Reaction 5: $2CO \Rightarrow C + CO_2 \quad \Delta H^0_{298K} = -172\,kJ/mol$

Carbon monoxide reduction:

- Reaction 6: $CO + H_2 \Rightarrow C + H_2O \quad \Delta H^0_{298K} = -131\,kJ/mol$

Thermodynamically, the higher the H/C ratio in the mix, the lower the possibilities for carbon formation, therefore in the membrane reactor in discussion, the extraction of H_2 could raise a problem from this point of view. Methane decomposition is likely to be the predominant mechanism for coking, because at the feed inlet, the hydrogen content is initially nil, and along the reactor in the vicinity of the membrane the lower H_2 partial pressure would also favour this reaction.

Equations [14.44]–[14.46] show the kinetics for Reactions 4–6 which are obtained experimentally (Hou and Hughes, 2001).

$$K_{\mathrm{eq}_4} = 4.161\times 10^{7} \times e\frac{-10614\,K}{T}\ (\mathrm{kPa}) \qquad [14.44]$$

$$K_{\mathrm{eq}_5} = 5.744\times 10^{-12} \times e\frac{20634\,K}{T}\left(\mathrm{kPa}^{-1}\right) \qquad [14.45]$$

$$K_{\mathrm{eq}_6} = 3.173\times 10^{-10} \times e\frac{16318\,K}{T}\left(\mathrm{kPa}^{-1}\right) \qquad [14.46]$$

Again one could analyse the equilibrium quotient for the reactions in consideration as follows:

$$\frac{K_4}{K_{\mathrm{eq}_4}} = \frac{p_{\mathrm{H}_2}^2}{p_{\mathrm{CH}_4}} \times \frac{1}{K_{\mathrm{eq}_4}} \qquad [14.47]$$

$$\frac{K_5}{K_{\mathrm{eq}_5}} = \frac{p_{\mathrm{CO}_2}}{p_{\mathrm{CO}}^2} \times \frac{1}{K_{\mathrm{eq}_5}} \qquad [14.48]$$

$$\frac{K_6}{K_{\mathrm{eq}_6}} = \frac{p_{\mathrm{H}_2\mathrm{O}}}{p_{\mathrm{CO}} \times p_{\mathrm{H}_2}} \times \frac{1}{K_{\mathrm{eq}_6}} \qquad [14.49]$$

Figure 14.11 plots the equilibrium quotient for Reactions 4 and 5 (the situation for Reaction 6 is very similar to the plot of Reaction 5), for different S/C and a case with high permeance. The figures take the data for the partial pressures in the vicinity to the membrane (r = 0.030 m), because in this region the occurrence of Reaction 4 is more critical than in the bulk of the reforming section. For Reaction 5 the equilibrium quotient does not have an evident preference in the radial direction, but the plot is also taken close to the membrane in order to study the possibility of blockage over the membrane surface due to carbon deposition.

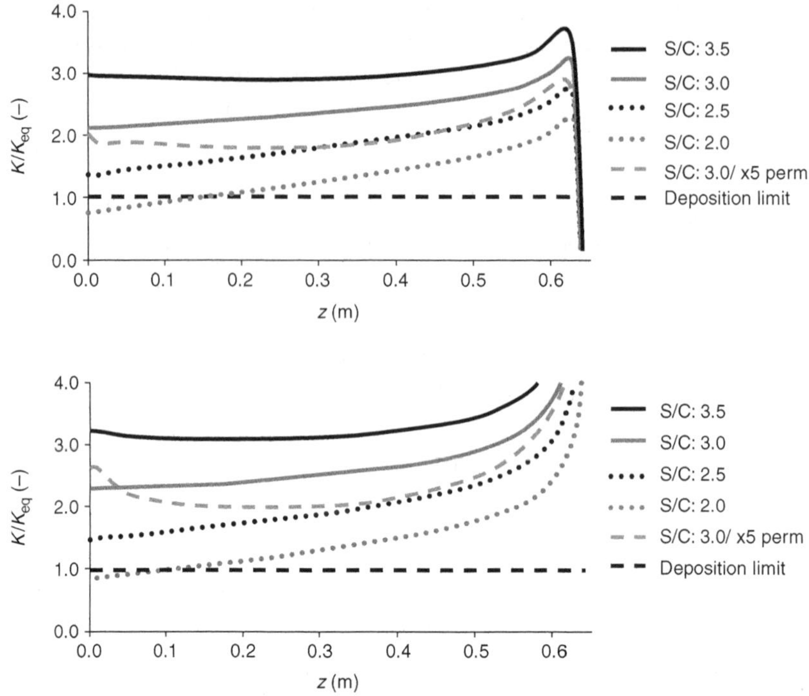

14.11 Carbon deposition study for different S/C and permeance through equilibrium quotient analysis. Carbon deposition through (a) CH_4 cracking and (b) Boudouard coking.

By means of the equilibrium quotient we can study the affinity of carbon formation in different cases; a quotient below 1 means that the reaction is thermodynamically driven to convert towards carbon formation. As expected, methane cracking is almost inevitable in the reactor inlet, and after some hydrogen is produced this effect is quickly bypassed. The equilibrium quotient for methane cracking, after an initial peak begins to fall due to the extraction of hydrogen from the reacting mix. With regards to Boudouard coking, the evolution of the quotient is towards lower values, due to the fact that even though the partial pressure of both CO and CO_2 rise, the quotient depends on the square of CO partial pressure. As expected, in either of the mechanisms studied, low S/C ratios favour coke formation. As already discussed, a system with very high permeance imposes stronger radial profiles for the species and in particular H_2, leading to higher affinity to carbon formation against the membrane. From the figure we can see that for the same S/C of 3.0, the equilibrium quotient is closer to the critical value for the case having higher permeance.

It must be added that carbon deposition cannot be entirely predicted by using equilibrium calculations; in practice, carbon deposition can occur before reaching the thermodynamic limit (due to poisons, real temperature gradients, etc.) (Rostrup-Nielsen, 1984; van Beurden, 2004). However, the analysis still provides reasonable guidelines or 'rules of thumb' that allow identifying those conditions which are critical for carbon formation, therefore providing aid for decision making about reactor design. An aspect of great importance is the selection of the catalyst, for instance, compared to Nickel-based catalysts, the use of noble metals reduces the likelihood of coking because they do not dissolve carbon.

From the analysis it is clear to see that some carbon formation is very likely to happen, at least at the reactor inlet by methane cracking. However, this effect would be bypassed by the adoption of a pre-reforming section which could be performed outside the membrane reactor using a small amount of catalyst or even at different temperature. The catalyst of a small pre-reforming section could be replaced periodically. Another solution, being that a small amount would be required, a noble based catalyst could be used for a pre-reformer, which in some cases could turn out to be economically advantageous since, besides the higher capital cost, it would avoid more frequent replacement.

14.4.5 Kinetics models

In this final section we point out the importance of considering the appropriate kinetic model when dealing with CFD simulations. It takes into consideration that every single existing model followed a campaign of experiments undertaken using some specified catalyst, with a specified metal content, density, surface area, support type, etc. Moreover, these models were calibrated for tests covering a certain range of pressure and temperature, outside of which the validity of the model can be put into question. Therefore, a certain model might or might not fit a different set of reactor, catalyst and operating conditions for a number of reasons. In order to be precise, a model is to be calibrated to the system subject to be modelled. This being said, from an engineering point of view it is best to consider those models which have been validated under conditions closest to the ones to simulate, and with that as a starting point, adjust by calibration for example by means of appropriate effectiveness factors (Xu and Froment, 1989b).

In this section we compare two of the most widely accepted kinetics models for steam methane reforming. For once the Xu and Froment (from now on X&F) (Xu and Froment, 1989a), which we used for all the above sections and Numaguchi and Kikuchi (from now on N&K) (Numaguchi and Kikuchi, 1988). In Table 14.7 we put into evidence the differences in the experimental conditions used in both references following what we have just highlighted on the asymmetry of diverse models.

Table 14.7 Experimental conditions for development of the models in two references

	Xu and Froment (1989)	Numaguchi and Kikuchi (1988)
Catalyst type	$Ni/MgAl_2O_4$	Ni/Al_2O_3
Nickel content (weight %)	15.2	8.7
Catalyst dimensions	0.18–0.25 mm (crushed from 10 mm ring-shaped)	5/8″ × 1/4″ cylindrical
Temperature range (K)	773–848	674–1160
Pressure range (kPa)	300–1500	120–2550
S/C	3.0–5.0	1.44–4.50

The kinetics for the X&F model are reported in Section 14.3. The kinetics given by the N&K model for Reactions 1–3 are as follows:

$$r_1^{\mathrm{N\&K}} = \frac{k_1^{\mathrm{N\&K}} \times \left(p_{\mathrm{CH_4}} \times p_{\mathrm{H_2O}} - \dfrac{p_{\mathrm{H_2}}^3 \times p_{\mathrm{CO}}}{K_{\mathrm{eq_1}}} \right)}{p_{\mathrm{H_2O}}^{1.596}} \qquad [14.50]$$

$$r_2^{\mathrm{N\&K}} = k_2^{\mathrm{N\&K}} \times \left(p_{\mathrm{CO}} \times p_{\mathrm{H_2O}} - \frac{p_{\mathrm{H_2}} \times p_{\mathrm{CO_2}}}{K_{\mathrm{eq_2}}} \right) \qquad [14.51]$$

$$r_3^{\mathrm{N\&K}} = k_3^{\mathrm{N\&K}} \times \left(p_{\mathrm{CH_4}} \times p_{\mathrm{H_2O}}^2 - \frac{p_{\mathrm{H_2}}^4 \times p_{\mathrm{CO_2}}}{K_{\mathrm{eq_3}}} \right) \qquad [14.52]$$

where:

$$k_1^{\mathrm{N\&K}} = 1.468 \times 10^5 \times e \frac{-1.069 \times 10^5 \,\mathrm{kJ/kmol}}{RT} \left(\frac{\mathrm{kmol}}{\mathrm{kPa}^{0.404}\,\mathrm{kg_{cat}}\,\mathrm{h}} \right) \qquad [14.53]$$

$$k_2^{\mathrm{N\&K}} = 8.820 \times e \frac{-5.453 \times 10^4 \,\mathrm{kJ/kmol}}{RT} \left(\frac{\mathrm{kmol}}{\mathrm{kPa}^{2}\,\mathrm{kg_{cat}}\,\mathrm{h}} \right) \qquad [14.54]$$

$$k_3^{\mathrm{N\&K}} = 0 \qquad [14.55]$$

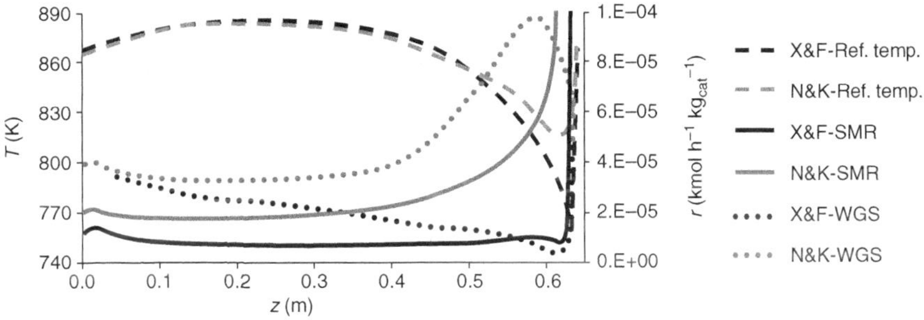

14.12 Comparison of the kinetic models of Xu and Froment with Numaguchi and Kikuchi: reforming temperature and reaction rates.

The results for the simulation set for conditions identical to the base case (Table 14.3) returned different results when using the two kinetic models. Figure 14.12 shows reforming temperature profiles and reaction rates considering SMR = $r_1 + r_2$ and WGS = $r_2 + r_2$. It is evident that X&F-SMR is much faster than N&K-SMR, explaining therefore the more accentuated initial temperature drop. The water–gas shift (WGS) also differs from one model to the other, for example X&F-WGS also peaks right at the inlet due to the very high rate of SMR producing CO, but N&K shows a softer evolution of the kinetic rates. The overall results showed X_{CH_4} and HRF of 80 and 63% for X&F and 76 and 58% for N&K, which are not minor differences.

The discrepancies between the two models are very noticeable in the section closer to the inlet of the reactor; however, in the final two-thirds the results seem to merge quite well; hence, if we were to model a membrane reactor where the feed has been pre-reformed (bypassing the discrepancy at the inlet), the results for both models would be much alike. In X&F in fact, the problem at the inlet is also present for the numerical resolution of the reaction rate, as was mentioned in Section 14.3.4.

In conclusion, the selection of the kinetic model is not an issue to underestimate, and its adoption must be taken with caution, expertise and knowledge of the system intended for emulation.

14.5 Conclusion and future trends

This chapter presented the application of bi-dimensional CFD modelling to MREFs integrated in the PEMFC-based micro-CHP system; the electric output of the systems was set in the range of 1–2 kW. The design of the MREF was carried out starting from the optimal HRF calculated from overall system simulation: the target HRF is between 60% and 70% and heat for sustaining the reforming reactions is supplied by a hot gas stream obtained from the combustion of the retentate leaving the reactor. The bi-dimensional CFD modelling was adopted to determine required membrane surface area for hydrogen permeation as well as the optimal reactor design conditions and the effect of different operating conditions on the overall hydrogen production and fuel conversion performance.

Simulations showed that the optimal MREF configuration is based on hot gases flowing co-currently with the reforming stream: in this configuration, the highest hot gases temperature perfectly matches the reaction heat demand leading to a nearly constant feed temperature. Moreover, the co-current configuration achieves a better control of the temperature over the membrane, reducing the risk of damage and loss of selectivity. It can be stated that the results support the feasibility of thermally coupling a MREF for hydrogen production with the hot exhaust gases resulting from the combustion of the exhaust of the fuel processor. Strengths of the 2D model in comparison with a 1D model become significant when dealing with high hydrogen fluxes and high membrane permeances, and for the analysis of carbon formation/deposition thermodynamics.

Finally, it can be stated that the proposed 2D CFD model showed improved performance prediction of SMR membrane reactors compared to previous simulations; however, the adoption of adequate kinetic models is necessary and must be considered carefully.

14.6 Acknowledgements

This work is related to research activities supported by Fondazione Silvio Tronchetti Provera within the framework of a PhD program and by Italian Ministry of Economic Development under the 'Industria 2015' project. The authors also wish to thank Dr Riccardo Mereu for his helpful ideas and suggestions for the CFD programming. The authors also acknowledge all published sources which are referred to.

14.7 References

ANSYS Inc. 2010. Fluent v6.3.26. Available at: www.ansys.com.

Berkheimer G D and Buxbaum R, 'Hydrogen pumping with palladium membranes', *J. Vac. Sci. Technol., A*, 1985, **3**, 412–416.

Branan C R, *Rules of Thumb for Chemical Engineers 3rd ed.* Gulf Professional Publishing, 2002.

Catalano J, Ciacinti M and Sarti G, 'Influence of water vapor on hydrogen permeation through 2.5 μm Pd–Ag membranes', *Int. J. Hydrogen Energy*, 2011, **36**, 8658–8673.

Chang H F, Pai W J, Chen Y J and Lin W H, 'Autothermal reforming of methane for producing high-purity hydrogen in a Pd/Ag membrane reactor', *Int. J. Hydrogen Energy*, 2010, **35**, 12986–12992.

Chen Y, Wang Y, Xu H and Xiong G, 'Efficient production of hydrogen from natural gas steam reforming in palladium membrane reactor', *Appl. Catal., B*, 2008, **81**, 283–294.

De Falco M, Di Paola L and Marrelli L, 'Heat transfer and hydrogen permeability in modelling industrial membrane reactors for methane steam reforming', *Int. J. Hydrogen Energy*, 2007, **32**, 2902–2913.

De Falco M, Di Paola L, Marrelli L and Nardella P, 'Simulation of large-scale membrane reformers by a two-dimensional model', *Chem. Eng. J.*, 2007, **128**, 115–125.

De Wasch A P and Froment G F, 'Heat transfer in packed beds', *Chem. Eng. Sci.*, 1972, **27**, 567–576.

Dittmeyer R, 'Membrane reactors for hydrogenation and dehydrogenation processes based on supported palladium', *J. Mol. Catal. A: Chem.*, 2001, **173**, 135–184.

Elnashaie S S E H and Elshishini S S, *Modelling, simulation, and optimization of industrial fixed bed catalytic reactors*, London, Gordon and Breach Science Publishers, 1993.

Ergun S, 'Fluid flow through packed columns', *Chem. Eng. Prog.*, 1952, **48**, 89–94.

Gallucci F, Paturzo L, Famà A and Basile A, 'Experimental study of the methane steam reforming reaction in a dense Pd/Ag membrane reactor', *Ind. Eng. Chem. Res*, 2004, **43**, 928–933.

Gallucci F, van Sint Annaland M and Kuipers J A M, 'Theoretical comparison of packed bed and fluidized bed membrane reactors for methane reforming', *Int. J. Hydrogen Energy*, 2010, **35**, 7142–7150.

Hou K and Hughes R, 'The kinetics of methane steam reforming over a Ni/α-Al2O catalyst', *Chem. Eng. J.*, 2001, **82**, 311–328.

Incropera F, DeWitt D, Bergman T, Lavine A, *Fundamentals of Heat and Mass Transfer 6th ed*. John Wiley & Sons. 2006.

Lin Y M, Liu S L, Chuang C H and Chu Y T, 'Effect of incipient removal of hydrogen through palladium membrane on the conversion of methane steam reforming. Experimental and modeling', *Catal. Today*, 2003, **82**, 127–139.

Madia G S, Barbieri G and Drioli E, 'Theoretical and experimental analysis of methane steam reforming in a membrane reactor', *Can. J. Chem. Eng.*, 1999, **77**, 698–706.

Merk H, 'The macroscopic equations for simultaneous heat and mass transfer in isotropic, continuous and closed systems', *Appl. Sci. Res.*, 1959, **8**, 73–99.

Numaguchi T and Kikuchi K, 'Intrinsic kinetics and design simulation in a complex reaction network: Steam Methane Reforming', *Chem. Eng. Sci.*, 1988, **43**, 2295–2301.

Oklany J, 'A simulative comparison of dense and microporous membrane reactors for the steam reforming of methane', *Appl. Catal., A*, 1998, **170**, 13–22.

Oyama S T and Hacarlioglu P, 'The boundary between simple and complex descriptions of membrane reactors: The transition between 1-D and 2-D analysis', *J. Membr. Sci.*, 2009, **337**, 188–199.

Patel K and Sunol A, 'Modeling and simulation of methane steam reforming in a thermally coupled membrane reactor', *Int. J. Hydrogen Energy*, 2007, **32**, 2344–2358.

Pizzi D, Worth R, Giacinti Baschetti M, Sarti G C and Noda K, 'Hydrogen permeability of 2.5μm palladium–silver membranes deposited on ceramic supports', *J. Membrane Sci.*, 2008, **325**, 446–453.
Roses L, Bonalumi D, Campanari S, Iora P, Manzolini G, 'Simulation Comparison of PEMFC micro-cogeneration units with conventional and innovative fuel processing', *ASME 2010 Eight International Fuel Cell Science, Engineering and Technology Conference proceedings*, New York, 2010a.
Roses L, Manzolini G and Campanari S, 'CFD simulation of Pd-based membrane reformer when thermally coupled within a fuel cell micro-CHP system', *Int. J. Hydrogen Energy*, 2010b, **35**, 12668–12679.
Roses L, Gallucci F, Manzolini G, Campanari S and van Sint Annaland M, 'Comparison between fixed bed and fluidized bed membrane reactor configurations for PEM based micro-cogeneration systems', *Chem. Eng. J.*, 2011, **171**, 1415–1427.
Rostrup-Nielsen J, 'Catalytic steam reforming', in Anderson and Boudart, *Catalysis Science and Technology*, Springer, 1–117, 1984.
Salemme L, Menna L, Simeone M and Volpicelli G, 'Energy efficiency of membrane-based fuel processors – PEM fuel cell systems', *Int. J. Hydrogen Energy*, 2010, **35**, 3712–3720.
Taylor R, Krishna R, *Multicomponent Mass Transfer*, John Wiley & Sons, 1993.
Tucho W M, Venvik H J, Stange M, Walmsley J C, Holmestad R, Bredesen R, 'Effects of thermal activation on hydrogen permeation properties of thin, self-supported Pd/Ag membranes', *Sep. Purif. Technol.*, 2009, **68**, 403–410.
Twigg M, *Catalyst Handbook 2nd ed.* Manson Publishing, 1996.
van Beurden P, '*On the catalytic aspects of steam-methane reforming: A literature survey*', 2004
Xu J and Froment G F, 'Methane steam reforming, methanation and water-gas shift: I. Intrinsic kinetics', *AIChE J.*, 1989, **35**, 88–96.
Xu J and Froment G F, 'Methane steam reforming: II. Diffusional limitations and reactor simulation', *AIChE J.*, 1989, **35**, 97–103.

14.8 Appendix: nomenclature

14.8.1 Notation

$Ac_{p,i,j}$ coefficients in polynomial expression for specific heat of component i

$A_{\lambda,i,j}$ coefficients in polynomial expression for thermal conductivity of component i

$A_{\mu,i,j}$ coefficients in polynomial expression for viscosity of component i

$c_{p,i}$ specific heat of the component i $\left(\frac{\text{kJ}}{\text{kmol K}} \text{ or } \frac{\text{kJ}}{\text{kg K}}\right)$

$c_{p,\text{mix}}$ gas mixture specific heat $\left(\frac{\text{kJ}}{\text{kg K}}\right)$

D_{ij}	binary diffusion coefficients
d_p	equivalent particle diameter (m)
d_{eq}	equivalent diameter of flow section (m)
e	emissivity of the catalyst surface
E_a	apparent activation energy $\left(\frac{\text{kJ}}{\text{kmol}}\right)$
F_i	i component gas flow-rate $\left(\frac{\text{kmol}}{\text{h}}\right)$
HRF	hydrogen recovery factor
i_{sw}	sweep gas ratio
J_i	i component flux through the membrane $\left(\frac{\text{kmol}}{\text{h m}^2}\right)$
K_i	adsorption equilibrium constant of the component i
k_j	rate constant of reaction j (depends on the reaction)
K_{eq_j}	equilibrium constant of reaction j
L	reactor length (m)
M_i	molecular weight of the component i $\left(\frac{\text{kg}}{\text{kmol}}\right)$
M_m	average molecular weight of the gas mixture $\left(\frac{\text{kg}}{\text{kmol}}\right)$
$\breve{p}_i(z)$	cross-sectional averaged partial pressure of component i at axial coordinate 'z'
p	pressure (kPa)
Δp	pressure drop along the reactor (kPa)
p_i	i-component partial pressure (kPa)
P_i	i component permeability coefficient $\left(\frac{\text{kmol}\cdot\text{m}}{\text{h m}^2\,\text{kPa}^n}\right)$
P_i^0	i component permeability pre-exponential factor $\left(\frac{\text{kmol}\cdot\text{m}}{\text{h m}^2\,\text{kPa}^n}\right)$
Pr	Prandtl number $\left(\frac{c_p\mu_g}{k_g}\right)$
PR_i	i component permeation ratio
r_j	reaction rate of reaction j
r	radial coordinate (m)

R universal gas constant $\left(\frac{\text{kJ}}{\text{kmol K}}\right)$

Re_p Reynolds number referred to catalyst particle diameter $\left(\frac{Gd_p}{\mu_g}\right)$

S/C steam to carbon ratio

T temperature (K)

u_s gas superficial velocity (m / s)

x_i mass fraction of the component i

X_i i-component conversion

y_i molar fraction of the component i

z axial coordinate (m)

Greek symbols

δ membrane thickness (m)

ε void fraction of packing

λ_{er} effective radial thermal conductivity $\left(\frac{\text{kJ}}{\text{h m K}}\right)$

λ_g gas phase thermal conductivity $\left(\frac{\text{kJ}}{\text{h m K}}\right)$

λ_i thermal conductivity of the component i $\left(\frac{\text{kJ}}{\text{h m K}}\right)$

λ_s thermal conductivity of packing material $\left(\frac{\text{kJ}}{\text{h m K}}\right)$

μ_g gas mixture viscosity $\left(\frac{\text{kg}}{\text{h m}}\right)$

μ_i viscosity of the component i $\left(\frac{\text{kg}}{\text{m h}}\right)$

ρ_B catalytic bed density (kg/m^3)

ρ_g gas density (kg/m^3)

ρ_s catalyst density (kg/m^3)

Subscripts and superscripts

i component in mixture

in inlet to reactor

inner inner limit of the radial section

j	reaction number
out	outlet from reactor
outer	outer limit of the radial section
p	permeate side
N&K	Numaguchi and Kikuchi kinetic model
r	reforming side
X&F	Xu and Froment kinetic model

14.8.2 Abbreviations

CFD	computational fluid dynamics
CHP	combined heat and power
MREF	membrane reformer
NG	natural gas
PEMFC	polymer electrolyte membrane fuel cell
SMR	steam methane reforming
WGS	water–gas shift
WHSV	weight hourly space velocity

15

Computational fluid dynamics (CFD) analysis of membrane reactors: modelling of membrane bioreactors for municipal wastewater treatment

Y. WANG, T. D. WAITE and
G. L. LESLIE, University of New South Wales, Australia

DOI: 10.1533/9780857097330.3.532

Abstract: Computational methods provide invaluable insight to the analysis of complex two-phase and three-phase flow in municipal-scale membrane bioreactors. This capacity to simulate air, liquid and solid movement is an invaluable contribution to the design process. The following chapter provides an overview of membrane bioreactor (MBR) design, computational fluid dynamics (CFD) theory and modelling techniques for two-phase and three-phase flow. Topics include the effect of reactor geometry and membrane configuration on turbulence, the accumulation of solids, energy consumption and the efficiency of aeration. Emphasis is placed on the importance of model calibration and validation and the outlook for future work.

Key words: membrane bioreactor (MBR), municipal wastewater treatment, CFD (computational fluid dynamics), hydrodynamics, MBR design.

15.1 Introduction

The membrane bioreactor (MBR) is a viable process for the treatment and reuse of wastewater at the municipal scale. The market value of MBR technology was estimated to be approximately US$217 million in 2005, rising at an average annual growth rate of between 9.5 and 12%, and is estimated to reach US$500 million by 2013 (Judd, 2011). During this time the capacity of MBRs has expanded from a typical 10 000 m^3/day to over 100 000 m^3/day, including an installation in Shiyan, Hubei province in China commissioned in 2009 with a capacity of 120 000 m^3/day (Judd, 2011).

Communities have selected MBRs as an alternative to conventional wastewater treatment plants, in which either suspended or fixed film biological systems convert soluble nitrogenous, carbonaceous and phosphorous nutrients into *settleable* biomass that is removed by gravity separation. In contrast, MBRs use a microporous membrane in lieu of a clarifier to separate biomass from the treated effluent (Fig. 15.1) (Leslie, 2001). Thus, as

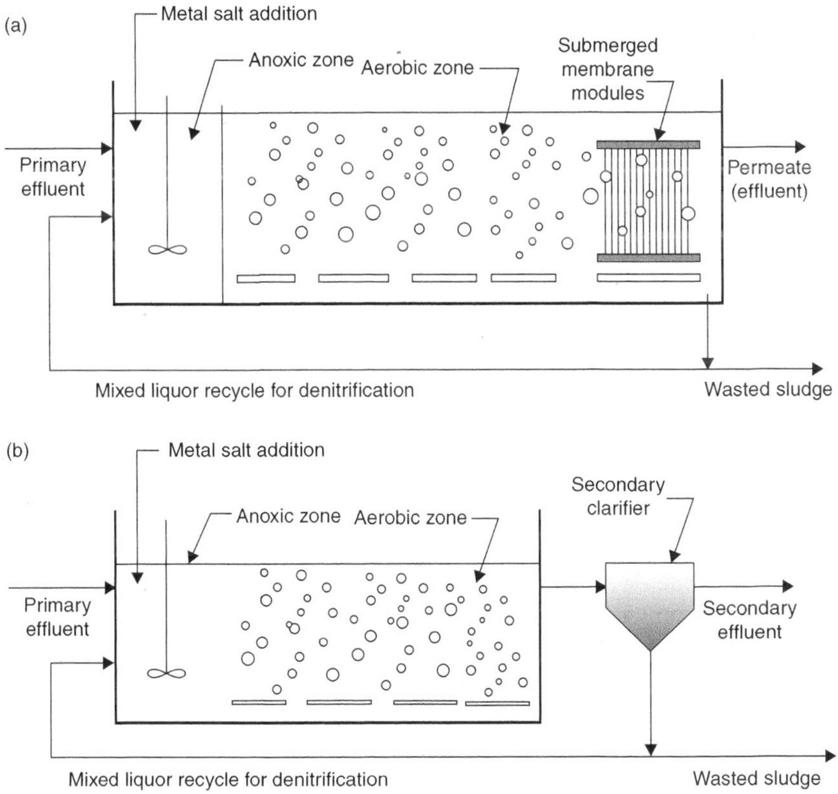

15.1 Typical schematic representation of activated sludge process: (a) membrane bioreactor; (b) conventional activated sludge process.

it is not necessary to produce a *settleable* biomass (which typically limits the mixed liquor suspended solids (MLSS) concentration in the reactor to 3–5 g/L), MBRs can be designed with *filterable* biomass concentrations of up to 20 g/L (Judd, 2006), though MBR MLSS concentrations typically range in practice from 8–15 g/L. This reduces the space requirements for MBRs as a result of improved reactor efficiency (higher biomass per unit volume) and the elimination of the clarifier. Moreover, because the pore size of the microporous membrane (typically < 0.2 microns) is smaller than bacterial and protozoan pathogens, it is possible to discharge the treated effluent to very sensitive environments such as bathing water and systems that recharge ground water (Leslie, 2001).

MBR technology is an active area for academic and industrial research in view of the many factors affecting the performance of the process. One area of particular development has been the use of numerical modelling techniques, such as CFD, as an aid to the design and optimisation of this technology. The following section provides a brief introduction to MBR technology, a comprehensive summary of the aspects of MBRs in which CFD models have been applied, and a description of the experimental techniques that may be used to generate data for model calibration and validation.

15.2 Design of the membrane bioreactor (MBR)

For a given wastewater, the parameters affecting the MBR design can be categorised into three groups: (i) biological factors, (ii) membrane factors and (iii) hydrodynamic factors (Manem and Sanderson, 1996) (Fig. 15.2). Biological treatment efficiency relies on the utilisation by the microbial community of the major wastewater nutrients carbon, nitrogen and phosphorous. Microbial growth, which depends on factors such as dissolved oxygen (DO) concentration, pH and temperature, is slow and therefore requires long hydraulic retention times (HRT) and large reactor volumes (i.e., hydrodynamic factors). Use of an MBR rather than a conventional activated sludge process enables a shortening of the HRT by retaining the solids in the reactor (resulting in a longer solids retention time (SRT)).

In comparison to the above hydrodynamic factors, the micro-organisms also affect membrane filtration performance. The separation of liquid and solids using microfiltration (MF) or ultrafiltration (UF) membranes is driven by trans-membrane pressure (TMP). The permeate flux, that is, the quantity of water passing through a unit area of membrane per unit time, is related to TMP by Darcy's equation:

$$J = \frac{\text{TMP}}{\mu R_t} \qquad [15.1]$$

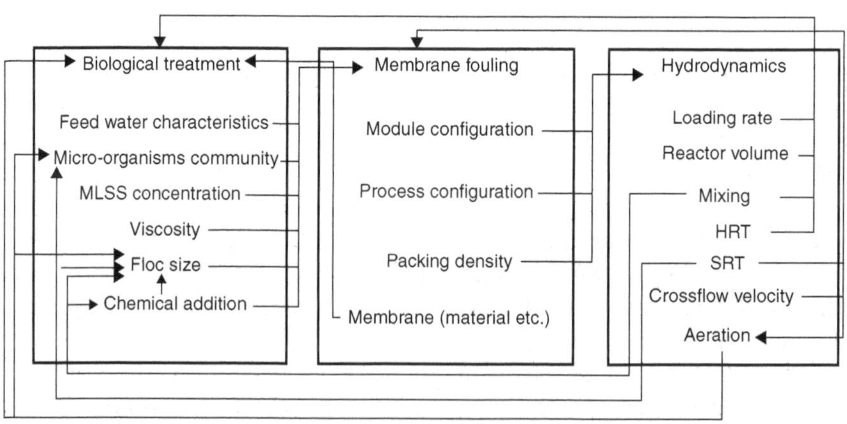

15.2 Interactive relationships between the MBR performance attributes. MLSS, mixed liquor suspended solids.

where J is the permeate flux, μ is viscosity of the permeate (effluent) and R_t is the total resistance, which is made up of the resistance of the membrane itself and the resistance of a fouling layer formed from the biomass on the membrane surface and/or inside the membrane pores.

During the separation process, particles may deposit on the membrane surface or in membrane pores, causing the TMP to increase (or flux to decline) and affecting the quality of the water produced (membrane fouling). The nature of the feedwater, concentration and structure of the mixed liquor, and nature and concentration of soluble microbial products (SMPs) are all recognised to influence fouling propensity (Judd, 2011).

Both the removal of nutrients and membrane filtration performance are largely determined by the hydrodynamics of the bioreactor. In conventional biological wastewater treatment, it is recognised that hydrodynamic factors such as the reactor type (i.e., completely mixed tank reactors (CSTRs) or plug flow reactors (PFRs)) and the mixing conditions in the reactor can affect both the efficiency of nutrient removal in the bioreactor and the settling of the sludge (Metcalf and Eddy, 1991). Good mixing promotes the transfer of substrates and heat to the micro-organisms and ensures the effective use of the entire reactor volume (Metcalf and Eddy, 1991).

Hydrodynamics are also important to the membrane filtration process. MBRs can be operated with the membrane being placed either external to the bioreactor (side-stream MBR) or immersed in the reactor (submerged MBR) (Fig. 15.3). The submerged configuration is usually preferred for medium- to large-scale domestic wastewater treatment, to the 'pressurised' side-stream configuration because of the lower energy requirements (Judd, 2011). The submerged MBRs can be further configured as 'inside submerged' and 'outside submerged' (Fig. 15.4). The membrane modules are set up directly in an aerated biological tank for the 'inside submerged' configuration, while the membranes are submerged in a separate tank dedicated to filtration for the 'outside submerged' configuration with the sludge being circulated between the biological and the filtration zones. Membranes used in municipal MBR plants can be configured as hollow fibre (HF), flat sheet (FS) and multi-tube (MT) modules (Fig. 15.5). HF and FS membranes are mainly used in submerged MBR systems, while side-stream MBRs use either FS or MT membranes. The HF membranes can be mounted either horizontally or vertically. A key difference between the HF and FS configuration is the greater packing density that can be achieved with the HF configuration compared with the FS configuration. Another key difference is how the distance and flow path between adjacent membrane vary in FS and HF configurations. FS configurations are rigid and have a fixed separation distance creating well-defined flow paths, while in HF configurations the membranes are allowed to move which create variable distance between fibres and highly variable flow paths (Buetehorn *et al.,* 2009; Drews *et al.,*

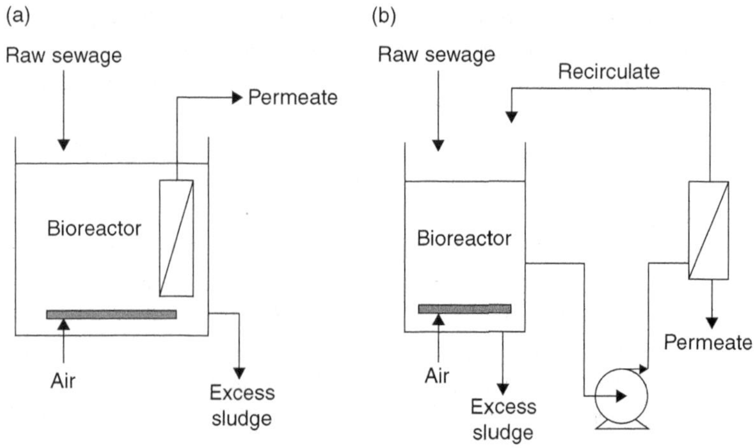

15.3 Diagrams of MBR configurations: (a) submerged MBR; (b) side-stream MBR.

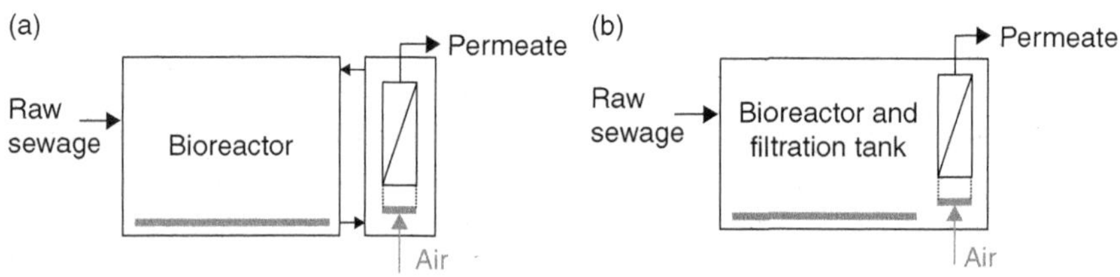

15.4 Inside submerged versus outside submerged configurations: (a) inside submerged; (b) outside submerged.

(a) (b) (c)

15.5 Different membrane modules used in MBRs: (a) FS; (b) HF; (c) MT.

2010). The membranes can be operated at different crossflow velocities (CFV). All of these various configurations combine to give a unique hydrodynamic condition in the reactor and result in different filtration behaviour and fouling propensity.

Another important component in the hydrodynamic category exerting a critical impact on the biological activity and membrane performance is the management of oxygen levels in different parts of the bioreactor. In the aerobic zone, adequate DO levels are essential for nitrification and the formation of nitrification organisms. However, carryover of DO into the anoxic zone interferes with the conversion of nitrite to nitrogen. Consequently, modelling DO in MBRs is a very important part of the design of the biological process. Similarly, the DO units will impact the oxidation-reduction potential, which may influence the form and behaviour of redox-active metals such as iron which are used widely in facilitating phosphorous removal (Waite *et al.*, 1999).

In addition to facilitating an understanding of the effects of aeration on the biological and chemical processes, modelling of aeration is also important in understanding the hydrodynamics of the liquid and solid phases in the reactor. MBRs generally use fine bubble diffusers for supplying oxygen to the micro-organisms and coarse bubble diffusers to provide air for fouling control. In each case, changes in the aeration area, airflow rate and intensity affect bubble size and frequency may lead to different biomass behaviour (e.g., floc size and viscosity). The feedwater characteristics (e.g., the level of surfactants in the wastewater) in turn affect the bubble size, shape and stability (Judd, 2011). Aeration also promotes the mixing in the reactor and is required for scouring of the membrane. The shear force produced by the air bubbles increases back transport, prevents large particles depositing on the membrane surface and promotes mass transfer of liquid through the membranes. Aeration however is the largest contributor to the energy consumption and the operating cost of an MBR process (Cote *et al.*, 2004; Brannock *et al.*, 2010a) (Fig. 15.6). Consequently, modelling the effect of aeration is an important part of optimising energy consumption.

In summary, the design of an MBR is a complex process. It requires design of both the biological process and the membrane modules and requires an understanding of the effect of hydrodynamics and aeration systems on both organism growth and filtration aspects. Notwithstanding this, there is a need for a method of design that takes into account all of these performance attributes. A powerful technique for assessing reactor hydrodynamics is the use of CFD modelling. CFD is used to generate a solution to the Navier-Stokes equation for a given reactor design and is able to be coupled with various numerical models accounting for membrane filtration and organism growth (Fig. 15.7).

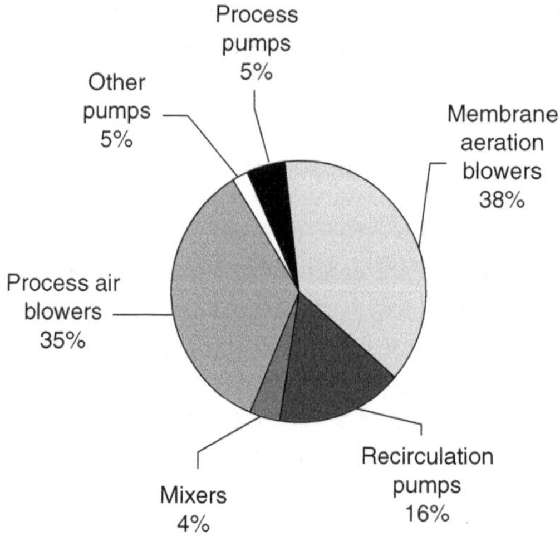

15.6 Representative energy demand for a 5.7 MLD (mega–litre per day) MBR (Judd, 2006).

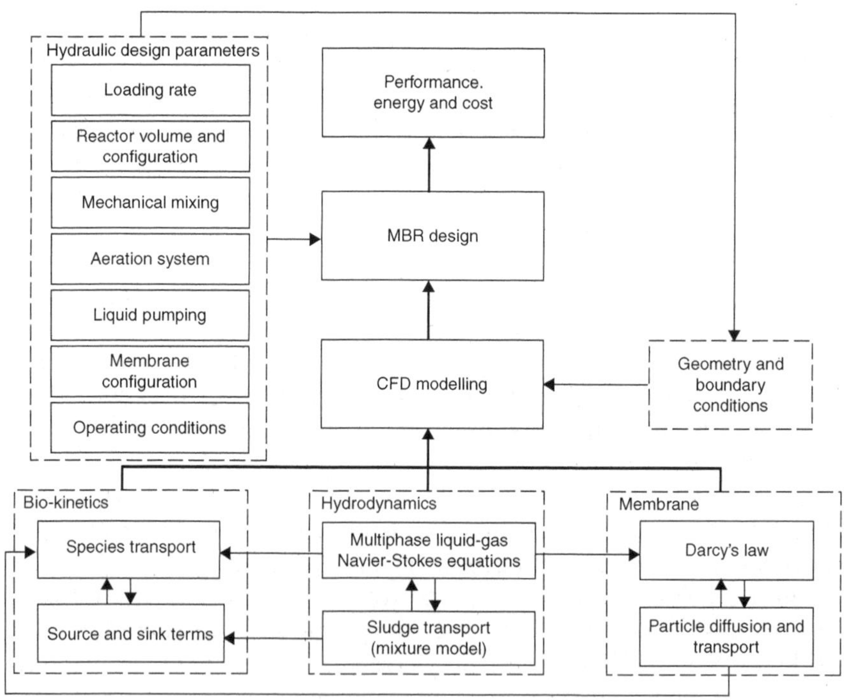

15.7 MBR CFD model components.

15.3 Computational fluid dynamics (CFD)

CFD uses numerical methods to solve the equations describing the motion of a fluid in three-dimensions. The following partial differential equations represent the mathematical statements of the conservation laws of physics (Versteeg *et al.*, 1995):

Mass conservation (or continuity) equation: the mass of a fluid is conserved (Equation [15.2]), that is,

$$\frac{\partial \rho}{\partial t}+\frac{\partial}{\partial x_i}(\rho u_i)=0 \quad [15.2]$$

Momentum conservation equation: the rate of change of momentum equals the sum of the forces on a fluid particle (Newton's second law, Equation [15.3]), that is,

$$\frac{\partial \rho}{\partial t}(\rho u_i)+\frac{\partial}{\partial x_j}(\rho u_i u_j)=-\frac{\partial p}{\partial x_i}+\frac{\partial \tau_{ij}}{\partial x_j}+\rho f_i \quad [15.3]$$

Energy conservation equation: the rate of change of energy is equivalent to the sum of the rate of heat addition to and the rate of work done on a fluid particle (First law of thermodynamics, Equation [15.4]), that is,

$$\frac{\partial}{\partial t}(\rho e)+\frac{\partial}{\partial x_j}(\rho e u_j)=-\rho\frac{\partial u_j}{\partial x_j}+\frac{\partial}{\partial x_j}\left(k\frac{\partial T}{\partial x_j}\right)+S_e \quad [15.4]$$

These equations can be generalised to:

$$\frac{\partial \rho\phi}{\partial t}+\nabla\bullet\left(\rho\phi\tilde{V}\right)=\nabla\bullet(\Gamma\nabla\phi)+S_\phi \quad [15.5]$$

or in Cartesian tensor form rather than the above vector form:

$$\underbrace{\frac{\partial}{\partial t}(\rho\phi)}_{\text{Accumulation term}}+\underbrace{\frac{\partial}{\partial x_j}(\rho\phi u_j)}_{\text{Convective term}}=\underbrace{\frac{\partial}{\partial x_j}\left(\Gamma\frac{\partial\phi}{\partial x_j}\right)}_{\text{Diffusive term}}+\underbrace{S}_{\text{Source term}} \quad [15.6]$$

where ϕ is the flow variable, t is time (s), ρ is the fluid density (kg/m^3), x_j is the length in dimension j, u_j is the velocity (m/s) in the jth direction, Γ is the diffusive coefficient (m^2/s) and S is the source or sink term of the flow variable.

Note that the mass conservation (or continuity) equation (Equation [15.2]) is equivalent to the general conservation equation with ϕ assigned to be unity. These equations are non-linear and cannot be solved analytically in the vast majority of practical cases (Patankar, 1980; Versteeg and Malalasekera, 1995). The equations must be linearised and solved over many small control volumes (the computational mesh). For determination of the flow-field, CFD simulations require input of geometry, boundary conditions and fluid properties.

The mixed liquor can be modelled as an incompressible fluid. Therefore, the continuity equation becomes:

$$\frac{\partial u}{\partial x}+\frac{\partial v}{\partial y}+\frac{\partial w}{\partial z}=0 \qquad [15.7]$$

The momentum equations (Navier-Stokes equation) for the incompressible fluid are:

$$\rho\left(\frac{\partial u}{\partial t}+u\frac{\partial u}{\partial x}+v\frac{\partial u}{\partial y}+w\frac{\partial u}{\partial z}\right)=-\frac{\partial p}{\partial x}+\mu\left(\frac{\partial^2 u}{\partial x^2}+\frac{\partial^2 u}{\partial y^2}+\frac{\partial^2 u}{\partial z^2}\right)+\rho f_x$$

$$\rho\left(\frac{\partial v}{\partial t}+u\frac{\partial v}{\partial x}+v\frac{\partial v}{\partial y}+w\frac{\partial v}{\partial z}\right)=-\frac{\partial p}{\partial x}+\mu\left(\frac{\partial^2 v}{\partial x^2}+\frac{\partial^2 v}{\partial y^2}+\frac{\partial^2 v}{\partial z^2}\right)+\rho f_y$$

$$[15.8]$$

$$\rho\left(\frac{\partial w}{\partial t}+u\frac{\partial w}{\partial x}+v\frac{\partial w}{\partial y}+w\frac{\partial w}{\partial z}\right)=-\frac{\partial p}{\partial x}+\mu\left(\frac{\partial^2 w}{\partial x^2}+\frac{\partial^2 w}{\partial y^2}+\frac{\partial^2 w}{\partial z^2}\right)+\rho f_z$$

The energy equation only needs to be solved if the problem involves heat transfer. Therefore, the energy equation can be ignored in the MBR systems that have small variations in temperature.

Thus, for solving the three-dimensional velocity distribution of continuous fluid, the continuity equation (Equation [15.7]) and the x, y and z direction momentum balance equations (Equation [15.8]) form a set of equations with four unknowns (ρ, u, v, w).

The improvements in both CFD tools and computer power that have occurred in recent years have resulted in application of CFD as a design tool in a variety of wastewater treatment systems including wastewater ponds (Wood *et al.,* 1995; 1998; Marshall 1999; Peterson *et al.,* 2000), secondary clarifiers (Devantier and Larock, 1987; Krebs, 1991; McCorquodale *et al.,* 1991; Brouckaert *et al.,* 1998), flocculation basins (Crossley *et al.,* 2001; Haarhoff and van der Walt, 2001; Baek *et al.,* 2005; Bridgeman *et al.,* 2010;

Samaras *et al.*, 2010), and anoxic and anaerobic vessels (Hunze, 1996; Karama *et al.*, 1999; Brannock, 2003). In addition, CFD techniques have been applied to membrane processes (Ghidossi *et al.*, 2006) and bubble column reactors (interactions between air and liquid flow) (Joshi, 2001).

However, CFD modelling of MBR processes potentially requires the simulation of a 3D fluid field in combination with multi-phase and turbulent flows and complex module configurations and must also account for the effects of rheology and viscosity of the mixed liquor on the fluid flow. As such, application of CFD to MBR systems is particularly difficult and requires substantial computational effort.

15.4 CFD modelling for MBR applications

15.4.1 Overview

Due to the complexity of the MBR process, CFD has been used principally to study particular aspects of an MBR system with the major types of modelling best described as (Nopens *et al.*, 2008):

1. Macro-scale: modelling of entire MBR with flow and mixing characterisation within the reactor, scale-up and optimisation.
2. Meso-scale: modelling of membrane modules and aeration system.
3. Micro-scale: modelling of interaction of bubbles on membrane, filtration and fouling.
4. Biology: coupling of CFD and biomass activities.

Macro-scale simulations have been focused on the flow structures of the entire full-scale plants (Saalbach and Hunze, 2008) and the macro-mixing of full-scale MBR plants using different membrane modules (i.e., FS vs. HF) (Brannock *et al.*, 2010b). Biological transformations have also been included in CFD models by coupling the activated sludge models (Henze *et al.*, 1987) in order to evaluate the effect of mixing on nutrient removal (Brannock *et al.*, 2010c). For models at the micro-scale, research has been focused on the impact of bubble frequency, size and shape on liquid velocities and shear stress on the membrane surface (Taha *et al.*, 2006; Ndinisa *et al.*, 2006b; Prieske *et al.*, 2007; Buetehorn, 2010; Drews *et al.*, 2010; Martinelli *et al.*, 2010). However, review of the literature indicates that there is no clear conclusion reached as yet with regards to the most effective bubbling regime in reducing membrane fouling. There are contradictory results with regards to the optimal aeration pattern (gas flowrate) and its ability to reduce fouling. Moreover, the filtration process is not modelled by CFD as the small suction that is present in the MBR system is assumed to minimally affect the hydrodynamics of the overall system. A summary of CFD models that have been applied to the MBR process is provided in Table 15.1.

Table 15.1 Summary of CFD models that have been applied to the MBR process

Scale	Purpose of modelling	Software	Numerical models	Calibration and Validation techniques	References
Macro-scale	The impact of energy demand on the macro-mixing (RTD) of full-scale MBR plants	FLUENT® 6.3	• Eulerian-Eulerian model to model two-phase (water-air) • Standard k-ε to model turbulence	Tracer studies	Brannock *et al.* (2010b)
	Inclusion of Activated Sludge Model No.1 to model the nutrient removal and the effect of macro-mixing regimes on nutrient removal	FLUENT® 6.3	• Eulerian-Eulerian model to model two-phase (water-air) • Standard k-ε to model turbulence	Benchmarked against European 'Co-operation in the field of Scientific and Technical Research' (COST) simulation benchmark (COST 2002)	Brannock *et al.* (2010c)
Meso- to macro-scale	Liquid and air velocity distribution surrounding FS and HF membrane modules	FLUENT® 6.3	• Algebraic-Slip-Mixture model to model the interactions between water, air and sludge • RNG k-ε to model turbulence • The activated sludge was modelled as substances concentration in the water phase • The membrane modules were modelled as porous media	On-site liquid velocity measurements to calibrate the coefficients in porous media model	Saalbach and Hunze (2008)

Comparison of modelling results of membrane filtration zone in pilot and full-scale MBRs	FLUENT® 6.3	• Realizable k-ε model to model turbulence • Eulerian-Eulerian model to model two-phase (water-air) flow • Porous media model using correlations from tube banks to represent the resistance of membrane module to mixed liquor circulation • Empirical equation of MLSS concentration dependent viscosity	Potential methods: PIV; high-speed camera to measure bubble size distribution and movement (Liu *et al.*, 2010)	Kang *et al.* (2008)
Hydrodynamic descriptions (liquid and air) of membrane filtration zone in a full-scale MBR plant	FLUENT® 6.3	• Standard k-ε model to model turbulence • Eulerian-Eulerian model to model two-phase (water-air) flow • Porous media model using experimentally determined correlations to represent the resistance of membrane module to liquid flow in three-dimensions	Measurement of friction losses due to HF bundles at different liquid velocities, viscosities and flow direction	Wang *et al.* (2010a)

(*Continued*)

Table 15.1 Continued

Scale	Purpose of modelling	Software	Numerical models	Calibration and validation techniques	References
Micro-scale	The effect of aeration flow rate on liquid circulation velocity of membrane tank	ANSYS CFX10	• Eulerian-Eulerian model to model two-phase (water-air) flow • Grace Drag model to account for the effect of bubble deformation and gas hold-up on the resistance coefficient • SST model to model the turbulence of water (continuous) phase while zero-equation model to model the turbulence of air (dispersed) phase	Optical analysis using video imaging	Prieske *et al.* (2007)
	Shear-stress distribution on the FS membrane surface at different air flowrates, bubble size and baffles	ANSYS CFX5.6	• Eulerian-Eulerian model to model two-phase (water-air)	Video imaging to capture the bubble pattern	Ndinisa *et al.* (2006)
	The movement of a Taylor bubble through a tubular membrane and the wall shear stress under different conditions	FLUENT® 5.4	• RNG k-ε model to model turbulence • VOF model to model the gas–liquid interface	N/A	Taha *et al.* (2006)
	Movement of bubbles and shear-stress distributions near the membrane surface at different air velocities of a side-stream MBR	FLUENT® 6.3	• Inclusion of Herschel–Buckley with Papanastasiou's adaption to model the rheology of wastewater sludge in the water phase	Measurement of sludge rheology to calibrate viscosity values in the model	Rios *et al.* (2007b)

		• VOF model to model the gas–liquid interface		
Impact of bubble size, FS membrane spacing and aeration intensity on wall shear stresses and on mixed liquor particle deposition	ANSYS FLUENT/CFX 11	• VOF model to model two-phase (water-air) flow • Inclusion of Power-law function to describe the rheology of wastewater sludge	N/A	Drews *et al.* (2010)
Quantify liquid circulation and shear stresses along the membranes and their impact on cake resistance	FLUENT®	• Eulerian-Eulerian model and VOF model to model two-phase (water-air) flow • 2-Dimensional unsteady state • Standard k-e model to model turbulence of both phases	2-Phase flow PIV to observe particle cake deposition	Martinelli *et al.* (2010)
Impact of fibre arrangement, air bubbling and flow channels on the hydrodynamic conditions	ANSYS CFX 11.0/12.0	• VOF model to model two-phase (water-air)flow • RNG k-e model to model turbulence	X-ray CT to map the instantaneous fibre displacement Direct observation with video camera to observe the impact of the distance from the header of the aerator, aeration rate and packing density on fibre movement	Buetehorn (2010)

15.4.2 Techniques for modelling turbulence

For turbulent flow, the fluctuations of velocity and other flow properties are random and have a three-dimensional spatial character. Turbulent flows are characterised by eddies with a wide range of length and time scales. The time-dependent three-dimensional Navier-Stokes equations (Equation [15.8]) are able to provide all the information for a given turbulent flow. It is possible, in theory, to directly resolve the whole spectrum of turbulence scales using direct numerical solution (DNS) without the need to resort to modelling. However, DNS is not feasible for practical engineering problems due to the limitations in computing power. The computing cost required in applying DNS to resolving the entire range of turbulence scales is proportional to Re_t^3, where Re_t is the turbulent Reynolds number. Therefore, two alternative methods, Reynolds averaging Navier-Stokes (RANS) and large eddy simulation (LES), are commonly used to approximate the effects of turbulence on the flow-field so that the small-scale turbulent fluctuations do not have to be directly simulated. For the LES approach, the Navier-Stokes equations are filtered to compute larger eddies while the smaller eddies are modelled. In the RANS turbulence models, the instantaneous velocity is replaced by the mean velocity and six additional stresses – three normal stresses, $-\rho\overline{u'^2}, -\rho\overline{v'^2}, -\rho\overline{w'^2}$ and three shear stresses, $-\rho\overline{u'v'}, -\rho\overline{v'w'}, -\rho\overline{v'w'}$, are generated. The six additional terms lead to un-closed equations and therefore the main task of turbulence modelling is to use turbulence models which relate the unknown Reynolds stress terms to flow properties thereby enabling development of a closed set of equations.

As a result, the selection of turbulence models in the existing commercial CFD packages is challenging, with researchers typically using different models for different purposes as described below.

Standard k-ε model

The standard *k*-ε model is the most established and the most widely validated turbulence model (Versteeg *et al.*, 1995). It can provide robust, economic and reasonably accurate solutions for a wide range of turbulent flows. The standard *k*-ε turbulence model is based on solution of the turbulence kinetic energy *k* and its rate of dissipation ε. Ndinisa *et al.* (2006) modelled the impact of bubbles on the shear-stress distribution on the membrane surface and used standard *k*-ε models to model the liquid phase. Brannock *et al.* (2010b) used it to model the macro-mixing of the entire full-scale MBR plants and validated the results through field measured residence time distributions. Wang *et al.* (2010a) expanded this work by including a porous media model to account for the pressure loss induced by the HF membrane bundles. Martinelli *et al.* (2010) modelled the liquid phase in fine bubble flow using standard k-ε models.

RNG k-ε model

The RNG k-ε model was derived using a rigorous statistical technique (ReNormalisation Group Theory) rather than experiments. One extra term, R_ε, appears in the ε equation which can significantly improve the accuracy for modelling rapidly strained flows. The accuracy for modelling swirling flows is also enhanced by coupling the effect of swirl on turbulence in the model. The Prandtl numbers in the RNG k-ε model are evaluated through an analytical formula while the standard k-ε model uses user-specified, constant values. Therefore, the RNG is more accurate and reliable for a wider class of flows than the standard k-ε model but it also consumes 10–15% more CPU time than the standard k-ε model due to the extra terms and functions and a greater degree of non-linearity (ANSYS, 2010). A number of researchers have chosen the RNG k-ε model to model the liquid phase in the MBR process, ranging from the liquid flow structures of the MBR tanks (Saalbach and Hunze, 2008), movement of a Taylor bubble through a tubular membrane, wall shear stress under different conditions (Taha *et al.*, 2006) and impact of fibre arrangement on the hydrodynamics of the system (Buetehorn, 2010).

Realizable k-ε model

The Realizable k-ε model ensures the positivity of normal stresses ('realizable') by making the empirical constants of k-ε turbulence model, C_μ, a function of the mean flow (mean deformation) and the turbulence (k, ε) while the Boussinesq theory used in the standard and RNG k-ε model allows for negative normal stresses. The Realizable k-ε model is more accurate in the prediction of the spreading rate of both planar and round jets. Kang *et al.* (2008) used a Realizable k-ε model to simulate the hydrodynamics in the membrane filtration zone of pilot and full-scale MBRs.

k-ω model

Rather than finding ε, the k-ω model solves the transport equation for the specific dissipation rate, ω, described as a frequency characteristic of the turbulent decay process under its self-interaction or, alternatively, can be thought of as the ratio of ε to k (Wilcox, 1998). The k-ω model predicts free shear flow spreading rates that are in close agreement with measurements for far wakes, mixing layers, and plane, round, and radial jets, and is thus applicable to wall-bounded flows and free shear flows (ANSYS, 2010). The k-ω model has not been used to model MBR systems so far.

Shear-stress transport (SST) model

The SST model uses the standard k-ε model in the bulk flow and incorporates the transport of turbulent shear stress while using the k-ω model in the boundary layer. This model can therefore be used over a wider range of operating parameters than the standard k-ω model, such as the flow around curved bodies and flows with separating boundary layers. Prieske *et al.* (2007) examined the relationship between circulation velocity using the SST model and aeration flow rate in a pilot-scale MBR.

The proper selection of turbulence models is important; however, the turbulence models are all based on approximation and empirical constants and each of them has advantages and disadvantages for different applications. Therefore, the selection of models is case dependent. The modelling results should always be validated in order to confirm the appropriateness and accuracy of the chosen model.

15.4.3 Techniques for modelling the interactions of water and air

Modelling of the interactions between water and air in MBR systems is important as aeration plays a key role in the mass transfer, particle size distribution and back transport of particles from the membrane surface. In addition, the behaviour of bubbles (shape and size, Fig. 15.8) is recognised to be affected by the injection method, sparger type and aeration rate. Table 15.2 summarises the models for the evaluation of effects of air bubbling in MBRs.

Currently there are two approaches for the numerical calculation of multiphase flows: the Euler-Lagrange approach and the Euler-Euler approach. The Euler-Lagrange approach solves the time-averaged Navier-Stokes

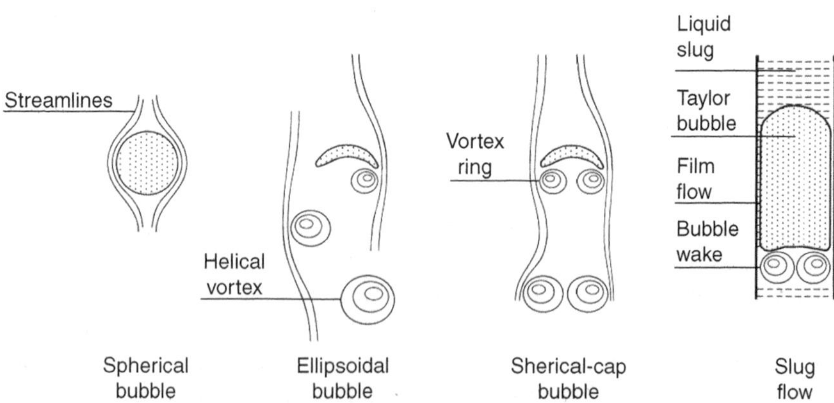

15.8 Schematic representation of air-liquid flow pattern with different shapes of bubbles. (Adapted from Cui *et al.*, 2003; Buetehorn, 2010.)

Table 15.2 Summary of CFD modelling of air bubbling in MBR

Membrane configuration	Parameters studied	Behaviour of bubbles	Findings from models	References
Tubular	• Air flow rate (using controlled pulse injection rather than uncontrolled gas sparging)	• Slug flow	• Local shear was computed. At fixed bubble frequency, the wall shear rate and permeate flux are highest when the membrane module was inclined at 45° from horizontal	Taha *et al.* (2006)
Flat sheet	• Nozzle size: 0.5–2.0 mm • Aeration flow rate: 2–8 L/min	• Number of bubbles: 11–554 • Bubble diameter: 1.88–9.92 mm	• Bubble size on air and water velocity: – The water velocity ranges from 0.35 to 0.58 m/s for bubble sizes between 2–15 mm – Bubble sizes of 4–5 mm is most popular for the aeration flow rate of 4 L/min • Effect of shear-stress distribution on membrane surface – Shear stress increased 31% when the air flowrate increased from 2 to 8 L/min (using bubble diameter of 5 mm) – The distribution of shear stress across the membrane surface increased with bubble size	Ndinisa *et al.* (2006b)
Flat sheet	• Aeration flow rate: 0.6 m^3/h and 3.45 m^3/h • Space between membrane plates	• Bubble size 1, 2, 3 mm was modelled	• Gas hold-up increased from 0.01 to 0.05 as the bubble diameter decreased from 3 to 1 mm • Larger bubbles increased the liquid circulation velocities and therefore are expected to be more efficient for air scouring of membrane surface due to the increase on the drag and lift forces on the particles	Prieske *et al.* (2007)
Tubular	• Air velocity: 0–1 m/s (liquid velocity was kept as a constant of 0.5 m/s)	• Bubble size and shape	• Taylor shape bubbles with a round nose and a cylindrical body were observed at the outlet • Bubbles have different sizes due to the acceleration and coalescence with preceding bubbles • At 1 m/s of air velocity, slug flow changes to churn flow • There are oscillations in the shear stress on the membrane surface due to the continuous passing of bubbles near the membrane wall	Rios *et al.* (2007)

(*Continued*)

Table 15.2 Continued

Membrane configuration	Parameters studied	Behaviour of bubbles	Findings from models	References
Flat sheet	• Single bubble movement between membrane plates • Distance between membrane plates (3–11 mm)	• Bubble size: equivalent spherical diameter ranging from 3–24 mm	• 5 mm bubbles in 5 mm channels appear to be less clogged due to the increase in shear stress • Drag and lift forces on single particle: large particles can be removed by the lift force while smaller particles are transported to the membrane and form cake layer	Drews *et al.*, (2010)
HF (tightly potted)	• Aeration type • Aeration flowrate	• Spherical cap bubbles with diameter of 3 mm or fine bubbles	• Fine bubbles were found to be evenly distributed in the whole reactor while spherical cap bubbles concentrated on the membrane • Wall shear stress was found to be less than 0.25 Pa • The liquid flow towards membrane surface increased with air flowrate, which could increase the particle transport to the membrane surface and caused higher particle concentration at membrane surface and hence higher fouling resistance	Martinelli *et al.* (2010)
HF	• Fibre arrangement	• Deformation of bubbles	• Deformation of bubbles (non-circular shape) was found due to the flow channel confinement • Higher mixture velocities were observed in the flow channels without fibres • The deformation of bubbles was affected by module design such as air sparger and the local fibre arrangement	Buetehorn (2010)

equations for the continuous phase while the equations for the dispersed phase are solved by tracking a large number of particles, bubbles, or droplets through the calculated flow-field. This approach assumes that the dispersed second phase occupies a low volume fraction and therefore would be inappropriate for the modelling of MBRs in which the volume fraction of the second phase (air bubbles) is unlikely to be negligible. Therefore, the Euler-Euler approach is selected for the CFD modelling of MBR systems, with the volume fractions of the dispersed phase assumed to be continuous functions of space and time and their sum equal to one. FLUENT® provides three different multi-phase models: the volume of fluid model (VOF), the mixture model and the Eulerian model (ANSYS, 2010).

Eulerian model

The Eulerian model is commonly used for bubble columns (Joshi, 2001; Sokolichin *et al.*, 2004) and gas–liquid wastewater treatment reactors (Le Moullec *et al.*, 2008) that have high numbers of bubbles despite the low gas fraction. The Eulerian model is the model that has been most commonly adapted in MBR simulations (Ndinisa *et al.*, 2006; Prieske *et al.*, 2007; Kang *et al.*, 2008; Martinelli *et al.*, 2010; Wang *et al.*, 2010a; Brannock *et al.*, 2010b). Kang *et al.* (2008) found that the Eulerian model was effective in tracking the dispersed phase while Ndinisa *et al.* (2006) found that it could be applied to dispersed particles smaller than the grid size by predefining a uniform particle size.

Mixture model

The mixture model is based on the solution of a single mixture momentum equation for all phases, which significantly reduces computational effort. Saalbach and Hunze (2008) used the Algebraic-Slip-Mixture model to simulate the interactions of three phases: water, air and sludge in pilot and full-scale plants. The mixture model could account for the slip velocities of the dispersed phase and the continuous phase relative to the mixture.

Volume of fluid (VOF) model

The VOF model solves momentum equations for the continuous phase while the dispersed phase follows directly from the closure condition of the volume fraction for the incompressible flow. All variables and properties of the fluid are computed as cell-average weighted volume fractions. The VOF models have been found to be suitable for the impact of air bubbles such as bubble movement (Rios *et al.*, 2007b), bubble size and intensity, FS membrane spacing (Drews *et al.*, 2010), and aeration intensity and fibre arrangement (Buetehorn, 2010) on the shear stress on the membrane surface and the cake formation. Martinelli *et al.* (2010) used a Eulerian model for fine mono-dispersed bubbles of 3 mm diameter while the VOF model was used for coarse bubble with the tracking of the gas–liquid interface and a geo-reconstructed scheme in the laminar flow. These simulation efforts have confirmed that aeration promotes the movement of fibres and increases turbulence and wall shear stress. The wall shear stress has found to be affected by the bubble frequency, size, shape and intensity. Moreover, high air flow rates could result in fibre breakage. An optimum value of the air flow rate has not been determined although this is possible (Drews *et al.*, 2010).

It has been found that VOF models are more suitable for small-scale simulations, while for the modelling of the large-scale systems the Eulerian and Mixture models are more feasible since the VOF models requires much longer simulation times (Nopens *et al.*, 2008).

15.4.4 Modelling of sludge rheology

Sludge rheology may play an important role in CFD modelling of the MBR process as it determines the hydraulics and transport phenomena near the membranes (Rios *et al.*, 2007a). It has been found that activated sludge behaves as a non-Newtonian fluid and therefore the relationship between the shear stress (τ) and the shear rate ($\dot{\gamma}$) exhibits a power-law (Table 15.3) rather than a linear relationship (Rosenberger *et al.*, 2002; Rios *et al.*, 2007a).

Rios *et al.*, (2007a) investigated the rheology of activated sludge from the aerobic zone of a lab-scale side-stream MBR with MLSS concentration of 8 g/L. The sludge was concentrated or diluted to obtain 11 different concentrations and correlated with shear rate-stress. The Herschel-Buckley rheology model with Papanastasiou's (H-B-P) adaption was found to exhibit the best fit to the experimental data for MLSS concentrations between 6 and 16 g/L and shear rates between 0.05 and 1600 s^{-1}.

Table 15.3 Non-Newtonian shear stress-rate correlations

Fluid behaviour	Shear-stress models
Shear-thinning (pseudo plastic) ($n < 1$) Shear-thickening (dilatant) ($n > 1$)	$\tau = k\dot{\gamma}^n$
Herschel-Buckley ($0 < n < \infty$)	$\tau = \tau_0 + k\dot{\gamma}^n$
Bingham plastic ($n = 1$)	$\tau = \tau_0 + k\dot{\gamma}$
Herschel-Buckley with Papanastasiou's adaption ($0 < n < \infty$)	$\tau = \tau_0\left(1 - e^{-m\dot{\gamma}}\right) + k\dot{\gamma}^n$

Source: Rios *et al.* (2007a).

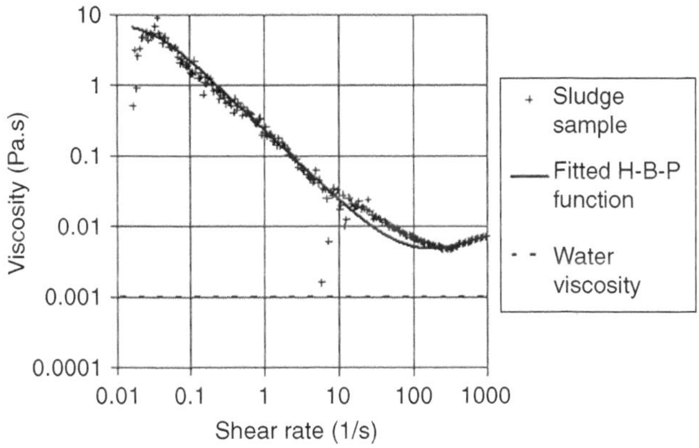

15.9 Shear rates versus viscosity for the pilot plant MBR sludge with MLSS of 8 g/L (Brannock *et al.*, 2010b).

Brannock *et al.* (2010b) investigated the rheology of a pilot-scale MBR with an MLSS concentration of 8 g/L. The fitted H-B-P model showed good agreement with the experimentally measured data which exhibited large deviation from the viscosity of water (Fig. 15.9).

The derived H-B-P model was then applied to a full-scale HF MBR and compared with the CFD simulated results using the viscosity of water. Minor differences were observed from the simulated residence time distribution profiles (Brannock *et al.*, 2010b). The impact of sludge settling on the models was also investigated. Using the densimetric Froude Number, which is a ratio between inertial forces and gravimetric forces, the positive buoyancy forces from rising bubbles were found to be at least ten times greater than the negative buoyancy forces from sludge settling, hence the effect of sludge settling can be ignored.

Buetehorn *et al.*, (2008) evaluated the rheology of solutions of carboxymethyl cellulose (CMC), glycerol and xanthan gum containing different concentrations of silica particles and found that both CMC and glycerol solutions show Newtonian characteristics for shear rates between 0 and 1000 1/s while xanthan -gum could well mimic the properties of sludge for MLSS concentrations between 10 and 16 g/L. Therefore, xanthan-gum solution could be used for validation purposes.

Although mixed liquor is a non-Newtonian fluid, it might not be necessary to include the impact of sludge rheology in all CFD simulations. The sludge rheology would have a large impact on the particle size distribution, shear forces and mass transfer and therefore would be essential to micro-scale simulations. However, for meso- to macro-scale simulations, it is unlikely that the accuracy of the results would be affected significantly by the sludge rheology. Furthermore, the inclusion of additional models representing sludge viscosities is computationally expensive (Nopens *et al.*, 2007).

15.4.5 Modelling the head loss caused by membrane module

In the CFD modelling of membrane filtration process, membranes are usually modelled as a porous wall while the flow within a membrane is usually solved using both Navier-Stokes and Darcy equations (Ghidossi *et al.*, 2006). A porous media model is widely used for determining the pressure loss during flow through packed beds, filter papers, perforated plates, flow distributors and tube banks (ANSYS, 2010). A momentum source term is added to the governing momentum equations, which creates a pressure drop that is proportional to the fluid velocity:

$$S_i = -\left(\underbrace{\sum_{j=1}^{3} D_{ij}\mu v_j}_{\text{Viscous loss term}} + \underbrace{\sum_{j=1}^{3} C_{ij}\rho v_{\text{mag}} v_j}_{\text{Inertial loss term}} \right) \qquad [15.9]$$

where S_i is the source term for the ith (x, y, or z) momentum equation, and D and C are prescribed matrices.

In laminar flows through porous media, the pressure drop is typically proportional to velocity. Ignoring convective acceleration and diffusion, the porous media model then reduces to Darcy's law:

$$\nabla p = -\frac{\mu}{\alpha}\vec{v} \qquad [15.10]$$

where α is the permeability.

At high flow velocities, the inertial resistance factor, C_{2ij}, can be viewed as a loss per unit length along the flow direction, thereby allowing the pressure drop to be specified as a function of dynamic head.

As noted earlier, membrane filtration in any scale of MBR plants has not been modelled using CFD techniques because the modelling of the small suctions involved is too computationally costly given that high mesh resolution is required with each single membrane element needing to be solved by Navier-Stokes equations coupled to Darcy's law. However, since we are interested in the flow resistance caused by the HF array or the FS plate and the impact of this resistance on hydrodynamic profile, the whole membrane module can be modelled as a porous medium with macroscopic characteristics enabling computational effort to be reasonably constrained. Therefore, investigators have attempted to transfer the effects of the membrane modules on the flow-field to a porous zone (Kang *et al.,* 2008; Saalbach and Hunze, 2008; Buetehorn, 2010; Wang *et al.,* 2010a) with the effect of small suction assumed, as noted earlier, to not affect the hydrodynamics of the overall system (Saalbach and Hunze, 2008; Wang *et al.,* 2010a).

The first attempt to model the flow resistance caused by the membrane module using the porous media approach was undertaken by Saalback and Hunze (2008). They examined the membrane zones of two wastewater treatment plants (WWTP), one with HF membranes and the other with FS membranes. The flow around the membranes was modelled as zones of porous media. The resistance values required by the porous media model were determined by velocity measurements at the real plants. However, the paper does not report on the values of the coefficients that were used in the porous media model. Kang *et al.* (2008) observed different hydrodynamic behaviour of HF membranes in a pilot system (which involved a single membrane module) and a larger system (which contained160 HF membrane modules). The empirical correlations used for modelling tube banks were applied to the porous media model to represent the resistance caused by the membrane module. In a more recent study, Wang *et al.* (2010a) incorporated the porous media model in a CFD model in order to describe the flow behaviour around the HF membrane bundle of a full-scale MBR. The inertial resistance factor used in the porous media model was determined experimentally by measuring the pressure drops across the membrane bundle for various liquid velocities with the flow direction both perpendicular and parallel to the membrane bundle. They found that the empirical correlations used for modelling the tube banks significantly underestimated the flow resistance induced by the fibres, particularly for the flow direction perpendicular to the fibre bundle. The porous media approach was found to be able to provide more accurate predictions of the hydrodynamics in the membrane filtration zone. Buetehorn *et al.* (2010) also modelled the HF bundle as a porous medium. However, instead of modelling the bundle as a homogenous porous zone, the friction factor was calibrated as a function of Reynolds number of 'bubble-entrained' liquid flow through the bundle and local porosity of the module.

So far, the porous media approach has been shown to be effective in representing the resistance caused by the membrane module and in describing their impact on the overall hydrodynamics however assumptions inherent in the approach including that filtration has a negligible effect on the hydrodynamics and that single phase modelling is appropriate (i.e., that air bubbles have no effect in the porous zone) may limit its accuracy. These issues highlight the importance of experimental validation of the model. Various approaches to validation are discussed in the next section.

15.5 Model calibration and validation techniques

Calibration and validation are two important aspects of any computational work that involves assumptions and numerical solution. Calibration involves the specification of boundary conditions required in the CFD models. Validation is the last step to be performed and is vital to confirming that the results represent a reasonable solution from both conceptual and numerical perspectives. The following section summarises the current experimental techniques for MBR CFD model calibration and validation.

15.5.1 Velocity measurements in the fluid field

Laser Doppler velocimetry (LDV), acoustic Doppler velocimetry (ADV) and particle image velocimetry (PIV) have been widely used to measure the liquid velocities in the MBR for model calibration purposes. Tacke *et al.* (2008) measured the liquid velocities of selected cross-sections in a MBR using ADV. A 34% glycerol solution was used to model the activated sludge with MLSS concentration of 12 g/L. In comparison to Doppler velocimetry, PIV measures the whole velocity field by taking two images shortly after each other and calculating the distance that individual particles travel within this time. From the known time difference and the measured displacement, the velocity is calculated. Liu *et al.* (2010) measured the liquid velocities in pure water and water–air two-phase flow, respectively. Martinelli *et al.* (2010) used PIV to acquire the images of bubbles and deduced liquid velocities of a fluid field using baker's yeast suspensions as model solution. Note that none of the above studies was carried out in a real activated sludge system or in a real MBR. This is because the above techniques can only be applied in transparent solutions and therefore cannot be used in the sludge systems. Moreover, the accuracy of results is usually affected by disturbances such as the presence of bubbles.

A more feasible method for velocity measurement involves the use of an electromagnetic current meter. Brannock (2003) measured the three-dimensional liquid velocities at different positions within an anoxic reactor using an electromagnetic current meter, which is not affected by the physical properties of the fluid or the presence of bubbles and solids.

15.5.2 Determination of physical properties of activated sludge

A rotational rheometer has been commonly used to measure the rheology of activated sludge (Rosenberger *et al.*, 2002; Rios *et al.*, 2007a; Brannock *et al.*, 2010b). Rheograms can then be developed based on the measurement results and the correlations obtained can be used to represent the viscosities of the sludge in the CFD models (Rios *et al.*, 2007b).

15.5.3 Video imaging

Direct observation using a high-speed video camera has been widely used by many researchers to observe flow patterns including the behaviour of bubbles. Ndinisa *et al.* (2006) used Nikon Coolpix995 digital camera to record the two-phase flow pattern in an MBR and obtained information on the total number of bubbles, the surface area of each bubble, the diameter and perimeter of the bubbles by analysing the images. Buetehorn (2010) applied an approach established by Wicaksana *et al.* (2005) in which a Sony DCR-TRV50E video camera was used to investigate the impact of the distance between the aerator header and membranes, the aeration rate and membrane packing density on fibre movement. The position of the fibre at different conditions was deduced from the recorded images. Liu *et al.* (2010) observed the air bubble size and movement using a high-speed camera operating at a recording speed of 500 frames per second, resolution of 1024 × 1024 pixels and shutter speed of 1/500 s. Direct observation provides an effective way to study the bubble behaviour; however, the accuracy of the results are highly dependent on the image processing techniques used.

15.5.4 Nuclear magnetic resonance (NMR) imaging

NMR imaging allows the monitoring of the composition of a sample by measuring the distribution of mobile protons in any slice of the specimen (Buetehorn *et al.*, 2011). The microfiltration process (permeate flow and cake growth) of model solutions (water or silica suspensions) has been visualised using NMR imaging (Buetehorn *et al.*, 2011).

15.5.5 X-ray computer tomography (CT) scan

An X-ray CT scanner consists of an X-ray source combined with a detector located on the opposite side of the specimen. The sample may be scanned from many directions and a three-dimensional map of the specimen is formed via image processing (Buetehorn, 2010). Buetehorn (2010) performed CT scans on a bundle of HFs of 0.85 m in length. The instantaneous displacement of fibres along the fibre length (from 0 to 0.75 m) was mapped showing that the fibres were arranged regularly in the lower part of the module and became less regular further from the base. The porosity used for the porous medium approach (Buetehorn, 2010) was based on the CT scan results.

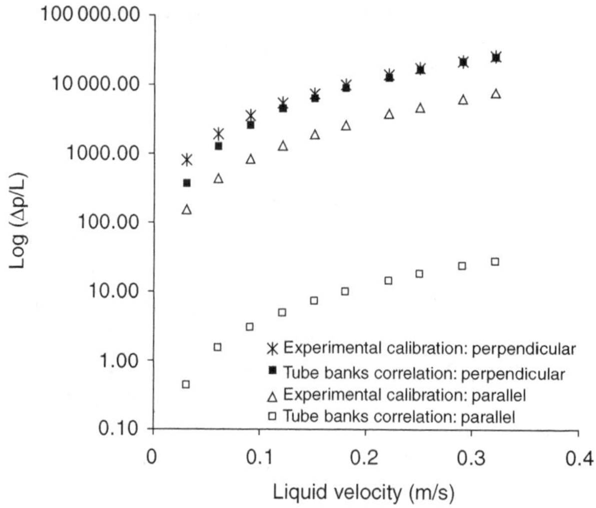

15.10 Comparison of the logarithmic scale of pressure drop obtained from experimental calibration and empirical correlations for tube banks.

15.5.6 Pressure drop measurements

Wang *et al.*, (2010a) calibrated the inertial resistance factor, C_{2ij}, used in the porous media model in a bench scale set-up using membrane bundles with the same packing density as the Siemens Memcor Memjet® B10R HF membranes that are used in the full-scale plant being modelled. The pressure drops across the membrane bundle for flow directions perpendicular and parallel to the membrane bundle and at different fluid viscosities were measured for various liquid velocities. The empirical correlations used for modelling the pressure drop caused by tube banks were found to underestimate the pressure drop caused by the HF bundles (Fig. 15.10).

Similarly, Buetehorn (2010) measured the pressure loss of a HF bundle for different parallel flow velocities using pure water. The pressure loss without fibres was also measured so that pressure loss due to pipes etc. could be subtracted.

15.5.7 Residence time distributions (tracer study)

One method that can be used to characterise mixing is based on the concept of residence time distributions (RTD). The mixing energy applied, as well as the bioreactor and membrane configuration, affects the output response of the bioreactor. The experimental determination of RTD typically involves the injection of an inert tracer. Brannock *et al.* (2010a) conducted tracer studies on a 2 MLD HF MBR in Sydney and a 5 MLD FS

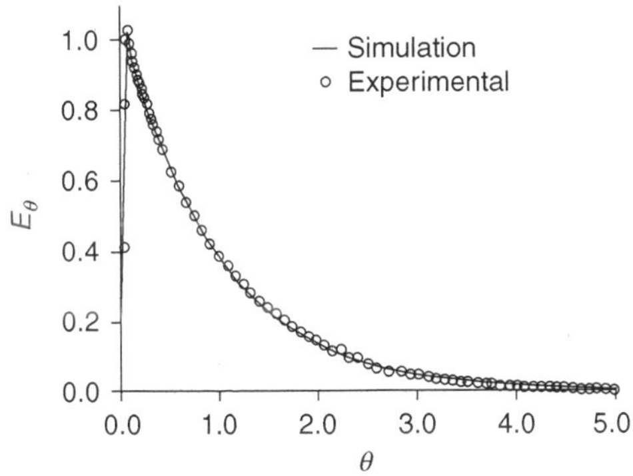

15.11 Comparison of experimentally measured and CFD simulated dimensionless residence time distribution (E_θ) versus dimensionless time (θ).

MBR in South Australia by injecting a pulse input of lithium chloride (LiCl) at the MBR inlet (post-screening) with the tracer response being measured in the permeate in order to validate CFD models that had been used to simulate the macro-mixing regimes of these two plants. The experimentally determined RTD profiles showed good agreement with the simulated RTDs (Fig. 15.11).

The selection of experimental methods for model calibration and validation should be based on the objectives of the study. The techniques used for meso- and macro-scale issues might not be suitable for micro-scale considerations. More advanced methods also need to be developed for the specific media of interest in MBR systems (i.e., water–air–sludge combinations).

15.6 Future trends and conclusions

Application of CFD to describe air bubbling is an area of intense activity, with the effects of air flowrate, intensity, bubble shape, diameter and frequency of particular interest. The major purpose of modelling the behaviour of bubbles is to evaluate the shear stress on the membrane wall. However, the interactions between the particles and bubbles are still not clear. For example, how are the floc shape and size and EPS formation affected by air bubbling? Both flocs and extracellular polymeric substances (or extracellular polysaccharides, EPS) contribute to formation of the cake layer and may block the membrane pores, with the extent of fouling potentially affected by the size of any particulates present. Models for cake layer build up and the removal by shear are still not well developed. Moreover, it is unclear how accurately the viscosity of the sludge should be modelled. The viscosity of the sludge could affect the extent of shear experienced by the flocs but the inclusion of viscosity models could be computationally expensive. The coupling of biological and chemical reactions with CFD models also requires huge computational resources but the link between flocs shape and size, and shear stress could be established. In addition, filtration has not been modelled and its impact on overall hydrodynamics has been neglected to date. More advanced experimental techniques need to be developed to validate this assumption.

When considering the interactions between particles and air bubbles, the effects of coagulation on MBR performance should not be ignored. The aggregation of particles is recognised to depend on the particle surface charge and coagulant concentration (Wiesner *et al.,* 1989; Hlavacek and Remy, 1995; Waite *et al.,* 1999). An increase in aggregate size could reduce pore blocking and decrease the specific cake resistance (Hlavacek and Remy, 1995; Waite *et al.,* 1999). CFD techniques have been used to model the aggregation and breakup of flocs in stirred tanks (Prat and Ducoste, 2006; Moussa *et al.,* 2007; Soos *et al.,* 2007; 2008; Ehrl *et al.,* 2008), in turbulent (Wang *et al.,* 2005a) and laminar (Wang *et al.,* 2005b) Taylor-Couette flow, and the effects of turbulent shear rate on aggregate size have been examined (Marchisio *et al.,* 2006). However, the impact of air bubbles on the aggregation and breakup of particles has not been modelled. The deformation of the particles might change the cake formation process and the properties of the cake. Population balance modelling (PBM) could be used to model the coalescence and breakage of bubbles and flocs (Kim and Kramer, 2006; Nopens *et al.,* 2008). In addition, the effect of mixing on the coagulant species transformation could also be of interest, given that the extent of mixing may determine the concentration of coagulant species at any location in the reactor. A recent study suggested that the addition of iron-based coagulants could lead to a variety of reactions in the MBR (Wang and Waite, 2010) with the rate of many of these reactions expected to be concentration dependent (Fig. 15.12). CFD has been used to model the effect of mixing (and the effect of different types of mixers) on the coagulation process (Farrow *et al.,* 2000; Jones *et al.,* 2000; Korpijarvi *et al.,* 2000; Craig *et al.,* 2002; Park *et al.,* 2003; Byun *et al.,* 2005; Bridgeman *et al.,* 2010; Vadasarukkai and Gagnon, 2010). The model could be further developed to gain the insight into the effect of local mixing (residence times and flow paths) on the transport and transformation of coagulant species in MBR systems.

For the meso to macro-scale, further studies should be carried out on the design of aeration systems and operational mode with consideration of factors such as the best backwashing and relaxation periods and the optimal aeration intervals required to minimise energy consumption of the process. In addition, more feasible methods should be developed for the validation of the porous media models (Wang, 2010b).

Much effort has been applied in recent years to using 'state-of-the-art' CFD methods to simulate the hydrodynamic behaviour, to scale-up and to

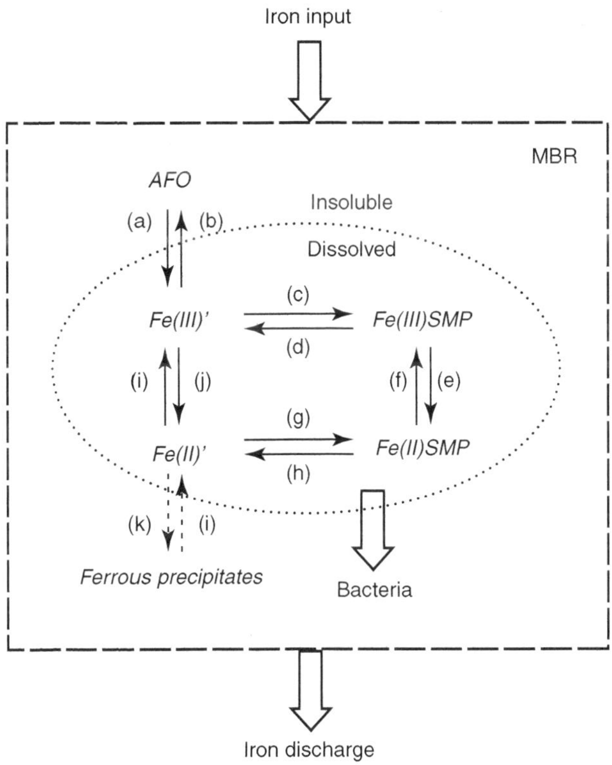

15.12 Schematic of key chemical transformations that may occur on the addition of iron-based coagulant to a MBR (Wang and Waite, 2010). Addition of inorganic Fe(III) salts will result in either precipitation as amorphous ferric oxide (AFO) (reaction b), complexation by soluble microbial products (SMPs) (reaction c) or reduction to Fe(II) (reaction j). Fe(III) in both inorganic (as AFO) and SMP-bound forms (as Fe(III)SMP) may undergo reduction (especially in the low Eh (<-0.2 mV) anoxic region (reactions e, a and j). Any Fe(II) present may be oxidized (especially in the aerated oxic compartment) (reactions i and f). Fe(II) may form insoluble precipitates such as ferrous phosphate or vivianite (reaction k).

optimise performance of MBRs. Despite this, much remains to be done to improve the available models and the modelling tools in order to produce a more accurate design tool for the MBR process.

A comprehensive MBR CFD model would be one that enables prediction of the impact of biological and chemical transformation kinetics, membrane filtration properties and fouling and hydrodynamic regimes on MBR performance, and which enables prediction of the energy usage of the plant as a function of design and operation choices. While some progress has been made towards developing such a model, a substantial amount of computational development and experimental validation is required before it is a reality.

15.7 Acknowledgement

The authors acknowledge the financial support from Australian Research Council through ARC Linkage grants LP100100056 and Sydney Water for the access to different water treatment plants. Special thanks to Mr Han Tng for proofreading.

15.8 References

ANSYS (2010). *ANSYS 13.0 Fluent User's Guide*, Canonsburg, PA.

Baek, H K, Park, N S, Kim, J S, Lee, S J and Shin, H S (2005), 'Examination of three-dimensional flow characteristics in the distribution channel to the flocculation basin using computational fluid dynamics simulation', *Journal of Water Supply Research and technology-AQUA*, **54**(6): 349–354.

Brannock, M (2003), *Computational Fluid Dynamics Tools for the Design of Mixed Anoxic Wastewater Treatment Vessels*, Department of Environmental Engineering, University of Queensland thesis, Brisbane, Queensland.

Brannock, M, Howes, T, Johns, M, De Clercq, B, Keller, J. (2003), *Computational Fluid Dynamics Tools for the Design of Mixed Anoxic Wastewater Treatment Vessels*, School of Engineering, University of Queesland thesis, Brisbane.

Brannock, M, Leslie, G L, Wang, Y and Buetehorn, S (2010c), 'Optimising mixing and nutrient removal in membrane bioreactors: CFD modelling and experimental validation', *Desalination*, **250**(2): 815–818.

Brannock, M, Wang, Y and Leslie, G (2010a), 'Evaluation of Full-Scale Membrane Bioreactor Mixing Performance & the Effect of Membrane Configuration', *Journal of Membrane Science*, **350**: 101–108.

Brannock, M, Wang, Y and Leslie, G (2010b), 'Mixing Characterisation of Full-Scale Membrane Bioreactors: CFD Modelling with Experimental Validation', *Water Research*, **44**: 3181–3191.

Bridgeman, J, Jefferson, B and Parsons, S A (2010), 'The development and application of CFD models for water treatment flocculators', *Advances in Engineering Software*, **41**(1): 99–109.

Brouckaert, C J, Fortmann, N E and A., B C (1998), 'A computational fluid dynamic investigation of a secondary clarifier'. *Proceedings of WISA '98*, Cape Town.

Buetehorn, S (2010). *Experimental and Numerical Investigation of the Hydrodynamics of Microfiltration Processes Using a Multi-Scale Approach*, RWTH Aachen University, PhD thesis, Aachen, Germany.

Buetehorn, S, Brannock, M, Le-Clech, P, Leslie, G L, Volmering, D, Vossenkaul, K, Wintgens, T and Melin, T (2009), 'Observation of cake layer formation and removal on microporous hollow-fibre membranes', *Desalination and Water Treatment*, **9**: 82–85.

Buetehorn, S, Carstensen, F, Wintgens, T, Melin, T, Volmering, D and Vossenkaul, K (2008). 'Permeate flux decline in cross-flow microfiltration at constant pressure'. *12th Aachener Membran Kolloquium (AMK)*, 29–30 Oct. 2008, Aachen, Germany.

Buetehorn, S, Utiu, L, Kuppers, M, Blumich, B, W, Intgens, T, Wessling, M and Melin, T (2011), 'NMR imaging of local cumulative permeate flux and local cake growth in submerged microfiltration processes', *Journal of Membrane Science*, **371**: 52–64.

Byun, S, Oh, J, Lee, B Y and Lee, S (2005), 'Improvement of coagulation efficiency using instantaneous flash mixer (IFM) for water treatment', *Colloids and Surfaces A physicochemical and Engineering Aspects*, **268**(1–3): 104–110.

Cote, P, Masini, M and Mourato, D (2004), 'Comparison of membrane options for water reuse and reclamation', *Desalination*, **167**: 1–11.

Craig, K, De Traversay, C, Bowen, B, Essemiani, K, Levecq, C and Naylor, R (2002). 'Hydraulic study and optimisation of water treatment processes using numerical simulation'. *3rd World Water Congress: Drinking Water Treatment*, Melbourne, Australia, IWA.

Crossley, I A, Valade, M T and Shawcross, J (2001), 'Using lessons learned and advanced methods to design a 1,500 ML/day DAF water treatment plant', *Water Science and Technology*, **43**(8): 35–41.

Cui, Z F, Chang, S and Fane, A G (2003), 'The use of gas bubbling to enhance membrane processes', *Journal of Membrane Science*, **221**: 1–35.

Devantier, B A and Larock, B E (1987), 'Modeling Sediment-Induced Density Currents in Sedimentation Basins', *Journal of Hydraulic Engineering-Asce*, **113**(1): 80–94.

Drews, A, Prieske, H, Meyer, E-L, Senger, G and Kraume, M (2010), 'Advantageous and detrimental effects of air sparging in membrane filtration: Bubble movement, exerted shear and particle classification', *Desaliantion*, **250**: 1083–1086.

Ehrl, L, Soos, M and Morbidelli, M (2008), 'Dependence of aggregate strength, structure, and light scattering properties on primary particle size under turbulent conditions in stirred tank', *Langmuir*, **24**(7): 3070–3081.

`Farrow, J B, Fawell, P D, Johnston, R R M, Nguyen, T B, Rudman, M, Simic, K and Swift, J D (2000), 'Recent developments in techniques and methodologies for improving thickener performance', *Chemical Engineering Journal*, **80**(1–3): 149–155.

Ghidossi, R, Veyret, D and Moulin, P (2006), 'Computational fluid dynamics applied to membranes: state of the art and opportunities', *Chemical Engineering and Processing*, **45**: 437–454.

Haarhoff, J and vander Walt, J J (2001), 'Towards optimal design parameters for around-the-end hydraulic flocculators', *Journal of Water Supply Research and technology-AQUA*, **50**(3): 149–159.

Henze, M, Grady, C. P. L., Jr, Gujer, W., Marais, G. v. R, Matsuo, T. (1987). *Activated Sludge Model No. 1*, IAWPRC, London.

Hlavacek, M and Remy, J F (1995), 'Simple relationships among zeta-potential, particle size distribution, and cake specific resistance for colloid suspensions coagulated with ferric chloride', *Separation Science and Technology*, **30**(4): 549–563.

Hunze, M (1996). *Numerische Modellierung reaktiver Strömungen in oberflächenbelüfteten*, Institute of fluid dynamics and computational applications in civil engineering, University of Hannover thesis, Hannover.

Jones, S C, Amirtharajah, A, Sotiropoulos, F and Skeens, B M (2000). *Using static mixers to mix coagulants: CFD modelling and pilot-plant experiments*. 9th Inernational Gothenburg Symposium. H. H. Hahn, E. Hoffmann and H. Odegaard. Istanbul, Turkey, Springer-Verlag, Berlin. Chemical Water and Wastewater Treatment VI: 79–88.

Joshi, J B (2001), 'Computational flow modelling and design of bubble column reactors', *Chemical Engineering Science*, **56**(21–22): 5893–5933.

Judd, S (2006). *The MBR Book, Principles and Applications of Membrane Bioreactors for Water and Wastewater Treatment*, 1st ed. London, Elsevier.

Judd, S (2011). *The MBR Book, Principles and Applications of Membrane Bioreactors for Water and Wastewater Treatment*, 2nd ed. London, Elsevier.

Kang, C, Hua, J, Lou, J, Liu, W and Jordan, E (2008), 'Bridging the gap between membrane bio-reactor (MBR) pilot and plant studies', *Journal of Membrane Science*, **325**: 861–871.

Karama, A B, Onyejekwe, O. O, Brouckaert, C. J, Buckley, C. A. (1999), 'The use of Computational Fluid Dynamics (CFD). Technique for evaluating the efficiency of an activated sludge reactor', *Water Science and Technology*, **39**: 329–332.

Kim, J and Kramer, T A (2006), 'Improved orthokinetic coagulation model for fractal colloids: Aggregation and breakup', *Chemical Engineering Science*, **61**(1): 43–53.

Korpijarvi, J, Laine, E and Alhstedt, H, Eds. (2000). *Using CFD in the study of mixing in coagulation and flocculation*. Chemical Water and Wastewater Treatment VI. Berlin, Springer-Verlag.

Krebs, P (1991), 'The hydraulics of final settling tanks', *Water Science and Technology*, **23**(4–6): 1037–1046.

Le Moullec, Y, Potier, O, Gentric, C and Pierre Leclerc, J (2008), 'Flow field and residence time distribution simulation of a cross-flow gas-liquid wastewater treatment reactor using CFD', *Chemical Engineering Science*, **63**(9): 2436–2449.

Leslie, G L (2001), 'Membrane bioreactors (MBR) for municipal wastewater treatment – an Australian prospective', *International Survey of Membrane Bioreactors, Report Prepared for South Australian Water Corporation, Brief No. 21/01 August 2001*.

Liu, N, Zhang, O, Chin, G-L, Ong, E-H, Lou, J, Kang, C-W, Liu, W and Jordan, E (2010), 'Experimental investigation of hydrodynamic behavior in a real membrane bio-reactor unit', *Journal of membrane Science*, **353**: 122–134.

Manem, J and Sanderson, R (1996). *Chapter 17: Membrane Bioreactors in 'Water Treatment Membrane Processes'*, McGraw-Hill, New York.

Marchisio, D L, Soos, M, Sefcik, J and Morbidelli, M (2006), 'Role of turbulent shear rate distribution in aggregation and breakage processes', *AIChE Journal*, **52**(1): 158–173.

Marshall, P R (1999). *Computational Fluid Dynamics Modelling of an Aerated Wastewater Pond – Development and Validation*, Dept. of Chemical Engineering, University of Queensland. thesis, (St Lucia, Qld.).

Martinelli, L, Guigui, C and Line, A (2010), 'Characterisation of hydrodynamics induced by air injection related to membrane fouling behaviour', *Desalination*, **250**: 587–591.

McCorquodale, J A, Yuen, E M, Vitasovic, Z and Samstag, R (1991), 'Numerical simulation of unsteady conditions in clarifiers', *Water Quality Research Journal of Canada*, **26**(2): 201–222.

Metcalf and Eddy (1991). *Wastewater Engineering, Treatment, Disposal, and Reuse*, McGraw-Hill, New York.

Moussa, A S, Soos, M, Sefcik, J and Morbidelli, M (2007), 'Effect of solid volume fraction on aggregation and breakage in colloidal suspensions in batch and continuous stirred tanks', *Langmuir*, **23**(4): 1664–1673.

Ndinisa, N, Fane, A, Wiley, D and Fletcher, D (2006), 'Fouling Control in a Submerged Flat Sheet Membrane System: Part II – Phase Flow Characterization and CFD Simulations', *Separation Science and Technology*, **41**(7): 1411–1445.

Nopens, I, Rios, N and Lesjean, B (2008). *Minutes of the '2nd Workshop on CFD modelling for MBR applications: From the fibre to the plant'*. Ghent, Belgium.

Nopens, I, Rios, N, Lesjean, B and Duc, E N C (2007). *Minutes of the '1st Workshop on CFD modelling for MBR applications: From the fibre to the plant'*. Berlin, Germany.

Park, N S, Park, H and Kim, J S (2003), 'Examining the effect of hydraulic turbulence in a rapid mixer on turbidity removal with CFD simulation and PIV analysis', *Journal of Water Supply Research and technology-AQUA*, **52**(2): 95–108.

Patankar, S V (1980). *Numerical heat transfer and fluid flow*, Hemisphere Pub. Corp. New York: McGraw-Hill, Washington.

Peterson, E L, Harris, J A and Wadhwa, L C (2000), 'CFD modelling pond dynamic processes', *Aquacultural Engineering*, **23**(1–3): 61–93.

Prat, O P and Ducoste, J J (2006), 'Modeling spatial distribution of floc size in turbulent processes using the quadrature method of moment and computational fluid dynamics', *Chemical Engineering Science*, **61**(1): 75–86.

Prieske, H, Drews, A and Kraume, M (2007), 'Prediction of the circulation velocity in a membrane bioreactor'. *IWA 4th International Membrane Conference*, 15–17 May 2007, Harrogate, UK, IWA.

Rios, N, Nopens, I, de Schepper, V, Jiang, T, Verstraete, W and Vanrolleghem, P (2007a). 'A rheological model for activated sludge in a side-stream MBR '. *IWA 4th International Membrane Conference*, 15–17 May 2007, Harrogate, UK, IWA.

Rios, N, Nopens, I and Vanrolleghem, P (2007b), 'Hydrodynamic CFD simulation of a two-phase flow in a single tube of an ultrafiltration membrane for a side-stream membrane bioreactor'. *2nd IWA National Young Water Professionals Conference*, Berlin, Germany, IWA Publishing.

Rosenberger, S, Kubin, K and Kraume, M (2002), 'Rheology of activated sludge in membrane bioreactors', *Engineering in life sciences*, **2**(9): 269–275.

Saalbach, J and Hunze, M (2008), 'Flow structures in MBR-tanks', *Water Science and Technology*, **57**(5): 699–705.

Samaras, K, Zouboulis, A, Karapantsios, T and Kostoglou, M (2010), 'A CFD-based simulation study of a large scale flocculation tank for potable water treatment', *Chemical Engineering Journal*, **162**(1): 208–216.

Sokolichin, A, Eigenberger, G and Lapin, A (2004), 'Simulation of buoyancy driven bubbly flow: Established simplifications and open questions', *Aiche Journal*, **50**(1): 24–45.

Soos, M, Moussa, A S, Ehrl, L, Sefcik, J, Wu, H and Morbidelli, M (2008), 'Effect of shear rate on aggregate size and morphology investigated under turbulent conditions in stirred tank', *Journal of Colloid and Interface Science*, **319**(2): 577–589.

Soos, M, Wang, L, Fox, R O, Sefcik, J and Morbidelli, M (2007), 'Population balance modeling of aggregation and breakage in turbulent Taylor-Couette flow', *Journal of Colloid and Interface Science*, **307**(2): 433–446.

Tacke, D, Pinnekamp, J, Prieske, H and Kraume, M (2008), 'Membrane bioreactor aeration: investigation of the velocity flow pattern', *Water Science and Technology*, **57**(4): 559–565.

Taha, T, Cheong, W L, Field, R W and Cui, Z F (2006), 'Gas-sparged ultrafiltration using horizontal and inclined tubular membranes – A CFD study', *Journal of Membrane Science*, **279**(1–2): 487–494.

Vadasarukkai, Y S and Gagnon, G A (2010), 'Determination of conventional velocity gradient (G) using CFD technique for a pilot-scale flocculation system', *Journal of Water Supply Research and technology-AQUA*, **59**(8): 459–470.

Versteeg, H K and Malalasekera, W (1995). *An introduction to computational fluid dynamics: the finite volume method*, Longman Scientific & Technical, Harlow, England.

Waite, T D, Schafer, A I, Fane, A G and Heuer, A (1999), 'Colloid fouling of ultrafiltration membranes: impact of aggregate structure and size ', *Journal of Colloid and Interface Science*, **212**: 264–274.

Wang, L, Marchisio, G L, Vigil, R D and Fox, R O (2005b), 'CFD simulation of shear-induced aggregation and breakage in laminar Taylor-Couette flow', *Journal of Colloid and Interface Science*, **282**(2): 380–396.

Wang, L, Vigil, R D and Fox, R O (2005a), 'CFD simulation of shear-induced aggregation and breakage in turbulent Taylor-Couette flow', *Journal of Colloid and Interface Science*, **285**(1): 167–178.

Wang, X-M and Waite, T D (2010), 'Iron speciation and iron species transformation in activated sludge membrane bioreactors', *Water Research*, **44**: 3511–3521.

Wang, Y (2010b). *Evaluation of Membrane Bioreactor Mixing Performance via Computational Fluid Dynamics Modelling*, UNESCO Centre for Membrane Science & Technology, School of Chemical Engineering, The University of New South Wales, PhD thesis, Sydney, Australia.

Wang, Y, Brannock, M, Cox, S and Leslie, G (2010a), 'CFD simulations of membrane filtration zone in a submerged hollow fibre membrane bioreactor using a porous media approach', *Journal of Membrane Science*, **363**: 57–66.

Wicaksana, F, Fane, A G and Chen, V (2005), 'The relationship between critical flux and fibre movement induced by bubbling in a submerged hollow fibre system', *Water Science and Technology*, **271**: 186–195.

Wiesner, M R, Clark, M and Mallevialle, J (1989), 'Membrane filtration of coagulated suspensions', *Journal of Environmental Engineering*, **115**(1): 20–40.

Wilcox, D C (1998). *Turbulence Modeling for CFD*, DCW Industries, Inc., La Canada, California.

Wood, M G, Greenfield, P F, Howes, T, Johns, M R and Keller, J (1995), 'Computational fluid dynamic modeling of wastewater ponds to improve design', *Water Sci. Technol.*, **31**(12, Waste Stabilisation Ponds and the Reuse of Pond Effluents): 111–118.

Wood, M G, Howes, T, Keller, J and Johns, M R (1998), 'Two dimensional computational fluid dynamic models for waste stabilisation ponds', *Water Research*, **32**(3): 958–963.

15.9 Appendix: nomenclature

15.9.1 Notation

C_{2ij}	inertial resistance factor
D_{ij}	viscous resistance factor
E_θ	dimensionless residence time distribution
e	energy (in energy balance equation)
J	permeate flux
f_i	body force in ith direction
k	turbulent kinetic energy
k	fitted parameters for shear-stress models (in Table 15.3)
L	length
m	fitted parameters for shear-stress models
n	fitted parameters for shear-stress models
p	pressure
Δp	pressure drop
R_t	total resistance
S	source or sink term of flow variable
T	temperature
t	time
u_i	fluid velocity in the ith direction (or u, v, w is the fluid velocity in x-, y-, z- direction respectively, while u', v', w' is the three-dimensional fluctuating velocity in Reynolds averaging)
v_{mag}	maximum velocity in the channel
x_i	length in dimension i (i.e., x, y, z in three-dimensional problem)

Greek symbols

α	permeability
ε	turbulent kinetic dissipation
ϕ	flow variable
Γ	diffusive coefficient
$\dot{\gamma}$	shear rate
θ	dimensionless time
μ	dynamic viscosity
ρ	density of fluid
τ	shear stress
τ	fitted parameters for shear-stress models (in Table 15.3)
τ_{ij}	viscous stress tensor
ω	specific dissipation rate

15.9.2 Abbreviations

3D	three-dimensional
ADV	acoustic doppler velocimetry
AFO	amorphous ferric oxyhydroxide
ASM	activated sludge model
CFD	computational fluid dynamics
CFV	crossflow velocity
CMC	carboxymethyl cellulose
COST	European Co-operation in the field of Scientific and Technical Research
CSTR	completely mixed reactor (continuous stirred-tank reactor)
CT X-ray	computer tomography scan
DNS	direct numerical simulation
DO	dissolved oxygen
EPS	extracellular polymeric substances
FS	flat sheet
H-B-P	Hershel–Bulkley model with Papanastasiou's adaption
HF	hollow fibre
HRT	hydraulic retention time
LDV	laser Doppler velocimetry
LES	large eddy simulation
MBR	membrane bioreactor
MF	microfiltration
MLD	mega-litre per day
MLSS	mixed liquor suspended solids
MT	multi-tube
NMR	nuclear magnetic resonance
PBM	population balance modelling
PFR	plug flow reactor
PIV	particle image velocimetry
RANS	Reynolds-averaged Navier-Stokes
RNG	renormalised group
RTD	residence time distribution
SMP	soluble microbial product
SRT	sludge retention time
SST	shear-stress transport (model)
TMP	trans-membrane pressure
UF	ultrafiltration
VOF	volume of fluid (model)

16
Models of membrane reactors based on artificial neural networks and hybrid approaches

S. CURCIO and G. IORIO, University of Calabria, Italy

DOI: 10.1533/9780857097330.3.569

Abstract: Mathematical modeling is widely used for the design and control of industrial processes and, in particular, membrane processes; the approach is based on the formulation of transport models to analyze the transport phenomena occurring in the membrane module as well as within the membrane. However, analysis of such complex behaviors is rather onerous and time consuming for practical purposes and some of the interactions between the fluid and the membrane structure or related to the actual kinetics, in the case of membrane reactors, are not yet completely understood. For these reasons, several simplified approaches have been proposed in the literature to describe the behavior of real membrane systems.

In contrast to mathematical modeling, a model based on artificial neural networks (ANNs) does not make use of any transport equation that could help to determine, on the basis of fundamental principles, the mutual relationships existing between the inputs and the outputs. ANNs, in fact, represent a data-driven method capable of learning from examples and are a particularly useful tool when difficult-to-interpret phenomena are to be described since no *a priori* knowledge is required. However, a neural model is still rather complicated. A good trade-off between a theoretical and a 'pure' neural network approach is represented by hybrid modeling – a combination of both theoretical and 'pure' neural network models – which allows the behavior of complex systems to be predicted in a more efficient way. Hybrid model predictions are indeed given as a combination of both theoretical and 'pure' neural network models, together concurring at the obtainment of system responses. The main advantage of a hybrid system is the possibility of describing some well-assessed phenomena by means of a fundamental approach. Other phenomena, that could be very difficult to interpret, are, instead, described by straightforward 'cause–effect' models, as represented, for instance, by ANN.

The present chapter describes how various kinds of advanced models, based either on ANNs or on a hybrid approach, can be used to predict the behavior of some typical membrane processes and to implement efficient control systems.

Key words: advanced modeling, control, simulation.

16.1 Introduction

Mathematical modeling represents an effective support for the design and control of industrial processes and, in particular, membrane processes. Different approaches can be used to develop reliable models aimed at investigating how membrane system responses may change, with time, under the influence of both external disturbances and manipulated variables.

Fundamental or theoretical modeling is based on the formulation of transport models analyzing the transport phenomena occurring in the membrane module as well as within the membrane. An exhaustive analysis of such a complex behavior, however, is rather onerous and time consuming for practical purposes, since the resulting system of nonlinear partial differential equations can only be solved by means of numerical methods. Moreover, some of the interactions between the fluid and the membrane structure or related to the actual kinetics, in the case of membrane reactors, are not yet completely understood and, therefore, are very difficult to interpret by proper mathematical relationships. For these reasons, several simplified approaches have been proposed in the literature to describe the behavior of real membrane systems.

On the other hand, a model based on ANNs does not make use of any transport equation that could help to determine, on the basis of fundamental principles, the relationships existing between the inputs and the outputs. ANNs, in fact, are a data-driven method capable of learning from examples and representing a particularly useful tool when difficult-to-interpret phenomena need to be described since no a prior*i* knowledge is required. However, a neural model is, generally, rather complicated, since it requires a large number of connections and, therefore, a large number of parameters to be estimated. Moreover, it is worthwhile observing that since extrapolation based on ANN predictions is an unreliable procedure, it is often necessary to perform many different experiments in order to train the network in as wide as possible range of practical situations.

A good trade-off between a theoretical and a 'pure' neural network approach is represented by hybrid modeling, which allows prediction of the behavior of complex systems in a more efficient way. Hybrid model predictions are indeed given as a combination of both theoretical and 'pure' neural network models, concurring in the obtaining of system responses. The main advantage of a hybrid system lies in the possibility of describing some well-assessed phenomena by means of a fundamental approach. Some others, that could be very difficult to interpret are, instead, described by straightforward 'cause-effect' models, such as ANN.

In the present chapter, it will be shown how various kinds of advanced models, based either on ANNs or on a hybrid approach, can be used to predict the behavior of a typical membrane process and to implement efficient control systems.

16.2 Fundamentals of artificial neural networks

ANNs are composed of several simple interconnected computational elements called neurons or nodes, which operate in parallel. These elements are inspired by the biological nervous system, whose functions are an attempt to somehow reproduce. The simplest kind of node is the so-called single input neuron, characterized by a scalar input (i), multiplied by a scalar weight (w), which is fed into a transfer function (F) that eventually produces a scalar output (o). In some cases, the weighted input ($i \times w$) is summed to a second scalar, known as the bias (b); the actual network input ($n = i \times w + b$) is then fed into a transfer function, which eventually produces an output. Both w and b can be adjusted, during the so-called training step, so as to allow the network to perform a particular task or exhibit a desired behavior. Any kind of differentiable mathematical function can be used as neuron activation function, but the most common, due to their forecasting capability, are the sigmoid logistic function, the hyperbolic tangent function and the linear function. A single neuron may also receive P different inputs ($i[1]$, $i[2]$, ..., $i[P]$), each of them multiplied by the corresponding weights so as to get a multiple input neuron, which is characterized, as in the previous case, by a net input (n), given by the sum of the bias b and the dot product of the row vector w and the column vector i, a transfer function (F) and an output (o). Actually, two or more such multiple input neurons may be combined in a layer and a specific network might contain one or more such layers. Each neuron, therefore, receives input signals from the related units, elaborates these stimuli by an activation or transfer function and generates an output signal that is then transferred to other neurons. In a multilayer network the layer which produces the network output(s) is called the output layer; all the other layers are called hidden layers. An ANN is a data-driven method, capable of learning from examples and capture the functional relationships existing between input(s) and the output(s). Even if the prediction of each single neuron were imperfect and bias-affected, the outcome of the interconnection(s) among the neurons is via a computational tool capable of learning from examples and thus providing accurate predictions, even with examples never processed before (Reilly and Cooper, 1990). This feature makes an ANN a particularly useful tool for modeling phenomena that are difficult to describe by an equation-based approach, since no a priori knowledge is actually required.

The architecture of the network actually consists of description of the number of layers composing the network, of the number of neurons and of the types of transfer function in each layer, of the connections established between each of the layers and each of the neurons and, finally, of the network inputs. The neural network architecture strongly determines the reliability of neural network prediction and, also, the computational effort required to properly train the network with a given set of experimental data. Generally, the more neurons there are in a hidden layer, the more powerful is the network, but the computational effort is higher. The number of input and output variables is problem-dependent since it is related, respectively, to the number of independent and dependent variables of the problem under consideration. On the other hand, the number of layers and the number of neurons in each layer is the result of an optimization process. A lower number of hidden neurons may result in under-fitting the problem. Correspondingly, the exploitation of too many hidden neurons may result in so-called over-fitting, that is, the incapability of the network to generalize to new situations that different from the examples memorized during the training. Even if several methods were proposed to achieve the final network architecture (Roy *et al.*, 1993; Murata *et al.*, 1994; Wang *et al.*, 1994), a general procedure is not yet available and the network structure is usually determined according to heuristic guidelines and to trial-and-error procedures.

To complete the network definition, it is, however, necessary to perform the so-called neural network training. During the training phase the network learns how to correlate the inputs to the output variables. More specifically, the network is submitted to a certain number of input and output data and, according to an error minimization algorithm, it changes the values of the weights, which can be considered as the key elements in which the network knowledge is stored (Anthony and Bartlett, 1999). Only a certain number of the available experimental points are exploited during the training phase; the remaining experimental data are used during a post-training analysis, called the test phase. During the test phase, the neural network is called on to predict the output values corresponding to an input combination never seen before and, therefore, to a set of experimental points not belonging to the training set. Even if the test phase is a post-training analysis only, it usually plays a fundamental role in the ANN architecture definition; in fact, the convergence of the above-mentioned trial-and-error procedure is usually considered achieved with reference to a performance index based on test points. The test phase of the network is usually carried out in the definition domain in which the training procedure has been achieved. As a matter of fact, the forecasting capability of the neural networks outside this definition range cannot be guaranteed; due to the intrinsic black-box nature of neural network models, the validity domain does indeed strictly depend on the range of data used in model definition (van Can *et al.*, 1998). Another kind of post-training analysis of neural network performance is the so-called validation phase. Similarly to the test phase, the network is called on to predict the experimental points excluded from both the training and the test sets;

unlike the test phase, the evaluation of model prediction reliability does not influence the definition of neural network architecture.

16.2.1 Multilayer perceptron network

The perceptron is a particular type of ANN created by Rosenblatt (Rosenblatt, 1961). One of the simplest forms of perceptron is represented by a single-layer network whose weights and biases are continuously adjusted during neural network training so as to obtain a correct target vector when presented with the corresponding input vector. Perceptrons are especially exploited to solve simple problems in pattern classification, and they suffer from several limitations: (a) the output values of a perceptron can take on only one of two values (0 or 1); (b) only linearly separable sets of vectors can actually be classified by a perceptron (Minsky and Papert, 1969). Due to these intrinsic limitations, single-layer networks are generally replaced by multilayer ANNs, in which the input signal is propagated through different processing layers before the network output is calculated. A multilayer perceptron (MLP) network is a supervised network consisting of at least, three layers: two or more hidden layers (the first one is sometimes called the input layer) and an output layer which produces the final output(s) of the network. Figure 16.1 shows a typical MLP structure consisting of three layers. Each layer is composed of one or more nodes; each node has input connections only with the nodes of the previous layer,

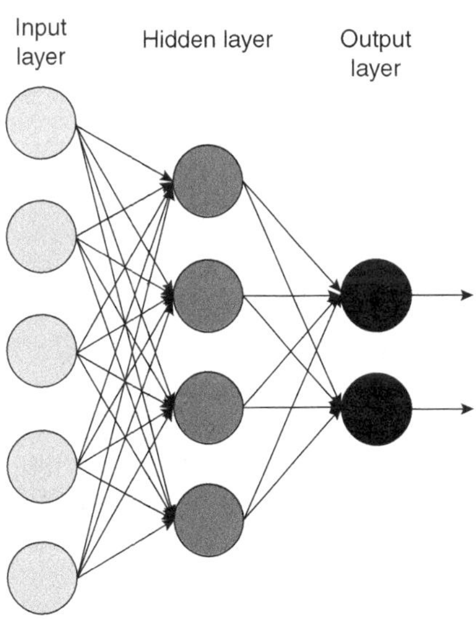

16.1 Multilayer perceptron (MLP) network with three layers.

and the node's output is transmitted only to the neurons belonging to the next layer. Each node has an associated weight vector and, usually, a bias, both transforming its input vector and a transfer function which allows the neuron's non-linear response to be obtained. These weights and the biases, therefore, represent the ANN model parameters which are to be adjusted to allow the network to produce the desired outputs. This feature makes ANN a parametric model. The final output of the network is compared with a target vector, generally represented by a set of available experimental data; the difference, known as the error, between the network output and the desired target is used to update the weights and biases during network training by means of a proper learning algorithm aimed at minimizing the error (McClelland, 2011).

Learning paradigms can be subdivided into three types, which are supervised learning, unsupervised learning and reinforcement learning (Kartalopoulos, 1996). The supervised learning paradigm is comparable to learning with a teacher providing some supervision and continuous feedback on the quality of solutions. An unsupervised learning paradigm only needs an input data-set; no target output is actually necessary; for successful learning, although unsupervised, the neural network may need some feature-selecting guidelines. A reinforcement learning paradigm may be regarded as a special case of supervised learning; however, instead of getting information on how close the actual output is to the desired output, only an indication on whether the output is right or wrong is made available. Therefore, during training, there is no indication of whether the output response is moving in the right direction or of how similar it is to the desired output.

Back propagation is one of the most successful algorithms exploited to train a network which is aimed at either approximating a function, or associating input vectors with specific output vectors or classifying input vectors in an appropriate way as defined by ANN designer (Rojas, 1996). Learning with back propagation is often exploited for supervised learning networks. Several algorithms, most of which are based on optimization theory, have been formulated for training ANNs. The general procedure, however, can be summarized as follows (Zhou and Therdthai, 2010). The learning process begins by randomly selecting, from all the available experimental data, a training vector. The weights associated with each neuron are guessed during the so-called network initialization. The transfer functions are then assigned to both the neurons in the hidden layers and the neurons in the output layer. The error associated with each neuron belonging to the output layer and, afterwards, the errors across the hidden layer, are calculated. The obtained errors are eventually exploited to update the connection weights of each neuron in both the output and the hidden layers and the overall procedure is repeated until a properly defined performance criterion is achieved.

Generally, learning stops when the difference (error) between the network output and the target vector is decreased to an acceptable value. Other types of performance criteria can be defined; one of the most common states that the learning process is complete when the sum of square errors of all the training patterns (i.e., the number of data points belonging to the training data-set) across the units belonging to the output layer is minimized. The simplest way of formulating a back propagation algorithm is to change the weights in the direction in which the selected performance criterion lowers more rapidly; in gradient descent (or steepest descent) learning rule, for instance, the gradient of the performance function is measured at a particular point in the weights space and this information is then used to properly adapt the ANN's weights. Besides gradient descent, other kinds of algorithm have been proposed to achieve neural network learning. With the conjugate gradient algorithm the minimization is performed along conjugate directions; this algorithm actually exhibited faster convergence for quadratic functions. The Levenberg–Marquardt (LM) algorithm is a method that allows minimizing nonlinear functions without the necessity of computing the second derivative Hessian matrix. Although slower than other techniques, the LM algorithm is rather robust and is demonstrated to be capable of achieving a solution even with poor initial guesses. Bayesian Regularization (MacKay, 1992) aims at improving an ANN's generalization capability and at preventing data over-fitting provided by some types of ANNs.

Generalization can be defined as the ability to train with one data-set and then successfully classify independent test sets. Although continued training will increase the training set accuracy, it has been proved that test set accuracy decreases after a certain point. This is attributable to over-fitting, which occurs when a model, generally an excessively complex one with too many parameters relative to the number of observations, begins to memorize training data rather than learning to generalize from trend. The Bayesian Regularization approach modifies the usually used objective function, such as the sum of squares errors, by including an additional term, which is the sum of squares of the network weights. In this way, a new performance criterion, given by the linear combination of the sum of squares errors and of the sum of squares weights, is obtained. It is assumed that the weights and biases of the network are random variables following Gaussian distributions and that the parameters to be optimized in order to achieve convergence are related to the unknown variances associated with these distributions.

16.3 An overview of hybrid modeling

A reasonable trade-off between the theoretical and the empirical approach is represented by hybrid modeling, leading to a so-called 'grey-box' model

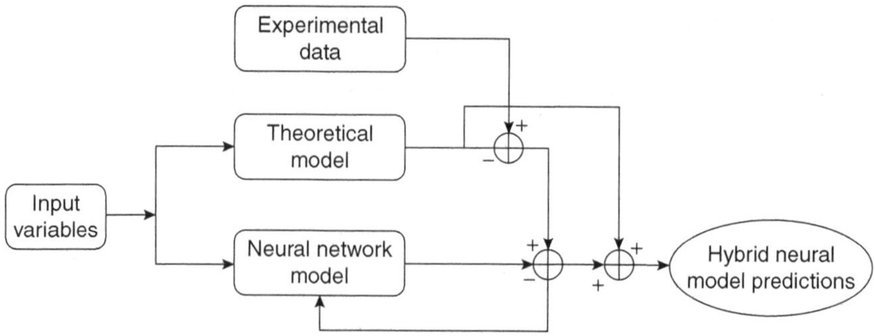

16.2 Hybrid neural model based on a parallel architecture.

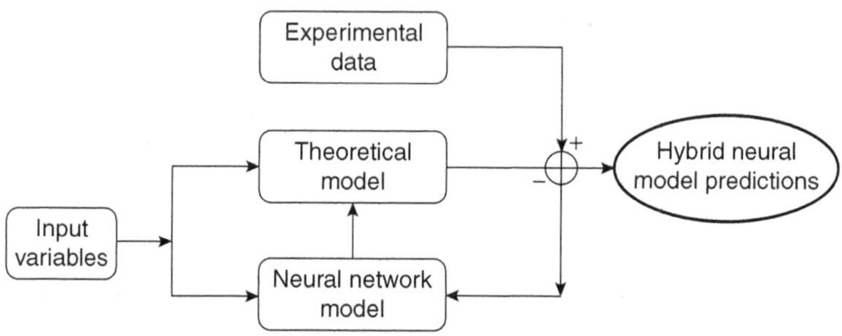

16.3 Hybrid neural model based on a serial architecture.

capable of good performance in terms of data interpolation and extrapolation (Saraceno *et al.*, 2010). Hybrid model predictions are given as a combination of both a theoretical and a 'pure' neural network approach, together concurring in the obtaining of system responses. The main advantage of hybrid modeling regards the possibility of describing some well-assessed phenomena by means of a theoretical approach, leaving the analysis of other aspects, very difficult to interpret and describe in a fundamental way, to rather simple 'cause-effect' models (Psichogios and Ungar, 1992; Agarwal, 1997; Curcio *et al.*, 2009). Two kinds of hybrid neural models (HNMs) can be generally defined depending on the interactions existing between the neural and the theoretical blocks. In a model based on a parallel architecture (Fig. 16.2), the inaccuracy in the predicted value with respect to the fundamental part is minimized by the addition of residuals calculated by the neural network. In a model based on a serial architecture (Fig. 16.3) a process variable, which is difficult to measure, is estimated by a neural network and then fed to the theoretical block as an input. Finally, the output coming from the fundamental part is checked with the experimental value for convergence.

16.3.1 Exploitation of ANNs and hybrid modeling in membrane processes

The interest of the scientific and academic community in the applications of ANNs to membrane technology is progressively increasing. The predictions of neural models indeed exhibited a rather good agreement with the available experimental data also when new conditions, never exploited during the training process, were used. Ou and Achenie (2005) presented a neural network modeling approach applicable to proton exchange membrane (PEM) fuel cells. The authors tested different models showing that the ANNs were capable of simulating the effects of Pt loading on the performance of the PEM fuel. ANNs were also used to predict the performance of a membrane sequencing batch reactor (MSBR) treating hypersaline oily wastewater (Pendashteh *et al.*, 2011). A cascade-forward network was used to model a submerged membrane bioreactor receiving cheese whey as input; the bioreactor behavior was simulated over a period of 100 days changing the solids retention times (Çinar *et al.*, 2006). ANNs were successfully employed also to model the permeate flux decay of both ultrafiltration and microfiltration as a function of process operating variables. Flux declines during crossflow microfiltration of a mixture containing phosphate and fly ash were modeled, showing that ANN accurately predicted the permeate flux as a function of the trans-membrane pressure, the feed solution and the membrane type (Aydiner *et al.*, 2005).

Chellam (2005) derived and validated ANN models for membrane fouling induced by polydispersed feed suspensions during crossflow microfiltration. Rai *et al.* (2005) used ANN models to describe the permeate flux and permeate concentration (total soluble solid) profiles during the ultrafiltration of different fruit juices. ANNs were also developed to model crossflow ultrafiltration of milk, thus allowing a prediction of the time-dependent rate of both permeate flux decay and membrane rejection (Razavi *et al.*, 2003). A well-trained ANN may represent the basis for the implementation of an optimal control strategy that could be actuated by means of a specific environment in which to develop an application that interfaces measurement and control hardware. An ANN control system, based on a proper integration of two different computational environments (specifically, MATLAB, where the ANN architecture was obtained and LabVIEW which actuated the neural network predictions), was presented (Curcio *et al.*, 2005). Classical feedback controllers were proposed in membrane technology to decide either when to perform the membrane backwash (Smith *et al.*, 2006) or the duration of backwash process (Smith *et al.*, 2005). Both proportional–integral (PI) and proportional–integral–derivative (PID) control algorithms were adopted in water desalination to regulate the process flow rates and to adjust the operating pressure so as to achieve a desired rate of clean water production (Alatiqi *et al.*, 1999). However, the exploitation of neural-based feedback controllers is not so common. Neural-based controllers may, in principle, offer several advantages, since a properly developed neural model providing reliable predictions of the actual system behavior is definitely essential for the implementation of high-performance control systems (Brown and Harris, 1994; Åström and Wittenmark, 1989; Brown and Harris, 1994). The utilization of modern optimization techniques and the formulation of reliable models of the actual dynamics of the system under study make it possible to implement high-efficiency feedback controllers, whose action is based on the actual transfer function of the plant, thus optimizing the true closed-loop performance (Toscano, 2005).

With reference to three case studies, it will be shown how various kinds of advanced models, based either on ANNs or on a hybrid approach, can be used to predict the behavior of some typical membrane process and to implement efficient control systems.

16.4 Case study: prediction of permeate flux decay during ultrafiltration performed in pulsating conditions by a neural model

The aim of this section is to present an innovative methodology modeling the permeate flux decay in membrane separation processes. An ANN was developed on the basis of the experimental results collected during the ultrafiltration of bovine serum albumin (BSA) solutions under pulsating conditions (Curcio *et al.*, 2005).

The experiments necessary to achieve the neural network model were performed filtering 1.0% (w/w) BSA aqueous solutions through a poly-ethersulfone ultrafiltration membrane with a molecular weight cut-off of 20 kDa. The membrane was assembled in a flat geometry lab-scale module, consisting of two different parts: the cover containing both the entry and the exit sections, and the support where the flat sheet membrane was positioned. The channel length was 0.25 m, its height was 0.003 m and the channel width was 0.03 m. All the experiments were performed changing, during each trial, the applied trans-membrane pressure (TMP) that was progressively increased by steps of 0.5 bar, from 0.5 to 2.0 bar; the operating temperature was maintained at 45°C. A computer program running under LabVIEW allowed monitoring of the main operating variables and controlling the pulse generation on both feed flow rate and TMP. With respect to a traditional ultrafiltration plant, an on–off electric valve (EV) was positioned on a 'slave' retentate line. When the EV valve was kept closed during the whole experiment, ultrafiltration was performed in the so-called constant conditions, that is, in a traditional way. On the other hand, when the EV was periodically opened and closed, pulsatile conditions were applied. In particular, opening the EV caused a two-fold effect: TMP abruptly dropped from the operating value to practically zero, while the feed flow rate increased significantly, in the tested cases by a factor of about 2. Therefore, a generation of both negative pulses for TMP and positive pulses for Q was obtained according to a square-wave-like profile, so that each real filtration step was followed, several times, by disruption of the concentration polarization layer formed on membrane surface. The graphical user interface running under LabVIEW allowed control of both the duration of each square-wave upstream pressure release, that is, pulse time (t_{pul}) and the duration of upstream operating pressure restoration, that is, operating time (t_{op}). Three different values for operating time, that is, $t_{op} = 60$ s, $t_{op} = 90$ s and $t_{op} = 120$ s were used, at a fixed pulse duration value, $t_{pul} = 10$ s, to train and subsequently test the developed neural network model.

The ultrafiltration of BSA in pulsating conditions was modeled accounting for the influence of three variables, that is, the operating time (t_{op}), the sampling time (t) and the inlet flow rate (Q), on permeate flux decay. Actually, the dependence on the applied TMP, that could be a further input for the network, was neglected as it related to the normalized permeate flux (J_n), no longer affected by TMP and defined as the ratio between the actual BSA permeate flux and that measured with pure water, ultrafiltered at the same values of temperature, Q and TMP. The above assumption, excluding TMP from the list of significant inputs, could be justified by the consideration that the intrinsically unsteady-state operations of the proposed device (alternation of pressure imposition and release) drastically reduced the formation of cake on the active membrane side, whose compressibility could non-linearly affect the relationship between TMP and permeate flow rate. The present neural network model consisted of three different inputs, t, Q, t_{op} and one single output, J_n. A multilayer perceptron (MLP) feed-forward architecture was eventually developed by MATLAB Neural Network Toolbox in order to obtain the non-linear relationships existing between the ANN output and the operating variables, that is, the ANN inputs. Bayesian Regularization was exploited to improve the generalization of the neural model.

Different sets of experimental data collected during BSA ultrafiltration, performed at different values of both feed flow rate and operating time, were adopted. The original database, consisting of 364 experimental values, was equally partitioned into test and training data. In particular, a total of 182 samples was deliberately and randomly excluded from the original data-set in order to test the neural network capability of predicting J_n values measured in conditions not exploited during training, but belonging to the same training range. On the basis of the above-described considerations, a neural model was finally identified according to a trial-and-error procedure, that allowed determination of the final network architecture once both the post-training and post-simulation analyses had given a maximum relative

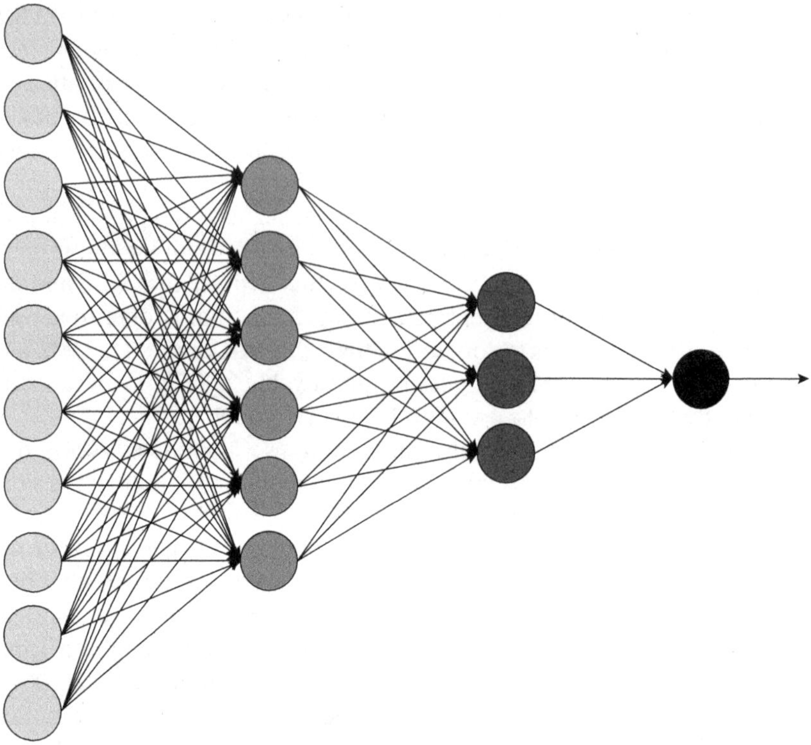

16.4 Multilayer perceptron feed-forward network predicting the decay of permeate flux in pulsating conditions (10 neurons in the first hidden layer (input layer), 6 neurons in the second hidden layer, 3 neurons in the third hidden layer, 1 neuron in the output layer).

error lower, respectively, than 5% and 10%. The neural model (Fig. 16.4) had a pyramidal architecture with four different layers:

- three hidden layers, the first of which was actually the input layer, having, respectively, ten, six and three neurons with a logistic sigmoid activation function;
- one output layer, corresponding to the normalized permeate flux, with one neuron and a logistic sigmoid activation function.

Figure 16.5 shows a typical post-training analysis in which the dynamic evolution of normalized permeate flux decay, J_n, calculated by the developed neural model was compared to the corresponding flux decay measured during the experiments. A very good agreement between these two data sets was observed, since the maximum relative error never exceeded 3%.

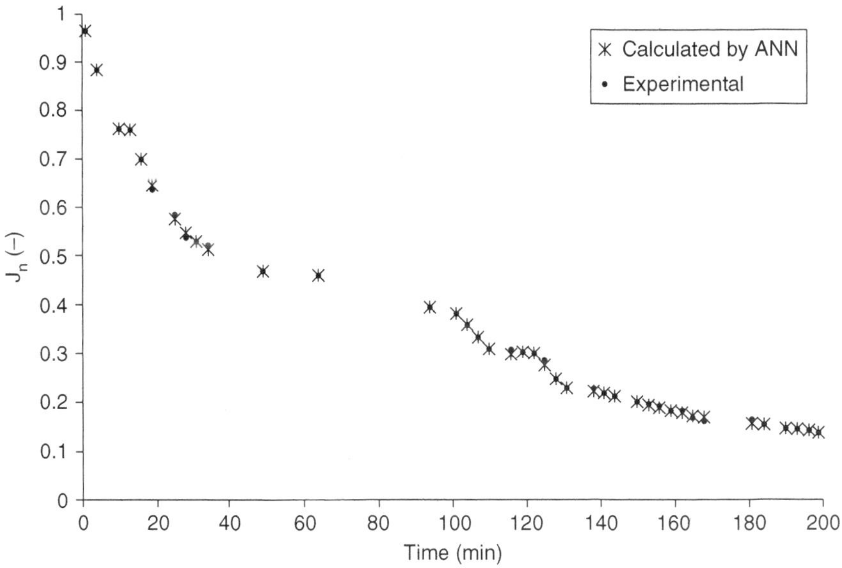

16.5 Post-training analysis. Normalized permeate flux decay (Q = 2 L/min, t_{op} = 90 s).

Figure 16.6 shows both post-training and post-simulation analyses obtained with reference to a different operating time. Also in this case, good agreement could be observed between neural network predictions and the real experimental values, the maximum relative error, calculated with respect to the sampling data not adopted for training, indeed being less than 5%.

16.5 Case study: prediction of permeate flux decay during ultrafiltration performed in pulsating conditions by a hybrid neural model

The 'pure' ANNs approach presented in the previous section poses, however, some problems. The main criticism concerns the fact that the developed ANN model did not make use of any equation that could have helped to determine, on the basis of fundamental principles, the mutual relationships between the output and the inputs. Since no fundamental equation was adopted, the resulting ANN model was rather complicated and required a large number of neurons, each of them associated to several weights and biases, and, therefore, a large number of parameters that had to be estimated. The possibility of combining a fundamental model with an ANN could, in principle, improve the model's capability to generalize to new situations. Moreover, it is expected that the resulting model would require a

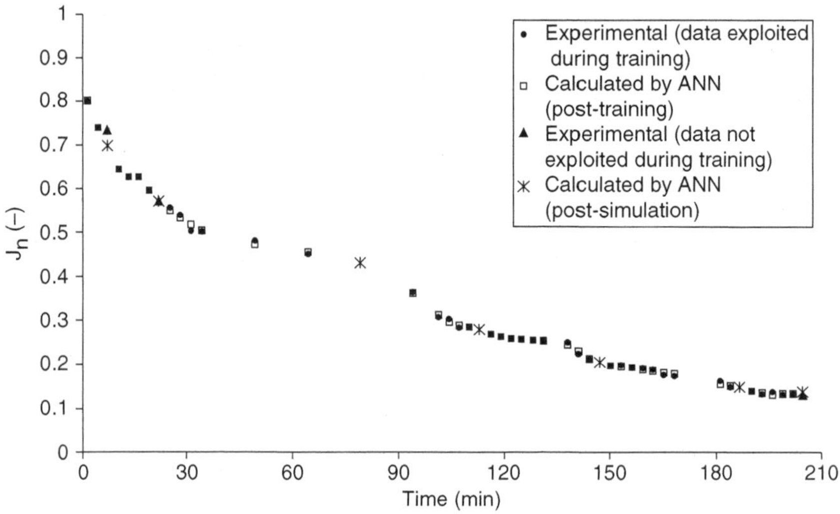

16.6 Post-training and post-simulation analyses. Normalized permeate flux decay (Q = 2 L/min, t_{op} = 120 s).

lower number both of experimental data and of parameters, thus reducing the computational effort. The previous considerations actually represent the basis for the development of a hybrid model (HNM), that is, a model whose predictions are given as a combination of both a fundamental and a 'pure' neural network approach, concurring together in the obtaining system responses (Curcio *et al.*, 2009).

To develop the present HNM aimed at predicting the performance of BSA ultrafiltration in pulsating conditions, both a theoretical and a neural model were, therefore, formulated and properly coupled. As far as the development of the theoretical model is concerned, the lab-scale membrane unit used to collect the experimental data was exactly reproduced and represented as a 2D straight rectangular channel.

To describe the velocity field of BSA solutions entering the membrane module, the momentum and the continuity equations were formulated as:

$$-\nabla \times \eta\left(\nabla \mathbf{u} + (V\mathbf{u})^T\right) + \rho(\mathbf{u} \times V)\mathbf{u} + \nabla p + \frac{\partial \mathbf{u}}{\partial t} = 0 \qquad [16.1]$$

$$\nabla \times \mathbf{u} = 0 \qquad [16.2]$$

and coupled to the Brinkman and continuity equations aimed at calculating the flow field within the membrane:

$$-\eta_1 \nabla^2 \mathbf{u}_1 + \frac{\eta_1}{k(t)} \mathbf{u}_1 + \nabla p_1 + \rho_1 \frac{\partial \mathbf{u}_1}{\partial t} = 0 \qquad [16.3]$$

$$\nabla \times \mathbf{u}_1 = 0 \qquad [16.4]$$

In the previous equations, ρ and η were the density and the viscosity of BSA solutions, respectively, **u** the velocity vector in the membrane module, and p the pressure within the module; ρ_1 and η_1 were, respectively, the viscosity and the density of the fluid (permeate) flowing within the membrane, $\mathbf{u}_1$ the velocity vector and p_1 the pressure in the membrane.

Both the outlet pressure and the feed velocity were varied in order to reproduce, as a function of t_{op} and t_{pul} values, negative pulses for TMP and positive pulses for Q, according to the already-described periodic square-wave profiles. At the interface of the channel and the membrane, it was assumed that the components of the velocity vectors were continuous, while, on the basis of the osmotic pressure model, the pressure p_1, evaluated on the 'membrane-side', was equal to the pressure p, calculated on the 'channel-side', decreased by the difference of osmotic pressure, $\Delta\Pi\ (x,t)$, between the membrane surface and the permeate:

$$\Delta\Pi(x,t) = \alpha\left[C_w(x,t) - C_p\right] + \beta\left[C_w(x,t)^2 - C_p^2\right] + \gamma\left[C_w(x,t)^3 - C_p^3\right]\ldots\ldots \qquad [16.5]$$

$\Delta\Pi\ (x,t)$ actually depended on BSA concentration build-up on the membrane wall, $C_w(x, t)$, and on the BSA concentration in the permeate; the membrane under study exhibited total rejection towards BSA: Cp, therefore, was set equal to zero.

The calculation of rejected species concentration was performed by solving, in the module channel only, the convection-diffusion equation:

$$\nabla \times \left(-D_{AB} \nabla C + C\mathbf{u}\right) + \frac{\partial C}{\partial t} = 0 \qquad [16.6]$$

where D_{AB} is the diffusion coefficient of BSA in water and the expression within brackets represents the flux vector, accounting for both diffusive and convective contributions to mass transfer.

It is worth observing that the membrane permeability, $k(t)$, appearing in the Brinkman equation was actually a function of time since an additional resistance, R_{ad}, due to the adsorption of BSA on the membrane surface had to be accounted for. In particular, it was assumed that the resistance due to adsorption was a function of protein concentration on the membrane, that is, $R_{ad} = R_{ad}(C_a)$ (Ruiz-Bevia *et al.*, 1997). Several empirical relationships were proposed in the literature to express the dependence of R_{ad} on C_a; one of the most common, determined in the case of the ultrafiltration process performed in constant conditions, was a power law model depending on two parameters, K and n:

$$R_{\text{ad}} = KC_a^n \qquad [16.7]$$

To determine the amount of adsorbed BSA, C_a (kg/m^2), a material balance on the membrane surface was formulated too:

$$\left(-D_{AB}\nabla C + C\mathbf{u}\right)_w + \frac{\partial C_a(x,t)}{\partial t} = 0 \qquad [16.8]$$

where the subscript w indicates that the flux vector was evaluated at the membrane wall.

It is worth observing that both the dependence of osmotic pressure on the local concentration values and the variation of membrane permeability, as determined by the amount of adsorbed BSA, determined a mutual interaction between momentum and mass balance equations. The solution of the developed theoretical model could be attained only by means of a numerical method, which, however, necessitated preliminary knowledge of the actual relationship between R_{ad} and C_a, achievable, in the case of proteins ultrafiltration, by a proper (and cumbersome) experimental analysis. The neural network's ability to describe difficult-to-interpret phenomena, such as the actual relationship existing between the amount of adsorbed proteins and the additional resistance due to membrane fouling, was exploited to formulate a hybrid neural model. In the present case, the main advantage of hybrid modeling was actually in the possibility of describing some well-assessed phenomena, such as concentration polarization, by means of a fundamental theoretical approach. Some others, such as the complex interactions existing between the adsorbed solute(s) and the membrane surface, were expressed by an ANN, which made up for the limited knowledge of the complex physical phenomena occurring at the membrane-fluid interface, thus playing the role of parameter estimator.

A block diagram of the developed hybrid system is shown in Fig. 16.7. It consisted of both a theoretical and a neural block composing an iterative scheme. The theoretical (FEM) block simulated the ultrafiltration process behavior according to a hypothesized relationship of R_{ad} vs. C_a; in this way, an expression of permeate flux decay could be obtained and compared to a set of experimental data collected during BSA ultrafiltration under pulsating conditions. The difference between simulated and measured values

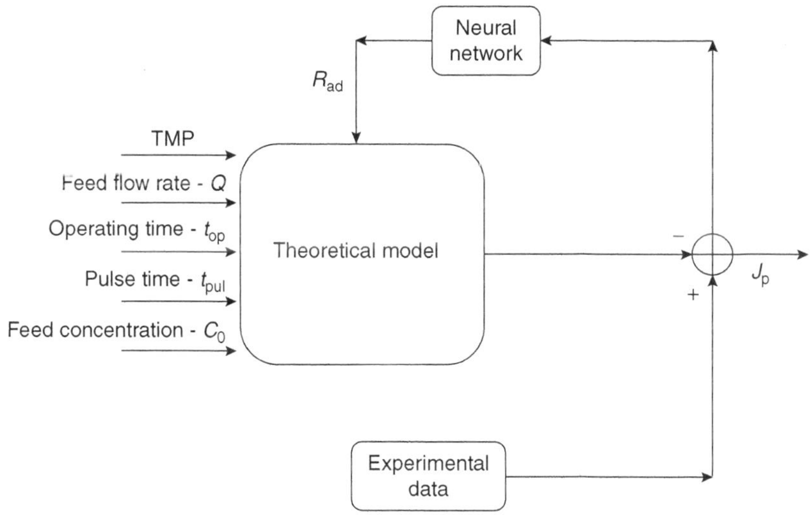

16.7 Schematic representation of the hybrid model (TMP: trans-membrane pressure; Q: feed flow rate; C_0: feed concentration of BSA solutions; t_{pul}: pulse time; t_{op}: operating time; $J_p(t)$: permeate flux decay).

was then fed to a rather simple neural model aimed at determining a new relationship between R_{ad} and C_a. The additional resistance calculated by the neural network was compared to that hypothesized at the beginning of iterative cycle; if the maximum relative error among the hypothesized values and the calculated ones was lower than 2%, the final results were obtained, otherwise a new iteration started using, as one of the inputs, the functionality between R_{ad} and C_a that the neural network had just determined. In the present work, a target data-set consisting of 318 experimental points was actually available. Of these, 212 points were used for network learning and 106 (one third of the total) to test neural network predictions. A very simple neural architecture was eventually obtained by a trial-and-error procedure; the neural network consisted of only one hidden layer with four neurons and two end-layers, with one neuron each, corresponding to the single input and the single output. Figure 16.8 shows the comparison among the network outputs and the target values in post-training and post-simulation analyses. A very good agreement was observed since, in both cases, the linear regression relating the targets to the network outputs had a slope nearly equal to unity, a nil intercept and a correlation coefficient almost equal to unity. Figure 16.9 shows the relationship that was finally achieved between R_{ad} and C_a. A sigmoid-like curve was obtained, proving that the exploitation of pulsating conditions resulted in a rather different behavior as compared to what is generally reported in the literature when BSA ultrafiltration is

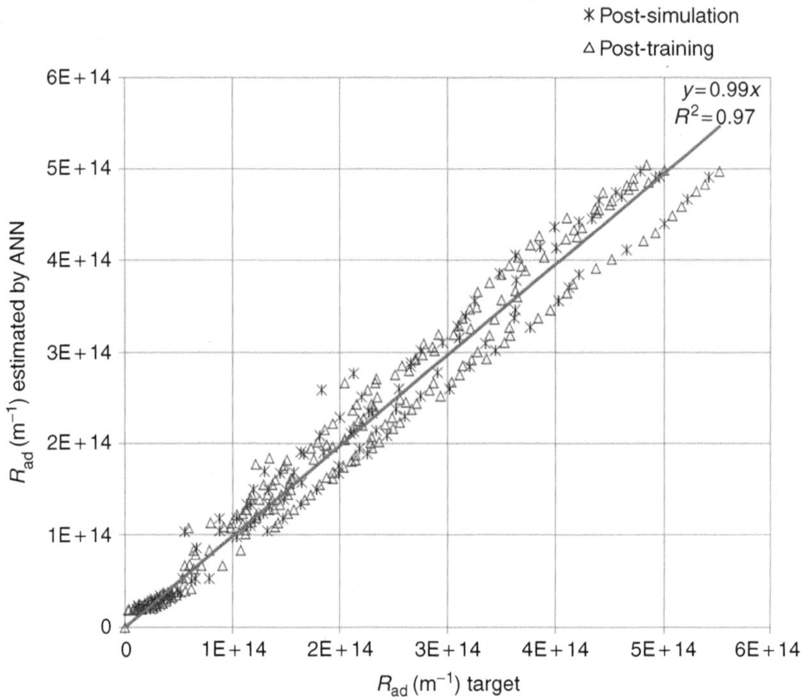

16.8 Comparison between target and calculated R_{ad} in both post-training and post-simulation analyses.

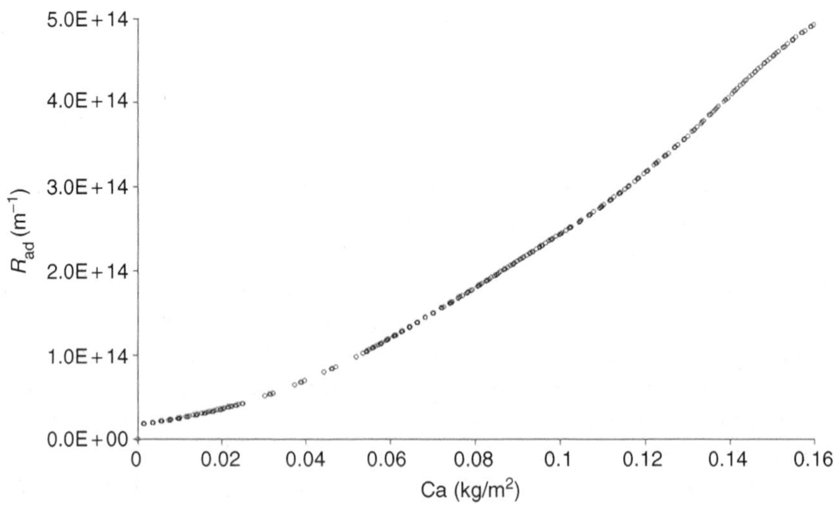

16.9 Dependence of R_{ad} upon Ca as estimated at the end of the iterative procedure.

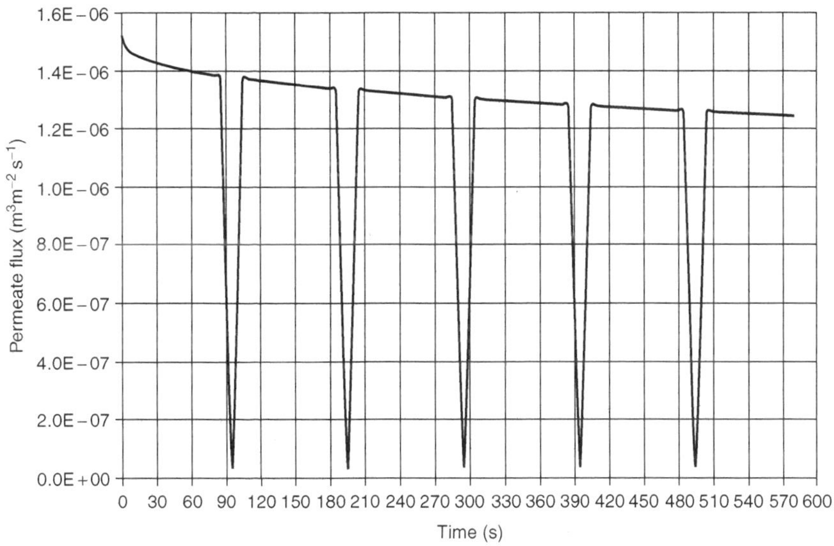

16.10 Permeate flux decay as estimated by the hybrid model (Q = 2 L/min, t_{op} = 90 s).

performed under constant conditions. Figure 16.10 shows the decay of permeate flux as calculated by the present hybrid neural model.

16.6 Case study: implementation of feedback control systems based on hybrid neural models

The previously-described hybrid neural model predicted rather well the actual behavior of BSA ultrafiltration under pulsating conditions; it was, therefore, worthwhile developing a feedback control system that, on the basis of HNM predictions, allowed suppression of the influence of external disturbance and optimization of the performance of membrane operations. The characteristic equation defining either a proportional-integral (PI) or a proportional-integral derivative (PID) controller was added to the already developed hybrid neural model in order to obtain true closed-loop responses, thus accomplishing a proper design and an accurate tuning of both controllers. Among the five inputs of the system (Fig. 16.7), the inlet concentration was considered the sole disturbance, since it influenced the time evolution of adsorbed BSA and, therefore, strongly affected permeate flux decay. TMP was, instead, chosen as the sole manipulated variable that, properly instructed by the control block, should allow maintaining the permeate flux as similar as possible to a desired set-point pattern (Curcio *et al.*, 2011). Feed flow rate (Q), pulse time (t_{pul}) and operating time (t_{op}) were, instead, simple inputs to the system and were fixed at definite values

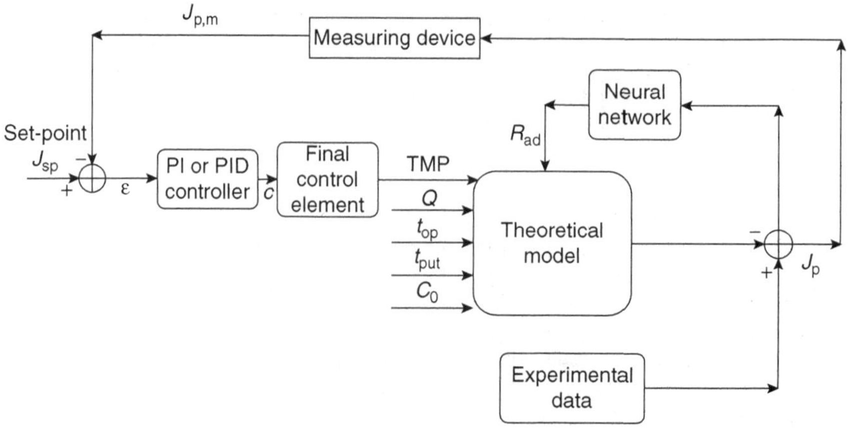

16.11 Schematic representation of the feedback hybrid neural control system.

before performing each simulation. The block diagram corresponding to the present feedback control system is shown in Fig. 16.11. The characteristic equations relating either the PI or the PID output, $c(t)$, to the deviation, $\varepsilon(t)$, between the set-point value and the actual permeate flux ($\varepsilon(t) = J_{p,SP}(t) - J_p(t)$), were taken into account in order to ascertain how the type of controller and its adjustable parameters affected the closed-loop responses of the controlled system:

$$\text{PI controller: } c(t) = K_c\varepsilon(t) + \frac{K_c}{\tau_I}\int_0^t \eta(t)\,\mathrm{d}t + c_s \qquad [16.9]$$

$$\text{PID controller: } c(t) = K_c\varepsilon(t) + \frac{K_c}{\tau_I}\int_0^t \varepsilon(t)\,\mathrm{d}t + K_c\tau_D\frac{\mathrm{d}\varepsilon(t) + c_s}{\mathrm{d}t} \qquad [16.10]$$

The dynamics of the measuring device were assumed negligible, and that the final control element was capable of instantly converting the signal produced by the controller; therefore, the manipulated variable, TMP(t), was related to the controller output by the following relationship:

$$\mathrm{TMP}(t) = a \times c(t) \qquad [16.11]$$

where a was a gain, actually representing the dynamics of the final control element, that allowed TMP(t) to assume the proper value within the feedback loop. Generally speaking, the main features of a PI controller are: it

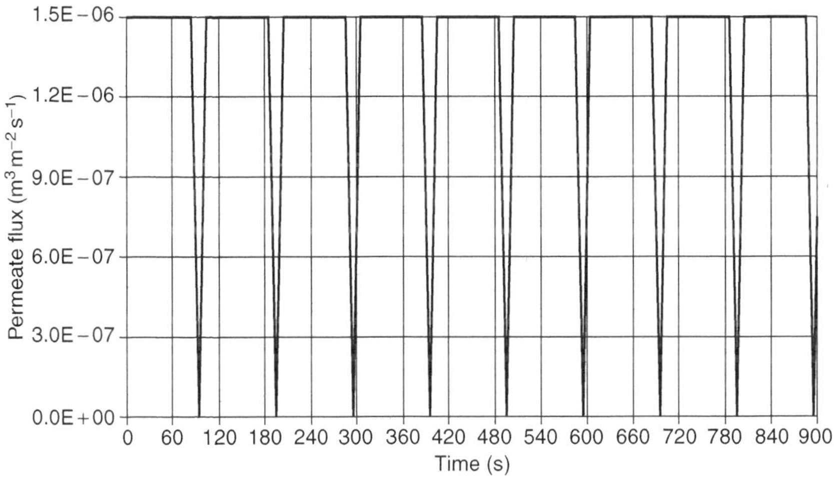

16.12 Case a: uniform set-point. The set-point, $J_{p,SP}$, is kept constant throughout the experiment.

allows elimination of the offset; as K_c increases, the response becomes faster but more oscillatory to set-point changes; moreover, the response is faster as the reset time, τ_I, decreases, for constant values of K_c. As far as the PID controller is concerned, its closed-loop response has the same qualitative dynamic characteristics as those resulting from PI control alone; however, the introduction of the derivative mode, accounted for by the parameter τ_D, brings a stabilizing effect to the system.

In the following, two different situations were actually considered in analyzing the closed-loop responses of the controlled systems when:

1. the set-point, that is, the permeate flux, $J_{p,SP}$, was kept constant throughout the experiment, except when each pulse occurred (Fig. 16.12). The control system, therefore, was called on to modify the manipulated variable so as to avoid or limit permeate flux decay;
2. the disturbance, that is, the feed concentration, C_0, did not change, whereas the set-point was modified according to a specific pattern (servo problem). Two positive steps and two negative steps occurred, at definite times, as far as $J_{p,SP}$ was concerned (Fig. 16.13).

On developing the present feedback control system it was necessary not only to choose the type of controller, but also to select the best values of its adjustable parameters. Controller tuning methods ensure that the obtained control system would be stable and would meet given objectives. All the methods proposed in the literature to tune a feedback controller were based on the knowledge about the process to be controlled. In the case of

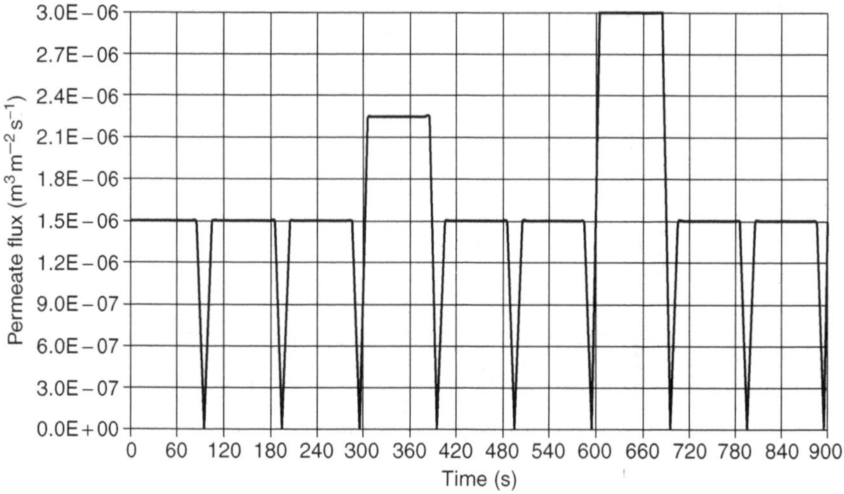

16.13 Case b: servo problem. The disturbance, C_0, does not change, whereas the set-point is modified according to a specific pattern.

BSA ultrafiltration performed in pulsating conditions, the knowledge of the complex process dynamics was definitely precise since it was provided by the previously-described HNM, which demonstrated offering very accurate predictions of the time evolution of permeate flux as a function of the significant process inputs. As a starting point for the selection of the best value for the adjustable parameters of the present control system, the Ziegler–Nichols (Z–N) tuning rule was adopted; then, the parameters were improved by the minimization of the integral of the time-weighted absolute error (ITAE) that allowed the final tuning of both developed feedback controllers to be achieved. The actual shape of the complete closed-loop response, from time t = 0 until a definite time, t_{fin}, considered as representative of system dynamics and chosen, in the present case, as equal to 15 m, was used for the formulation of a dynamic performance criterion based on the minimization of the ITAE, defined as (Stephanopoulos, 1984):

$$\mathrm{ITAE} = \int_0^{t_{\mathrm{fin}}} t\,|\,\varepsilon(t)\,|\,\mathrm{d}t \qquad [16.12]$$

With respect to other similar time-integral performance criteria, the ITAE criterion was chosen since it is able to suppress errors that might persist for long times. The minimization of ITAE was performed by available numerical algorithms implemented within the exploited simulation software.

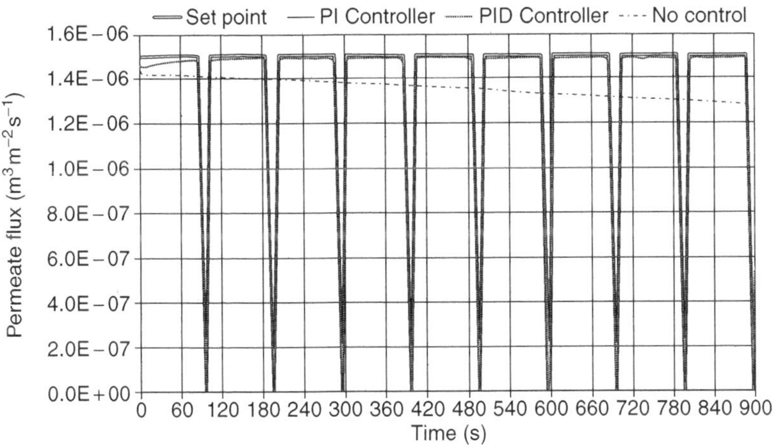

16.14 Closed-loop responses of PI and PID controllers and comparison to the uniform set-point (Case a).

The closed-loop responses of PI and PID controllers are shown in Fig. 16.14 and compared to both the uniform set-point pattern (case a) and to the response of the uncontrolled system. The input variables were set equal to: $Q = 4$ L/min, $t_{pul} = 10$ s, $t_{op} = 90$ s and were never changed further on, even though they represented the actual process inputs. It is worth observing that when no control action occurred, the permeate flux tended to progressively decay; any observable difference between the set-point pattern and the actual permeate flux, instead, tended to zero as soon as either a PI or a PID controller was utilized. Actually, the PI controller exhibited similar responses, which overlapped throughout the considered time horizon, to the PID controller. However, the PI controller was much simpler to operate and to tune than the PID. For this reason, the behavior of the PI controller was analyzed in more detail since it gave high-quality responses and required a lower computational effort to achieve a proper tuning compared to the PID. The effect of reset time, τ_I, on permeate flux decay is shown in Fig. 16.15 in the case of a servo problem. It can be observed that, for a constant value of $K_c = 2.2$, the control action was faster and more effective when τ_I was decreased from 44 to 7.33 s. When a lower value of τ_I is chosen, the actual trend of permeate flux decay reproduces rather well each required change of set-point pattern.

The developed control system exhibited good performance in all the tested conditions; however, it was necessary to perform a more rational choice of the controller parameters so as to obtain improved closed-loop responses of the present membrane system. The Ziegler–Nichols tuning technique was initially used. The value of the proportional gain providing an oscillatory response was equal to 4.508, whereas the oscillation period

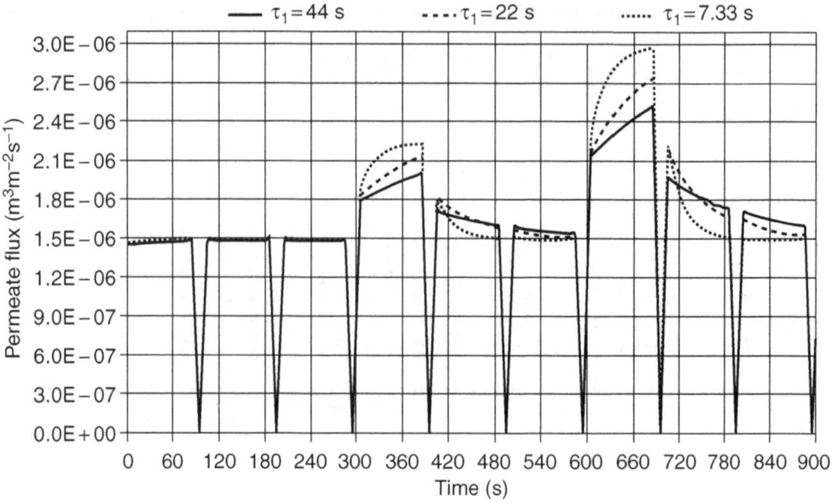

16.15 Effect of the reset time on the performance of PI controller (servo problem).

Table 16.1 Settings of both PI and PID controllers calculated by the Ziegler–Nichols tuning rule

Type of feedback controller	Proportional gain, K_c	Reset time, τ_I	Derivative time constant, τ_D	ITAE
PI	2.049	4.17	–	0.041
PID	2.652	2.5	0.625	0.044

was found equal to 5 s. According to the Z–N tuning rule, the values of the parameters were then determined for both a PI and a PID controller and reported in Table 16.1. The corresponding ITAEs, in the case in which both the controllers were called on to reproduce the set-point change shown in Fig. 16.13, were reported too. Starting from the calculated Z–N parameters and according to the performance criterion based on the minimization of the ITAE, a new set of optimized controller parameters was eventually obtained with reference to the previously-described servo problem. The minimization of the objective function, ITAE, was performed numerically repeating the simulations with different values of the controller parameters, which were changed in an iterative scheme, until a minimum was reached. Table 16.2 summarizes the obtained values of K_c, τ_I and τ_D referred to both a PI and a PID controller, together with the corresponding value of minimum ITAEs. It is worth observing that the controller parameters, as estimated by the present improved technique, were rather different

Table 16.2 Settings of both PI and PID controllers obtained by minimizing the ITAE

Type of feedback controller	Proportional gain, K_c	Reset time, τ_I	Derivative time constant, τ_D	ITAE
PI	2.724	1.911	–	0.0153
PID	2.928	2.272	0.033	0.0169

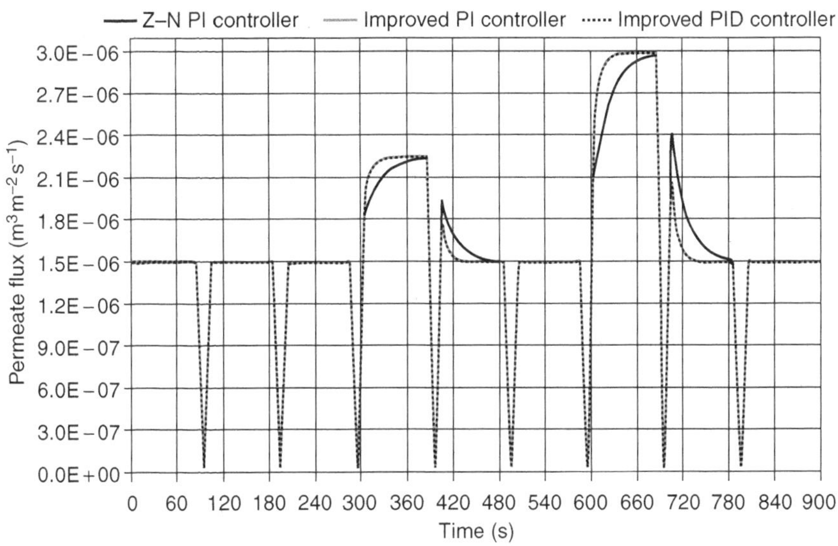

16.16 Comparison of the closed-loop responses obtained using the Z–N tuning rule and the proposed tuning rule for both PI and PID controllers (servo problem).

from those calculated according to the Z–N tuning rule. A considerably lower value of ITAE was also obtained, thus proving the effectiveness of the proposed tuning methodology. Figure 16.16 shows a comparison of the closed-loop responses obtained using either the Ziegler–Nichols tuning rule or the tuning rule based on the minimization of ITAE. It can be observed that the control action was faster and more effective when the improved controller parameters were used, since the actual trends of permeate flux decay reproduced very well each imposed change on the set-point. On the contrary, the performance of the Z–N PI controller was characterized by a much larger $\varepsilon(t)$, especially when the steps on set-point pattern actually occurred.

The methodology proposed in the present section had indeed several advantages, which could be successfully exploited during the implementation and the optimization of a feedback control system. These advantages can be briefly summarized as follows:

- the proposed control system was developed starting from the predictions of a reliable model that exhibited good performance over a wide range of process and operating conditions;
- no simplification, for example, linearization, approximation of the true process dynamics to a first-order plus time delay dynamics, etc., was performed to develop the present control system;
- the minimization of the ITAE allowed tuning the developed feedback controller on the basis of the actual transfer function of the ultrafiltration process, thus optimizing the true closed-loop performance;
- the proposed methodology, based on the formulation of a proper hybrid neural model and on the optimization of a definite performance criterion, was very general as it might be exploited to implement and optimize the performance of feedback controllers operating on different kinds of membrane processes.

16.7 Conclusions

In the present contribution, it has been shown that both ANNs and hybrid models definitely represent powerful computational tools, offering very reliable predictions of the actual behavior of membrane systems. The results obtained demonstrate, in particular, that the proper combination of a theoretical model with a straightforward neural model is able to widen the applicability of pure neural models beyond the training range, thus paving the way for the exploitation of the HNM for process optimization purposes and for the implementation of efficient on-line controllers operating on different kinds of membrane processes.

16.8 References

Agarwal M., Combining neural and conventional paradigms for modelling, prediction and control, *International Journal of Systems Science*, 1997, **28**, 65–81.

Alatiqi I., Ettouney H., El-Dessouky H., Process control in water desalination industry: an overview, *Desalination*, 1999, **126**, 15–32.

Anthony M., Bartlett P. L., *Neural Network Learning: Theoretical Foundations*, 1999, Cambridge University Press, Cambridge UK.

Åström K. J., Wittenmark B., *Adaptive Control*, 1989, Addison Wesley, Reading, MA.

Aydiner C., Demir I., Yildiz E., Modeling of flux decline in crossflow microfiltration using neural networks: the case of phosphate removal, *Journal of Membrane Science*, 2005, **248**(1–2), 53–62.

Brown M., Harris C., *Neurofuzzy adaptive modelling and control*, 1994, Prentice Hall International, UK.

Chellam S., Artificial neural network model for transient crossflow microfiltration of polydispersed suspensions. *Journal of Membrane Science*, 2005, **258**(1–2), 35–42

Çinar Ö., Hasar H., Kinaci C., Modeling of submerged membrane bioreactor treating cheese whey wastewater by artificial neural network, *Journal of Biotechnology*, 2006, **123**(2), 204–209

Curcio S., Calabrò V., Iorio G., Design and tuning of feedback controllers: effects on proteins ultrafiltration process modeled by a hybrid system, *Desalination and water treatment*, 2011, **34**(1–3), 295–303.

Curcio S., Scilingo G., Calabrò V., Iorio G., Ultrafiltration of BSA in pulsating conditions: an artificial neural networks approach, *Journal of Membrane Science*, 2005, **246**(2), 235–247.

Curcio, S., Calabrò, V., Iorio, G., Reduction and control of flux decline in cross-flow membrane processes modeled by artificial neural networks and hybrid systems. *Desalination*, 2009, **236**(1–3), 234–243.

Kartalopoulos, S.V. *Understanding Neural Networks and Fuzzy Logic*, 1996, IEEE Press, New York.

MacKay D. J. C., Bayesian interpolation, *Neural computation*, 1992, **4**(3), 415–447.

McClelland J. L., *Explorations in parallel distributed processing: a handbook of models, programs and exercises*, 2nd edition, 2011, Stanford University, CA.

Minsky M., Papert S., *Perceptrons: an introduction to computational geometry*, 1969, MIT Press, Cambridge, MA.

Murata N., Yoshizawa S., Amari S., Network information criterion determining the number of hidden units for an artificial neural network model, *IEEE transaction on Neural Networks*, 1994, **5**(6), 865–872.

Ou S., Achenie L.E.K., A hybrid neural network model for PEM fuel cells, *Journal of Power Sources*, 2005, **140**(2), 319–330.

Pendashteh A. R., Fakhru'l-Razi A., Chaibakhsh N., Abdullah L. C., Madaeni S. S., Abidin Z. Z., Modeling of membrane bioreactor treating hypersaline oily wastewater by artificial neural network, *Journal of Hazardous Materials*, 2011, **192**(2), 568–575

Psichogios D. D., Ungar L.H., A hybrid neural network-First principle approach to process modeling, *AIChE Journal*, 1992, **38**(10), 1499–1511.

Rai P., Majumdar G. C., Das Gupta S., De S., Modeling the performance of batch ultrafiltration of synthetic fruit juice and mosambi juice using artificial neural network, *Journal of Food Engineering*, 2005, **71**(3), 273–281.

Razavi M.A., Mousavi M., Mortazavi A., Dynamic prediction of milk ultrafiltration performance: a neural network approach, *Chemical Engineering Science*, 2003, **58**, 4185–4195.

Reilly D. L., Cooper L.N., An overview of neural networks: early models to real world systems. In: Zornetzer, S.F.; Davis, J. L.; Lau, C. (Eds), *An Introduction to neural and Electronic Networks*, 1990, Academic Press, New York.

Rojas R., *Neural Networks a systematic introduction*, 1996, Springer-Verlag Eds., Berlin.

Rosenblatt F., *Principles of neurodynamics*, 1961, Spartan Press, Washington D.C.

Roy A., Kim L.S., Mukhopadhyay S., A polynomial time algorithm for the construction and training of a class of multilayer perceptron, *Neural Networks*, 1993, **6**, 535–545.

Ruiz-Bevia F., Gomis-Yagues V., Fernandez-Sempere J., Fernandez-Torres M.J., An improved model with time-dependent adsorption for simulating protein ultrafiltration, *Chemical Engineering Science*, 1997, **52**(14), 2343–2352

Saraceno A., Curcio S., Calabrò V., Iorio G., A hybrid neural approach to model batch fermentation of 'ricotta cheese whey' to ethanol. *Computers and Chemical Engineering*, 2010, **34**(10), 1590–1596.

Smith P.J., Vigneswaran S., Ngo H.H., Ben-Aim R., Nguyen H., A new approach to backwash initiation in membrane systems, *Journal of Membrane Science*, 2006, **278**, 381–389.

Smith P.J., Vigneswaran S., Ngo H.H., Ben-Aim R., Nguyen H., Design of a generic control system for optimising backflush durations in a submerged membrane hybrid reactor, *Journal of Membrane Science*, 2005, **255**, 99–106.

Stephanopoulos G., *Chemical Process Control*, 1984, Prentice Hall, Englewood Cliffs.

Toscano R., A simple robust PI/PID controller design via numerical optimization approach, *Journal of Process Control*, 2005, **15**(1), 81–88.

van Can, H.J.L., Te Braake, H.A.B., Dubbelman, S., Hellinga, C., Luyben, K. C. A. M., Heijnen, J. J., Understanding and Applying the Extrapolation Properties of Serial Grey-Box models, *AIChE Journal*, 1998, **44**(5), 1071–1089.

Wang Z., Massimo C. D., Tham M.T., Morris A.J., A procedure for determining the topology of multilayer feed-forward neural networks, *Neural Network*, 1994, **7**(2), 291–300.

Zhou W., Therdthai N., Artificial Neural Networks in Food Processing, In: M. M. Farid (Ed.), *Mathematical Modeling of Food Processing*, 2010, CRC Press -Taylor & Francis Group, Broken Sound Parkway NW.

16.9 Appendix: nomenclature

16.9.1 Notation

a	gain
$c(t)$	feedback controller output
C	BSA concentration in the membrane module (mol m^{-3})
C_0	BSA concentration fed to the membrane module (mol m^{-3})
C_a	BSA concentration adsorbed on the membrane surface (kg m^{-2})
C_p	BSA concentration in the permeate (mol m^{-3})
D_{AB}	diffusion coefficient of BSA in water (m^2 s^{-1})
J_n	normalized permeate flux (m^3 m^{-2} s^{-1})
$J_{p,SP}$	set-point value of permeate flux (m^3 m^{-2} s^{-1})
J_p	permeate flux (m^3 m^{-2} s^{-1})
$k(t)$	membrane permeability (m^2)
K_c	proportional gain
p	pressure in the membrane module (Pa)
p_1	pressure in the membrane (Pa)
Q	feed flow rate (m^3 s^{-1})
R_{ad}	additional resistance due to proteins adsorption

t	time (s)
TMP	trans-membrane pressure (Pa)
t_{op}	operating time (s)
t_{pul}	pulse time (s)
$\mathbf{u}$	velocity vector in the membrane module (m s^{-1})
$\mathbf{u}_1$	velocity vector in the membrane (m s^{-1})

Greek symbols

∇	nabla operator
$\Delta\Pi$	difference of osmotic pressure (Pa)
α, β, γ	virial coefficients defining the difference of osmotic pressure
$\varepsilon(t)$	deviation between the set-point value and the actual permeate flux
η	viscosity of BSA solution (Pa s)
η_1	permeate viscosity (Pa s)
ρ	density of BSA solution (kg m^{-3})
ρ_1	permeate density (kg m^{-3})
τ_D	derivative time constant (s)
τ_I	reset time (s)

Subscripts and superscripts

T	transpose
w	evaluated at the membrane wall

17

Assessment of the key properties of materials used in membrane reactors by quantum computational approaches

G. DE LUCA, ITM-CNR, Italy

DOI: 10.1533/9780857097330.3.598

Abstract: The evaluation of key properties of materials using quantum mechanics (QM) methods is the aim of this chapter. The use of QM is necessary to calculate properties that depend on electron interactions or electron density polarization. Following the Introduction, which covers computational chemistry notions, some basic concepts concerning the Density Functional Theory (DFT) used in the presented calculations are illustrated, in addition to a brief review of intermolecular interactions. The chapter then reviews the assessment of some fundamental quantities, such as the adsorption energies of gases and hydrogen in nano-porous materials and on metallic surfaces, respectively. Finally, the calculation of hydrogen solubilization in metal alloys will be also presented.

Key words: Density Functional Theory (DFT), adsorption energies, silica nanotube, Nb–Mo alloys, hydrogen adsorption, solubilization.

17.1 Introduction

Computational approaches, based on accurate QM methodologies, generally require significant computational resources. However, the development of the high performance computing service (parallel supercomputers) and efficient computational algorithms have allowed QM calculations to be carried out that would have been unthinkable few years ago. This opportunity opens up new possibilities for research because QM calculations enable the evaluation of chemical–physical quantities that could not have been estimated by experimental tests, or only with remarkable inaccuracy. Furthermore, multi-scale simulations can now be developed, starting with the valuation of molecular and atomic features, by using QM, which must be used to evaluate properties related to the interactions involving electrons or electron density polarization (see Fig. 17.1). The quantities obtained at QM level can be used as inputs in subsequent molecular dynamics, Monte-Carlo (MC) or semi-empirical calculations to describe properties related to a huge number of molecules, atoms or in general elementary particles. In turn, the information provided by molecular dynamics and MC simulations can be utilized in meso-scale calculations (coarse grain MC or Lattice

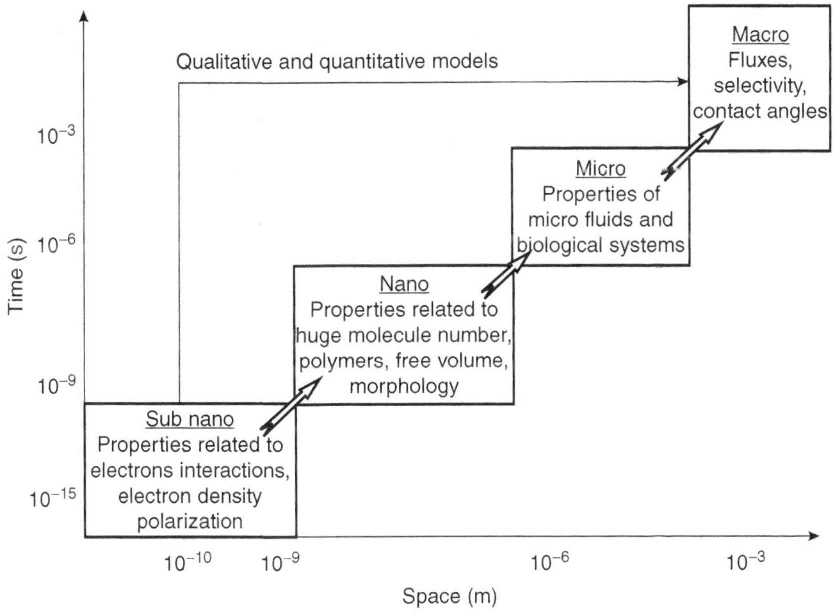

17.1 Properties and corresponding time-space scale.

Boltzmann) to reproduce properties in microscale (Fig. 17.1). Finally, the description of the macro properties can be carried out using information coming from the lower scales. It is interesting to note that the starting point for the highlighted multi-scale computational procedure is the QM calculations. Nevertheless, the QM results can be used 'directly' as inputs in qualitative or quantitative analytical mathematical models, which give the average macroscopic properties as shown in Fig. 17.1. Both computational procedures allow, at least theoretically, eliminating empirical values or empirical models obtained by the fitting of experimental data. However, using computational procedures starting from QM calculations, particular attention should be paid to the molecular models (structural chemical models). Molecular models are dependent on the properties to be assessed and consequently by the computational procedure. For example, QM methods are not applicable to the geometric optimization of macromolecules such as polymers or systems containing thousands of atoms (such as biological systems). MD methodologies or coarse grain MC can be used in these cases. However, the MD approaches cannot be used to study systems in which the breaking or/and formation of bonds occurs, such as in some phenomena occurring in membrane reactors. Thus, the choice of molecular model is crucial if QM approaches must be used. Even if quantum methodologies based on the Density Functional Theory (DFT) need computational efforts significantly lower than post-Hartree-Fock procedures, nevertheless large chemical models cannot be used currently. These structural models inevitably lead to the neglect of some aspects, though in some cases they may be irrelevant if the choice of molecular model is done correctly.

The last consideration concerns the study of catalysis in conjunction with permeability analysis. Since in catalysis the breaking and formation of bonds are involved, then QM approaches permit the description of the properties of catalysts appropriately, by means of the characterization of the transition states along the reaction paths. Generally, in membrane reactors (MRs), the catalysts can be found as solid surfaces or as nano-clusters. These latter can be deposited on membrane surfaces or dispersed in the membranes. Moreover, membranes can surround pellets of catalyst located in the lumen of the reactors. Other configurations can be found, but their description is obviously not the purpose of this chapter. As outlined above, QM calculations require the definition of a structural model of the catalysts which, for the most part and in any type of configuration, are modelled by infinite surfaces, slabs or by different types of atomic clusters. The membrane, that is the interface, is indirectly taken into account in all these structural models. A huge literature exists about QM studies on catalysis (periodic surfaces or clusters, etc.) (Thomas, 1989; Ertl, 1991) and many aspects, such as role of defect, nanostructured surfaces and cluster sizes have been investigated (Abbet *et al.*, 2000; Judai *et al.*, 2005). Probably, QM studies on nano-clusters, embedded in particular new materials as nanotubes, could still be a novel field of investigation. In addition, specific computational procedures, specifically tailored to carry out accurate QM studies on heterogeneous catalysts in the form of nanostructured surfaces and clusters have been widely developed. Briefly, these procedures can be divided into three main categories: 'single cluster' (van Santen and Neurock, 2001), 'embedded cluster' (Sierka and Sauer, 2000) and periodic methods (Dong *et al.*, 1998). Readers can easily find a wide range of literature – papers and several books – on this topic. The considerations expressed are useful to emphasize the fact that QM studies on catalysis in MRs could be treated independently from membrane permeability investigation. However, it is important to emphasize that the merging of the results obtained by these different investigations has to be done. For example, the kinetic constants characterizing a reaction path, related to a particular catalyst and evaluated by the QM approach, should be compared with the diffusion coefficients characterizing the permeability of the reagents and products through the membrane. Thus, a complete theoretical modelling must take into account both aspects.

Since QM studies on catalysis are numerous, this chapter will show how some QM results can be directly used to obtain information on macro-scale properties, following the scheme presented in Fig. 17.1 (Qualitative and quantitative mathematical models). In particular, the chapter is organized as follows. After this Introduction, in Section 17.2, some theoretical concepts are presented to better understand the results shown in the following sections. The description of the theoretical approaches, related to important issues, is given in Sections 17.3 and 17.4. The QM calculations of the molecular properties involved in the selected theoretical models are shown in the same sections. Finally, conclusions and future trends are presented in the last section.

17.2 Basic concepts of computational approaches

As stressed in the Introduction, when molecular and supra-molecular properties have to be calculated, the computational approach and structural models must be set. The choice of computational procedure strictly depends on the target property. For example, binding energies, molecular electrostatic properties, such as dipole, partial charges and electron density distribution, as well as orbital configurations must be evaluated by using QM approaches. The structural models not only depend on the target property but also on the computational time required to derive these properties. Thus, large structural models cannot be used in conjunction with accurate QM approaches, such as correlated Hartree-Fock (Szabo and Ostlund, 1994), due to the huge computational time required by these methods. DFT (Parr and Yang, 1989) instead allows obtaining different electronic properties in a relatively short computational time with respect to the mentioned methods. Thus, this theoretical approach can be considered a powerful tool to investigate large chemical systems.

17.2.1 The Density Functional Theory (DFT)

To better understand the results described in the following sections, the fundamental equations of this relative new quantum theory for a closed shell system are summarized below (Parr and Yang, 1989). The total energy of a electronic system is evaluated through the functional defined in the Equation [17.1].

$$E_{\mathrm{DFT}}\left[\rho(r)\right]=T_s\left[\rho(r)\right]+J\left[\rho(r)\right]+E_{\mathrm{xc}}\left[\rho(r)\right]+\int v(r)\rho(r)\mathrm{d}r \qquad [17.1]$$

This functional depends on the electron density of the quantum system, $\rho(r)$. In particular, the kinetic energy is described by the functional, $T_s[\rho(r)]$, related to an isoelectronic non-interacting system and $J[\rho(r)]$ describes the Coulomb electrostatic energy. $E_{xc}[\rho(r)]$ is the exchange and correlation functional, which takes into account the difference between the true kinetic energy and the kinetic energy of the isoelectronic non-interacting system; in addition to the difference between the true electron–electron and the Coulomb electrostatic energy ($E_{xc}[\rho(r)] = T[\rho(r)] - T_s[\rho(r)] + V_{ee}[\rho(r)] - J[\rho(r)]$). The sum of the functionals: $T_s[\rho(r)]$, $J[\rho(r)]$ and $E_{xc}[\rho(r)]$ defines the universal functional, $F_{HK}[\rho(r)]$, which depends only on the total number of electrons (N) of the system, since the last functional in Equation [17.1], that is $\int v(r)\rho(r)\mathrm{d}r$, is the energy due to the electrons–nuclei interaction. Although the exact form of $F_{HK}[\rho(r)]$ is not yet known, several approximations of this functional have been published. In particular, the approximations of the $F_{HK}[\rho(r)]$ are expressed as a function of the exchange and correlation functional, $E_{xc}[\rho]$ (Becke, 1993; Perdew, *et al.*, 1997; Xu and Goddard, 2004). In general, the external potential $v(r)$ in the functional $\int v(r)\rho(r)\mathrm{d}r$, takes into account the interactions among electrons and nuclei of molecule. It depends on the type and geometrical disposition of the nuclei, considered in fixed positions (Born-Oppenheimer approximation). A specific arrangement of the nuclei defines a particular potential $v(r)$. Nevertheless, it is worth noting that a particular arrangement does not necessarily provide the lowest total energy of the molecular system. Thus, the optimization of the molecular geometry is necessary to find the nuclei arrangement corresponding to the lowest molecular total energy, that is, the minimum. However, in addition to a global minimum, several molecular structures may have different local minima (conformational isomers). The research of all the conformational isomers may require high computational time. In this case, the combination of QM methods with other computational approaches is advised. The research of local and global minima is the necessary and first step to obtaining the chemical and physical properties of the system.

The external potential $v(r)$ can be substituted by an effective potential that replaces the core electrons of the system with an effective electrostatic potential (effective core potential, ECP); in this case only the electrons of the external orbitals are explicitly considered in the calculation (Pacios and Christiansen, 1985). This procedure decreases the computational time and allows to take into account relativistic effects, which are important when transition metals are present in the molecule. Finally, the electron density, $\rho(r)$, in the functional [17.1] is defined through a set of basic functions (atomic orbital basis set, $\phi_j(r)$) (Szabo and Ostlund, 1994) according to the linear combination reported in Equation [17.2]:

$$\rho(r) = \sum_i^N |\psi_i(r)|^2 \qquad [17.2]$$

where:

$$\psi_i(r) = \sum_{\substack{1 \le i \le N \\ i \le j \le K}} c_{i,j}\, \varphi_j(r)$$

A large basis set is necessary to obtain high accuracy and convergence of the non-linear Khon-Sham Equations [17.3].

$$\left[-\frac{1}{2}\nabla_i^2 + V_{\text{eff}}(r)\right]\psi_i(r) = \varepsilon_i \psi_i(r) \qquad [17.3]$$

in which the effective potential $V_{\text{eff}}(r)$ is defined as follows:

$$V_{\text{eff}}(r) = V(r)\frac{\partial J[\rho(r)]}{\partial \rho(r)} + \frac{\partial E_{xc}[\rho(r)]}{\partial \rho(r)} = v(r) + \int \frac{\rho(r')}{|r - r'|}\mathrm{d}r' + v_{xc}(r) \qquad [17.4]$$

The Khon-Sham equations allow the coefficients $c_{i,j}$ of the above linear combination to be obtained, therefore the electron density, $\rho(r)$ which as mentioned before allows the calculation of all the properties of the quantum system. The variational principle ensures that the minimum of the functional $E_{\text{DFT}}\,[\rho(r)]$, in the space of $\phi(r)$, is the best approximation of the ground state energy. Thus, the Khon-Sham equations are obtained by minimizing the energy functional Equation [17.1] (Parr and Yang, 1989). The Coulomb, exchange and correlation potentials, which are the last two terms in the relation Equation [17.4], can be evaluated by a grid of points or expanding these potentials using another set of functions, called the 'auxiliary' basis set. The large atomic orbital basis makes the calculations very time-consuming. Nevertheless, the calculation is faster if 'auxiliary' basis set is used instead of a grid of points to evaluate the Coulomb, exchange and correlation potentials. This brief overview of DFT highlights some critical issues that significantly affect the accuracy and the computational time of a density functional calculation:

- the choice of functional $E_{xc}(\rho)$;
- the definition of an ECP;
- the choice of an appropriate orbital basis set;
- the use of an appropriate grid or 'auxiliary' basis set to evaluate the Coulomb, exchange and correlation potentials.

All these points should be taken into account during a calculation to optimize both the accuracy and computational effort.

17.2.2 Computational approaches for the evaluation of non-covalent interactions

Some basic considerations on non-covalent bonds (supra-molecular chemistry) are given in this section. It is worth noting that non-covalent interactions (Rudkevich, 2004) control many phenomena that occur at the interfaces. Thus, hydrogen bonding and electrostatic interactions, viz. ion–dipole, ion–induced dipole, dipole–induced dipole and dispersion forces (London interactions) appear of crucial importance for the interpretation of many phenomena involving membranes. For example, hydrogen bonding and dipolar interactions are determinant in the absorption and adsorption of polar molecules, such as H_2O, CO_2, CO and apolar gas (N_2, O_2, H_2) in polymeric membranes (De Luca *et al.*, 2009b), and in porous inorganic materials such as zeolites and nanotubes (De Luca *et al.*, 2005). Furthermore, non-covalent interactions control the adsorption of gas molecules such as N_2, O_2 and H_2 on metallic surfaces and the adhesion between different polymers (blends). Although several studies within the last three decades on supra-molecular chemistry have led to significant successes, theoretical investigations concerning the influence on the membrane properties of the hydrogen bonds, van der Waals interactions (dipolar interactions) and dispersion force definitely deserve attention. In particular, non-covalent binding energies, bond lengths and angles, as well as the charge distribution on the atoms involved in the intermolecular interactions, are important properties to be assessed. So, a brief explanation of the calculation of binding energies related to non-covalent interactions (key property) will be given.

The calculation of the binding energies can be achieved through two main strategies: 'indirect' and 'direct'. Following the 'indirect' approach, the electrostatic features, such as dipole, quadruple moments, polarizability and charge distribution of the fragments forming the intermolecular adduct, are calculated by QM, and then these properties are used in classical equations describing the non-covalent interaction between the different fragments. Conversely, in the 'direct' approach, the entire supra-molecular adduct is considered as a single quantum system and the energy of its ground state is evaluated. Hence, the binding energy is calculated as the difference between the energy of the supra-molecular adduct and the sum of the energies of its single fragments. Once the electron density, $\rho(r)$, is obtained, the energy of the ground state of the complex can be calculated, in addition to other chemical–physical properties of the supra-molecule. Although the 'direct' method takes into account aspects that the 'indirect' procedure does not, the computational effort correlated with the 'direct' calculation may be a problem. To better understand the two computational approaches just described, two case studies are illustrated below.

The adhesion between two polymers is a problem of considerable interest in the membrane field. For example, a polymer can be mixed with another polymer in a common solvent in order to obtain a blend with novel properties. In this context, the choice of polymers to be used is very important. The non-covalent binding between the polymer chains is an important factor controlling the adhesion between the polymers, in addition to their ability to randomly intertwine due to polymeric flexibility. Thus, calculations, in the framework of DFT, can provide useful information concerning polymer–polymer non-covalent interactions in order to make an appropriate choice of polymers. As pointed out in the Introduction, structural models (polymeric fragments) of the target polymers have to be defined. Afterwards, optimization of their geometry must be performed. If various fragments of two polymers are considered, their optimizations cannot be carried out using QM methods since, for each polymer fragment, several conformations can exist. In this case, a multi-computational approach is required. Semi-empirical Hamiltonian can be used for the conformation analysis of each polymeric fragment. Then, for each polymer, the dipole moments of the monomer belonging to each conformers of a polymeric fragment will be evaluated at QM level. The final dipole moment for a specific monomer is therefore obtained as the weighted average of the dipoles of the monomer in the various conformations in the polymeric fragments considered. Thus, this average dipole is connected to the relative stability of each monomer in the corresponding polymeric chain (fragment). Finally, the comparison of the average dipole moments of different polymers permits us to make the appropriate choice of polymers (IMETI, 2010). The presented methodology has been successfully used to describe, qualitatively, the relative affinity of five polymers in order to find the best combination of two polymers (IMETI, 2010). It is important to note that this method does not consider the possibility that non-covalent interactions occur between specific chemical groups located on the polymeric chains, for example by means of hydrogen bonding. In this case, the 'direct' computational approach should be used.

To better illustrate the 'direct' computational approach and also the importance of hydrogen bonding hereafter, a case study is shown where the role of this bond is crucial. De Luca *et al.* (2009b) analysed the hydrogen-bonding interactions involved between an amphiphilic modifier (*guest*, N-ethyl-o,p-toluenesulfonamide) and polar functional groups of a block (ether/amide) polymer (*host*). In this system, the interactions occur among specific chemical groups of the *guest* and *host* via hydrogen bonding. As a consequence, supra-molecular adducts formed by *guest–host* models were defined. Then the 'direct' computational procedure reported above has been adopted. The calculations, both in vacuum and in solvent, have shown that the bonding energies *of host–host* and *guest–host* hydrogen bonds are similar (De Luca *et al.*, 2009b). Thus, when the *guest* is dissolved with the *host* in a common solvent, the *guest* prevents the formation of several hydrogen bonds among the polymer chains leaving polar functional groups free. This increases the concentration of the free amide and ester groups of the modified polymer with respect to the unmodified one. The macroscopic result was: the enhancement of the polymer hydrophilicity; the simultaneous reduction of the packing density of the polymer chains; and the increase of sorption of polar and polarizable molecules in the membrane. Increasing the concentration of the *guest* molecules, these bind with the free amide and ester functional groups of the polymer and this leads to a decrease of the membrane hydrophilicity and sorption of polar and polarizable penetrants. This study showed that the sorption of water, into modified polymer matrices having polar groups, is not linear with respect to the concentration of the modifiers. The mechanism, carried out in this work (De Luca *et al.*, 2009b), namely the competition between *host–host* and *guest–host* hydrogen bonding, can be generalized for other types of small *guests* dispersed in different polymeric media. In particular, the *guests* could be residue molecules of solvent used for the production of the membranes and entrapped in them (De Luca *et al.*, 2009a, p.60). Moreover, the proposed mechanism could be used to interpret the swelling and fouling of polymeric membranes. Finally, it can also be used to understand the increase of affinity to specific molecules of polymeric imprinted membranes prepared by means of a non-covalent molecular procedure (De Luca *et al.*, 2011). Summarizing, hydrogen-bonding, electrostatic (dipolar) and dispersion (London) interactions control several membrane macroscopic properties. In the following section, insights into the role of these non-covalent interactions in the adsorption of gases in inorganic porous materials will be provided.

17.3 Gas adsorption in porous nanostructured materials

Diffusion regimes describing transport in porous materials for which well-accepted diffusivity formulae exist can be briefly summarized into four types: gaseous diffusion, Knudsen diffusion, diffusion in liquid and atomic diffusion in solids (Xiao and Wei, 1992). Gaseous diffusion occurs when the ratio of the molecular mean free path to the channel sizes is much less than 1, whereas Knudsen diffusion arises when this ratio is equal or greater than 1. Instead, thermal vibrations and free volume govern the molecular diffusion in liquids and polymers. Finally, activated jumps control the transport of molecules and atoms in solid materials. In porous nanostructured materials, particular attention has to be paid to another transport mechanism, namely configurational diffusion (Xiao and Wei, 1992). Configurational diffusion becomes significant when pore sizes becomes comparable to molecular size.

For example, in zeolite pores, some gas molecules may not be as free as in Knudsen diffusion regimes in which the effect of the solid surface potential is negligible. In the configurational diffusion regime, molecules or atoms always feel the potential field of the pores but they are not covalently bonded as in the interstitial sites of a solid solution (see Section 17.4). The molecules vibrate with their lattice (the host), accumulating enough energy to jump to neighbouring sites. When the molecular concentration increases, interactions among molecules inside the nano-porous material should be taken into account. Membranes, formed by meso-porous unmodified zeolites with pore diameter of 4 nm, perform most gas separations according to Knudsen diffusion. However, the selectivity is usually not very high for various gas separations and it provides limited conversion enhancement in catalytic membrane applications (Burggraaf and Cot, 1996). Configurational diffusion, instead, yields economically interesting separation factors although this mechanism is limited to easily condensable gases and rather low pressure drops (Burggraaf and Cot, 1996; Bakker *et al.*, 1997). Therefore, for particular gas separations, configurational diffusion is preferred. In these cases, meso-porous membranes are replaced by different types of nano-porous membranes with pore diameters less than 2 nm, such as silica nanotubes, microporous zeolites and carbon nanotubes embedded in polymer matrices. It is worth noting that when nano-porous zeolites are used in nanostructured materials, the correct terminology is microporous materials (Kärger and Ruthven, 1992).

Krishna (2000), starting from the fundamental concepts of Maxwell J. C. and Stefan J., describes the configuration diffusion, also called surface diffusion, within zeolite crystals by means of the following set of equations, namely the generalized MS approach (Kapteijn *et al.*, 2000; Krishna, 2000). Nevertheless, this set of equations can also be used to describe diffusion in other nanostructured materials such as carbon nanotubes:

$$-\rho_z \frac{\theta_i}{RT} \nabla \Phi_i = \left[\sum_{j \neq i}^{n} \frac{1}{D_{ij}^{\mathrm{MS}}} \frac{C_j J_i - C_i J_j}{C_i^{\mathrm{sat}} C_j^{\mathrm{sat}}} + \frac{1}{D_i^{\mathrm{MS}}} \frac{J_i}{C_i^{\mathrm{sat}}} \right] \quad i = 1, 2, 3, \ldots, n \qquad [17.5]$$

where ρ_z is the density of the zeolite, Φ_i is the chemical potential of the gas i, which is the force acting on the species i that moves along the inner channels of the zeolite, while J_i are the macroscopic fluxes. θ_i is the fractional surface occupancy or coverage of the adsorbed molecule. It is defined as $\theta_i = C_i / C_i^{\mathrm{sat}}$ where C_i is the molecular coverage expressed in molecules of adsorbed gas per unit cell of zeolite, whereas C_i^{sat} is the saturation coverage of the species i. The diffusion coefficient D_i^{MS} appearing in the Equation [17.5] takes into account the interaction between the species i and the surface of the nanopores; instead the coefficient D_{ij}^{MS} reflects the sorbate–sorbate interactions, which is the friction exerted by the species j on the motion of the species i inside the pore. The D_i^{MS} and D_{ij}^{MS} coefficients are called the Maxwell–Stefan surface diffusivities. In particular, D_i^{MS} is also called corrected diffusivity (Kärger and Ruthven, 1992). The Maxwell and Stefan formulation for a single component (single gas) can be obtained by setting $n = 1$ in Equation [17.5]. Therefore, the gas flux can be obtained from Equation [17.5] assuming equilibrium between the adsorbed species and the bulk phase equations:

$$J_1 = -\rho_Z C_1^{sat} D_1^{MS} \Gamma_1 \nabla \theta_1 \qquad [17.6]$$

Using Equation [17.6], it is easy to verify that the familiar transport or phenomenological Fick coefficient can be obtained as the product of the corrected diffusivity (MS diffusivity) and the thermodynamic factor Γ, defined as $\Gamma = \theta_1 \frac{\partial \ln p_1}{\partial \theta_1}$,where p_1 equals the pressure of the considered gas:

$$D^{Fick} = D_1^{MS}(\theta_1)\Gamma \qquad [17.7]$$

The evaluation of the thermodynamic factor, Γ, requires the definition of a relationship between the pressure of the gas and its surface occupancy, θ_1, that is the isotherm. Generally, the Langmuir isotherm (mono- or multi-sites) is the most common function used (Burggraaf and Cot, 1996). The Langmuir isotherm is characterized by the adsorption parameter, b, and its pre-exponential parameter, b_o. In turn, b is defined as function of the isosteric heat or *adsorption enthalpy* at zero coverage of the target gas. Several studies showed that the corrected diffusivity can exhibit different types (at least five) of coverage dependencies. Krishna (2000) expressed this dependence through the following general relation:

$$D_1^{MS}(\theta_1) = D_1^{MS}(0) f(\theta_1) \qquad [17.8]$$

The term $D_1^{MS}(0)$ is the MS diffusivity in the limit of zero coverage, namely self-diffusivity. It only depends on the temperature and on the *activation energy* for the molecular diffusion inside the nanopores. The *activation energy* for diffusion is the energy necessary in order for the molecule to jump from one site to another one. At zero coverage limit, all three diffusivities, corrected, Fick and self-diffusivity, are equal. Finally, the macroscopic flux of a single gas, J_1, through a nano-porous material, for example silica nanotubes, can be obtained by the Equation [17.6], once defined the isotherm, the *activation energy* for diffusion and $f(\theta_1)$ function. Using the mathematical analytical model (Equation 17.6), only four molecular quantities must be calculated to achieve the flux J_1: the *adsorption enthalpy*, the *diffusional activation energy* in addition to the pre-exponential parameters appearing in the Langmuir parameter and self-diffusivity expression (Eyring equation). If the ratio of the effective diameter of gas molecule to the channel size of some zeolites is equal to 3 the potential curve, generally, shows a maximum in the middle of the nanopore (Burggraaf and Cot, 1996; Burggraaf, 1999). In this case, the *adsorption enthalpy* and the *diffusional activation energy* are similar, although they have different physical meaning (numerically almost equal). Furthermore, when the values of the *diffusional activation energy* of two gases are similar (for example near to 1) then the flux, J_1, is controlled by molecular adsorption along the nanopores. Thus, in these cases, the evaluation of the *adsorption enthalpy* would permit information on the macroscopic fluxes or on the ideal selectivity to be obtained (Fig. 17.1).

Finally, an example is presented below, showing how QM can be used to obtain the aforementioned *adsorption enthalpy* (De Luca *et al.*, 2001, 2002, 2005). Although the Equation [17.6] deals with diffusion of gases through three-dimensional zeolite pores, the example reported concerns the diffusion in a single nanotube of silica; nevertheless, the physical meaning of the adsorption energies is the same. Nanotubes of silica or titania are prepared by templating onto porous membranes that are commercially available as filters. These nanotubes of inorganic materials can be prepared through sol–gel chemistry using various membranes as templates, such as porous alumina membrane. Functional groups, anchored in the silica nanotubes, affect the transport through the membranes, walls decked with long hydrocarbon chains preferentially transport hydrophobic molecules, whereas those decorated with short polar groups behave in just the opposite way. Several silane groups have been anchored on the walls of silica nanotubules (Kshama *et al.*, 1997).

The hydroxyl groups of silica easily react with the silicon atoms of the silanes, while the end of these compounds can be used for chemical selectivity. These molecules offer the possibility of functionalizing the silica and also the carbon nanotubes easily for a wide variety of applications. Here, the functionalization of the border of a silica nanotubule (10-membered oxygen ring) is presented. The opening of the silica nanotube was defined using the bond distances (Si–Si, Si–O) and angles (Si–O–Si, Si–Si–Si) of the Silicalite-1 zeolite (Fig. 17.2a), while the number of Si atoms and dihedral angles was chosen to obtain a pore diameter almost equal to 1 nm. In

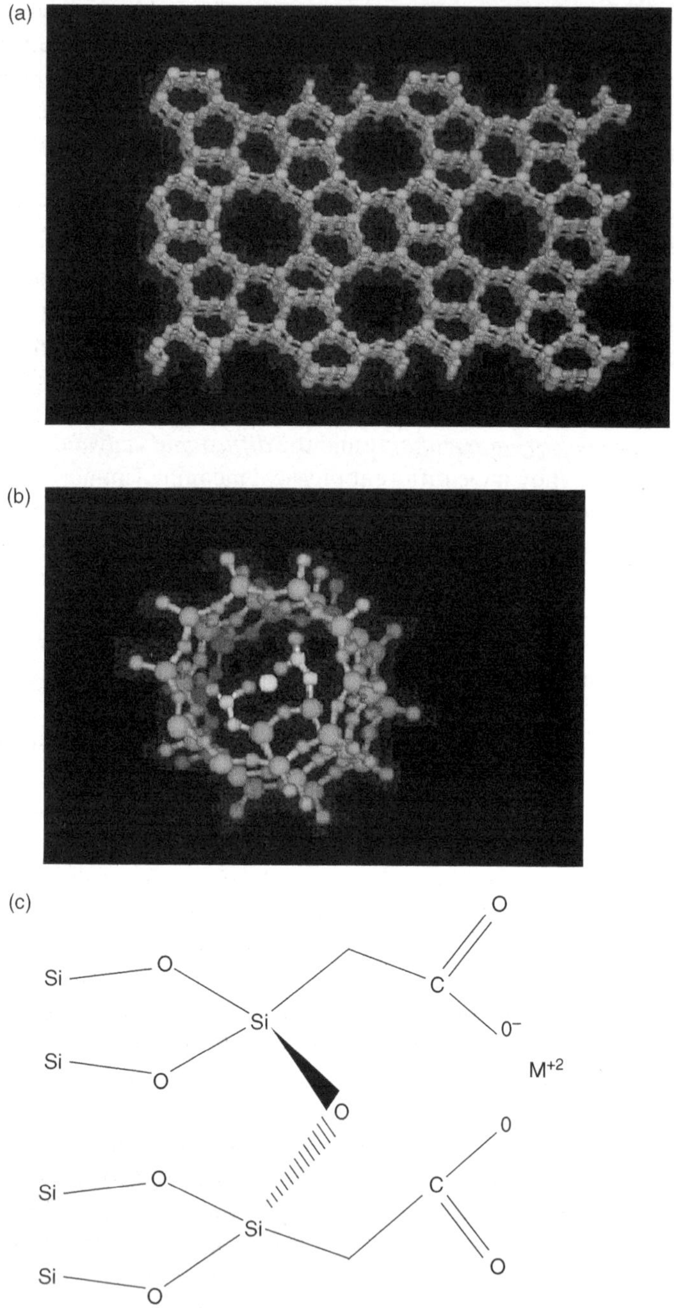

17.2 Chemical structure of the Silicalite-1 zeolite (a), schematic representation of a functionalized silica nanotube (b), chemical formula of silane functional group used in the silica nanotube (c).

Fig. 17.2a, the three-dimensional structure of the Silicalite-1 is reported in addition to the structural model of the functionalized silica nanotube. This model is used to calculate the aforementioned adsorption energies. The silane group, anchored as shown in Fig. 17.2b, ends with carboxylic groups ($^{-}OOCCH_2$–SiOSi–CH_2COO^{-}) that coordinate a specific cation, Ca^{+2} (Fig. 17.2b). The -CH_2–SiOSi–CH_2- group is necessary to connect the carboxylic moiety to the nanotube border (Fig. 17.2b). The number of -CH2- groups (Fig. 17.2c) was calibrated with respect to the diameter of the silica nanotube opening. Summarizing, the model system is composed of a silica ring with four hydroxyl groups directly bonded with a silane group that ends with carboxylic groups coordinating a calcium cation.

The adsorption energies for the CO, N_2, O_2, H_2 and CO_2 molecules were calculated by using two different quantum chemical approaches: Hartree-Fock and DFT. In the frame of DFT, the calculations were performed using the B3LYP (Becke, 1993) as exchange and correlation functional in conjunction with Gaussian type orbitals double-ζ basis set plus a polarization function (6-31G*). The two theoretical approaches have shown that the adsorption energy of the CO molecule is almost one order of magnitude larger than the H_2 one, whereas the heat of adsorption of N_2 and O_2 are similar and slightly lower compared to the CO adsorption energy. This result suggests that this nanostructure could be used to obtain a good separation of CO, N_2 and O_2 from H_2. However, the adsorption energy of CO_2 is similar to those of nitrogen and oxygen. Based on these considerations, the separation of CO, N_2 and O_2 would be difficult using this kind of inorganic nanostructure, since the *adsorption enthalpies* of these gases are similar. For an effective separation of these gases, the substitution of the Ca^{+2} cation with another cation should be investigated. The QM calculations show that the design of new silane groups is a promising strategy for achieving efficient separations of CO or CO_2 from gas mixtures containing hydrogen using very cheap materials. Studies on the functionalization of carbon nanotube with novel silane groups are in progress.

17.4 Adsorption and absorption of hydrogen and small gases

17.4.1 Adsorption of hydrogen and inhibitor gases on metallic surfaces

Dense palladium-alloy membranes are generally used for the production of ultrapure hydrogen, although their high cost prevents their widespread use. However, their properties, such as infinite selectivity to hydrogen, chemical and thermal resistance, allow the use of the Pd-based membranes in MRs. Theoretical studies on the adsorption of gases on palladium-alloy surfaces are a stimulating topic that could be developed reliably by quantum computational approaches. As with the previous example, in this section first a relationship among molecular observables and a macroscopic quantity is illustrated. The decrease of hydrogen permeation through Pd-alloy membranes is a serious problem because it affects the membranes' efficiency (Eriksson and Ekedahl., 1998; Amandusson *et al.*, 2000; Mitsui *et al.*, 2002). In general, this decrease can be due to the dilution and/or inhibition of hydrogen, in addition to the polarization effect due to the concentration gradient of the permeating molecules near the membrane surface (Eriksson and Ekedahl, 1998). The H_2 dilution in the feed causes the decrease of its partial pressure involving the reduction of the H_2 flux according to the Richardson equation. Since the inhibition of the H_2 is owing to reversible interactions between the non-permeating gas, mixed with H_2, and the Pd-alloy surfaces, it can be investigated reliably by using QM approaches. The adsorption energy of inert gases, constituting feed mixtures, is small, therefore in this case the decrease of H_2 flux is principally caused by the dilution of hydrogen. Instead, inhibitor gases, such as CO and H_2O that strongly adsorb on the metallic surface, produce both dilution and inhibition effects (Eriksson and Ekedahl, 1998; Li *et al.*, 2001).

The inhibition of hydrogen permeation caused by the adsorption of water, oxygen or carbon monoxide on Pd-alloys has been widely analysed (Amano *et al.*, 1990; Amandusson *et al.*, 2000; Li *et al*, 2001; Gielens *et al.*, 2006). Here, the inhibition owing to carbon monoxide is taken into account, however the presented analysis can also be extended to other inhibitor gases that do not react with the surface, as in the case of O_2 or H_2S. At molecular level, carbon monoxide inhibition can occur through this general mechanism (Eriksson and Ekedahl, 1998; Amandusson *et al.*, 2000). The molecules of CO are adsorbed on the metal surface sites on which hydrogen atoms are also adsorbed. In this way, the CO directly decreases H_2 adsorption reducing the membrane surface available for the hydrogen permeation. However, owing to the electron repulsion among the neighbouring molecules of CO, the adsorption of carbon monoxide does not occur on all the possible adsorption sites. Thus, among the CO molecules, free adsorption sites for hydrogen permeation can exist. Furthermore, the adsorbed CO molecules also change the path of the H_2 dissociation reaction, occurring before the atomic diffusion in the solid bulk. Thus, reduced hydrogen permeation will take place even after the complete saturation of the Pd-alloy surface (complete coverage, $\theta = 1$).

Summarizing, two different mechanisms are proposed to explain the inhibition effect of CO molecules: the direct blocking of the adsorption sites (reduction of the membrane surface for the H_2 permeation) and the increase of the transition state energy controlling the hydrogen dissociation, a preliminary step for the subsequent atom diffusion in the Pd-alloy (Eriksson and Ekedahl, 1998). Really, these two mechanisms should be considered separately only in two extreme cases: the site blocking in low surface coverage ($0 < \theta < \theta_{thr}$), and the increase of the transition state energy for the H_2 dissociation in the saturation limit ($\theta = 1$). For θ values higher than the threshold value, θ_{thr}, the two mechanisms should be coupled. With regard to this aspect, the slope of the Richardson flux equation with respect to the H_2 partial pressure difference between feed and permeate can be scaled by a factor, $g(T, p_{co})$, which takes into account the inhibition effect for different values of θ:

$$P_{H_2} = \left[P^0_{H_2} e^{-\frac{E_a}{kT}} \right] g(T, p_{co}) \qquad [17.9]$$

The temperature dependence of the hydrogen permeability can be expressed by the well-known Arrhenius expression: $\left[P^0_{H_2} \exp\left(-\frac{E_a}{kT} \right) \right]$, formally similar to the self-diffusivity coefficient mentioned in Section 15.3. However, in this case, the activation energy for diffusion, E_a, is referred to the jumps of the hydrogen atoms in the bulk of the Pd-alloy (the rate-determining step is the H diffusion according to the Richardson equation). Considering that θ depends on the partial pressure of the carbon monoxide, in addition to the temperature of the mixture, that is $\theta(T, p_{co})$, the scaling factor $g(T, p_{co})$ can be expressed as function of the surface coverage as follows:

$$g(\theta) = \begin{cases} (1-\theta), & 0<\theta\leq\theta_{thr} \\ \left[\omega(\theta)+(1-\theta)\right], & \theta_{thr}<\theta\leq 1 \end{cases} \qquad [17.10]$$

where:

$$\omega(\theta) = \begin{cases} 1, & 0<\theta\leq\theta_{thr} \\ \omega^{max}\theta, & \theta_{thr}<\theta<1 \\ \omega^{max}, & \theta=1 \end{cases} \qquad [17.11]$$

The key parameter ω^{max} depends on the temperature and chemical structure of the alloy surface. As described in the previous Section 17.3, $\theta(T, p_{co})$ can be expressed using a multi-site Langmuir isotherm. Thus, the surface coverage θ is defined by the Langmuir adsorption parameter, b_i, related to each CO adsorption site and to the pre-exponential parameter $b_{o,i}$. As for the gas adsorption on the silica nanotube, the *adsorption enthalpy* of carbon monoxide is again a fundamental parameter that should be evaluated to assess the inhibition effect.

QM calculations were carried out, in the frame of DFT, using the B3LYP functional in conjunction with the extended split-valence double-zeta (6-31G*) atomic orbital basis to obtain reliable adsorption energies. The LANL2DZ basis with 28 core electrons being replaced by the relativistic ECP of Hay and Wadt were used to describe the Pd atoms of the surface. In particular, the palladium surface was mimicked using different Pd clusters, in which the metal atoms were placed at a distance of 2.75 Å, which represents the X-ray distance found in the nearest palladium atoms present in the (100) faces of the Pd fcc crystal (Dong *et al.*, 1998). The adsorption energies, obtained for the carbon monoxide, were −27.51 (kcal/mol) for the Pd cluster in the singlet state and −23.12 (kcal/mol) for the Pd cluster in the triplet state. Moreover, in order to evaluate the tricky threshold value θ_{thr}, the geometry optimization of two neighbouring CO molecules on the palladium clusters with 20, 14 and 11 atoms was also performed. These calculations have shown that two CO molecules adsorbed on two H_2 neighbouring adsorption sites do not show stable geometries (minima) with the main molecular axis perpendicular to the Pd surface. The stable structures are twisted in the opposite direction to each other. These structures are less stable than the molecular arrangement showing larger intermolecular distances among neighbouring CO molecules. The maximum packing of the CO molecules on the alloy surface permits the θ_{thr} threshold to be evaluated. It is worth noting that once the arrangement of maximum packing of the CO molecules is established, the ω^{max} parameter may be evaluated. The novel proposed relationships (Equations [17.10] and [17.11]) have been formulated in order to have a direct and simple correlation between the decrease of the hydrogen permeance and the adsorption of the molecules on the alloy surface system. However, other types of relationships can also be formulated.

Finally, the activation energy for the diffusion, E_a, appearing in Equation [17.9] is another important parameter to be evaluated. However, the calculation of this quantity is not easy to obtain in the Born-Oppenheimer approximation. Quantum dynamics approaches, such as Car-Parinello, would be more appropriate in this case. It is worth remarking that, according to the Richardson model, the diffusion of hydrogen atoms in the alloy bulk is the rate-determining step. This diffusion occurs by means of consecutive jumps of hydrogen atoms between neighbouring sites (solid diffusion) (Xiao and Wei, 1992). Thus, knowledge of the absorption site energies in the solid is also great importance for a subsequent description of hydrogen diffusion using for example MC approaches. Furthermore, the calculation of the H absorption energies can be useful for a qualitative comparison of hydrogen solubility in different metal alloys.

17.4.2 Hydrogen solubilization in metal alloys

Hydrogen solubility is an important property useful for the design of new alloys alternative to the Pd-based membranes. Concerning this aspect, here, the calculation of the main interactions experimentated of hydrogen atoms in the unit cell of niobium–molybdenum and pure niobium are presented. This calculation is useful for understanding if the molybdenum atoms in the unit cell of niobium increase or decrease the solubility of hydrogen in the metal alloy. The use of alloys formed by metals different from palladium and presenting low cost, high hydrogen solubility and diffusion can be an attractive alternative to pure palladium membranes in many applications. Thus, the choice of metals with specific features becomes a problem of particular interest. Based on hydrogen permeability, solubility and its cost, the niobium appears to be the cheapest among transition metals. Various experimental works (Yagi *et al.*, 1989; Sumin *et al.*, 2001; Leu *et al.*, 2003) have shown that the hydrogen atoms in Nb–Mo alloys as in Nb occupy symmetrical positions around tetrahedral interstitial sites at 298 K, shown in Fig. 17.3. The hydrogen atoms can occupy several positions in the unit cell; however, the tetrahedral interstitial sites are experimentally considered the most probable. As a first approximation, the atoms of an H_2 molecule, absorbed in an Nb_n–Mo or Nb unit cell, can occupy any of the possible positions shown in Fig. 17.3, thereby creating a large number of possible atomic arrangements in the unit cell.

Various dispositions of hydrogen atoms have equal interatomic distances (equivalent atomic coordinates); specifically, there are equivalent atomic arrangements. However, even if the only non-equivalent positions are considered (corresponding to 2, 4 and 6 hydrogen atoms), the number of calculations necessary to evaluate all the trapping energies (see Equation [17.12]) is equally high. For this reason, it was decided to consider only those non-equivalent dispositions having an equal number of atoms on each side of the unit cell. These arrangements form a homogeneous distribution of hydrogen atoms, along the crystallographic planes, translating and rotating the unit cell in the directions of the crystallographic axes. A complete analysis of the trapping energies, evaluated using DFT (Basile and De Luca, 2005), showed that arrangements with larger H–H distances have lower energy. The trapping energies, corresponding to a particular hydrogen atomic arrangement in a unit cell of Nb–Mo and Nb, can be evaluated using the following equations, respectively:

$$\Delta E_{\mathrm{DFT}}^{\mathrm{NbMo}} = E_{\mathrm{DFT}}\left(\mathrm{Nb_8Mo} - 2n\mathrm{H}\right) - \left[E_{\mathrm{DFT}}\left(\mathrm{Nb_8Mo}\right) + nE_{\mathrm{DFT}}\left(\mathrm{H_2}\right)\right] + \Delta E_{\mathrm{DFT,BSSE}}^{\mathrm{NbMo}} \qquad [17.12]$$

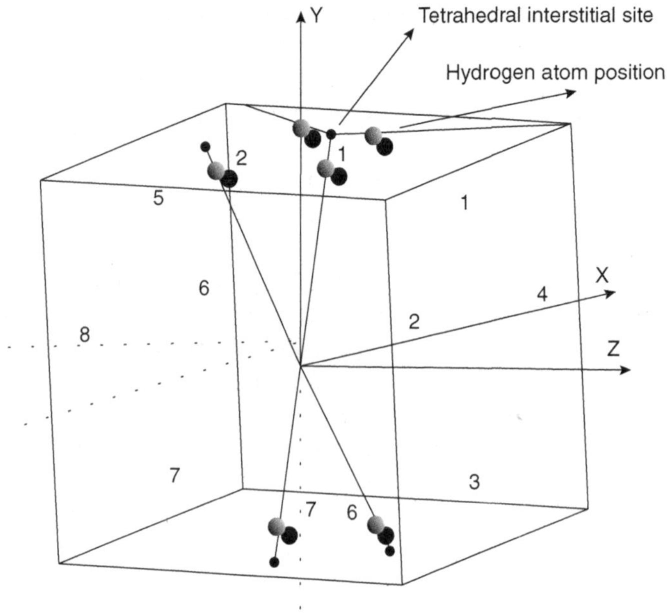

17.3 Nb–Mo and Nb unit cell with tetrahedral interstitial sites for the hydrogen atoms, indicated by the numbers 1 to 8.

$$\Delta E^{\mathrm{Nb}}_{\mathrm{DFT}} = E_{\mathrm{DFT}}\left(\mathrm{Nb}_9 - 2n\mathrm{H}\right) - \left[E_{\mathrm{DFT}}\left(\mathrm{Nb}_9\right) + nE_{\mathrm{DFT}}\left(\mathrm{H}_2\right)\right] + \Delta E^{\mathrm{Nb}}_{\mathrm{DFT,BSSE}} \quad [17.13]$$

The differences between the considered trapping energies, corresponding to seven non-equivalent arrangements of hydrogen atoms, are reported in Table 17.1. The errors due to different sizes of the basis set superposition error (BSSE) are considered almost equal. All the differences between the trapping energies, are negative, showing that the molybdenum atom, instead of niobium at the centre of the unit cell, stabilizes the hydrogen atoms. The trapping energy for the 1,3–6,8 atomic arrangement was evaluated using both an auxiliary basis set and point grid to evaluate the Coulomb, exchange and correlation potentials. The difference between the obtained energies is equal to 13 kcal/mol. Furthermore, the difference between the $E_{\mathrm{DFT}}(\mathrm{Nb}_n\mathrm{Mo}\text{-}2n\mathrm{H})$ energies of two equivalent arrangements, corresponding to three molecules of hydrogen in the unit cell ($n = 3$) that is H17-H48-H17 and H17-H17-H17, is about 20 kcal/mol. This difference has to be almost zero. This discrepancy can be attributed to several causes: the use of symmetry and auxiliary functions employed to describe the electron density; the ECP (LAND2Z) and functional B3LYP employed; and the orbital basis set used (Basile and De Luca, 2005).

Table 17.1 Differences between trapping energy corresponding to six non-equivalent arrangements of hydrogen atoms in Nb–Mo and Nb unit cell

Number of H_2 molecules	Tetrahedral interstitial non-equivalent position of hydrogen atoms	$\Delta E_{DFT}^{NbMo} - \Delta E_{DFT}^{Nb}$ (kcal / mol)
1	1–7	–46.6
	3–2	–49.6
2	1,3–6,8	–107.3
	1,3–6,8*	–120.0
	1,7–1,7	–64.6
3	1,7–1,7–3,5	–99.8
	2,6–2,6–3,5	–78.1

*Value obtained using a grid to evaluate the Coulomb potential, as well as the exchange and correlation potential.

However, while bearing in mind the errors related to the computational procedure, the marked differences between the reported trapping energies, ranging from 46.6 to 120.0 kcal/mol, provide a solid indication of the ability of molybdenum atoms to stabilize the hydrogen in the unit cell. Yagi (1989) and Sumin (2001) showed that no hydrogen precipitation is observed in Nb–Mo alloy with low Mo concentration. In their work, they concluded that the Mo atom leads to an increase of the hydrogen solubility limit, at least for low temperatures. Although the presented DFT calculations do not take into account the thermal vibrations in the trapping energy evaluation, they are in qualitative agreement with the above experimental observations referring to Nb–Mo alloys at low concentration of Mo and temperatures around 298 K. The corrections due to thermal vibrations of the hydrogen atoms in the pure Nb and Nb–Mo alloys can be carried out at QM level. Thus, this calculation can indicate whether the effect of the Mo atom can be extended to higher temperatures.

17.5 Conclusions and future trends

The evaluation of key properties of materials utilized in the preparation of MR by QM approaches is the aim of this chapter. QM should be used when the key properties depend on electronic interactions or electron density polarization. In this chapter, the main concepts on which the computational chemistry is based were illustrated, in addition to the fundamental equations of DFT (quantum theory). The chapter reviews the assessment of some fundamental *molecular* quantities, such as the binding energies in polymeric membranes, the energies of small gases adsorbed in inorganic nanostructured materials and the hydrogen trapping energies in Nb–Mo alloys.

The calculation of the binding energies in different *host–guest* polymeric systems has allowed us to hypothesize a competitive mechanism for the interpretation of their properties. In particular, the competition between the *host–host* and *guest–host* hydrogen-bonding and electrostatic interactions is at the root of some properties of the considered materials.

Using the generalized MS equations, it is possible to deduce that the flows of a single gas through a nano-porous material (zeolites, nanotubes) directly depend on the adsorption enthalpy of the gas under certain conditions. The QM calculation of the adsorption energies of the CO, N_2, O_2, H_2 and CO_2 on a silane group anchored on a silica nanotube was shown. The calculation has shown that the adsorption energies of the CO molecule are almost one order of magnitude larger than the H_2 one, whereas the heat of adsorption of N_2 and O_2 are similar and slightly lower as compared to the CO energy. This result suggests that modified silica nanotube or zeolites with the proposed silane could be used to obtain an efficient separation of CO, N_2 and O_2 from H_2. However, the separation of CO, N_2 and O_2, using this kind of inorganic nanostructure, will be difficult since the adsorption enthalpies of the gases are similar.

Concerning hydrogen adsorption on metallic surfaces, novel relationships (Equations [17.10] and [17.11]) have been formulated in order to evaluate the decrease of hydrogen permeance due to the inhibition of gases such as CO. These new relationships, unlike those reported in the literature, permit a direct and simple correlation between the decrease of the hydrogen permeance and the adsorption of the gas molecules on alloy surface system. The validation of these relationships is in progress.

Finally, the trapping energies of hydrogen in Nb–Mo and Nb alloys were also evaluated. All the differences between the trapping energies related to the Nb_8Mo and Nb_9 unit cell, respectively, are negative showing that the molybdenum atom, instead of niobium at the centre of the unit cell, stabilizes the hydrogen atoms. While bearing in mind the errors related to the computational procedure, the marked differences between the above trapping energies, ranging from 46.6 to 120.0 kcal/mol, provide a solid indication on the ability of the molybdenum alloys to stabilize the hydrogen as experimentally observed.

17.6 References

Abbet S, Sanchez A, Heiz U, Schneider W D, Ferrari A M, Pacchioni G and Rösch N (2000), 'Acetylene cyclotrimerization on supported size-selected Pdn clusters: one atom is enough!', J. Am. Chem. Soc., **122**, 3453–3457.

Amandusson H, Ekedahl L G and Dannetun H (2000), 'The effect of CO and O_2 on hydrogen permeation through a palladium membrane', Appl. Surf. Sci., **153**, 259–267.

Amano M, Nishimura C and Komaki M (1990), 'Effects of high concentration carbon monoxide and carbon dioxide on hydrogen permeation through palladium', Mater. Trans., J. I. M., **31**, 404–408.

Bakker W J W, van den Broeke L J P, Kapteijn F and Moulijn J A (1997), 'Temperature dependence of one-component permeation through a Silicalite-1 membrane', AIChE J., **43**, 2203–2214.

Basile A and De Luca G (2005), 'Studio di nuovi materiali da utilizzare come permeatori nel processo di produzione di idrogeno con il ciclo zolfo-iodio', final report FUS TN BB-R-014 – Allegato n. 2, Commessa Interna ENEA n. FUS 20/2004.

Becke A D (1993), 'Density-functional thermochemistry. III. The role of exact exchange', J. Chem. Phys., **98**, 5648.

Burggraaf A J (1999), 'Single gas permeation of thin zeolite (MFI) membranes: theory and analysis of experimental observations', J. Mem. Sci., **155**, 45–65.

Burggraaf A J and Cot L (1996), *Fundamentals of Inorganic Membrane Science and Technology*, Membrane Science Technology Series, 4, Amsterdam, Elsevier.

De Luca G, Donato L, Del Blanco S G, Tasselli F and Drioli E (2011), 'On the cause of controlling affinity to small molecules of imprinted polymeric membranes prepared by noncovalent approach: a computational and experimental investigation', J. Phys. Chem. B, **115**, 9345–9351.

De Luca G, Gugliuzza A and Drioli E (2009a), 'The role of hydrogen-bonding and weak interactions in competitive interacting processes influencing the polymeric membrane properties', Euromembrane The EMS Conference Montpellier – France, Abstracts, 60.

De Luca G, Gugliuzza A and Drioli E (2009b), 'Competitive hydrogen-bonding interactions in modified polymer membranes: a Density Functional Theory investigation', J. Phys. Chem. B, **113**, 5473–5477.

De Luca G, Tocci E and Drioli E (2005), 'Quantum and molecular mechanics calculations on modified silica nano ring', J. Mol. Struct., **739**, 163–172.

De Luca G, Pullumbi P and Russo N (2002), 'Impact of zeolites and other porous materials', The New Technologies at the Beginning of the New Millennium, Proceedings of Second International FEZA, Taormina –Italy.

De Luca G, Arbouznikov A, Goursot A and Pullumbi P (2001), 'Quantum chemical study of low-pressure adsorption in zeolitic materials', J. Phys. Chem. B, **105**, 4663–4668.

Dong W, Ledentu V, Sautet Ph, Eichler A and Hafner J (1998), 'Hydrogen adsorption on palladium: a comparative theoretical study of different surfaces', Surf. Sci., **411**, 123–136.

Eriksson M, and Ekedahl L G (1998), 'Real time measurements of hydrogen desorption and absorption during CO exposures of Pd hydrogen sticking and dissolution', Appl. Surf. Sci., **133**, 89–97.

Ertl G. (1991), Dynamics of reactions at surfaces, Adv. Catal., **45**, 1–69.

Gielens F C, Knibbeler R J J, Duysink P F J, Tong H D, Vorstman M A G and Keurentjes J F T (2006), 'Influence of steam and carbon dioxide on the hydrogen flux through thin Pd/Ag and Pd membranes', J. Membr. Sci., **279**, 176–185.

IMETI (2010), *Quantum mechanics evaluation of the affinity among polymers: internal report of the IMETI project.*

Judai K, Wörz A S, Abbet S, Antonietti J M, Heiz U, Del Vitto A, Giordano L and Pacchioni G (2005), 'Acetylene trimerization on Ag, Pd and Rh atoms deposited on MgO thin films', Phys. Chem. Chem. Phys., **7**, 955–962.

Kapteijn F, Moulijn J A and Krishna R, (2000), 'The generalized Maxwell-Stefan model for diffusion in zeolites: sorbate molecules with different saturation loadings', Chem. Eng. Sci., **55**, 2923–2930.

Kärger J and Ruthven D M (1992), *Diffusion in Zeolites and Other Microporous Solids*, New York, Wiley.

Krishna R (2000), 'Diffusion of binary in zeolites: molecular dynamics simulations versus Maxwell-Stefan theory', Chem. Phys. Lett., **326**, 477–484.

Kshama B, Jirage J, Hulteen C and Martin C R (1997), 'Nanotubule-based molecular-filtration membranes', Science, **278**(5338), 655–658.

Leu B, Wipf H, Coluzzi B, Biscarini A, Mazzolai G, Mazzolai F M and Sumin V V (2003) 'Low-temperature Snoek-type relaxation of hydrogen interstitial atoms in $Nb_{0.8}Mo_{0.2}$', J. Alloy Compd., 356–357, 322–325.

Li A, Liang W and Hughes R (2001), 'The effect of carbon monoxide and steam on the hydrogen permeability of a Pd/stainless steel membrane', J. Membrane Sci., **165**, 135–141.

Mitsui T, Rose M K, Fomin E, Ogletree D F and Salmeron M, (2002) 'Coadsorption and interaction of O and H on Pd(1 1 1)', Surf. Sci., **511**, 259–266.

Pacios F and Christiansen P A (1985), 'Ab initio relativistic effective potentials with spin-orbit operators. I. Li through Ar', J. Chem. Phys., **82**, 2664–2671.

Parr R G and Yang W (1989), *Density Functional Theory of Atoms and Molecules*, Oxford, Clarendon Press, New York, Oxford University Press.

Perdew J P, Burke K and Ernzerhof M (1997), 'Generalized gradient approximation made simple', Phys. Rev. Lett., **78**, 1396.

Rudkevich D M (2004), 'Emerging supramolecular chemistry of gases', Angew. Chem. Int. Ed., **43**, 558–571.

Sierka M and Sauer J (2000), 'Finding transition structures in extended systems: a strategy based on a combined quantum mechanics-empirical valence bond approach', J. Chem. Phys., **112**(16), 6983–6996.

Sumin V V, Wipf H, Coluzzi B, Biscarini A, Campanella R, Mazzolai G and Mazzolai F M (2001), 'A neutron-spectroscopy study of the local vibrations, the interstitial sites and the solubility limit of hydrogen in niobium-molybdenum alloys', J. Alloy. Compd., **316**, 189–192.

Szabo A and Ostlund N (1994), *Modern Quantum Chemistry Introduction to Advanced Electronic Structure Theory*, London, Macmillan Publishing CO. INC, Collier Macmillan Publishers.

Thomas J M (1989), *Advanced catalysts: interfaces in the physical and biological sciences*, Angewandte Chemie International Edition, **28**, 1079–1088.

van Santen R A and Neurock M (2001), *Encyclopedia of Catalysis*, New York, Wiley.

Xiao J and Wei J (1992), 'Diffusion mechanism of hydrocarbons in Zeolites I. Theory', Chem. Eng. Sci., **47**, 1123–1141.

Xu X and Goddard WA (2004), 'The X3LYP extended density functional for accurate descriptions of nonbond interactions, spin states, and thermochemical properties', PNAS, **101**(9), 2673–2677.

Yagi E, Nakamura S, Kano F and Kobayashi T (1989), 'Hydrogen trapping by solute atoms in Nb-Mo (3at. %) alloys as observed by the channeling method', Phys. Rev. B, **39**(1), 57–60.

17.7 Appendix: nomenclature

17.7.1 Notation

A	membrane area (m^2)
b	Langmuir adsorption parameter (kPa^{-1})
b_o	Langmuir pre-exponential parameter (kPa^{-1})
Bi	Biot number $\left(\frac{h_W \times \frac{d_t}{2}}{\lambda_{er}}\right)$
C_i	molecular coverage (adsorbed gas per unit cell)
C_i^{sat}	saturation coverage
$\tilde{c}_i$	dimensionless i component concentration
c_i	i component concentration $\left(\frac{kmol}{m^3}\right)$
$c_{p,i}$	specific heat of the component i $\left(\frac{kJ}{kmol \cdot K} \text{ or } \frac{kJ}{kg \cdot K}\right)$
$c_{p,mix}$	gas mixture specific heat $\left(\frac{kJ}{kg \cdot K}\right)$
D_i^{MS}	Maxwell–Stefan surface diffusivity (corrected diffusivity)
Da	Damköhler number
D_e	effective diffusivity in the solid particle $\left(\frac{m^2}{h}\right)$
D_{ez}	effective axial mass diffusivity $\left(\frac{m^2}{h}\right)$
D_{er}	effective radial mass diffusivity $\left(\frac{m^2}{h}\right)$
d_p	equivalent particle diameter (m)
d_t	internal tubular reactor diameter (m)
E_a	apparent activation energy $\left(\frac{kJ}{mol}\right)$

E_{DFT}	total energy of quantum system (a.u.)
E_{xc}	exchange and correlation energy functional (a.u.)
f	friction factor
F_{sweep}	sweeping gas flow-rate $\left(\frac{kmol}{h}\right)$
F_{TOT}	inlet total gas mixture flow-rate $\left(\frac{kmol}{h}\right)$
G	Gibbs free energy (kJ/mol)
g	scaling factor (–)
h_f	heat transport coefficient between gas and solid phase $\left(\frac{kJ}{h \cdot m^2 \cdot K}\right)$
$h_{i,perm}$	*i*-component enthalpy in permeation zone $\left(\frac{kJ}{mol}\right)$
$h_{i,reac}$	*i*-component enthalpy in reaction zone $\left(\frac{kJ}{mol}\right)$
$(-\Delta H_j)$	heat of reaction *j* $\left(\frac{kJ}{mol}\right)$
h_{perm}	forced convection heat transport coefficient in the permeation zone $\left(\frac{kJ}{h \cdot m^2 \cdot K}\right)$
h_W	heat transport coefficient near the tube wall $\left(\frac{kJ}{h \cdot m^2 \cdot K}\right)$
$h_{W,f}$	wall heat transport coefficient for the fluid phase $\left(\frac{kJ}{h \cdot m^2 \cdot K}\right)$
$h_{W,s}$	wall heat transport coefficient for the solid phase $\left(\frac{kJ}{h \cdot m^2 \cdot K}\right)$
J_i	*i*-component flux through the membrane $\left(\frac{kmol}{h \cdot m^2}\right)$
K_i	adsorption equilibrium constant of the component *i* (kPa^{-1})
k_j	rate constant of reaction *j* (depends on the reaction)
$K_{eq,j}$	equilibrium constant of reaction *j*
k_{mem}	membrane thermal conductivity $\left(\frac{kJ}{h \cdot m \cdot K}\right)$
k_{met}	tube wall conductivity $\left(\frac{kJ}{h \cdot m \cdot K}\right)$

L	reactor length (m)
M_i	molecular weight of the component i $\left(\frac{\text{kg}}{\text{kmol}}\right)$
M_m	average molecular weight of the gas mixture $\left(\frac{\text{kg}}{\text{kmol}}\right)$
p	pressure (kPa)
$\tilde{p}$	dimensionless pressure in the reaction zone
Δp	pressure drop along the reactor (KPa)
p_0	inlet pressure (kPa)
P_i	permeability coefficient $\left(\frac{\text{kmol}\cdot\text{m}}{\text{h}\cdot\text{m}^2\cdot\text{kPa}^n}\right)$
P_i^0	permeability pre-exponential factor $\left(\frac{\text{kmol}\cdot\text{m}}{\text{h}\cdot\text{m}^2\cdot\text{kPa}^n}\right)$
Pe_a	axial Péclet number
Pe_{cr}	critical Péclet number
Pe_{mr}	radial mass Péclet number $\left(\frac{u_z d_p}{D_{\text{er}}}\right)$
p_i	i-component partial pressure (kPa)
$p_{i,p}$	i-component partial pressure on the permeate (downstream) side (kPa)
$p_{i,r}$	i-component partial pressure on the retentate (upstream) side (kPa)
$p_{i,s}$	i-component partial pressure on the catalyst surface (kPa)
Pr	Prandtl number $\left(\frac{c_p \mu_g}{k_g}\right)$
Pr_{perm}	Prandtl number in the permeation (downstream) zone
q_m	heat flux through the membrane $\left(\frac{\text{kW}}{\text{m}^2}\right)$
q_r	heat flux through the walls $\left(\frac{\text{kW}}{\text{m}^2}\right)$
r	radial coordinate (m)
$\tilde{r}$	radial dimensionless coordinate
R	universal gas constant $\left(\frac{\text{kJ}}{\text{kmol}\cdot\text{K}}\right)$
Re	Reynolds number
Re_p	Reynolds number referred to catalyst particle diameter $\left(\frac{G d_p}{\mu_g}\right)$
Re_{perm}	Reynolds number in the permeation (downstream) zone

$r_{i,j}$	reaction rate of component i in reaction j $\left(\frac{\text{kmol}}{\text{h}\cdot\text{m}^2} \text{ or } \frac{\text{kmol}}{\text{h}\cdot\text{kg}_{\text{cat}}}\right)$
Sc	Schmidt coefficient $\left(\frac{\mu}{\rho_g D}\right)$
St	Stanton number
S/C	steam to carbon ratio
T	temperature (K)
T_{bi}	normal boiling temperature of the component i (K)
T_{in}	inlet temperature (K)
T_m	membrane temperature (K)
T_W	wall temperature (K)
U	overall heat transport coefficient $\left(\frac{\text{kJ}}{\text{h}\cdot\text{m}^2\cdot\text{K}}\right)$
u_s	gas superficial velocity (m/s)
V_p	pellet volume (m^3)
V_{reactor}	reactor volume (m^3)
$v(r)$	nuclei–electron potential
W/F	apparent residence time $\left(\frac{\text{kg}_{\text{cat}}\text{s}}{\text{mol}}\right)$
X_i	i-component conversion
y_i	molar fraction of the component i
Y_i	overall yield to species i
z	axial coordinate (m)
$\tilde{z}$	axial dimensionless coordinate

Greek symbols

δ	membrane thickness (m)
ε	void fraction of packing
η	effectiveness factor
η_j	effectiveness factor of reaction j
λ_e	effective thermal conductivity in the solid particle $\left(\frac{\text{kJ}}{\text{h}\cdot\text{m}\cdot\text{K}}\right)$
λ_{ea}	effective axial thermal conductivity $\left(\frac{\text{kJ}}{\text{h}\cdot\text{m}\cdot\text{K}}\right)$
λ_{er}	effective radial thermal conductivity $\left(\frac{\text{kJ}}{\text{h}\cdot\text{m}\cdot\text{K}}\right)$

λ_g	gas phase thermal conductivity $\left(\frac{kJ}{h \cdot m \cdot K}\right)$
λ_i	thermal conductivity of the component i $\left(\frac{kJ}{h \cdot m \cdot K}\right)$
λ_S	thermal conductivity of packing material $\left(\frac{kJ}{h \cdot m \cdot K}\right)$
ξ	radial coordinate inside the particle (m)
μ_g	gas mixture viscosity $\left(\frac{kg}{h \cdot m}\right)$
μ_i	viscosity of the component i $\left(\frac{kg}{m \cdot h}\right)$
θ_i	fractional surface occupancy or coverage
θ_{thr}	threshold fractional surface coverage
ρ_B	catalytic bed density $\left(\frac{kg}{m^3}\right)$
ρ_g	gas density $\left(\frac{kg}{m^3}\right)$
$\rho(r)$	electron density
ρ_s	catalyst density $\left(\frac{kg}{m^3}\right)$
ρ_z	density of the zeolite
Φ_i	chemical potential
$\breve{\phi}_i$	fugacity coefficient of species i in the gas mixture (kPa)
$\phi_j(r)$	spatial atomic orbital
ω^{max}	maximum value of scaling factor
Λ	mean free path (m)
Γ	thermodynamic correction factor (–)

Subscripts and superscripts

app	apparent
eff	effective parameter
g	gas translational
i	component in mixture
j	reaction number
s	surface
sat	saturation
SM	Stefan–Maxwell

17.7.2 Abbreviations

MR	membrane reactor
PEMFC	polymer electrolyte membrane fuel cell
QM	quantum mechanics
TR	traditional reactor (reactor without membranes)
WHSV	weight hourly space velocity

18

Non-equilibrium thermodynamics for the description of transport of heat and mass across a zeolite membrane

S. K. SCHNELL, and T. J. H. VLUGT, Delft University of Technology, The Netherlands and S. KJELSTRUP, Norwegian University of Science and Technology, Norway

DOI: 10.1533/9780857097330.3.627

Abstract: Non-equilibrium thermodynamics offers a better and more precise way to describe transport of heat and mass over membranes than the simple Fick and Fourier models. In this work, we derive the equations for single component transport (of *n*-butane) across a zeolite membrane. We use non-equilibrium molecular dynamics (NEMD) data to show how transport phenomena are related. Significant differences are discussed between this method, and methods invoking Fick's and Fourier's laws.

Key words: non-equilibrium thermodynamics, zeolite membrane, coupled transport of heat and mass.

18.1 Introduction

A precise description of membrane transport phenomena requires detailed attention to the surface of the membrane, as well as to the membrane itself (Kjelstrup and Bedeaux, 2008). Before or after transport through the membrane, the component to be transported can be more or less strongly adsorbed to the membrane interface. The transported component may also be involved in a chemical reaction at the interface. Such phenomena are usually closely related to the exchange of heat, meaning that there are sinks or sources for heat present at the membrane external surface. This surface is of molecular thickness, and it is therefore to be expected that the mass flux is affected by the heat flux through the membrane, and also that these fluxes vary with position in the membrane. Nevertheless, in chemical engineering the transport of heat and mass is usually described using Fourier's law and Fick's laws, which do not include such effects. We consider a series of layers, from a homogeneous gas-phase on the 'feed' side, to a homogeneous gas-phase on the 'permeate' side. The boundary layers are of macroscopic thickness. We will here distinguish such macroscopic boundary layers from a molecularly thin interface, and consider five regions of transport, namely the homogeneous phases of the macroscopic boundary layers, the interface layers between each homogeneous phase and the membrane phase, called the surface, and the membrane phase itself.

In the modelling of membrane separations, knowledge of chemical potentials and temperatures is required for the adjacent homogeneous phases and the membrane. The surface is often assumed to be at equilibrium, making these properties continuous at the surface. With large mass fluxes and/or large heat productions at the surface, these assumptions may no longer be correct and more traditional descriptions using Fick's and Fourier's laws will fail to capture the correct behaviour. In this case, one must use the theory of irreversible thermodynamics to correctly describe heat and mass transfer (de Groot and Mazur, 1984; Kjelstrup and Bedeaux, 2008).

The purpose of this chapter is to present such a description, to identify the effects and consequences of coupling heat and mass transport, and to study the commonly used assumption of equilibrium at the interface at steady state conditions. The simple transport of a single component through a membrane is used as an example. A typical alkane, *n*-butane, is chosen as the component to be transported and silicalite-1 (Baerlocher *et al.*, 2007) is used as membrane material. Membrane reactors using zeolites are well studied; see for example a recent review (McLeary *et al.*, 2006). Most importantly, all transport coefficients for heat and mass transfer are already available for this system; see the work of Inzoli *et al.* (Inzoli, 2008; Inzoli *et al.*, 2009). In this work, molecular dynamics simulations were used to obtain the transport coefficients. Equilibrium data needs to be supplied, in order to be able to define the driving forces for transport. To the best of our knowledge, this is the first time data obtained from molecular dynamics simulations have been used exclusively to accurately model the transport through a heterogeneous system under non-equilibrium conditions at the surface.

18.1.1 A zeolite membrane in equilibrium with a gas

A zeolite membrane separating two gas-phases is in the thermodynamic sense a heterogeneous system composed of five subsystems. This is illustrated in Fig. 18.1. Two of these, the two gas-phases, are homogeneous, three-dimensional systems. The membrane can also be regarded as a homogeneous phase on a scale in units of length of two zeolite unit cells (Schnell *et al.*, 2011) with gas absorbed in the various channels. The two external surfaces, or the transition zones of molecular thickness between gas and zeolite membrane, are the fourth and fifth thermodynamic systems. The surfaces are two-dimensional systems, defined according to Gibbs, with excess properties as state variables (Kjelstrup and Bedeaux, 2008). We will use the word surface or interface to indicate this type of thermodynamic system.

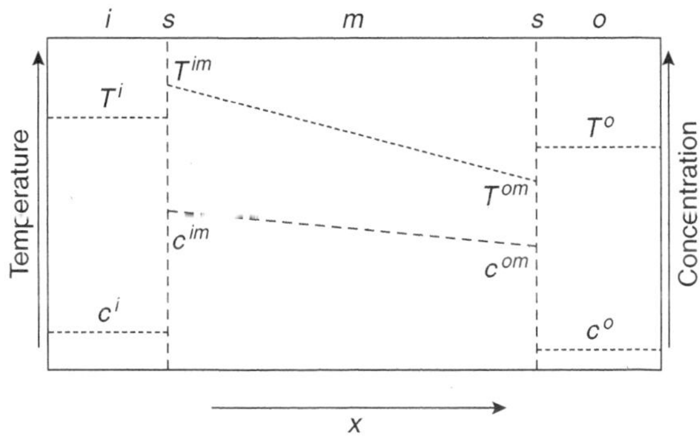

18.1 Schematic representation of the system studied here. At the surface, jumps in temperature, *T*, and concentration, *c*, of guest molecules occur. For simplicity, it is assumed that temperature and concentration gradients inside the membrane are constant. The membrane is denoted by *m*; *i* and o denote the pure gas-phases (also considered as the 'feed' and 'permeate' side, respectively). The surfaces are denoted by *s*. The superscripts im and om denote the temperature and concentration in the membrane on the *i*- and *o*-side.

Figure 18.2 shows the density profiles of *n*-butane in the zeolite and close to the surface at equilibrium. The local density of guest molecules in the zeolite strongly varies with position in the zeolite. In the work of Inzoli *et al.* (2008, 2009), the external surface has two different surface structures; one where the surface is cut between the sinusoidal channels and the other one across the sinusoidal channels. This shifts the density profile as seen in Fig. 18.2, and these represent two quite different surface structures. Inzoli *et al.* (2009) showed that for this system the details of the surface structure have a minor influence on the surface resistance to transport. In Fig. 18.3 we show a typical snapshot of a zeolite/gas interface from an equilibrium simulation at a pressure of 100 kPa, at 400 K. Clearly, *n*-butane molecules are adsorbed on the external surface of silicalite-1 and some molecules are adsorbed in the zeolite pores. The external surface in this case has a flat structure.

Equilibrium adsorption isotherms for gases and gas mixtures are in general well known for bulk zeolite systems, experimentally as well as computationally (Dunne *et al.*, 1996; Vlugt *et al.*, 1999; Dubbeldam *et al.*, 2004). We have used established procedures to obtain equilibrium adsorption isotherms data (Frenkel and Smit, 2002), which are needed to describe transport problems in our case. Thermodynamic adsorption data is obtained using Configurational-bias Monte-Carlo simulations in the grand-canonical ensemble (Frenkel and Smit, 2002). Interactions between atoms and molecules are described using a classical force field (Dubbeldam *et al.*, 2004). Figure 18.4 shows adsorption isotherms for *n*-butane adsorption in silicalite-1 at a

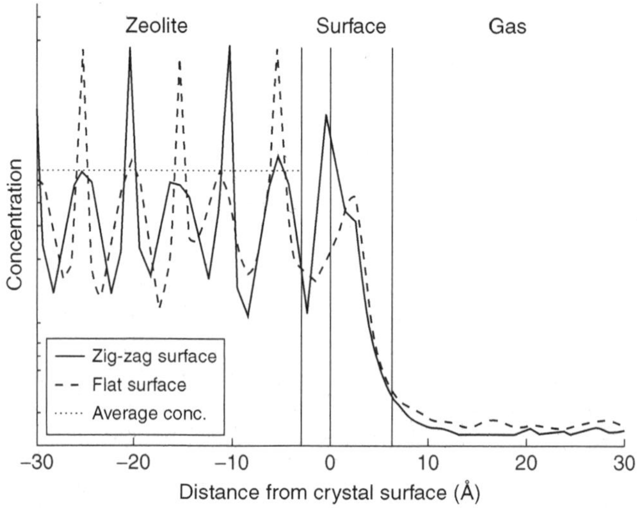

18.2 *n*-Butane local density near a gas–zeolite interface. The density profiles are given in arbitrary units, to illustrate the density fluctuations inside such a zeolite. Density profiles are shown for (1) a flat surface cut between the sinusoidal channels, and perpendicular to the straight channels and (2) a textured surface, where the cut is in the sinusoidal channels, and also here, perpendicular to the straight channels. For macroscopic applications, the average concentration over several unit cells is a more convenient measure for the concentration in the zeolite.

temperature from 320 K to 400 K, which is the range of temperatures represented by coefficients given by Inzoli *et al.* (Inzoli, 2008, 2009). In Fig. 18.4 we have plotted the loading of *n*-butane in the zeolite (in *molecules/nm^3*) as a function of the fugacity. In this work we take the fugacity coefficient to be one. The isotherms are well described with a Langmuir model (Langmuir, 1918) with a maximum loading of approximately 1.7 molecules/nm^3. The silicalite-1 oriented so that the straight channels face the gas-phase is one of the most interesting orientations for this crystal and effort have been put in place for producing membranes with only this crystallographic orientation; see for example the work of Lee and Jeong (2011).

18.1.2 Transport across a heterogeneous system

When a fluid with one or more components is transported across a selective membrane, concentration gradients arise in the layer immediately outside the membrane. Likewise, due to negative or positive enthalpies of adsorption in the membrane, temperature gradients will arise, depending on the ratio of the relative conductivity of the membrane to its surroundings (Inzoli *et al.*, 2009). In a pure gas, there is obviously no composition gradient, only

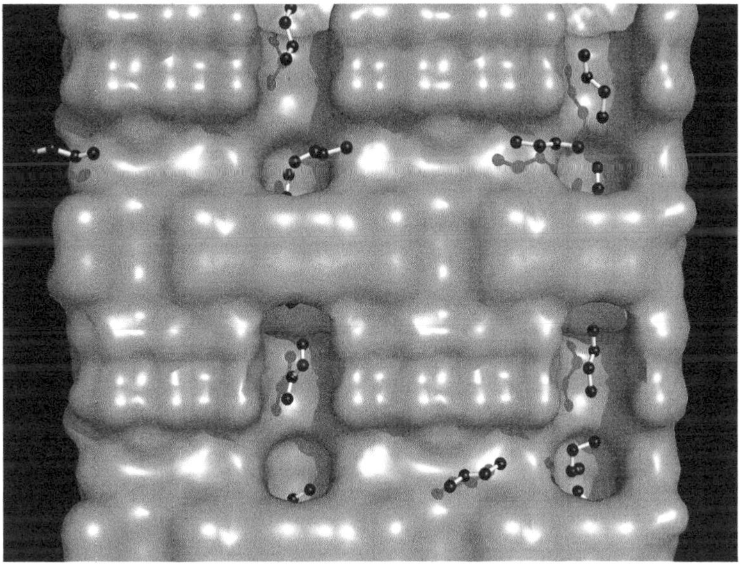

18.3 Typical snapshot of a silicalite-1 zeolite with gas–zeolite interfaces. Adsorption data are obtained from molecular simulations and this snapshot is taken from a Molecular Dynamic simulation in the canonical ensemble. This simulation is performed with an explicit gas–zeolite interface. The *n*-butane molecules are rendered as a string of four beads, using the united atom model, see Frenkel and Smit (2002), while the zeolite is shown as a grey surface. In this case, the *n*-butane molecules are located in the straight and sinusoidal channels. The gas-phases are located to the left and right of the zeolite crystal. The silicalite-1 straight channels are parallel to the paper plane.

a temperature gradient. With pores of molecular size, the zeolite becomes a component in the thermodynamic sense. A concentration gradient inside a zeolite can thus be maintained in the presence of a temperature gradient. We will investigate the stationary state transport of one component through the membrane.

Typical profiles of the intensive variables, such as chemical potential, μ and temperature, T, are illustrated in Fig. 18.1. At the interface during transport, a possibility arises to maintain a difference in chemical potential between the two phases adjacent to the interface; this means that the interface is not in equilibrium under such conditions, cf. a similar discussion of liquid–vapour interfaces (Ge *et al.*, 2007). Likewise, a temperature difference can be maintained across the interface. We will study the details of this situation, that is, what are the causes of such jumps, what do they physically mean and how can we take advantage of them? We will also examine the assumption of continuity of intensive variables at the interface: what does it mean in practice to have the correct non-equilibrium description of the

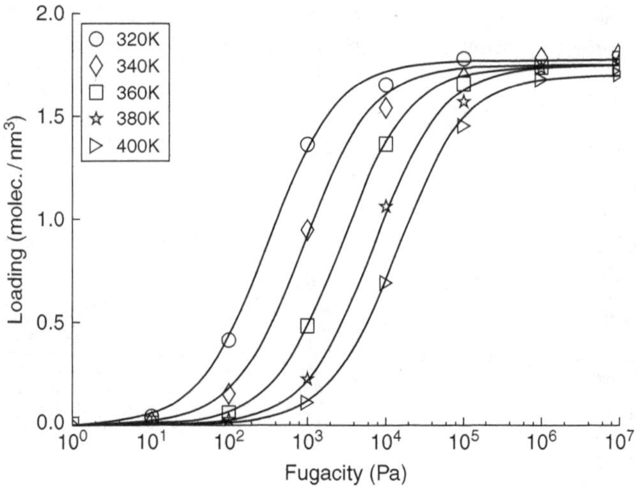

18.4 Adsorption isotherms of *n*-butane in a bulk zeolite crystal with periodic boundary conditions, without an interface. Therefore, no gas–zeolite interface was present here. The temperature range is selected to be in the range of the operating conditions of such a membrane system. In all cases, the isotherms are well described using a Langmuir single site isotherm (Langmuir, 1918).

surface? This work can be seen as a step in the direction of analysing multi-component systems using realistic transport coefficients.

18.2 Fluxes and forces from the second law and transport coefficients

18.2.1 Fluxes and forces from the second law

The theory of non-equilibrium thermodynamics is used to analyse the questions above. The sections below provide the essential steps in such a description. The detailed background and explanation of the theory, as set up for heterogeneous systems, can be found elsewhere (Kjelstrup and Bedeaux, 2008). The same symbols are used as this reference.

The equations for transport of heat and mass

In the theory of non-equilibrium thermodynamics, the fluxes and forces pertinent to the system are derived from the entropy production per volume or area element. The entropy production is formulated using the assumption that there is local equilibrium in the system. This assumption is surprisingly valid even for systems with very large gradients (Kjelstrup *et al.*, 2008). The expression for the entropy production follows by introducing mass- and energy-balances into the Gibbs equation. In the present case, the mass and energy-balances are simple, as the mass and total energy fluxes, J and J_q, respectively, are constant and directed along the x-axis only:

$$J^i = J^m = J^o = J \quad [18.1]$$

$$J_q^{\prime i} + J^i \mathrm{H}^i = J_q^{\prime m} + J^m \mathrm{H}^m = J_q^{\prime o} + J^o \mathrm{H}^o = J_q \quad [18.2]$$

Superscripts m, i and o denote that the location is the membrane, the i-side (feed) or o-side (permeate) surfaces; see Fig. 18.1. The flux of n-butane is denoted by J; the sensible heat flux is $J_q^{\prime}$ (Kjelstrup and Bedeaux, 2008) and H is the molar enthalpy of n-butane. The total energy flux is J_q. We distinguish between the sensible heat flux and the total heat flux. The sensible heat flux is the one that can be measured directly and is described by Fourier's law in a pure component. The total heat flux is the sensible heat flux and in addition the enthalpy carried by mass transport. The definitions of the various types of heat fluxes are described and explained in detail in Kjelstrup and Bedeaux, (2008). The entropy production and the corresponding force-flux relations for the zeolite membrane and its two surfaces are discussed in the next section.

The zeolite membrane

As discussed in Section 18.2, we can regard the zeolite membrane as a homogeneous phase. Strong evidence has been obtained that it is meaningful to consider thermodynamic variables in volume elements as small as a single unit cell, and also when the system is exposed to large gradients (Kjelstrup and Bedeaux, 2008; Schnell *et al.*, 2011). The entropy production σ is:

$$\sigma = J_q^{\prime m} \frac{\mathrm{d}}{\mathrm{d}x}\frac{1}{T} + J\left(-\frac{1}{T}\frac{\mathrm{d}}{\mathrm{d}x}\mu_T\right) \quad [18.3]$$

We define the fluxes into the membrane as positive and use the membrane as the frame of reference for all fluxes. The driving forces for transport of heat and mass are the gradients in inverse temperature T, and minus the gradient in chemical potential, μ_T, taken at constant temperature. The fluxes conjugate to these forces are those of n-butane and sensible heat. The force-flux relations that follow from this are:

$$\frac{\mathrm{d}}{\mathrm{d}x}\frac{1}{T} = r_{qq}^{m} J_q^{\prime m} + r_{qm}^{m} J \quad [18.4]$$

$$-\frac{1}{T}\frac{\mathrm{d}}{\mathrm{d}x}\mu_T = r_{mq}^{m} J_q^{\prime m} + r_{mm}^{m} J \tag{18.5}$$

In Equations [18.4] and [18.5] the resistivities are denoted by the symbol r, and the superscript denotes that these relate to the membrane, while the subscript identifies the type of resistivity. The diagonal resistivities, r_{qq} and r_{mm} can be related to the conductivity and diffusivity respectively, while the off-diagonal resistivities describe the coupling between heat and mass transfer. Also, the mass flux is denoted by J, and the sensible heat flux is denoted by $J_q^{\prime m}$. We have chosen to use the sensible heat flux in all of this work, especially since this is the one that is determined in experiments. The resistivities are related to the more commonly used thermal conductivity, λ, and Fick's diffusivity, D. The resistivity r_{qq} is related to the thermal conductivity by:

$$\lambda = \frac{1}{T^2 r_{qq}} \tag{18.6}$$

while r_{mm} is related to Fick's diffusivity by:

$$D = \frac{1}{T r_{mm}} \frac{\partial \mu_{b,T}}{\partial c_b} \tag{18.7}$$

The zeolite external surfaces

The entropy production of the surface, σ^s, is an excess property, given for stationary state transport of one component and heat as (Kjelstrup and Bedeaux, 2008):

$$\sigma^s = J_q^{\prime i} \Delta_{i,m} \frac{1}{T} + J\left(-\frac{1}{T^m}\Delta_{i,m}\mu_T\left(T^m\right)\right) \tag{18.8}$$

The driving forces are now expressed as differences instead of gradients. For instance, $\Delta_{i,m}$ indicates the difference between the phases m and i. The dimension of σ^s is J/K m^2. The force-flux relations for the surface are therefore:

$$\Delta_{i,m}\frac{1}{T} = r_{qq}^{s,i} J_q^{\prime i} + r_{qm}^{s,i} J \tag{18.9}$$

$$-\frac{1}{T^m}\Delta_{i,m}\mu_T\left(T^m\right)=r_{mq}^{s,i}J'^{i}_{mm}J \qquad [18.10]$$

Again, the superscript indicates that the resistivities are related to the surface, s. We use the superscript i to indicate that these values are related to the i-side of the membrane; see Fig. 18.1. The subscript m again means the membrane.

For the o-side of the interface we obtain similarly,

$$\Delta_{\mathrm{m,o}}\frac{1}{T}=r_{qq}^{s,o}J'^{o}_{q}+r_{qm}^{s,o}J \qquad [18.11]$$

$$-\frac{1}{T^m}\Delta_{m,o}\mu_T\left(T^m\right)=r_{mq}^{s,\mathrm{o}}J'^{\mathrm{o}}_{q}+r_{mm}^{s,\mathrm{o}}J \qquad [18.12]$$

Similarly for these equations, the superscript indicates which surface the value relates to. The values of $J_q^{'\mathrm{i}}$ and $J_q^{'\mathrm{o}}$ are not necessarily identical. The values of the resistivities for the surface are evaluated from the fluxes in the corresponding gas-phase.

In order to find the temperature and concentration profiles across the membrane, we solve Equations [18.2], [18.4], [18.5] and [18.9]–[18.12], simultaneously.

18.2.2 Transport coefficients and driving forces

Because of the Onsager symmetry relation, $r_{ij} = r_{ji}$, only three coefficients are needed in the Equations [18.4] and [18.5], and six in the set Equations [18.9]–[18.12]. The nine coefficients needed were already determined by Inzoli *et al.* (Inzoli *et al.*, 2008, 2009), and the values of the transport coefficients are shown in Fig. 18.5 (transport inside the membrane) and Fig. 18.6 (transport at the membrane surface). Altogether, the six equations and nine coefficients are sufficient to compute the profiles of the temperature and chemical potential across the heterogeneous system fully, for any set of inlet conditions on the i-side, or feed side.

The chemical potential driving force is taken at a specified temperature everywhere. This driving force in the membrane is

$$-\frac{1}{T}\frac{\mathrm{d}}{\mathrm{d}x}\mu_T=-R\frac{\mathrm{d}}{\mathrm{d}x}\ln a \qquad [18.13]$$

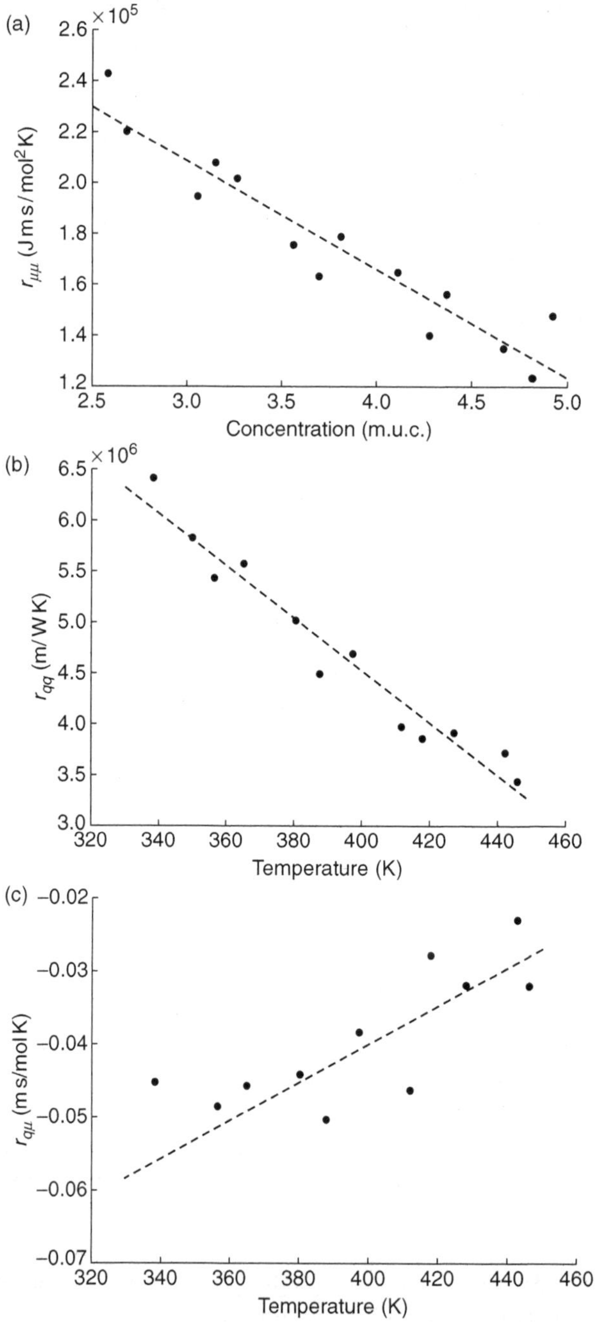

18.5 The resistance to (a) mass transfer, (b) heat transfer, and (c) the coupling between mass and heat transfer for *n*-butane in silicalite-1 zeolite. The resistance to mass transfer, $r_{\mu\mu}$, is given as a function of the concentration, while the resistance to heat transfer, r_{qq}, is given as a function of temperature. Note that the coupling coefficient, $r_{q\mu}$, is very small.

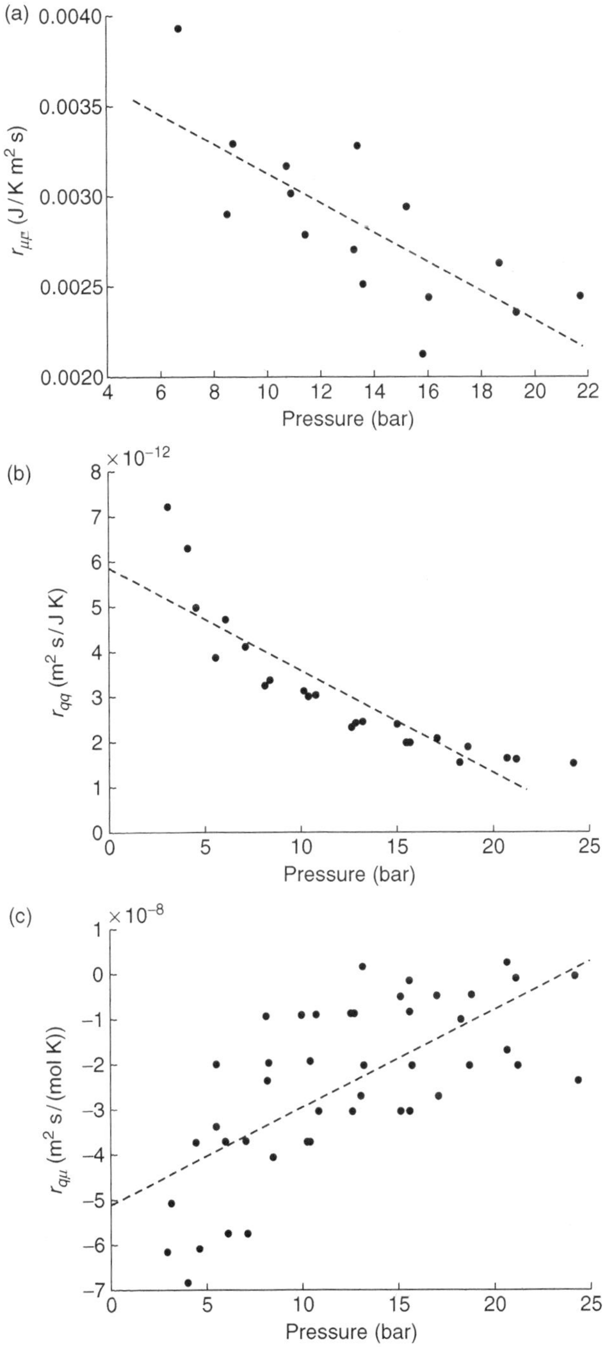

18.6 The resistance to (a) mass transfer, (b) heat transfer, and (c) the coupling between mass and heat transfer for *n*-butane at the external surface of silicalite-1 zeolite. All values are depending on the pressure in the facing gas-phase, and the temperature dependence is found to be negligible.

where $a = p/p^*(T)$ is the activity, defined by p, the vapour pressure of the component that would be in equilibrium with the gas in the membrane (assuming ideal gas behaviour), and p^*, the vapour pressure for saturation of the gas at temperature T. The driving force across the surface is likewise

$$-\frac{1}{T^m}\Delta_{m,o}\mu_T\left(T^m\right) = -R\ln\frac{p}{p^*\left(T^m\right)} \qquad [18.14]$$

These equations will be used to calculate jumps in concentration at the surface.

18.3 Case studies of heat and mass transport across the zeolite membrane

To examine the interaction of heat and mass transport across the zeolite membrane, we have chosen to present the profiles of concentration and temperature for the following conditions. Results are reported in tables or figures as indicated:

1. The heat and mass fluxes at constant and equal temperatures in the gas-phases; see Table 18.1 and Fig. 18.7.
2. The heat and mass fluxes at isobaric transport of n-butane; see Table 18.2 and Fig. 18.8.

Table 18.1 Heat and mass fluxes for various fugacity differences over the membrane

Δf(kPa)	J (mol/m²s)	J'^{i}_{q} (kJ/m²s)
2	0.05508	–2.40
4	0.11250	–4.89
6	0.17248	–7.50
8	0.23524	–10.23
10	0.30107	–13.10

Notes: The fugacity on the 'feed' side (denoted by i in the text and in Fig. 18.1) is 50 kPa, and the 'permeate' side (denoted by o) is $f_i - \Delta f$. The temperature is kept at 400 K on both sides, and the thickness of the membrane is 50 μm. The measurable heat flux is given for the i-side of the membrane.

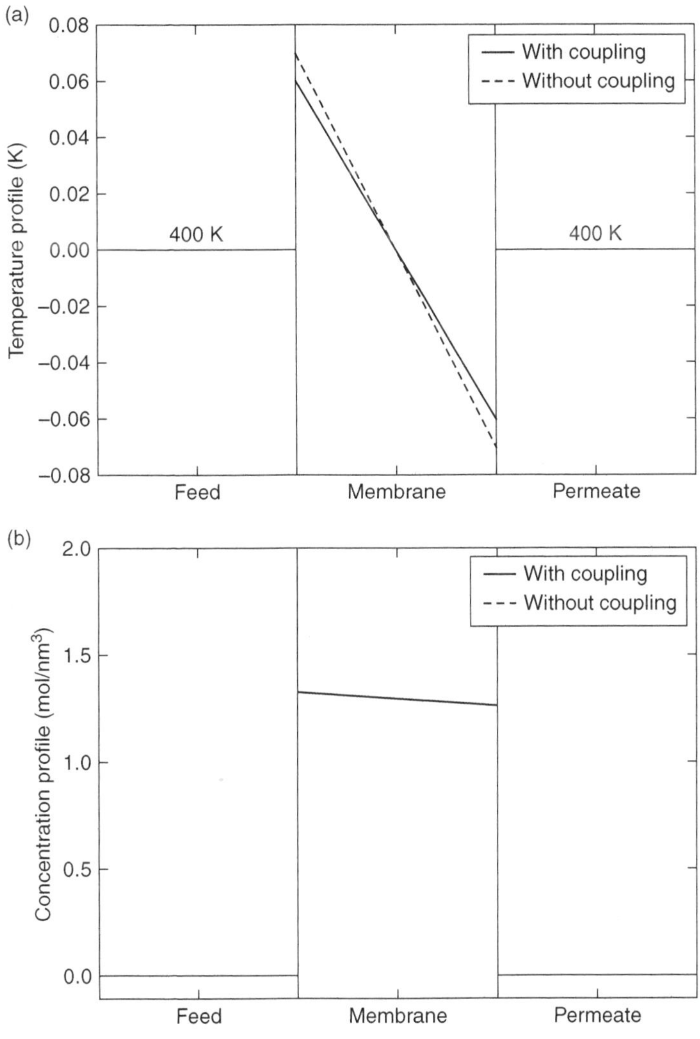

18.7 Concentration and temperature profiles in the zeolite membrane. On both sides of the membrane the temperature is 400 K; (a) shows the temperature difference, while (b) shows the concentration profile across the membrane. The fugacity difference between the two gas phases is 10 kPa. In both cases, the modelling is performed both with coupling of heat and mass transfer, and without coupling. For the temperature across the membrane, we observe that including the coupling of heat and mass transfer in the modelling leads to differences in the temperature profile, while the concentration profile is not affected.

Table 18.2 Heat and mass fluxes for non-isothermal conditions over the membrane

ΔT (K)	J (mol/m²s)	J'^{i}_{q} (kJ/m²s)
0.2	–0.00179	–5.18
0.4	–0.00358	–10.35
0.6	–0.00537	–15.51
0.8	–0.00715	–20.67
1.0	–0.00894	–25.83

Notes: In this table we present the mass flux, as well as the measurable heat flux for the 'feed' side (denoted by *i*) of the membrane. Here, ΔT is $T^o - T^i$, and $T^i = 400$ K. We denote the 'permeate' side by *o*. In all cases, the fugacity of *n*-butane is set to be $f^i = 100$ kPa on both sides of the membrane, and the membrane is kept at 50 μm thickness. In all cases, we have used coupling of heat and mass transport over both the membrane and the surfaces.

18.3.1 Solution procedure

There are different ways to treat this type of membrane system. Equations [18.4]–[18.12] are formulated such that we can calculate fluxes and temperature profiles on the membrane surface. For other formulations, we refer to the work of Kjelstrup and Bedeaux (2008). In this case, we have the resistivities as a function of pressure for the gas–zeolite interface, and the resistivities for transport in the membrane as a function of the local temperature and concentration. The cases of equal temperature difference over the membrane and isobaric transport, can be solved directly by solving Equations [18.4], [18.5] and [18.9]–[18.12]. In addition the total energy balance of Equation [18.2] is used. This leads to a system with eight equations and eight unknowns. The equations can be solved numerically. In this case, we wish to find the fluxes, J, J'^{m}_{q}, J'^{i}_{q}, and J'^{o}_{q}. In addition, we obtain the temperature and concentration profiles across the membrane.

The resistivities on the surfaces are determined from the pressure in the adjoining gas-phase. The surface resistivities were relatively independent of the temperature over the range of temperatures from 350 to 450 K. In Fig. 18.6 we have plotted the values of resistivities obtained from Inzoli *et al.* (2009).

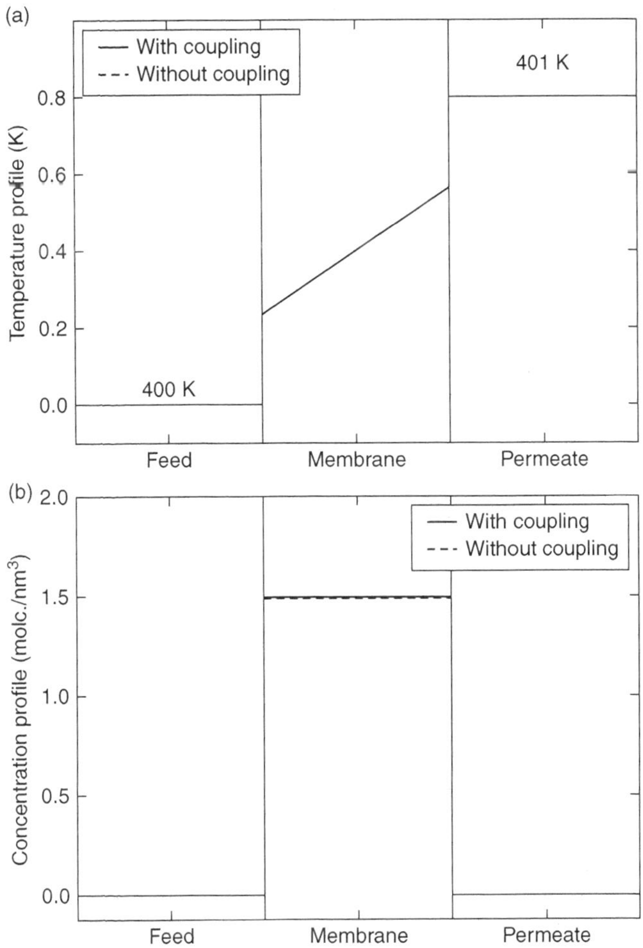

18.8 Concentration and temperature profiles at 400 K; (a) shows the temperature difference, while (b) shows the concentration profile across the membrane. The temperature difference between the two gas phases is kept at 0.8 K. In both cases, we considered the situations in which coupling between heat and mass transfer was included or excluded in the model. The effect of the coupling is negligible for this case.

The resistivities for the membrane are found from the concentration and temperature on the *i*-side of the membrane. All values for resistivities inside the membrane are taken from Inzoli *et al.* (2008); see Fig. 18.5. These values are obtained for a system with *b*-oriented silicalite-1, where the straight channels in the silicalite emerge on the gas–zeolite interface. All values in the work of Inzoli are obtained for such a system. Also, that model does not take defects, for example, pinholes or small fractures, into account. For practical applications, the zeolite membrane has to be fixed to some frame or suspended in a polymer matrix. These are effects that will clearly affect the transport, but are not taken into account here.

18.3.2 Results

In this section we present results obtained by solving the reported equations. There is a huge number of possible configurations of such a membrane setup; in this case, we have focused on a few basic setups and the choice of temperature is well within the range of temperatures used by Inzoli (Inzoli *et al.*, 2008, 2009).

Equal temperatures in gas-phases

The temperature difference from gas-phase to gas-phase is first set to zero and then setting the temperature equal to 400 K on both sides. This still produces a jump in temperature from the surface to the membrane. We set the membrane thickness to be 50 μm and the difference in pressure is taken to be from 2 to 10 kPa, in increments of 2 kPa.

In Fig. 18.7 we present the temperature and concentration profile for this case. The temperature profile has a jump across the surfaces on both sides of the membrane. There is also a jump in the chemical potential (not shown). The concentration in the different parts of the system is shown in Fig. 18.7b. The concentration in the gas-phases is low and the concentration in the zeolite membrane is relatively high.

In Table 18.1, we presented the transport of heat and mass through the membrane when there is no temperature difference, but there is a difference in pressure of *n*-butane. We clearly see how increasing the gradient in pressure changes the transport of *n*-butane through the membrane. The membrane thickness was 50 μm.

Thermally induced transport

In this case, we study the transport over the membrane, by setting the pressure on both sides of the membrane to a constant value, 50 kPa. The temperature is kept at 400 K on the *i*-side of the membrane in all the cases, while the temperature of the *o*-side is given by $T^i + \Delta T$. Concentration and temperature profiles for this case are presented in Fig. 18.8, and the resulting fluxes for different gradients in temperature are given in Table 18.2.

The temperature gradient strongly influences the transport of heat in the system. However, we see that also the contribution to mass transport can be quite substantial, especially at higher temperature differences. In this case, we see that heating the *o*-side will decrease the total transport in the membrane. Still, the most important contribution to mass transport is found in the pressure difference.

Enhancing membrane transport

In Table 18.1 we have presented the heat and mass transport over the membrane with a constant pressure difference and differences in temperature. It is obvious that increased transport can be obtained by increasing the temperature on the o-side of the membrane. The transport of heat with no temperature difference is caused by the coupling of heat transport with the mass transport. This is a quite significant contribution for such a small pressure difference.

Transport in the membrane

The transport inside the membrane can very well be described using the classical equations, that is, Fick's and Fourier's laws. Coupling of heat and mass is zero in pure systems, and small in two-component bulk phases such as silicalite-1 with *n*-butane. The coupling is found to have a small effect on the thermal gradient, as can be seen in Fig. 18.7a, while the concentration gradient seems to be more or less unaffected by the coupling. This can easily be seen from the isotherms; see Fig. 18.4, where all of the isotherms show saturation at the pressures and temperatures used in these calculations.

18.4 Conclusions and future trends

It has recently been clear that Fourier's and Fick's laws, plus the energy balance (Taylor and Krishna, 1993), are not sufficient when it comes to describing interface transport for zeolite membranes, not even for transfer of single components (Kjelstrup and Bedeaux, 2008). These equations alone are not sufficient to obey the requirement for Onsager symmetry in the interaction of the heat and mass flux. This requirement can be fulfilled only by invoking the theory of non-equilibrium thermodynamics. Going to this level of description necessarily means that the description becomes more complicated. The gain may be a better basis for further use, for optimization of yield and energy efficiency of membrane reactors.

In this chapter, we have presented results obtained for transport under non-equilibrium conditions, and shown how this method models transport across a zeolite membrane in agreement with the second law of thermodynamics. We clearly see changes in temperature at the surface of the zeolite, and this effect needs further study. In this case we have only presented transport of single components across the zeolite. In future work, the selectivity and separation purposes should be studied further.

18.5 Acknowledgement

The authors acknowledge financial support from NWO-CW through an ECHO grant, and computational resources was sponsored by the Stichting

Nationale Computerfaciliteiten (National Computing Facilities Foundation, NCF), with financial support from the Nederlandse Organisatie voor Wetenschappelijk onderzoek (Netherlands Organization for Scientific Research, NWO).

18.6 References

Baerlocher C, McCusker L B, and Olson D H (2007), *Atlas of Zeolite Framework Types*, Amsterdam, Elsevier.

Dubbeldam D, Calero S, Vlugt T J H, Krishna R, Maesen T L M, and Smit B (2004), 'United Atom Force Field for Alkanes in Nanoporous Materials', *J Phys Chem B*, **108**, 12301–12313.

Dunne J A, Mariwala R, Rao M, Sircar S, Gorte R J, and Myers A L (1996), 'Calorimetric heats of adsorption and adsorption isotherms. 1. O_2, N_2, Ar, CO_2, CH_4, C_2H_6, SF_6 on silicalite', *Langmuir*, **12**, 5888–5895.

Frenkel D and Smit B (2002), *Understanding Molecular Simulations, 2nd edition*, London, Academic Press.

Ge J, Kjelstrup S, Bedeaux D, Simon J M, and Rousseau B (2007), 'Transfer coefficients for evaporation of a system with a Lennard-Jones long-range spline potential', *Phys Rev E*, **75**, 061604.

de Groot S R, and Mazur P (1984), *Non-equilibrium thermodynamics*. New York, Dover Publications Inc.

Inzoli I (2008), *Coupled transport of heat and mass at the surface of and inside silicalite.* PhD thesis, Norwegian University of Science and Technology.

Inzoli I, Simon, J M, Bedeaux D, and Kjelstrup S (2008), 'Thermal diffusion and Partial Molar Enthalpy Variations of *n*-butane in Silicalite-1', *J Phys Chem B*, **112**, 14937–14951.

Inzoli I, Kjelstrup S, Bedeaux D, and Simon, J M (2009), 'Transport coefficients of *n*-butane into and through the surface of silicalite-1 from non-equilibrium molecular dynamics study', *Micropor Mesopor Mater*, **125**, 112–125.

Kjelstrup S and Bedeaux D (2008), *Non-equilibrium thermodynamics of heterogeneous systems*, Singapore, World Scientific.

Kjelstrup S, Bedeaux D, Inzoli I, and Simon, J M (2008), 'Criteria for validity of thermodynamic equations from non-equilibrium molecular dynamics simulations', *Energy*, **33**, 1185–1196.

Langmuir I (1918), 'The adsorption of gases on plane surfaces of glass, mica and platinum', *J Am Chem Soc*, **40**, 1361–1403.

Lee L and Jeong, H K (2011), 'Synthesis and gas permeation properties of highly *b*-oriented MFI silicalite-1 thin membranes with controlled microstructure', *Micropor Mesopor Mater*, **141**, 175–183.

McLeary E E, Jansen J C, and Kapteijn F (2006), 'Zeolite based films, membranes and membrane reactors: Progress and prospects', *Micropor Mesopor Mater*, **90**, 198–220.

Schnell S K, Liu X, Simon, J M, Bardow A, Bedeaux D, Vlugt, T J H, and Kjelstrup S (2011), 'Calculating thermodynamic properties from fluctuations at small scales', *J Phys Chem B*, **115**, 10911–10918, doi:10.1021/jp204347p.

Taylor R and Krishna R (1993), *Multicomponent Mass Transfer*, New York, John Wiley and Sons.

Vlugt T J H, Krishna R, and Smit B (1999), 'Molecular Simulations of Adsorption Isotherms for Linear and Branched Alkanes and their Mixtures in Silicalite', *J Phys Chem B*, **103**, 1102–1118.

18.7 Appendix: nomenclature

18.7.1 Notation

a	activity
c	concentration
D	Fick's diffusivity
H	molar enthalpy
J	mass flux
J_q	total heat flux
J_q'	measurable heat flux
p	pressure
r	resistivity
R	universal gas constant
T	temperature
λ	thermal conductivity
μ	chemical potential
σ	entropy production

Index